32nd UIT (Italian Union of Thermo-fluid-dynamics) Heat Transfer Conference 2014

Journal of Physics: Conference Series Volume 547

Pisa, Italy
23 – 25 June 2014

ISBN: 978-1-63439-661-5
ISSN: 1742-6588

Printed from e-media with permission by:

Curran Associates, Inc.
57 Morehouse Lane
Red Hook, NY 12571

Some format issues inherent in the e-media version may also appear in this print version.

Printed by Curran Associates, Inc. (2015)

For permission requests, please contact the Institute of Physics
at the address below.

Institute of Physics
Dirac House, Temple Back
Bristol BS1 6BE UK

Phone: 44 1 17 929 7481
Fax: 44 1 17 920 0979

techtracking@iop.org

Additional copies of this publication are available from:

Curran Associates, Inc.
57 Morehouse Lane
Red Hook, NY 12571 USA
Phone: 845-758-0400
Fax: 845-758-2634
Email: curran@proceedings.com
Web: www.proceedings.com

TABLE OF CONTENTS

INVITED LECTURE

HEAT TRANSFER AND EFFICIENCY IN ENERGY SYSTEMS, ENVIRONMENTAL TECHNOLOGIES, AND BUILDINGS

NATURAL, FORCED AND MIXED CONVECTION

CONDUCTION AND RADIATION

Journal of Physics: Conference Series **547** (2014) 011001 doi:10.1088/1742-6596/547/1/011001

32nd UIT (Italian Union of Thermo-Fluid Dynamics) Heat Transfer Conference

Pisa, Italy, 23-25 June 2014

PREFACE

The annual Conference of the "Unione Italiana di Termofluidodinamica" (UIT) aims to promote cooperation in the field of heat transfer and thermal sciences by bringing together scientists and engineers working in related areas.

The 32nd UIT Conference was held in Pisa, from the 23rd to the 25th of June, 2014 in the buildings of the School of Engineering, just a few months after the celebration of the 100th anniversary of the first Institution of the School of Engineering at the University of Pisa.

The response was very good, with more than 100 participants and 80 high-quality contributions from 208 authors on seven different heat transfer related topics: Heat transfer and efficiency in energy systems, environmental technologies, and buildings (25 papers); Micro and nano scale thermo-fluid dynamics (9 papers); Multi-phase fluid dynamics, heat transfer and interface phenomena (14 papers); Computational fluid dynamics and heat transfer (10 papers); Heat transfer in nuclear plants (8 papers); Natural, forced and mixed convection (10 papers) and Conduction and radiation (4 papers).

To encourage the debate, the Conference Program scheduled 16 oral sessions (44 papers), three ample poster sessions (36 papers) and four invited lectures given by experts in the various fields both from Industry and from University. Keynote Lectures were given by Dr. Roberto Parri (ENEL, Italy), Prof. Peter Stephan (TU Darmstadt, Germany), Prof. Bruno Panella (Politecnico di Torino), and Prof. Sara Rainieri (Universitá di Parma).

This special volume collects a selection of the scientific contributions discussed during this conference. A total of 46 contributions, two keynote lectures and 44 papers both from oral and poster sessions, have been selected for publication in this special issue, after a second accurate revision process. These works give a good overview of the state of the art of Italian research in the field of Heat Transfer related topics at the date.

The editors of the volume would like to sincerely thank the authors for presenting their works at the conference and in this special issue. Special thanks are also due to the Scientific Committee, to all the reviewers, and to all the authors for their accurate revision process of each paper for this special issue. Special thanks go to the Organizing Committee, chaired by Prof. Paolo Di Marco.

Walter Grassi (Chairman of the Scientific Committee), Alessandro Franco, Nicola Forgione, Daniele Testi – Editors of the Special Issue

Journal of Physics: Conference Series **547** (2014) 011001 doi:10.1088/1742-6596/547/1/011001

About U.I.T. (Italian Union of Thermo-Fluid Dynamics)

Journal of Physics: Conference Series **547** (2014) 011001 doi:10.1088/1742-6596/547/1/011001

UIT Conferences

I	1983, Trieste
II	1984, Bologna
III	1985, Palermo
IV	1986, Genova
V	1987, Torino
VI	1988, Bari
VII	1989, Firenze
VIII	1990, Ancona
IX	1991, Pisa
X	1992, Genova
XI	1993, Milano
XII	1994, L'Aquila
XIII	1995, Bologna
XIV	1996, Roma
XV	1997, Torino
XVI	1998, Siena
XVII	1999, Ferrara
XVIII	2000, Cernobbio (CO)
XIX	2001, Modena
XX	2002, Maratea (PZ)
XXI	2003, Udine
XXII	2004, Genova
XXIII	2005, Parma
XXIV	2006, Napoli
XXV	2007, Trieste
XXVI	2008, Palermo
XXVII	2009, Reggio Emilia
XXVIII	2010, Brescia
XXIX	2011, Torino
XXX	2012, Bologna
XXXI	2013, Moltrasio (CO)
XXXII	2014, Pisa

Journal of Physics: Conference Series **547** (2014) 011001 doi:10.1088/1742-6596/547/1/011001

UIT Heat Transfer Conference, Pisa 2014 – Scientific Committee

Giovanni Sebastiano Barozzi	Università degli Studi di Modena e Reggio Emilia, Italy
Gian Paolo Beretta	Università degli Studi di Brescia, Italy
Romano Borchiellini	Politecnico di Torino, Italy
Gian Piero Celata	Centro Ricerche ENEA Casaccia, Roma
Giovanni Cesini	Università Politecnica delle Marche, Italy
Michele Ciofalo	Università degli Studi di Palermo, Italy
Elvio Cossali	Università degli Studi di Bergamo, Italy
Andrea De Lieto Vollaro	Università di Roma "La Sapienza", Italy
Mario De Salve	Politecnico di Torino, Italy
Stefano Del Giudice	Università degli Studi di Udine, Italy
Walter Grassi (Chairman)	Università di Pisa, Italy
Oronzio Manca	Seconda Università degli Studi di Napoli, Italy
Mario Misale	Università degli Studi di Genova, Italy
Gian Luca Morini	Università degli Studi di Bologna, Italy
Vincenzo Naso	Università degli Studi di Napoli "Federico II", Italy
Alfonso Niro	Politecnico di Milano, Italy
Giorgio Pagliarini	Università degli Studi di Parma, Italy
Luisa Rossetto	Università di Padova, Italy
Peter Stephan	Technische Universität Darmstadt, Germany
Lounes Tadrist	Université de Provence Aix-Marseille I, France
John Richard Thome	École Polytechnique Fédérale de Lausanne, Switzerland

Organizing Committee

Walter Ambrosini	DICI, Università di Pisa
Carlo Bartoli	DESTEC, Università di Pisa
Claudio Casarosa	DESTEC, Università di Pisa
Paolo Di Marco (Chairman)	DESTEC, Università di Pisa
Sauro Filippeschi	DESTEC, Università di Pisa
Nicola Forgione	DICI, Università di Pisa
Alessandro Franco	DESTEC, Università di Pisa
Walter Grassi	DESTEC, Università di Pisa
Francesco Leccese	DESTEC, Università di Pisa
Daniele Testi	DESTEC, Università di Pisa

Keynote Lecturers

Bruno Panella	Politecnico di Torino, Italy
Roberto Parri	Enel Green Power, Larderello (PI), Italy
Sara Rainieri	Università degli Studi di Parma, Italy
Peter Stephan	Technische Universität Darmstadt, Germany

Conference Topics

A. Heat transfer and efficiency in energy systems, environmental technologies, and buildings
B. Micro and nano scale thermo-fluid dynamics
C. Multi-phase fluid dynamics, heat transfer and interface phenomena
D. Computational fluid dynamics and heat transfer
E. Heat transfer in nuclear plants
F. Natural, forced and mixed convection
G. Conduction and radiation

Journal of Physics: Conference Series **547** (2014) 011001 doi:10.1088/1742-6596/547/1/011001

Conference Program

Monday, June 23rd, 2014		
09:00 – 09:30	Registration [Polo Etruria, room F06]	
09:30 – 11:00	**Opening Ceremony** [Aula Magna Pacinotti] Commemoration of Prof. Sergio Faggiani Presentation of UIT 2013 Prize "Sergio Faggiani"	
11:00 – 11:15	Coffee break [Polo Etruria]	
11:15 – 12:00	**Invited Lecture 1: R. Parri** [Polo Etruria - Room F09] *Chairman: C. Casarosa*	
12:00 – 13:00	**Topic A, Session 1** [Room F09] *Chairman: V. Naso*	**Topic B, Session 1** [Room F07] *Chairman: G.L. Morini*
13:00 – 14:30	Lunch [Polo Etruria]	
14:30 – 15:15	**Invited Lecture 2: P. Stephan** [Room F09] *Chairman: W. Grassi*	
15:15 – 16:35	**Topic C, Session 1** [Room F09] *Chairman: G.P. Celata*	**Topic D, Session 1** [Room F07] *Chairman: S. Del Giudice*
16:35 – 16:50	Coffee break [Polo Etruria]	
16:50 – 18:10	**Topic A, Session 2** [Room F09] *Chairman: A.M. Lezzi*	**Topic B, Session 2** [Room F07] *Chairman: P. Stephan*

Tuesday, June 24th, 2014		
09:00 – 09:45	**Invited Lecture 3: B. Panella** [Room F09] *Chairman: W. Ambrosini*	
09:45 – 10:45	**Topic E, Session 1** [Room F09] *Chairman: M. De Salve*	**Topic D, Session 2** [Room F07] *Chairman: N. Forgione*
10:45 – 11:00	Coffee break [Polo Etruria]	
11:00 – 11:50	**Topic A, Poster Session** [Room F08] *Chairman: A. Niro*	
11:50 – 13:10	**Topic A, Session 3** [Room F09] *Chairman: G. Pagliarini*	**Topic F, Session 1** [Room F07] *Chairman: G.S. Barozzi*
13:00 – 14:30	Lunch [Polo Etruria]	
14:30 – 15:10	**Topic G, Session 1** [Room F09] *Chairman: S. Rainieri*	
15:10 – 16:00	**Topics B-D-E-G, Poster Session** [Room F08] *Chairman: B. Panella*	
16:00 – 16:30	Coffee break [Polo Etruria]	
16:30 – 18:00	**UIT 2014 Council** [Room F09]	
20:30 – 23:00	**Gala Dinner** [Chiostro della Chiesa di Santa Maria del Carmine]	

Wednesday, June 25th, 2014		
09:00 – 09:45	**Invited Lecture 4: S. Rainieri** [Room F09] *Chairman: P. Tartarini*	
09:45 – 11:05	**Topic F, Session 2** [Room F09] *Chairman: A. Mariani*	**Topic C, Session 2** [Room F07] *Chairman: D. Del Col*
11:05 – 11:30	Coffee break [Polo Etruria]	
11:30 – 12:20	**Topics C-F, Poster Session** [Room F08] *Chairman: L. Rossetto*	
12:20 – 13:00	**Conference Closure** [Room F08]	

Peer review statement

All papers published in this volume of *Journal of Physics: Conference Series* have been peer reviewed through processes administered by the proceedings Editors. Reviews were conducted by expert referees to the professional and scientific standards expected of a proceedings journal published by IOP Publishing.

Journal of Physics: Conference Series **547** (2014) 012001 doi:10.1088/1742-6596/547/1/012001

Two-Phase flow instrumentation for nuclear accidents simulation

G Monni, M De Salve and B Panella
Energy Department, Politecnico di Torino, Corso Duca degli Abruzzi, 24, 10129 Torino, Italy

E-mail: bruno.panella@polito.it

Abstract
The paper presents the research work performed at the Energy Department of the Politecnico di Torino, concerning the development of two-phase flow instrumentation and of models, based on the analysis of experimental data, that are able to interpret the measurement signals. The study has been performed with particular reference to the design of power plants, such as nuclear water reactors, where the two-phase flow thermal fluid dynamics must be accurately modeled and predicted. In two-phase flow typically a set of different measurement instruments (Spool Piece - SP) must be installed in order to evaluate the mass flow rate of the phases in a large range of flow conditions (flow patterns, pressures and temperatures); moreover, an interpretative model of the SP need to be developed and experimentally verified. The investigated meters are: Turbine, Venturi, Impedance Probes, Concave sensors, Wire mesh sensor, Electrical Capacitance Probe. Different instrument combinations have been tested, and the performance of each one has been analyzed.

1. Introduction

The design of power plants, involving two-phase flow, such as nuclear water reactors, requires the ability to model and predict the flow thermal fluid dynamics, with a required degree of accuracy, while on the other hand accurate experimental data need to be used in order to validate the simulation tools usually adopted during the design phase. For example, the design of nuclear reactors requires to carry out integral tests, used to simulate the thermo-hydraulics of a PWR in a scaled-down test facility and for safety codes validation (like RELAP and TRAC), and separate effect tests, where the local phenomena of the flow are investigated using sensor characterized by a sufficiently high time and space resolution.

Due to the complex nature of the two-phase flow, the measurement of the flow parameters, such us phases distribution, velocities, interfaces evolution, require the application of special instrumentation an d the development of interpretative models of the experimental signals based on both the physics of the sensor and of the flow. Successful application of a measuring system/ device in a two-phase test facility does not guarantee its applicability in every flow or loop conditions. Intrusive and nonintrusive measurement techniques of two-phase flows, with special application to the determination of mass flow rates, have been developed, in particular to investigate Design Basis Accidents like loss of coolant accidents (LOCA), where the measurement of the mass flow rate of the two phases was required to analyze the accident evolution. Nowadays new experimental campaigns are required for the licensing of the new advanced reactors, like the Small Modular Reactors, so that the development of reliable instruments is required for the measurement of the fundamental parameters of the flow, under as many as possible two-phase flow conditions. As an example, the SPES3 experimental facility [1], able to simulate the innovative small and medium size PWR nuclear reactors, is being built and will be operated by the SIET Company.

An accurate accident analysis requires the measurement of the mixture mass flow rate and instruments and methodologies to evaluate different two-phase flow parameters have to be developed. Typically a set of different measurement instruments (Spool Piece - SP) must be installed in order to evaluate the mass flow rate of the phases and a model of the SP, depending on the geometry and on the SP orientation, is required. The selection of the instruments strongly depends on the experimental

Journal of Physics: Conference Series **547** (2014) 012001 doi:10.1088/1742-6596/547/1/012001
conditions: pressure, temperature and phases velocities. In the present paper the experimental result and the models developed for different instruments and different SP is presented. Based on a bibliographic analysis of the available instruments, suitable to be installed in the SPES3 simulation facility [2], a number of instruments have been experimentally and theoretically studied: Turbine [3], Drag disk [3], Venturi [4], Impedance Probes, Concave sensors, Wire mesh sensor (WMS) [5], Electrical Capacitance Probe (ECP) [6], designed by SIET S.P.A. Their measurement characteristics have been investigated, taking in account the range of measurement, dynamic response, installation requirements, materials/electrical compatibility with pressure and temperature conditions, flow velocity compatibility. The selected devices can operate in the whole range of conditions expected in the break lines of the experimental facility: void fractions (from 0 to 1), pressure (up to 155 bar), temperature (higher than 200°C), velocity (gas velocity up to 200 m/s). Moreover the flow pattern is likely to be also time dependent and the measurement is complicated by the fact that during the transient the fluid becomes a non homogeneous mixture of water and steam.

Different instrument combinations have been tested, with air and water at low pressure and temperature, and the performance of each one has been analyzed in terms of estimation of the mass flow rate of the two phases, identifying their advantage and drawbacks, and estimating the phases mass flow rate measurement accuracy for each SP configuration. All details, concerning the two- phase flow instrumentation and the general criteria for the instruments selection, can be found in ref. [7].

2. Two-Phase Flow Investigation with WMS

The Wire-Mesh Sensors (WMS), based on the measurement of the local instantaneous conductivity of the two-phase mixture, are used for a high-speed visualization of a gas–liquid flow as well as for the measurement of void fraction profiles, bubble size distributions and gas velocity distributions [8,9,10]. The WMS, able to measure directly the conductivity between pairs of crossing wires, can be used to detect the presence of the phases in the different measuring points, and allows an high space and time resolution. In the present work, the WMS has been used to investigate the air-water flow in a horizontal Plexiglas pipeline (inner diameter D=19.5 mm), for several flow patterns. The local, chordal, cross-sectional void fraction are estimated from the sensor data. The flow evolution in time and space is analyzed and the characteristics for all the flow patterns are derived by means of a statistical analysis. The adopted sensor (Fig. 1) has been manufactured by Teletronic Rossendorf GmbH [11]. It consists of two planes of parallel wire grids (16x16) that are placed across the channel at a short distance from each other (1.5 mm); the wires of both planes cross under an angle of 90°. The sensor has been designed to cover the cross section of a channel having a 19.5 mm inner diameter; the wires have a diameter D_{wire} of 70 μm and a pitch p equal to 1.3 mm, so only the 5.4% of the pipe section is occupied by the sensor.

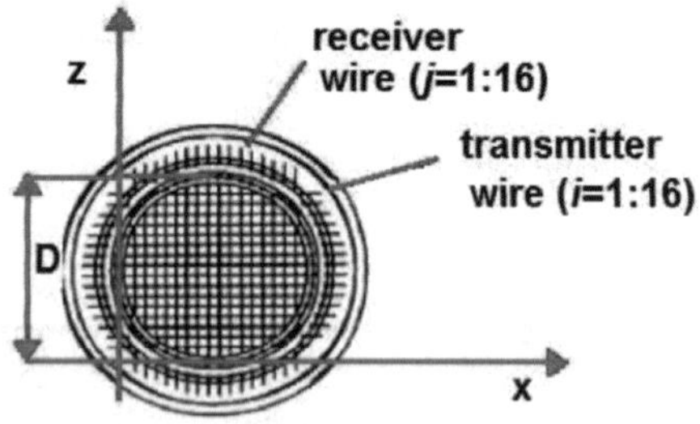

Fig. 1. Scheme of the WMS.

The measuring grid allows a spatial resolution of the order of the pitch length (1.3 mm), that combined with the high time resolution (up to 10.000 frames/s) allow the investigation of the local phenomena of the flow and their time evolution. The output is a 3-D matrix $V^*(i,j,k)$, where the indexes i and j are related to the space position of the mesh points and k is the time index, having a value between 1 and $T_T f_{acq}$, where T_T is the total observation time and f_{acq} is the acquisition frequency.
The indexes i and j refer to transmitting wires and to receiving wires respectively. The developed signal processing scheme is structured to obtain the desired two-phase flow parameters. First of all the signal is normalized taking into account the single phase reference matrices, $V_g(i,j)$ and $V_l(i,j)$:

$$V^*(i,j,k)=\frac{V(i,j,k)-V_l(i,j)}{V_g(i,j)-V_l(i,j)} \quad (1)$$

The signal normalization can be considered as an approximation of the local void fraction value, if a linear relationship between conductivity and void fraction and a reference area equal to the square of the wire pitch p are assumed. From the evaluated local void fraction, the instantaneous chordal profiles are derived. The vertical profiles in the centre line of the pipe (index j=8) are used to evaluate the characteristic times and the characteristic shape of the void fraction profiles for the different flow conditions. The local instantaneous void fraction α has been obtained with an acquisition frequency f_{acq} equal to 1250 Hz for a total observation time T_T equal to 20 s. Further details can be found in ref. [12]. The local void fraction in each point of the sensor grid is characterized by the signal histogram: the analysis of the normalized histograms, that are evaluated for the total observation time, allows us to define if a deterministic or a stochastic behavior characterizes the flow. The parameter σ/α_{mean}, where σ is the standard deviation of the signal and α_{mean} is the average void fraction is used as an index of the fluctuation of the signal. For non intermittent flows, the standard deviation value of the local void fraction is small, and the flow can be characterized with well defined void fraction profiles. If the probability density function PDF is not a Gaussian curve and a quasi periodic behavior can be defined for the macroscopic structures of the flow, the characteristic times and the characteristic profiles have to be derived. In order to analyze fast phenomena in terms of average values, a moving average of the void fraction time values has been performed using a time interval τ. Once the averaging time interval is defined, every average profile has been characterized from the statistical point of view by the root mean square profiles; while the variation in time of the chordal profile has been analyzed considering the evolution of the parameter $M(t)$ defined as the sum along the chord of the difference between the moving average local void fraction evaluated at the time t and at the time t-Δt:

$$M(t)=\sum_{i=1}^{16}(<\alpha(i,j,t)>-<\alpha(i,j,t-\Delta t)>) \quad (2)$$

t is equal to k/f_{aca} and Δt is equal to $1/f_{aca}$. Starting from the evolution in time of the $M(t)$ parameter it is possible to define the time interval in which the profile is constant and the time interval in which is evolving. The chordal profiles at the selected time are then analyzed and used to characterize the flow pattern in terms of time evolution and characteristic shape and parameters. A detailed description of the methodology can be found in ref. [12].

The experimental facility (Fig. 2) consists of the feed water loop (tap water with conductivity of about 620 μS is used), the feed air loop and the test section. The liquid flow rate is measured with a ±0.5% r.v. accuracy value. The air flow rate is measured by means of different rotameters for the different ranges with a ±2% f.s.v. accuracy value.

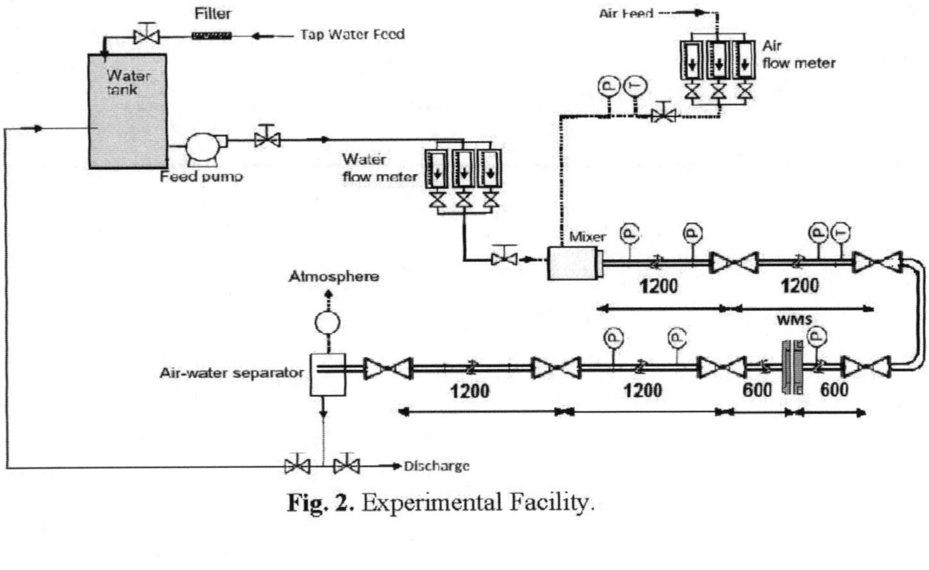

Fig. 2. Experimental Facility.

The test section consists of a 19.5 mm diameter and about 6 m long horizontal pipe. The WMS is located between two Plexiglas pipes having a length of 600 mm. Experiments have been performed at water temperature of about 20 °C. The flow quality ranges from 0 to 0.73 and the superficial velocity ranges from 0.14 to 31.94 m/s for air and from 0.019 to 2.62 m/s for water. The pressure ranges from atmospheric pressure to 3.7 bar depending on the experimental conditions. The typical observed flow patterns are stratified flow, intermittent flow (slug and plug) and non symmetric annular flow. The comparison between the observed flow patterns and the prediction by the Mandhane's map is reported in Fig.3; the flows are identified by superficial velocities (J_g, J_l). Some runs have been performed in transition regimes on the ground of the visual observation and maps results.

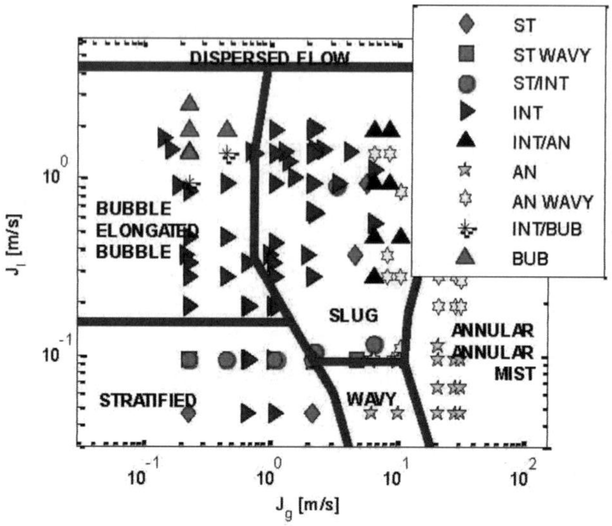

Fig. 3. Comparison between Mandhane's Map flow patterns prediction and visual observation. ST=Stratified, INT=Intermittent, AN=Annular, BUB=Bubble

To obtain a characteristic shape of the void fraction profile that can be related to the superficial flow velocities, the moving average profiles have been evaluated. In Fig. 4 the time evolution of the parameter *M(t)*, that is obtained using a moving average time τ of 0.4 s, is shown for different air and water superficial velocities.

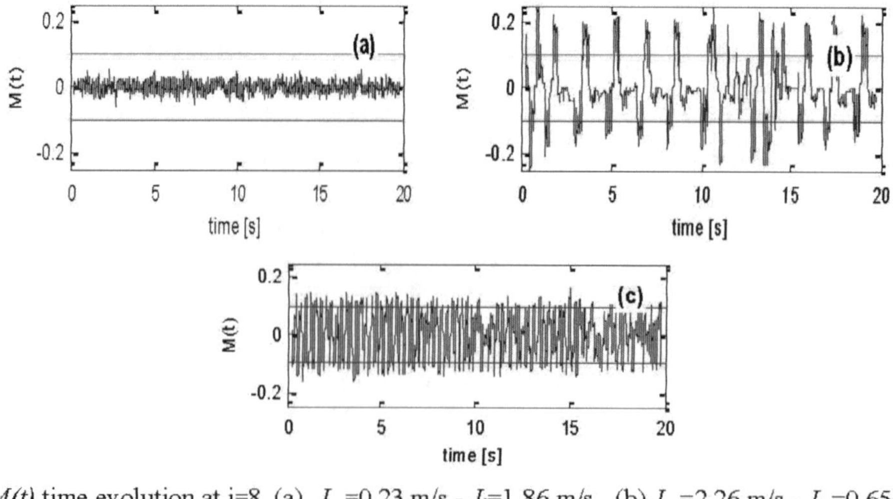

Fig. 4. *M(t)* time evolution at j=8. (a) J_g =0.23 m/s - J_l=1.86 m/s, (b) J_g =2.26 m/s - J_l =0.65 m/s, (c) J_g =8.73 m/s - J_l =1.86 m/s.

Journal of Physics: Conference Series **547** (2014) 012001 doi:10.1088/1742-6596/547/1/012001

The optimum time interval, used for the moving average profiles, has been selected as the minimum averaging time that allows us to maximize the time period in which *M(t)* is zero. If the amplitude is lower than a threshold value the flow can be characterized by the mean profile and the correspondent standard deviation (Fig. 4 (a)), and it is classified as "non intermittent" or "non periodic profile" flow; while for higher amplitudes, (*M(t)* higher than 0.1 in absolute value) a number of profiles depending on the time history have to be analyzed. In this case the flow evolution can be divided in two categories, namely "quasi periodic profile" (Fig. 4 (b)) , and "non periodic profile with high noise" (Fig. 4 (c)): the profile is classified as "quasi periodic" if there is a time period in which the profile doesn't change significantly, while it is classified as "non periodic with high noise" if the absolute value of *M(t)* is higher than the threshold value and the local flow is characterized by a chaotic behavior. For the "high noise" profile, in which the value of the *M(t)* is continuously changing, the time evolution is characterized by the mean profile, that is evaluated adopting the observation time of 20 s, and by the relative standard deviation. For the "quasi periodic" profile at least two characteristic profiles with the related time are derived considering the time periods in which *M(t)* is near zero and the time periods in which the flow is evolving and *M(t)* is higher than zero. For classification purpose and considering the time interval that has been adopted to perform the moving average, the limit value has been considered of the order of 0.5 s: the profile is considered as "quasi periodic" if the profile doesn't change significantly for a period higher than 0.5 s. From the analysis of the parameter *M(t)* time evolution, the flow characteristic profiles are derived and qualified in terms of shape and flow parameters. As an example, in Figs. 5 the characteristic void fraction chordal profile is shown for two different flows, that are characterized by low amplitude values of *M(t)*. In Fig. 5a the typical profile of an horizontal bubbly flow is shown, while in Fig.5b the flow can be classified as annular. Considering the flow of Fig. 5a, the parameters that have to be defined in order to classify it are: the height of the stratified region, the mean value of the void fraction in the bubbles region, the mean value of the void fraction and the related standard deviation in the stratified region, and finally the mean value of the void fraction in the upper part of the pipe, that is close to the wall.

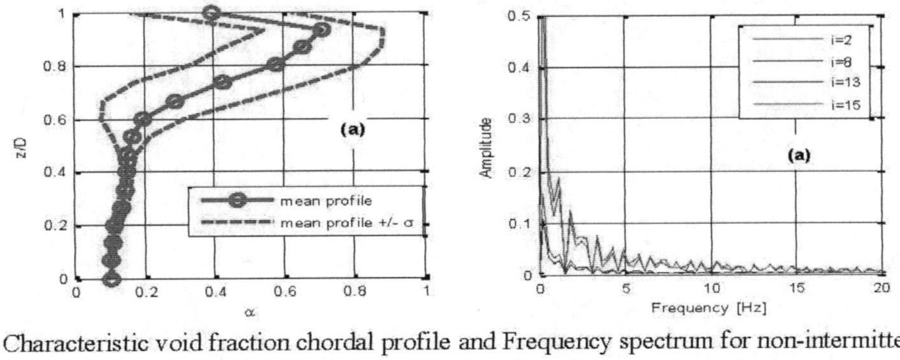

Fig. 5a. Characteristic void fraction chordal profile and Frequency spectrum for non-intermittent flow. (a) J_g =0.23 m/s – J_l =1.86 m/s, (b) J_g=29.3 m/s – J_l =0.05 m/s.

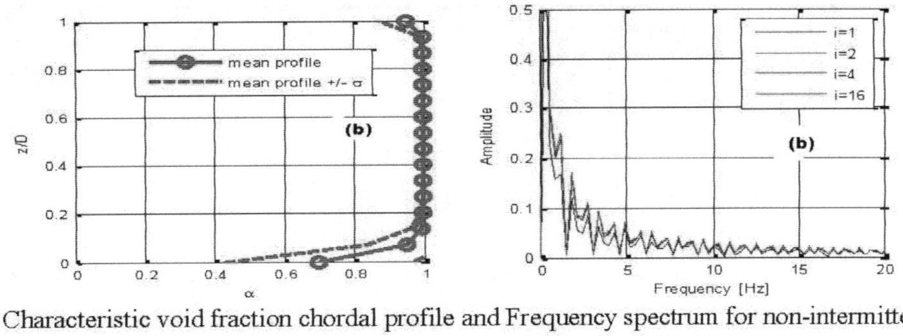

Fig. 5b. Characteristic void fraction chordal profile and Frequency spectrum for non-intermittent flow at j=8. (a) J_g=0.23 m/s – J_l =1.86 m/s, (b) J_g=29.3 m/s – J_l=0.05 m/s.

At higher water velocity the evolution from bubble flow to annular flow occurs; the air flow rate increase causes an increase of the void fraction and a reduction of the liquid film thickness both in the upper and lower region of the pipe. If the liquid velocity increases the flow tends to homogenize with a void fraction at the channel bottom higher than zero while the maximum void fraction value decreases. By increasing the air velocity the mean chordal void fraction increases and the profile flattens evolving in annular flow. The transition to the annular flow seems to coincide with a shift of the void fraction maximum value towards the centre of the pipe. The pictures in Fig. 6 show that the flow is symmetric with respect to the vertical pipe axis and the profile of the void fraction is strongly affected by the gravity effect. The flow can be classified as plug flow and it is characterized by a deterministic time evolution of the void fraction, with the lower part of the pipe (up to j=6) always covered by the liquid phase, while in the upper part of the pipe, large bubble of gas (plugs) are spaced out by slugs of liquid. The void distribution for the flows classified as bubbly is quite different depending on the liquid velocity. The comparison of the chordal profiles at different superficial velocities combinations of water and air, is pointed out in Fig. 7 showing the effect of the superficial velocities on the vertical profiles. The developed methodology for the analysis of the Wire Mesh Sensor signals allows the identification of the flow pattern and the flow characterization in terms of void fraction profiles and characteristic times. From the analysis of the time evolution of the local void fraction and of the parameter $M(t)$ it is possible to characterize the flow in terms of the void fraction profile shape and other parameters like phases distribution, liquid level, mean void fraction in the different pipe regions, characteristic frequencies. The evolution of the void fraction profiles, related to the superficial velocity of the two-phases (J_g and J_l), and the flow evolution in time and space show that such methodology is useful to identify and characterize the two-phase flow patterns. All the test results are reported in refs. [7] and [12].

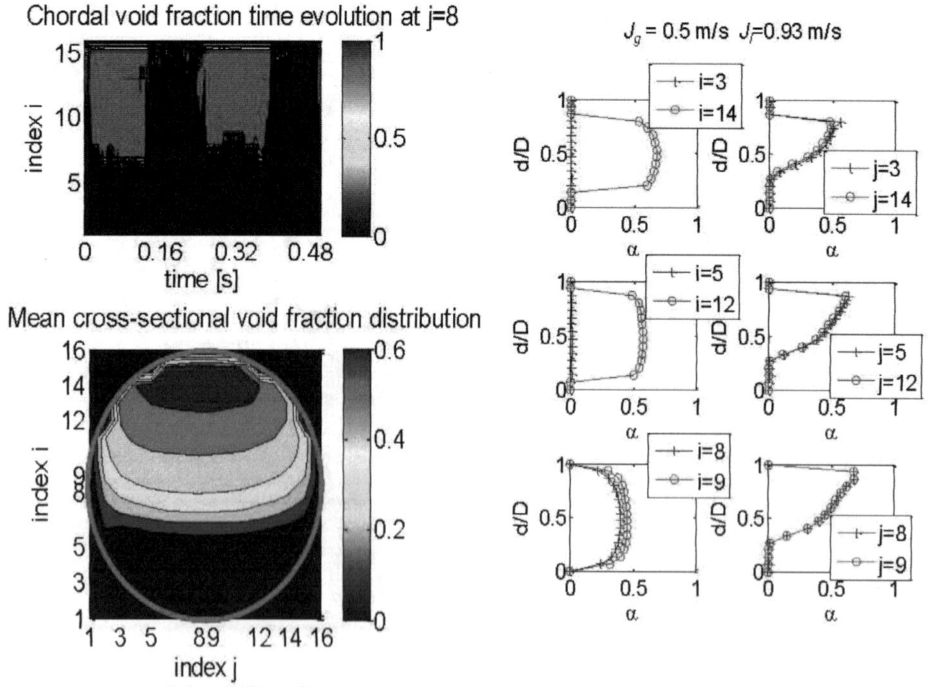

Fig. 6. Plug flow example: J_l=0.93 m/s and J_g=0.5 m/s.

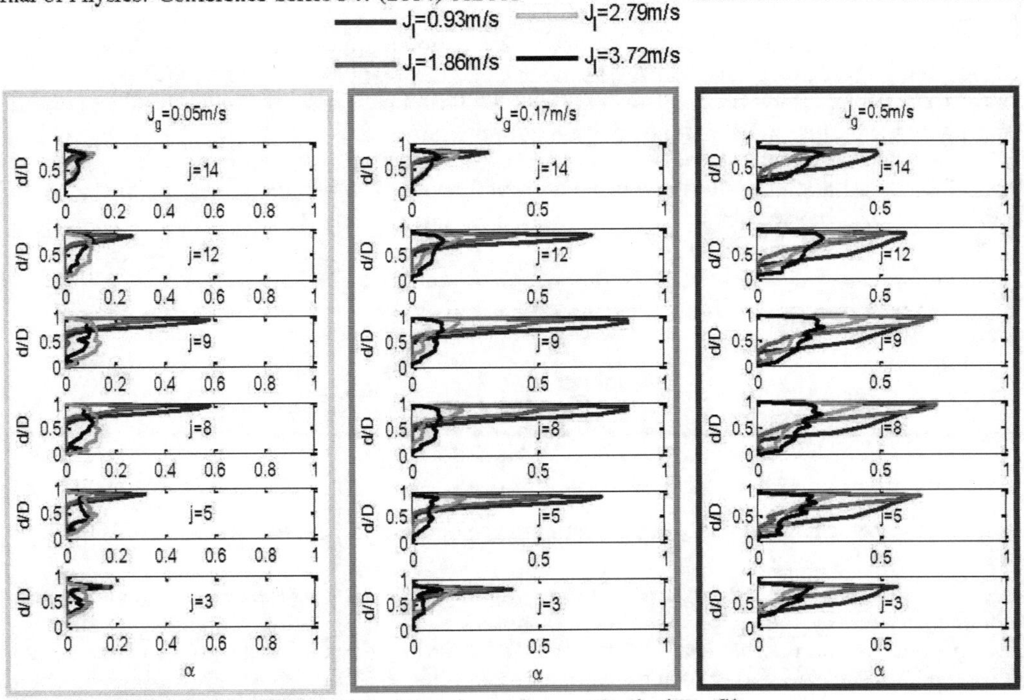

Fig. 7. Velocities dependence: Vertical Profiles.

3. Two-Phase Flow Measurement for Nuclear applications in Simulation Facilities

3.1 Spool Piece: Turbine-Drag Disk-Impedance Probe

The SP consisting of a turbine flow meter (TFM), a drag disk (DD) has been one of the most analyzed instruments combination during the first experimental tests for nuclear reactors [13] .

In the present study a SP made up of three instruments, turbine, drag-disk and a concave electrodes impedance probe, has been analyzed. The test section where the SP is installed is part of the facility shown in Fig. 8: the main components of the facility and the instruments of the SP (1- Impedance probe, 2- Turbine and 3- Drag Disk in the picture below it), are schematically reported in Fig. 8.

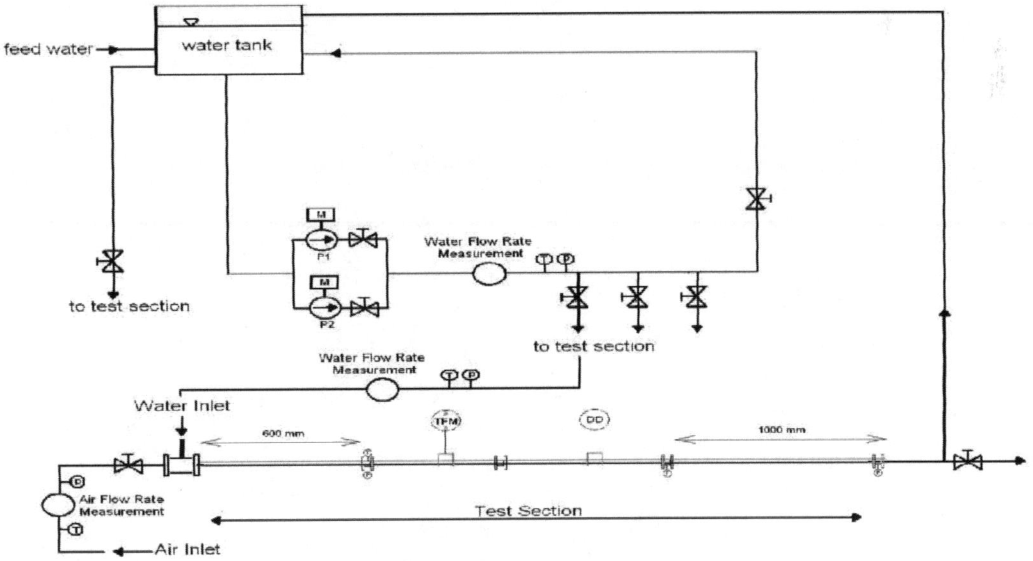

Fig. 8. Experimental facility and test section picture.

The turbine flow meter and the drag disk have been calibrated under single-phase flow, and then the response of the instruments of the SP have been analyzed in two-phase flow between 0.055 and 0.3 kg/s and 0.005 to 0.013 kg/s for water and air respectively. The dependence of turbine flow meter and drag disk signals on the mass flow rate of the phases are reported in [15] .

The impedance probe (IP) installed in the SP (fig.9), consists of two concave externally mounted electrodes having a length of 166 mm and a width of 18 mm. The two electrodes are installed at 180° along the axial length of the pipe.

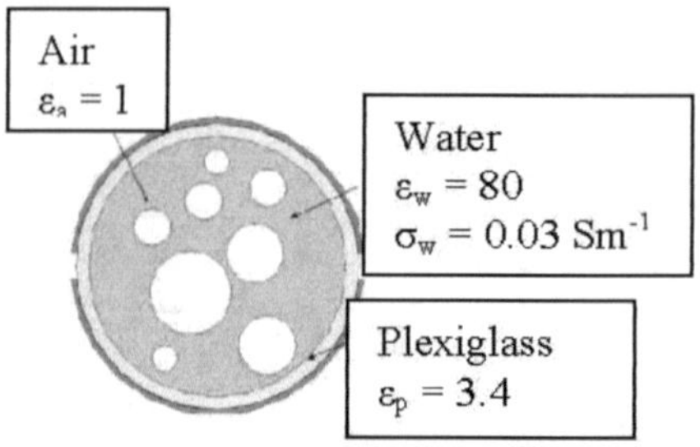

Fig. 9. Scheme of the IP cross-section and electrical properties of the materials.

The signals are acquired for 30 s, using a sample frequency of 1 kHz. At each run the volumetric void fraction is measured by means of two Quick Closing Valves (QCV) after the signal acquisition. The volumetric void fraction is measured 5-10 times at the same flow condition in order to have a sufficient statistics. The tests provided data points for $0.2<\alpha<0.9$. The observed flow patterns have been related with the measured void fractions: bubbly and slug flows are confined to low-range flow quality with void fractions ranging between 0.2-0.45 and 0.45-0.7 respectively, while annular flow has been observed for void fraction ranging between 0.7 and 0.9. The impedance probe analysis has been performed comparing the signal with the experimental volumetric void fraction measured by means of the QCV, dividing the data into groups depending on the flow pattern (Figs. 10 and 11).

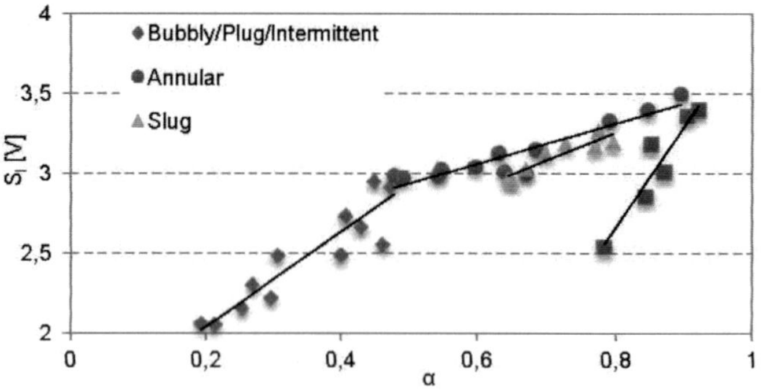

Fig.10. Impedance probe signal vs. void fraction.

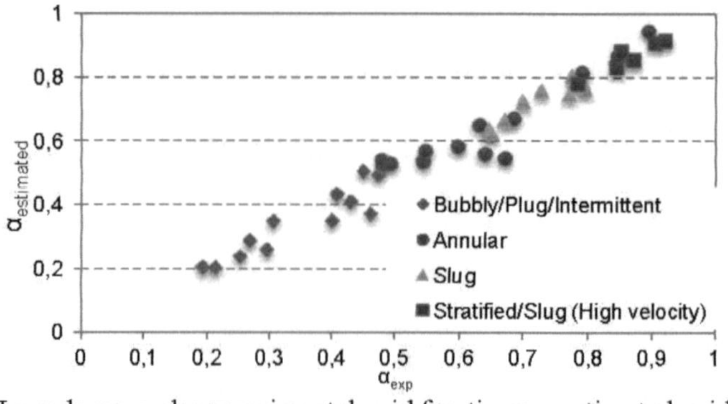

Fig. 11. Impedance probe experimental void fraction vs. estimated void fraction.

The Aya's model [14] has been used to calculate the mass flow rate of the two phases, adopting in this case a turbine coefficient equals to the water single-phase flow coefficient, and evaluating the void fraction by means of the developed model of the impedance sensor. The parametric analysis, performed by the Aya's model, has shown that, the knowledge of the flow pattern, and then of the void fraction, is required to evaluate the two-phase flow parameters. So the SP has been completed with the third instrument able to measure the void fraction of the flow (an impedance sensor), whose signal has been analyzed as a function of the flow parameters and compared with the experimental volumetric void fraction measured by means of the QCV. The analysis shows that a linear relationship between void fraction and impedance sensor can be used, but the knowledge of the flow pattern is needed for a proper signal interpretation. The experimental data, corresponding to different flow patterns from bubbly to annular flow, have been used to get an operating map of the SP, able to evaluate the mass flow rates of the phases in the mixture, with the relative errors, and to predict the flow pattern. The measurement of the void fraction allows the use of the Aya's model in all the experimented range, obtaining the mass flow rate of the phases with an accuracy of 15% in the 75% of the cases. Fig. 12 shows that most of the points calculated with the model of Aya are in the range of 15% of error, which means that the data has been obtained correctly. It can be concluded that the results of a SP, consisting of a turbine flow meter, a drag disk and an impedance concave probe, under annular and stratified annular air-water two-phase flow, have allowed the characterization of the flow in terms of flow pattern and mass flow rates of the phases, with an accuracy of 15%, in the experimented range. Further details can be found in ref. [15].

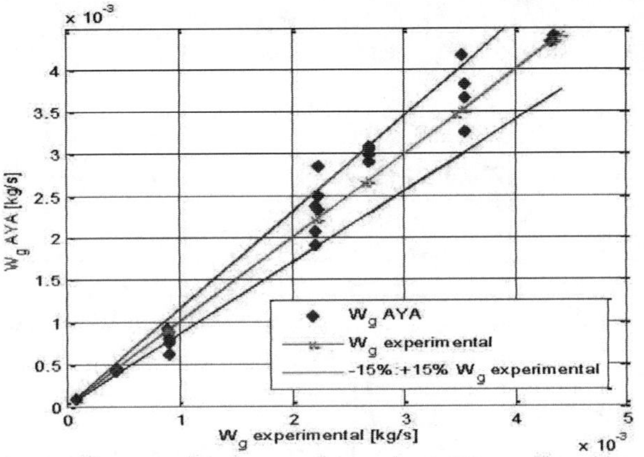

Fig. 12. Air mass flow rate (W_g) experimental vs. W_g predicted by Aya's model.

3.2 Spool Piece: Venturi Flow Meter (VFM) and WMS

In the second analyzed SP, the Wire Mesh Sensor (WMS), presented in section 2 and a classical Venturi Flow Meter (VFM) have been installed in horizontal configuration and analyzed under air-water two-phase flow. Compared with single-phase flow, the two-phase flow VFM pressure drops increase due to the interaction between the gas and the liquid phase; for this reason the value of the effective two-phase mass flow rate must be estimated using available correlations derived from experimental data or semi-empirical models.

The test section described previously has been modified with the insertion of the second instrument (Fig. 13). It consists of a 19.5 mm diameter and 7 m long pipe. The SP is installed at L/D = 192 from the entrance; the WMS is installed between two Plexiglas pipes having a length of 600 mm, while the VFM is installed between two Plexiglas pipes having a length of 500 mm and 490 mm upstream and downstream respectively. The experimental test section is equipped with two quick closing valves (QCV) that allow the measurement of the volumetric void fraction in a length of 1300 mm. The signals have been acquired using a frequency f = 1250 Hz for all the instruments for a total acquisition time of 20 s. Instruments characterization in single-phase flow of air and water has been performed. The flow quality ranges from 0 to 0.73 and the superficial velocity from 0.14 to 32 m/s for air and from 0.019 to 2.62 m/s for water.

Journal of Physics: Conference Series **547** (2014) 012001 doi:10.1088/1742-6596/547/1/012001

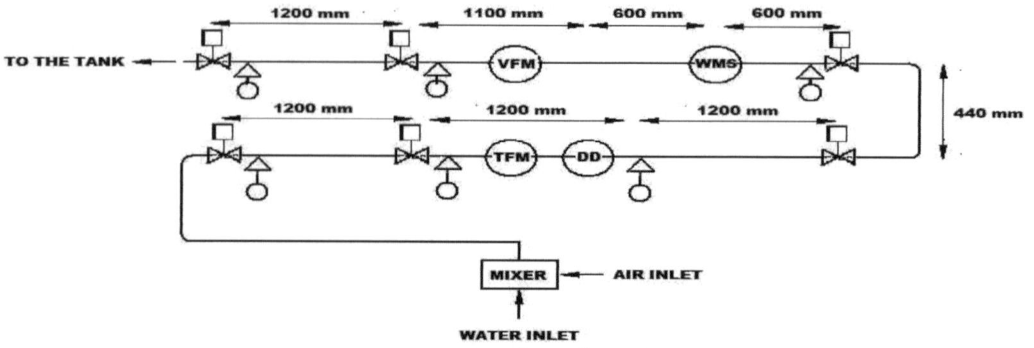

Fig. 13. Test section schematic.

According to Baker's Map [16] the typical observed flow patterns were stratified flow, intermittent flow (slug and plug) and annular flow. The WMS output signal $V(i,j,k)$ is analyzed in order to obtain the mean cross sectional value of the void fraction, that is used to model the response of the VFM. The local instantaneous void fraction (eq.1) has been obtained with an acquisition frequency f_{acq} equal to 1250 Hz for a total observation time T_T equal to 20 s, so that the value of k ranges from 1 to k_T, where k_T is the total number of measured frames, equals to $f_{acq} \cdot T_T$.

The averaged time cross-sectional void fraction is obtained by averaging over the observation time and over the measuring section (A_p is the pipe cross section):

$$\langle \alpha \rangle = \frac{1}{k_T} \sum_{k=1}^{k_T} \frac{1}{A_p} \sum_{i=1}^{16} \sum_{j=1}^{16} \alpha(i,j,k) \tag{3}$$

The WMS experimental void fraction has been then compared with different literature correlations (Lockhart-Martinelli [17], Baroczy [18], Chisholm [19] and Wallis [20]): the Lockhart-Martinelli and Wallis correlations give a good estimation of the experimental data at flow qualities higher than 10^{-2}. In order to estimate the momentum density, the void fraction value must to be known; then the output of the WMS becomes an input for the VFM signal interpretation. The two-phase flow discharge coefficient is then evaluated, based on the experimental data [7].

The SP model consists of a set of equations that, based on the measurement of the VFM pressure drop, and on the measurement of the void fraction in the WMS, together with the measurement of pressure and temperature of the fluid, allows the estimation of the mass flow rate of the phases in the test section:

$$\begin{cases} x = f_1(\alpha) \\ J_l = f_2(\alpha, \Delta p_{VFM}, p) \\ J_g = f_3(\alpha, \Delta p_{VFM}, p) \end{cases} \tag{4}$$

An iterative approach is adopted for the estimation of the two mass flow rate of the phases. The model input are the void fraction, the pressure drop in the VFM, and the properties of air and water that are evaluated at the working pressure and temperature. An initial guess value for the flow quality is introduced. Using a x-α correlation the void fraction is calculated and the flow quality value is iterated till the error between the void fraction calculated and the void fraction measured by the WMS is lower than a threshold value. When the required accuracy is reached, the two-phase flow Reynolds number and the discharge coefficient are calculated in a new iterative loop, where for the first iteration the discharge coefficient for the liquid is assumed. When the convergence is got ($W_{TP,i}$–$W_{TP,i-1}$<10^{-4}) the mass flow rate are calculated for the two phases. For the x-α correlation to be used in the calculation, the Lockhart-Martinelli correlation has been modified, based on present experimental data.

$$\begin{cases} \alpha = \left[1 + 0.2 \cdot \left(\frac{1-x}{x}\right)^{0.6} \cdot \left(\frac{\rho_g}{\rho_l}\right)^{0.36} \cdot \left(\frac{\mu_l}{\mu_g}\right)^{0.07}\right]^{-1} & x < 1 \cdot 10^{-3} \\ \alpha = \left[1 + 0.28 \cdot \left(\frac{1-x}{x}\right)^{0.6} \cdot \left(\frac{\rho_g}{\rho_l}\right)^{0.36} \cdot \left(\frac{\mu_l}{\mu_g}\right)^{0.07}\right]^{-1} & 10^{-3} \le x \le 6 \cdot 10^{-2} \\ \alpha = \left[1 + 0.45 \cdot \left(\frac{1-x}{x}\right)^{0.6} \cdot \left(\frac{\rho_g}{\rho_l}\right)^{0.36} \cdot \left(\frac{\mu_l}{\mu_g}\right)^{0.07}\right]^{-1} & x > 6 \cdot 10^{-2} \end{cases} \quad (5)$$

The comparison between the original correlation, the modified correlation and the void fraction value obtained from the WMS signal analysis is shown in Fig. 14.

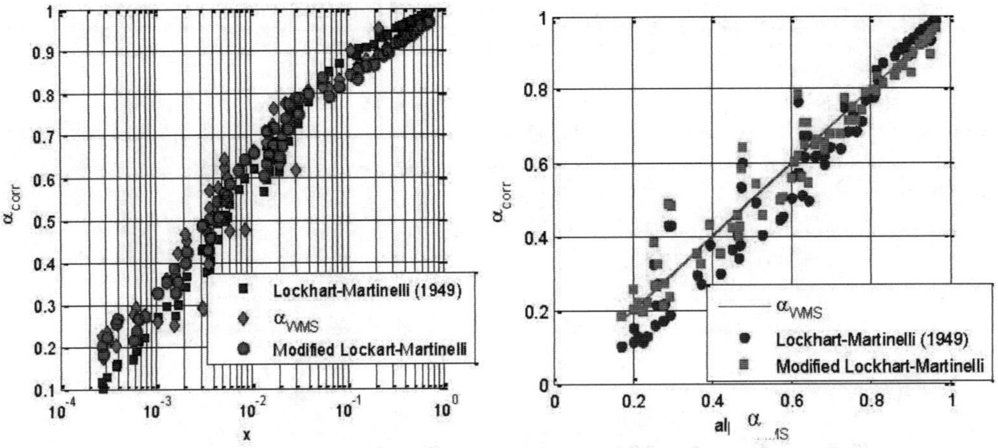

Fig. 14. Comparison between WMS void fraction and correlations.

The developed SP Model has been applied to the experimental data in order to derive the mass flow rate of the phases. The flow quality is estimated with an error lower than 10% for the 72% of the runs. The estimated mass flow rates are shown in Fig. 15. The mass flow rate has been estimated with an error lower than 10% in the 73.3% of the cases and the estimation error is considerably lower for the flow characterized by higher values of quality and void fraction. In this range, with reference to facility SPES3 test conditions, the error in the mass flow rate estimation is always lower than 10% and 20% for water and air respectively.

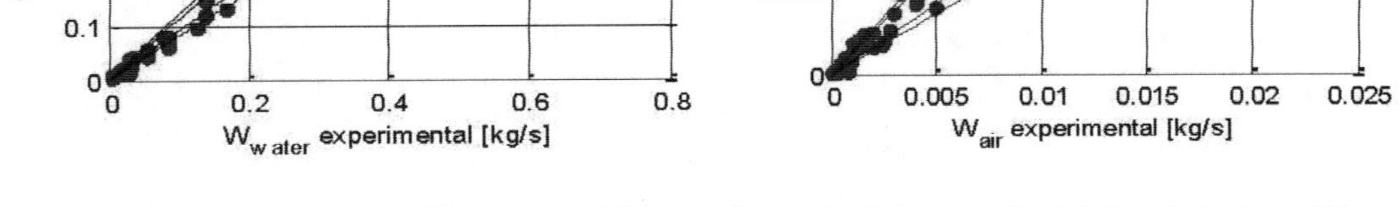

Fig. 15. Experimental mass flow rate vs. SP mass flow rate for water phase (a) and air phase (b).

3.3 Electrical Capacitance Probe (ECP)

The electrical sensor designed at the SIET laboratories [21, 22, 23], for the measurement of the void fraction in the SPES-3 break lines, has been developed in order to be inserted in a SP for the measurement of the mass flow rate of the phases. The range of interest of the flow parameters in the SPES3 facility, in steady state and transient conditions, has been investigated by means of the RELAP-5 code [24]: the simulations show that in a relative long period of the transients, the two-phase flow corresponds to an annular flow pattern, characterized by very high void fraction and high flow velocity, so the selected instruments of the Spool Piece must be sensitive to the flow in such conditions. The sensor (ECP) consists of 10 electrodes: 9 external and one internal (Fig. 16 a). The internal and the external diameters of the Plexiglas pipe, where the probe is mounted, are 80 mm and 90 mm respectively. The external electrodes (steel stripes of 400 mm length and 5 mm width) are spaced with 22.5° angle only in half of the pipe circumference, as the symmetry of the vertical flow has been verified. The angle corresponds to an external chord of 17.56 mm and an internal chord of 15.6 mm. Details can be found in ref. [21]. The main purpose of the sensor is the evaluation of the effect of the angular electrode distance on the output signal.

The experimental test section (Fig. 16 b) consists of the feed water and the feed air loops equipped with instruments to measure the single-phase flow parameters (flow meter, temperature and pressure). Details can be found in ref. [23]. The first ECP qualification step is the characterization of the sensor for single-phase conditions [23].

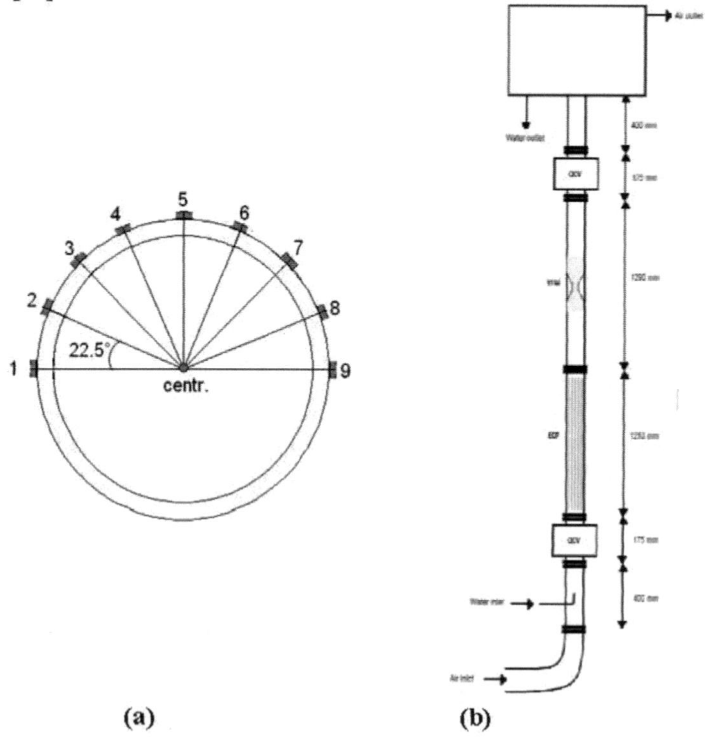

(a) (b)

Fig. 16. Schematic of the ECP (a) and of the test section (b).

In order to evaluate the response of the sensor at high void fractions, corresponding to annular flow, preliminary tests at atmospheric pressure have been carried out with a mixture of air and demineralized water introduced in the test section. The mass flow rate ranges from 0.094 to 0.15 kg/s for air and from 0.002 to 0.021 kg/s for water, corresponding to a void fraction higher than 95%. In order to take into account the single-phase signal variations, the two-phase flow measured values are normalized as follows:

$$V_{ij}^* = \frac{RMS_{TP-ij} - RMS_{l-ij}}{RMS_{g-ij} - RMS_{l-ij}} \quad (6)$$

where the subscript *ij* identifies the measuring electrodes combination. As an example of the preliminary results, the average signals measured between electrodes at 22.5°, 45°, 90° and 180° are presented in Fig. 17 as a function of the volumetric void fraction measured by the QCV technique.
V_{ij}^* , evaluated by means of eq. (6), increases with θ at constant void fraction, and approaches the air value at higher angular distance.

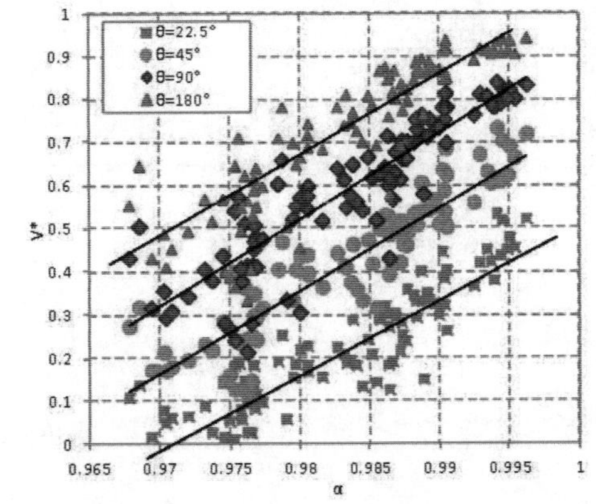

Fig. 17. Mean signal in the external electrodes as a function of the volumetric void fraction α.

Due to the symmetry of the flow, the annular flow regime in a vertical channel can be analyzed as a liquid film region and a core region with entrained droplets. The liquid film is characterized in terms of film thickness, frequency and amplitudes of the waves at the liquid-gas interface, while the core region is characterized in terms of the mean void fraction value.

Adopting the Ishii correlation [24] for the fraction of the liquid flux flowing as droplets entrained in the vapor core, the liquid film thickness is shown in Fig. 18.

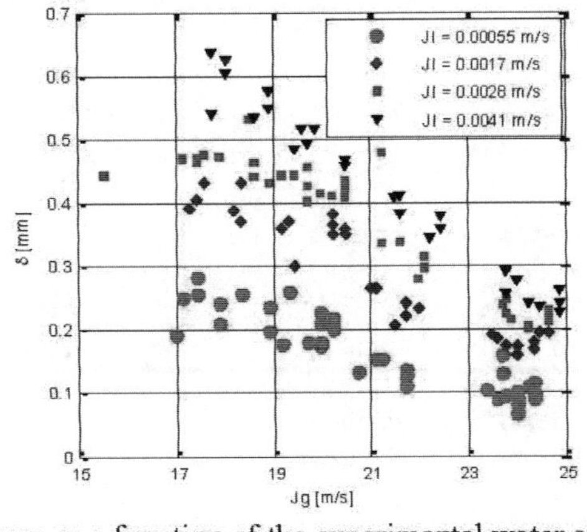

Fig. 18. Mean film thickness as a function of the experimental water and air superficial velocities.

In the second experimental campaign, when downstream of the ECP a VFM has been installed between two straight pipes of 1290 mm upstream and downstream (Fig.18 b), the air superficial velocity range is between 14 and 18 m/s while the water superficial velocity ranges from 0.0008 to 0.005 m/s; the flow pattern is annular and the corresponding void fraction is higher than 0.97, while the flow quality ranges from 0.78 to 0.96. The ECP response has been studied by means of a FEM analysis, able to evaluate the sensitivity of the sensor to the void fraction variation in the measurement volume. The FEM

Journal of Physics: Conference Series 547 (2014) 012001 doi:10.1088/1742-6596/547/1/012001

analysis allows the evaluation of the capacitance associated with the different measuring electrodes, for the different phases distributions, by means of the electrical field resolution in the control volume. Three different phases distribution are simulated: homogeneous, ideal annular flow, annular flow with entrainment (thin and smooth liquid film at the pipe wall and a homogeneous mixture in the core region). In the last case the void fraction value depends on the liquid film thickness and on the dielectric constant of the mixture in the core region. The numerical analysis highlights that the electrodes response strongly depends on the phases distribution.

The signal obtained under two-phase flow conditions is normalized as reported in eq. 6.

As an example in Fig. 19 the comparison between experimental and numerical data is shown for the external electrodes characterized by a measuring distance of 22.5° and 180°, showing the dependence on the void fraction and on the liquid film thickness (C^* is the normalized electrical capacitance).

The comparison shows a qualitative good agreement of the experimental data with the numerical predictions and that the electrode, characterized by a measuring distance of 180°, shows a very low data dispersion and a linear dependence on the mean cross sectional void fraction that highlights a similar behavior under homogeneous and annular flow.

The performed analysis allows us to conclude that the signal related to the electrodes at 180° can be used to directly estimate the average cross sectional void fraction, due to the low sensitivity to the phases distribution, while the measurement at 22.5° can be useful for the liquid film thickness and the core phases distribution.

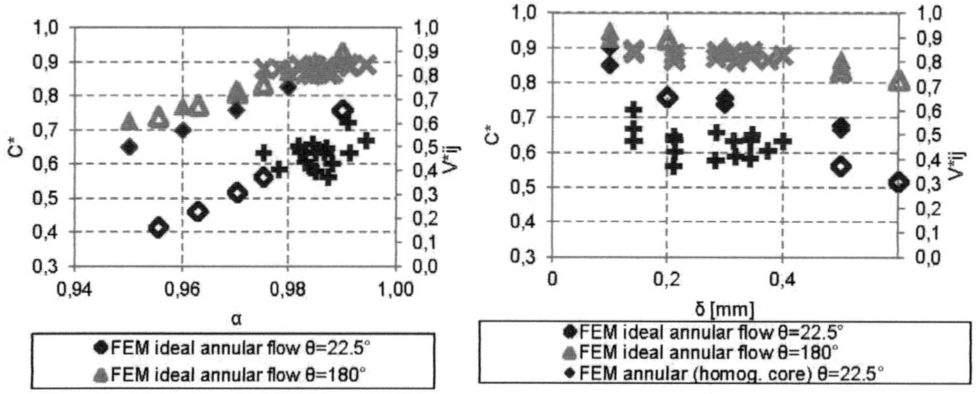

(a) (b)

Fig. 19. Comparison between experimental and numerical ECP external electrodes results at 22.5° and 180°. Void fraction dependence (a) and film thickness dependence (b).

In Fig. 20 the comparison between numerical and experimental results for the central electrode is reported. Also in this case the comparison shows a good agreement between the experimental measurement and the simulations corresponding to the ideal annular flow. Although the sensitivity of the Electrical Capacitance Probe (ECP) in single-phase flow is low, two-phase flow void fraction variations lower than 1% has been detected.

The presence of the central electrode allows the evaluation of the average cross-section void fraction also in annular flow, where the liquid film is the principal cause of the sensor low sensitivity. The signal measured in the central electrode is linear with the void fraction for values higher than 98% and it is quite sensitive to the flow pattern (flow patterns, from rather regular to more disturbed annular flow, have been clearly characterized).

The variation of the signal measured in the external electrodes has been related to the average liquid film thickness that is evaluated by means of the Ishii's model [24]. The sensitivity of the sensor must be increased, in order to make the instrument suitable to be used for void fraction measurement with a good accuracy.

The FEM numerical model, compared with the experimental data, shows a good agreement and the possibility to use it to identify the possible sensor modifications finalized to increase the sensitivity under annular flow conditions.

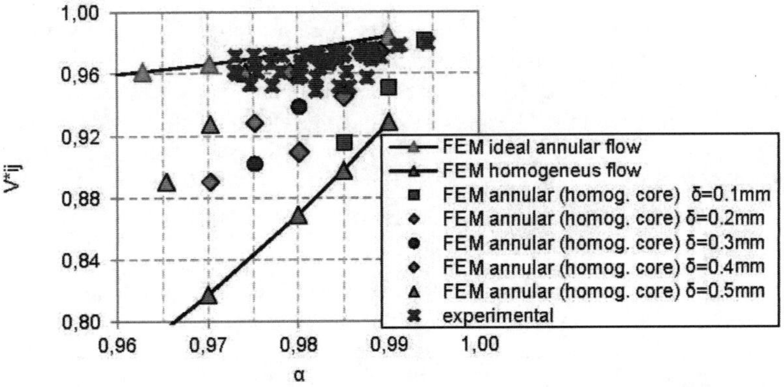

Fig. 20. Comparison between experimental and numerical ECP central electrode results as a function of the void fraction.

3.4 Venturi flow meter in vertical annular flow

Two Spool Pieces have been previously presented for horizontal flow, consisting of a Turbine Flow Meter and a Drag Disk and of a Venturi Flow Meter and a Wire Mesh Sensor [26]: if the void fraction, and then the flow pattern, can be correctly identified, the two instruments can be used to estimate the mass flow rate of the two phases with a good accuracy. A similar approach has been applied in a vertical test section, where a new Venturi Flow Meter has been coupled with the Electrical Capacitance Probe (Fig. 18 a). The VFM (80 mm maximum diameter, 340 mm total length) has been home designed and has been constructed in Plexiglas in order to allow the visualization of the flow (Fig. 21).

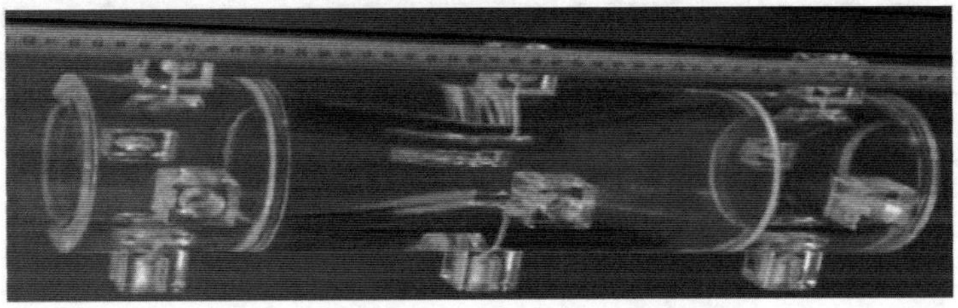

Fig. 21. Venturi Flow Meter picture.

The air superficial velocity range is between 14 and 18 m/s while the water superficial velocity range is between 0.0008 and 0.005 m/s, to have an annular flow pattern, and the corresponding void fraction is higher than 0.97, while the flow quality ranges from 0.78 to 0.96. The Venturi has been designed to operate in both fluid direction (symmetrical flow meter), with the angle of the convergent section equal to the divergent section angle (θ=21°), in order to be able to work in reverse flow conditions. Compared with the Herschel model the present VFM is characterized by a higher divergent angle (usually set to 15°), with higher irreversible pressure drops, due to the flow separation. In single-phase flow the discharge coefficient value takes into account the phenomena of the wall flow detachment at the inlet of the divergent section.

According to the well known approach, the frictional pressure gradient of a two-phase mixture flowing in a pipe can be correlated to the single-phase pressure drop by means of the two-phase flow multiplier [27], that is expressed by empirical or semi-empirical correlations (Chisholm [28], Murdock [29]). The mixture void fraction is measured by means of the QCV technique. The acquisition time is equal to 30 s with an acquisition frequency of 1250 Hz. The pressure drop values across the Venturi, reported in ref. [7], show the dependence on the superficial velocities of the two phases and the small effect due to the liquid flow rate. The flow field is characterized and complicated by the continuous deposition and entrainment of liquid droplets along the Venturi length and by the presence of waves on the liquid film surface. On the ground of the previous considerations, the pressure drop in the VFM can be expressed as a function of the single-phase pressure drop with a correction factor (a two-phase flow multiplier) that can take into account the presence and the effect of the liquid phase at different flow qualities. The two-phase flow multiplier can been written as (χ is the famous Martinelli parameter, that is function of the flow quality):

$$\phi^2_{g,mod} = 7.8 \cdot \chi + 1 \tag{7}$$

where the two constants have been obtained as best-fit of the present experimental data.
The two-phase flow multiplier values, evaluated by the correlation (7), are compared with the correlations of Murdock, Chisholm and the present test data (Fig. 22): the classical correlations under predict the two-phase multiplier ϕ^2_g.

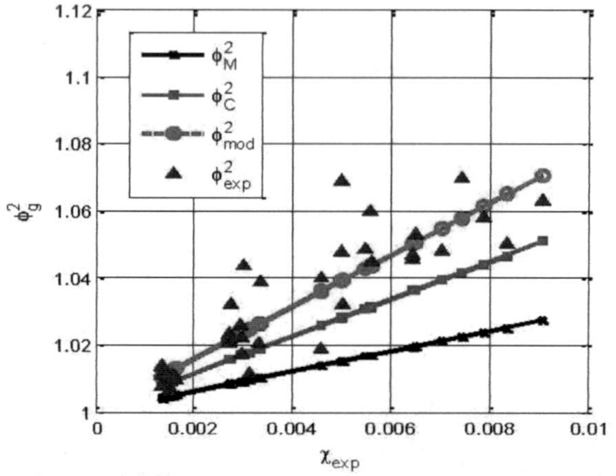

Fig. 22. Two-phase flow multiplier: comparison between experimental data and correlations.

In Fig. 23 the pressure drops evaluated with the new correlation are compared with the experimental values, showing a good agreement and a calculation accuracy better than 5%.

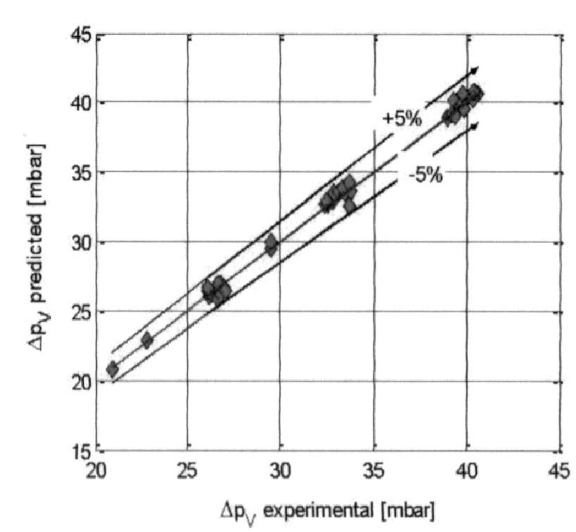

Fig. 23. VFM pressures drops: comparison between experimental data and predicted values.

The irreversible pressure losses, measured between the Venturi inlet and outlet sections have been correlated with the flow parameters. The analysis of this component allows us to understand better the effect of the dispersed phase in the two- phase pressure drops. Fig. 24 shows the dependence of the irreversible pressure loss on the superficial velocities of the phases. While, due to the liquid phase presence, the inlet-throat pressure drop increases of about 10%, if compared to the single-phase flow, the irreversible pressure loss increases from about 20% to 100% depending on the liquid flow rate.

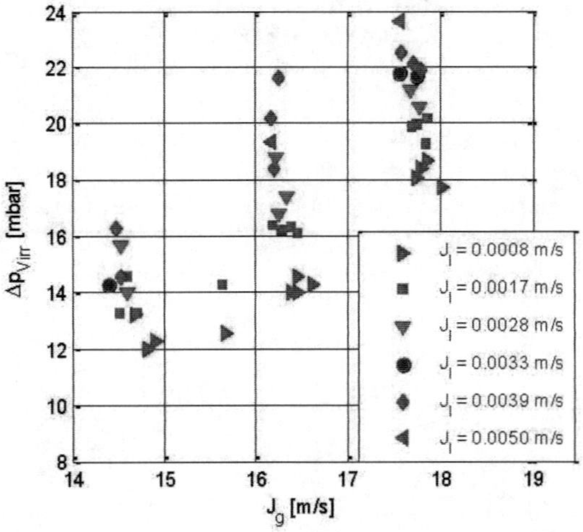

Fig. 24. VFM irreversible pressure loss as a function of the superficial velocities of air and water.

The previous analysis allows the derivation of a correlation able to describe the irreversible pressure loss change as a function of the flow rate of the two phases. The proposed correlation, developed for the present tested conditions, expresses the irreversible pressure loss as a function of the gas superficial velocity and of the ratio between the liquid and the gas superficial velocities, highlighting the effect of the dispersed phase:

$$\Delta p_{irr} = k_1 \cdot \left(\rho_g J_g^{k_2}\right) \cdot \left(J_l / J_g\right)^{k_3} + k_4 \tag{8}$$

The following constants have been obtained from the best fit of the present experimental data: k_1 = 0.2096, k_2 = 2, k_3 = 0.13, k_4= -2.9786.

The comparison between the predicted irreversible pressure loss and the test data is shown in Fig.25: the proposed correlation allows the estimation of the VFM irreversible pressure loss, in annular flow, at high void fraction, with an accuracy of 5%.

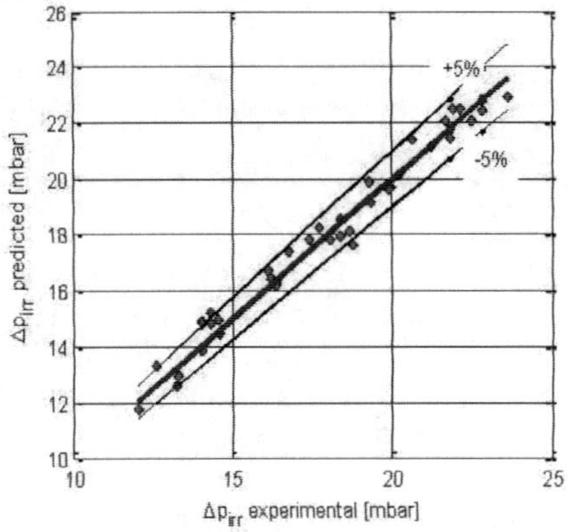

Fig. 25. Comparison between predicted (eq.8) and experimental VFM irreversible pressure losses.

In single-phase flow the VFM allows the estimation of the mass flow rate from the pressure drop across the pipe restriction, but in presence of two phases the direct correlation between pressure drops and mass flow rate is not possible, so additional information and a model able to interpret the signals are required [26,30]. In limited application ranges, if the flow pattern does not change, as in the present experimental conditions, it is possible to derive the essential information from the signals of a VFM. The closure equation of the model in this case is based on the measurement of the irreversible pressure losses and on the proposed correlation (8).

By means of an iterative approach the flow quality and the mass flow rate of the two phases have been evaluated: in the first iteration the air flow rate is guessed by considering a single-phase flow in the VFM, while starting from the second iteration this value is corrected using the estimated value of the flow quality, by using the correlation (7); an initial guess value of this parameter is used to evaluate the VFM pressure drops, and the iterative loop goes on until the estimated pressure drop reaches the experimental value. Then the mass flow rates of the two phases are estimated by using the flow quality value and the VFM pressure losses correlation (8). The reference signals have been obtained as the mean values of the 30 s acquisition time. In Fig. 26 the estimated mass flow rates of air and water are compared with the values measured by the rotameters: the air mass flow rate is estimated with an accuracy of 2%, while a lower accuracy characterizes the liquid flow rate prediction (the relative error is lower than 30% for all conditions). All the details are given in ref. [7].

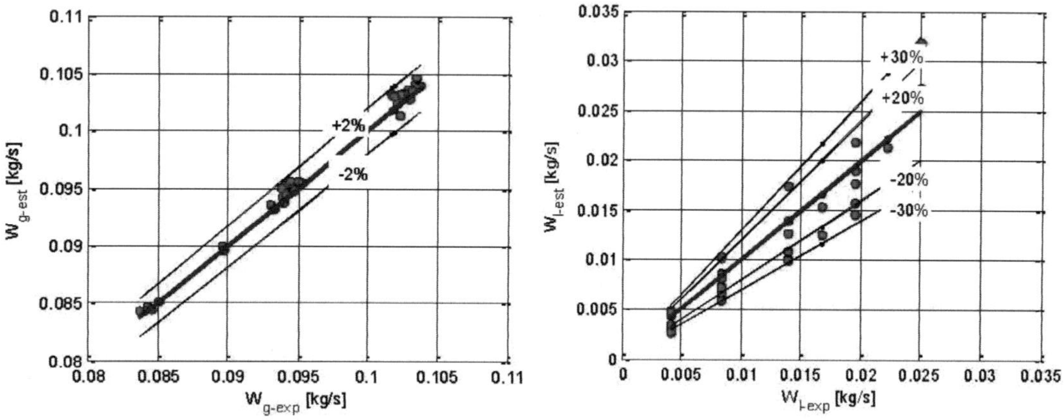

Fig. 26. Comparison between VFM experimental and estimated mass flow rates.

4. Conclusions

The performance of different measurements instruments has been evaluated in two-phase flow applications (like the simulation of nuclear plants accidents) and interpretative models for each one, as well as, experimental correlations for the quality - void fraction and for the pressure drops, have been developed and tested.

The developed WMS signal analysis allows to evaluate the relation between the evolution of the void fraction profiles and the superficial velocity of the two-phases (Jg and Jl). The flow evolution in time and space has been analyzed and discussed, showing that such methodology is useful to identify and characterize, in terms of void fraction characteristic profiles, characteristic frequencies and interface evolution, the two-phase flow patterns.

By coupling different instruments in Spool Pieces (suitable to be installed in industrial facilities) and by the development of models for the instruments, the mass flow rate of the two phases for different flow conditions has be estimated. The proposed methods allow the possibility to measure the mass flow rate of the phases by using three, two or one instruments with different degree of accuracy.

Advantage and drawbacks of the different instrument combinations have been highlighted, and the accuracy achievable for each SP configuration has been estimated. Moreover the experimental analysis performed with the different instruments, at different flow conditions and geometrical configurations, has allowed the investigation and the characterization of the air-water flow in a wide range of flow patterns, and new correlations able to estimate two-phase flow parameters, such as void fraction and two-phase flow pressure drops, have been found.

Future work will be addressed to investigate the performance of the different SP in transient conditions with fixed periodic oscillations. Tests with a SP consisting of a Venturi flow meter and a an Electrical Capacitance Probe in horizontal flow across 80 mm diameter pipes are being carried out too.

Ackowledgements
The present research activity has been supported by ENEA and by the Ministry of Economic Development. A special thanks to R. Costantino and G. Vannelli for their support in the laboratory.

Nomenclature

C VFM discharge coefficient
C^* normalized electrical capacitance
F signal acquisition frequency
J superficial velocity
Re Reynolds number
V* WMS normalized signal
W mass flow rate
x flow quality
α void fraction
Δp differential pressure
ρ density
Φ two- phase multiplier
χ Martinelli parameter

Subscripts

g gas
i WMS i-th vertical index
irr irreversible
j WMS j-th horizontal index
k WMS time index
l liquid
TP Two-Phase

References

[1] Carelli M D, Conway L E, Oriani L, Petrović B, Lombardi C V, Ricotti M E, Barroso A C O, Collado J M, Cinotti L, Todreas N E, Grgić D, Moraes M M, Boroughs R D, Ninokata H, Ingersoll D T, Oriolo F, The Design and Safety Features of the IRIS Reactor, *Nucl. Eng. Design*, 230, pp. 151-167 (2004).

[2] Bertani C, De Salve M, Malandrone M, Monni G, Panella B, State-of-Art and selection of techniques in multiphase flow measurement, Report RdS/2010/67 ENEA (2010).

[3] De Salve M, Monni G, Panella B, Turbine Flow Meter and Drag-Disk in Horizontal Air Water Flow, UIT2011, Torino (Italia), 20-22 Giugno 2011. pp. 107-112, (2011)

[4] De Salve M, Monni G, Panella B, A Model for a Spool Piece made up of Venturi and Void fraction Flow Meter in Horizontal Flow, *ANS TRANS.*, vol. 108 , pp. 1013-1016, ISSN 0003-018X (2013).

[5] De Salve M, Monni G, Panella B, Horizontal Air-Water Flow Analysis with Wire Mesh Sensor. JoP, vol. 395, 012179, ISSN 1742-6588 (2012).

[6] Monni G, De Salve M, Panella B, Randaccio C, Electrical Capacitance Probe Characterization in Vertical Annular Two-Phase Flow. In: *SCI TECHNOL NUCL INSTAL*, vol. 2013, Article ID 568287, pp. 1-12, ISSN 1687-6075 (2013).

[7] Monni G, PhD Thesis, Politecnico di Torino (2014).

[8] Prasser H M, Krepper E, Lucas D, Evolution of the two-phase flow in a vertical tube-decomposition of gas fraction profiles according to bubble size classes using wire-mesh sensors, *Int. J. Therm. Sci.*, 41 17–28 (2002).

[9] Prasser H M, Böttger A, Zschau J, A new electrode-mesh tomograph for gas–liquid flows, *Flow Meas. Instr.*, 1998, 9 111–9 (1998).

[10] Pietruske H, Prasser H M, Wire-mesh sensors for high-resolving two-phase flow studies at high pressures and temperatures, *Flow Meas. Instr.*, 2007 18 87-94 (2007).

[11] Teletronic Rossendorf GmbH, Wire Mesh Sensor System, WMS200 Manual, Version 1.2, December 9, 2010

[12] De Salve M, Monni G, Panella B, Horizontal Two-Phase Flow Pattern Recognition, 8th World Conf. on Exp. Heat Transfer, Fluid Mech. and Thermodyn., Lisbon, Portugal, June 16-20 (2013), *Exp. Therm. Fluid Sc., In Press, Corrected Proof, Available online 26 April 2014.*

[13] Averill and Goodrich, Design and performance of a drag Disc and Turbine transducer. For LOFT Experimental Program (1979).

[14] Aya I, A model to calculate mass flow rate and other quantities of two-phase flow in a pipe with densitometer , a drag disk and a turbine meter, ORNL-TM-4759 (1975).

[15] De Salve M, Monni G, Panella B. Analisi delle prestazioni di uno *Spool Piece* in deflusso orizzontale bifase aria-acqua . SPOOL PIECE costituto da TURBINA e DRAG DISK, (in Italian), Report ENEA, Luglio (2011).

[16] Baker C R, Flow Measurement Handbook, Cambridge University Press (2000).

[17] Lockhart R W and Martinelli R C, Proposed correlation of data for isothermal two- phase two-component flow in pipe, *Chem. Eng. Prog.* 45 39-48 (1949).

[18] Baroczy C J, A systematic correlation for two-phase pressure drop, AIChE reprint 37 presented at 8th National Heat Transfer Conference, Los Angeles (1965).

[19] Chisholm D and Rooney D H 1974 Research note: Two-phase flow through sharp-edged orifices, *J. Mech. Eng. Sci.* 16 353-55 (1962).

[20] Wallis G B, One-Dimensional Two-Phase Flow, McGraw Hill, 2nd Edition, New York, 1979.

[21] Achilli A, Greco M, Progettazione di una sonda capacitiva per misurazione del grado di vuoto medio di sezione di miscele bifase acqua-vapore, (in Italian), Report RdS/2010/x ENEA (2010).

[22] De Salve M, Monni G, Panella B, Caratterizzazione di una Sonda capacitiva in deflusso verticale ascendente aria-acqua, (in Italian), Report ENEA, RdS/2011/121 (2011).

[23] Monni G, De Salve M, Panella B, Randaccio C, Electrical Capacitance Probe Characterization in Vertical Annular Two-Phase Flow, *Sc. Techn. Nucl. Installations*, Vol. 2013, Article ID 568287, 12 pages, ISSN: 1687-6075, doi:10.1155/4262 (2013).

[24] Ishii M, Mishima K, Two-fluid model and Hydrodynamic constitutive relations, *Nucl. Eng. Des.,* Vol. 82, pp. 107-126 (1984).

[25] Ferri R, Congiu C, SPES3-IRIS facility RELAP5 base case transient analyses for design support, SIET document 01 489 RT 09 Rev.0 (2009).

[26] De Salve M, Monni G, Panella B, A Model for a Spool Piece made up of Venturi and Void fraction Flow Meter in Horizontal Flow, *Trans. ANS*, Vol. 108 1013-1016, Atlanta, Georgia, June 16–20 (2013).

[27] Lockhart R W and Martinelli R C, Proposed correlation of data for isothermal two- phase two-component flow in pipe *Chem. Eng. Prog.* 45 39-48 (1949).

[28] Chisholm D, Two-phase flow through sharp edged orifices. *J. Mech. Eng. Sci.*, 19(3), 128–30 (1977).

[29] Murdock J W, Two-phase Flow Measurement with Orifices, *ASME J. Basic Eng.*, December, 419-433 (1962).

[30] Zhang H J, Yue W T, Huang Z H I, Investigation of oil–air two-phase mass flow rate measurement using venturi and void fraction sensor, J. Zhejiang University Science 6A (6) (2005).

Passive techniques for the enhancement of convective heat transfer in single phase duct flow

S. Rainieri[1,2], **F. Bozzoli**[1], **L. Cattani**[1]
Department of Industrial Engineering, University of Parma, Parco Area delle Scienze 181/A I-43124 Parma, Italy.

SITEIA.PARMA Interdepartmental Centre, University of Parma, Parco Area delle Scienze 181/A, I-43124 Parma, Italy.

E-mail: sara.rainieri@unipr.it

Abstract. This review presents the main results of the experimental campaign on passive techniques for the enhancement of forced convective single phase heat transfer in ducts, performed in the last years at the Laboratory of the Industrial Engineering Department of the University of Parma by the Applied Physics research group. The research was mainly focused on two passive techniques, widely adopted for the thermal processing of medium and high viscosity fluids, based on wall corrugation and on wall curvature. The innovative compound heat transfer enhancement technique that couples together the effect of wall curvature and of wall corrugation has been investigated as well. The research has been mainly focused on understanding the causal relationship between the heat transfer surface modification and the convection enhancement phenomenon, by accounting the effect of the fluid Prandtl number. The pressure loss penalties were also evaluated. The principal results are presented and discussed.

1. Introduction

The enhancement of convective heat transfer is an everlasting issue within applied research, since it can ultimately provide an effective way to optimize industrial thermal processes by reducing, at the same time, the size of the equipment and then the energy and materials costs [1].

This issue becomes particularly important in industrial applications in which the thermal processing of medium and high viscosity fluids is required, such as in the food, chemical, pharmaceutical and cosmetics industries.

In fact, often, in these conditions, the momentum transfer mechanism is necessarily laminar and therefore the efficiency of the heat transfer apparatuses in which the fluids are conveyed is inevitably penalized.

In these industrial fields, engineers have then been striving for techniques producing enhanced heat transfer coefficients, accompanied by reduced pumping power requirements. This is witnessed also by the huge amount of papers available within the world scientific literature of the last decades regarding heat transfer augmentation and by the growing numbers of registered patents related to heat transfer enhancement technology or devices (in 2001 Bergles estimated a steady-state rate of literature generation of about 400 papers and reports per year [2]).

According to the classification suggested by Bergles (see for instance [3]), the techniques for the enhancement of forced convection encompass the optimal use of active devices, propelled by mechanical aids or electrostatic fields (see for instance [4]), and passive devices, mostly based on some suitable conformation of the interface surface (see for instance [2,5]), or else a combination of both active and passive techniques [1].

Passive techniques are more attractive because no power is required to facilitate the enhancement.

Among them, treated surfaces, rough surfaces, displaced enhancement devices, swirl-flow devices, surface-tension devices, coiled tubes, or flow additives [1–2] are found. Moreover the simultaneous use of two or more of these techniques has been investigated in literature in order to verify if it is able to produce an enhancement larger than that produced by only one technique. This simultaneous use is termed compound enhancement [1]. Regarding compound techniques, it must be noted that they offer a promising tool for further enhancing the performances of thermal apparatuses, since the heat transfer coefficients are expected to be increased above any of the several techniques acting alone [6].

Another possible classification of heat transfer enhancement techniques into different generations and based on the stage of development of the heat transfer technology was further suggested by Bergles [2]: according to this approach, the first generation heat transfer technology is represented by the plain surfaces, while the fourth is represented by compound techniques.

The effectiveness of all of the above referred passive techniques is often well assessed in literature and performance evaluation criteria under both first and second principle approach have also been developed in order to compare the thermal performance benefits achievable through their use with the inevitably correlated pressure drop penalties (see for instance [7,8]).

By focusing on passive heat transfer enhancement techniques, it is important to remind that the main mechanisms on which they rely are: periodic disruption of the development of the boundary layers, generation of swirling/rotating and/or secondary flows, promotion of boundary-layer separation and attachment/reattachment phenomena, promotion of an early departure from the laminar flow and increase of the turbulent intensity.

In the experimental investigations described in the following paragraphs, two passive techniques widely adopted for the thermal processing of medium and high viscosity fluids and based on wall corrugation and/or wall curvature will in particular be analyzed, by reviewing the main results of the experimental campaign performed in the last years by the Applied Physics research group at the Laboratory of the Industrial Engineering Department of the University of Parma.

2. The enhanced tubes tested, experimental conditions and data processing

The investigation was focused on straight corrugated wall tubes and on helically coiled tubes having both smooth and corrugated wall.

2.1 Wall corrugated tubes

The introduction of roughened wall surfaces has revealed a successful method for enhancing heat transfer in tubes and, therefore, for reducing the heat exchangers' size for a given heat duty requirement. Among this kind of techniques, the most successful was found in single and multistart spirally wall corrugated tubes [5,9]. They are characterized by an internal helical ridging corresponding to and external helical grooving, obtained by embossing a smooth tube, generally made of stainless steel.

The enhancement effect associated to the wall corrugation is generally ascribed to the periodic interruption of the boundary layers development, to the increase in heat transfer area, to the generation of swirling and/or secondary flows and, above all, to the promotion of the transition to the turbulent flow regime [1,5].

The enhanced sections tested in the experimental investigations, were generated by stainless steel type AISI 304 tubes having a wall thickness of 1 mm. Both spirally (with single and cross helical profile) and transversely wall corrugated tubes were considered. Some representative enhanced tubes are shown in figure 1.

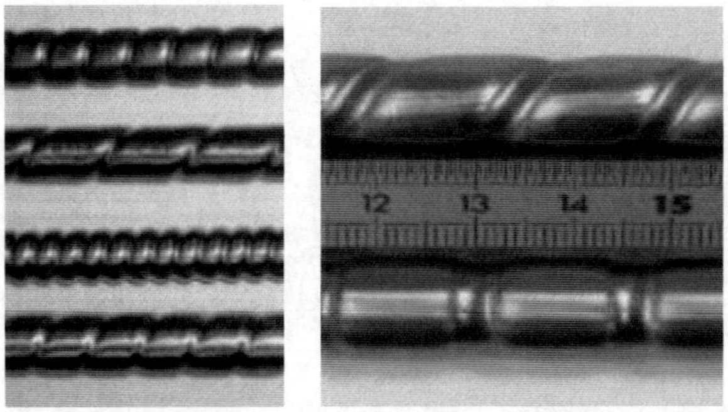

Figure 1. Representative spirally and transversely corrugated pipes.

2.2 Helically coiled tubes

In the experimental campaign hereby reviewed, both smooth and corrugated wall helically coiled tubes made of stainless steel type AISI 304, having a wall thickness of 1 mm were tested. They were characterized by coils following a helical profile along the tube's axis with different coil radius. The corresponding dimensionless curvature δ (the ratio of the tube diameter to the helix diameter) ranged from 0.031 to 0.056.

Two helically coiled pipes (one having smooth wall and the other having corrugated wall), representative of the ones tested in the experimental campaign, are shown in figure 2.

It is well known in literature that the enhancement of heat and mass transfer in coiled pipes with respect to the straight tube behaviour, is due to the distortion of the velocity and temperature profiles induced by the centrifugal force that acts with different intensity over the tube's cross section. The centrifugal force produces the onset of secondary flow structures, appearing as twin counter rotating vortices in the cross-sectional plane [10], that are responsible for a further enhancement of the forced convection mechanism. By neglecting all other effects related to geometry, except the centrifugal acceleration term, Dean derived a simplified momentum balance equation (known as ''Dean approximation'') in which the only dimensionless independent parameter is the Dean number, that couples together inertial with centrifugal effects [11], defined as $Dn = \delta^{1/2}Re$, where δ is the dimensionless curvature and *Re* is the Reynolds number.

Figure 2. Representative smooth and corrugated helically coiled pipes.

2.3 Experimental conditions and data processing

The enhanced sections were tested in heating condition obtained by Joulean dissipation directly within the tube wall. The experimental setup allowed investigating the heat transfer performance of the tubes under the thermal entrance problem.

As working fluids, both glycerol and ethylene glycol, that both show a Newtonian flow behaviour, and mixtures of these two fluids and water were used. Some tests were also performed by adopting real fluid foods, such as orange concentrate, that behaves has as non-Newtonian pseudoplastic fluid [9].

The directly measured quantities were: the electric power dissipated within the tube's wall, the fluid mass flow rate, the bulk fluid temperature at the inlet and outlet sections, the outer wall temperature (both by traditional contact sensors and the infrared thermographic technique) and the pressure drop along the test section. Further details about the experimental apparatus and the data acquisition system can be found in [5,12-16].

The derived quantities were: the average wall heat flux (obtained by the value of the heat power dissipated by Joule effect into the wall) the internal wall temperature (obtained by solving the steady state heat conduction problem with heat generation in the tube wall), the bulk temperature at any location in the heat transfer section (obtained by the energy balance of the system).

The heat transfer performance of the enhanced sections was then quantified, for different corrugation profiles, by evaluating the Nusselt number (both the local value along the axial coordinate and the average value over the heated section) and the friction factor (average value along the test section) by varying Reynolds, Dean and Prandtl numbers. The experimental condition enabled to test the enhanced sections in the Reynolds, Dean and Prandtl number ranges 4÷2000, 10-300 and 30-15000 respectively.

It must be noted that only if the axial and peripheral heat conduction effects within the wall can be considered negligible, the Joulean heating condition well approximates the thermal boundary condition of uniform wall heat flux at the fluid wall interface. If these assumptions cannot be made, a conjugate heat transfer approach is instead needed [17]. Regarding the experimental conditions hereby reviewed, being wall thermal conductivity relatively low and being the wall thin, the axial heat conduction effects were expected to be negligible. Regarding the peripheral heat conduction effects within the wall instead, they were expected to be negligible in straight tubes, while they couldn't be overlooked in coiled pipes, due to the secondary flow pattern.

In particular, in helically coiled tubes, being the fluid temperature profile strongly asymmetrical with respect to the tube's axis, the peripheral heat conduction in the wall caused a heat flux redistribution and the convective heat transfer coefficient was then expected to vary significantly along the periphery of the tube's cross section. In particular it was expected that the convective heat transfer coefficient presented higher values at the outer bend side of the wall surface than at the inner bend side due to the secondary flow pattern. It must be reminded that this irregular distribution may be critical in some industrial applications, such as in those that involve a thermal process. For instance, in food pasteurisation, the irregular temperature field induced by the wall curvature could reduce the bacteria heat-killing or could locally overheat the product.

From the value of the heat power dissipated by Joule effect into the wall, the circumferentially averaged values of the wall heat flux was derived, while for estimating its distribution along the wall periphery a different approach was needed.

At some specific cross section in the downstream region of the heated section, the local convective heat transfer coefficient and the local Nusselt number as a function of the angular coordinate along the wall periphery were then derived by processing, under the Inverse Heat Conduction Problem (IHCP) approach, the external wall temperature distribution acquired by means of an infrared camera [14,15]. This procedure was based on the inverse solution of the energy equation within the solid domain in which the external wall temperature distribution was experimentally measured and in which the heat flux distribution at the internal wall-fluid interface was assumed to be unknown.

To cope with the difficulty embedded in IHCPs, many techniques have been suggested in literature, and the most well-known are: function specification methods [18,19], iterative methods [20,21], methods based on filtering proprieties [22,23] and regularisation techniques [24]. In the processing of the data acquired in the experimental campaign hereby reviewed, the Tickhonov regularization methods and the filtering methods were adopted.

Among the regularisation techniques, Tikhonov regularization method is perhaps the most common: it promotes the construction of stable approximate solutions to the original IHCP by solving a well-posed problem via the minimisation of an objective function [24]. The objective function is generally expressed by the sum of the squared difference between the measured and the estimated temperature discrete data and of a regularisation parameter times a term that expresses the smoothness of the unknown quantity. The success of this approach relies on a proper choice of the regularization parameter, and this didn't reveal to be an easy task. In the processing of the data acquired in the experimental campaign hereby reviewed, the Tickhonov regularization methods was implemented with the aid of the most common approaches for choosing the regularization parameter, i.e. the classical L-curve method and the fixed-point iteration technique [14,15].

Another approach to solve the above described IHCP and that revealed very promising is the filtering technique [25]. This approach avoids the formulation of complex algorithms, since the wanted information (heat transfer coefficient or heat sources distribution) can be derived by directly solving the heat conduction equation that uses as input data a de-noised temperature field obtained by filtering the raw infrared map.

In order to estimate the local convective heat transfer coefficient along the wall periphery of the coiled section, the filtering technique approach, based on the Gaussian filter, was successfully applied to the infrared temperature maps acquired on the external wall surface [25].

2.4 Enhancement efficiency

The performances of the enhanced geometries were also suitably described by the following dimensionless quantities: the average friction factor enhancement, the heat transfer enhancement and the overall enhancement efficiency defined as follows:

$$\varepsilon_f = \frac{f}{f_0} \tag{1}$$

$$\varepsilon_H = \frac{Nu}{Nu_0} \tag{2}$$

$$\eta = \frac{\varepsilon_H}{\varepsilon_f} \tag{3}$$

where Nu and f are the Nusselt number and Darcy friction factor averaged along the test section respectively and where the subscript 0 refers to the plain geometry, i.e. the straight smooth wall tube behaviour.

In the data reduction the average bulk temperature between the inlet and outlet sections was used for evaluating all fluid properties, unless otherwise specified. Moreover, the heat transfer area was taken equal to the envelope cylinder surface area.

For the highest viscosity fluid (glycerol) in order to take into account the effect of the viscosity variation with temperature, which was expected to activate modifications within the boundary layers which alter the heat transfer mechanism in comparison to the constant property problem, the approach suggested by Sieder and Tate was adopted [26].

3. Results

3.1 Corrugated Wall Tubes

In the investigation both Straight tubes having Helical Corrugation (SHC) and Straight tubes having Transverse Corrugation (STC) were tested. The results reported in [5,27,28] clearly showed that the effect of the wall corrugation is negligible for Reynolds number values lower than a critical value (of the order of 500). In these conditions, the helical corrugation induced significant swirl components to which, however, an as much significant heat transfer enhancement was not associated [5].

Therefore for the spirally corrugated tubes as the ones exemplified in figure 1 the main augmentation mechanism was mainly ascribed to the turbulence promotion. Moreover the data pointed out a stronger ability of the transverse corrugation and of the cross helical corrugation with respect to the single helical corrugation in promoting the onset of instabilities within the flow [5,27]. A significant effect on the onset of instabilities within the flow was also proved to be related to the effect of the variation of the fluid thermophysical properties, mainly viscosity, with temperature. Moreover for the helical corrugation, the critical local Reynolds number was proved to be proportional to the dimensionless corrugation pitch. Some representative results are reported in figure 3 and in figure 4. The Nusselt number values averaged over the heated length, reported in figure 3 for SHC tubes with corrugation depth equal to 1 mm, confirmed that by increasing the Reynolds number the data depart from the smooth wall tube solution and approach a behaviour that was reported for the turbulent heat transfer in spirally fluted tubes [27]. The transition occurred at different Reynolds numbers according to the corrugation profile and, in any case, well below the critical Reynolds number value expected for smooth wall tubes. The experimental results for the Darcy friction factor, shown in figure 4 were instead by far below the turbulent flow data predicted for the spirally fluted tubes, even in the Reynolds number range where the flow should approach this flow regime, as suggested by the heat transfer results.

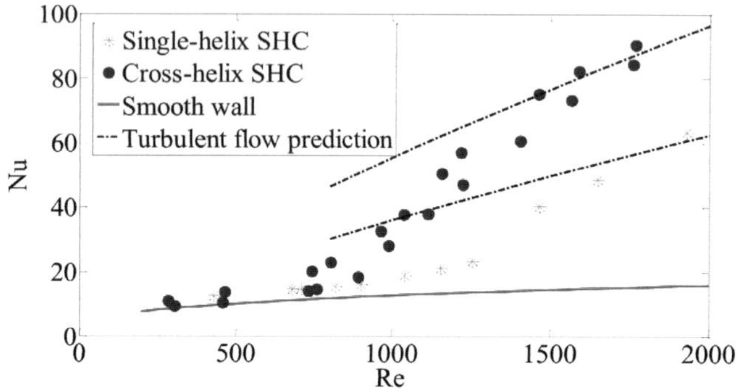

Figure 3. Averaged Nusselt number versus Reynolds number for the corrugated wall tubes and comparison with the smooth wall solution and with the turbulent flow data predicted for spirally fluted tubes [27] ($Pr \cong 30$).

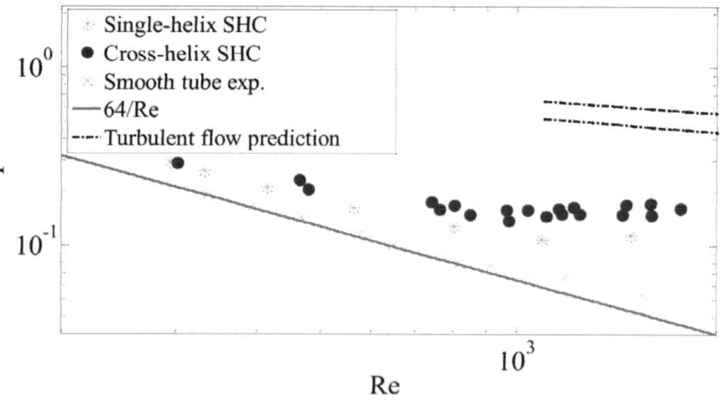

Figure 4. Averaged friction factor versus Reynolds number for the corrugated wall tubes and comparison with the smooth wall solution and with the turbulent flow data predicted for spirally fluted tubes [27].

3.2 Helically coiled Smooth Wall (HSW) and Helically coiled Corrugated Wall tubes (HCW)

Due to the secondary fluid pattern, the heat transfer enhancement registered for the HSW sections over the straight smooth wall behaviour, was significant, already for very low Reynolds number values [12]. In these conditions the HCW sections were instead less effective and the beneficial effect due to the pipe's curvature was even destroyed by the wall corrugation [12]. By increasing the Reynolds number to about 100 the effect of the wall curvature prevailed, while a further increase of the Reynolds number to values up to 1000 showed that in this flow regime the enhancement effect due to the wall corrugation superimposes to the one due to the curvature, by further augmenting the Nusselt number [5,27]. This behaviour was interpreted as a consequence of a departure from the steady laminar flow induced by the wall corrugation, as it was observed in wall corrugated straight tubes.

Some representative results are reported in figures 5 and 6. The data confirm that both wall curvature and wall corrugation are responsible for an additional convective transport mechanism that increases both the heat transfer and the pressure drop when compared to that in a straight tube. The same conclusion is pointed out by the synoptic plot of figure 7 in which the enhancement efficiency is reported vs. the Reynolds number for both the smooth and wall corrugated helically coiled tubes for a representative curvature ratio value. The data confirmed that, above the critical condition, the wall corrugation, as superimposed to the wall curvature, brings an additional heat transfer enhancement that is combined with a modest expense of the additional pressure-losses. In figure 7 the data regarding the SHC tubes behaviour obtained for a different Prandtl number fluid are reported, too. The results show that the simultaneous use of the two passive heat transfer enhancement techniques (wall curvature compounded with wall corrugation) is particularly effective if compared to the performance achievable through their separate use.

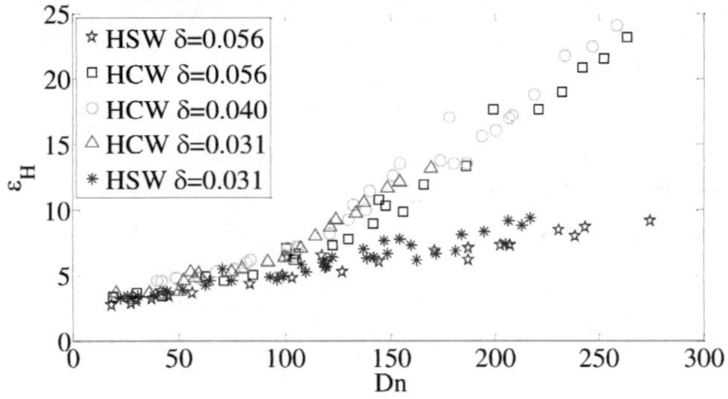

Figure 5. Heat transfer enhancement vs. Dean number for the helically coiled tubes ($Pr \cong 200$).

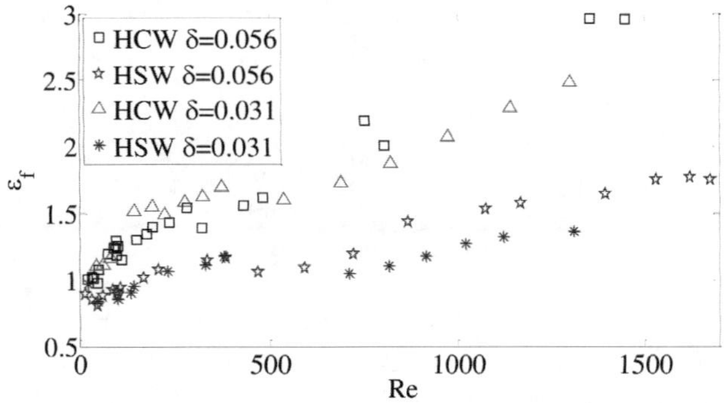

Figure 6. Friction factor enhancement vs. Reynolds number for the helically coiled tubes.

3.3 Convective heat transfer coefficient distribution

The data so far discussed referred to Nusselt number averaged along the wall circumference and these data are certainly important for the design of helically coiled pipes heat exchangers. Although this, the knowledge of the distribution of the convective heat transfer coefficient along the wall periphery is often a crucial and necessary information whose importance cannot be overlooked. For instance, in food pasteurization and sterilization a non-uniform heating could result in a failure to achieve a safe process. While for straight sections the heat transfer coefficient variations along the wall periphery are expected to be limited, in coiled tubes this phenomenon cannot be disregarded. In particular, the uneven distribution of the convective heat transfer coefficient along circumferential angular coordinate at the fluid-wall interface could negatively affect the residence time distribution thus impacting on the safety of the thermal process. The few data available in literature that face this problem mainly adopt the numerical approach and most of them consider low Prandtl number fluids [29].

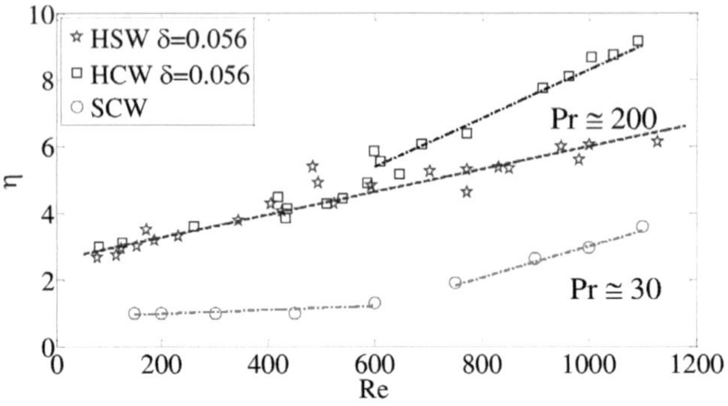

Figure 7. Enhancement efficiency vs. Reynolds number for the HSW and HCW tubes with δ=0.056 and comparison with SCW tube behaviour.

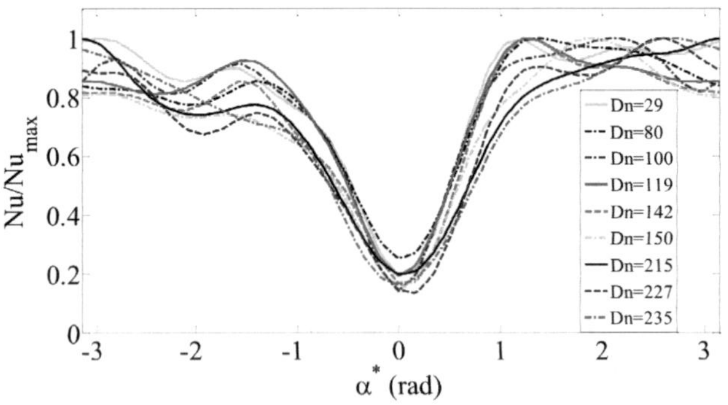

Figure 8. Normalized local Nusselt number for various Dean numbers along the wall perimeter ($Pr \cong 200$).

The research work performed in the last two years by the Applied Physics research group of the Industrial Engineering Department of the University of Parma tried to fill this literature gap, by optimizing an experimental procedure aimed to estimate the local convective heat transfer coefficient in coiled tubes. So far, the investigation was focused on the laminar flow regime in coiled tubes having smooth wall.

The temperature distribution maps on the external coil wall, acquired by means of the infrared thermographic technique at some representative pipe's cross sections, were employed as input data of the linear IHCP in the wall under a solution approach based on both the Tikhonov regularization method [15] and the filtering technique [25].

Some representative results obtained with the Tickhonov regularization method with the support of the fixed-point iteration technique to determine the proper regularization parameter are reported in figure 8. The data confirm a strong uneven distribution of the convective heat transfer coefficient along the wall perimeter: at the outer bend side of the coil it is about five times that one at the inner bend side. Within the experimental uncertainty, the Nusselt number profile along the curvilinear coordinate, normalized with its maximum value, was almost independent on Dean number, by confirming the data of Jayakumar et al. [30] for turbulent heat transfer in helical pipes. Similar results were obtained by adopting the filtering approach that resulted also a powerful tool, alternative to the Tikhonov Regularization approach, to obtain the convective heat transfer coefficient distribution by starting from the same data, i.e. the infrared temperature maps acquired on the external tube wall [25].

4. Conclusions

The main results of the experimental campaign on passive techniques for the enhancement of forced convective heat transfer in ducts performed in the last years by the Applied Physics research group at the Laboratory of the Industrial Engineering Department of the University of Parma were presented.

Two passive techniques, widely adopted for the thermal processing of medium and high viscosity fluids and based on wall corrugation and/or wall curvature, were in particular discussed.

The main results can be summarized as follows:

-the solution based on the corrugations of the heat transfer surface, such as rough wall tubes, mainly relies on the promotion of the boundary layer transition to the turbulent flow regime. Therefore, this technique is very effective for medium viscosity fluids, i.e. fluids that are able to reach Reynolds numbers of the order of at least a few hundred while this solution is almost ineffective for highly viscous fluids.

-the solution based on modifications of the heat transfer surface that produce a swirl, or helical type flow appears very interesting instead also in the conditions in which the flow persists in the laminar regime, also in case of very low Reynolds numbers. In fact, although it is well known that the curvature of the wall stabilizes the flow by delaying the transition to the turbulent flow, the effect of the secondary flow due to the centrifugal force that acts on the fluid flowing along the curvilinear trajectory, has a positive effect on the convective heat transfer coefficient.

-the simultaneous use of the two passive heat transfer enhancement techniques (wall curvature compounded with wall corrugation) is particularly effective if compared to the performance achievable through their separate use for Dean numbers higher than about 120, while it produces a negative effect for highly viscous fluids which flow under very low Reynolds numbers (of the order of 10).

-the heat transfer augmentation due to curvature is accompanied by a side effect related to the unevenness of the convective heat transfer coefficient distribution along the wall perimeter: in the Dean and Prandtl number range hereby investigated, the convective heat transfer coefficient at the outer bend side of the coil is about five times that one at the inner bend side, and the normalized Nusselt number distribution shows a shape that is almost independent on Dean number.

The here discussed results highlight that, as expected, within passive heat transfer enhancement, the proper solution to obtain a significant benefit strongly depends on the fluid flow regime: below a critical condition a given augmentation technique could be ineffective or even counterproductive.

Further efforts are needed in order to establish a fruitful connection between the scientific research outcomes such as the ones here reviewed and the technological innovation process which the process industry nowadays requires.

Acknowledgments
This work was partially supported by Emilia-Romagna Region (POR FSE 2007-2013).
The Author thanks the whole research group of the Applied Physics area of the Industrial Engineering Department of the University of Parma.

5. References

[1] Webb R and Kim NH 2005 *Principles of Enhanced Heat Transfer* 2nd ed. (New York: Taylor & Francis)
[2] Bergles AE 2002 *Exp. Therm. Fluid Sci.* **26** 335
[3] Bergles AE *J. Heat Transf.* 1997 **119** 8
[4] Rainieri S, Bozzoli F, Mordacci M and Pagliarini G 2012 *Heat Transfer Eng.* **33**(13) 1120
[5] Rainieri S and Pagliarini G 2002 *Int. J. Heat Mass Tran.* **45** 4525
[6] Kuppan T 2000 Heat Exchanger Design Handbook (New York: Marcel Dekker, Inc.).
[7] Bejan A 1978 *Int. J. Heat Mass Tran.* **21** 655
[8] Fan J F, Ding W K. Zhang J F He Y L and Tao W Q 2009 *Int. J. Heat Mass Tran.* **52** (1-2) 33
[9] Rozzi S, Massini R, Paciello G, Pagliarini G, Rainieri S and. Trifirò A 2007 *J. Food Eng.* **79** (1) 249
[10] Rainieri S, Bozzoli F, Schiavi L and Pagliarini G 2011 *Int. J. Numer. Method H.* **21**(5) 559
[11] Dean WR 1927 *Philos. Mag. J. Sci.* **4** (20) 208
[12] Rainieri S, Bozzoli F and Pagliarini G 2012 *Int. J. Heat Mass Tran.* **55** 498
[13] Rainieri S, Bozzoli F, Cattani L and Pagliarini G 2013 *Int. J. Heat Mass Tran.* **59** 353
[14] Bozzoli F, Cattani L, Rainieri S and Pagliarini G 2013 *Exp. Therm. Fluid Sci.* In press
[15] Bozzoli F, Cattani L, Rainieri S, Viloche Bazán FS and Borges LS 2014 *Int. J. Heat Mass Tran.* **72** 352
[16] Rainieri S, Bozzoli F, Cattani L and Pagliarini G 2012 *Journal of Physics: Conference Series* **395** 012032
[17] Barozzi GS and Pagliarini G 1984 *Int. J. Heat Mass Tran.* **27** (12) 2321
[18] Beck JV and Arnold K 1977 *Parameter Estimation in Engineering and Science* (New York: Wiley Interscience)
[19] Beck JV, Balckwell B, St.Clair Jr CR 1985 *Inverse Heat Conduction – Ill-posed problems* (New York: John Wiley & Sons Inc.)
[20] Alifanov OM 1994 *Inverse Heat Transfer Problem* (Berlin: Springer)
[21] Orlande HRB, Fudym O and Maillet D 2011 *Thermal Measurements and Inverse Techniques* (Boca Raton: CRC Press, Taylor & Francis Group)
[22] Murio DA, The Mollification Method and the Numerical Solution of Ill-Posed Problems (New York: John Wiley and Sons)
[23] Bozzoli F, Pagliarini G and Rainieri S 2013 *Exp. Therm. Fluid Sci.* **44** 858
[24] Tichonov AN and Arsenin VY 1977 *Solution of Ill-posed Problems* (Washington: Winston & Sons)
[25] Bozzoli F, Cattani L, Pagliarini G and Rainieri S *Proc. 12th International Workshop on Advanced Infrared Technology and Applications*, Torino, September 10 - 13, 2013.
[26] Incropera FP and DeWitt DP 1996 *Fundamentals of Heat and Mass Transfer* fourth ed. (John Wiley and Sons)
[27] Rainieri S, Farina A and Pagliarini G *Proc. 2nd European Thermal-Sciences and 14th UIT National Heat Transfer Conference*, Roma 1996 **1** 203
[28] Rainieri S and Pagliarini G 1997 *Heat Technol.* **15** (2) 69
[29] Di Liberto M and Ciofalo M 2013 *Int. J. Heat Mass Tran.* **59** 112
[30] Jayakumar JS Mahajani SM Mandal JC Kannan NI and Vijayan PK 2010 *Comput. Chem. Eng.* **34** 430

Journal of Physics: Conference Series **547** (2014) 012003 doi:10.1088/1742-6596/547/1/012003

Porous filtering media comparison through wet and dry sampling of fixed bed gasification products

G Allesina[1], S Pedrazzi[1], L Montermini[2], L Giorgini [3], G Bortolani[2], P Tartarini[1]

[1] University of Modena and Reggio Emilia, Department of Engineering 'Enzo Ferrari', Via Vignolese 905, 41125 Modena, Italy

[2] Curti Divisione Energia - Via Emilia Est, 181/D - 41121 Modena, Italy

[3] University of Bologna, Department of Industrial Chemistry "Toso Montanari", Viale Risorgimento 4 - 40136 Bologna, Italy

E-mail: giulio.allesina@unimore.it

Abstract.
The syngas produced by fixed bed gasifiers contains high quantities of particulate and tars. This issue, together with its high temperature, avoids its direct exploitation without a proper cleaning and cooling process. In fact, when the syngas produced by gasification is used in an Internal Combustion engine (IC), the higher the content of tars and particulate, the higher the risk to damage the engine is. If these compounds are not properly removed, the engine may fail to run. A way to avoid engine fails is to intensify the maintenance schedule, but these stops will reduce the system profitability. From a clean syngas does not only follow higher performance of the generator, but also less pollutants in the atmosphere. When is not possible to work on the gasification reactions, the filter plays the most important role in the engine safeguard process. This work is aimed at developing and comparing different porous filters for biomass gasifiers power plants. A drum filter was developed and tested filling it with different filtering media available on the market. As a starting point, the filter was implemented in a Power Pallet 10 kW gasifier produced by the California-based company "ALL Power Labs". The original filter was replaced with different porous biomasses, such as woodchips and corn cobs. Finally, a synthetic zeolites medium was tested and compared with the biological media previously used. The Tar Sampling Protocol (TSP) and a modified "dry" method using the Silica Gel material were applied to evaluate the tars, particulate and water amount in the syngas after the filtration process. Advantages and disadvantages of every filtering media chosen were reported and discussed.

1. Introduction

The increasing cost of conventional energy sources is a matter of serious concern worldwide [1]. The unceasing increasing of greenhouse gas emissions enhances the global warming causing a global climate change. In addition, the energy demand doubled in developed countries in the last years and this is mostly fulfilled through the use of conventional fossil fuels [2]. Between the renewable technologies which can relieve this condition, fixed bed biomass gasification is probably the most promising technology due to its high potential for rural power generation and for its neutral carbon balance [3].

Fixed bed biomass gasification is a thermochemical conversion composed of four different steps: drying, pyrolysis, oxidation and reduction [4, 5]. This thermal process generates a combustible gas, known as syngas, composed by CO, CO_2, N_2, H_2 and CH_4. Syngas can run IC engines for generating electricity and heat reducing greenhouse gas emissions [4]. In addition, small-scale district power generation enables off-grid villages to get access to electricity for the first time. A small scale gasifier ensures a complete permeability of the territory and it supplies energy to small users or isolated villages in developing countries. Furthermore, the plant cost is low compared to other renewable energy plants of same size [6].

In order to meet the requirements of IC engines, the syngas from fixed bed reactors must to be properly clean. In particular, an IC engine can be operated successfully with particulate content below 50 mg/Nm^3 and tars content below 100 mg/Nm^3 [5].

There are many methods and processes for hot gas cleaning process, i.e. the cyclones or other gravitational methods separate the particulate from the gas [7]. These systems are not highly efficient and they can separate only big solid particles, besides they require big spaces and they may need to be heated in order to avoid the condensation of water and tars on their inner surfaces. More efficient cleaning systems are other dry methods like electrostatic filters or wet method such as scrubber with oil or water as working fluid. Above all these methods, bio-filters are the only kind of filters where the filtering bed material can be gasified. Moreover, the cost of installation and maintenance is low [7]. For these economical and technical reasons, a dry bio-filtration system with a woodchips filter bed is the most appropriate solution especially for small size gasifiers [8].

This work aims to select filtering bed materials which ensure a good filtration of the syngas produced by a 10 kW$_{el}$ APL Power Pallet [9]. Woodchips, corncobs and synthetic zeolites were tested as filter media and the amounts of particulate, tars and water in the filtered syngas were evaluated and discussed.

2. Material and methods

2.1. Experimental setup and sampling procedure

The facility used in this work is depicted in Fig. 1. The device "A" is the APL downdraft gasifier, at the bottom of the reactor there is the first pressure port "P reactor". "B" is the drum filter equipped with another pressure port "P filter". During the tests, the filtered gas is burnt into a torch while a little amount of it is analyzed in the device "C" which is the Tar Sampling Protocol apparatus or the Silica Gel apparatus sketched in Figures 3 and 4. During each test, after about 30 minutes of reactor warmup, 30 minutes of Tar Sampling Protocol was carried out setting a syngas flow rate of about 1 Nm^3/h. Subsequently, the Tar Sampling Protocol apparatus was substituted with the Silica Gel one for other 30 minutes test maintaining fixed the flow rate. After that, a tedlar bag was filled with cleaned syngas in order to analyze it into a gas cromatographer.

The drum filter used in this work is basically a closed cylinder containing: two grates, two sponges and the chosen porous media. Under and above the filtering media there is some space for the gas to spread and slow down. The filter was filled with different materials then tested with the methods previously described. The media choice occurred on the basis of several criteria, such as material availability and quality/price ratio. The choice fells on the following three illustrated in Fig. 2:

- **Woodchips G10 W5:** it is currently the most used as syngas biofilter bed. The average dimension of the chips about 10 mm and the moisture is below 5%.
- **Corncobs:** chosen for their high surface area in relation to their size. The properties of the corncobs adopted in this work are reported in Reference [6].
- **Composite filter**: it is a material based on synthetic zeolites.

At the beginning, the height of the filtering bed was the same for each material. However, tests shown that the syngas residence time inside the filter depends on the materials used as

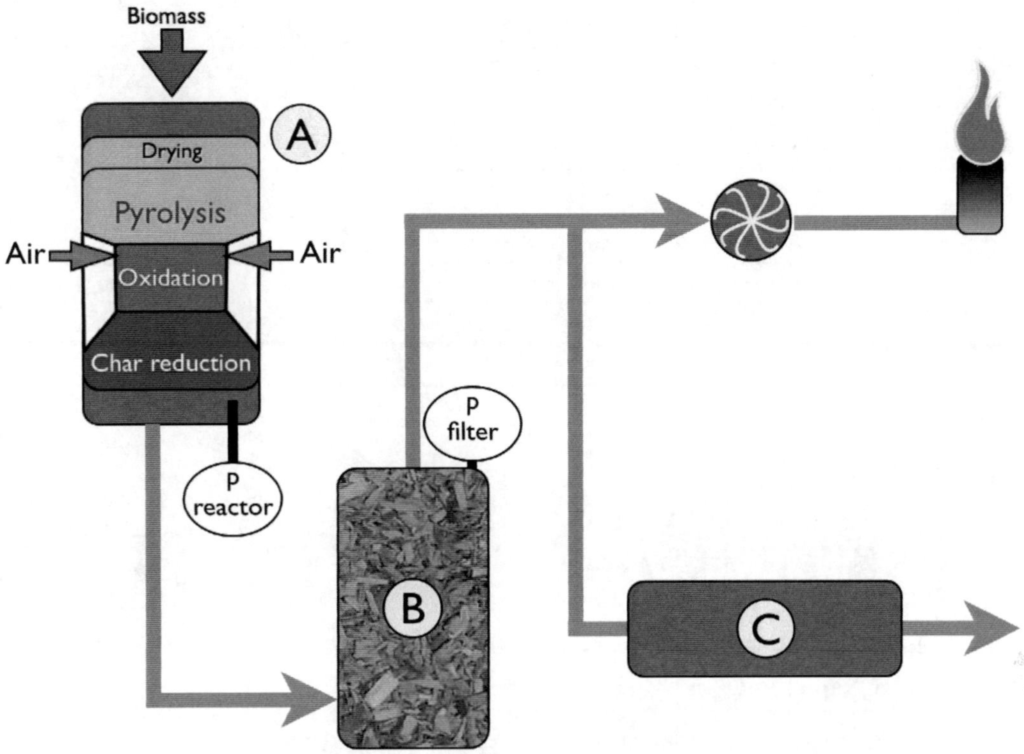

Figure 1. Facility adopted in this work

Figure 2. Filter porous media

result of their different porosity. Therefore, the bed height was varied to ensure a fixed value of differential pressure between the input and the output of the filter during all the tests. A metal mesh with 1 mm holes was placed at the bottom of the filter to support even the finer material. The filter material was laid on the grate and covered by two sponges with different porosity. The sponge placed below has a porosity of 65 dpi and the sponge placed above has a porosity of 45 dpi. A second metal mesh ensured that the height of the bed during the test was constant. Physical properties of the filter media are reported in Tab. 1. Furthermore, heights and differential pressures measured during the tests are reported in Tab. 2.

Table 1. Physical properties of the porous media

Filtering materials	True density [g/cm^3]	Bulk density [g/cm^3]	Porosity [%]	Moisture [%wt.]
Woodchips	1.450	0.218	85.97	10
Corncobs	1.360	0.107	92.13	9.4
Zeolites	2.910	0.902	69.00	$\cong 0$

Table 2. Parameters of the filter drum

Filtering materials	Height [m]	Superficial area [m^2]	Volume [m^3]	P reactor [kPa]	P filter [kPa]
Woodchips	0.330	0.113	0.0373	6.23	6.97
Corncobs	0.475	0.113	0.0537	6.23	6.97
Zeolites	0.100	0.113	0.0113	6.23	6.97

2.2. Physical and chemical methods

All tests performed were carried out after a 30 minutes start up. During the tests, were taken several gas samplings in order to analyze syngas quality and composition. In particular, the syngas was analyzed through the following chemical and physical methods:

- Tar Sampling Protocol;
- Silica Gel analysis;
- Gas Chromatographic analysis.

In order to analyze tar, particulate and water content in the syngas and to compare the filtering properties of the three different porous materials, the Tar Sampling Protocol was adopted [10]. The method consists in a series of 6 impingement bottles in which the syngas flows and where the first one acts like a moisture collector in which water is condensed. After the moisture collector, the gas flows through a series of 4 impinger bottles filled of approximately 70 ml of isopropyl alcohol that is the solvent used to dilute the tars and particulate of the syngas. The last bottle is intentionally empty to ensure the collection of final condensates. The heat released by gas cooling and tar condensation is removed through an external bath, that consists in a mixture of dry ice and glycol. Six standard glass impingers of 250 ml each are used while the external bath is maintained at a constant temperature of -18 ^oC as shows in Fig. 3.

The whole system is maintained in depression by a pump positioned downstream which draw the syngas after the drum filter. In the sampling line it is also implemented a rotameter with adjustable valve that allows the measurement and the regulation of the syngas flow rate [10]. After tars dilution in the impingement bottles, tars content was approximated proportional to the concentration of six compounds in the isopropyl alcohol according to Basu and Milne [5, 11]:

- Benzene
- Toluene
- Naphthalene
- Xylene
- Styrene
- Ethylbenzene

As suggested by [12, 13], these compounds can be detected with gas chromatography-flame ionization detection (GC-FID), gas chromatography-mass spectrometry (GC-MS), or gas chromatography-Fourier transform infrared detection (GC-FTIR). In this work the gas

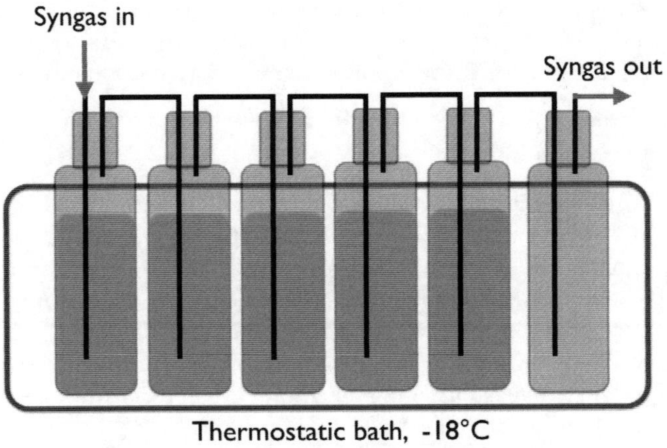

Figure 3. Tar Sampling protocol apparatus

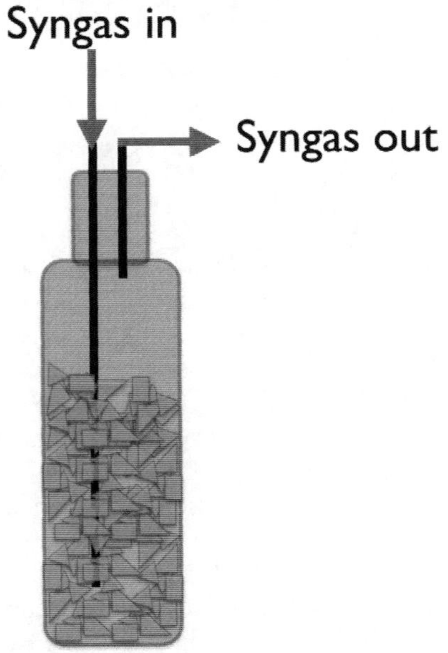

Figure 4. Silica gel apparatus

chromatography-mass spectrometry (GC-MS) was used successfully through the adoption of a specific column for hydrocarbons detection.

In addition to the Tar Sampling Protocol, another tests with Silica Gel was carried out in order to validate the results. The analysis was performed with the same procedure described for the Tar Sampling Protocol using only one impingement bottle filled with about 70 ml of Silica Gel instead of isopropyl alcohol as shown in Fig. 4. This method is carried out at room temperature and it is already adopted in literature coupled with a charcoal filter [14]. In this case the condensates amount is evaluated weighting the gel before and after the tests.

Journal of Physics: Conference Series **547** (2014) 012003 doi:10.1088/1742-6596/547/1/012003

During each test, a bag of syngas was sampled from the top of the biofilter and then it was analyzed by a double coloumn portable gas chromatographer using argon as gas-carrier. This analysis allows to determine the composition of the gas and therefore to monitor the gasification process in order to validate the assumption of constant particulate content and tars production in all the tests.

3. Results

In this section, results concerning Gas Chromatograpy, Tar Sampling Protocol and Silica Gel analyses are reported and discussed. Each filtering medium was tested under steady state conditions of gasification. Tab. 3 summarizes the results of the Gas Chromatograpic analyses in order to check the reliability of the test. The composition of the syngas is affected by the presence of little amount of O_2. This is symptomatic of some air leakage inside the syngas flow lines as result of holes in the seals. However, the amounts of combustible gases CO, CH_4 and H_2 are high, therefore the gasification is efficient and the syngas produced is adequate to run an IC engine.

Table 3. Results of the syngas Gas Chromatograpic analyses

	Woodchips	Corncobs	Zeolites
N_2 fraction [% vol.]	60.5	62.8	60.0
O_2 fraction [% vol.]	9.5	9.2	11.7
CH_4 fraction [% vol.]	2.7	4.0	5.1
H_2 fraction [% vol.]	10.6	8.9	8.0
CO_2 fraction [% vol.]	6.7	5.2	5.8
CO fraction [% vol.]	10.0	8.3	11.0
Higher Heating Value [MJ/Nm^3]	3.69	3.77	4.44

The results of the Tar Sampling Protocol analyses are reported in Tab. 4. The concentration of water and particulate in the syngas were obtained from the difference between the mass of matter collected in the impinger bottles and the mass of tars detected by the GC-MS. In every test, the great part of tars (about 80 %) is benzene, while about 15% is toulene as shown in Tab. 4. The total amount of tars, particulate and water (TPW) collected in the impingement bottles gives the filtering performance of the porous medium. In particular, corncobs reveals the greater performance. The volumetric tar amount in the syngas filtered by corncobs is the lower (1.32 g/Nm^3), therefore this feedstock is able to adsorb water and particulate very efficiently when compared to woodchips and zeolites. This behaviour is related to the high porosity of corncobs compared to the other porous media. In addition, corncobs can be further gasified after their use as filtering medium, instead zeolites need to be rigenerated or disposed.

Tab. 5 summarizes the results of the Silica Gel analyses. The total amount of tars, water and particulate is similar to the value obtained by the Tar Sampling Protocol. The corncobs results confirmed the great performance of this feedstock which is able to reduce at 6.59 g/Nm^3 the T&P&W volumetric amount in the syngas. The values referred to zeolites and wood chips are two and three times grater than the corncobs one. Fig. 5 shows the filtering performance comparison of the porous media analyzed in terms of T&P&W volumetric amount in the syngas and Fig. 6 depicted the comparison in terms of tars volumetric amount.

4. Conclusion

This work discusses the comparison of three different porous materials used as filter in a small scale gasification power plant. Results show that all the filters investigated reduced the tar

Table 4. Results of the Tar Sampling Protocol analyses

	Woodchips	Corncobs	Zeolites
Duration [min]	30	30	30
Average Syngas Flow [Nm^3/h]	0.96	0.96	0.96
T&P&W amount [g]	10.58	2.83	7.48
T&P&W volumetric amount [g/Nm^3]	20.3	5.43	14.4
P&W amount [g]	9.55	2.14	6.65
P&W volumetric amount [g/Nm^3]	20.31	4.11	12.76
T amount [g]	1.03	0.69	0.83
T volumetric amount [g/Nm^3]	1.97	1.32	1.60
Benzene fraction [%]	79.2	80.9	79.1
Toluene fraction [%]	15.2	16.4	16.8
Ethylbenzene fraction [%]	0.52	0.24	0.49
Xylene fraction [%]	0.56	0.18	0.64
Stirene fraction [%]	1.64	1.07	1.24
Naphthalene fraction [%]	2.84	1.20	1.73

Legend: T = tars; P = particulate; W = water.

Table 5. Results of the Silica Gel analyses

	Woodchips	Corncobs	Zeolites
Duration [min]	30	30	30
Average Syngas Flow [Nm^3/h]	0.96	0.96	0.96
T&P&W amount [g]	10.2	3.43	7.65
T&P&W volumetric amount [g/Nm^3]	19.6	6.59	14.7

Legend: T = tars; P = particulate; W = water.

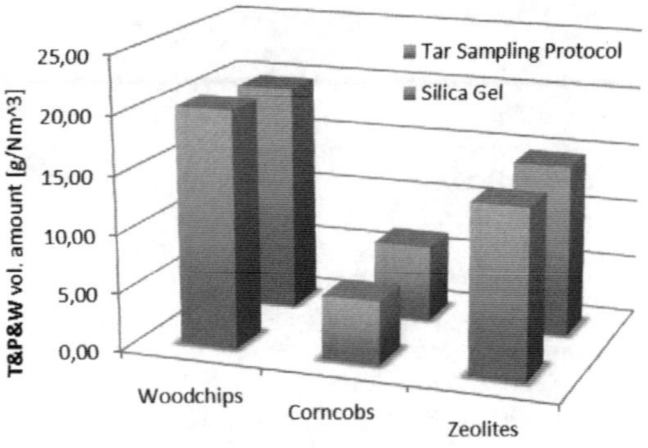

Figure 5. Porous media T&P&W content comparison

Journal of Physics: Conference Series **547** (2014) 012003 doi:10.1088/1742-6596/547/1/012003

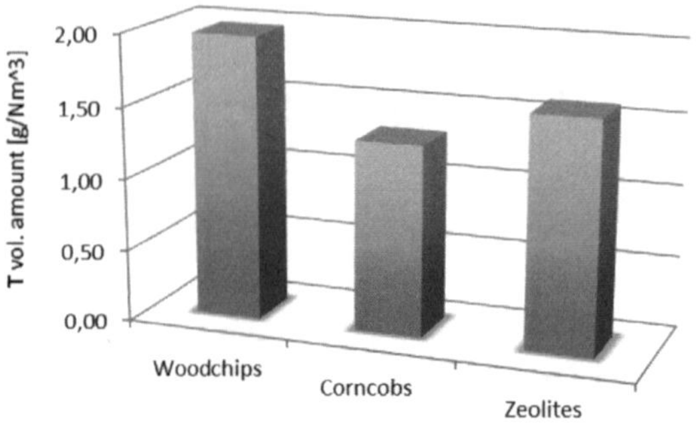

Figure 6. Porous media tar content comparison

content under 2 g/Nm^3. Furthermore, even if the woodchips are widely used as biofilter material, the corncobs resulted to be almost twice better in terms of water, particulate and tars absorption. On the other hand, the zeolites filter investigated in this work gave results halfway between the woodchips and the corncobs.

Corncobs, as well as the other parts of the corn stover, are a by-product with marginal applications. This work opens the possibility of finding new purposes for this material. It can efficiently substitute or complement woodchips in the drum filters. Even though other filtering solutions may produce a cleaner gas, biofilters remain a simple and inexpensive solution that founds a perfect match with small scale gasification systems. In these micro power plants, the simplicity and handiness are more desirable than high perfomance. Moreover, the BEELab is investigating the behavior of corncobs as fuel in downdraft reactors, this solution will lead to a filtering system that has no disposed materials.

5. References

[1] ENEA. La strategia comunitaria in tema di biomasse. Technical report, ENEA, 2010.

[2] IEA. Technology roadmap - bioenergy for heat and power. Technical report, IEA, 2012.

[3] Simone Pedrazzi. *Modeling and optimization of advanced systems for electrical energy production from wood biomass.* PhD thesis, HIGH MECHANICS AND AUTOMOTIVE DESIGN & TECHNOLOGY, University of Modena and Reggio Emilia, Dep. of Engineering 'Enzo Ferrari', 2013.

[4] Thomas B. Reed and Agua Das. *Handbook of Biomass Downdraft Gasifier Engine Systems.* The biomass energy foundation press, 1988.

[5] Prabir Basu. *Biomass Gasification and Pyrolysis: Practical Design and Theory.* Academic Press, Elsevier, 2010.

[6] Giulio Allesina, Simone Pedrazzi, Federico Sgarbi, Elisa Pompeo, Camilla Roberti, Vincenzo Cristiano, and Paolo Tartarini. Appapproach sustainable development through energy management, the case of foto, cameroon. In *VIII Congresso Nazionale AIGE, Reggio Emilia, June 9, 10 2014*, 2014.

[7] Lopamudra Devi, Krzysztof J Ptasinski, and Frans J.J.G Janssen. A review of the primary measures for tar elimination in biomass gasification processes. *Biomass and Bioenergy*, 24(2):125–140, 2003.

[8] Giulio Allesina. *Experimental and analytical investigation of downdraft stratified gasifiers.* PhD thesis, HIGH MECHANICS AND AUTOMOTIVE DESIGN & TECHNOLOGY, University of Modena and Reggio Emilia, Dep. of Engineering 'Enzo Ferrari', 2012.

[9] All-Power-Labs. Gasifier experimental kit, 2013.

[10] J.P.A. Neeft, H.A.M. Knoef, U. Zielke, K. Sjöström, P. Hasler, P.A. Simell, M.A. Dorrington, L. Thomas, N. Abatzoglou, S. Deutch, C.Greil, G.J. Buffinga, C. Brage, and M. Suomalainen. Guideline for sampling and analysis of tar and particles in biomass producer gases. Technical report, European Commission (DGXII) Netherlands Agency for Energy and the Environment (NOVEM) Swiss Federal Office of Education and Science Danish Energy Agency (Energistyrelsen) US Department of Energy (DoE) National Resources Canada, 2002.

[11] T.A. Milne, R.J. Evans, and N. Abatzoglou. *Biomass gasifier" tars": their nature, formation, and conversion.* National Renewable Energy Laboratory Golden, CO, 1998.

[12] B.M. EI-Haj, A.M. Al-Amri, M.H. Hassan, R.K. Bin-Khadem, and A.A. Al-Hadi. A gc-ms method for the detection of toluene and ethylbenzene in volatile substance abuse. *Journal of Analytical Toxicology*, 24:390–394, 2000.

[13] R.J. Flanagan, P.J. Streete, and J.D. Ramsey. Volatile substance abuse, practical guidelines for analytical investigation of suspected cases and interpretation of results. Technical Report 5, United Nation Office on Drugs and Crime, 1997.

[14] Thana Phuphuakrat, Tomoaki Namioka, and Kunio Yoshikawa. Tar removal from biomass pyrolysis gas in two-step function of decomposition and adsorption. *Applied Energy*, 87(7):2203 – 2211, 2010.

Journal of Physics: Conference Series **547** (2014) 012004 doi:10.1088/1742-6596/547/1/012004

A Design of Experiments (DOE) approach to optimise temperature measurement accuracy in Solid Oxide Fuel Cell (SOFC)

F Barari[1, 2], R Morgan[1] and P Barnard[2]

[1]University of Brighton, Cockcroft Building, Lewes Road, Brighton, BN2 4GJ, UK
[2]Ceres Power Limited, Viking House, Foundry Lane, Horsham RH13 5PX, UK
E-mail: farzad.barari@cerespower.com

Abstract. In SOFC, accurately measuring the hot-gas temperature is challenging due to low gas velocity, high wall temperature, complex flow geometries and relatively small pipe diameter. Improper use of low cost thermometry system such as standard Type K thermocouples (TC) may introduce large measurement error. The error could have a negative effect on the thermal management of the SOFC systems and consequential reduction in efficiency. In order to study the factors affecting the accuracy of the temperature measurement system, a mathematical model of a TC inside a pipe was defined and numerically solved. The model calculated the difference between the actual and the measured gas temperature inside the pipe. A statistical Design of Experiment (DOE) approach was applied to the modelling data to compute the interaction effect between variables and investigate the significance of each variable on the measurement errors. In this study a full factorial DOE design with six variables (wall temperature, gas temperature, TC length, TC diameter and TC emissivity) at two levels was carried out. Four different scenarios, two sets of TC length (6 – 10.5 mm and 17 – 22 mm) and two different sets of temperature range (550 – 650 °C and 750 – 850 °C), were proposed. DOE analysis was done for each scenario and results were compared to identify key parameters affecting the accuracy of a particular temperature reading.

1. Introduction

Accurate measurement of the gas temperature is vital for thermal management of SOFC system. Temperature measurement error could have a detrimental effect on system overall efficiency, reduce system life and may even cause a complete system failure. In addition, running the system outside of the material safe operating temperature range may generate toxic gas or hazardous material. An expensive sophisticated thermometry system may improve the system performance but this is not a practical option for manufacturing this system at a competitive price in high volume. Challenges arose when a low cost, robust and accurate thermometry system is needed for SOFC systems.

A K-type TC is a simple and inexpensive thermometry sensor with a reasonable accuracy. It is widely used in industry for high temperature measurement. However, a common mistake is to simply place a TC into a small diameter pipe to measure a gas temperature. The problem exacerbates when it is used in high temperature dynamic system like SOFC. A common problem of measuring hot-gas temperature inside a pipe is ignoring the effects of thermal radiation emitted by the surrounding as well as the heat conduction from the wall via the TC stem [1]. This may introduce systematic temperature measurement errors. The effect of thermal conduction and radiation on TC measurements accuracy has been investigated several times mainly in gas turbine, combustors and fire studies [2-7]. However, the dominant effect in this type of measurement has not been clearly identified.

In this study, a one-dimensional mathematical model was developed to investigate the effect of heat conduction, convection and radiation on temperature measurement with a thermocouple attached to a wall. Since there were six major factors influencing the temperature measurement, Design of Experiment (DOE) method was used to determine the effect of critical factors and their interaction. Four different temperature measurement scenarios were proposed. A full factorial DOE was designed at two levels for each scenario and the results were analysed and plotted by Minitab software.

2. Mathematical model

The aim of measuring gas temperature is to measure the gas thermal energy via convection. However, heat transfers from the adjacent components via radiation and conduction could affect the temperature measurement. In fact, the temperature measurement is a result of heat balance on the TC stem [1]. For instance, if the wall temperature is higher than the gas temperature, heat transfers from the wall to the tip via conduction and the wall radiates heat to the TC stem and tip.

In this study the TC is considered to be a cylindrical probe immersed to a gas stream inside a pipe. It was assumed that the gas steam is uniform in terms of velocity (non-viscous fluid) and temperature across the pipe. The pipe wall was assumed to be at a constant temperature (no thermal gradient). There was no thermal resistance at the junction between the TC and the wall. A perfect thermal contact between the TC and the wall is not realistic assumption. However, the actual thermal resistance is depends on several factors such as fitting martial, contact force between the TC fitting and TC, method of attaching fitting to the wall and wall material. Therefore, a perfect thermal contact assumption was simplified the analysis and reduced the ambiguity regarding the thermal conductivity between the TC and wall.

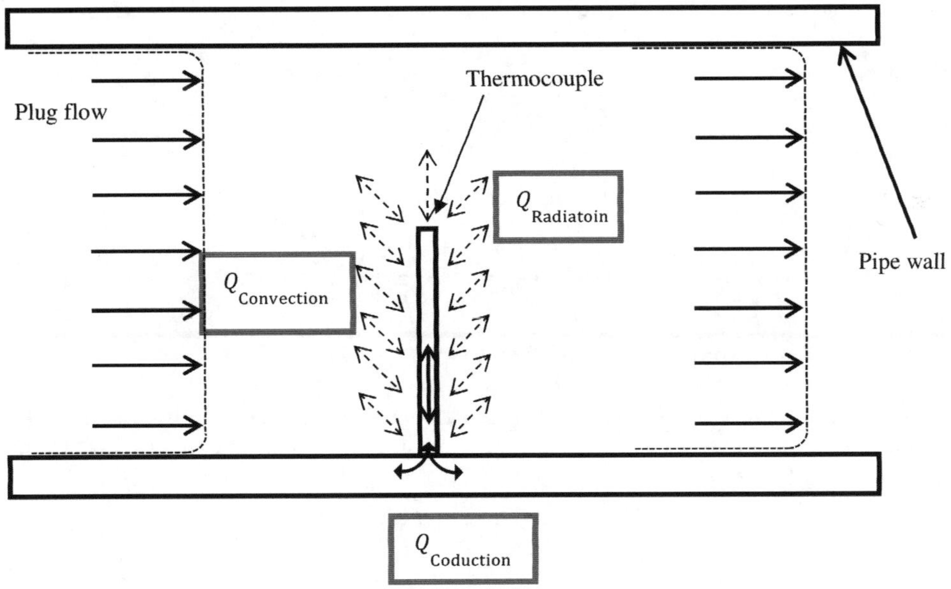

Figure 1. Schematic view of a thermocouple in a pipe subjected to a gas stream.

Journal of Physics: Conference Series **547** (2014) 012004 doi:10.1088/1742-6596/547/1/012004

A 1-D energy balance equation describes the heat balance for this system. It can be seen from equation 1 that the left term is the heat that accumulate on the TC. On the right-hand side are the three terms that transfers heat to/from the TC.

$$\underbrace{\rho_{TC} C_{TC} \frac{d\, T_{TC}(x)}{dt}}_{heat\ accumulation\ on\ TC} = \underbrace{k_{TC} \frac{d^2 T_{TC}(x)}{dx^2}}_{Conduction} - \underbrace{\frac{4}{D_{TC}} h \left(T_{TC}(x) - T_g\right)}_{Convection} - \underbrace{\frac{4}{D_{TC}} F\, \sigma \left(T_{TC}^4(x) - T_W^4\right)}_{Radiation} \quad (1)$$

where ρ is the density, k is the thermal conductivity, t is time, T is temperature D_{TC} is the TC diameter, h is convection heat transfer coefficient, is σ is the Stefan-Boltzmann constant and F is radiation transfer factor.

It can be seen from the equation 2 that the Nusselt number (Nu) and gas thermal conductivity (k_g) should be calculated in order to estimate h.

$$h = Nu \frac{k_g}{D_{TC}} \quad (2)$$

The Nusselt number was calculated based on a cross flow over a cylinder (equation 3) [8]. This equation is valid for Reynolds numbers (Re_D) ≤ 4000 and Prandtl numbers (Pe_D) $Pr \geq 0.2$

$$Nu = 0.3 + \frac{0.62 Re_D^{1/2} Pr^{1/2}}{\left[1 + \left(\frac{0.4}{Pr}\right)^{2/3}\right]^{1/4}} \left[1 + \left(\frac{Re_D}{282000}\right)^{5/8}\right]^{4/5} \quad (3)$$

The gas thermal conductivity was calculated by fitting a line to Kadoya and Matsunaga experimental data [9]:

$$k_{air} = Tg\, 6 \times 10^{-5} + 0.0077 \quad (4)$$

Gas viscosity (μ) was calculated with Sutherland's law:

$$\mu = \mu_{ref} \left(\frac{T}{T_{ref}}\right)^{3/2} \frac{T_{ref} + S}{T + S} \quad (5)$$

where S is Sutherland's constant which is 120 for air.

It was assumed that the TC was a small grey body enclosed in a larger body (pipe) with hypothetically homogeneous temperature (wall temperature). As a result, the transfer factor (F), can be simplified to the TC emissivity [8]:

$$F = \frac{1}{\frac{1}{\varepsilon_{TC}} + \frac{A_{TC}}{A_w}\left(\frac{1}{\varepsilon_{TC}} - 1\right)} \quad (6)$$

$$A_w \gg A_{TC} \therefore F \cong \varepsilon_{TC}$$

Following boundary conditions were used to solve equation 1:

$$T_{TC}(0) = T_w \tag{7}$$

$$\frac{dT_{TC}}{dx} = 0 \ @ \ x = L_{TC} \tag{8}$$

3. Numerical solution:

The exact solution for a steady state version of equation 1 is available in literature [10]. Since SOFC systems are dynamic systems and pipes temperature varies at different electric loads, it is important to know the transient time to accurately control the systems. For a transient condition, analytical solution is more complicated and numerical solution is more favourable than the analytical one. The transient effect on temperature measurement accuracy is dependent on several factors such as start-up, shutdown and electric load profile of SOFC systems. Therefore it is not practical to add this factor to DOE analysis and this effect is not considered in this paper.

The equation was solved by the implicit Backward-Time Centred-Space (BTCS) method. In this method the time derivative part of the equation was replaced by the first-order backward-time finite difference approximations and the distance derivative part was replaced by the second-order centred-space approximation and they were evaluated at the solution time level n+1.

$$\frac{d^2T(x)}{dx^2} \cong \frac{T_{i+1}^{l+1} - 2T_i^{l+1} + T_{i-1}^{l+1}}{\Delta x^2} \tag{9}$$

$$\frac{d\,T(x)}{dt} = \frac{T_i^{l+1} - T_i^l}{\Delta t} \tag{10}$$

Substituting equation 9 and 10 in equation 1 yields equation 11:

$$\rho_{TC} C_{TC} \frac{T_i^{l+1} - T_i^l}{\Delta t} = k_{TC} \frac{T_{i+1}^{l+1} - 2T_i^{l+1} + T_{i-1}^{l+1}}{\Delta x^2} - \frac{4}{D_{TC}} h\left(T_i^l - T_g\right) - \frac{4}{D_{TC}} \sigma \left({T_i^l}^4 - {T_w}^4\right) \tag{11}$$

After introducing λ, a and b, equation 11 can be rewrite as equation 12.

$$\lambda = \frac{k_{TC}\Delta t}{\rho_{TC} C_{TC} \Delta x^2} \qquad a = \frac{4\,\Delta t\, h}{\rho_{TC} C_{TC} D_{TC}} \qquad b = \frac{4\,\Delta t\, \varepsilon_{TC}}{\rho_{TC} C_{TC} D_{TC}}$$

$$(1 + 2\lambda)T_i^{l+1} - \lambda T_{i-1}^{l+1} - \lambda T_{i+1}^{l+1} = -aT_i^l - b{T_i^l}^4 - aT_g - bT_w^4 \tag{12}$$

$$\underbrace{\begin{bmatrix} 1+2\lambda & -\lambda & \cdots & 0 \\ -\lambda & 1+2\lambda & -\lambda & \vdots \\ \vdots & -\lambda & \ddots & -\lambda \\ 0 & \cdots & -\lambda & 1+2\lambda \end{bmatrix}}_{A} \underbrace{\begin{bmatrix} T_1^1 \\ \vdots \\ \vdots \\ T_i^{l+1} \end{bmatrix}}_{X} = \underbrace{\begin{bmatrix} -aT_1^0 - b{T_1^0}^4 - aT_g - bT_w^4 \\ \vdots \\ \vdots \\ -aT_i^l - b{T_i^l}^4 - aT_g - bT_w^4 \end{bmatrix}}_{B} \tag{13}$$

All terms on the right-hand side of equation 12 are known. Hence, the equation 12 forms a tridiagonal linear system. TC stem temperature (matrix *X*) can be calculated by multiply matrix *B* to inverse matrix of *A*.

The physical domain of thermocouple stem was discretised into 100 nodes and a code was written in Matlab to solve the above matrix. Shown in figure 2 is an example of the predicted temperature profile alongside the thermocouple stem generated by the model. The figure shows the difference between the indicated temperatures (thermocouple tip temperature) and the actual gas temperature (the red line in figure 2). In this example, a 0.5 mm diameter (K type) thermocouple with stainless steel 310 sheath (thermal conductivity 13 W/m.K) was placed at the centre of a 15 mm (internal diameter of 12 mm and 1.5 mm wall thickness) pipe. The TC was attached to the tube wall and had a perfect thermal contact with the wall. The TC length was 6 mm. The TC emissivity was set at 0.1 and the air velocity was 6 m/s. The air and wall temperature was 650 °C and 550 °C respectively. The graph shows that the TC tip temperature was at 634.6 °C but the actual air temperature was at 650 °C. In other word, the temperature measurement in this condition had 15.4 °C error. The modelling results were compared with the results reported by [1] (1) and there were good agreement between the reported results.

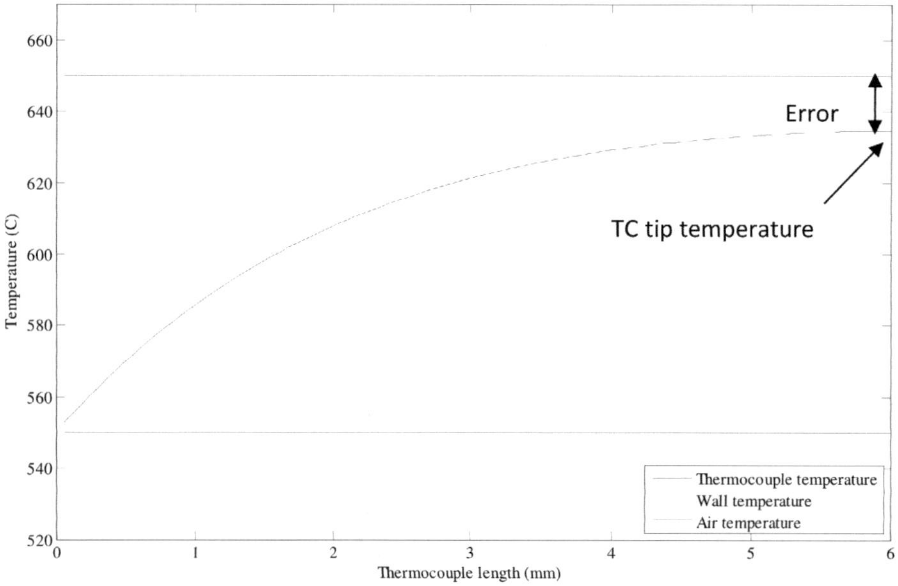

Figure 2 An example of temperature variation in a TC stem.

4. Design of Experiment

Design of Experiment (DOE) technique was performed to identify the key variables which affect the temperature measurement process and to determine at what levels these factors must be kept to reduce the measurement error. The results of the experiment were analysed and plotted using Minitab software.

In order to see the effect of different variables, four different scenarios and therefore four DOEs were considered. Figure 3 shows the first two scenarios. For the first scenario, a TC attached to a 15 mm (1.5 mm wall thickness) pipe with 6 mm exposed length inside the pipe. For the second scenario, a TC attached to a 25 mm (2 mm wall thickness) pipe with 10.5 mm exposed length. In figure 4, the TC fitting moved 10 mm away from the pipe. Therefore, for the third and fourth scenarios, TC exposed length were 17 mm and 22 mm for 15 mm and 25 mm pipes respectively.

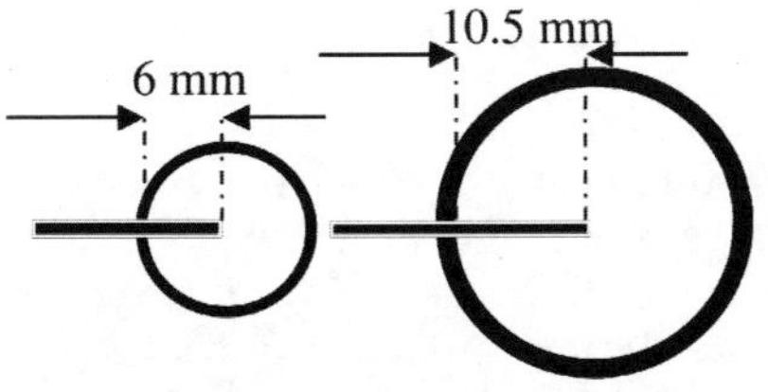

Figure 3. TCs attached to a 15 mm and 25mm pipe.

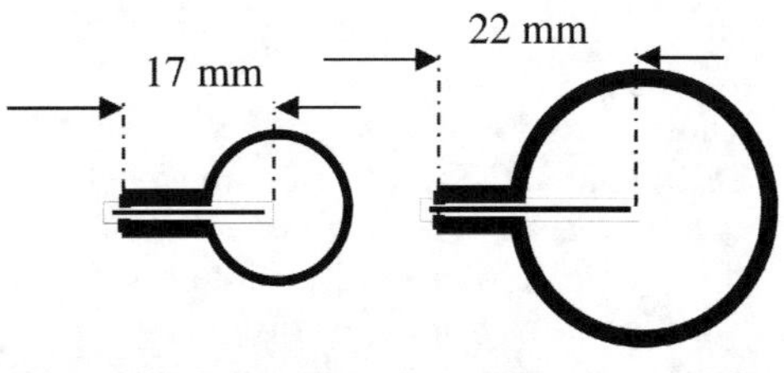

Figure 4. TCs attached to a 15 mm and 25mm pipe with 10mm extension.

The following tables (Table 1, 2) show the four scenarios and the factors/levels considered in this study. Two temperature levels, Low and high, were considered. The lower range, 550 °C – 650 °C, represents a nominal operating temperature range for metal-supported SOFC modules. Tabulated emissivity for stainless steel 316 is available in literature [11] but it varies depending on the condition of the surface, wave length, the exposed temperature and time. In general it could be 0.5 - 0.7 depends on finished surface and surface oxidation. One should bear in mind that TC emissivity increases by increasing temperature due to oxidisation. It was reported that the polished stainless steel at room temperature has emissivity of 0.16 however by increasing the temperature to 1027 °C the emissivity could increases to 0.8 [8]. However, the emissivity could be kept at lower value by coating the TC surface. Therefore in this study a range of 0.1 to 0.7 was adapted to study the effect of emissivity in temperature measurement error. In addition to the above factors, two different TC diameters, 0.5 mm and 1.5 mm were studied. Thermocouple diameter could affect the heat transfers to thermocouples via conductivity (cross section area), radiation (effective surface area) and convection (exposed surface area).

Table 1. List of factors considered and their level for low thermocouple length

		Scenario 1		Scenario 3	
Factor	Labels	Low-level	High-level	Low-level	High-level
T_W (°C)	A	550	650	750	850
T_G (°C)	B	550	650	750	850
V_g (m/s)	C	6	40	6	40
ε	D	0.1	0.7	0.1	0.7
L_{TC}	E	6	10.5	6	10.5
D_{TC} (mm)	F	0.5	1.5	0.5	1.5

Table 2. List of factors considered and their level for high thermocouple length

		Scenario 2		Scenario 4	
Factor	Labels	Low-level	High-level	Low-level	High-level
T_W (°C)	A	550	650	750	850
T_G (°C)	B	550	650	750	850
V_g (m/s)	C	6	40	6	40
ε	D	0.1	0.7	0.1	0.7
L_{TC}	E	17	22	17	22
D_{TC} (mm)	F	0.5	1.5	0.5	1.5

Full factorial (two levels or one degree of freedom) design with six factors was considered for each scenario. The design matrix had 64 rows (2^6) and 6 columns. The design matrix then imported to the model and the estimated thermocouple errors (the temperature difference between the gas and the TC tip) were modelled for each 64 runs. The response matrix (estimated thermocouple errors) then exported to Minitab software to run the DOE analysis.

5. Results and discussion

Having obtained the results from the model, the first step was to compute the effect of factors and their interaction effects. The main effect is the difference in average response due to changes in the level factor. Shown in figure 5 is a graph of the main effects of the single variable affecting temperature measurement error in scenario 1. This plot demonstrates the factors influencing the temperature measurement error and it also shows the level of strength of each factor. In order to plot these graphs the mean response (measurement error) at each factor level was computed and then connected by a straight line. The reference line was drawn based on the overall mean of the response. The slope of the line indicates the effect of that factor on the results. If the slope is close to zero, then it means that there is no major effect presents. For instance in figure 5, the gas and the wall temperature had little effect on temperature measurement error. However, the graph shows that less error predicted for low wall temperature and high gas temperature. The reason for this behaviour was higher wall temperature causes a larger error due to conduction and radiation effect from the wall. The graph shows that TC diameter was the most important factor which affects the measurement error followed by TC length, Gas velocity and TC emissivity. The graph shows that increasing the TC diameter increases the measurement error due to having higher surface area for radiation and larger cross section area for higher thermal conductivity. It can be seen from the graph that increasing the TC length decreases the error due to lowering the TC stem effect (heat transfer through from the wall through the TC stem to the TC tip). The result also shows that increasing the gas velocity reduces the error. The actual gas temperature is measured by measuring the gas thermal energy transferred from/to the TC's tip by the convection and any other heat transfers from the adjacent components such as pipe wall would influence the measurement accuracy. Therefore, at high gas velocity the effect of convection is more dominant which reduces the effect of radiation and conduction effect on temperature measurement.

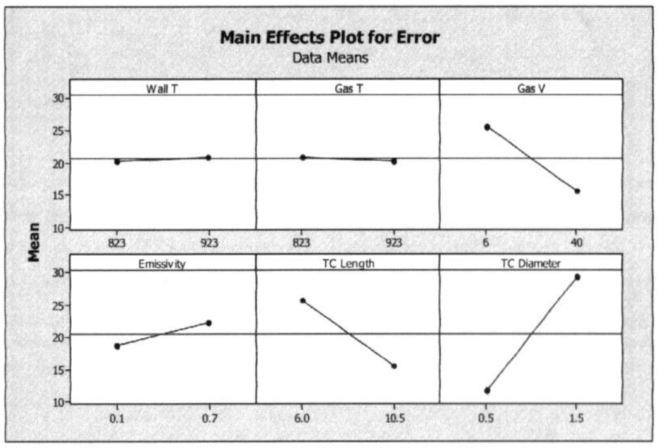

Figure 5. Main effects plot of the significant effects for scenario 1(temperature in K).

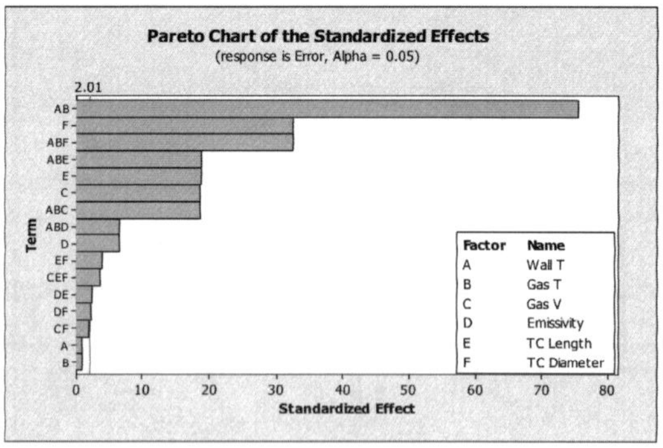

Figure 6. Pareto plot of factor effects for scenario 1.

Figure 6 shows the Pareto plot of the main effects and their combination. To generate this graph, the single and combined factors were analysed and the initial plot was produced. Then, the statistically insignificant factors and the combined factors, which were more than three single factors, were eliminated from analysis and process was iterated again. The final Pareto graph, which can be seen in Figure 6, was plotted to investigate significant of the factors and their interactions. The factors which were lower than the red line (5% statistically significant level) were statistically insignificant. It can be seen for the graph that all the single factors excluding wall and gas temperature were significant. The graph also shows that the wall and gas temperature were insignificant factors; however, their combine factor (AB) was the highest. The next combined significant factor was ABF followed by ABE. This indicates the importance of factor F and E in this scenario. The single factor C was the next single important factor after E followed by the combined factor ABC and ABD. This shows that the emissivity factor, D, was the least significant single factor for this scenario.

Shown in figure 7 and 8 are the DOE analysis for scenario 2. In this scenario the length of the TC inside the pipe increased by 10 mm. Both graphs show that after increasing the TC length, the D factor was the most significant single factor followed by F, C and E. This shows that although the temperature range was similar to scenario 1, the emissivity factor became the most significant factor by increasing the TC length. In other words, by increasing the TC length, the error due to TC stem effect was reduced and the error due to radiative heat became a dominant factor in this scenario. Since the TC diameter also related to radiate heat transfer, it became the next significant factor. The graph shows that increasing the gas velocity reduces the error. By increasing the gas velocity, the Nusselt number also increases (equation 3) and consequently increases the heat transfer rate to the TC. This makes the convection more dominant than the other heat transfer mechanisms and reduces the temperature measurement error. The velocity effect can also be seen in figure 8 which shows that C, ABC, CF and CD were the significant factor next to D.

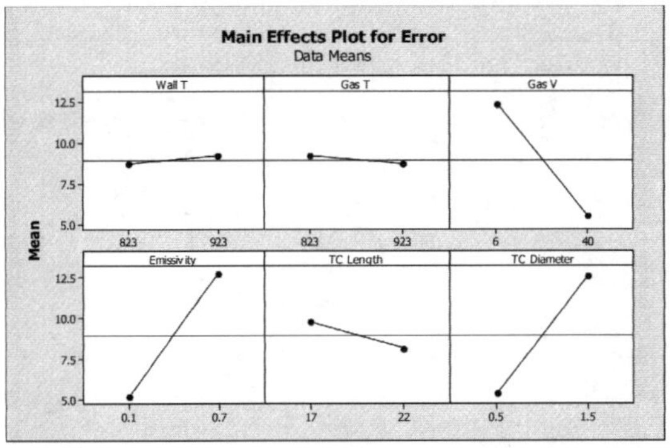

Figure 7. Main effects plot of the significant effects for scenario 2 (temperature is in K).

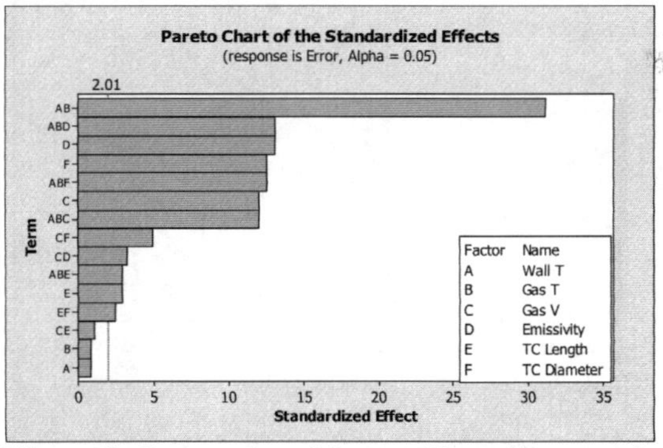

Figure 8. Pareto plot of factor effects for scenario 2.

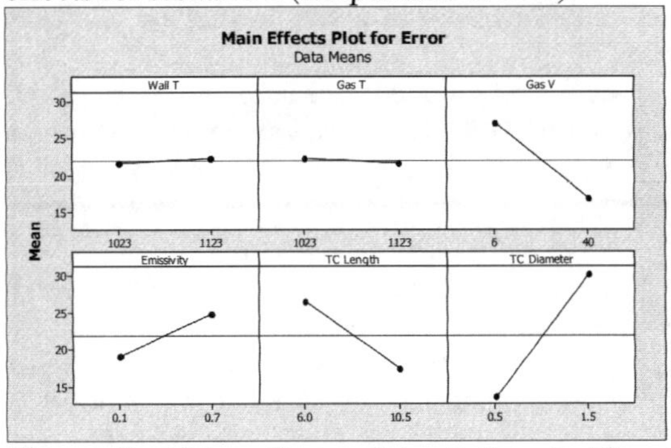

Figure 9. Main effects plot of the significant effects for scenario 3 (temperature is in K).

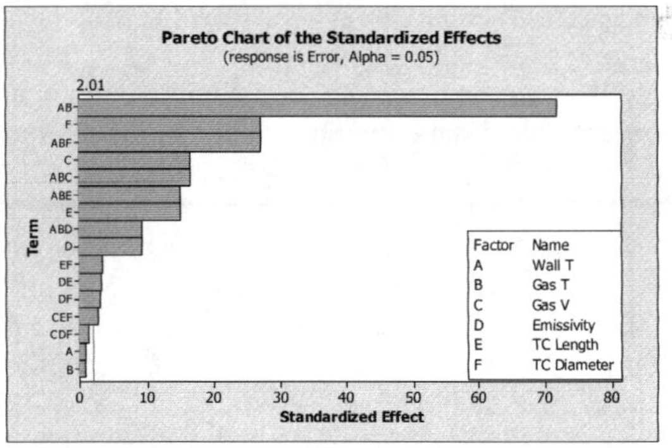

Figure 10. Pareto plot of factor effects for scenario 3.

Figure 9 and 10 show the DOE results for scenario 3. Although in scenario 3 the temperature range was 100 °C higher than scenario 1, D was the least single factor effect. This means that the TC stem effect (combination of E and F) was dominant source of error in both scenarios (1 and 3). Despite the similarity between scenario 1 and 3, the errors (Standard Effect) for various factors were higher in scenario 3 than scenario 1.

The DOE results for scenario 4 were depicted in figure 11 and 12. The results show that D was the most significant single factor and E was the least significant single factor affecting the temperature measurement error. This was similar to the results were obtained from scenario 2. The gas velocity, C, was the second highest significant single factor in scenario 4 but in scenario 2 F was the second highest significant factor. Despite this difference in scenario 1 and 4, the difference between C and F were not significant for both scenarios. This means that D, C and F were significant factors affecting the temperature measurement regardless of the operating temperature when the TC length was longer that scenario 1 and 3.

As discussed earlier, some of the assumptions such as thermal conductivity (between and the TC and fitting), TC emissivity and flow profile are not fully representative of the real world. Therefore, the DOE and modelling results require experimental validation. A test rig has been manufactured and the results and comparison with the DOE model will be reported in a future paper.

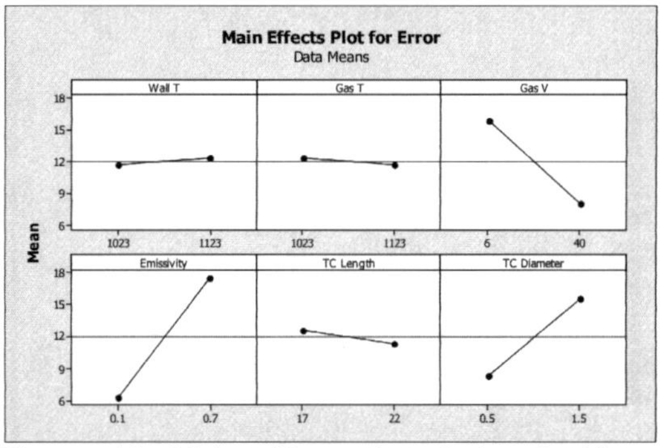

Figure 11. Main effects plot of the significant effects for scenario 4 (temperature is in K).

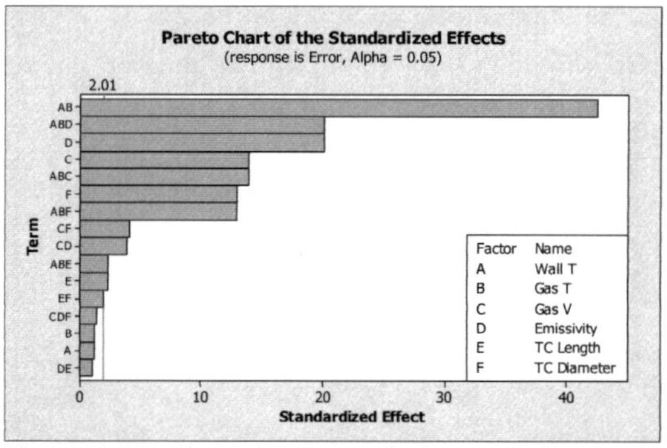

Figure 12. Pareto plot of factor effects for scenario 4.

6. Concluding remarks

A one-dimensional mathematical model was developed and numerically solved to investigate the temperature measurement error for a thermocouple placed at a centre of a pipe to measure the flow temperature. DOE method was applied to the modelling results to investigate the key variables affecting the temperature measurement. Four temperature measurement scenarios were proposed based on conditions similar to those found in metal-supported SOFC. One scenario was the TC length was equal to the half of the pipe internal diameter and the second scenario was extending the TC length for 10mm. Both scenarios were analysed at two different temperature ranges. One for SOFC stack temperature (550 °C – 650 °C) and the other was for the gas temperature after the burner (750 °C – 850°C). In this study a full factorial design at two levels for six variables was considered for each scenario. The DOE results for each scenario were reported and analysed. The results showed that for short TC length, TC diameter and TC length were the most significant factors even at higher temperature range. By extending the TC length, TC emissivity and gas velocity became significant factors even at lower temperature range.

Acknowledgements
The authors wish to thank Knowledge Transfer Partnership (KTP), Technology Strategy Board (TSB), University of Brighton and Ceres Power Ltd for their funding of this project.

References

[1] Tsikonis L, Van herle J and Favrat D 2012 The Error in Gas Temperature Measurements with Thermocouples: Application on an SOFC System Heat Exchanger *Fuel Cells* **12** 32-40

[2] Villafañe L and Paniagua G 2013 Aero-thermal analysis of shielded fine wire thermocouple probes *Int.l J. of Thermal Sci* **65** 214-223

[3] Yilmaz N 2011 Detailed multiphysics modeling and validation of thermocouple readings in fires *J. of Fire Sci* **29**(5) 443-464

[4] Roberts I L and Coney J 2011 Estimation of radiation losses from sheathed thermocouples *Appl Thermal Eng.* **31** 2262-2270

[5] Blevins L G and Pitts W M 1999 Modeling of bare and aspirated thermocouples in compartment fres *Fire Safety J.* **33** 239-259

[6] Rizika J W and Rohsenow W M 1952 Thermocouple Thermal Error *Eng. and Pro. Development* **44**(5) 1168-1171

[7] Yilmaz N 2012 Effect of thermocouple insertion depth in multiple directions on temperature measurements *J. of Fire Sci* **30**(3) 201-210.

[8] Incropera F P, Dewitt D P, Bergman T L and Lavine A S 2013 *Principles of Heat and Mass Transfer* 7th ed. (Singapore: John Wiley & Sons) p796

[9] Kadoya K, Matsunaga N and Nagashima A 1985 Viscosity and thermal conductivity of dry air in the gaseous phase *J. Phys. Chem. Ref. Data* **14** 947-9770

[10] Asadi M and Haghighi Khoshkho R 2013 Temperature Distribution along a Constant Cross Sectional Area Fin *International Journal of Mechanics and Applications* **3** 131-137

[11] Cao G, Weber S J, Martin S O, Sridharan K, Anderson M H and Allen T R 2013 Spectral emissivity of candidate alloys for very high temperature reactorsin high temperature air environment *J. of Nuclear Materials* **441**(1–3) 667-673

Thermal characterization of intumescent fire retardant paints

L. Calabrese[1], F. Bozzoli[1], G. Bochicchio[2], B. Tessadri[2] , S. Rainieri[1*], G. Pagliarini[1]

[1]Department of Industrial Engineering - University of Parma
Parco Area delle Scienze 181/A, 43124 Parma, Italy.
[2]National Research Council of Italy (CNR), Trees and Timber Institute – Fire Lab. Via Biasi, 75, 38010 San Michele all'Adige, Italy.

E-mail address of the corresponding Author (Sara Rainieri): sara.rainieri@unipr.it

Abstract Intumescent coatings are now the dominant passive fire protection materials used in industrial and commercial buildings. The coatings, which usually are composed of inorganic components contained in a polymer matrix, are inert at low temperatures and at higher temperatures, they expand and degrade to provide a charred layer of low conductivity materials. The charred layer, which acts as thermal barrier, will prevent heat transfer to underlying substrate. The thermal properties of intumescent paints are often unknown and difficult to be estimated since they vary significantly during the expansion process; for this reason the fire resistance validation of a commercial coatings is based on expensive, large-scale methods where each commercial coating-beam configuration has to be tested one by one. Adopting, instead, approaches based on a thermal modelling of the intumescent paint coating could provide an helpful tool to make easier the test procedure and to support the design of fire resistant structures as well. The present investigation is focused on the assessment of a methodology intended to the restoration of the equivalent thermal conductivity of the intumescent layer produced under the action of a cone calorimetric apparatus. The estimation procedure is based on the inverse heat conduction problem approach, where the temperature values measured at some locations inside the layer during the expansion process are used as input known data. The results point out that the equivalent thermal conductivity reached by the intumescent material at the end of the expansion process significantly depends on the temperature while the initial thickness of the paint does not seem to have much effect.

1. Introduction

Steel structures are widely used in building construction due to their high mechanical strength, ductility and execution times. However, due to deterioration of mechanical properties with temperature, it is essential that the whole structure and its elements have the necessary fire resistance to prevent a collapse in case of fire. One measure intended to the design of structures with the required fire resistance is to use higher cross-sectional elements or fire-resistant steels. An increase of the fire resistance time can be achieved by applying passive fire protection materials that insulate the steel structure from the effects of the elevated temperatures that are generated during a fire. They can be divided into two categories: non-reactive materials, of which the most common types are incombustible boards and spray of a cooling fluid, and reactive materials, of which the intumescent coatings are an example. They are available as solvent or water-based systems, which are applied with a thickness of up to approximately 3 mm. The time for the steel component to reach a critical point of behavior confers the rating to the coating. Typically this occurs when the steel component reaches about 550°C, the temperature at which steel shows a considerable loss in strength. The advantages of this kind of protection include the reduced overweight compared to other materials, off-site application and a good surface finishing, which can be improved with the application of decorative films[1]. Intumescent systems consist generally of three compounds: an acid source (ammonium phosphate-APP), a carbon source (pentaerythriol - PER) and a blowing agent (melamine). These compounds are linked each other by a polymer matrix which plays a capital role in the intumescent behavior. Phosphorous polymeric binders are often used as flame retardancy. The paint layer evolves under the effect of the heating source in a complex way [2], with an interplay of physical and chemical processes that occur in a suitable sequence (in series or in parallel) [3,4].

Exposed to a heating flux, the behaviour of the intumescent coating can be described by successive steps. Initially under the effect of the thermal flux, the ablation of a part of the intumescent paint then occurs, gases are released and they cause the swelling of a part of the paint. Subsequently, a charring zone appears, which grows with time, until being stabilized in stationary state. The ablative layer goes on regressing and the expanding layer swells. The non reactive char layer grows to the detriment of the reactive layer and the ablative layer. Finally, the original intumescent paint is totally ablated, the paint stops to swell and just a stable char layer stays [2].

The fire resistance of a commercial coating is usually tested experimentally, by using large-scale tests (see, for instance, EN 13381-4 and EN 1363-1 standards [5,6]). Such tests are expensive and time-consuming, with a large number of tests required to cover the range of steel configurations and protection thicknesses typically required in the construction field.

Adopting a modelling approach based on thermal properties of the intumescent paint could, instead, provide a helpful tool to make easier the test procedure and to support in the design of fire resistant structure, as well. Therefore, the knowledge of the equivalent thermal conductivity of the intumescent layer is therefore an important task, for the purpose of predicting the behavior of metal structures subject to the action of fire, when protected by intumescent paints. Some researchers [7-11] have proposed to estimate the equivalent thermal conductivity by solving the one-dimensional non-steady inverse heat conduction problem in the intumescent layer. In this procedure, the directly measured parameters are the temperature-time response of the metallic element on which the intumescent paint is smeared and the flame temperature [10] or the heat flux density produced by a cone calorimeter apparatus [8,9]. However, the determination of the net heat flux absorbed by the free surface of the intumescent layer is problematic, because of the uncertainty in parameters which characterize its thermal interaction with the surrounding environment, such as the coefficients of absorption, reflection and emission of the surface, the view factor towards the environment and its temperature, the coefficient of convection through which the surrounding air removes the heat from the surface of the intumescent layer. Others [9,12,13] have attempted an estimate of the equivalent thermal conductivity through the inversion of simple formulas, as the ones proposed in the technical regulations aimed at the determination of the temperature limit which is reached by metallic structures protected by thermally insulating coatings. However, the estimated equivalent thermal conductivity values are sometimes questionable (see, for example,[7,9,12]), since they are much lower than the thermal conductivity of air at the same temperature.

All these approaches, do not guarantee to achieve the sought result with a sufficient accuracy, due to the lack of reliable data on the thermal phenomena that occur at the free boundary of the intumescent layer. To overcome this difficulty, in this paper the temperature inside the intumescent layer is included among the variables being measured. In this way, the formulation of the inverse problem no longer requires the definition of the thermal conditions at the free surface of the intumescent layer, which is subjected to the heat flux, while the effective thermal conductivity of the intumescent layer becomes the only unknown parameter of the protective coating to be estimated through the solution of the inverse heat conduction problem.

2. Experimental setup and estimation procedure

The considered intumescent paint consists of a polymeric PVC resin with additives: titanium dioxide, melamine and 2-butoxyethanol. The equivalent thermal conductivity of the intumescent layer was measured for two different initial thicknesses of paint, i.e. 0.8 mm (test specimen S08) and 1.2 mm (test specimen S12), evenly distributed on the face of a square steel plate with a side of 100 mm and 8 mm thick. In these conditions, the specimens were subjected to a heat flux of nominal 50 kW/m^2, produced by a cone calorimeter apparatus placed at the distance of 80 mm from its surface. The other surfaces of the specimen, both the rear and the sides, was thermally insulated with 2 cm of a calcium silicate board. The whole system was then placed in a metal box open at the top, figure 1.

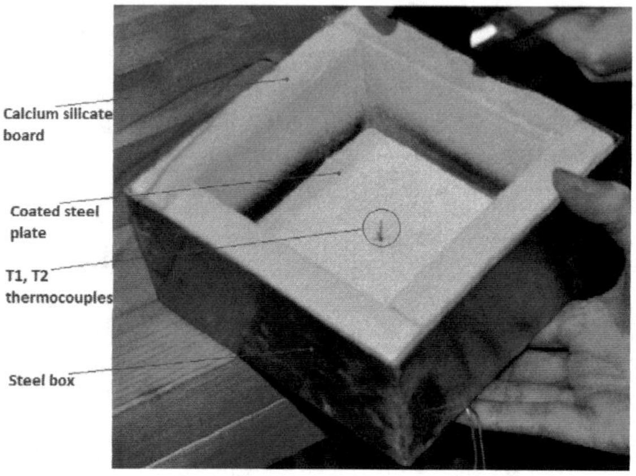

Figure 1. Specimen under test.

The steel plate temperature, T_s, was measured by two K-type thermocouples, thermocouples *TS*, placed in contact with the rear face. The temperature in the intumescent layer was measured by two additional K-type thermocouples, which are protected by a metal sheath of circular cross section with a diameter of 1 mm, thermocouples *T1* and *T2*, respectively.

These thermocouples were installed through a small hole drilled in the metal plate, normally to the coated surface, with the axes at a distance of 2 mm. These arrangement prevent to possible displacements of the thermocouple during the swelling of the material under test. However it makes the exact identification of the position of the sensing element of the thermocouple problematic. In this respect, on the basis of the technical data provided by the manufacturer, it was assumed that the sensing element of the thermocouple is located at 0.5 mm from the apex of the protective metal sheath. With this assumption, the distance of the sensing element from the upper surface of the steel plate is given in table 1.

Table 1. Distance (mm) of the sensing element of the thermocouples from the metal surface of the specimen.

Specimen	Thermocouple *T1* [mm]	Thermocouple *T2* [mm]
S08	5.5	13.3
S12	8.8	15.6

After a sufficiently long time interval, thermocouples *T1* and *T2* are incorporated in the intumescent material and they begin to detect its temperature. In consideration of the relatively high thermal conductivity of the material (steel) and of the small thickness of the metal plate under test, the lumped parameter thermal model was adopted to describe its temperature behavior when exposed to the heat flux. This assumption allows to estimate transmitted to the metal plate through the protective layer, by using the energy balance of the plate itself. In fact, by denoting with q_s and q_d the heat absorbed by the plate and the heat lost through its back face, respectively, with reference to the unit surface area of the substrate plate, its energy balance gives:

$$q = q_s + q_d = \rho_s \, c_s \, s \, \frac{dT_s}{dt} + K\,(T - T_a) \qquad (1)$$

where ρ_s, c_s, s are, respectively, density, specific heat and thickness of the plate; *K* is the overall heat transfer coefficient which accounts for the heat lost from the back face of the specimen towards the environment, whose temperature is T_a. Anderson et al.[10] note that, once the intumescent layer has developed, *q* is approximately a constant. This circumstance implies that the thermal regime through the intumescent layer is almost steady and, as a consequence, the equivalent thermal conductivity of the layer could be estimated through direct application of Fourier's law for a simple plane situation:

$$\lambda_{eq} = \frac{q \cdot d}{\Delta T} \qquad (2)$$

where ΔT is the temperature difference between two planes parallel to the face of the specimen, *d* away.

Unfortunately, with the present experimental arrangement the application of eq. 2 is not allowed, since it is expected that the thermocouples installed in the intumescent material perturb the temperature field. In fact, a non-negligible heat flux towards the base plate is established along the metal sheath of the thermocouple, which tends to depress the measured temperature, compared to the value detectable in its absence. For this reason in this paper, the inverse problem of heat conduction in the intumescent layer was formulated as shown in figure 2.

Journal of Physics: Conference Series **547** (2014) 012005 doi:10.1088/1742-6596/547/1/012005

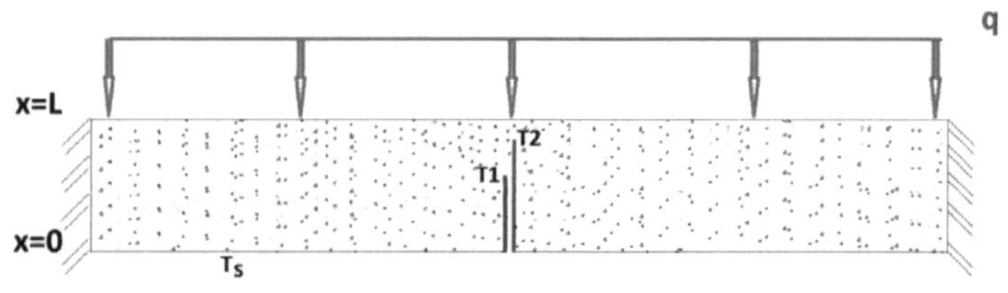

Figure 2. Thermal schematization of inverse heat conduction problem.

At x = 0 the temperature is known, as measured by the thermocouples, *TS*, placed in contact with the metal plate. Thickness *L* is set in an arbitrary way, in order to be sufficient to consider the temperature for x = L to be uniform. At x = L a Neumann boundary condition is adopted and the imposed heat flux value is estimated by eq. 1. The thermocouples were modeled in the inverse problem following the manufacturer specification. At x = L the temperature is unknown, while it is known at points 1 and 2, as measured by the thermocouples *T1* and *T2*, respectively. The side surfaces of the intumescent layer was considered to be adiabatic. The presence of thermocouples *T1* and *T2*, whose metal sheath may significantly perturb the temperature field, makes the thermal problem three-dimensional. In view of the slow variation of the thermal conditions which follow the development of the intumescent layer, the conductive thermal problem was solved in steady state.

In the cone calorimetric tests it is expected that the temperature across the intumescent layer varies by several hundreds of Kelvin. According to the analysis of Staggs [4], the corresponding variation of the equivalent thermal conductivity of the expanded intumescent char should then be very significant. Nevertheless, in the present investigation a uniform equivalent thermal conductivity is assumed, as representative of a first approximation estimate.

The equivalent thermal conductivity was then estimated by requiring that the following objective function have a minimum [14]:

$$F(\lambda_{eq}) = [T_{measured} - T_{computed}(\lambda_{eq})]_1^2 + [T_{measured} - T_{computed}(\lambda_{eq})]_2^2 \quad (3)$$

Where $T_{measured}$ is the temperature value measured by thermocouples *T1* and *T2*, while $T_{computed}(\lambda_{eq})$ is the corresponding temperature value given by the numerical model as a function of the equivalent thermal conductivity.

3. Experimental results

The tests under the cone calorimeter were continued for approximately 4000 s (S08 and S12). Figure 3 shows the temperature trend of the metal plate, obtained as the average value of the readings of the two thermocouples arranged in contact with the rear surface of the specimen.

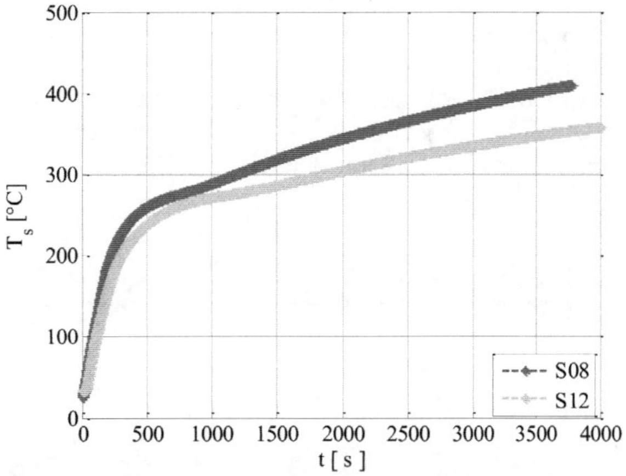

Figure 3. Steel plate temperature histories acquired in the cone calorimeter test.

In the time interval considered for estimating the thermal conductivity of the intumescent layer, the two thermocouples to provided temperature values which are coincident within 1 or 2 °C, for the samples S08 and S12, respectively. As expected, the plate temperature increases monotonically, but at a varying velocity, which continues to decrease till the end of the test time. A similar behavior was detected by Anderson et al. [10]. To the opinion of these authors, the decrease of the heating velocity even after the completion of intumescence is due to the increase of the heat loss, q_d, towards the environment as the steel plate temperature increases.
The heat absorbed by the steel plate, q_s, was then calculated from the temperature time response curve, with the properties of steel specified in table 2.

Table 2. Thermal conductivity, density and specific heat for steel.

λ_s [W/m·K]	ρ_s [kg/m^3]	c_s [J/kg · K]
45	7850	425+0.7773 T-0.00169 T^2+0.0000022 T^3

The heat lost through the back face of the specimen, q_d, was instead computed in steady state (eq. 1). By neglecting the thermal resistance of the metal box wall, the overall heat transfer coefficient, *K*, was defined as a series of two heat transfer resistances:

$$K = \frac{1}{R_{cs} + R_{cr}} \tag{4}$$

The first one is the thermal resistance of the calcium silicate board:

$$R_{cs} = \frac{s}{\lambda_{cs}} \tag{5}$$

while the second one is the thermal resistance from the external wall of the metallic box to the environment, whose temperature T_a was 25°C. It consists of two thermal resistances in a parallel configuration, due to heat convection and thermal radiation:

$$R_{cr} = \frac{1}{h_c + h_r} \tag{6}$$

The coefficient of radiation and of convection was computed as follows:

$$h_r = \sigma_0\, a\, (T + T_a)(T^2 + T_a^2) \tag{7}$$

$$h_c = \frac{\lambda}{l\,(\mathrm{Gr}\cdot\mathrm{Pr})^{\frac{1}{3}}} \tag{8}$$

The thermal properties of air was computed at film temperature. For the wall of the metallic box the absorption coefficient a=0.7 was assumed. The overall heat transfer coefficient slightly depends from the temperature of both the box wall and the environment. In the present measurements its averaged value was 7.2 and 7.0 W/m^2K, for specimens S08 and S12, respectively.
For both samples, the heat flux, q, that reaches the metal plate is shown in figure 4 and 5, together with its components, q_s and q_d.

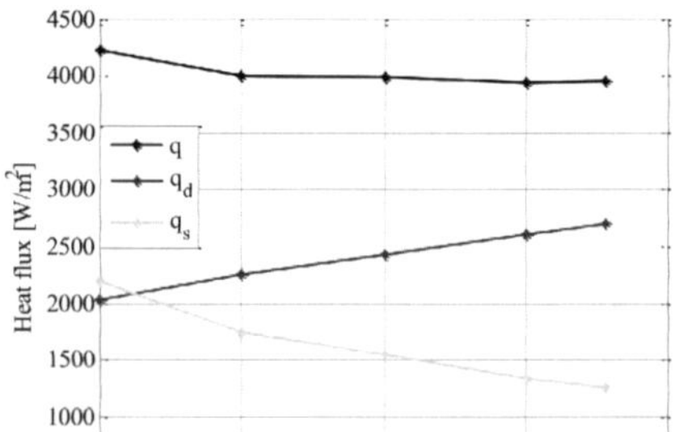

Figure 4 . Heat fluxes for sample S08.

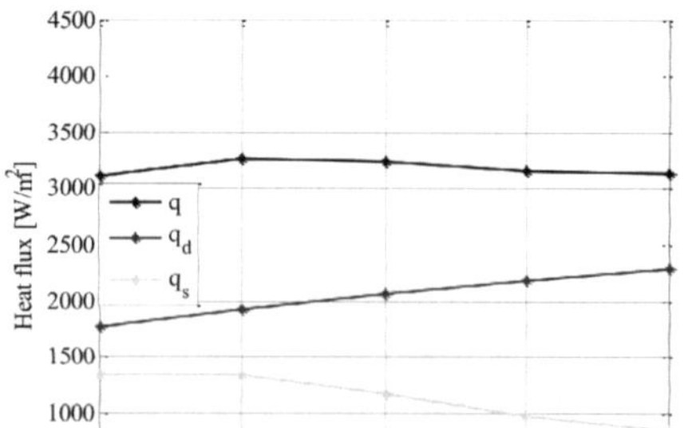

Figure 5. Heat fluxes for sample S12.

The graphs of figure 4 and 5 confirm that, after completion of intumescence, the heat flux through the intumescent layer is just constant with time. In particular, from t = 2000 s, q remains constant within ±1 and ±2% for specimens S08 and S12, respectively. Instead, q_s and q_d are not steady. In particular,

while q_s decreases with time, q_d increases, since the temperature difference between the metal plate and the environment increases.
To the maximum heat flux through the metal plate it corresponds a maximum temperature difference, in the direction normal to the coated surface, of about 0.6°C, less than 0.2% of the maximum temperature difference which the plate experiences during the whole heating process. Therefore, the assumption of lumped parameters which underlies eq. 1, can be considered verified with sufficient accuracy for the purposes of the present analysis.
The assumption of negligible heat losses in the experimental test aimed to the estimation of the equivalent thermal conductivity is at the base of the procedure suggested by many researchers, see, for instance, Anderson et al. [10], Mesquita et al. [9]. Disregard the heat losses in the energy balance equation of the steel plate (eq. (1)) may result in an underestimation of the heat flux through the intumescent layer and, as a consequence, it may result in a proportional underestimate of the equivalent thermal conductivity. To the assumption of negligible heat losses may be due some unrealistically low values of the equivalent thermal conductivity which are reported in the technical literature on the subject of intumescent protective coatings.
The temperature time response by thermocouples *T1* and *T2* is shown in figures 6 and 7, for specimens S08 and S12, respectively.

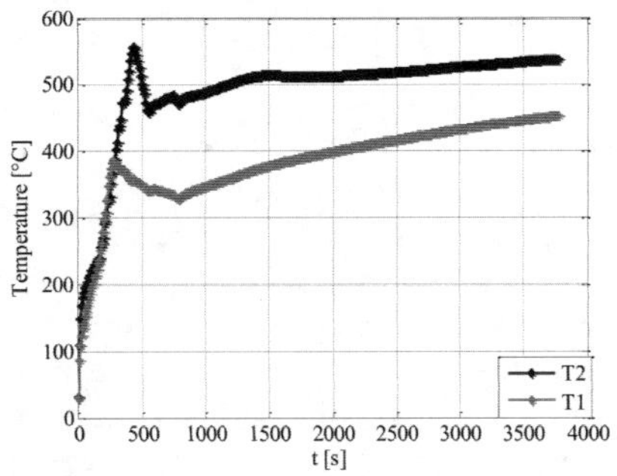

Figure 6. Temperature detected by thermocouples T1 and T2 during the intumescence process for sample S08.

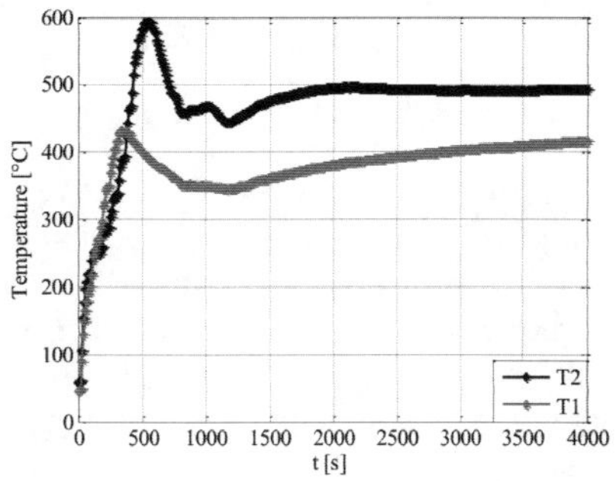

Figure 7. Temperature detected by thermocouples T1 and T2 during the intumescence process for sample S12.

The maximum value of temperature is likely reached when the intumescent layer incorporates the thermocouple, which previously is directly exposed to the heat flux emitted by the cone calorimeter. The graphs show that the intumescent layer had already reached the thickness of almost 14 mm (about 35% of the final thickness (40 mm)) at t = 500 s, for specimen S08, while, for specimen S12, the thickness of 16 mm (about 27% of the final thickness (59 mm)) was reached at t = 750 s.

4. Equivalent thermal conductivity estimation

The steady thermal problem schematized in figure 2 was solved by the numerical finite elements method, implemented inside the Comsol Multiphysics environment. The objective function (eq. 3) subject to the given constraints was then minimized through the Matlab Optimization Toolbox® by using Nelder-Mead minimization algorithm. The results, in terms of equivalent thermal conductivity, obtained for the two specimens are shown in figure 8, as a function of the temperature of the substrate metal plate.

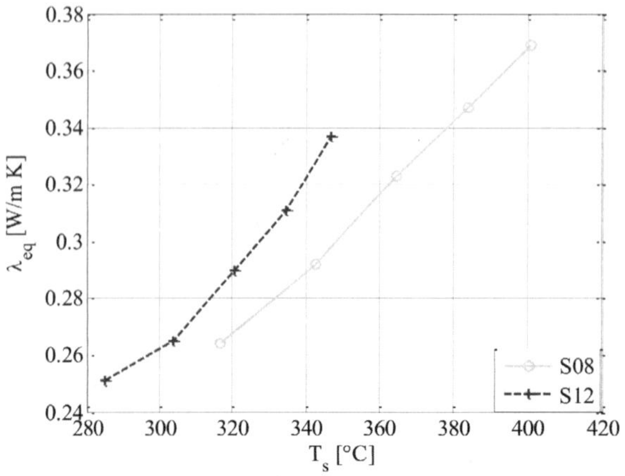

Figure 8. Equivalent thermal conductivity, obtained for the two specimens versus the steel temperature.

As expected, the data show a great effect of temperature. An increase in the plate temperature of hundred Kelvin produces an increase of about fifty per cent in the effective thermal conductivity. Therefore, a proper data reduction procedure should account for the temperature variation across the intumescent layer, since a temperature variation of several hundreds of Kelvin is expected between the free surface of the protective layer and the surface in contact with the metal plate.

Regarding the effect of the initial paint thickness, the results presented in figure 8 show a systematically higher equivalent thermal conductivity for specimen S12. However, this behavior might be due to the above cited dependence on temperature. Indeed, in spite of the higher plate temperature for specimen S08 as shown by figure 3, it is expected that the average temperature across the layer thickness is higher for specimen S12 than for specimen S08. This circumstance derives from the estimated heat flow rate through the layer, q, and from the layer thickness, as measured at the end of the test. The average difference between the mean temperature through each layer is about between 25°C to 30°C. Shifting the curve for specimen S12 on the right of this temperature interval reduce the difference between the estimated λ_{eq} values. As a consequence, the initial thickness of the paint seems to have a quite negligible effect on the effective thermal conductivity of the intumescent layer.

5. Conclusion

A method for estimating the equivalent thermal conductivity of the char layer generated by the intumescence process has been presented and assessed through its application to two specimens, with different initial paint thickness. The method is based on the heat conduction inverse problem approach, with temperature measured in two different point of the intumescent layer as inputs known data.

The results have confirmed that the equivalent thermal conductivity highly depends on the temperature, while paint thickness does not seem to have great importance.

6. Acknowledgements

The authors thank Renato Leveghi from National Research Council of Italy (CNR), Trees and Timber Institute for the help in the experimental activity.

References

[1] Wilkie C A, Morgan A B, 2012 The Fire Retardancy of Polimeric Materials, Second edition, *CRC Press*.

[2] Rodier S, Autrique L, Perez L, Ramdani N, Serra J J, 2005 Fire protective coatings evaluation: tests and modelling based on a nonlinear partial differential equations system, *Imacs -World Congress.* Paper T6-R-00-0406

[3] Griffin G J, 2010 The modelling of heat transfer across intumescent polymer coatings , *Journal of Fire Sciences*. **28**

[4] Staggs J E J, 2010 Thermal conductivity estimates of intumescent chars by direct numerical simulation, *Fire Safety Journal*. **45** 228-237

[5] UNI EN 13381-4, 2013 Test methods for determining the contribution to the fire resistance of structural members - Part 4: Applied passive protection to steel members

[6] UNI EN 1363-1, 2001 Fire resistance tests General requirements

[7] Mesquita L, Piloto P, Vaz M, 2009 Efficiency evaluation of intumescent coatings used for fire protection:comparison between numerical method and experimental results, *Integrity Reliability and Failure*. **Chapter XVIII** 537-538

[8] Bartholmai M, Schartel B, 2007 Assessing the performance of intumescent coatings using benchscaled cone calorimeter and finite difference, *Fire and Materials*. **31** 187-205

[9] Mesquita L, Piloto P, Vaz M, 2007 Experimental study of intumescent fire protection coatings, *Fire Retardant Coatings II, Berlin, Germany.*

[10] Anderson C E, Ketchum D E, Mountain W P, 1988 Thermal Conductivity of Intumescent Chars, *Journal of Fire Sciences*. **6** 390-410

[11] Bartholmai M,Schriever R, Schartel B, 2003 Influence of external heat flux and coatings using cone calorimeter and numerical analysis, *Fire and Materials*. **27** 151-162

[12] Xianghe D,Wang Y, Bailey C, 2010 A simple method to predict temperatures in steel joints with partial intumescent coating fire protection, *Fire Technology*. **46** 19-35

[13] Li G Q, Lou G B, Zhang C,Wang L L, Wang Y C, 2012 Assess the fire resistance of intumescent coatings by equivalent constant thermal resistance, *Fire Technology*. **48** 529-546

[14] Beck J V, Blackwell B, Clair JR. C R ST , Inverse heat conduction, Ill-posed problem, New York etc., J. Wiley & Sons 1985.

Analysis of thermodynamic losses in ground source heat pumps and their influence on overall system performance

C Casarosa, P Conti, A Franco, W Grassi, D Testi

Department of Energy, Systems, Territory and Constructions Engineering (DESTEC), University of Pisa, Largo Lucio Lazzarino – 56122 Pisa, Italy

E-mail: paolo.conti@for.unipi.it

Abstract. The present work aims at identifying the relative influence of GSHP subsystems (viz. ground source, earth heat exchangers, heat pump unit, pumping devices) on the overall efficiency and the limits to which technological improvements should be pushed (because, beyond these limits, only minor benefits may be achieved). To this end, an analysis of thermodynamic losses is conducted for a case study, followed by a sensitivity analysis on the heat pump unit thermal performance. Primary energy consumptions of nine configurations with different combinations of ideal and real subsystems are compared. The completely ideal system is used as the reference to normalize energy consumptions and obtain a dimensionless efficiency parameter. The results show that – when a proper design methodology is employed – the performance of the borehole heat exchangers slightly affects the overall efficiency. On the contrary, the thermal response of the ground and the thermal and hydraulic performances of the heat pump unit are key factors. Finally, a sensitivity analysis is conducted by increasing the heating and cooling efficiencies of the heat pump device.

1. Introduction

Ground source heat pump (GSHP) systems are globally recognized as one of the most promising technologies in terms of economic and energy savings. However, despite the aroused interest, operative performances can be lower than expected [1–4], possibly reducing the attractiveness of GSHPs with respect to air heat pumps and condensing boilers.

As well known, GSHPs involve different subsystems: ground source, ground-coupled heat exchangers, ground-coupled heat pump unit (GHP), and pumping devices. Remarkable benefits can be achieved through the application of proper sizing and management strategies aimed at optimizing the synergy among GSHP subsystems [5–9]. Further performance enhancement could be obtained by reducing all the possible causes of losses in the system. This work aims to find which component mainly affects the overall performance, in order to evaluate the corresponding room for improvement and – through a sensitivity analysis on technological performances – identify appropriate strategies for GSHP development.

Thermal inefficiencies are typically investigated by means of the "*second-law analysis*" or "*exergy analysis*" (see, for instance, [10]). *Exergy efficiency* (ψ) is the corresponding index of performance: its value, given by the ratio of the desired exergy output to the exergy used, is generally associated with a measure of the relative deviation between a real system and the corresponding ideal one.

Several works have involved exergy analyses of ground heat pumps (see, for instance, [11–15]). However, it is worth recalling that the results of this type of analysis are strongly dependent on the choice of the reference state, especially when the operative temperatures of the system are close to it [12,16]. Heat pump applications are usually investigated using external air as reference state [11–14,17], therefore, this choice is pivotal.

In the present work, an alternative approach has been followed: instead of analyzing exergy fluxes, ideal subsystems are simulated, comparing primary energy consumptions. The GSHP system comprises four subsystems, viz. ground source, borehole heat exchangers (BHEs), ground-coupled heat pump unit (GHP), and pumping devices; each of them has an ideal reference configuration. Combining ideal and real subsystems, nine different configurations, listed in table 1, are obtained.

Table 1. Analyzed configurations in terms of real and ideal subsystems.

Configuration	Ground source	BHEs	GHP	Hydraulic losses
#1	Real	Real	Real	Real
#2	Ideal	Real	Real	Real
#3	Real	Ideal	Real	Real
#4	Real	Real	Ideal	Real
#5	Real	Real	Real	Ideal
#6 = #2 + #4	Ideal	Real	Ideal	Real
#7 = #3 + #4	Real	Ideal	Ideal	Real
#8 = #4 + #5	Real	Real	Ideal	Ideal
#9 = #2 + #3 + #4 + #5	Ideal	Ideal	Ideal	Ideal

Global design and control strategy have been optimized in every configuration, thus, final energy consumptions depend only on the performance of the various subsystems. Considering an ideal subsystem, it is possible to quantify the maximum benefits that can be achieved improving the technological level of each component. The optimization procedure has been presented in [6,7] and it is shortly recalled in Sections 2 and 3.

A dimensionless efficiency parameter (ε) is used to normalize and compare the energetic performances of the different simulated configurations. ε is based on the *"task efficiency"* definition provided by Moran [10]. In the present case, it reads:

$$\varepsilon = \frac{En_p^*}{En_p} \tag{1}$$

Where En_p^* is the theoretical minimum primary energy consumption, obtained by a loss-free system (Configuration #9), and En_p is the actual primary energy consumption of the system.

2. GSHP systems modelling

An introductory scheme for GSHPs simulation has been outlined in [5] and fully developed in [6]. A quasi-steady-state approach is adopted: in each time step, a full set of equations (2), made of each subsystem model, is solved. The two coefficients f_H and f_C represent the control strategy: their value correspond to the fraction of building load delivered by the geothermal heat pump in heating and cooling mode, respectively. The system comprises:

- equation (2.1), which imposes that the total heat exchanged between the BHE field and the ground (Q_g) is equal to the heat transferred in the evaporator/condenser ($Q_{E/C}$);
- equation. (2.2), which is the energy balance for the ground loop in the evaporator/condenser;
- equation (2.3), representing the heat pump unit; the function F correlates the HP performance to the operative conditions; in this work, generators performances are evaluated according to the current Italian technical standard [18–20], including the electric energy needed for pumping;
- equation (2.4), representing the BHE field; the function E correlates Q_g to the ground temperature at the borehole surface (T_g), to the BHEs characteristics, and to the ground-coupled loop operative parameters (flow rate and temperature); in this work, the classical ε-*NTU* method for heat exchangers analysis has been used;
- equation (2.5), representing the ground source; the function S correlates T_g to heat fluxes and thermo-physical properties; in this work, the finite line source model (FLS)[21] has been used together with time and space superposition technique [7,21];
- equation (2.6), which is the share of thermal load at the GHP for heating (f_H) and cooling (f_C);
- equation (2.7), representing the back-up system; similarly to the GHP, the performance of the back-up generator is influenced by its capacity ratio and by the temperature of the end-user loop; the function B characterizes the employed back-up technology;
- equation (2.8), which imposes that the building thermal load (Q_l), is given by the sum of the thermal energies delivered/removed by GSHP and back-up generators.

$$\begin{cases} Q_{E/C} = Q_g & (2.1) \\ Q_{E/C} = \dot{m}_w c_w \left| T_{win} - T_{wout} \right| \tau & (2.2) \\ Q_{E/C} = F(T_{win}, T_l, CR_{H/C}) & (2.3) \\ Q_g = E(\dot{m}_w, T_{win}, R_b, H, N_{BHE}) & (2.4) \\ T_g = S(T_g{}^0, Fo_g, Q_g) & (2.5) \\ f_H = \frac{Q_E}{Q_l}\left(\frac{COP}{COP-1}\right) \quad f_C = \frac{Q_C}{Q_l}\left(\frac{EER}{EER+1}\right) & (2.6) \\ Q_{E/C} = B(T_l, CR_{H/C}) & (2.7) \\ Q_{bk} = (1 - f_{H/C}) \cdot Q_l & (2.8) \end{cases}$$

3. Design and control optimization algorithm

As mentioned in Section 1, both the equipment sizing and control strategy are optimized for each configuration in order to highlight intrinsic inefficiencies. The overall optimization procedure involves two types of control variables: the ones related to the sizing of the earth heat exchangers (viz. number and depth of BHEs and flow rate in the ground-coupled loop) and the ones related to the control strategy (capacity ratio of GHP unit).

The optimization problem was formulated as a "multistage decision problem" [22]. A full discussion on problem formulation and resolution strategy can be found in [7]; here we shortly recall the main aspects. The problem reads:

$$E_p^{tot}(\mathbf{U}) = \min_{\mathbf{U}} \sum_{n=1}^{N} R(\mathbf{u}^n, \mathbf{x}^n) \qquad (3)$$

where:

- $\mathbf{u}^n = (u_1{}^n, u_2{}^n, ...) = (N_{BHE}, H, \dot{m}_w, CR^n_{H/C})$ is the vector of control variables at the nth stage; design variables have the same value at each stage;
- $\mathbf{x}^n = (x_1{}^n, x_2{}^n, ...)$ is the vector of state variables at the nth stage;
- $E_p^{tot}(\mathbf{U})$ is the objective function; in this work, we considered primary energy consumption;
- $\mathbf{U}$ is the set containing all the $\mathbf{u}^n$;
- $R(\mathbf{u}^n, \mathbf{x}^n)$ is the so-called "return function"; it represents the contribution of the nth stage to the total objective function;
- $f(\mathbf{u}^n, \mathbf{x}^n)$ is the mathematical model for GSHP simulation described by set of equations (2); f relates the state variables of a stage to the control and state variables of the previous stage.

The constraints of the optimization variables are:

- the depth of a single borehole (H) cannot exceed 100 m;
- the flow rate ($\dot{m}_w$) in the ducts must be high enough to guarantee a fully turbulent regime ($Re_D > 6000$) and a fluid velocity higher than 0.3 m/s;
- the supply temperature of the ground-coupled loop cannot be lower than 3°C or exceed 35°C, in order to avoid water freezing in the pipes or overheating of the ground.

4. Description of the analyzed configurations

4.1 Configuration #1: Benchmark

The benchmark configuration is based on the test case illustrated in [7]. A ground-coupled vertical heat exchanger heat pump system has been simulated during 10 years of operational life, a convenient period for evaluating the effects of possible long-term ground temperature drifts. The back-up generators are a condensing boiler and an air/water cooler. Heating and cooling loads are imposed, as shown in table 2, according to a numerical example given in [23], referring to a typical medium-scale office in the Mediterranean climate. The characteristics of ground, BHEs and generators employed in the simulation are reported in tables 3 and 4.

4.2 Configuration #2: Ideal ground source

This configuration considers an ideal ground source with an infinite thermal capacity. In this way, BHEs surface temperatures remain always constant: equation (2.e) is replaced by: $T_g^n = T_g^0$.

4.3 Configuration #3: Ideal borehole heat exchangers

In this case, ideal BHEs are employed: borehole thermal resistance (R_b) is considered null and heat exchanger effectiveness is set equal to 1. equation (2.d) is replaced by: $Q_g = \dot{m}_w c_w \left(T_{w_{in}} - T_g\right)\tau$.

4.4 Configuration #4: Ideal ground-coupled heat pump unit

The real GHP unit is replaced by a totally reversible thermodynamic cycle. All the HP components are considered ideal: operating and secondary fluids can exchange heat without temperature difference (no external irreversibilities) and compression and expansion processes are isentropic (no internal irreversibilities). This ideal device can deliver any thermal load without power limitations and with no penalizations due to low capacity ratios ($CR_{H/C}$).

4.5 Configuration #5: Ideal head loss

In this configuration, no distributed or lumped losses are present; thus, pumping energy is null.

Table 2. Monthly heating and cooling loads (GWh) of the tested office building.

	JAN	FEB	MAR	APR	MAY	JUN	JUL	AUG	SEP	OCT	NOV	DEC	TOT
Heating demand[a]	16.1	11.7	6.9	1.4	0	0	0	0	0	1.4	8.3	13.9	**59.7**
Cooling demand[b]	0	0	0	0	7.5	14.4	17.2	17.2	6.9	0	0	0	**63.3**

[a] Supply / return temperature of the building end-user loop: 45 / 40 °C.
[b] Supply / return temperature of the building end-user loop: 7 / 12 °C.

Table 3. Ground thermal properties and BHE characteristics.

Property	**Value**
Ground thermal conductivity [W/(m K)]	1.7
Ground thermal diffusivity [mm^2/s]	0.68
BHE diameter [m]	0.15
BHE configuration	Double U
Spacing between boreholes [m]	10
Grouting thermal conductivity [W/(m K)]	1.7
BHE pipe diameter (inner – outer) [m]	0.0262 – 0.032
U shank spacing [m]	0.094
Pipe thermal conductivity [W/(m K)]	0.35
BHE thermal resistance R_b [m K/W]	0.062

Table 4. Declared capacity (DC) of the ground heat pump and back-up generators.

Ground-coupled unit (water/water)		**Condensing boiler**	**Air/water unit**
Heating DC[a]	Cooling DC[b]	Heating DC	Cooling DC[b]
24.7 kW	22.9 kW	33.5 kW	59.2 kW

[a] Outdoor Heat Exchanger: Inlet 0°C / Outlet -3°C; Indoor Heat Exchanger: Inlet 40°C / Outlet 45°C [24].
[b] Outdoor Heat Exchanger: Inlet 30°C / Outlet 35°C; Indoor Heat Exchanger: Inlet 12°C / Outlet 7°C [24].

5. Results and discussion

Optimal control strategies and design variables, together with the main performance indices, are reported for each configuration in tables 5 and 6.

In Configuration #1, the $f_{H/C}$ sequence (control strategy) is given by the optimal synergy among GHP unit and back-up generators (condensing boiler and air/water heat pump). In heating mode, the geothermal solution performs better than the boiler, except during the transitional months (April and October), when the building load is below the control range of the GHP unit. $f_H = 0.9$ in January is due to the constraint imposed on the supply temperature of the ground-coupled loop ($T_{w_{in}} \geq 3$°C): indeed, the optimal BHEs number and depth (7 x 100 m) resulting from the best trade-off between heat transfer performance and pumping energy, is not sufficient to exchange all the heat required to match the building heating load. For the same reasons, during the cooling season, the BHEs field is not able to match the total cooling load, therefore, the air unit integration is always required. Nevertheless, the optimal solution does not split the cooling load between air and ground sources, but it finds more convenient operating the sole air unit during the hottest months (July and August) and the sole GHP during the others (June and September).

In Configuration #2, the ground temperature remains constant; hence, a reduced number of BHEs is sufficient to meet the total heating load.

Table 5. Optimal control strategies.

Configuration	**JAN**	**FEB**	**MAR**	**APR**	**MAY**	**JUN**	**JUL**	**AUG**	**SEP**	**OCT**	**NOV**	**DEC**
#1	0.9	1	1	0	1	1	0	0	1	0	1	1
#2	1	1	1	0	1	1	0.7	0.7	1	0	1	1
#3	1	1	1	0	1	1	0	0	1	0	1	1
#4	0.9	1	1	1	1	1	1	1	1	1	1	1
#5	1	1	1	0	1	1	0.7	0.7	1	0	1	1
#6 = #2 + #4	1	1	1	1	1	1	1	1	1	1	1	1
#7 = #3 + #4	0.9	1	1	1	1	1	1	1	1	1	1	1
#8 = #4 + #5	1	1	1	1	1	1	1	1	1	1	1	1
#9 = #2 + #3 + #4 + #5	1	1	1	1	1	1	1	1	1	1	1	1

Table 6. Optimal design variables and performance indices.

Configuration	E_p^{tot} [MWh]	ε [-]	N_{BHE} [-]	H [m]	$\dot{m}_w$ [kg/s]	f_H [-]	f_C [-]	COP [-]	EER [-]	CR_H [-]	CR_C [-]	η [-]	EER_{air} [-]	CR_{air} [-]
#1	856	0.19	7	100	2.39	0.93	0.46	3.94	3.75	0.98	1	1.09	3.18	0.94
#2	721	0.22	4	100	1.36	0.95	0.84	4.83	4.70	1	0.84	1.09	2.18	0.28
#3	843	0.19	7	100	2.39	0.95	0.46	3.98	3.84	1	1	1.09	3.18	0.94
#4	375	0.42	7	100	2.39	0.97	1	6.41	12.0	0.98	1	1.09	-	-
#5	713	0.22	n.a. (∞)	100	n.a. (∞)	0.95	0.84	5.00	4.67	1	0.84	1.09	2.18	0.28
#6 = #2 + #4	214	0.74	4	100	1.36	1	1	9.08	31.9	1	1		-	-
#7 = #3 + #4	358	0.44	7	100	2.39	0.99	1	6.50	12.7	1	1	1.09	-	-
#8 = #4 + #5	214	0.75	n.a. (∞)	100	n.a. (∞)	1	1	9.49	28.7	0.98	1	1.09	-	-
#9 = #2 + #3 + #4 + #5	159	1	n.a	n.a	n.a	1	1	10.8	77.5	1	1		-	-

The ideal properties of the ground allow to operate the GHP also during the hottest months: $f_C = 0.7$ corresponds to the optimal capacity ratios for the actual air and ground temperatures. Energy consumption is slightly reduced and the corresponding ε value increases from 0.19 to 0.22.

In Configuration #3, the heating load is fully delivered by the GSHP, thanks to the enhanced heat transfer performance of the BHEs. However, in cooling mode, we deal with the same situation illustrated for Configuration #1; therefore, E_p^{tot} value does not decrease significantly.

In Configuration #4, the high performances of the ideal GHP allow to match the total building load. As in Configuration #1, $f_H = 0.9$ in January is due to the heat transfer effectiveness of the BHEs and to the constraints on $T_{w_{in}}$. The ε value reaches 0.42 and the primary energy consumption is notably reduced.

In Configuration #5, hydraulic head losses are neglected. Both N_{BHE} and $\dot{m}_w$ tend to infinity; thus, temperature alteration of the ground results negligible. The behavior is very similar to the one of Configuration #2: the small difference between the two E_p^{tot} values is due to the pumping energy.

In summary, for the first five configurations, where only the effect of single subsystems is investigated, the greatest improvement of the system performance is obtained by replacing the GHP unit with an ideal one (Configuration #4); on the contrary, the other components slightly affect the overall performance, even when loss-free.

The results of Configurations #6, #7, and #8 show that, when an ideal heat pump is present, it can be advantageous to improve the other subsystems, too. In Configurations #6 and #8, when an ideal heat pump is coupled with an ideal ground and with a system free of head losses, ε reaches, respectively, 0.74 and 0.75.

In conclusion, a technological development of GSHP components does not produce adequate benefits unless the efficiency of the heat pump unit is concurrently augmented. Conversely, the equivalence of ε values in Configurations #6 and #8 suggests that reducing both thermal and hydraulic losses of the heat pump is a possible way of obtaining high performances, even in the presence of a soil with unfavorable thermal properties.

6. Sensitivity analysis

In the previous section, it has been shown that the GHP is the key element for the improvement of the whole GSHP system. In this section, a sensitivity analysis is carried out on system performances, increasing the GHP second-law efficiencies in heating and cooling modes ($\eta_{H/C}^{II}$). The aim is to find a preferable path and practical upper limits for technological development of the heat pump device. $\eta_{H/C}^{II}$ is the ratio of actual COP/EER of the unit and the coefficient of performance of a theoretical loss-free heat pump, operating at the same temperatures of the sources (COP^*/EER^*).

$$\eta_H^{II} = \frac{COP}{COP^*} \quad (4.1) \qquad \eta_C^{II} = \frac{EER}{EER^*} \quad (4.2)$$

Optimal design and control strategies for minimum primary energy consumption are evaluated for different combinations of $\eta_{H,nom}^{II}$ and $\eta_{C,nom}^{II}$ values, starting from the nominal values and moving towards the ideal ones (see table 7).

Table 7. Nominal second-law efficiencies of GHP unit used for the sensitivity analysis (in parentheses, the efficiency increase with respect to the benchmark configuration).

$\eta_{H,nom}^{II}$	0.555[a]	0.6 (+8%)	0.7 (+26%)	0.8 (+44%)	0.9 (+62%)	1 (+80%)	
$\eta_{C,nom}^{II}$	0.397[a]	0.5 (+26 %)	0.6 (+51%)	0.7 (+76%)	0.8 (+102%)	0.9 (+127%)	1 (+152%)

[a] Current value (benchmark configuration) in rating conditions [24].

In every case, Minimum energy consumptions are obtained with 7 BHEs. The relative energy savings (percentage savings with respect to the benchmark value) are depicted in the contour plot of figure 1, as a function of second-law efficiencies in heating and cooling modes. A saturation trend can be observed for the system performance, which reaches its maximum (37% savings) for unitary values of $\eta^{II}_{H,nom}$ and $\eta^{II}_{C,nom}$.

The continuous red line represents the shortest path from the benchmark to the maximum, but its practical meaning is poor, as it is impossible to eliminate all the inefficiencies with a single technological leap. The dotted blue line, instead, is obtained by following the shortest paths between each consecutive iso-line (with steps of 5% savings) and shows a more realistic technological development strategy, based on step-by-step evolutions.

A practical indication that can be derived from the graph is that heating and cooling efficiencies should increase concurrently, but with a small, though significant, preference for heating mode improvement. This can be explained by the higher values of f_H with respect to f_C, with a greater weight associated to the heating performance.

It is worth recalling that – although the suggested development path and the outlined conclusions, strictly speaking, are valid only for the analyzed case study – the proposed method is generally applicable to any other building system, even selecting different objective functions.

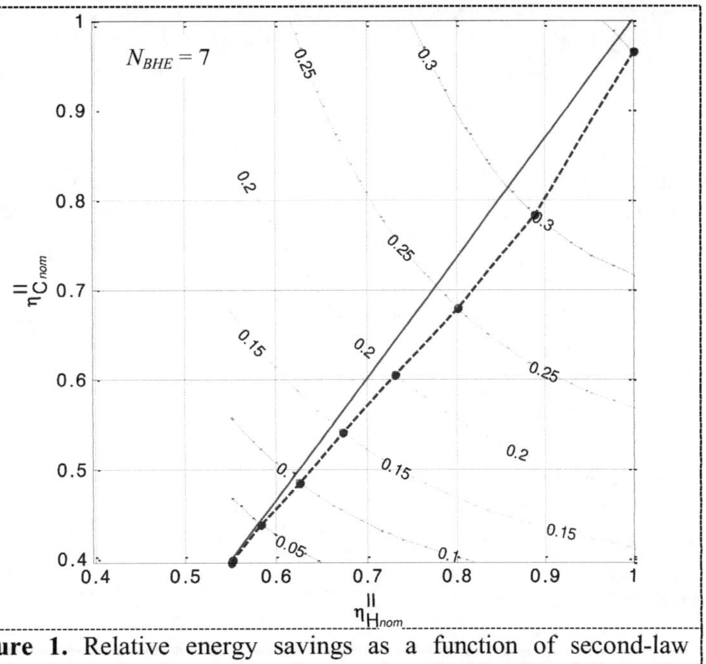

Figure 1. Relative energy savings as a function of second-law efficiencies in heating and cooling modes; the blue line follows the shortest path between two consecutive iso-lines.

7. Conclusions

In this work, the thermodynamic losses of a GSHP case study have been analyzed. Primary energy consumptions of nine configurations with different combinations of real and ideal subsystems have been compared, identifying their relative influence on the overall performance of the system. Furthermore, the use of theoretical loss-free components, together with optimized sizing and control strategies, allows to calculate the maximum energy savings achievable through the development of each subsystem.

The results reveal the possibility of an inherent hierarchical approach to the development of the subsystems. Specifically, the ground-coupled heat pump unit is the key element on which

Journal of Physics: Conference Series **547** (2014) 012006 doi:10.1088/1742-6596/547/1/012006

technological development should be focused. Increasing GHP performances also allows to enhance the positive effects given by other components: e.g. the task efficiency is 42% in Configuration #4 (ideal heat pump unit) and 22% in Configuration #5 (no head losses); combining the two ideal subsystems (Configuration #8), ε reaches 0.75.

Focusing on the heat pump device, a sensitivity analysis is performed on its heating and cooling efficiencies, aimed at finding the best path of technological development. The results show a saturation trend of the system performance, but with different behaviors in heating and cooling modes, suggesting a small but significant preference for the promotion of the heating efficiency.

It has to be stressed that these conclusions do not have a general value, but depend on the particular case under exam. This notwithstanding, the proposed methodological approach can be applied to any other GSHP system (e.g. in other climatic conditions or on larger or smaller scale buildings).

As for future developments, they are mainly required for the hydraulic design of the evaporator/condenser and control capacity of the heat pump unit. The head losses in the evaporator/condenser have to be lowered through an optimized hydraulic design or with a proper layout of the ground-coupled loop. Besides, the penalization effects due to low capacity ratios should be reduced. To do this, the control range of the GHP unit should be as wide as possible or smaller capacity units should be installed, compatibly with the available economic budget.

This work deals only with energy savings. Economic and thermoeconomic aspects will be included in future works, in order to appropriately take into account in the optimization procedure also installation and technological development costs. In this way, a graph similar to the one of figure 1, but with iso-lines of thermoeconomic savings, should show a maximum, and the optimal way to reach it.

Nomenclature

Symbols and Acronyms

BHE	Borehole heat exchanger
c	Specific heat [J/(kg K]
COP	Coefficient of performance of the heat pump unit in heating mode
$CR_{H/C}$	Capacity ratio in heating/cooling mode
EER	Coefficient of performance of the heat pump unit in cooling mode
En_p	Primary energy consumption [MWh]
$f_{H/C}$	GSHP share of building load in heating/cooling mode
Fo_{BHE}	Fourier number at borehole surface
GHP	Ground-coupled heat pump unit
H	Borehole depth [m]
$\dot{m}$	Mass flow rate [kg/s]
N_{BHE}	Number of boreholes
$\dot{q}$	Heat flow per unit length [W/m]
Q	Thermal energy [MWh]
$\dot{Q}$	Thermal power [W]
R	Return function
R_b	Borehole thermal resistance [m K/W]
R_{BHE}	Borehole radius [m]
T	Temperature (K or °C)
$\mathbf{u}^n$	Vector of control variables
$\mathbf{U}$	Set containing all the u^n
$\mathbf{x}^n$	Vector of state variables

Greek Letters

α	Thermal diffusivity [m^2/s]
ε	Task efficiency (see equation 1)
η	Boiler efficiency
$\eta^{II}_{H/C}$	Second-law efficiency in heating/cooling mode
τ	Reference time scale [h]
ψ	Exergy efficiency

Subscripts

air	Air unit
bk	Back-up generator
C	Condenser
DC	Declared capacity
E	Evaporator
g	Ground
in	Inlet/supply
l	Building thermal load
nom	Nominal
out	Outlet/return
W	Water circulating in the ground-coupled loop

Superscripts

0	Initial time
$*$	Ideal conditions

References

[1] Bakirci K 2010 *Energy* **35** 3088–96
[2] Karabacak R, Güven Acar Ş, Kumsar H, Gökgöz A, Kaya M and Tülek Y 2011 *Int. J. Refrig.* **34** 454–65
[3] Montagud C, Corberán J M, Montero Á and Urchueguía J F 2011 *Energy Build.* **43** 3618–26
[4] Yu X, Wang R Z and Zhai X Q 2011 *Energy* **36** 1309–18
[5] Conti P, Grassi W and Testi D 2013 Proposal of a Holistic Design Procedure for Ground Source Heat Pump Systems *Proc. European Geothermal Congress 2013 (Pisa, IT)*
[6] Conti P, Grassi W and Testi D 2014 A Design Method for Ground Source Heat Pump Systems Based on Optimal Year-Round Performance - Part 1: Model Definition and Discussion *Submitt. to Appl. Energy*
[7] Conti P, Grassi W and Testi D 2014 A Design Method for Ground Source Heat Pump Systems Based on Optimal Year-Round Performance - Part 2: Conduction of a Case Study *Submitt. to Appl. Energy*
[8] Conti P, Grassi W and Testi D 2015 Proposal of Technical Guidelines for Optimal Design of Ground-Source Heat Pump Systems *Proc. World Geothermal Congress 2015 (Melbourne, AU)*
[9] Robert F and Gosselin L 2013 *Appl. Therm. Eng.* **61** 481–91
[10] Moran M J 1989 *Availability Analysis - A guide to efficient energy use (Corrected Edition)* ed ASME Press (New York: ASME Press)
[11] Bi Y, Wang X, Liu Y, Zhang H and Chen L 2009 *Appl. Energy* **86** 2560–5
[12] Lohani S P 2010 *Energy* **35** 3323–31
[13] Hepbasli A and Akdemir O 2004 *Energy Convers. Manag.* **45** 737–53
[14] Li R, Ooka R and Shukuya M 2014 *Energy Build.* **75** 447–55
[15] Ally M R, Munk J D, Baxter V D and Gehl A C 2013 *Int. J. Refrig.* **36** 1417–30
[16] Rosen M A and Dincer I 2004 *Int. J. Therm. Sci.* **43** 121–33
[17] Lohani S P and Schmidt D 2010 *Renew. Energy* **35** 1275–82
[18] UNI 2008 *UNI/TS 11300-2*
[19] UNI 2010 *UNI/TS 11300-3*
[20] UNI 2012 *UNI/TS 11300-4*
[21] Ingersoll L R, Zobel O J and Ingersoll A C 1954 *Heat conduction with engineering, geological and other applications* (New York: McGraw-Hill)
[22] Rao S S 1996 *Engineering Optimization: Theory and Practice* (Hoboken: John Wiley & Sons, Inc.)
[23] UNI 2012 *UNI 11466*
[24] UNI 2013 *UNI EN 14511-2*

Thermal characterization of micro thermoelectric coolers: an analytical study

G De Aloysio[1] and F de Monte[2]
[1]Department of Information Engineering, Computer Science and Mathematics, University of L'Aquila, Via Vetoio, 67100 L'Aquila, ITALY
[2] Department of Industrial and Information Engineering and Economics, University of L'Aquila, Via Gronchi 18, 67100 L'Aquila, ITALY

E-mail: giulia.dealoysio@graduate.univaq.it

Abstract. Aim of this paper is to provide a complete characterization of the transient thermal behaviour of thermoelectric (TE) micro-coolers by solving the parabolic heat diffusion equation, through two different analytical methods, the Laplace Transform (LT) and the Separation of Variables (SOV). The results of the performed analysis show that a numerical inversion of the Laplace Transform is necessary, because the analytical-based Riemann-sum approximation does not yield temperature values at very early times. Once the temperature distribution is known in both the semiconductors of the p-n junction, the heat fluxes and the coefficient of performance (COP) of the cooler may be obtained. Then, by applying the proposed procedure to an existing micro-TE cooler, it is observed that the cooling load and the COP reach their maximum values, 2.9 W and 2, respectively, at different times, 3.4 ms vs. 1.7 ms. When a steady-state is reached, the micro-system is characterized by a cooling load of about 1.9 W with an efficiency of nearly 0.5, both in agreement with experimental and numerical data.

1. Introduction

Thermoelectric devices are solid state energy converters which can convert electricity into thermal energy for cooling purposes or thermal energy into electricity for power generation.

Based on semiconductor materials, these systems offer many advantages compared with conventional fluid-based devices:

- High reliability due to the absence of moving parts (no extra-maintenance).
- Absence of refrigerants which are harmful for the environment.
- No waste matter in the conversion process.

As the interest in micro- and nanotechnology has increased in the last years, miniature thermoelectric systems attract a lot of attention, in particular the micro-TEC devices because they offer effective cooling in micro-processors and in other microelectronic applications [1,2].

A TEC is composed of pairs of thermoelements, p-type and n-type semiconductors, and copper interconnectors at the top and at the bottom of the semiconductors. These thermoelectric elements are connected electrically in series and thermally in parallel. If an electric current is applied, the charged carriers in the material (electrons and holes) diffuse from the cold side to the hot side, leaving behind their oppositely charged and immobile nuclei [3], so that a temperature difference is produced. This temperature difference drives the phonons drift back to the cold side, so that heat conduction arises at equilibrium. Many attempts have been made to improve the performances of TEC, such as the design of nanostructured thermoelectric materials, nanocomposites, for instance alloys of Bi_2Te_3, as they are the best thermoelectric materials at room temperature.

However, it is important to understand the thermal behaviour of these devices in order to improve their performance. Most of the existing models focus on thermal behaviour under steady state conditions, whereas only a few papers [1,3,4,5,7,8] analyse the transient thermal behaviour of micro-thermoelectric devices. The dynamic characteristics play an important role in operation of micro thermoelectric coolers, especially during start-up and shut down and in case of variable operational parameters. In the last years different complete three-dimensional models [3,6-9] have been proposed to investigate the dynamic behaviour of miniature TECs, but despite their completeness, they provide solutions by means of numerical methods.

Starting from the available in the literature models [3-10], the aim of this paper is to provide an analytical study of the transient behaviour of micro- thermoelectric coolers, including the Thomson effect. In order to obtain the temperature distributions for both the semiconductors, the governing equations of the semiconductors (i.e. the parabolic heat diffusion equations with boundary conditions of first kind and initial uniform condition) are solved through the Laplace Transform (LT) and the Separation of Variables (SOV) Methods. The temperature solution is then implemented in MATLAB® ambient. In particular a comparison among the SOV Method, the analytical and a numerical inversion of Laplace Transform is presented. Once the temperature distributions are known, the heat fluxes and the COP of the micro cooler may be obtained. The influence of the Thomson effect on the performance of the micro-TEC is analysed under steady state conditions.

2. Mathematical formulation

A TEC module (TECM) is composed of many thermoelectric pairs embedded inside it. By considering the periodicity of the thermal characteristics, only one pair is considered for the analysis, as shown in figure 1.

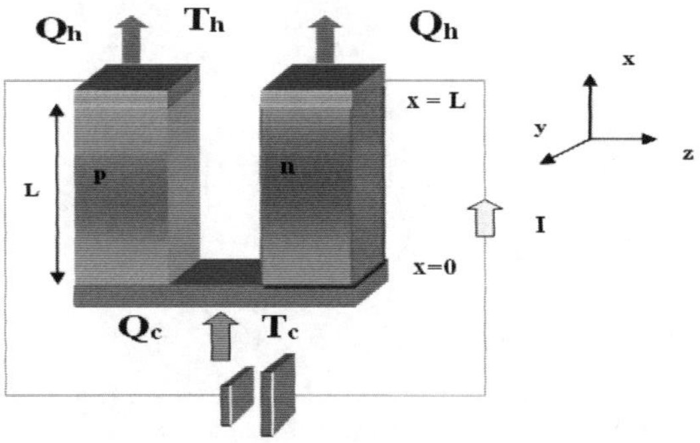

Figure 1. Simplified schema of a thermoelectric cooler.

As previously said, the working principle of these devices is the Peltier effect. When an applied current flows into the system, a cooling load Q_c is absorbed from the reservoir at T_c and an heating load Q_h is transferred to the reservoir at T_h. In order to perform the thermal behavior of the system, three additional effects have to be considered: the Fourier effect, the Joule effect and the Thomson effect. The following different assumptions underlie the development of the analytical model

- The material of the thermoelectric pair is homogeneous and isotropic so that the thermal conductivity and electric resistivity may be considered constant.
- Thermoelectric pellets are insulated thermally and electrically, except at the contact between the junctions and the two reservoirs which have fixed temperatures, T_h and T_c, respectively. The side surfaces of the TEC are considered adiabatic.
- The cross-sectional areas A_p and A_n and the lengths L_p and L_n of the thermoelements are respectively equal.
- Initially both semiconductors have a room temperature T_0 fixed at 298 K.
- The applied current is constant and uniform.

In this way, the transient thermal behaviour of both semiconductors (p and n-type) can be described by means of the 1-D heat flow governing equation along the *x*-direction, as shown in [4,5].

$$k_i \frac{\partial^2 T_i}{\partial x^2} \mp \gamma_i \frac{I}{A_i}\frac{\partial T_i}{\partial x} + \frac{I^2 R_i}{L_i A_i} = \rho c \frac{\partial T_i}{\partial t} \tag{2.1}$$

where the subscript i stands for n and p respectively. In detail, the plus sign is used for the n-type thermoelement, whereas the minus one is for the p-element. By using the dimensionless parameters introduced in [4,5] and here defined in appendix A, the temperature distributions may be obtained by solving the following dimensionless equations:

- Parabolic heat diffusion equations

$$\frac{\partial^2 \vartheta_n}{\partial \xi^2} + B_1 \frac{\partial \vartheta_n}{\partial \xi} + B_2 = \frac{\partial \vartheta_n}{\partial \tau} \tag{2.2}$$

$$\alpha_r \frac{\partial^2 \vartheta_p}{\partial \xi^2} - D_1 \frac{\partial \vartheta_p}{\partial \xi} + D_2 = \frac{\partial \vartheta_p}{\partial \tau} \tag{2.3}$$

- Boundary Conditions

$$\begin{aligned} \vartheta_n(\tau,0) &= \vartheta_p(\tau,0) = 1 \\ \vartheta_n(\tau,1) &= \vartheta_p(\tau,L_r) = h \end{aligned} \tag{2.4}$$

- Initial Conditions

$$\vartheta_n(0,\xi) = \vartheta_p(0,\xi) = 0 \tag{2.5}$$

where the dimensionless coefficients B_i and D_i (with i = 1 or 2) are defined as follows.

$$B_1 = \frac{I\gamma_n L_n}{A_n k_n} \tag{2.6}$$

$$B_2 = \frac{I^2 L_n R_n}{A_n k_n (T_c - T_0)} \tag{2.7}$$

$$D_1 = \frac{I\gamma_p L_n}{A_p k_n Cr} \tag{2.8}$$

$$D_2 = \mathrm{R_p I^2 L^2{}_n}\left[\mathrm{A_p k_n L_p (T_c - T_0) Cr}\right]^{-1} \tag{2.9}$$

3. Solution of the governing equations

The temperature solutions have been carried out by using the well-established methods of Separation of Variables and the Laplace Transform as shown in the next subsections.

3.1. SOV method

Following the Ozisik's approach [11], the temperature solutions in both domains p and n are expressed as sum of a steady state part and a complementary transient part that is the solution of the associated transient homogeneous problem. We have:

$$\vartheta_n(\xi,\tau) = \frac{\exp(-B_1) - h - \frac{B_2}{B_1}}{\exp(-B_1) - 1} + \frac{h + \frac{B_2}{B_1} - 1}{\exp(-B_1) - 1}\exp(-B_1\xi) - \frac{B_2}{B_1}\xi + \sum_{m_n=1}^{\infty} c_{m_n}\sin(\beta_{m_n}\xi)\exp(-\beta_{m_n}^2\tau) \times \exp\left(\frac{(-B_1)}{2}\xi - \frac{(-B_1)^2}{4}\tau\right) \tag{3.1}$$

$$\vartheta_p(\xi,\tau) = \frac{\exp\left(\frac{D_1}{\alpha_r}L_r\right) - h + \frac{D_2}{D_1}L_r}{\exp\left(\frac{D_1}{\alpha_r}L_r\right) - 1} + \frac{h - \frac{D_2}{D_1}L_r - 1}{\exp\left(\frac{D_1}{\alpha_r}L_r\right) - 1}\exp\left(\frac{D_1}{\alpha_r}\xi\right) + \frac{D_2}{D_1}\xi + \sum_{m_p=1}^{\infty} c_{m_p}\sin(\beta_{m_p}\xi)\exp(-\alpha_r\beta_{m_p}^2\tau)\exp\left(\frac{D_1}{2\alpha_r}\xi - \frac{(D_1)^2}{4\alpha_r}\tau\right) \tag{3.2}$$

Where and $\beta_{m_n} = m\pi$ $\beta_{m_p} = \frac{m\pi}{L_r}$ are the m eigenvalues of the problem and c_{m_n} and c_{m_p} are the integration constants depending on the initial conditions.

3.2. LT method

By applying Laplace transform to equations (2.2) and (2.3) as proposed in [4,5], we obtain

$$\Theta_n(s,\xi) = C_{1n}(s)\exp(l_{1n}(s)\xi) + C_{2n}(s)\exp(l_{2n}(s)\xi) + \frac{B_2}{s^2} \tag{3.3}$$

$$\Theta_p(s,\xi) = C_{1p}(s)\exp(l_{1p}(s)\xi) + C_{2p}(s)\exp(l_{2p}(s)\xi) + \frac{D_2}{s^2} \tag{3.4}$$

$$\begin{aligned} &\Theta_n(s,0) = \Theta_p(s,0) = \frac{1}{s} \\ &\Theta_n(s,1) = \Theta_p(s,L_r) = \frac{h}{s} \\ &\Theta_n(0,\xi) = \Theta_p(0,\xi) = 0 \end{aligned} \tag{3.5}$$

where $\Theta_j(s,\xi)=L[\vartheta(\xi,\tau)]$ with j= n, p. Two different approaches are used to solve the inverse Laplace Transform (ILT, for short), the analytical-based Riemann-sum approximation [4,5] and a numerical MATLAB® function [12]. By means of the former method obtain

$$\vartheta(\xi,\tau)=\frac{\exp(\varphi\tau)}{\tau}\left[\frac{1}{2}\Theta(\xi,\varphi)+\mathrm{Re}\sum_{n=1}^{N}\Theta\left(\xi,\varphi+\frac{i\pi n}{\tau}\right)(-1)^n\right] \tag{3.6}$$

where for a rapid convergence of the Riemann sum, φ has to be such that $\varphi\tau\approx4.7$[4,5].

Note that i is the imaginary unit and Re denotes the real part of the complex summation.

3.3. Performance of the micro-TEC

Once the temperature distributions are known, the dimensionless cooling and the heating loads may be obtained as:

$$q_c=\mathrm{G}-\frac{\partial\vartheta_n(\tau,0)}{\partial\xi}-\mathrm{k_r A_r}\frac{\partial\vartheta_p(\tau,0)}{\partial\xi} \tag{3.7}$$

$$q_h=\mathrm{H}-\frac{\partial\vartheta_n(\tau,1)}{\partial\xi}-\mathrm{k_r A_r}\frac{\partial\vartheta_p(\tau,L_r)}{\partial\xi} \tag{3.8}$$

where the coefficients G and H are the dimensionless contributions due to the Peltier effect. Therefore the COP (coefficient of performance) is given by:

$$COP=\frac{q_c}{q_h-q_c} \tag{3.9}$$

4. Results and discussion

A module consisting of 8 embedded thermoelectric pairs [10] has been considered for the simulations. The values of the semiconductor properties and their geometrical characteristics are listed in tables 1 and 2. The steady-state temperature profiles in presence and in absence of the Thomson effect are plotted in figures 2 where T_h is set at 300 K and T_c at 290 K with an electric current of 3.8A.

Table 1. Semiconductor elements properties and their geometrical sizes.

Parameter	p-type	n-type	Reference
k ($Wm^{-1}K^{-1}$)	1.472	1.643	[10]
A (m^2)	2.5×10^{-7}	2.5×10^{-7}	[10]
L (m)	5×10^{-4}	5×10^{-4}	[10]
ρ (kg m^{-3})	7740	7740	[9]
c (J kg^{-1}K)	187	187	[3]
r_e (Ω m)	8.826×10^{-6}	8.239×10^{-6}	[10]

Table 2. Seebeck coefficients [10], [13].

	p-type $(Bi_{0.25}Sb_{0.75})_2Te_3$	n-type $Bi_2(Te_{0.94}Se_{0.06})_3$
σ (VK^{-1})	$-6.95\times10^{-10}(T-300)^2+3.42\times10^{-7}(T-300)+2.207\times10^{-4}$	$1.04\times10^{-9}(T-300)^2-1.25\times10^{-7}(T-300)-2.23\times10^{-4}$

The steady state temperature profiles for both semiconductors are slightly lower than the results obtained numerically in [10], with a maximum temperature difference of 6.6 K between the two pellets compared with the 7-8 K obtained in [10]. This is due to the different solution method and to the different evaluation of the Seebeck and Thomson coefficients.

With reference to the steady state cooling loads in presence and in absence of the Thomson effect (figure 3), the maximum value of Q_c is 1.926 W with Thomson effect and 1.774 W without Thomson effect.

As regards the steady state COP (figure 4), a slightly higher difference with the results in [10] may be underlined. In fact the COP values with and without Thomson effect shown in [10] seem to be coincident (4.84 at 0.292 A), whereas, in the present paper the maximum value for the steady state COP in presence of Thomson effect is 4.84 at 0.3 A and nearly 4.4 in absence of Thomson effect.

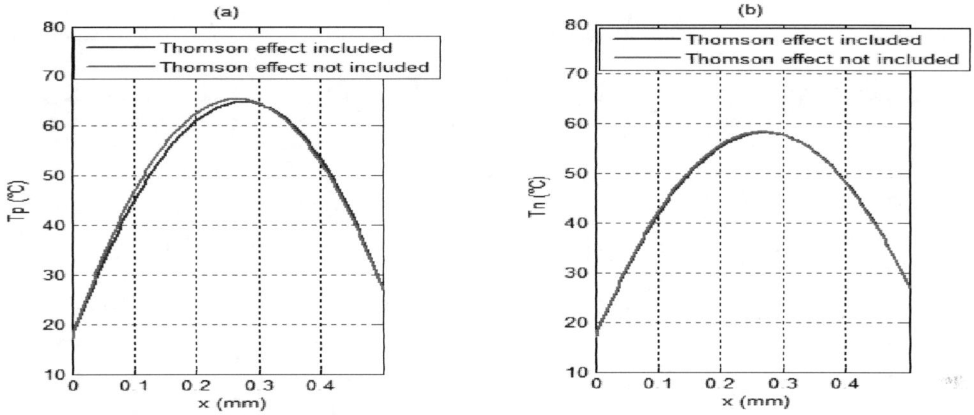

Figure 2. Steady-state T profiles in the semiconductors a) p-type and b) n-type.

The transient temperature profiles are shown in figure 5, (a) and (b), including the Thomson effect with an electric current of 3.8 A. Note that the temperature profiles for times t < 2 ms are plotted only by using the SOV and the numerical ILT methods, as the analytical ILT does not converge for dimensionless times shorter than 0.01 (to which corresponds a dimensional of 2ms). During the simulations we have taken N= 2000 in the Riemann sum, defined in equation (3.6). For dimensionless times higher than 0.01 no difference between the three techniques arises.

A more accurate analysis is required for the transient performances of the device, i.e. the cooling load Q_c and the COP, shown in figures 6 and 7, respectively. As regards Q_c, the different values obtained are listed in table 3. It can be seen that at very early times, less than 2 ms, the AILT does not yield physically- consistent results, whereas the other two techniques may also be used only for time values less than 2 ms, but greater than 0.1 ms. This is due to the imposed non-realistic time-independent boundary conditions of the first kind. For time values higher than 2 ms the three techniques give the same results. The maximum Q_c value is of 2.89 W and occurs at 3.4 ms.

Journal of Physics: Conference Series **547** (2014) 012007 doi:10.1088/1742-6596/547/1/012007

Different COP profiles are shown in figure 7. The maximum COP value of about 2 obtained with the different techniques occurs after 1.66 ms and was obtained by using the SOV and the NILT methods, as the AILT (based on the Riemann sum Approximation) is numerically unstable at early times

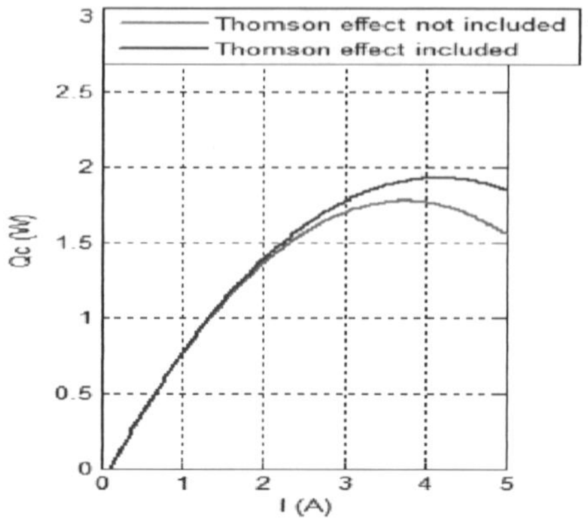

Figure 3. Steady state cooling load.

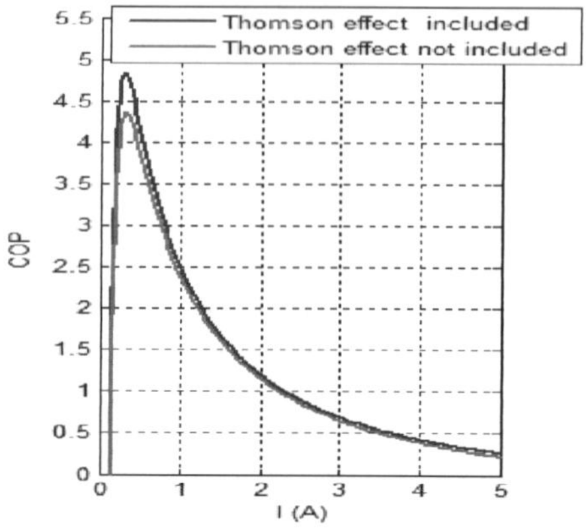

Figure 4. Steady state COP.

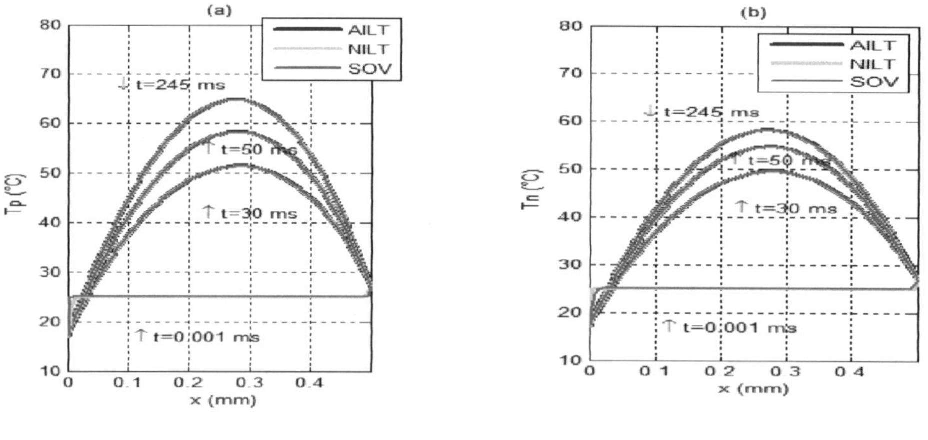

Figure 5. Transient T profiles in the semiconductors a) p-type and (b) n-type. AILT stands for analytical inverse Laplace Transform, NILT stands for numerical inverse Laplace Transform.

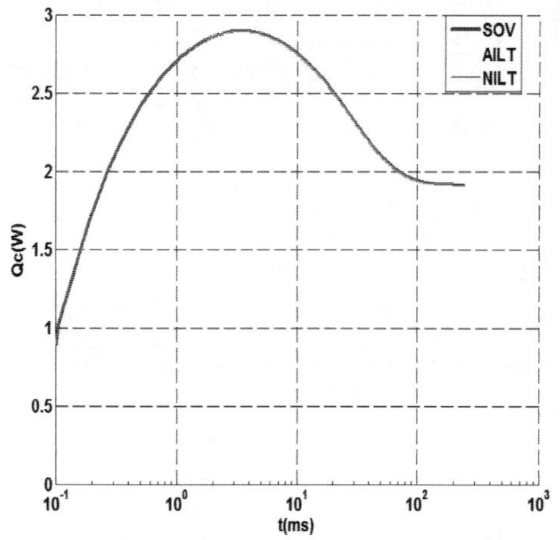

Figure 6. Transient Q_c profiles with the three techniques.

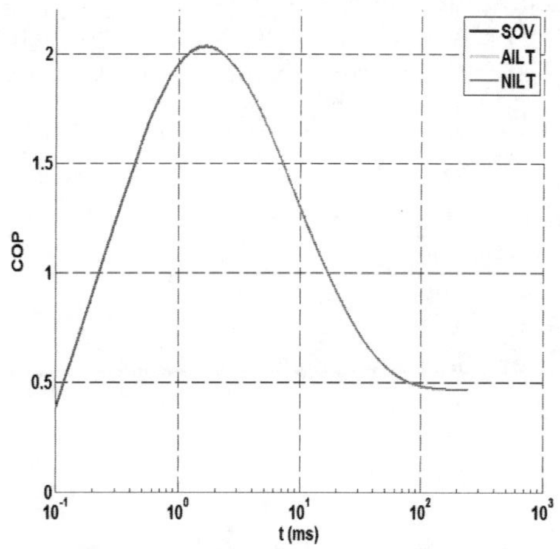

Figure 7. Transient COP profiles with the three techniques.

Table 3. Comparison among the Q_c values obtained with the different techniques.

t (ms)	Qc_SOV (W)	Qc_NILT (W)	Qc_AILT (W)	COP_SOV	COP_NILT	COP_AILT
0.100	0.911	0.912	-	0.365	0.365	-
0.200	1.722	1.722	-	0.891	0.891	-
1.000	2.710	2.710	-	1.955	1.954	-
2.000	2.860	2.860	2.860	2.021	2.020	2.020
3.000	2.893	2.893	2.893	1.928	1.928	1.928
4.000	2.892	2.892	2.892	1.805	1.803	1.803
5.000	2.877	2.877	2.877	1.697	1.696	1.696
10.000	2.715	2.715	2.715	1.302	1.300	1.300
15.000	2.626	2.626	2.626	1.070	1.069	1.069
20.000	2.510	2.510	2.510	0.918	0.916	0.916
30.000	2.324	2.323	2.323	0.735	0.734	0.734
50.000	2.106	2.105	2.105	0.577	0.576	0.577
100.000	1.941	1.941	1.941	0.484	0.483	0.483

The investigation of the hyperbolic heat diffusion equation by using Green's functions [14] for a more realistic description of the micro-TEC transient behaviour at early times remains a subject of the future research.

5. Conclusion

Three different approaches for the thermal transient modelling of a micro-thermoelectric cooler have been presented and compared. The Inverse Laplace Transform performed analytically by Riemann sum approximation does not allow its transient thermal behaviour to be modelled for very short times (less than 1.719 ms). As regard the cooling load and COP, the SOV and NILT cannot be used for times less than 0.1 ms, due to the simplifying assumption of time-independent boundary

Journal of Physics: Conference Series **547** (2014) 012007 doi:10.1088/1742-6596/547/1/012007

conditions. An important remark is for the COP and Q_c that, during the transient behaviour, reach a maximum value at different times.

A comparison of the results with data available in the literature (and obtained both experimentally and numerically - by FEM) indicates the good accuracy of the proposed solution methods under steady state conditions.

Appendix A. Nomenclature

A = cross-sectional area
A_r = cross-sectional area ratio= A_p/A_n
c = specific heat
Cr = dimensionless heat capacity = $\rho_p c_p/\rho_n c_n$
h = temperature ratio= $(T_h-T_0)/(T_c-T_0)$
I = electric current
J = current density
k = thermal conductivity
k_r= thermal conductivity ratio= k_p/k_n
L = length
$L_r = L_p/L_n$ = Length ratio
Q_c = Cooling load
Q_h = Heating load
q_c = Dimensionless cooling load = $Q_cL_n/(k_nA_n(T_c-T_0))$
q_h = Dimensionless heating load = $Q_hL_n/(k_nA_n(T_c-T_0))$
R = electric resistance
r_e = electric resistivity
t = time
T = temperature
T_0 = room temperature
x = space coordinate

Greek symbols

α = thermal diffusivity
$\alpha_r = \alpha_p/\alpha_n$= dimensionless thermal diffusivity
γ = Thomson coefficient, defined as $\gamma = T d\sigma / dT$
ϑ = dimensionless T= $(T-T_0)/(T_c-T_0)$
ξ = dimensionless space = x/L
ρ = density
σ = Seebeck coefficient.
τ = dimensionless time = $\alpha t L^{-2}$

Subscripts

c=cold
h = hot
n = semiconductor n
p = semiconductor p

References

[1] Yang R, Chen G, Kumar A R, Snyder G J and Fleurial J P 2005 Transient cooling of thermoelectric coolers and its applications for microdevices *Energy Conv. Manag.* **46** 1407-21

[2] Zhao D and Tan G 2014 A review of thermoelectric cooling: Materials, modeling and applications *Appl. Therm. Eng.* **66** 15-24

[3] Cheng C H, Huang S Y and Cheng T C 2010 A three-dimensional theoretical model for predicting transient thermal behaviour of thermoelectric coolers *Int. J. Heat and Mass Transfer* **53** 2001-11

[4] Naji M, Alata M A and Al Nimr M 2003 Transient behaviour of a thermoelectric device *Proc. Inst. Mechan.Eng. A J. Power Energy* **17** 615-621

[5] Naji M, Alata M A and Al Nimr M 2003 Transient behaviour of a thermoelectric device under the hyperbolic heat conduction model *Int. J. Thermophys.* **24** 1753-68

[6] Wang X D, Huang Y X, Cheng C H, Lin D T W and Kang C H 2012 A three dimensional numerical modeling of thermoelectric device with consideration of coupling of temperature field and electric potential field *Energy* **47** 488-497

[7] Meng J H, Wang X D and Zhang X X, 2013 Transient modeling and dynamic characteristics of thermoelectric cooler *Appl. Energy* **108** 340-348

[8] Cheng C H and Huang S Y 2012 Development of a non-uniform-current model for predicting transient thermal behavior of thermoelectric coolers *Appl. Energy* **100** 326-335

[9] Zhu W, Deng Y,Wang Y and Wang A 2013 Finite element analysis of miniature thermoelectric coolers with high cooling performance and short response time *Microelectron. J.* **44** 860-868

[10] Chen W H, Liao C Y and Hung C I 2012 A numerical study on the performance of miniature thermoelectric cooler affected by Thomson effect *Appl. Energy* **89** 464-473

[11] Özisik M N 1993 *Heat Conduction* (New York: Wiley)

[12] Hollenbeck K J 1998 INVLAP.M: A matlab function for numerical inversion of Laplace transforms by the de Hoog algorithm http://www.isva.dtu.dk/staff/karl/invlap.htm

[13] Yamashita O 2009 Effect of linear and non–linear components in the temperature dependences of thermoelectric properties on the cooling performance *Appl. Energy* **86** 1746-56

[14] Haji-Sheikh A, de Monte F and Beck J V 2013 Temperature solution in thin films using thermal wave Green's function solution equation *Int. J. Heat and Mass Transfer* **62** 78-86

Journal of Physics: Conference Series **547** (2014) 012008 doi:10.1088/1742-6596/547/1/012008

Borehole modelling: a comparison between a steady-state model and a novel dynamic model in a real ON/OFF GSHP operation

M De Rosa[1], F Ruiz-Calvo[2], J M Corberán[2], C Montagud[2] and L A Tagliafico[1]

[1] DIME/TEC - Division of Thermal Energy and Environmental Conditioning, University of Genoa. Via All'Opera Pia 16145 Genoa, Italy.

[2] Instituto de Ingeniería Energética, Universitat Politècnica de València.
Camino de Vera sn 46022 Valencia, Spain. Phone: 34-96-3879910. Fax: 34-963877272

E-mail: mattia.derosa@unige.it

Abstract. The correct design and optimization of complex energy systems requires the ability to reproduce the dynamic thermal behavior of each system component. In ground source heat pump (GSHP) systems, modelling the borehole heat exchangers (BHE) dynamic response is especially relevant in the development of control strategies for energy optimization purposes. Over the last years, several models have been developed but most of them are based on steady-state approaches, which makes them unsuitable for short-term simulation purposes. In fact, in order to accurately predict the evolution of the fluid temperatures due to the ON/OFF cycles of the heat pump, it is essential to correctly characterize the dynamic response of BHE for very short time periods. The aim of the present paper is to compare the performance of an analytical steady-state model, available in TRNSYS environment (Type 557), with a novel short-term dynamic model. The new dynamic model is based on the thermal-network approach coupled with a vertical discretization of the borehole which takes into account both the advection due to the fluid circulating along the U-tube, and the heat transfer in the borehole and in the ground. These two approaches were compared against experimental data collected from a real GSHP system installed at the Universitat Politècnica de València. The analysis was performed comparing the outlet temperature profiles predicted by both models during daily standard ON/OFF operating conditions, both in heating and cooling mode, and the between both approaches were highlighted. Finally, the obtained results have been discussed focusing on the potential impact that the differences found in the prediction of the temperature evolution could have in design and optimization of GSHP systems.

1. Introduction

Ground source heat pump (GSHP) systems use the ground as a heat source in winter and as a heat sink in summer, taking an advantage of the more stable temperature of the ground, which allows achieving a greater efficiency and, hence, reducing the CO_2 emissions for heating and cooling applications.

Generally, the heat exchange takes place in a ground source heat exchanger (GSHE) and different configurations can be adopted (a detailed review of GSHE systems can be found in [1]). Among those, one of the most commonly used is the vertical borehole field, consisting on a certain number of borehole heat exchangers (BHE) drilled in the ground, inside which one or more U-tubes are located.

The heat transfer phenomena are complex and occur over a wide spatial and timescales ranges. Therefore, several assumptions are usually made in order to simplify the phenomena, with the aim to realize simplified models which permit to perform fast calculations. In fact, the ability to predict the BHE thermal performance can contribute to find the best solutions to increase the energy efficiency of the whole system.

In the last years, several approaches have been proposed in order to reproduce the thermal behavior of different BHE configurations (a complete review is reported by Yang et al. [2]). The most widely used approach is based on the line and cylindrical heat source theory [3, 4, 5], which models only the heat transfer between the borehole wall and the surrounding soil, neglecting the heat transfer inside the borehole. Eskilson [6] proposed a model based on the use of non-dimensional temperature response factors, called g-functions, that represent the temperature response to a constant heat injection pulse, for a certain time step. Thus, the actual thermal load can be subdivided into a series of constant step loads and then the temperature response can be obtained by superimposing the single response at each load step. This approach is valid only for a time scale greater than 3-6 hours for a typical borehole [7].

The thermal network model represents another simple approach in which the thermo-electrical analogy is adopted to represent the heat transfer phenomena inside the borehole. The basic thermal network is the steady-state delta network [7] in which one temperature node is located on each pipe of the U tube and another one is located on the borehole wall. The thermal network approach can also be used for modelling the behavior of the ground around the borehole, from the undisturbed ground temperature to the borehole wall. Moreover, it is possible to overcome the typical steady-state approach by adding thermal capacitances to each temperature node ([8, 9, 10]), allowing to reproduce the dynamic of the system.

Usually, GSHP systems for heating and cooling in buildings are governed by an ON/OFF control algorithm. In these cases, the thermal load is injected/extracted to the ground in short heat pulses of about 20 minutes approximately (depending on the instantaneous thermal energy demand). In this context, the short-term behavior of BHE represents a crucial topic for evaluating and optimizing the thermal performance of GSHP, especially in terms of system control criteria and heat pump performance (which are strongly affected by the temperature evolution of the working fluid). Therefore, there is a need for software able to reproduce the short term dynamic behavior of the fluid circulation along the BHE. The present paper reports the comparison between two different approaches for modelling a single U-tube for short-term periods:

- the typical DST model provided in Trnsys (Type 557) [11] based on the explicit finite difference method (FDM) coupled with an analytical steady-state solution.
- a novel short-term BHE dynamic model, called Borehole-to-Ground (B2G) model, based on the delta network approach and implemented in the Trnsys environment. This model has been validated against experimental data provided by different system installations in previous authors works.[1,2].

The analyses have been conducted comparing the BHE outlet temperature profiles obtained by the two different codes with real experimental ones collected during a step-test (7 hours of heat injection) and, moreover, during real ON/OFF operation of the GSHP system installed at the Universitat Politècnica de València,

[1] Ruiz-Calvo, F., De Rosa, M., Acuña, J., Corberán, J.M., Montagud, C. 2014. Experimental validation of a short-term Borehole-to-Ground (B2G) dynamic model. Unpublished results.

[2] De Rosa, M., Ruiz-Calvo, F., Corberán, J.M., Montagud, C., Tagliafico, L.A. 2014. A novel TRNSYS type for short-term borehole heat exchanger simulation: B2G model. Unpublished results.

Journal of Physics: Conference Series **547** (2014) 012008 doi:10.1088/1742-6596/547/1/012008

2. Experimental Plant

The experimental data used in the present work were obtained from an operating GSHP system installed at the Universitat Politècnica de València (UPVLC), Spain. The experimental plant was built on 2004 and started its operation on February 2005, in the framework of a European Fifth Framework project called GeoCool [12]. GeoCool plant consists of a reversible water-to-water heat pump, coupled to a ground source heat exchanger (GSHE). The system was designed for air conditioning a total area of approximately 250 m^2 comprising a set of offices and general purpose rooms.

Figure 1 shows a schematic diagram of the experimental plant, where it can be observed that the system mainly consists of two hydraulic loops coupled by the heat pump. The internal (building) circuit consists of a total of 12 parallel connected fancoils, a circulation pump, and a water storage tank. The external circuit includes the ground source heat exchanger and a circulation pump. The fluid used in both circuts is water. A more detailed description of the installation and the particular conditions of its operation is provided in [13].

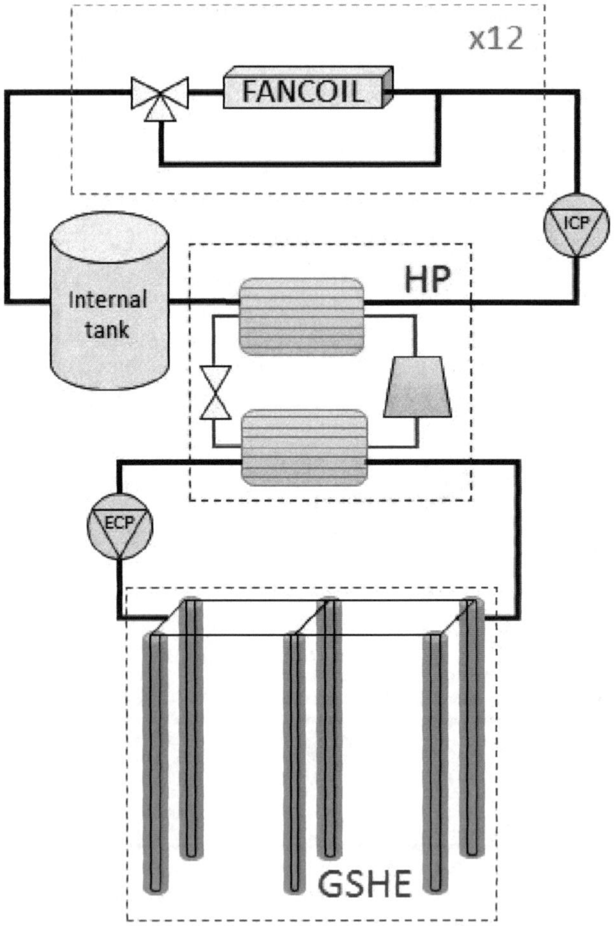

Figure 1. Experimental plant schematic diagram.

The heat pump installed for GeoCool project is a one-step reversible heat pump. The adjustment of the heat pump thermal capacity to the actual thermal load is achieved by an ON/OFF control strategy. This particular mode of operation produces a characteristic temperature evolution along the day, which affects the system performance and the heat pump thermal efficiency. Further improvements on the control algorithm have been implemented recently in order to optimize the overall performance of the plant [14].

The GSHE of the GeoCool plant consists of 6 vertical boreholes, drilled in the ground near the building. The boreholes are arranged in a 2x3 rectangular grid, spaced 3 m from each other, and connected in parallel. Each borehole has a diameter of 150 mm and a depth of 50 m. The water flows throughout the borehole by a U-tube, being the inner and external diameters of 25.4 mm and 32 mm respectively, and a center-to-center distance (shank spacing) of 70 mm. At the top of the U-tube, there are two temperature sensors, which measure the temperature of the water entering and exiting from each BHE. Further details about the design of the GSHE and its characteristics can be found in [15].

3. BHE Models

3.1. B2G Model

A short-term BHE dynamic model, called Borehole-to-Ground (B2G) model, has been developed at Universitat Politècnica de València with the aim of correctly predicting the short-term behavior of U-tube boreholes in terms of water temperature throughout the pipe. The model is based on the thermal network approach considering a discretized domain of the U-tube and the ground in vertical direction. Five thermal capacitances and six thermal resistances are taken into account at each node depth (5C6R-n model, where n is the number of the nodes), considering the thermal properties of the ground, the grout and the pipes and neglecting the vertical heat conduction (Figure 2b).

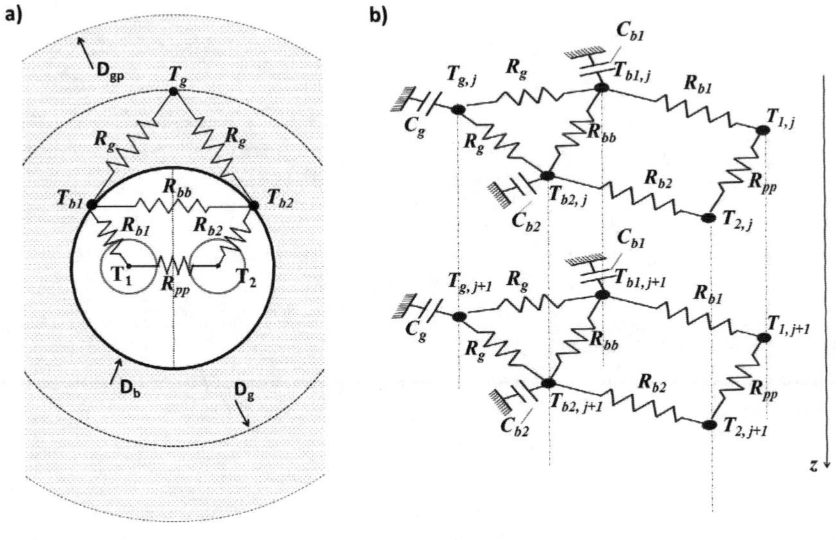

Figure 2. Thermal network model adopted in the present work: a) 2D model; b) 3D model.

The model divides the grout into two separate capacitance nodes according to the pipes location, and it situates the delta network between those nodes and the one located in the surrounding ground. The U-tube is represented by two nodes, one for the downward side and the other for the upward side. Each fluid node exchanges heat with the correspondent borehole node and with the adjacent fluid node. Moreover, the fluid advection is also taken into account in the energy balance. The balance equations in transient state for all nodes typology correspond to Eqs. 1, 2, and 3.

$$\frac{\partial T_1(z)}{\partial t} = -v\frac{\partial T_1(z)}{\partial z} - \frac{1}{C_f}\left(\frac{T_1(z) - T_{b1}(z)}{R_{b1}} + \frac{T_1(z) - T_2(z)}{R_{pp}}\right) \tag{1}$$

$$C_{b1}\frac{\partial T_{b1}(z)}{\partial t} = \frac{T_1(z) - T_{b1}(z)}{R_{b1}} + \frac{T_{b1}(z) - T_{b2}(z)}{R_{bb}} - \frac{T_{b1}(z) - T_g(z)}{R_g} \tag{2}$$

$$C_g\frac{\partial T_g(z)}{\partial t} = \frac{T_{b1}(z) - T_g(z)}{R_g} + \frac{T_{b2}(z) - T_g(z)}{R_g} \tag{3}$$

The B2G parameters are the thermal resistances and capacitances of the different nodes of the thermal network. These parameters can be determined taking into account the borehole geometrical characteristics and thermophysical properties of each material. In particular:

- the thermal capacitances C_{b1} and C_{b2} are calculated considering the volume of each grout zone;
- the thermal resistance R_{b1} and R_{b2}, supposed to be equal, are calculated starting from the overall borehole thermal resistance, which can be obtained by standard step-test procedures, assuming an equivalent diameter and considering that the borehole nodes have been located at the borehole diameter [10];
- the resistance between the two borehole nodes R_{bb} is obtained assuming a one-dimensional heat transfer, as shown in Eq. 4, in which D_b and $D_{p,e}$ are the borehole and the pipe external diameters, while W is the shank spacing;

$$R_{bb} = \frac{W}{k_b(D_b - D_{p,e})\mathrm{d}z} \tag{4}$$

- the thermal resistance between the pipe nodes R_{pp}, which is quite complex to determine due to the two-dimensional heat transfer, is estimated considering a one-dimensional linear heat conduction between them and the convection inside the pipes.
- the ground thermal capacitance C_g is calculated considering the radial penetration depth D_{gp} of the borehole which is an adjustment parameter that depends on the considered heat injection/extraction time and on the ground thermal diffusivity [6]. The ground node is located at D_g, which is calculated as the average between the borehole diameter, D_b, and the penetration diameter, D_{pg};
- the thermal resistances R_g between each borehole node and the ground node are calculated considering a cylindrical conductive heat transfer between them.

3.2. TRNSYS Tess Model - Type 557

TRNSYS type 557 implements a vertical BHE model based on the *Duct Ground Heat Storage Model* (DST) developed by Hellström [11]. DST model is focused on the characterization of the temperature profile of the ground around the BHE. The storage region, a cylindrical volume of the ground around the BHE, is discretized using a two-dimensional mesh adopting radial and vertical coordinates. The temperature of each point in the storage region of the ground is calculated with the superposition of a global solution, a steady-flux solution, and a local solution.

The temperature of the outlet fluid of the BHE is calculated by means of a steady-state heat balance, as shown in Eq.5 in which c_f is the heat capacity of the fluid, $\dot{m}_f$ is the flow rate,

$T_f(s,t)$ is the fluid temperature along the pipe, α is the heat transfer coefficient between the fluid and the surrounding ground with temperature T_a.

$$c_f \dot{m}_f \frac{\partial T}{\partial s} + \alpha(T_f - T_a) = 0 \quad (5)$$

Solving this equation by integration along the tube, the outlet BHE fluid temperature can be calculated from Eq.6, in which β is a damping factor defined by equation 7.

$$T_{f,out} = \beta T_{f,in} + (1-\beta)T_a \quad (6)$$

$$\beta = e^{-\frac{\alpha L}{c_f \dot{m}_f}} \quad (7)$$

As it can be observed in Eq.6 and in Eq.7, the outlet fluid temperature approaches the ground temperature (T_a) when the flow rate tends to zero, while it tends to the inlet temperature ($T_{f,in}$) as the flow rate increases.

Generally, the DST model produces an accurate and detailed prediction of the ground temperature evolution of a BHE. However, it is based on steady-state assumptions, which could affect the performance of the model for very short time steps like the ones existing in ON/OFF GSHP systems for heating and cooling. Moreover, in the prediction of the outlet fluid temperature, advection is not taken into account.

Differences between the results produced by the DST model based on a steady-state approach and the dynamic B2G model are highlighted when confronting both models with experimental data from a real installation operating under ON/OFF conditions. Section 4 presents a comparison of the performance of both models using experimental data from GeoCool Plant.

4. Results

The performance of the B2G model and of the TRNSYS type 557 has been compared, considering experimental data measurements from the GSHP system installed at UPVLC as reference. Both models use as input the inlet water temperature and the water mass flow rate, taken directly from the experimental measurements, and provide the outlet water temperature as an output.

B2G parameters have been estimated using the theoretical approach described in section 3.1. The same parameters have also been used for the TRNSYS model in order to ensure a fair comparison.

The comparison has been conducted comparing the simulated BHE outlet water temperatures provided by B2G and TRNSYS with the experimental measurements. Two different operating conditions were adopted for the comparison. First, the performance of both models has been compared considering experimental data from a step-test in carried out on the 4th of November 2013, in order to validate the transient response of the models. Afterwards, the comparison has been extended to two typical ON/OFF operation conditions for two different days, one in heating and the other in cooling mode. The ability of each model to reproduce the characteristic temperature evolution during ON/OFF cycles has been described and commented.

4.1. Step-test

The step-test was carried out in cooling mode, forcing the heat pump to remain switched on so that the condensing heat was injected into the ground loop during a continuous period of 7 hours approximately. After this time, the heat pump switched off, according to the schedule of the installation, producing a recovery step of 12 hours, also useful for model validation purposes. As previously stated, the input data for both models are the inlet water temperature and the mass flow rate. The simulated outlet water temperature profiles obtained by each approach have been compared against the experimental data. Both inlet and outlet experimental temperatures are measured at the top of the BHE.

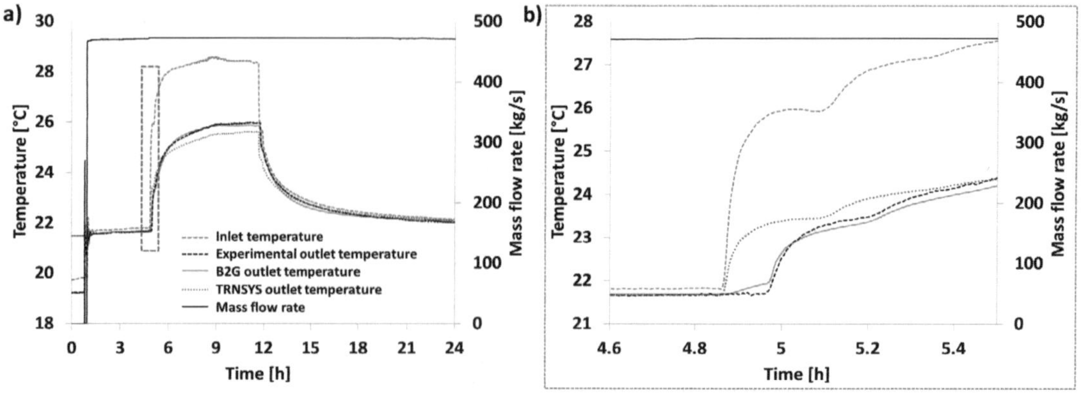

Figure 3. Step-test results: a) general response comparison, b) short-term response comparison (first hours).

Figure 3 shows the comparison results for the step-test simulation. It can be observed that the TRNSYS model predicts an outlet water temperature profile slightly lower than the experimental measurements (Figure 3a), with a difference of around 0.5°C. Although this little discrepancy is perfectly acceptable for simulation purposes, it could be corrected by slightly adjusting the model parameters. On the other hand, B2G results show a very good agreement with the experimental measurements.

Focusing on the short-term temperature evolution shown in figure 3b, it can be observed that the temperature increase at the outlet of the BHE occurs a few minutes after the correspondent increase on the inlet water temperature. This corresponds to the necessary time for the water to travel along the U-tube. B2G response perfectly reproduces this delay, since the advection terms are taken into account in this model. However, TRNSYS model only produces a steady-state response for the outlet water temperature, without taking into account the displacement time of the water inside the pipes. Therefore, the outlet water temperature calculated by the TRNSYS model increases exactly at the same time than the inlet water temperature. In the step-test conditions simulation, this could be easily corrected by adding some kind of delay to the TRNSYS model outlet water temperature, e.g. by means of a loss-less pipe. However, when using this model for a typical day simulation, with ON/OFF cycling, the steady-state approach may introduce non-negligible deviations from reality.

4.2. Typical performance

In order to compare the performance of a typical day, two representative days have been selected, one for heating mode and one for cooling mode.

Figure 4 shows the simulation results for the heating mode operation day (17/02/2010); in particular, Figure 4a presents an overview of the BHE water temperature evolution along the day, while Figure 4b shows an augmented section of the previous one, allowing the detailed study of the temperature evolution along some heat pump cycles.

In order to study the performance of both models, it is necessary to understand the characteristic experimental water temperature evolution that is produced by the ON/OFF cycling of the heat pump. For this purpose, critical points (A-E) have been identified in Figure 4b. First of all, during the OFF periods (from A to B), the evolution of the outlet water temperature depends on the environmental conditions rather than on the BHE performance, since there is no mass flow rate and the water is stopped at the top of the BHE. So, the

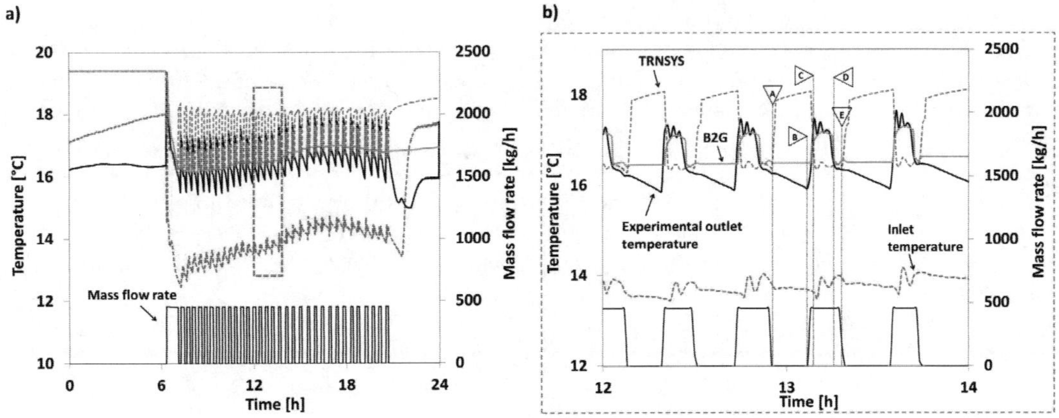

Figure 4. Comparison of both models performance for a typical day in heating mode: a) simulation results for the whole day, b) comparison for two hours of simulation (17/02/2010).

temperature evolution during this period is out of the scope of the studied models and it is not taken into account for the comparison.

On the other hand, at the start of the ON periods (from B to C), there is a sudden increase of the water temperature, due to the displacement of the water that remains inside the borehole during the OFF period, whose temperature tends to the ground one. Water entering the borehole at the start of the ON period takes about 7 minutes to reach the end of the U-tube (point D). At this momentm, a temperature decrease can be observed at the outlet water temperature, which remains fairly constant until the heat pump switches off again (point E).

Regarding the response of the studied models, it can be observed that B2G correctly reproduces the outlet water temperature evolution, according to the experimental measurements. All the phenomena described before and related to the water temperature evolution are correctly modelled by the B2G thermal network approach. The differences observed between B2G results and experimental measurements can be attributed to measurements uncertainty and vertical heat transfer effects that are out of the scope of the model.

The TRNSYS model produces a very different output. On one hand, during the OFF periods, since there is no mass flow rate, the calculated outlet water temperature tends to the ground temperature (Eq. 6), which is about 19.4°C. Although this has no further influence on the results of the ON cycle, when integrated in a global model, the BHE model has to be connected to other components which would receive an incorrect water temperature value during the OFF cycle, consequently producing errors in the global model results.

On the other hand, during the ON periods, the temperature evolution is different from the experimental measurements. The values of the temperature match the experimental measurements corresponding to the entering water temperature at the moment it reaches the end of the U-tube (point D), but without taking into account the displacement of the water inside the U-pipe. In this case, delaying the response of the TRNSYS type is not enough to solve the problem: this solution would correct the position of the temperature values but, since the volume of water inside the U-pipe is not taken into account in the TRNSYS model, it would be impossible for this model to correctly reproduce the temperature increasing happening at the start of the ON period (from B to C).

Results for the cooling mode comparison are shown in Figure 5. The results of the simulation and the comparison between the two models with the experimental measurements are analogous to the ones observed in the heating mode comparison.

The differences observed in the response of both models may not be highly relevant from a long-time evolution point of view. However, instantaneous water temperatures are directly related to the heat pump efficiency and capacity. Moreover, control algorithms that might be used in the installation usually act according to the instantaneous temperature measurements. Therefore, for a short-term analysis of the system, it will be necessary to use a model capable of reproducing the instantaneous temperature response. Results presented in this section have shown that TRNSYS model would not be useful for this purpose. On the contrary, B2G produces an accurate enough prediction of the BHE outlet water temperature and it is useful for short-term simulations.

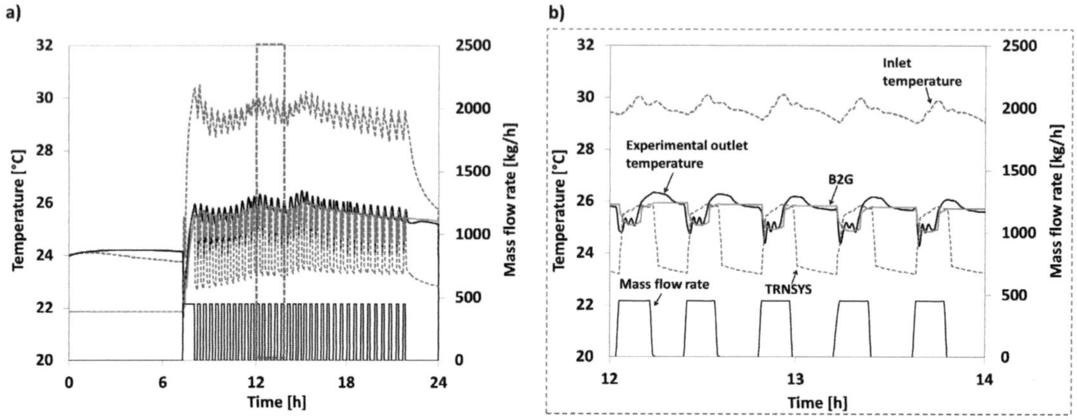

Figure 5. Comparison of both models performance for a typical day in heating mode (17/09/2010).

5. Conclusions

The aim of this work was to compare the performance of two different BHE models, regarding their ability to predict the outlet water temperature of the borehole. A novel BHE model called B2G was presented. This model is based on a dynamic approach using a thermal network, and has been validated in previous works. B2G model was compared to the one already existing in the TRNSYS software (type 557), based on a steady-state approach. For the comparison, experimental data from a real installation located at UPVLC was used.

Two different operating conditions were simulated: (i) a step-test and (ii) typical operating conditions of a GSHP installation for heating and cooling mode. The response of both models for the step-test simulation was very similar. However, it was found that the TRNSYS model could not reproduce the temperature delay corresponding to the time that the water takes to circulate through the U-tube of the borehole. On the other hand, in (ii), real operating conditions were simulated, which included several ON/OFF cycles. In this case, B2G proved to be able to reproduce the water temperature evolution with a great accuracy, while the TRNSYS model wasn't able to provide a satisfactory result, due to the steady-state approach on which it is based.

The differences observed between both models predictions highlight the importance of using a detailed dynamic approach when modeling a BHE, especially at ON/OFF operating conditions. In fact, assuming a steady-state approach could lead to inaccurate outlet water temperature predictions, which may introduce errors in calculating the heat pump efficiency or in the application of temperature control algorithms.

Acknowledgments
The present work has been supported by the FP7 European project Advanced ground source heat pump systems for heating and cooling in Mediterranean climate (GROUND-MED).

References
[1] Florides G, Kalogirou S, 2007 Ground heat exchangers - A review of systems, models and applications, *Renew. Energ.* **32** 2461-2478
[2] Yang H, Cui P, Fang Z 2010 Vertical-borehole ground-coupled heat pumps: A review of models and systems, *Appl. Energ.* **87** 16-27
[3] Ingersoll L R, Plass H J 1948 Theory of the ground pipe source for the heat pump, *ASHRAE Trans.* **54** 339-348
[4] Ingersoll L R, Adler F T, Plass H J 1950 Theory of earth heat exchangers for the heat pump, *ASHRAE Trans.* **56** 16788
[5] Carslaw H S, Jaeger J C 1946 *Conduction of heat in solids* (Oxford UK: Claremore Press)
[6] Eskilson P 1987 *Thermal analysis of heat extraction boreholes* PhD Thesis (University of Lund, Sweden)
[7] Eskilson P, Claesson J 1988 Simulation model for thermally interacting heat extraction boreholes, *Numer. Heat Transfer* **13** 149-65
[8] Bauer D, Heidemann W, Müller-Steinhagen H, Diersch H-J G 2011 Thermal resistance and capacity models for borehole heat exchangers, *Int. J. Energ. Res.* **35** 312-320
[9] Bauer D, Heidemann W, Diersch H-J G 2011 Transient 3D analysis of borehole heat exchanger modeling, *Geothermics* **40** 250-260
[10] Pasquier P, Marcotte D 2012 Short-term simulation of ground heat exchanger with an improved TRCM *Renew. Energ.* **46** 92-99
[11] Hellström G 1989 *Duct Ground Heat Storage. Manual for Computer Code* (Department of Mathematical Physics, University of Lund)
[12] GeoCool project (EU 5th Framework Programme, NNE5-2001-00847), http://cordis.europa.eu/projects/rcn/86940_en.html
[13] Ruiz-Calvo F, Montagud C 2014 Reference data sets for validating GSHP system models and analyzing performance parameters based on a five-year operation period, *Geothermics* **51** 417-428
[14] Montagud C, Corbern J M, Montero Á 2014 In situ optimization methodology for the water circulation pumps frequency of ground source heat pump systems, *Energ. Buildings* **68** 42-53
[15] Magraner T 2010 *Validación Experimental de los Métodos de Diseño de Instalaciones de Bomba de Calor Acoplada al Terreno* PhD thesis (UPVLC)

Journal of Physics: Conference Series **547** (2014) 012009 doi:10.1088/1742-6596/547/1/012009

Environmental monitoring of a Sardinian earthen dwelling during the summer season

G Desogus[1], S Di Benedetto[1], W Grassi[2] and D Testi[2]

[1] DICAAR, Department of Civil and Environmental Engineering and Architecture, University of Cagliari, Italy
[2] DESTEC, Department of Energy, Systems, Territory and Constructions Engineering, University of Pisa, Italy

E-mail: s.dibenedetto@unica.it

Abstract. Increasing interest in earth architecture has led to the development of new international norms regarding these structures. Although Italy has no specific legislation for this building type, both national laws for the safeguard of rural architecture and regional norms regarding the conservation of historical centers have considerably slowed down the pace of their destruction. This is particularly true for Sardinia, which maintains a conspicuous heritage of "raw earth" architecture, mostly in the old town centers of the Campidano plain and in its adjacent valley. Due to the current legislation on energy efficiency in buildings, it has become essential – particularly for the Sardinian region – to define guidelines for the improvement of energy efficiency for this existing building heritage and identify the best parameters for their energetic classification. Currently, these constructions are heavily penalized by the gap that persists between the requirements of current energy balance evaluations, calculated upon heating and domestic hot water energy demands, and the actual year-round energy performance, which also includes the summer season. Moreover, this building type has a low lifecycle environmental impact, but this aspect is not properly "rewarded" by Italian regulations. The study proposed herein firstly took into account the simulation of the thermal transient characteristics of the adobe wall (brick made of clay, earth and straw, forged with wooden molds and sun dried). Analytical calculations were performed using a transient model, assuming sinusoidal behavior of all the parameters acting on the system. The results showed a high thermal inertia of the material and a good ability in dampening the external thermal wave. Next, we conducted an internal and external environmental monitoring of an existing earthen residential building in Sardinia ("Casa Mancosu", Serramanna, VS), which provided the experimental data for the evaluation of the whole building thermo-physical behavior. The measurements were taken during the 2010 summer season; the dwelling was not cooled by an air conditioning system. Thermal comfort analyses based on these experimental data indicate that the roof is the "weak" component, creating local discomfort due to radiant asymmetry. The described methodology is expected to be applicable also to the many buildings of this geographical area similar to the examined one.

1. Introduction

1.1 Earth architecture: the state of the Legislations

Current Italian norms referring to earth architecture lay out regulations for the conservation of rural architecture (Law 378/2003) and the enhancement of value of existing earthen architectures (Regional Law 2/2006, Piemonte, and Regional Law 17/1997, Abruzzo). These three laws have attempted to fill the legislative void for the conspicuous heritage of raw earth architecture, favoring the safeguard of this architectural patrimony in Italy and particularly in Sardinia, where the adobe (from the Arab *al-tub*, brick) building technique is very common. The sole interest of these legislative texts is the safeguard and conservation of buildings of rural and historical importance: a conservationist slant that neither encourages new constructions nor incentivizes the renovation of existing earth buildings in a territory where this is a well rooted building technique. The renewed national and international interest in earth architecture is not only tied to criteria of bio-architecture but is dictated first and foremost by its thermo-physical characteristics that meet the demands of a Mediterranean climate, particularly

Journal of Physics: Conference Series **547** (2014) 012009 doi:10.1088/1742-6596/547/1/012009

during the summer season [1-2].

Although new laws have been proposed in Italy in the last few years [3-5], there has been no positive outcome due to the absence of specific references to the construction technique and of technical elements to support the use of raw earth buildings rather than those built using other materials [6]. On the international front, many countries have legislated on the matter (see Figure 1). A picture on the legislative situation regarding earth construction in the world is given in [7]. This study has evidenced how 79% of norms analyzed worldwide focus on the construction technique most prevalent in the respective territories. Indeed in many countries adobe masonry is considered a weight-bearing structure, whereas Italian legislature does not include raw earth among construction materials [8], thus discouraging research and experimentation of building types associated to these techniques.

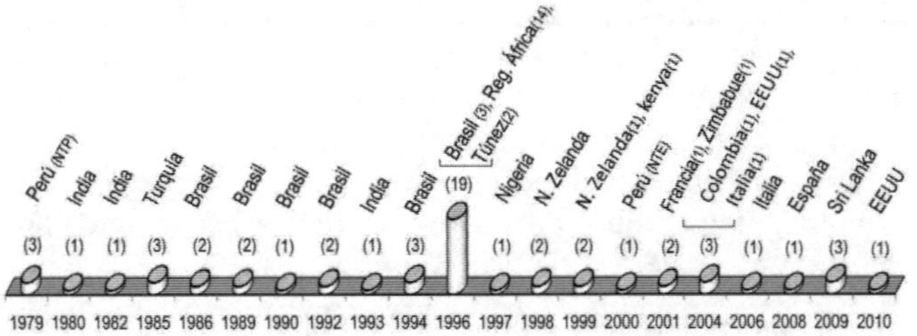

Figure 1. History of laws concerning earth constructions published worldwide [7].

The building type is also unfavorably treated by the Italian normative context, which is still lacking in norms regarding energetic classification of buildings in summertime conditions. The time has come to investigate which types of buildings best respond to Mediterranean climatic conditions, to limit summertime energy consumption. The scientific results collected through environmental monitoring of the buildings should form the basis of the passive requisites to be introduced in current national laws regarding energy efficiency.

With [9] Italy has acknowledged the European Guideline 2002/91/EC, substituted by the current 2010/31/EU, which promotes the improvement of energy efficiency performance in buildings within the Union, taking into account local external climatic conditions. As far as the summertime aspect is concerned, with the introduction of [10], Italian legislation has begun to define the performance limits for summertime cooling, introducing parameters to determine the cooling energy requirement of the building opaque envelope and outlining specific requisites on its periodic thermal transmittance (*Yie*) and surface mass (*Ms*). These parameters are also included in the evaluation dictated by the National Guidelines [11], with energy classes from I to IV, according to the degree of quality of energy performance reached.

These prescriptions make it possible to determine adequate thermal performance in summertime conditions, so the dampening and attenuation of the external thermal wave can become a controlled value even in the planning phase of the buildings in question.

To date, however, summertime performance does not contribute to the definition of a global energetic efficiency class of the building, but is reported by means of separate indicators in the energy efficiency certification document, named "Attestazione di Prestazione Energetica" (APE). Given Italy's climate conditions, it is important to obviate this shortcoming and to reward, through the APE certification, those buildings which maintain suitable levels of comfort with a reduced use of air cooling and dehumidification systems [12].

1.2 Objectives

This article presents the case study of an adobe building monitored in Sardinia during the summer season. Firstly, the article tackles, using the methodology described in the paragraph below, the effective thermal inertia of an adobe wall, based on its physical characteristics, tightly connected to the time constant of the building. Useful indications in terms of comfort are expected, so that minimum

Journal of Physics: Conference Series **547** (2014) 012009 doi:10.1088/1742-6596/547/1/012009

passive requisites can be suggested for introduction in the legislation on building summertime performance.

Environmental monitoring of the private residential building, "Casa Mancosu" [13], built in adobe and situated in the historical town center of Serramanna (VS), Italy, has provided all the necessary data to apply the technical standards ISO 7730:2005 [14] and ISO 7726:1998 [15], needed for the evaluation of thermal comfort levels in the monitored indoor environments. The analysis of thermal comfort according to these norms allows to classify the environment, according to 3 comfort classes, and identify the inefficient components of the building, responsible for creating conditions of local discomfort.

2 Analytical evaluation of the thermal behavior of adobe masonry

2.1 Set up and resolution of the problem

The initial study of the thermal behavior of an adobe monolayer wall in dynamic regime was tackled assuming that the indoor temperature ($T_{w,indoor}$) is measured, considering outdoor conditions, air temperature ($T_{a,out}$) and solar radiation (W), on an hourly basis [16].

In line with the measured values, let us consider the law of temporal variation of outdoor temperature by approximating it to a sinusoidal curve of period (P) equal to 24 hours:

$$T_{a,out} = T_{a,mean\ out} + \theta cos(2\pi t/P) \tag{1}$$

Solar radiation is also assumed to follow a sinusoidal trend.

Given the Fourier equation for conduction, simplified for one-dimensional conditions in space:

$$\frac{\partial T}{\partial t} = \alpha \frac{\partial^2 T}{\partial x^2} \tag{2}$$

we simplify the study of the temperature variation inside the masonry, assuming that the latter is given by the product of the functions *A(x)* dependent on space and *B(t)* dependent on time, according to the technique of separation of variables:

$$T(x,t) = A(x)B(t) \tag{3}$$

To calculate the constants, which we will call *M* and *N*, the following boundary conditions were set:

$$T(x=0) = T_{w,indoor} \tag{4}$$

$$\left(\frac{\partial T}{\partial x}\right)_{x=d} = -\frac{1}{\lambda}\left[aW + h_c\left(T_{a,out} - T_{w,out}\right) + h_r\left(T_{sky} - T_{w,out}\right)\right] \tag{5}$$

From the resolution of the system, we obtain:

$$T(x,t) = \left(Me^{\sqrt{\frac{i\omega}{\alpha}}x} + Ne^{-\sqrt{\frac{i\omega}{\alpha}}x}\right)e^{i\omega t} \tag{6}$$

where the constants *M* and *N* depend on known quantities:

$$M = \frac{T_{w,indoor}\sqrt{\frac{i\omega}{\alpha}}e^{-\sqrt{\frac{i\omega}{\alpha}}d} - \frac{aW}{\lambda} + \frac{h_c}{\lambda}\left(T_{w,indoor}e^{-\sqrt{\frac{i\omega}{\alpha}}d} - T_{a,out}\right) + \frac{h_r}{\lambda}\left(T_{w,indoor}e^{-\sqrt{\frac{i\omega}{\alpha}}d} - T_{sky}\right)}{\sqrt{\frac{i\omega}{\alpha}}e^{\sqrt{\frac{i\omega}{\alpha}}d} + \sqrt{\frac{i\omega}{\alpha}}e^{-\sqrt{\frac{i\omega}{\alpha}}d} - \frac{h_c}{\lambda}\left(e^{\sqrt{\frac{i\omega}{\alpha}}d} - e^{-\sqrt{\frac{i\omega}{\alpha}}d}\right) - \frac{h_r}{\lambda}\left(e^{\sqrt{\frac{i\omega}{\alpha}}d} - e^{-\sqrt{\frac{i\omega}{\alpha}}d}\right)} \tag{7}$$

$$N = \frac{T_{w,indoor}\sqrt{\frac{i\omega}{\alpha}}e^{\sqrt{\frac{i\omega}{\alpha}}d} + \frac{\alpha W}{\lambda} + \frac{h_c}{\lambda}\left(T_{a,out} - T_{w,indoor}e^{\sqrt{\frac{i\omega}{\alpha}}d}\right) + \frac{h_r}{\lambda}\left(T_{sky} - T_{w,indoor}e^{\sqrt{\frac{i\omega}{\alpha}}d}\right)}{\sqrt{\frac{i\omega}{\alpha}}e^{\sqrt{\frac{i\omega}{\alpha}}d} + \sqrt{\frac{i\omega}{\alpha}}e^{-\sqrt{\frac{i\omega}{\alpha}}d} - \frac{h_c}{\lambda}\left(e^{\sqrt{\frac{i\omega}{\alpha}}d} - e^{-\sqrt{\frac{i\omega}{\alpha}}d}\right) - \frac{h_r}{\lambda}\left(e^{\sqrt{\frac{i\omega}{\alpha}}d} - e^{-\sqrt{\frac{i\omega}{\alpha}}d}\right)} \tag{8}$$

Function (6) represents the solution to the Fourier equation in periodic dynamic regime. Let us add now the steady-state solution:

$$< T > = c_1 x + c_2 \tag{9}$$

where c_1 and c_2 depend on the mean values of the oscillating functions and on the wall thermal transmittance (U-value):

$$c_1 = \frac{T_{w,out} - T_{w,indoor}}{d} = \frac{U}{\lambda}\left(T_{a,mean\ out} - T_{a,indoor}\right) \tag{10}$$

$$c_2 = T_{w,indoor} \tag{11}$$

The general solution will therefore be given by:

$$\Theta(x,t) = \left(Me^{\sqrt{\frac{i\omega}{\alpha}}x} + Ne^{-\sqrt{\frac{i\omega}{\alpha}}x}\right)e^{i\omega t} + c_1 x + c_2 \tag{12}$$

The real part of the complex function (12) represents the temperature at different locations within the wall (variable x) and at different times (variable t).

The main limit of this calculation lies in the assumption of sinusoidal trend for the external climatic variables, thus an approximation of the real profiles.

2.2 Application to the adobe wall

The application of the analytical solution (12) enables us to confirm the good inertial behavior of adobe. Figure 2 depicts the temporal evolution of temperature at different positions within a 40 cm thick adobe masonry wall (position x=0 refers to the internal wall surface, x=0.4 m to the external one). The utilized thermo-physical properties of the material are reported in Table 1, together with the corresponding source.

Table 1. Thermo-physical properties of adobe.

Property	*Unit*	*Value*
Mass density	kg/m^3	1842 [17]
Thermal conductivity (λ)	W/(m K)	0.663 [17]
Specific heat	J/(kg K)	1000 [18]
Thermal diffusivity (α)	mm^2/s	0.3599

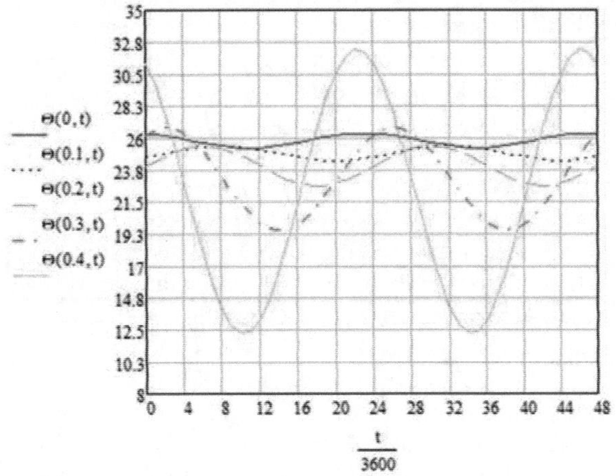

Figure 2. Temporal evolution of temperature inside an adobe masonry wall.

As expected, it can be noted that the external thermal wave increasingly lags out of phase and attenuates as we advance in the masonry towards the inside of the building; at a depth of 30 cm, the temperature is already almost constant, with oscillations of approximately 1 K, notwithstanding elevated thermal excursions (around 20 K).

From Figure 2, we can deduce that the phase displacement is above 12 hours, corresponding to a Class I performance according to the National Guidelines [11]. Applying the standard ISO 13786 [19], we obtain the dynamic thermal characteristics reported in Table 2.

Table 2. Dynamic thermal characteristics of adobe-made walls, according to [19].

Parameter	*Unit*	*Value with d=30 cm*	*Value with d=35 cm*	*Value with d=40 cm*
Periodic thermal trasmittance (*Yie*)	$W/(m^2\ K)$	0.347	0.210	0.127
Decrement factor (f_d)		0.216	0.146	0.098
Phase shift (φ)	H	11.0	12.9	14.8
Surface mass (*Ms*)	kg/m^2	553	645	737

All these data suggest that an environment with at least 30 cm of adobe masonry can provide satisfactory thermal comfort conditions in summertime if all the other components of the building envelope behave efficiently. This result is confirmed by a large quantity of data present in the literature, amongst which it is worth mentioning [20], an interesting analysis on the thermodynamic behavior of a raw earth structures.

3 Environmental monitoring of the case study

3.1 Analysis of the hygrothermal comfort of an earth building

The environmental monitoring of a real case study allowed us to perform an analysis of hygrothermal comfort in an indoor environment built using the adobe technique. The analysis was conducted using the classic Fanger's method [21], although – strictly speaking – this is quantitatively correct only for spaces in which the use is passive, in other words where there is no intervention by occupants on the building system to improve their level of comfort. The results will therefore be taken as simple qualitative indicators to shed light on the most critical elements of the structure and point towards possible interventions for energetic-environmental retrofit.

The monitoring of "Casa Mancosu" was conducted using two microclimatic control units (fitted with hot filament anemometer, global thermometric probe, and psychrometer) and appropriate temperature and humidity sensors, in accordance with the standard [15]. In addition to the measurements prescribed by this norm, all the temperatures of the internal vertical walls, floors and ceilings enclosing the spaces were also measured, using contact thermal resistances. The measurement chain was completed by a portable computer and a data logger. The data necessary to ascertain the level of thermal satisfaction are the indoor air temperature ($T_{a,indoor}$), the relative humidity of indoor air (RH_i), the indoor air velocity (v_{ar}), the mean radiant temperature (T_r), the type of clothing worn by the occupants (I_{cl}), and the metabolism according to the performed activity (M^*).

The experimental data were recorded at 10-minute intervals by the microclimatic units (positioned in the most significant places, according to frequent occupant activities), subdividing the 24-hour period into 4 subgroups of 6 hours for the living room (floor area: 24.5 m^2) and 6 subgroups of 4 hours for the bedroom (floor area: 14.7 m^2). This way of grouping the data was established following the positive verification of the cyclical temperature variations indicated in the standard [14], which specifies that if the peek-to-peek variation is lower than 1 K and the rate of temperature variation in response to a step input is less than 2K/h, the methods for steady-state regime can be applied.

The case-study building has a residential use and the monitored period was the summer season; for this reason, the following values were assumed: M^*=70 W/m^2 (or 1.2 met) and I_{cl}=0.5 clo.

The method prescribed by the technical standard is applicable, as the following conditions subsist: 46 $W/m^2 < M^* < 232\ W/m^2$ (0.8 met$<M^*<$4 met), $0<I_{cl}<0.310\ m^2K/W$ ($0<I_{cl}<2$ clo), $10°C<T_{a,indoor}<40°C$, $0<v_{ar}<1$ m/s, and $0<p_a<2700$ Pa.

The performed calculations of the global values *PMV* (Predicted Mean Vote index) and *PPD* (Predicted Percentage of Dissatisfied people) were validated by means of the test specified in [14].

Analytical determination of thermal comfort must be completed by the analysis of phenomena associated to local discomfort, that is to say drafts (*DR*), vertical air temperature variation ($\Delta T_{a,v}$), hot/cold floors (T_{pav}), radiant asymmetry (*AR*). The analysis of the latter characteristic, in particular, is useful to highlight shortcomings in the thermal behavior of the construction elements of the building.

Local discomfort due to radiant asymmetry is caused by an excessive temperature difference between opposite surfaces (roof or walls), too hot or too cold. To determine the degree of this discomfort, it is necessary to calculate the difference in radiant temperature along the flat plane between opposite and parallel surfaces (ΔT_{pr}).

According to [15], T_{pr} is calculated as follows:

$$T_{pr} = T_1^4 * F_{p-1} + T_2^4 * F_{p-2} + \cdots + T_N^4 * F_{p-N} \tag{13}$$

This relationship is valid if all surfaces forming the environment are black bodies. However, the calculation is significant, because the materials used in the case-study building present emissivity values (at infrared wavelengths) close to one, as most of the construction materials.

The building was subject to the verification of all the indices of local discomfort, both in the living room and in the bedroom. In particular, to verify the degree of radiant asymmetry, the most critical thermal situation was chosen. For the living room, where a large glass surface contrasts an opaque indoor surface, the calculation along the horizontal plane (wall-wall) is reported; on the contrary, for the bedroom, the vertical direction was chosen (floor-roof), because the monitoring concerned the summertime period and – again in accordance with [14] – the percentage of local thermal discomfort caused by the "hot roof" situation is greater than that caused by a "hot wall".

3.2 Thermal comfort analysis results

The analysis of thermal comfort in "Casa Mancosu" relates to sedentary activity: in the living room area we refer to a person sitting in a state of relaxation; the bedroom, on the other hand, has been divided in two sections: one related to a person in the "study" position and one related to the "sleeping" position, as suggested by [22]. This distinction between daytime and nighttime allows a more complete outlook of comfort conditions throughout the whole day.

As shown in Figure 3, the living room performs with a good comfort class during occupancy, with PPD above 15% for only 2.2% of the recorded data. Considering the summertime monitoring, the construction elements making up this envelope and its favorable south-east orientation do not call for energy optimization intervention.

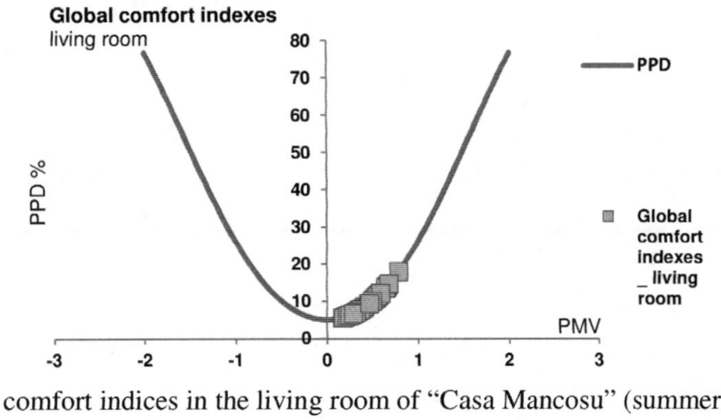

Figure 3. Global comfort indices in the living room of "Casa Mancosu" (summertime monitoring).

The bedroom, on the other hand, does not fit the conditions of thermal comfort. The major discomfort – according to calculations made in line with the technical standard – is caused by the roof covering, which has a negative effect particularly during the sleeping hours, as evidenced by the comparison of Figures 4 and 5. Indeed, the graph in Figure 5 shows a higher dissatisfaction percentage, together with an increase of the higher PPD values, from 30% to over 40%, highlighting the importance of better comfort conditions in the hours dedicated to sleep.

Adobe's high thermal inertia guarantees almost constant temperatures in indoor environments, while local discomfort is caused by other weak components of the construction, such as the roof covering and windows lacking external sun screens, as evident by indoor air and surface temperature measurements (see Figures 6 and 7).

Figure 8 illustrates the results of verifications of discomfort due to radiant asymmetry for "hot wall" in the living room (the negative values mark the absence of discomfort conditions). Figure 9 shows instead the verifications of radiant asymmetry in the bedroom for the "hot wall" and "hot ceiling": in this case, discomfort caused by "hot wall" is present, though not significant, whereas the incidence of discomfort caused by the roof, combined with the global PMV and PPD values, places this environment in borderline comfort conditions or in total discomfort, according to [14].

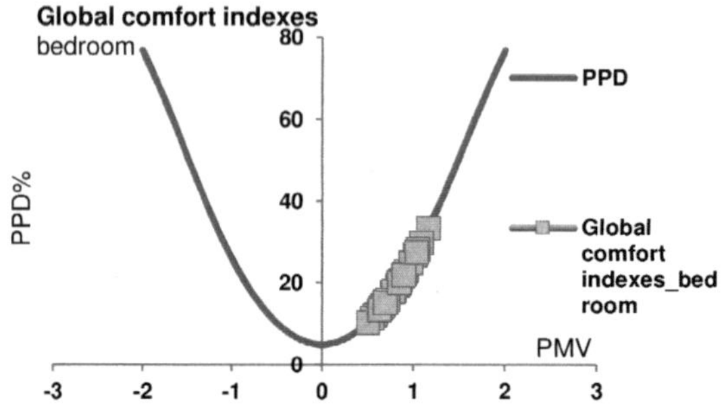

Figure 4. Global comfort indices for the bedroom of "Casa Mancosu" (daytime monitoring).

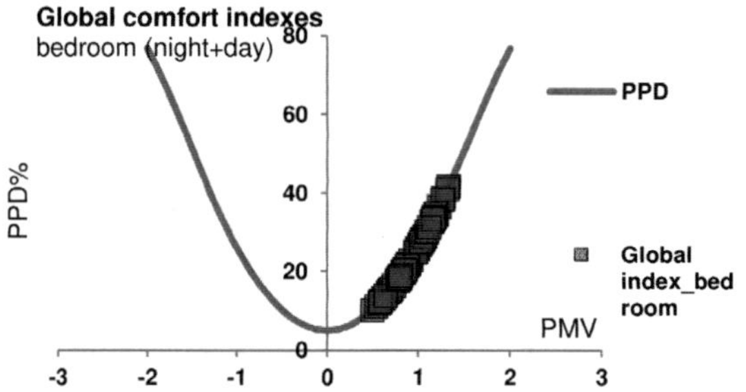

Figure 5. Global comfort indices for the bedroom of "Casa Mancosu" (24-hour monitoring).

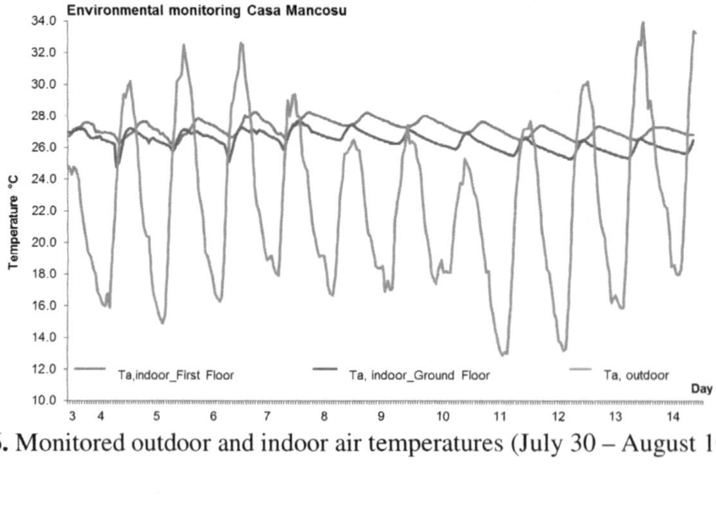

Figure 6. Monitored outdoor and indoor air temperatures (July 30 – August 10, 2010).

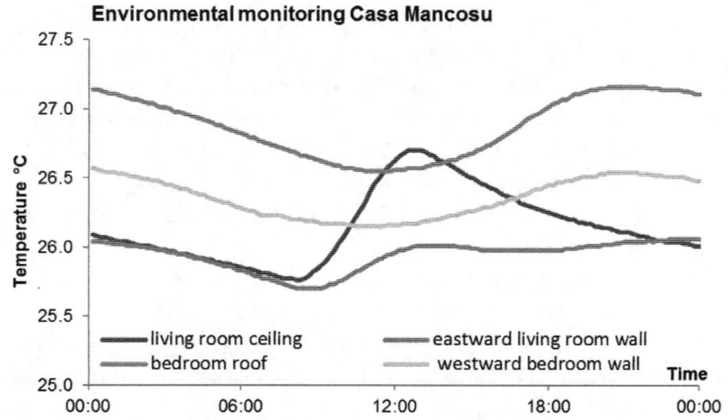

Figure 7. Monitored indoor surface temperatures (August 8, 2010).

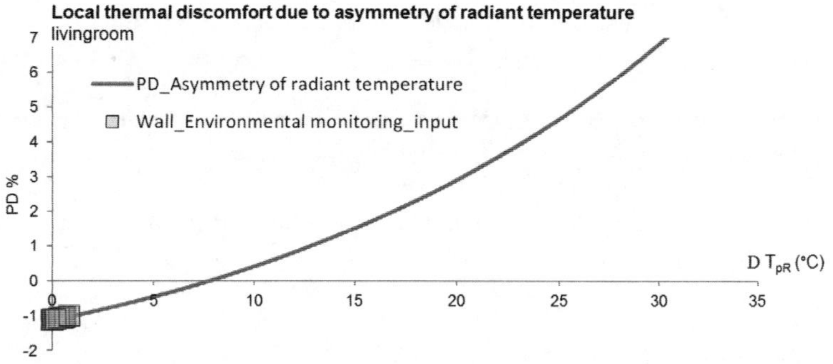

Figure 8. Local thermal discomfort due to asymmetry of radiant temperature in the living room ("hot wall" case).

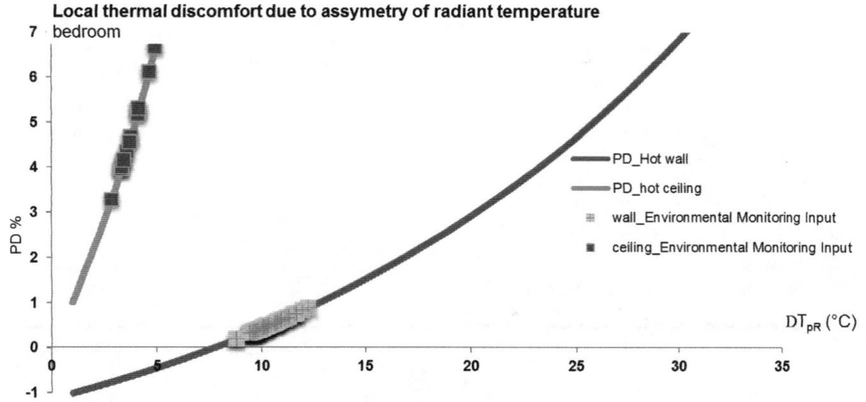

Figure 9. Local thermal discomfort due to asymmetry of radiant temperature in the bedroom ("hot wall" and "hot ceiling").

4. Conclusions and future developments

In the present paper, we employed methodologies capable of evaluating the thermodynamic behavior of a raw earth construction and the hygrothermal comfort perceived within the adobe private residential building "Casa Mancosu". The results confirm the high thermal inertia given by the thermo-physical properties of the adobe material. We verified the level of thermal comfort of the case-study building, calculated according to current international standards, using the experimental data from the environmental summertime monitoring. More specifically, the adobe-made building, without mechanical cooling, has proven to be able to maintain satisfactory levels of thermal comfort and internal temperatures close to the design set points of air conditioning systems. On the other hand, conditions of thermal discomfort were recorded in the presence of inefficient construction elements. Specifically, the presented graphs highlight how the roof of "Casa Mancosu" – which does not reach the periodic thermal trasmittance currently prescribed by Italian legislation – causes discomfort due to radiant asymmetry between ceiling and floor and a high predicted percentage of dissatisfied people (above 15% for as many as 83% of the recorded data).

As a future development, we suggest to perform a full dynamic analysis of the building thermal evolution, which could be properly validated by means of the monitoring data and used to simulate possible retrofit actions – such as roof insulation and venting – in a frame of reliability.

As for the evaluation of comfort levels, we suggest further analyses using the adaptive method [2], due to the mentioned limits of the Fanger's method.

Finally, a comparison between cooling energy needs obtained by monthly-based methods (implemented by current technical standards [23]) and monitoring results is expected to shed some light on the inaccuracy of quasi-steady-state energy calculations for the summer period.

Acknowledgments

The Energy Efficiency Laboratory of "Sardegna Ricerche" is gratefully acknowledged for providing us with the experimental equipment, as well as the owners of "Casa Mancosu", for letting us mount the instruments and perform the environmental monitoring of their house and for responding to our interviews on the building usage. Finally, we would like to thank Mr. Davide Della Vista, technician of the University of Pisa, for his support on the data processing activity.

List of symbols

a	external wall radiative absorption coefficient
d	wall thickness [m]
F_p	view factor
h_c	external convective heat transfer coefficient [W/(m^2 K)]
h_r	external radiative heat transfer coefficient [W/(m^2 K)]
I_{cl}	clothing insulation [m^2 K/W, clo]
M^*	metabolic rate [W/m^2, met]
P	period of the external air temperature wave [s]
p_a	vapor pressure [Pa]
PMV	Predicted Mean Vote index
PPD	Predicted Percentage of Dissatisfied people index
RH_i	indoor air relative humidity [%]
t	time [s]
$T_{a,indoor}$	indoor air temperature [°C]
$T_{a,out}$	external air temperature [°C]
$T_{a,mean\ out}$	mean external air temperature [°C]
T_{pr}	radiant temperature along the flat plane
T_r	mean radiant temperature [°C]
T_{sky}	equivalent sky temperature [°C]
$T_{w,indoor}$	indoor surface temperature [°C]
$T_{w,out}$	outdoor surface temperature [°C]
U	wall thermal trasmittance [W/(m^2 K)]
v_{ar}	indoor air velocity [m/s]
W	solar radiation [W/m^2]
x	spatial coordinate [m]

Greek symbols

α	wall thermal diffusivity [m^2/s]
λ	wall thermal conductivity [W/(m K)]
θ	semi-amplitude of the external air temperature wave [K]
Θ	wall temperature field [°C]
ω	angular frequency of the external air temperature wave [rad/s]

References

[1] Secci W 2005 Dalla tutela alla nuova costruzione: normativa e standardizzazione *Proc. Costruire in Terra Cruda Oggi (Novi Ligure, IT)* ed Edicom Edizioni
[2] Desogus G, Di Benedetto S and Ricciu R 2014 Adaptive Thermal Comfort in Earth Architecture *Submitt. to Build. Environ.*
[3] Italian Parliament 2004 *Proposta di Legge Cossu (XIV Legislatura)*
[4] Italian Parliament 2007 *Disegno di Legge 1349 (XV Legislatura)*
[5] Italian Parliament 2009 *Proposta di Legge Schirru (XVI Legislatura)*
[6] Galli D 2004 *Parere della Camera alla Legge Cossu (Commissione XI Lavoro)* Italian Parliament
[7] Cid J, Mazzaron F R and Canas I 2011 *Informes de la Construccion* **63**/523 159
[8] Italian Parliament 1974 *Legge 64/1974*
[9] Italian Parliament 2005 *D.Lgs. 192/2005*
[10] Italian Parliament 2009 *D.P.R. 59/2009*
[11] Italian Parliament 2009 *D.M. 26/06/2009*
[12] Prada A, Baratieri M, Baggio P, Romagnoni P, Gasparella A and Bettanini E 2008 Consumi energetici e certificazione energetica: situazione estiva ed invernale in Italia e in Europa *Proc. 63rd ATI National Congress (Palermo, IT)*
[13] Di Benedetto S 2011 *La Sostenibilità Ambientale degli Edifici – Diagnosi Energetica delle Architetture di Terra: Analisi del Comportamento Termico in Regime Dinamico di una Muratura in Adobe e Misura del Benessere Termico di un Edificio Realizzato con l'Utilizzo del Mattone Crudo* Ph.D. Thesis in Electrical and Thermal Energetics University of Pisa
[14] ISO 2005 *ISO 7730* International Organization for Standardization
[15] ISO 1998 *ISO 7726* International Organization for Standardization
[16] ARPAS, Dipartimento Specialistico Regionale IdroMeteoClimatico (Sassari, IT)
[17] Istituto Giordano, Sezione di Trasmissione del Calore (Bellaria, IT)
[18] Achenza M and Sanna U 2008 *Il Manuale Tematico della Terra Cruda* Regione Autonoma della Sardegna (Cagliari, IT)
[19] ISO 2007 *ISO 13786* International Organization for Standardization
[20] Wessling F 1975 Transient Thermal Response of Adobe *Adobe News 6 (Albuquerque, NM)*
[21] Fanger P O 1970 *Thermal Comfort Analysis and Applications in Environmental Engineering* (McGraw-Hill: New York)
[22] Lin Z and Deng S 2008 *Build. Environ.* **43**/5 905
[23] ISO 2008 *ISO 13790* International Organization for Standardization

Theoretical analysis of screened heat pipes for medium and high temperature solar applications

P Di Marco, S Filippeschi, A Franco, D Jafari*

Università di Pisa, DESTEC, Largo Lucio Lazzarino 2, 56122 Pisa, Italy

E-mail: d.jafari@studenti.unipi.it

Abstract. A mathematical model is applied to study the cylindrical heat pipes (HPs) behaviour when it is exposed to higher heat input at the evaporator for solar collector applications. The steady state analytical model includes two-dimensional heat conduction in the wall, the liquid flow in the wick and vapour hydrodynamics, and can be used to evaluate the working limits and to optimize the HP. The results of the analytical model are compared with numerical and experimental results available in literature, with good agreement. The effects of heat transfer coefficient, power input, evaporator length, pipe diameter, wick thickness and effective pore radius on the vapour temperature, maximum pressure drop and maximum heat transfer capability (HTC) of the HP are studied. The analysis shows that wick thickness plays an important role in the enhancement of HTC. Results show that it is possible to improve HTC of a HP by selecting the appropriate wick thickness, effective pore radius, and evaporator length. The parametric investigations are aimed to determine working limits and thermal performance of HP for medium temperature solar collector application.

1. Introduction

The HP is a two-phase device that transfers heat from a hot zone to a cold zone using capillary forces generated by a wick and a working fluid. The thermal performance of a HP can be characterized by its overall thermal resistance and its maximum power. These characteristics depend mainly on the capillary structure, which is usually made of grooves, meshes, sintered powder or a combination of them.

The interest in the use of HPs is recognized in many industrial applications such as control of satellites and spacecraft in aerospace, the cooling of electronic components and solar applications. Interesting connection can be evidenced in the literature between solar collector and HPs. The HPs that can be applied are based both on the conventional scheme in which the capillary action permits to return heat to the evaporator or according to the scheme of the wickless HP or closed loop two-phase thermosyphon working in gravity mode [1,2] or antigravity mode [3]. Recently, there has been a rising interest in the use of HP technology in the solar collector. Chun et al. [4] experimentally studied a particular system using different working fluids. Joudi and Witwit [5] used various lengths copper pipe with water as a working fluid. Mathioulakis and Belessiotis [6] presented the energy behaviour of a solar collector employing a HP filled with ethanol. Riffat et al. [7] investigated thermal performance of a thin membrane of HP solar collector. Azad [8] developed a theoretical model for analysing HP parameters. Taoufik et al. [9] developed the thermal efficiency of a flat plate HP solar collector. Finally, Aghbalou et al. [10] used HP in parabolic solar collector.

As a result of the interest in HP technology, several analytical and numerical analyses have been developed by many investigators. There has been a considerable research in the area of HP modelling from simple state thermodynamic analyses [11] and transient one dimensional (1-D) [12] to multidimensional [13-17].
As a result of the interest in HP technology, several analytical and numerical analyses have been developed by many investigators. Analytical modelling has been extensively carried on by researchers, with particular concern for fundamental analyses related to hydrodynamic and heat transfer processes [18-20]. Several models have been developed to predict the pressure drop that occurs in HPs; researchers used the Darcy's law [15, 16, 20] and non-Darcian transport [14, 21]. Evaporation and condensation at liquid–vapour interface have been described using the conjugate heat transfer model [14], and liquid–vapour coupling by means of interfacial velocity [16, 20]. In the case of thermal resistances of the wall and wick, researchers neglected conduction in the wall [18, 19, 21], considered heat conduction in radial direction [14], or in the axial direction [16], or the two-dimensional [20].
The specific objective of this paper is to investigate the HP for solar collectors, considering in particular the temperature range between 60 and 160°C, temperature range of specific interest for solar cooling and industrial process heat. The mathematical model presented in this paper is based on Shabgard and Faghri's model [16] and constituted a useful tool to perform optimization and parametric studies as well as to evaluate the HP capillary limits. The two dimensional steady state analytical model implemented in a MATLAB platform that couples axisymmetric heat conduction in the wall to liquid flow in the wick and vapour flow in the core. The results of the present model were compared to a numerical and experimental results reported in [13], and to an analytical model [14]. A very good agreement was observed.
Although several studies reveal the large potential of HP for solar collector, little attention has been paid to study the HP behaviour when it is exposed to high heat inputs at the evaporator. This paper presents a parametric study of the effect of different operation parameters on the HP performance. The effects of the wick thickness, the outside diameter (OD), the evaporator length, the power input, and the cooling fluid heat transfer coefficient of the condenser are studied. The motivation was to determine what are the effects on the heat transport behaviour of the screen wick structure, whether an optimum wick structure exists and how the value for maximum heat removal for a wick structure is determined for medium to high temperature solar application.

2. Mathematical model

The HP model is shown in figure 1a: it consists of the wall, the porous wick, and the vapour space in the radial direction as well as three sections in the axial direction (evaporator, adiabatic and condenser). As schematically shown in the thermal network, two-dimensional heat conduction in the wall is coupled with the 1-D heat conduction in the wick (radial). According to the electrical analogy (figure 1b), the radial and the axial heat conduction can be modelled as thermal resistances.

$$R_{radial} = \frac{\ln {}^{r_2}\!/_{r_1}}{2\pi k L} \tag{1}$$

$$R_{axial} = \frac{L}{kA} \tag{2}$$

Where r_1 and r_2 are the radial locations of the inner and outer surfaces of the element, respectively, k is the conductivity of the material, A is the cross section, $A = \pi(r_2^2 - r_1^2)$ and L is the length of the given zone. In case of solid wall k is the thermal conductivity of the pure material; concerning the wick, the effective conductivity (k_{eff}) takes into account the thermal conductivity of both liquid and metal.

$$k_{eff} = \frac{k_l\left[(k_l + k_w) - (1-\varepsilon)(k_l - k_w)\right]}{\left[(k_l + k_w) + (1-\varepsilon)(k_l - k_w)\right]} \tag{3}$$

where k_l and k_w are the thermal conductivities liquid and wick material, respectively and ε is the wick porosity. As schematically evidenced in figure 1b, there are two possible pathways for the conductive heat transfer through the HP, from the evaporator to the condenser. The thermal resistance of path 1 and 2, neglecting radial thermal resistance of the wall, can be described as

$$\text{Path 1: } R_{W,e} + R_{W,c} \tag{4}$$

$$\text{Path 2: } R_{3P,e} + R_{1P,a} + \frac{R_{2P,a}(R_{1W,a} + R_{2W,a})}{R_{2P,a} + R_{1W,a} + R_{2W,a}} + R_{3P,a} + R_{3P,c} \tag{5}$$

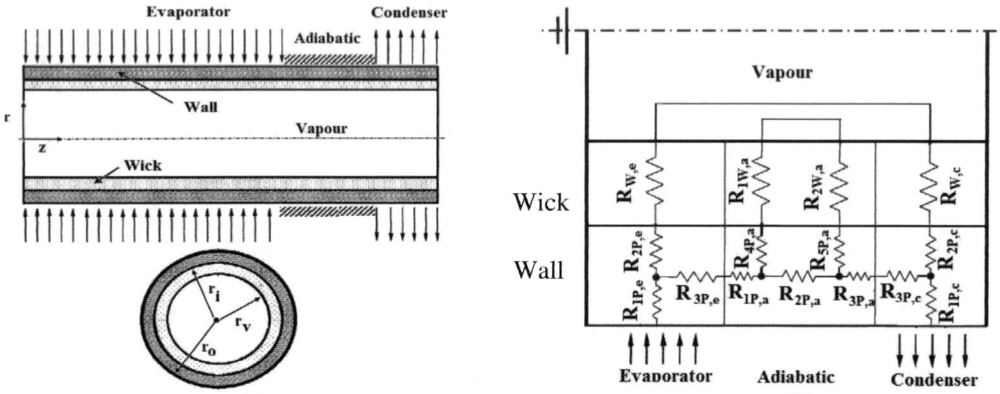

(a) (b)

Figure 1. (a) Schematic view of the cylindrical HP. (b) Network scheme for solid model (P, Pipe(wall); W, wick; e, evaporator; a, adiabatic; c, condenser)

The assumptions in the analysis are steady-state, incompressible and laminar flow, a saturated wick, constant properties and saturation temperature, and linear temperature profile across the thin wick structure. The conservation of mass, momentum, and energy equation is defined as follows.

- Heat conduction in the wall

The energy equation for two-dimensional (figure 1) and steady-state heat conduction in the wall with constant properties can be written as (θ is difference between the wall temperature and vapour temperature)

$$\frac{\partial^2\theta}{\partial r^2} + \frac{1}{r}\frac{\partial\theta}{\partial r} + \frac{\partial^2\theta}{\partial z^2} = 0 \tag{6}$$

- Liquid flow in wick

Continuity equation and Darcy's law [22] is used for the liquid flow in the porous wick:

$$r\frac{\partial u_l}{\partial z} + \frac{\partial}{\partial r}(rv_l) = 0 \tag{7}$$

$$u_l = -\frac{K}{\mu_l}\frac{dP_l}{dz} \tag{8}$$

where u_l is axial and v_l is radial liquid velocity, K is medium permeability, μ_l is liquid dynamic viscosity, and P_l is liquid pressure.

- Vapour flow analysis

A parabolic velocity profile is used for vapour flow within the HP. The conservation equations for mass and momentum are used to obtain vapour velocity and pressure in the core:

$$r\frac{\partial u_v}{\partial z}+\frac{\partial}{\partial r}(rv_v)=0 \tag{9}$$

$$\rho_v\left(u_v\frac{\partial u_v}{\partial z}+v_v\frac{\partial u_v}{\partial r}\right)=-\frac{\partial P_v}{\partial z}+\mu_v\left(\frac{\partial u_v}{\partial r^2}+\frac{1}{r}\frac{\partial u_r}{\partial r}\right) \tag{10}$$

where ρ_v is density, u_v is axial velocity, v_v is radial velocity, P_v is pressure, and μ_v is vapour dynamic viscosity.

- Liquid-vapour interface

At the interface the vapour velocity is related to the liquid velocity by a mass balance:

$$\rho_v v_{v,i}=\rho_l v_{l,i} \tag{11}$$

Boundary condition specifications at the outer pipe (r_o), wall-wick (r_i), wick-vapour interface (r_v) as well as at center axis and both ends of HP, are provided in table 1. A MATLAB platform is used to discretize the governing equations. In order to validate the model, the analytical results are compared with the experimental data reported in literature [13, 14]. The physical dimensions of the HP are considered as 890 mm in length (600 mm evaporator length, 200 mm condenser length and 900 mm adiabatic length) and OD= 19.1 mm. Figures 2 and 3 show a comparison of the current model with the literature data for the axial temperature distributions and pressure drop of the HP, respectively. As can be seen, an excellent match has been found with all the data reported.

Table 1. boundary condition for heat pipe analysis

$r=r_o$	$r=r_i$	$r=r_v$	r=0	z=0, L
$q''_e=q''_c=-k_{wall}\frac{\partial\theta}{\partial r}$	$\frac{\partial\theta}{\partial r}=h\theta\ (h=\frac{k_{wick}}{k_{wall}r_i\log(r_i/r_v)})$; $v_l=0$	$v_l=\frac{q''\vert_{r_i}}{\rho_l h_{fg}}(\frac{r_i}{r_v})\ q''\vert_{r_i}=-k_{wall}\frac{\partial\theta}{\partial r})$; $u_v=0$	$\frac{\partial u_v}{\partial r}=0$	$\frac{\partial\theta}{\partial z}=0$; $u_v=0$

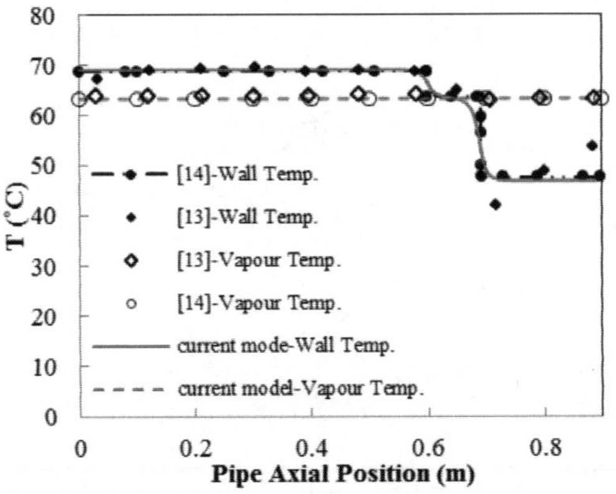

Figure 2. Comparison of the axial distribution of the wall and vapour temperatures.

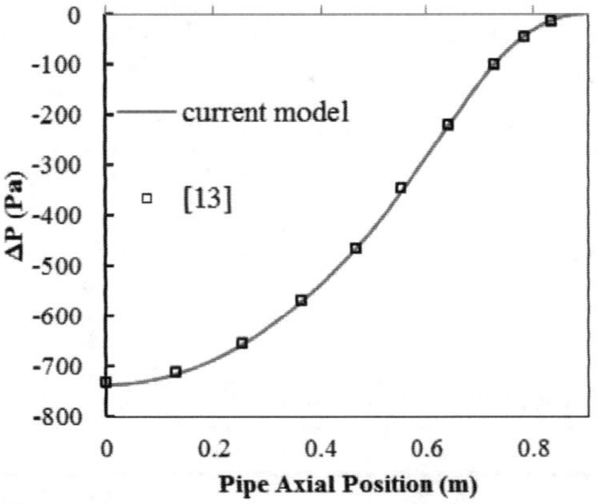

Figure 3. Comparison of the axial liquid pressure drop.

2.1. HP limits

HPs working are strongly dependent on their operational limitations. The operating limitations can be divided into two main categories [23]: the first ones linked to the liquid flow (entrainment, capillary, boiling) and the other ones linked to the vapour flow (viscous and sonic). The most important design consideration is the amount of power that the HP is capable of transferring at a given operational temperature and size (i.e.; length and external radius). The viscous, entrainment and sonic limits are generally much greater in comparison with the capillary limit and so do not pose any constraint [24].

In fact, for the selected temperature range (60-160°C), the high vapour pressure result in both a small vapour velocity and pressure drop. Therefore, the viscous, sonic, and entrainment limits are typically not reached, allowing the vapour to flow to the condenser section. In the present configuration, only the capillary and boiling limits are affecting the HP performance.

The capillary limit consist in the fact that, for a HP to operate properly, the net pressure drop must be greater than the capillary pressure which is derived from the Laplace-Young equation ($\Delta P_c = 2\sigma_l / r_c$) where σ_l is liquid surface tension and r_c ($=1/2N$) is effective capillary radius of the evaporator wick where N is the screen mesh number.

The other major limitation for the HP is the boiling limit. At higher applied heat flux, nucleate boiling may appear in the wick structure. At steady state operations, an expression for the heat flux beyond which bubble growth will occur may be developed by [25]:

$$Q_b = \frac{2\pi l_{eff} k_{eff} T_v}{\rho_v h_{fg} \ln\left(r_i / r_v \right)} \left(\frac{2\sigma_l}{r_n} - \Delta P_c \right) \tag{12}$$

where l_{eff} is the effective length and r_n is the critical nucleation site radius, which according to [22] ranges from 0.1 to 25.0 μm for conventional metallic case materials.

3. Results and discussion

There are several important factors affecting HP performance: the working fluid, the wick structure, materials, the dimensions and input parameters. In this study the effects of wick thickness, effective pore radius of wick, diameter, evaporator length, power input and heat transfer coefficient of the cooling fluid are analysed to achieve limitations of HP. The effects of these parameters are studied for a copper wrapped screen wick (150 mesh). All the arrangements were investigated at different heat fluxes from 500 to 20000 W/m^2, heat transfer coefficient of 200-1800 W/m^2K and cooling water temperature of 60-160 °C. Table 2 shows further values of the design parameters chosen to perform the present analysis, as well as the combinations of these parameters that were analysed. The design parameters shown in the table 3 are fixed for all investigations.

Table 2. Levels of the chosen parameters.

Level	*1*	2	*3*	4
l_e [m]	2	1.55	1	0.5
OD [mm]	12.7	20	30	-
t_{wick} [mm]	1	0.75	0.5	0.25
r_c [m]	2.4×10^{-4}	7.9×10^{-5}	3.2×10^{-5}	-

Table 3. Fix parameters.

Wick thermal conductivity [W/mK]	1.97
Wick permeability [m^2]	1.5×10^{-5}
Wall thickness [mm]	0.5
Condenser length [m]	0.17
Adiabatic length [m]	0.04

3.1. Effect of the wick thickness, outside diameter and evaporator length

The selection of the wick for a HP depends on many factors, several of which are closely linked to the properties of the working fluid, the maximum capillary length and the wick permeability. Another feature of the wick that must be optimized is its thickness.

The relationship between wick thickness and vapour temperature are presented in figure 4a, and the effects on the pressure drop for different evaporator lengths and diameters are shown in figures 4b, 4c, and 4d, respectively. As shown in figure 4a, when the wick thickness decreases from 1 mm to 0.25 mm, the vapour temperature decreases because of decreasing thermal resistance of the wick region. It also shows that the vapour temperature is weakly affected by evaporator length, as shown only in the case of OD = 12.7mm. Conversely, the pipe diameter affects the results: the vapour temperature for

OD=30 mm is about 15°C lower than the temperature for OD=12.7 mm, and its variation with increasing wick thickness is more sensible for the lower outside diameters.

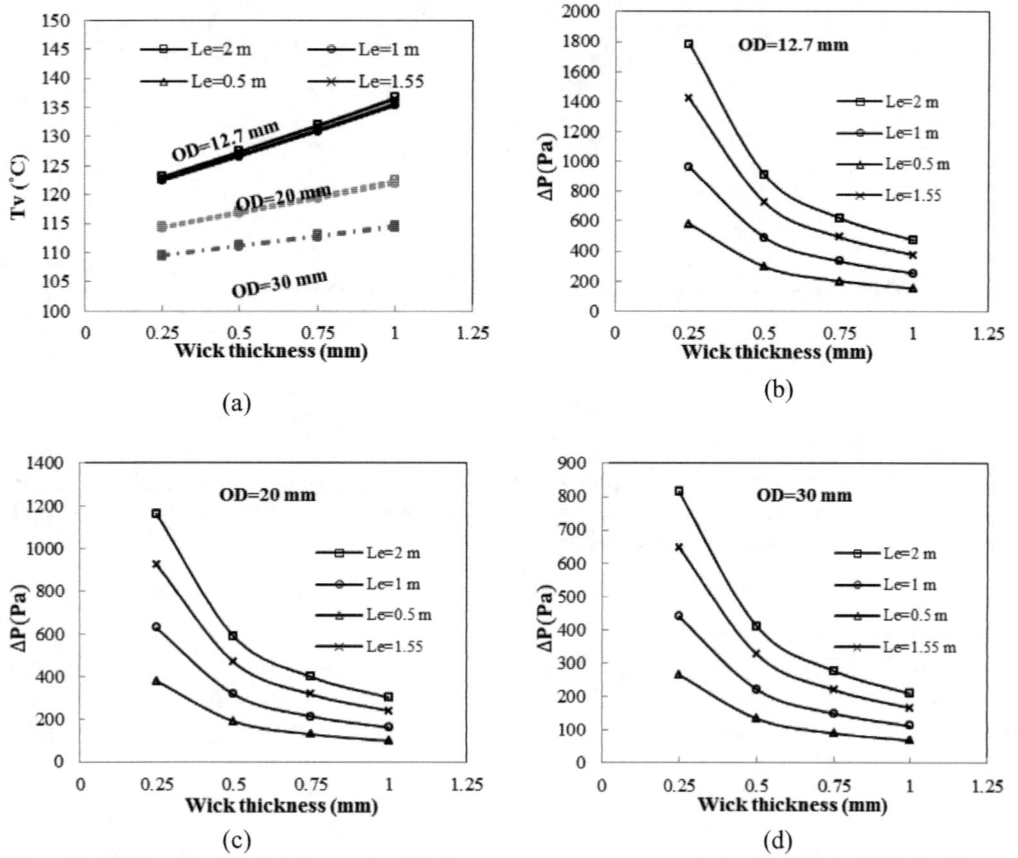

Figure 4. Effect of wick thickness on (a) the vapour temperature (T_v), (b), (c) and (d) the Pressure drop for OD=12.7 mm, OD=20 mm and OD=30 mm respectively for different evaporator lengths.

As indicated in figures 4b-d, the increase in the wick thickness reduces the liquid flow resistance and the pressure drop of liquid decrease too. In addition, when the evaporator length decreases from 2 m to 0.5 m, the slope of pressure drop vs. increasing wick thickness is reduced. It can be found that the evaporator length does not have a significant effect on the vapour temperature, but it does have a significant effect on the pressure drop.

Figure 4d shows also that the decrease of wick thickness, from 1 to 0.25 mm, increases the pressure drop up to 73% for different ODs. From figure 4a it can be seen that the same decrease of the wick thickness reduces the vapour temperature by approximately 10%, 6.5%, and 4% for the HP with OD of 12.7, 20 and 30 mm, respectively.

Figures 4b-d show that when the OD increases, the pressure drop decreases. These results indicate that, when using the HP for medium temperature, it is better to decrease the wick thickness to obtain a lower temperature and HP size. At the same time, we must pay attention to the pressure difference to prevent overcoming the capillary pressure, especially for higher power input.

3.2. Effect of the Heat Transfer Coefficient of the Cooling Fluid

The effects of the variation of heat transfer coefficient of the cooling fluid on the vapour temperature and pressure drop are shown in figures 5a and 5b, respectively. As can be seen, when the heat transfer coefficient increases, the vapour temperature decreases and the pressure drop increases. The influence of the heat transfer coefficient on the HP performance is less sensible for the higher values of it (after 800W/m^2 K). An increase in the heat transfer coefficient increases the non-uniform distribution of the

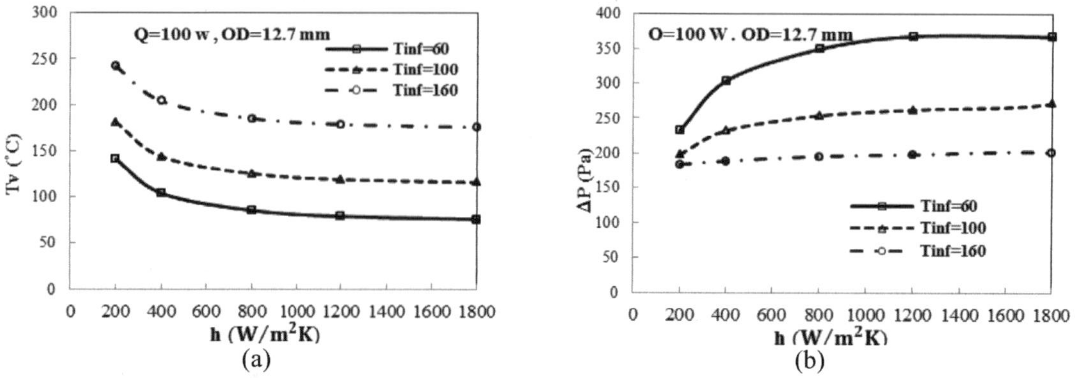

Figure 5. Effect of heat transfer coefficient (h) on a) vapour temperature and b) Pressure drop for different water cooling temperatures (T_{inf}).

heat flux at the wick-vapour interface, and hence, the maximum pressure difference of wick increases and leads to an increase of the total pressure difference. Further, figure 5b shows that increasing heat transfer coefficients does not have a great effect on the pressure drop for cooling water temperature of 160°C, compared with its effects for lower water temperatures.

3.3. Maximum heat transfer capability (HTC)

For the cooling heat transfer coefficient of 1600 W/m^2 K, the relationship between the maximum HTC and wick thickness is presented in figure 6 for two values of the cooling water temperatures (100°C and 160°C), different outside diameters (12.7, 20 and 30 mm), different evaporator lengths (2, 1.55, 1 and 0.5 m), and different effective pore radii (2.4×10^{-4} m, 7.9×10^{-5} m and 3.2×10^{-5} m).

As discussed previously, achieving thin wick is limited by the liquid pressure drop and superheat for the onset of bubble nucleation. As shown in figure 6, in some cases maximum HTC increases with reducing wick thickness and in some cases decreases. This depends on the fact that the boiling limit is less than the capillary limit, or vice-versa.

For instance, in the case of OD=12.7 mm, evaporator length of 2 m and cooling water temperature of 100°C, as shown in figure 6a, the capillary limit is less than the boiling limit for the wick effective pore radius 2.4×10^{-4} m, therefore HTC decrease with decrease in wick thickness. If effective pore radius changed to 7.9×10^{-5} m, for the wick thickness of 1 and 0.75 mm boiling limit is reached before capillary limit, so HTC increases with a reduction in wick thickness, and the contrary holds for wick thickness of 0.5 and 0.25 mm, because the capillary limit is reached first. A further reduction of effective pore radius leads to improve HTC in the cases that are limited by the capillary pumping.

By increasing OD, when the vapour radius increases, the vapour flow resistance decreases, but the conductive thermal resistance increases, and the size of the HP increases. The effect of the vapour core radius on the HTC is shown in figure 6. As can be seen, when the vapour radius decreases, the maximum HTC does not incur in a significant change, even though the conduction resistance decreases. For example for water cooling temperature of 100°C, evaporator length of 1 m, wick thickness of 0.5 and effective pore radius of 7.9×10^{-5} m, maximum HTC for OD=12.7, 20 and 30 mm is about 8000 W/m^2. Thus, the overall conductive resistance in the vapour region does not have a considerable effect on the HTC and the heat is transferred mainly by phase change phenomena.

In the analysed range of parameters, the capillary limit was higher than the boiling limit for the wick thickness of 1 and 0.75 mm and for high water temperature (160°C), but in some cases, for wick thickness of 0.5 and 0.25 mm, the capillary limit was reached before boiling limit (OD=12, 20 mm and $L_e = 2$, 1.55 and 1 m). Also, for evaporator length of 0.5 m, in all cases capillary limit was not reached before boiling limit. Maximum HTC for the water cooling temperature of 100 and 160 °C are about 12000 and 5000 W/m^2, respectively for the selected configuration as shown in figure 6.

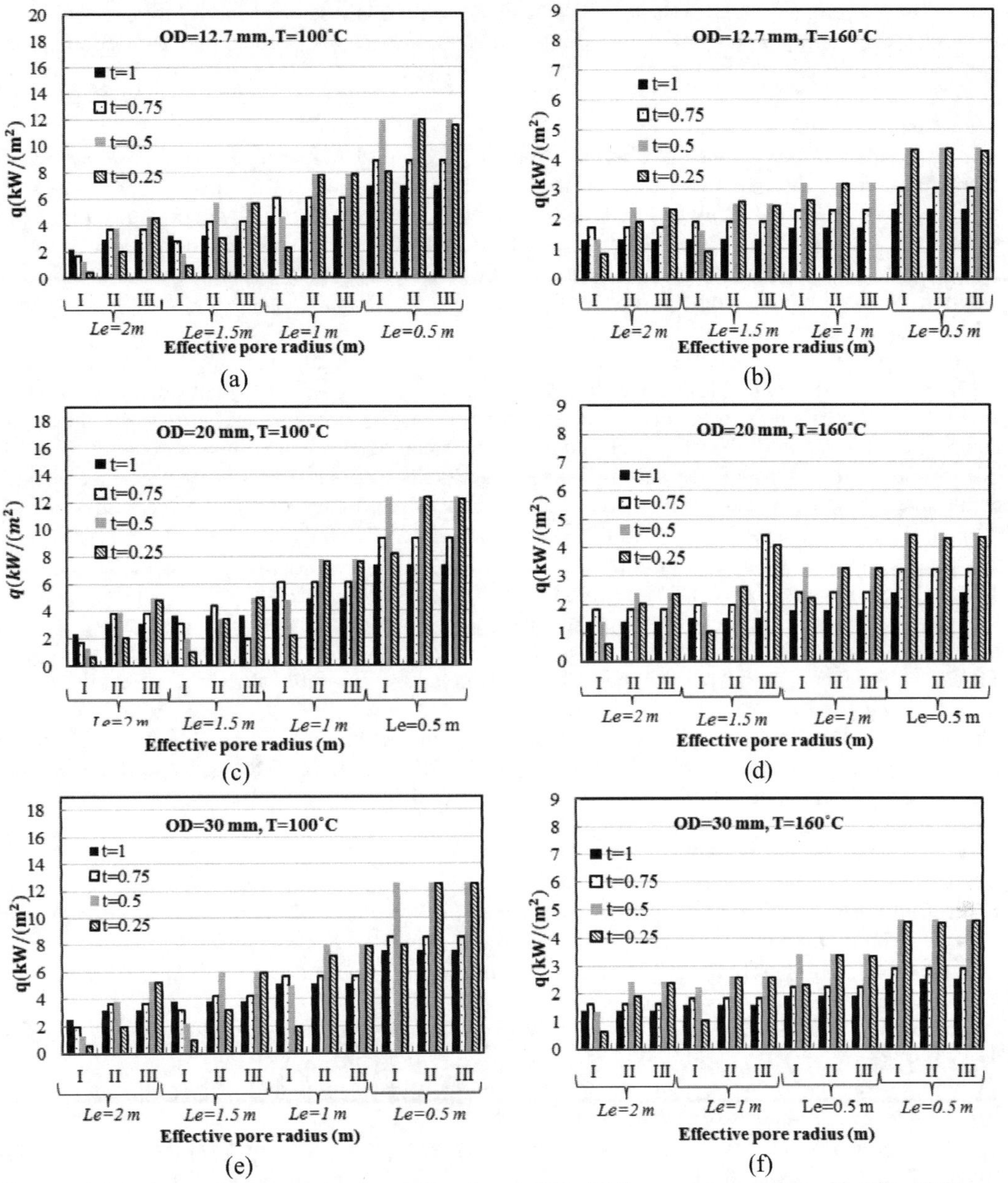

Figure 6. Maximum heat transfer capability for a) OD=12.7 mm, $T_{\text{inf}} = 100^{\circ}C$, b) OD=12.7 mm, $T_{\text{inf}} = 160^{\circ}C$, c) OD=20 mm, $T_{\text{inf}} = 100^{\circ}C$, d) OD=20mm, $T_{\text{inf}} = 160^{\circ}C$, e) OD=30 mm, $T_{\text{inf}} = 100^{\circ}C$, f) OD=30 mm, $T_{\text{inf}} = 160^{\circ}C$ [evaporator length, m (L_e); wick thickness, mm (t); effective pore radius, I=2.4×10^{-4} m, II=7.9×10^{-5} m and III=3.2×10^{-5} m].

Results presented here can effectively direct the design of high heat flux solar device, avoiding capillary and boiling limits, by selecting an appropriate configuration. Results indicate that the optimum wick thickness to achieve maximum heat flux depends on the effective pore radius and evaporator length. Based on the results of Figure 6, it is concluded that, to reach high input power without limitation, it is advisable to decrease the wick thickness, as it is consistent with the results of Kempers et al. [26], and decreases the pore radius to the smallest possible value. For small wick thicknesses the capillary limit becomes dominant, and the results predict that shorter evaporator lengths yield larger heat fluxes.

4. Conclusions

The heat conduction resistance, the boiling limit and capillary pumping pressure become primary factors determining the maximum HTC in the HP. This paper provides a sensitivity analysis of the issues encountered in the design of HP with specific interest for medium temperature application, typically for solar cooling and industrial process heat. A two dimensional model in the wall coupling the hydrodynamic and thermal models was used in this parametric study to predict the overall HTC. A validation of the numerical model was performed. The results show good agreement between the model results and those found in the literature.

The effects of heat transfer coefficient, power input, evaporator length, pipe diameter, wick thickness and effective pore radius on the vapour temperature, maximum pressure drop and maximum HTC are studied. The results show that the evaporator length has no a significant effect on the vapour temperature in HP for different power inputs. The effect of the heat transfer coefficient on HP performance decreases with increasing its value. The results also show that when the wick thickness increases, the vapour temperature increases, and the maximum pressure difference decreases.

The analysis shows that wick thickness plays an important role in the enhancement of HTC. There is an optimum wick thickness that minimizes the increment of temperatures which independent from the heating power. Results show that it is possible to improve HTC by selecting the appropriate wick thickness, effective pore radius and pipe diameter and evaporator length. By decreasing the wick thickness (until an optimum is reached), because of the decrease thermal resistance of the wick region, HTC can be enhanced. Results also indicate that the maximum dry-out heat flux significantly depends on the wick thickness.

Therefore, an optimal region of HP's lengths and wick characteristic must be chosen to enhance the heat transfer and the heat removal capability especially when the heat fluxes increase. Results of this investigation will assist in optimizing the heat transfer performance of HPs which may contribute to improved HP solar collector design.

References

[1] Rittidech S and Wannapakne S 2007 *Experimental study of the performance of a solar collector by closed-end oscillating heat pipe (CEOHP)* Appl. Therm. Eng. **27** 1978-1985

[2] Franco A and Filippeschi S 2013 *Experimental analysis of Closed Loop Two Phase Thermosyphon (CLTPT) for energy systems* Exp. Thermal Fluid Sci. **51** 302–311

[3] Filippeschi S 2006 On *periodic two-phase thermosyphons operating against gravity* Int. J. Thermal Sci. **45** 124–137

[4] Chun W, Kang YH, Kwak HY and Lee YS 1999 *An experimental study of the utilization of heat pipes for solar water heaters* Appl. Therm. Eng. **19(8)** 807–17

[5] Joudi KA and Witwit AM 2000 *Improvements of gravity assisted wickless heat pipes* Energy Convers. Manage. **41** 2041-2061

[6] Mathioulakis E and Belessiotis V 2002 *A new heat-pipe type solar domestic hot water system* Solar Energy **72** 13–20

[7] Riffat S, Zhao X and Doherty PS 2005 *Developing a theoretical model to investigate thermal performance of a thin membrane heat-pipe solar collector* Appl. Therm. Eng. **25** 899–915

[8] Azad E 2008 *Theoretical and experimental investigation of heat pipe solar collector* Exp. Thermal Fluid Sci. **32** 1666-1672

[9] Brahim T, Foued M and Abdelmajid J 2013 *Parametric Study of a Flat Plate Wick Assisted Heat Pipe Solar Collector* ASME J. Solar Energy Eng. **135** 031016-1

[10] Aghbalou F, Mimet A, Badia F, Illa J, El Bouardi A and Bougard j 2004 *Heat and mass transfer during adsorption of ammonia in a cylindrical adsorbent bed: thermal performance study of a combined parabolic solar collector, water heat pipe and adsorber generator assembly* Appl. Therm. Eng. **24** 2537–2555
[11] Zuo ZJ and Faghri A 1997 *A Network Thermodynamic Analysis of the Heat Pipe* Int. J. Heat Mass Transfer **41(11)** 1473–1484
[12] Xuan Y, Hong Y and Qiang L 2004 *Investigation of Transient Behaviours of Flat Heat Pipes* Exp. Thermal Fluid Sci. **28** 249-255
[13] Tournier JM and El-Genk MS 1994 *A heat pipe transient analysis model* Int. J. Heat Mass Transfer **37** 753-762
[14] Zhu N and Vafai K 1999 *Analysis of cylindrical heat pipes incorporating the effects of liquid e vapour coupling and non-Darcian transport-a closed form solution* Int. J. Heat Mass Transfer **42** 3405-3418
[15] Kaya T and Goldak J 2007 *Three-dimensional numerical analysis of heat and mass transfer in heat pipe* Heat Mass Transfer **43** 775–785
[16] Shabgard H and Faghri A 2011 *Performance characteristics of cylindrical heat pipes with multiple heat sources* Appl. Therm. Eng. **31** 3410–3418
[17] Brahim T and Jemni A 2014 *Effect of the Heat Pipe Adiabatic Region* ASME J. Heat Transfer **136** 042901-1
[18] Vafai K and Wang W 1992 *Analysis of flow and heat transfer characteristics of an asymmetrical flat plate heat pipe* Int. J. Heat Mass Transfer **35** 2087-2099
[19] Vafai K, Zhu N and Wang W 1995 *Analysis of assymetrical disk-shaped and flat plate heat pipes* ASME J. Heat Transfer **117** 209-218
[20] Aghvami M and Faghri A 2011 *Analysis of flat heat pipes with various heating and cooling configurations* Appl. Therm. Eng. **31** 2645-2655
[21] Zhu N and Vafai K 1996 *The effects of liquid e vapour coupling and non-Darcian transport on asymmetrical disk-shaped heat pipes* Int. J. Heat Mass Transfer **39** 2095-2113
[22] Reay DA and Kew PA 2006 *Heat Pipes, Theory Design and Applications* (New York: Elsevier)
[23] Bertossi R, Romestant C, Ayel V and Bertin V 2012 *A theoretical study and review on the operational limitations due to vapour flow in heat pipes* Front. Heat Pipes **3** 023001
[24] Faghri A 1995 *Heat Pipe Science and Technology* (Washington, DC: Taylor and Francis)
[25] Peterson GP 1994 *Heat Pipes: Modelling, Testing and Applications* (USA: John Willey & Sons)
[26] Kempers R, Ewing D and Ching CY 2006 *Effect of number of mesh layers and fluid loading on the performance of screen mesh wicked heat pipes* Appl. Therm. Eng. **26** 589–595

Thermal analysis of the building envelope of lightweight temporary housing

Fabio Fantozzi, Pietro Galbiati, Francesco Leccese, Giacomo Salvadori, Michele Rocca
Dept. of Energy Engineering, Systems, Territory and Constructions
University of Pisa, School of Engineering, Largo Lucio Lazzarino, Pisa, IT.

E-mail: f.leccese@ing.unipi.it

Abstract. In the last few years, to meet the need of build efficient homes in a short time and with maximum constructive simplification, lightweight prefabricated building systems have proved to be particularly suitable, especially in geographical areas which must deal with emergency situations (i.e., temporary housing). In this paper the winter and summer thermal performance of a building prototype, realised with modular steel framed elements, have been studied, in both winter and summer conditions. Special attention has been paid to the optimisation of the dynamic thermal performance of the multi-layered envelope structures. The dynamic thermal behaviour of the outer wall, analysed and discussed in details in the paper, shows that it is possible to improve the performance of lightweight walls by using an optimised stratigraphy characterised by an opportune sequence of resistive and capacitive layers. The influence of inner structures (partitions, floor and roof) on the building thermal behaviour has also analyzed trough the introduction of room performance indices appropriately defined. The results of the study have been discussed with special reference to the requirements fixed by the Energy Performance Buildings European Directive (EPBDs) and the resulting implementation in Italian Legislation.

1. Introduction

In the last few years, lightweight prefabricated building systems have proved to be particularly suitable in geographical areas which must deal with emergency situations such as earthquakes, weather-climate events, situations of conflict, and others (see Figure 1).

In the past a considerable part of post-disaster temporary housing programs have been unsustainable and culturally inadequate as a result of unsuccessful strategies, misunderstandings about users' real needs and misconceptions in dealing with local conditions and resources [1].

Disaster-affected families who have lost their homes need a private and secure place to restart their daily activities as soon as possible after the disaster, yet temporary housing programmes tend to be overly expensive, too late and responsible for undesirable impacts on the urban environment [2].

Housing reconstruction programmes play a decisive role on the disaster recovery and providing temporary housing is a crucial step of these programmes. During the reconstruction of permanent housing, it allows victims to have a private and secure place to return to their normal life. It has been widely used after the largest scale disasters but it has also been greatly criticized, mainly for being unsustainable and culturally inadequate [1].

The selection of temporary housing to ensure sufficient levels of comfort is very important also for the psychological effects on the earthquake survivor. A research on a comparison of people assigned to containers, converted into mobile homes, vs. wooden dachas vs. a control group, that had not lost their homes in the earthquake, have shown that container inhabitants reported greater discomfort, felt more dominated by the situation and, importantly, reported more psychological stress symptoms than those living in dachas or in regular homes. Again, it may surprise that the well-being of people living in dachas was closer to that of control participants than to fellow earthquake victims living in containers. In particular, people in dachas did not reliably differ from control participants in terms of feeling dominated or in terms of psychological stress symptoms, but they did express greater general discomfort [3].

Figure 1. Examples of temporary housing units in earthquake areas: (left) L'Aquila, 2009; Haiti, 2010; Japan, 2012 (right).

It may be stated that these buildings require shorter construction times and simple construction steps, and they should not be mistakenly considered as "temporary housing." On the contrary, they must ensure high safety standards (against fires, earthquakes, etc.) as well as high levels of living comfort (thermal, acoustic, etc.), even in remote areas lacking services and utilities, such as electricity or gas supply.

This study is focused on the analysis of the dynamic thermal behaviour of a new prefabricated building system; the EN ISO 13786 method has been followed, implementing a spreadsheet calculation file [4]. In particular, the parameters: periodic thermal transmittance Y_{ie} ($Wm^{-2}K^{-1}$), decrement factor f_a and time shift φ (h) have been analyzed. The aim of the research was to study the dynamic thermal characteristics of the building envelope in order to enhance its performance by optimizing the stratigraphy of the outer multi-layered walls. The choice of the stratigraphy which optimizes the dynamic thermal characteristics has been performed using the method of lumped parameters [5].

It also aims to explain that the inner structures (partitions, floor and roof) play a role in the summer thermal performance considering the passive behaviour of the building. The analysis of the influence of inner structures on the energy performance of the room was established through the introduction of performance indices, similar to those above, but appropriately redefined [6].

2. Description of the new building system

The study has been carried out (with both computational simulations and on-site measurements) on a building prototype [6]. The new building system consists of modular steel framed elements (pre-assembled at the factory), which are shipped to the construction site including fixtures and equipment (see Figure 2).

The main features that characterize the new building system can be summarized in the following aspects: shorter construction times, construction flexibility, good seismic performance, energy saving, acoustic insulation, workplace safety, installation without scaffolding, easy maintenance and low costs.

The system involves the construction of a foundation slab designed according to the building loads and to the ground properties, taking into account of seismic aspects. The outer wall are anchored on the edge of the foundation. The modular elements of walls and floors (which constitute the supporting structure) are joint whit bolts. The new building system allows to create different solutions of ventilated facade covering whit stone, ceramic, aluminium, plastic materials with various surface finishes. The building typologies that the systems includes are multi-storey residential buildings, emergency housing units, commercial/office and industrial building.

Figure 2. Images of the new lightweight prefabricated building system (courtesy of HOMLEG, www.homleg.it).

In Table 1 are shown the stratigraphies of the outer walls, horizontal structures (floor) and inner partitions used in the standard solution [6], currently under production in Italy (Tuscany Region). In particular in Table 1, for each layers, are shown: thickness (d), thermal conductivity (λ), density (ρ) and thermal capacity (c).

3. Winter and summer thermal performance of the building envelope

The evaluation of the winter and summer thermal performance was initially carried out on the outer wall (W0) without the external coating layer (Ref.: Outer wall, layers 1 and 2, Table 1).

The winter behaviour of the outer wall has been evaluated by calculating the overall thermal tramittance U, amounting to 0.29 $Wm^{-2}K^{-1}$ ($R_{Tot}=1/U$=3.45 m^2KW^{-1}). In Italy, as European Directive EPBD implementation [7-9], limit values (U_{lim}) of the overall thermal transmittance for the outer wall of buildings have been fixed for each of the six climates zone (A÷F) in which the Italian territory is slip up. In particular, the more stringent limit values (Italian colder climate, zone F) are: U_{lim}=0.33 $Wm^{-2}K^{-1}$ (for vertical walls). It can be observed that the overall thermal trasmittance of the W0 wall is always less than the limit values set for each climate zone [7].

As regards the summer thermal behaviour, has been calculated the periodic thermal transmittance Y_{ie} ($Wm^{-2}K^{-1}$), the decrement factor f_a. and the time shift φ (h). In Italy, for the municipalities in which the maximum intensity of the solar radiation on a horizontal plane is higher than 290 Wm^{-2}, the following requisite have been imposed [8]: for the vertical opaque walls (facing South, South-West and South East), the value of the surface mass M_s should be higher than $(M_s)_{lim}$=230 kgm^{-2}, otherwise the value of the periodic thermal trasmittance Y_{ie} should be lower than $(Y_{ie})_{lim}$=0.12 $Wm^{-2}K^{-1}$.

On the national guidelines for energy certification of buildings [9], in order to rate the "summer performance level" of the opaque outer walls the following parameters are used: decrement factor f_a and time shift φ (see Table 2). In [9] it is also specified that, in cases of values of f_a and φ do not belong to the same performance level, the assignation of the building summer performance level is based on the value of the time shift.

The W0 wall exhibits: Y_{ie}=0.121 $Wm^{-2}K^{-1}$, f_a=0.41 and φ=6h49' (see Table 3). The W0 wall does not fulfil the limits imposed on the summer thermal behaviour and results in performance level IV (see Table 2).

Journal of Physics: Conference Series **547** (2014) 012011 doi:10.1088/1742-6596/547/1/012011

In order to achieve better summer performance, it was decided to gradually fill (in step of 1cm) the air cavity (Ref.: Outer wall, layer 5, Table 1) with insulating material (i.e. rock wool). Consequently 15 different wall stratigraphies (from W1 to W15, see Table 3) has been examined.

In Table 3 the calculation results, for W0 (outer wall in the standard solution) and for the subsequent 15 stratigraphies (from W1 to W15), are shown. All the dynamic thermal characteristics has been calculated with respect to a reference period (24h) of the oscillation of the thermal field. It can be observed that it is sufficient the addition of a centimeter of insulating material to fulfil the limit imposed on periodic thermal trasmittance, however the W1 wall remain in performance level IV (φ=7h23', W1, see Table 3).

Table 1. Stratigraphy of the main components of the new lightweight prefabricated building.

			d (cm)	λ (Wm^{-1}K^{-1})	ρ (kgm^{-3})	***c*** (JK^{-1}kg^{-1})
Outer wall	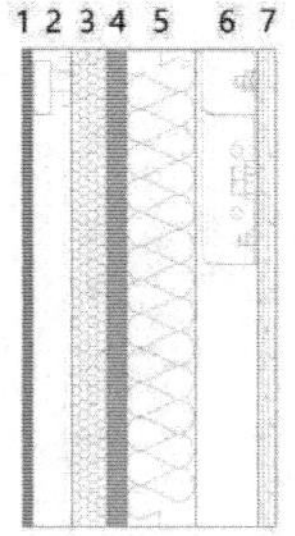	*Ext*				
		1- Outer coating	-	-	-	-
		2- Air cavity	4.5	0	1	1004
		3- Expanded polystyrene	8	0.033	20	1450
		4- OSB panels	1.8	0.13	650	1700
		5- Air cavity	15	0	1	1004
		6- OSB panels	5.4	0.13	650	1700
		7- Plasterboard sheet	1.2	0.21	900	837
		Int				
Floor	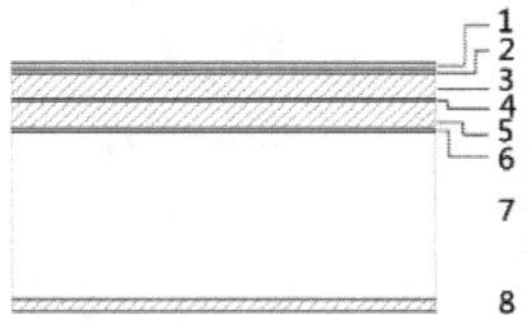	*Top*				
		1- Tile, cork	0.8	0.045	130	1764
		2- Silicone-mastic	0.5	-	-	-
		3- Plywood	3	0.13	500	2092
		4- Neoprene	0.3	-	-	-
		5- Plywood	3	0.13	500	2092
		6- Steel	0.5	52	7800	460
		7- Air cavity	19.5	0	1	1004
		8- Plywood	1.4	0.13	500	2092
		Down				
Partition	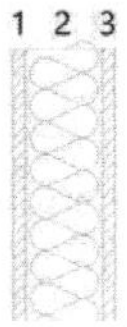	1- Plasterboard sheet	1.3	0.21	900	2300
		2- Rock wool	8	0.041	70	1030
		3- Plasterboard sheet	1.3	0.21	900	2300

Table 2. "Summer performance level" for the opaque building envelope.

Level	**I**	**II**	**III**	**IV**	**V**
Description	Excellent	Good	Medium	Sufficient	Poor
Time Shift φ (h)	$\varphi > 12$	$10 < \varphi \leq 12$	$8 < \varphi \leq 10$	$6 < \varphi \leq 8$	$\varphi \leq 6$
Decrement Factor f_a (-)	$f_a < 0.15$	$0.15 \leq f_a < 0.3$	$0.3 \leq f_a < 0.4$	$0.4 \leq f_a < 0.6$	$f_a \geq 0.6$

From Table 3 it can be observed that 3 cm of insulating material are necessary to achieve the performance level III (φ = 8h21', W3) and 9 cm of insulating material are necessary to achieve the performance level II (φ =10h10', W9). The most insulated solution (W15) exhibits the following thermal properties (see Table 3): U=0.146 Wm^{-2}K^{-1} (R_{Tot}=1/U=6.85 m^{2}KW^{-1}), Y_{ie}=0.026 Wm^{-2}K^{-1},

f_a=0.175 and φ=11h38'. Despite of the excellent thermal behaviour, $U<<U_{lim}$, $Y_{ie}<<(Y_{ie})_{lim}$, the W15 wall achieves a summer performance level rated as good (level II), being: $10<\varphi\leq12$h.
To further improve the dynamic thermal behaviour of the outer walls the Authors have proceeded with an optimization of the stratigraphy of the W15 wall [10-12].

Table 3. Thermal characteristics of the 16 stratigraphies analyzed: thermal trasmittance U, periodic thermal trasmittance Y_{ie}, time shift φ and decrement factor f_a.

Stratigraphy	**Thickness insulating material** (cm)	U (Wm^{-2}K^{-1})	Y_{ie} (Wm^{-2}K^{-1})	φ	f_a
W0	0	0.295	0.121	6h49'	0.410
W1	1	0.276	0.111	7h23'	0.403
W2	2	0.258	0.095	7h55'	0.370
W3	3	0.243	0.083	8h21'	0.340
W4	4	0.229	0.072	8h43'	0.314
W5	5	0.217	0.063	9h03'	0.292
W6	6	0.206	0.056	9h21'	0.273
W7	7	0.196	0.050	9h38'	0.256
W8	8	0.187	0.045	9h54'	0.242
W9	9	0.179	0.041	10h10'	0.228
W10	10	0.172	0.037	10h26'	0.217
W11	11	0.165	0.034	10h42'	0.207
W12	12	0.158	0.031	10h58'	0.197
W13	13	0.153	0.029	11h14'	0.188
W14	14	0.147	0.026	11h29'	0.180
W15	15	0.146	0.026	11h38'	0.175

4. Optimization of outer multi-layered walls

The problem of how to define the stratigraphic pattern of a wall provided with a thermal resistance R_{Tot} (including surface thermal resistances) and a thermal capacity C_{Tot} fitting for minimizing the periodic thermal transmittance Y_{ie} (i.e. the decrement factor f_a) has been solved by using the method of lumped parameters [10-12].
This method will be solved by considering that the definition "wall stratigraphy" refers to how many purely resistive layers provided with a thermal resistance r_s and how many purely capacitive layers provided with a thermal capacity c_s constitute that specific wall on the basis of a given layer sequence order. A $2n+1$ layered wall, consisting of n capacitive layers and $n+1$ resistive layers, can be outlined as follows (with $R=\Sigma r_s$ and $C=\Sigma c_s$):

$$[INT]\ [r_n]\ [c_n]\ [r_{n-1}]\ldots\ldots[r_1]\ [c_1]\ [r_0]\ [EXT]$$

Given that the different resistive and capacitive layers are respectively represented by triangular matrices of the following type ($j=\sqrt{-1}$ and ω angular frequency of oscillation of external field):

$$\begin{pmatrix} 1 & r_s \\ 0 & 1 \end{pmatrix}, \quad \begin{pmatrix} 1 & 0 \\ j\omega c_s & 1 \end{pmatrix}$$

For $n=\infty$, we obtain a typical homogeneous wall (i.e. a wall with uniformly distributed capacity C and thermal resistance R).
The problem of defining the wall stratigraphic pattern fitting for minimizing Y_{ie} is characterised by the following non-dimensional parameter:

$$\gamma = \omega R C$$

which corresponds to the product of an outer thermal field angular frequency ω and the wall's time constant *RC*. The obtained results can be summarized as follows.

For $\gamma < 18$, the optimal symmetry configuration is that obtained with $n=1$, corresponding to a three-layered wall, whose capacitive layer (*C*) is placed between two identical resistive layers (*R/2*), of the following type:

$$[INT]\ [R/2]\ [c]\ [R/2]\ [EXT] \qquad (\overline{T}_1)$$

Such a configuration is characterized by:

$$Y_{ie} = 1/R\sqrt{1+\gamma^2/16} \quad , \qquad \varphi = (P/2\pi)\arctan(\gamma/4)$$

For $18 < \gamma < 42$, the optimal symmetry structure is that obtained with $n=2$ (five-layered wall), of the following type:

$$[INT]\ [r_0]\ [c_1]\ [r_1]\ [c_1]\ [r_0]\ [EXT] \qquad (T_2)$$

For $42 < \gamma < 76$ the optimal solution is a T_3 symmetry structure obtained with $n=3$ (seven-layered wall), for $76 < \gamma < 100$ the optimal solution is a T_4 symmetry structure obtained with $n=4$ (nine-layered wall), and so on.

For each case with $n>1$, the optimal resistance and capacity values corresponding to the different layers are dependent on the γ value. It has been proved that entirely symmetric walls $\overline{T}_n$, consisting of n capacities and $n+1$ resistances whose values are all identical, with $c_n=C/n$ and $r_n=R/(n+1)$, approximate with a very good accuracy the behaviour of the above defined optimal configurations T_n, within the respective γ intervals.
It must be observed that whatever real wall can be outlined as a lumped-parameter wall provided with a sufficiently high number of layers. The above presented analysis shows that, for a given γ, the optimal stratigraphy is characterized by a low n value, which excludes the possibility of it being other than a lumped-parameter stratigraphy. The lumped parameter model is therefore not limitative.
Even the problem of defining the wall stratigraphic pattern fitting for maximizing the time shift φ is characterized by a non-dimensional parameter γ, the solution to this problem is very similar to the solution previously obtained with regard to minimizing Y_{ie}, if we except that the transition from one optimal stratigraphic pattern to another takes place for considerably lower γ values. The analysis turns out to be very simple when its spectrum is limited to entirely symmetric type $\overline{T}_n$ walls.
If such is the case: for $\gamma < 3.5$, a type $\overline{T}_1$ wall gives an advantageous solution, with $n=1$ (three-layered wall), for $3.5 < \gamma < 6.7$, a type $\overline{T}_2$ wall gives an advantageous solution, with $n=2$ (five-layered wall), for $6.7 < \gamma < 10.1$, a type $\overline{T}_3$ wall gives an advantageous solution, with $n=3$ (seven-layered wall) and so on.
In the case of high γ values, the number of layers which are needed to maximize the time shift φ is found to be in its turn very high, whereas the relevant structure approximates a homogeneous wall's structure, with an uniform distribution of *R* (including surface thermal resistances) and *C*.

5.Stratigraphic pattern optimization of the outer wall

In this analysis has been performed for the W15 wall without the innermost coating layer (Ref.: Outer wall, layer 7, Table 1), hereinafter named W15* (see Table 5). In Table 4, the thermal properties of the W15* wall, in particular the thermal resistance *R* (m^2KW^{-1}), the thermal capacity *C* ($Jkg^{-1}K^{-1}$) and the thermal diffusivity α ($m^2\mu s^{-1}$) of the different layers, have been reported. In Table 4 the parameters p_1 and p_2 have been also reported. The parameters p_1 and p_2 are defined as: $p_1=C/C_{Tot}$, $p_2=R/R_{Tot}$, where

C_{Tot} (kJm^{-2}K^{-1}) is the overall thermal capacity of the wall and R_{Tot} (m^{2}KW^{-1}) is the overall thermal resistance of the wall. In the analysed case (W15*) C_{Tot} and R_{Tot} are respectively: 92.7 kJm^{-2}K^{-1} and 6.81 m^{2}KW^{-1}.

Table 4. Thermal properties of the W15* wall (see also Table 1).

		EXT	Expanded polystyrene	OSB panel	Insulating material (Table 3, W15)	OSB panel	INT
R	m^{2}KW^{-1}	0.04	2.424	0.138	3.659	0.415	0.13
C	kJm^{-2}K^{-1}		2.3	19.9	10.8	59.7	
α	m^{2}μs^{-1}		1.14	0.12	0.57	0.12	
p_1			0.025	0.215	0.117	0.644	
p_2			0.365	0.021	0.551	0.063	

In order to understand the behaviour of the resistive layers (for example, a layer of insulating material), the value assumed by p_1 must be sufficiently low; similarly the capacitive layers should be evaluated considering the value assumed by p_2 [13-14].

In view of these aspects a new stratigraphy (Type 1) has been analysed. The Type 1 wall, keeping the same alternation of materials capacitive-resistive-capacitive-resistive of W15* and the same thicknesses of the different layers, is composed only of OSB panel (capacitive layers) and expanded polystyrene (resistive layers). The optimization of the dynamic thermal behaviour of the Type 1 is then processed by changing the sequence of layers from the inside outwards, see Figure 3.

Starting from the Type 1, three other stratigraphy have been designed in order to show the variation of the dynamic thermal characteristics of the opaque envelope of the new building system examined (see Figure 3 and Table 5). The Type 2 has been obtained without vary the thicknesses of the individual layers and by shifting toward the centre the capacitive layers (OSB panel) and outwards the resistive layers (expanded polystyrene). The Type 3 has been obtained by keeping the order of the layers of the Type 2 but making it symmetrical (the two layers of expanded polystyrene have the same thickness equal to 11.5 cm). The Type 4 has been obtained by searching the configuration that minimize the periodic thermal transmittance Y_{ie} through the use of the method of lumped parameters applied to the Type 3. The results of the optimization is a seven layers stratigraphy with alternated resistive and capacitive layers that shown the following thermal properties: ω=7.27·10^{-5}, γ =48.24 (42<γ<76), C_{Tot}= 86.23 kJm^{-2}K^{-1} and R_{Tot}=7.69 m^{2}KW^{-1}. The resistive layers (expanded polystyrene) are 4 of thickness 5.8 cm and the capacitive layers (OSB panel) are 3 of thickness 2.4 cm. A further fractionation of the stratigraphy does not allow appreciable improvements of the dynamic thermal performance of the wall.

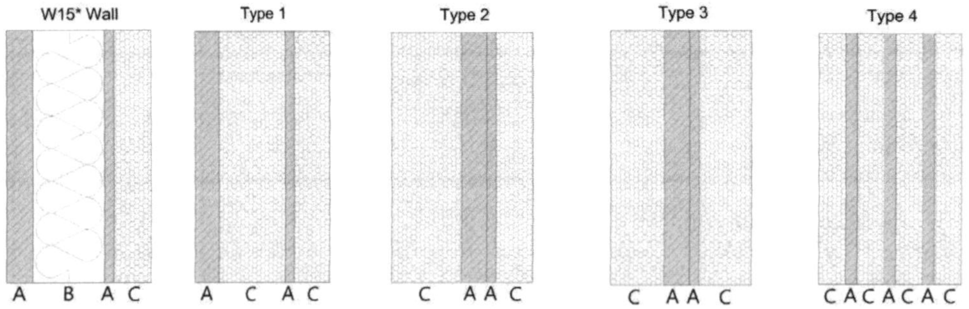

Figure 3. Different type of walls studied (A: OSB Panels, B: insulating material, C: expanded polystyrene).

From the results shown in Table 5, it can be observed that the different analysed solutions obviously have the same overall thickness (30.2 cm) and a surface mass M_s (kgm^{-2}) very low (always less than 60 kgm^{-2}). Of course the Type 1, 2, 3 and 4 walls, that are designed with the same materials and the same overall thickness, have the same thermal transmittance (U=0.130 Wm^{-2}K^{-1}) that is considerably lower than the limit value fixed for the climate zone more burdensome by the Italy legislation.
From the results reported in Table 5, it can be also observed that the Type 4 wall, optimized with the method of lumped parameters, shows a significant improvement of the time shift with respect to all the other proposed solutions. In particular the Type 4 wall, with φ=13h55', is characterised by a summer thermal performance level rated as excellent (level I).

Table 5. Thermal characteristics of W15* wall and the new Type 1..4 walls.

	d (cm)	***M$_s$*** (kgm^{-2})	***U*** (Wm^{-2}K^{-1})	***R$_{Tot}$*** (m^{2}KW^{-1})	***C$_{Tot}$*** (kJm^{-2}K^{-1})	***Y$_{ie}$*** (Wm^{-2}K^{-1})	***fa***	***φ***
W15*	30.2	59	0.147	6.81	92.69	0.029	0.20	11h02'
Type 1	30.2	51	0.130	7.69	86.23	0.028	0.22	10h07'
Type 2	30.2	51	0.130	7.69	86.23	0.012	0.09	8h51'
Type 3	30.2	51	0.130	7.69	86.23	0.011	0.08	8h47'
Type 4	30.2	51	0.130	7.69	86.23	0.006	0.05	13h55'

6. Thermal influence of the inner structures

The thermal comfort of a building depends not only on the outer structures (roof and façades) but also on the inner structures, this is especially evident when the passive behaviour of the building is studied, that is when the response of the building to variations of the external temperature is analysed in the absence of an air conditioning system. The question involved is one of great relevance. It must be noted that a building which has been the object of an effective thermal planning so as to show an excellent passive behaviour, can provide, in the summer time, a satisfactory comfort level even when no air-conditioning system is installed or, at the most, when the role of that system is limited, with clear savings in energy consumption. For an overview on how far inner structures contribute to buildings' thermal comfort, readers are referred to [10-12].
In order to perform this analysis the performance indices, decrement factor η and time shift ψ as defined in [5-6], can be used. The dynamic thermal insulation of the building is much higher (and therefore the internal conditions are much less bound to the external ones) the smaller is η and ψ. In the present paper a type room, with square plane (side length of 5m) and a useful height of 3 m, has been studied. We have assumed thermal transmission as uni-dimensional and any effects brought about by thermal bridges will be neglected. Under such conditions the surface passing, at a given temperature, through the wall's midspan, can be outlined as being adiabatic. As a consequence, the only sector being affected by thermal problems within the room is half of the wall giving onto the room itself.
To study the influence of the inner structures (partitions, floor and roof) and of the outer wall on the performance indices (η and ψ), four different configurations of the type room (see Table 8) have been considered. The configuration R0 is composed by the inner structures described in Table 1 (floors and partitions) and the outer wall W15*; the configuration R1 is the same that R0 except for the inner structures that are optimized (see description below and see Table 6); the configuration R2 is the same that R0 except for the outer wall that is Type 4, and finally the configuration R3 that is composed by optimised inner structures and the outer wall Type 4.
The optimized partitions is defined as a type of partition with a thermal capacity greater than the existing one. The optimised partition walls have been obtained using also the method of lumped parameters and alternating layers of hardwood panels and wood-cement panels (see Figure 4) with thermal properties shown in Table 6. In Table 6 are also shown the values of the coefficients κ_1 and κ_2, which represent the thermal capacity on both sides of the partitions.

From the results shown in Table 7, it is observed that the attenuation factor η is progressively reduced from $2{,}81\cdot 10^{-3}$ to $0{,}33\cdot 10^{-3}$ passing from the configuration R0 to configuration R3, with an overall reduction of 88%. The behaviour of the time shift ψ is, on the contrary, more complex. The transition from configuration R0 to configuration R1 (optimized inner structures) produces a reduction in the time shift from 14h36' to 13h32'. The configurations R2 (optimized outer wall) and R3 (optimized partitions and optimized outer wall) exhibits higher values of time shift; being the R2 configuration the best, with an increase of ψ (compared to the R0 configuration) equal to 20%.

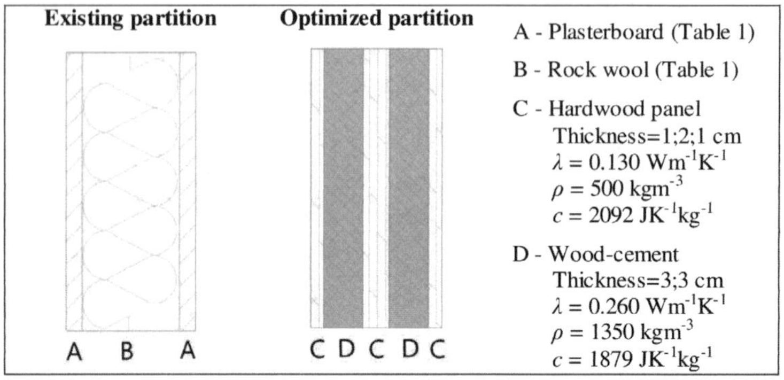

Figure 4: Stratigraphies of existing (standard solution) and optimized partition.

Table 6. Thermal properties of existing partition (standard solution) and optimized partition.

Partition	***Thickness*** (cm)	***U*** (Wm^{-2}K^{-1})	M_s (kgm^{-2})	κ_1 (kJm^{-2}K^{-1})	κ_2 (kJm^{-2}K^{-1})
Exsisting	10.46	0.43	29	12.5	12.5
Optimized	10	1.25	101	50.5	50.5

Table 7. Summary of the four configuration of the type room analyzed.

R0 "Standard solution"	**R 1**	**R 2**	**R 3**
Existing partition Outer wall **W15***	**Optimized** partition Outer wall **W15***	**Existing** partition Outer wall **Type 4**	**Optimized** partition Outer wall **Type 4**
η =2.81·10^{-3} ψ =14h36'	η =1.5·10^{-3} ψ =13h32'	η =0.6·10^{-3} ψ =17h28'	η =0.33·10^{-3} ψ =16h57'

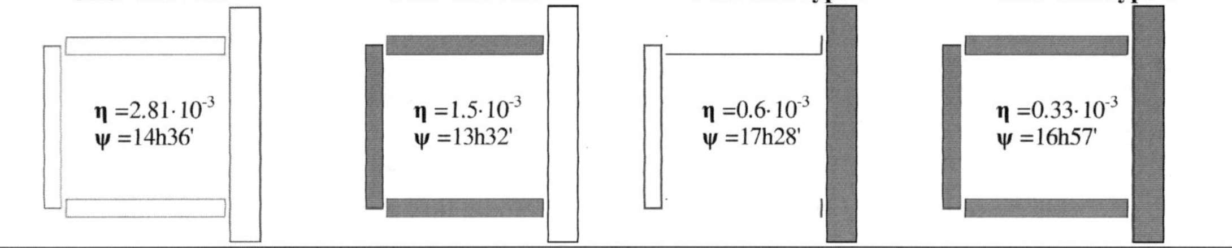

7. Conclusive remarks

The study presented in this paper has been designed in order to combine the well-know benefit of the prefabricated buildings with the energy high standard required by the market and the current construction sensivity.

The results show that it is possible to improve the dynamic thermal performance of the outer walls (lightweight prefabricated walls), by using an optimised stratigraphy characterised by an opportune sequence of resistive and capacitive layers.

The method of lumped parameters proves to be useful in deriving the stratigraphy which optimizes the dynamic thermal behaviour of the wall, minimizing the periodic thermal transmittance Y_{ie} and maximizing the time shift φ. The method also highlights how the inner partitions (optimized by

increasing the thermal capacity) help to keep the indoor air temperature constant against external temperature variations.
The calculation results make explicit the "weight" that the effects of thermal properties of each component on the determination of overall values of decrement factor and time shift for the type room. It is also clear that, with keeping attention on the simplification of the present case, the partitions play a significant role in the overall energy balance.

References

[1] Felix D., Branco J.M., Feio A., 2013. Temporary housing after disaster: A state of the art survey. *Habitat International*, **40**, pp. 136-141.

[2] Johnson C., 2007. Impacts of prefabricated temporary housing after disasters: 1999 earthquake on Turkey. *Habitat International*, **31**, pp. 36-52.

[3] Caia G., Ventimiglia F., Maass A., 2010. Container vs. dacha: The psychological effects of temporary housing characteristics on earthquake survivors. *Journal of Environmental Psychology*, **30**, pp. 60-66.

[4] EN ISO 13786, 2007, Thermal performance of building components – Dynamic thermal characteristics – Calculation methods.

[5] Leccese F, Fantozzi F, Salvadori G, Rocca M, 2014, Thermal performance of outer and inner multi-layered walls in buildings, *32rd UIT Heat Transfer Conference*, Pisa (I), 23-25 June 2014, ETS (Pisa), ISBN: 978-8846739971, pp.1-8.

[6] Fantozzi F., Galbiati P., Leccese F., Salvadori G., Rocca M., Maragno F., 2014. Dynamic thermal analysis of new lightweight prefabricated building system, *32rd UIT Heat Transfer Conference*, Pisa (I), 23-25 June 2014, ETS (Pisa), ISBN: 978-8846739971, pp.1-9.

[7] D.Lgs.vo n.192 del 19 agosto 2005, Attuazione della direttiva 2002/91/CE relativa al rendimento energetico nell'edilizia (modificato dal D.Lgs.vo n.311 del 29 dicembre 2006).

[8] D.P.R. n.59 del 2 aprile 2009, Regolamento di attuazione dell'articolo 4, comma 1, lettere a) e b) del D.Lgs.vo n.192 del 19 agosto 2005, concernente attuazione della direttiva 2002/91/CE sul rendimento energetico in edilizia.

[9] D.M. Sviluppo Economico del 26 giugno 2009, Linee guida nazionali per la certificazione energetica degli edifici.

[10] Ciampi M, Fantozzi F, Leccese F, Tuoni G, 2003, On the optimization of building envelope thermal performance – Multi-layered walls design to minimize heating and cooling plant intervention in the case of time varying external temperature fields, *Civil Engineering and Environmental Systems*, Taylor & Francis, Vol. **20 (4)**, pp.231-254.

[11] Ciampi M, Leccese F, Tuoni G, 2004, Multi-layered walls design to optimize building-plant interaction, *Int. J. of Thermal Science*, Elsevier, Vol. **43 (4)**, pp.417-429.

[12] Ciampi M, Leccese F, Tuoni G, 2005, On the thermal design of the external walls in buildings. *8th REHVA World Congress*, Lausanne (CH),9-12 October 2005, ISBN: 978-3-033-00585-3, CD-Rom, pp.1-8.

[13] Ciampi M, Fantozzi F, Leccese F, Tuoni G, 2001, A Criterion for the Optimization of Multi-layered Walls, *7th RHEVA World Congress*, Napoli (I), 15-18 September 2001, AICARR Milano (I), CD-Rom, pp.1-15.

[14] Ciampi M, Leccese F, Tuoni G, 1999, Optimization of multi-layered wall and building-systems interaction (in italian), *17th UIT Heat Transfer Conference*, Ferrara (I), 30 June-2 July 1999, ETS (Pisa), Vol. II, pp.503-514.

Journal of Physics: Conference Series **547** (2014) 012012 doi:10.1088/1742-6596/547/1/012012

Thermodynamic and heat transfer analysis of LNG energy recovery for power production

A Franco, C Casarosa
University of Pisa, Department of Energy, Systems, Territory and Constructions Engineering, Largo Lucio Lazzarino 56126 PISA

E-mail: alessandro.franco@ing.unipi.it

Abstract. An important option to transport the gas is to convert it into liquid natural gas (LNG) and convey it using insulated LNG tankers. At receiving terminals, the LNG is off-loaded into storage tanks and then pumped at the required pressure and vaporized for final transmission to the pipeline. The LNG production process consumes a considerable amount of energy, while the cold availability, as also known as cold energy, has been stored in LNG. At a receiving terminal, LNG needs to be evaporated into gas at environmental temperature before fed into the gas distribution system. Seawater is commonly used for the regasification process of the LNG. In the present paper, after a general analysis of the perspectives of the various thermodynamic schemes proposed for power production from the regasification, a detailed analysis of enhanced direct expansion system is carried out in order to identify the upper level of the energy that can be recovered. The analysis outlines that power production typical of optimized ORC plant configurations (120 kJ/kg) can be obtained with direct expansion solutions.

1 Introduction

A viable way to transport natural gas is to convert it into liquid natural gas (LNG) and convey it using insulated LNG tankers. At receiving terminals, the LNG is off-loaded into storage tanks and then pumped from storage at the required pressure and vaporized for final transmission to the distribution system. The LNG production process consumes a considerable amount of energy, while the cold availability, as also known as cold energy, has been stored in LNG. At a receiving terminal, LNG needs to be evaporated into gas at environmental temperature before fed into the gas distribution system. Seawater is commonly used for the regasification process of the LNG and such process needs about 800 kJ/kg of heat energy. The topic of LNG cold energy recovery both for power production and for different energetic uses is considered in the literature since the late 90s [1-2], mainly under the impulse of the Japanese experience, which is surely one of the most important in the world concerning LNG facilities.

Most LNG terminals regasify the liquid using the thermal energy of seawater or the warm seawater effluent from a power plant, destroying in this way all physical exergy. In the liquefying process, a large amount of mechanical energy is consumed in refrigeration process, so LNG contains a significant amount of cold energy (i.e. cryogenic exergy). If LNG is used as a fuel in a combined system, the waste heat of exhaust gases and the cold energy of LNG can be used at the same time. The

authors intend to focus the attention on the problem of cold the energy contained in LNG: several Thermodynamics schemes can be proposed for energy production [3-5].
The regasification process essentially consists of two operations: pumping up of the liquid gas up to the pressure of the distribution grid; heating of the natural gas up to the distribution temperature (typically in the range between 0 and 20 °C) using a heat exchanger. From a thermodynamic viewpoint, re-heating represents a net loss of available energy, which causes degradation of overall energy efficiency of the conversion chain.
The cold energy stored by the LNG could be recovered rather than directly taken off by seawater. The differences among the various regasification processes are the mode of heat transfer and the type of the loop: this can be an "open loop" (the fluid change) or "closed loop" (the operating fluid is always the same and there is a real thermodynamic cycle). Moreover a fundamental difference is the hot source: this can be at environmental temperature (basically the seawater) or at higher temperature.
Other LNG cold energy utilization ways can be air separation, material freezing, intake air cooling, dry ice production and refrigeration in chemical industry [6].
The cryogenic power generation is the most interesting option. There are several ways using the energy given off by LNG regasification to complete thermodynamic cycle to generate power, but they basically belong to three particular options. The methods are object of analysis in the literature: direct expansion cycle schemes [3-4]; Organic Rankine Cycle (ORC) with intervening media or more complex cascading Rankine cycle configurations [7-8]. While the direct expansion cycle directly uses LNG as working fluid, ORC uses seawater as the primary heat source and LNG as the heat sink with an auxiliary working fluid (usually a low boiling hydrocarbon) for power production in turbine.
In recent options including Brayton cycle with perfect gas cycle [9-12], different non conventional configurations that stands under the name of Combined cycle [13-15] and alternative thermodynamic systems like those based on Stirling cycle have been proposed too [16].
Considering the important handling capacity of typical regasification plants, the cold energy that can be available at LNG receiving terminal, often of the order of magnitude of 50-100 kg/s of natural gas, the potential for practical applications of the LNG cold energy should be further explored.
As previously discussed, in the literature by using the cooling capacity of LNG several thermodynamics schemes are proposed employing conventional and non conventional conversion cycles and different heat sources: typically sensible heat of the seawater is used as energy input, but often a high temperature heat source is used (in the range 60-1300 °C) [17]. For this reason it is not easy to have a clear comparison of the various available power production options.
In the present paper, after a general analysis of the perspectives of the various possible plant configurations to obtain power from the regasification a detailed analysis of the perspectives of direct expansion cycle is performed and a sensitivity analysis with respect to the various plant options and the various operating variables is carried out.

2. LNG Cold Energy Recovery and Power Generation: general considerations

LNG is produced by cryogenic refrigeration of natural gas at about -162 °C at atmospheric pressure. Liquefying natural gas is a high energy consumption process, and it is estimated that producing one kg of LNG, assuming the composition of CH_4 and considering an higher pressure of the process of 55 bar (the critical pressure of CH_4 is 46 bar) a compression work of about 800-860 kJ/kg is necessary (considering a compression efficiency in the range between 0.8 and 0.85).
If the real configurations of liquefaction plants are considered, the energy consumption is sensibly higher than the value considered before. Quiang and co-authors in [18] considers an amount of energy required of 850 kWh/kg correspondent to about 3 MJ/kg. Gerasimov et al. in [19] proposed a plant in which the amount of energy consumed is of about 700-800 kWh/kg (2.5-2.8 MJ/kg). In a textbook on Natural Gas, Medici considers a refrigeration cycle considering a ternary mixture of refrigerants, identifying the level of 1.9 MJ/kg as a possible minimum for the energy required for compression, [20]. Assuming real configurations, it is possible to identify the level of 2900 kJ/kg as a realistic value for the energy consumed in the process, and this is mainly electrical energy.

Considering that, and according to the advanced liquefaction processes about 2900 kJ/kg are consumed in the liquefaction process, the larger amount, about 2070 kJ/kg is dissipated as heat, but the remaining, estimating in the order of magnitude of 830 kJ/kg, called "cold energy" are stored in the LNG. It can be simply estimated with theoretical considerations that when the LNG is regasified to an ambient temperature of 20 °C, an interesting amount of the energy required for the liquefaction process could be theoretically recovered: this corresponds approximately to the value of 0.23 kWh per kg of LNG. Considering the typical annual handling capacity of various LNG receiving terminals (usually estimated in some million of tons for each year), the cold energy that can be available is an amount that cannot be ignored. It is clear that the potential for practical applications of the LNG cold energy should be further explored in order to define optimized solutions. The output pressure of natural gas required for LNG vaporizing terminals varies according to the pipeline requirements. The final pressure varies from 25-30 bar in case of combined cycle stations up to 80 bar typical for long distance distribution (table 1). According to the distribution requirements, an evaluation of the maximum available power (specific exergy difference, Δex, based on ambient temperature of 15 °C) can be given.

Table 1. Required pressure for several uses of NG and corresponding theoretical work available

Application	p [bar]	T [°C]	h [kJ/kg]	s [kJ/kg K]	Δex [kJ/kg]
LNG initial condition	1	-161.5	442.80	8.529	-
Combined cycle stations	27	15	573.40	9.761	531.00
Local distribution	35	15	564.70	9.604	494.46
Long-distance pipelines	70	15	525.50	9.143	400.82
Maximum pressure	150	15	442.80	8.529	306.60

Considering that the final pressure must be much higher that the atmospheric one, and that according to the required pressure the T-s diagram for LNG has very different configurations (figure 1), several possible options for the recovery can be considered and it is difficult to define the best available technology. Hence if options based on a simple Rankine cycle or Brayton cycle are probably better for pressure up to 30 bar, for higher values, the direct expansion options can be reconsidered. In order to use LNG cold energy to generate electricity, several different generation processes can be designed, in the following section the various options will be briefly reviewed.

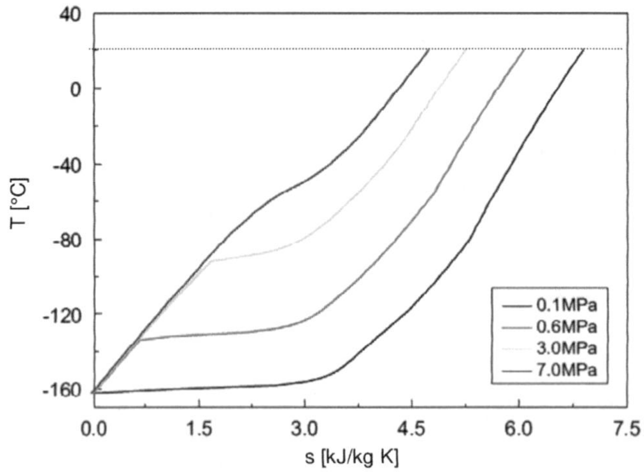

Figure 1. T-s diagram for LNG

3. The various options for electricity generation by LNG cold energy

The electricity generation by using the cold energy of LNG is possible by means of a variety of thermodynamic schemes. The more considered ones in the literature are the following: Rankine cycle, direct expansion cycle; gas expansion cycle (Brayton cycle) and combined cycles. In some papers options considering possible combination of the different options are also proposed such as Stirling cycle, but in general they represent more complex solutions.

3.1. Direct Expansion Cycle

The direct expansion conversion is basically the simplest configuration for power production. In this cycle, LNG liquid is firstly compressed up to a pressure higher than the user's need then heated and regasified through evaporator by seawater and use the vapour to drive the turbine-generator. LNG direct expansion cycle is simple and suitable for small regasification station which supply low pressure natural gas. Nevertheless as previously discussed, in most cases the gasified LNG is requested at supercritical conditions, therefore it can be dispatched to long distance pipelines and consequently the maximum pressure level is over 100 bar. After expansion in the turbine, where power is generated, its pressure decreases to the gas-supplying pressure. Considering conventional schemes, where the maximum pressure of natural gas is 150 bar and the final pressure in the pipeline is on the order of 80 bar (typical of long-distance distribution pipelines) the amount of net power that can be generated by expansion turbine can be easily estimated in the order of 20 kJ/kg, being about 60 kJ/kg the difference of enthalpy in the turbine and about 40 kJ/kg the specific enthalpy required for the pumps: a very little amount with respect to the cold energy. The value can be somewhat higher for lower values of the final pressure required for the pipelines and low mass flow rates. Moreover some authors state that if the high pressure after gas expansion is needed, direct expansion cycle is not suitable to be used. The expansion cycle can be selected in different ways in order to maximize the power extracted. The simplified scheme of figure 2 can be improved in order to increase the output power considering the imposed boundary condition of the requested pressure.

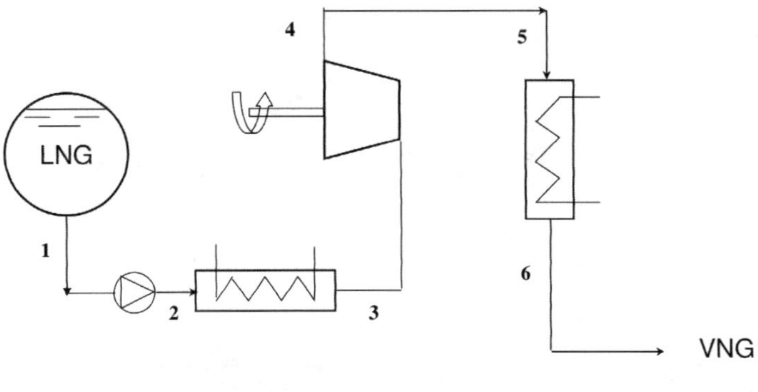

Figure 2. Direct expansion cycle

3.2. *Rankine Cycle with Intervening Media*

One of the most common options for power production is the use of LNG for cooling the condenser of a Rankine cycle which exploits sensible heat of the seawater as energy input. In the version of Rankine cycle, an auxiliary fluid (as for example propane R22, R23, R13B1 and, for the coldest applications, methane) is used for expansion within the turbine; LNG is used as low temperature source for the condenser. The auxiliary fluid is condensed by LNG and then pumped to evaporator heated by seawater and finally passed in turbine to drive generator to complete whole cycle (figure 3). As LNG need not expand to work, the send out pump can be set to the pipeline pressure level. In some cases, in order to reach the environment temperature, it is necessary to use additional heater. Even if with the Rankine cycle the high temperature can be selected with a certain grade of freedom, depending on the heat source available and the practical limit is only represented by the thermal stability of the organic fluid, the real limitation of the simple Rankine cycle stands in the fact that the gasified LNG is often required at a quite high pressure (more than 30 bars) so that, considering the thermodynamic scheme of figure 4 the available cooling capacity in non-isothermal, which implies a not perfect match with the condenser of the Rankine cycle. The conventional values for the energy recovery obtained with Rankine cycle can be estimated in the range between 40 and 120 kJ/kg, using conventional schemes [7-8]. For practical LNG cold power generation, ORC is most commonly used, in some cases using different configurations like a binary mixture as working fluids and combined with a vapour absorption process or using a cascading mode the output power can overcome 200 kJ/kg [21].

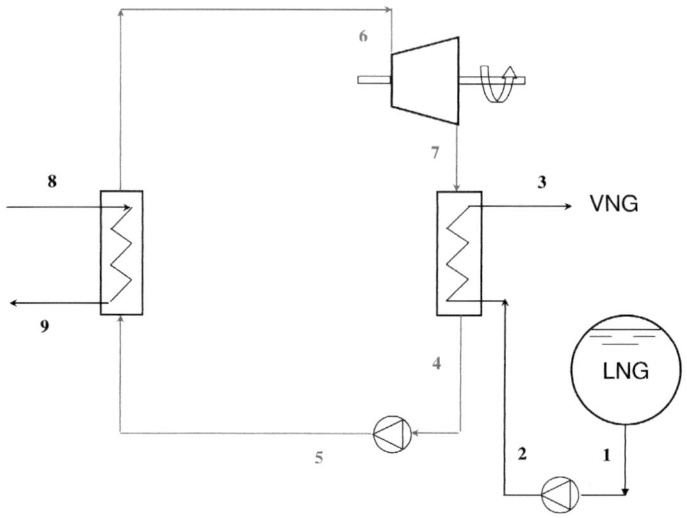

Figure 3. Rankine Cycle With Intervening Media

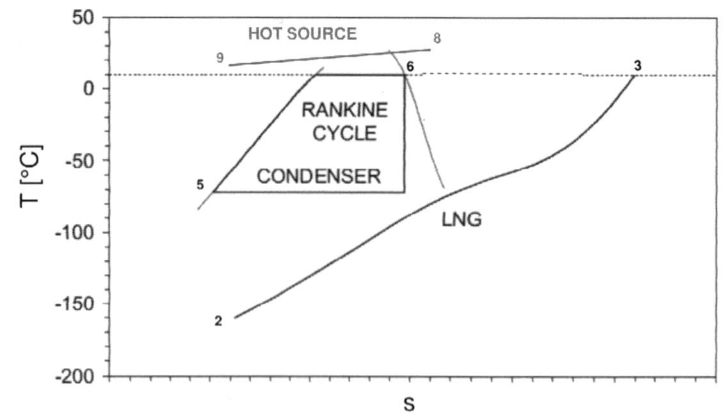

Figure 4. The cooling capacity of supercritical LNG in relation to Rankine cycle

3.3. Brayton cycle

From a conceptual point of view another simple system for obtaining work by the cold energy is the use of a fluid in a Brayton cycle. Since the level of cold in a LNG flow is thermodynamically predetermined, working fluids must be selected with a critical point which fits the LNG thermal capacity, i.e. 5-15 °C higher than the LNG temperature. Some possible operating fluids are reported in table 2. However ordinary Brayton cycles considering real gas conditions, exhibit a modest efficiency [9]; the problem is that the theoretical conditions are relative to real gas cycles and the cooling capacity of LNG is only partially used: a better exploitation is obtained from perfect gas cycles, selecting fluid such as argon, higher temperature of the hot source (figure 5) or a more complex configuration with a cascading Rankine cycle.

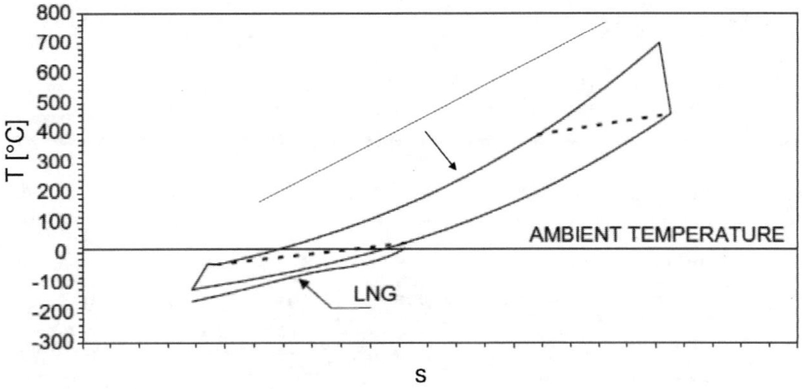

Figure 5. The cooling capacity of supercritical LNG in relation to high temperature Brayton cycle

Table 2. Working fluids for cryogenic Brayton gas cycles [9]

Fluid	Molecular mass [g/mol]	Critical temperature [K]	Critical pressure [bar]
N_2	28	126.2	33.98
Air	28.96	132.52	37.66
Ar	39.948	150.86	48.98
O_2	32	154.58	50.43
CH_4	16.043	190.56	45.99

4. Analysis and optimization of direct expansion recovery configurations

As briefly discussed in the previous section, the basic direct expansion configuration appears to be not convenient from the perspective of energy recovery even if it is the more simplified and basic configuration. In the conventional schemes the direct expansion in only obtained with a single pressure level by means of pumping the LNG at a pressure well higher than the required one with a simple expansion. However, as well known from the Thermodynamics, for obtaining higher energy productions, it is necessary to divide the mass flow rates at different pressure levels. For this reason a more convenient configuration involves three different pressure levels, in which the maximum pressure is higher than the pressure required for the pipeline (80 bar). The configuration considered is represented in figure 6 with the corresponding thermodynamic diagram reported in figure 7. In this case four different pressure levels are considered with the only boundary condition imposed by the value of the maximum pressure fixed at 150 bar. This particular configuration can be optimized modifying the values of the two intermediate pressure levels: for each value of the pressure different mass flow rates can be defined. In order to evaluate optimal configuration, a mathematical model can be written: it consists of mass balance (Eqs. 1-5) and energy balance equations (Eqs. 6-12):

$$m_1 = m_2 = m_{8*} = m_8 = m* \quad (1)$$

$$m_2 + m_{11} = m_3 \quad (2)$$

$$m_4 + m_9 = m_5 \tag{3}$$

$$m_7 - m_8 = m_9 + m_{10} \tag{4}$$

$$m_{11} = m_3 - m^* \tag{5}$$

$$m_2 h_2 + m_{11} h_{11} = m_3 h_3 \tag{6}$$

$$m_4 h_4 + m_9 h_9 = m_5 h_5 \tag{7}$$

$$m^* h_2 + (m_3 - m^*) h_{11} = m_3 h_3 \tag{8}$$

From the previous equations it is possible to derive the corresponding equation connecting the values of the various mass flow rate, identified with reference to the schemes of figures 6 and 7 as m_3, m_5, m_7 and m_{10} as a function of the mass flow rate directed to the pipeline m*, that permits to evaluate the power that can be obtained with high and low pressure turbine and consequently:

$$m_3 = \frac{(h_{11} - h_2)}{(h_{11} - h_3)} \cdot m^* \tag{9}$$

$$m_5 = m_3 \cdot \frac{(h_9 - h_4)}{(h_9 - h_5)} \cdot m^* \quad \rightarrow \quad m_5 = m_6 = m_7 = \frac{(h_{11} - h_2)}{(h_{11} - h_3)} \cdot \frac{(h_9 - h_4)}{(h_9 - h_5)} \cdot m^* \tag{10}$$

$$m_{10} = \frac{(h_3 - h_2)}{(h_{11} - h_3)} \cdot m^* \tag{11}$$

Considering the various configurations analyzed, the better recovery conditions, corresponding to the three following pressure levels: 4 bar, 35 bar and 150 bar are reported in table 3. The data are obtained considering isentropic efficiency of 0.85 and 0.9 for the high and low pressure turbine.

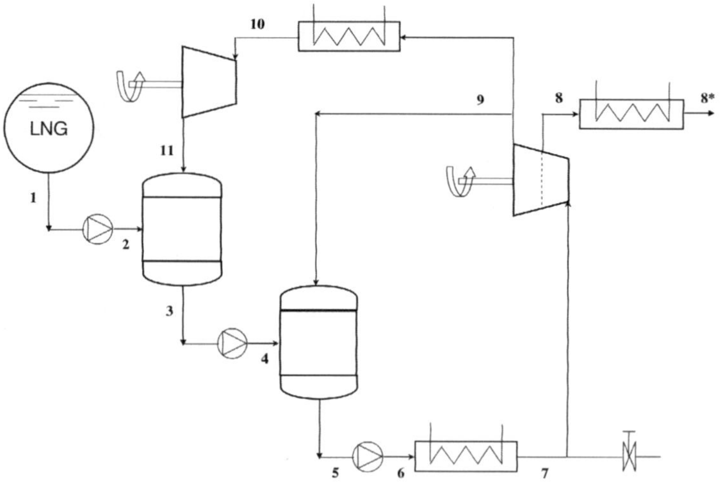

Figure 6. Direct expansion cycle configuration analyzed

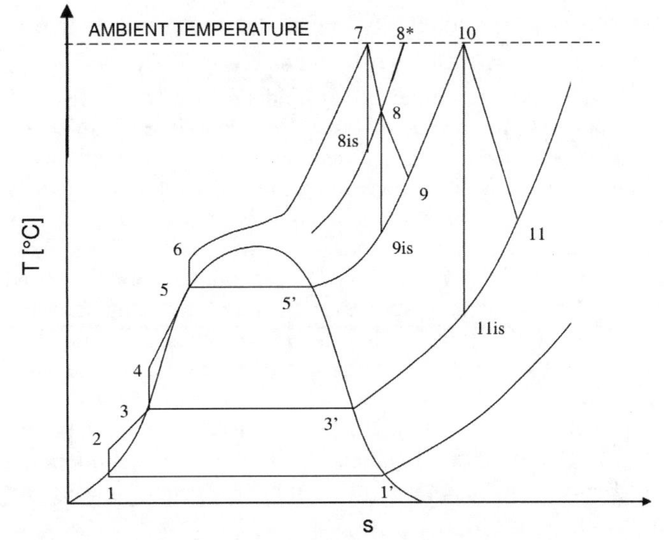

Figure 7. Direct expansion cycle considered: schematic T-s diagram

According to the model described before, the amount of power that can be produced as the sum of the two different contributions related to the high pressure and low pressure turbine respectively:

$$W_{HP} = m_7 \cdot (h_7 - h_8) + (m_7 - m^*) \cdot (h_8 - h_9) \quad (12)$$

$$W_{LP} = m_{10} \cdot (h_{10} - h_{11}) \quad (13)$$

Table 3. Data referred to the various Thermodynamic states of the direct expansion cycle

State	T [°C]	P [MPa]	v [m³/kg]	h [kJ/kg]	s [kJ/kgK]	x
1	-161.5	0.1013	0.002367	-286.5	4.943	0
1'	-161.5	0.1013	0.5501	223.8	9.504	1
2	-161.4	0.4	0.002366	-285.7	4.934	
3	-141.7	0.4	0.002552	-215.5	5.513	0
3'	-141.7	0.4	0.1543	252.7	9.075	1
4	-141	3.5	0.002532	-209.3	5.500	
5	-91.22	3.5	0.003738	11.79	6.897	0
5'	-91.22	3.5	0.01477	239.9	8.151	1
6	-90	15	0.003037	-10.43	6.569	
7	15	15	0.007871	442.8	8.529	
8	-25	8	0.01176	387.2	8.569	
9'is	-75.38	3.5	0.02104	318.7	8.569	
9	-72.48	3.5	0.02187	329	8.620	
8*	15	8	0.01689	542.9	9.142	
9is	-77.5	3.5	0.02041	310.8	8.529	
10	15	3.5	0.03978	564.7	9.604	
11is	-107.8	0.4	0.2044	330.5	9.603	
11	-97.2	0.4	0.2193	353.9	9.740	

From the data of table 3, obtained by means of the utility CATT (Computer Aided Thermodynamics Tables) it is possible to calculate the pumping power related to the various pressure increases, P_{1-2}, P_{3-4} and P_{5-6}; mainly the last value is quite relevant with respect to the specific expansion enthalpy. Considering the data of table 3 and the mathematical model assumed, the detailed results reported in table 4 can be obtained. Analyzing the results of table 4 it is possible to understand that, considering configurations like the one with the three pressure level considered, levels of output power of the order of magnitude of 120 kJ/kg, values typical for conventional ORC cycles and corresponding to a second law efficiency of about 28%. Assuming a LNG mass flow rate, typical of various regasification terminals (about 70 kg/s corresponding to a nominal production rate of about $3*10^9$ m^3 of natural gas) an output power of about 8.37 MW can be obtained. Using similar boundary conditions used for the basic cycle of figure 6 (final pressure of the lever required for long distribution pipelines and maximum pressure of 150 bar, working on the thermodynamic cycle with the introduction of internal recovery heat exchangers permitting further heat recovery and optimizing the various mass flow rate ratios, it is possible to reach a specific power production approaching 160 kJ/kg, with a Second Law efficiency of the recovery cycle quite close to 0.4. Even if it seems that the additional energy recovery gives origin to a higher complexity of the LNG recovery plant, further improvements of the direct expansion exposed in the figures 6 and 7 appears to be suitable for more detailed analysis.

Table 4. Details of the direct expansion optimized recovery configuration

Mass flow ratio	Power ratio	Values
m_3 / m^*		1.1233
m_5 / m^*		1.9062
m_{10} / m^*		0.1232
	W_{HP} / m^*	158.72 kJ/kg
	W_{LP} / m^*	25.98 kJ/kg
	W_{gross} / m^*	184.70 kJ/kg
	P_{1-2} / m^*	0.707 kJ/kg
	P_{3-4} / m^*	8.886 kJ/kg
	P_{5-6} / m^*	55.504 kJ/kg
	P / m^*	65.097 kJ/kg
	W_{net} / m^*	119.616 kJ/kg
	Q_{6-7} / m^*	863.94 kJ/kg
	Q_{9-10} / m^*	29.06 kJ/kg

5. Conclusions

In the present paper the problem of energy production for LNG cold energy has been analyzed with the specific aim to evaluate the perspectives of simple direct expansion configurations. After a preliminary analysis of the various conventional options, a specific three pressure level configuration has been analyzed and tested. Considering an upper limit for the higher pressure (150 bar) and a boundary condition imposed by the pipeline at 80 bar, a potential of power production of about 120 kJ/kg have been estimated: this value being correspondent to the typical values declared for optimized systems based on ORC. Analyzing more complex configurations that include recovery heat exchangers, it seems possible to obtain further increase of the specific power production.

As concluding remark it is possible to state that though if the analysis reported in the present paper is based on some theoretical assumptions (e.g. the availability of high pressure turbines operating with natural gas at a temperature well below the environmental temperature and ideal heat exchangers configurations), the results that can be obtained with a Thermodynamic optimization of the cycle are very interesting if compared with the results obtained in some simplified analysis available in the literature about direct LNG expansion, based on single pressure configurations.

References

[1] Liu H and You L 1999 Characteristics and applications of the cold heat exergy of liquefied natural gas, Energy Conversion & Management **40** 1515-25.

[2] Kim T S and Ro S T 2000 Power augmentation of combined cycle power plants using cold energy of liquefied natural gas, Energy **25** 841–56.

[3] Zhang F and Yu X 2009 LNG Cold Energy Recovery and Power Generation, Proc. of Power Energy Engineering Conference, APPEEC Asia Pacific, 27-31 March, 2009.

[4] Dispenza C Dispenza G La Rocca V and Panno G 2009 Exergy recovery during LNG regasification: Electric energy production – Part one App. Thermal Engineering **29** 380–87.

[5] Zhang L and Tang Q 2012 Comparison of different power generation processes by LNG cold energy, 2012 ASRI Conference on Power and Energy Systems, AASRI Procedia **2** 31-38.

[6] Bisio G and Tagliafico L 2002 On the recovery of LNG physical exergy by means of a simple cycle or a complex system, Exergy, an International Journal, **2** 34-50.

[7] Wang J Yan Z Wang M and Dai Y 2013 Thermodynamic analysis and optimization of an ammonia-water power system with LNG as its heat sink, Energy **50** 513-22.

[8] Pu L Qu Z Bai Y Qi D Song K and Yi P 2014 Thermal performance analysis of intermediate fluid vaporizer for liquefied natural gas, App. Thermal Engineering **65** 564-74.

[9] Angelino G and Invernizzi C M 2011 The role of real gas Brayton cycles for the use of liquid natural gas physical exergy, App. Thermal Engineering **31** 827-33.

[10] Zhang N and Lior N 2006 A novel near-zero CO_2 emission thermal cycle with LNG cryogenic exergy utilization, Energy **31** 1666–79.

[11] Liu M Lior N Zhang N and Han W 2009 Thermoeconomic analysis of a novel zero-CO_2-emission high-efficiency power cycle using LNG coldness, Energy Conversion and Management **50** 2768–81.

[12] Shi X and Che D 2009 A combined power cycle utilizing low-temperature waste heat and LNG cold energy, Energy Conversion and Management **50** 567–75.

[13] Gómez M R Ferreiro Garcia R Romero Gómez J and Carril J C 2014 Thermodynamic analysis of a Brayton cycle and Rankine cycle arranged in series exploiting the cold exergy of LNG (liquefied natural gas), Energy **66** 927-37.

[14] Querol E Gonzalez-Regueral B García-Torrent J and Ramos A 2011 Available power generation cycles to be coupled with the liquid natural gas (LNG) vaporization process in a Spanish LNG terminal, Applied Energy **88** 2382–90.

[15] Lu T and Wang K S 2009 Analysis and optimization of a cascading power cycle with liquefied natural gas (LNG) cold energy recovery, App. Thermal Engineering **29** 1478–84.

[16] Dong H Zhao L Zhang S Wang A and Cai J 2013 Using cryogenic exergy of liquefied natural gas for electricity production with the Stirling cycle, Energy **63** 10-18.

[17] Choi I H Lee S Seo Y and Chang D 2013 Analysis and optimization of cascade Rankine cycle for liquefied natural gas cold energy recovery, Energy **61** 179-95.

[18] Qiang W Yanzhong L and Jiang W 2004 Analysis of power cycle based on cold energy of liquefied natural gas and low-grade heat source, App. Thermal Engineering **24** 539–548.

[19] Gerasimov V E Kuz'menko I F Peredel'skii V A and Darbinyan R V 2004 Introduction of technologies and equipment for production, storage, transportation, and use of LNG, Chemical and petroleum engineering **40** 31-35.

[20] Medici M 1974 Natural Gas Industry, Newnes-Butterworth UK.

[21] Liu Y and Guo K 2011 A novel cryogenic power cycle for LNG cold energy recovery Energy **36** 2828-33.

Spectrophotometric evaluation of optical performances of polarizing technologies for smart window applications

N. Levati, L. Vitali, D. Fustinoni and A. Niro

Politecnico di Milano, Department of Energy,
Campus Bovisa, Via Lambruschini 4, 20156 Milano, Italy

E-mail: alfonso.niro@polimi.it

Abstract. In recent years, window-integrated solar protection systems are used and studied as a promising energy saving technology, both for cold and hot climates. In particular, smart windows, whose optical proprieties in the solar wavelength range can somehow be controlled, show interesting results, especially in reducing the air conditioning power consumption. With the improvement of nanolithography techniques as well as with the possibility of designing polarization intervals, coupled polarizing films show a good potential as a dynamic and wavelength-selective shading technology. In this paper, UV-Vis-NIR spectrophotometric measurements are carried out on two polarizing technologies, Polaroid crystalline polarizer and Wire Grid broadband polarizer, in single- and double- film layout, to evaluate their optical performances, i.e. spectral transmittance, reflectance and absorptivity. The solar radiation glazing factors, according to the standard UNI EN 410, are calculated. The measured data are also analyzed in detail to emphasize the optical peculiarities of the materials under study that do not stand out from the standard parameters, as well as the specific problems that arise in spectrophotometric evaluations of polarizing films.

1. Introduction

Through the history of thermal and lightning control for human comfort in indoor spaces, transparent windows have been coupled with a large variety of shading solutions. Since 1959, with the invention of the float (Pilkington) method, that allowed industrial production of glasses with uniform thickness and very flat and parallel surfaces, and in the following years thanks to the advancements in metallic and polymeric materials engineering, many research projects have been carried out on window-integrated solar protection systems.

Today, three main types of such systems can be identified. *Static/Passive* glasses have fixed optical properties, i.e. spectral transmittance, reflectance and absorptivity, determined by the use of purposely engineered superficial thin films or by multiple layer deposition. This category includes various kinds of low-emissivity (Low-E) glasses and selective coatings (solar control glasses). *Dynamic/Passive* glasses show a change in optical proprieties due to incoming solar radiation, like thermocromic (TC) and photocromatic (PC) glasses, while *Dynamic/Active* glasses require an external electric stimulus, that usually allows a greater flexibility of use and the optimization of daylighting and energy saving. Examples are Electrocromic (EC), Liquid Crystal glasses (LCD) and Suspended Particle Devices (SPD).

In recent years, researchers, standardization bodies, governement bodies and industrial companies are closely following the development and performance improvement of the aforementioned technologies. Among recent publications, the paper by Jelle [1] reports a comprehensive comparison of thermo-optical performances of several smart windows technologies, as well as a deep reference section. The report by Lee et al. [2] focuses on an experimental and numerical evaluation of the reduction of Heating, Ventilation and Air Conditioning (HVAC) power consumption given by the use of TC and EC smart windows in a test building, showing promising results in the case of EC windows, also studied by Sbar [3]. A previous study on the control strategy for active windows was carried out by Jonsson and Roos [4]. In Italy, spectrophotometric measurements were performed by Asdrubali et al. [5] on glass and plastic based window panes.

Fewer literature can be found on polarizing filters as a possible technology for a new concept of smart windows, that can combine the benefits of a passive approach (no energy is required to maintain a certain optical configuration) with the reconfiguration capability of active technologies. The recent development of nanolitographic techniques can lead to the design of the polarization spectral range, and to special films suited for both applications and climate zones.

In this paper, UV-Vis-NIR spectrophotometric measurements are carried out to measure the global optical performances, i.e., transmittance, reflectance and absorptivity parameters according to the standard UNI EN 410 [6], of two polarizing technologies, namely crystalline polarizer and Wire Grid broadband polarizer, in single- and double- film layout. The measured data are analyzed in detail to emphasize the optical peculiarities of the materials under study that do not stand out from the standard parameters, and to evaluate any issues that can arise in spectrophotometric measurements of polarizer filters.

2. Experiments

2.1. Theory and definitions

Let us consider an ideal linearly polarized beam, whose each wavelength component is totally polarized, with total intensity I_0, incident on an ideal polarizer filter; the total intensity of the transmitted radiation I is, according to Malus' law (1):

$$I = I_0 \cos^2 \varphi_l \tag{1}$$

where φ_l is the angle between the initial polarization orientation of the beam and the polarizer axis. When a beam of unpolarized radiation, like sunlight, is transmitted by a single polarizing film, its intensity is reduced by the average of $\cos^2(\varphi)$ within the 0-π interval, i.e., by one half. Thus, Malus' law applied in the case of two ideal polarizing filters crossed by unpolarized radiation can be written as (2):

$$I = \frac{1}{2} I_0 \cos^2 \varphi \tag{2}$$

where φ is the angle between the axis of the two polarizers. While being applicable to limited ranges and for certain polarizing technologies, this physical principle is not adequate to predict the behaviour of wavelength-selective polarizers, that must be experimentally evaluated.

On the other hand, the standard UNI EN 410 defines some global parameters to evaluate the performances of luminous and solar characteristics of glazing for glass in buildings. Light (visible) transmittance τ_v and reflectance ρ_v are defined by equations (3-4)

$$\tau_v = \frac{\sum_{\lambda=380\,nm}^{780\,nm} D_\lambda \tau(\lambda) V(\lambda)\, \Delta\lambda}{\sum_{\lambda=380\,nm}^{780\,nm} D_\lambda V(\lambda)\, \Delta\lambda} \tag{3}$$

$$\rho_v = \frac{\sum_{\lambda=380\,nm}^{780\,nm} D_\lambda \rho(\lambda) V(\lambda)\, \Delta\lambda}{\sum_{\lambda=380\,nm}^{780\,nm} D_\lambda V(\lambda)\, \Delta\lambda} \tag{4}$$

where D_λ is the relative spectral distribution of illuminant D65 and $V(\lambda)$ is the spectral luminous efficiency for photopic vision defining the standard observer for photometry or, in other words, the sensibility curve of an average human eye (figure 1).

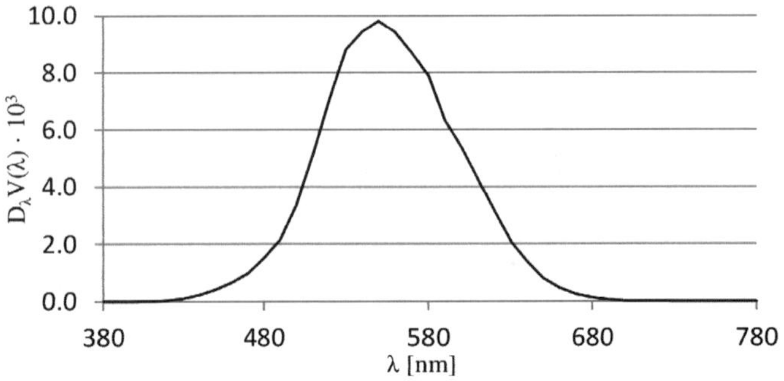

Figure 1. Visible parameters weight factor vs. wavelength

Solar direct transmittance τ_e and reflectance ρ_e are defined by equations (5-6)

$$\tau_e = \frac{\sum_{\lambda=300\,nm}^{2500\,nm} S_\lambda \tau(\lambda)\, \Delta\lambda}{\sum_{\lambda=300\,nm}^{2500\,nm} S_\lambda\, \Delta\lambda} \tag{5}$$

$$\rho_e = \frac{\sum_{\lambda=300\,nm}^{2500\,nm} S_\lambda \rho(\lambda)\, \Delta\lambda}{\sum_{\lambda=300\,nm}^{2500\,nm} S_\lambda\, \Delta\lambda} \tag{6}$$

where S_λ is the relative spectral distribution of the solar radiation (figure 2)

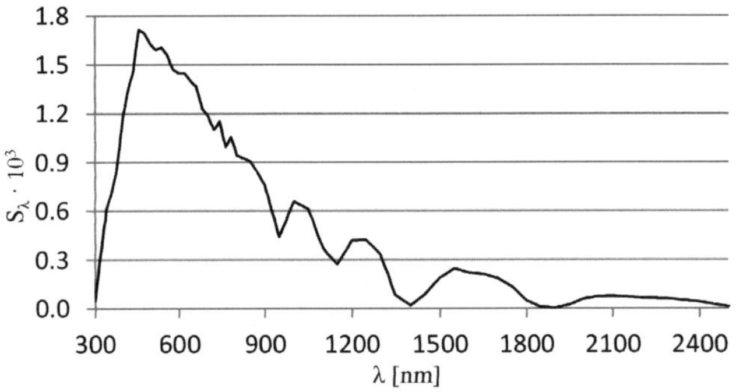

Figure 2. Solar parameters weight factor vs. wavelength

UV transmittance is defined by equation (7). Although the UV range is conventionally limited between 280 and 315 nm for the UV-B band, and from 315 to 380 nm for the UV-A band, our analysis starts at 300 nm, since shorter wavelengths do not reach ground level because of atmospheric absorption [7].

$$\tau_{uv} = \frac{\sum_{\lambda=280\,nm}^{380\,nm} U_\lambda \tau(\lambda)\, \Delta\lambda}{\sum_{\lambda=280\,nm}^{380\,nm} U_\lambda\, \Delta\lambda} \tag{7}$$

These parameters are directly calculated from spectral transmittance τ(λ) and reflectance ρ(λ) measurements. Visible and solar absorptivity, i.e., α_v and α_e, are instead derived from the relation between the three radiative proprieties, both spectral and total, equations (8-9).

$$\alpha(\lambda) = 1 - (\tau(\lambda) + \rho(\lambda)) \tag{8}$$

$$\alpha_e = 1 - (\tau_e + \rho_e) \tag{9}$$

2.2. Experimental setup

Two different polarizing materials are analyzed, in single-film (polarizer only) and double-film (polarizer/analyzer) configurations. One is a Polaroid polymeric-crystalline filter: each film is 0.13 mm thick and it is composed by an inner layer of polyvinyl-alcohol (PVA) with uniformly oriented iodoquinine sulphate crystals, laminated between two external layers of triacetate-cellulose (TAC). The other is a wire grid (WG) anisotropic polarizer, whose polarizing element is a layer of 100 nm-high aluminium nanowires deposited on a TAC substrate with a pitch of 100 nm, as schematically shown in figure 3, for a total film thickness of 0.08 mm.

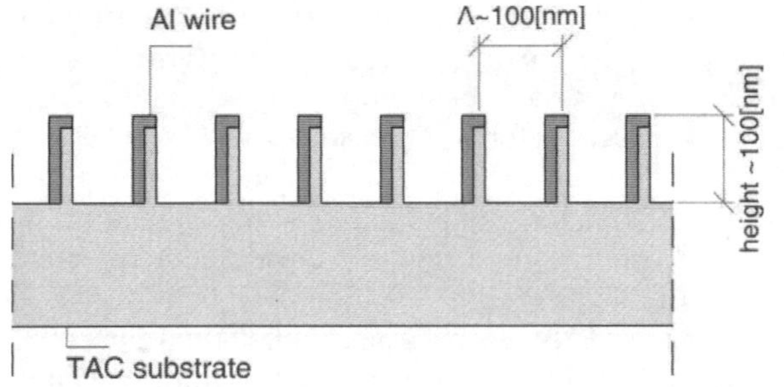

Figure 3. Scheme of the wire grid polarizer film

The samples, measuring 45 mm x 45 mm, are glued on an octagonal wood frame with a central aperture, as shown in figures 4 and 5. Henceforth, φ is the angle between the axis of the two polarizers, θ is the angle of the sample orientation with respect to the vertical, for both single- and double-film trial configurations.

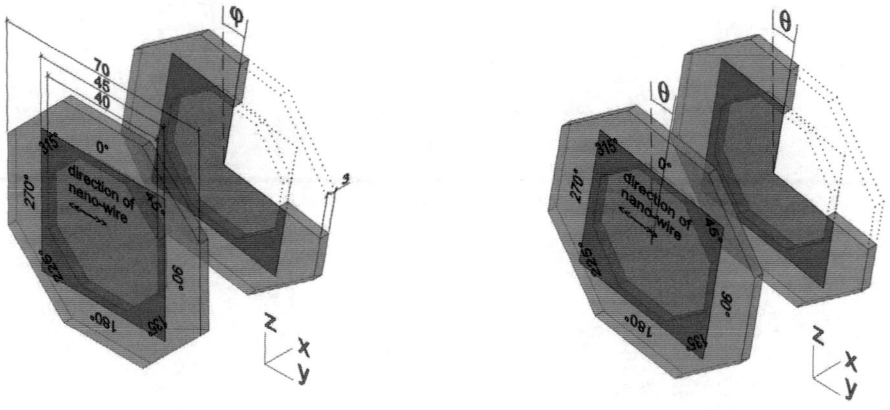

(a) (b)

Figure 4. 3D drawing of the frame-mounted WG samples.

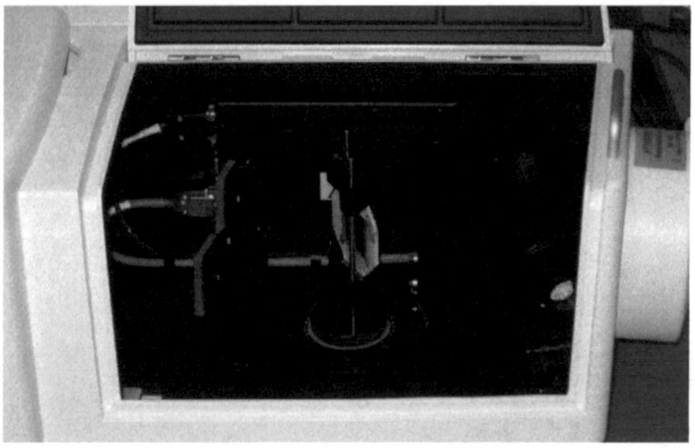

Figure 5. Photo of the sample (center) and integrating sphere (right) in the integrating sphere accessory.

The instrument used in this work is a Perkin Elmer Lambda 950 UV-Vis-NIR, double beam, double monochromator spectrophotometer. The radiation sources are a deuterium lamp for the UV region, and a tungsten lamp for the Vis/NIR range. The instrument is equipped with a 150 mm spectralon integrating sphere mounting a Photomultiplier Tube (PMT) sensor for the UV-Vis range, and a Gallium Arsenide (InGaAs) sensor for NIR measurements; the switch between the two detectors occurs at 860 nm. The total spectral range is between 200 and 2500 nm with a wavelength accuracy of ±0.08 nm in the UV/Vis range, and ±0.30 nm in the NIR range. The instrument have been calibrated with a certified Spectralon sample; the measurement errors, both in reflectance and transmittance mode, are less than 0.11% in the UV-Vis range, and less than 0.23% in the NIR range. In this work, the measurement range is between 300 and 2500 nm with 10 nm intervals. A Hanle-type common beam depolarizer, placed behind the ray chopper, minimizes the unwanted polarization of both test and reference beams caused by the instrument optics [8], even though some residual polarization effects are still noticeable in sections *3.1.1* and *3.2.1*.

The optical scheme of the integrating sphere is shown in figure 6. For transmittance measurements, the sample is placed in a rotating sample holder shown in figure 5, while for reflectance measurements the sample is placed in the reflectance sample holder located in the backside of the sphere, with a fixed 8° inclination relative to the incoming beam (quasi-parallel). It's worth remembering that reflectance measurements encompass both specular and diffuse components.

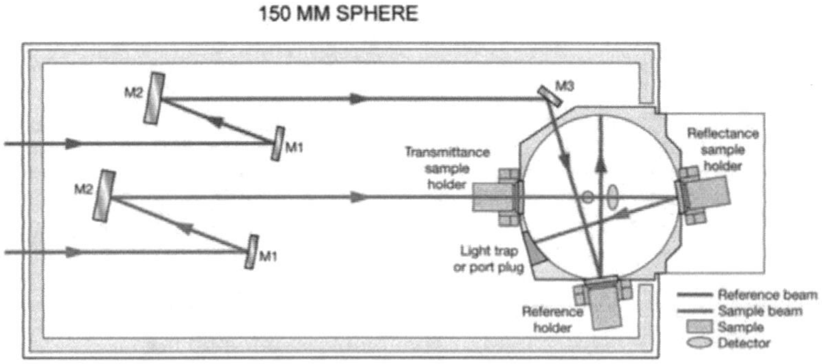

Figure 6. Optical scheme of the Integrating Sphere accessory (adapted from [9])

Data are collected with a PC using a Perkin Elmer dedicated software (UV Winlab) and analyzed with a custom made datasheet.

2.3. Preliminary analysis: scattering effects

As stated in EN 410, "the use of an integrating sphere is necessary when light scattering materials are tested", and "the size of the sphere and its aperture shall be large enough to collect all possible scattered light". In order to test the adequacy of the employed integrating sphere, the WG single-film specimen spectral transmittance has been measured by placing the specimen both close to the aperture and in the rotating sample holder, mounted 6 cm apart. As shown in figure 7, the two measurements differ at the most of 0,3% in the UV-Vis range, and of 1% in the NIR region; thus, for our purposes, scattering phenomena can be neglected.

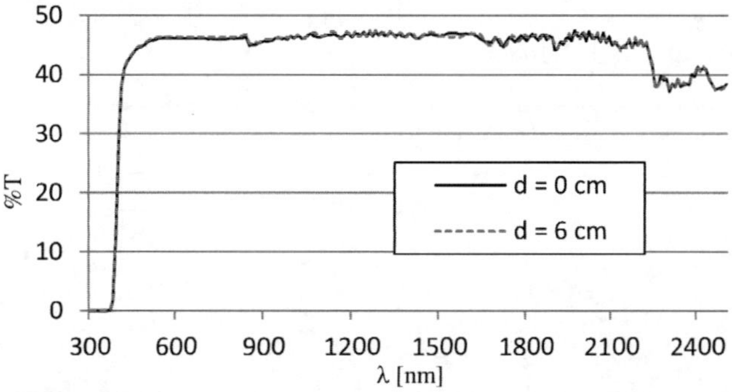

Figure 7. Single WG film, placed at the entrance of the integrating sphere (d = 0 cm) and on the rotating sample holder (d = 6 cm)

3. Results

3.1. Crystalline polarizer (Polaroid)

3.1.1. Single film. A single film of Polaroid has been preliminary tested in four orientations, namely θ = 0°, 45°, 90°, 135°. As shown in figure 8, only the visible part of the spectrum shows a partial agreement to Malus' law (equation 2); in fact, UV radiation is completely blocked, while the specimen is 90% transparent in most of the NIR region, amd hence only the visible portion of the spectrum is effectively polarized. Transmittance values show a little difference with the orientation, about 3%, again only in the visible range, between 0°/45° and 90°/135°, probably because of residual beam polarization combined with the polarized light dependence on the PMT sensor [10].

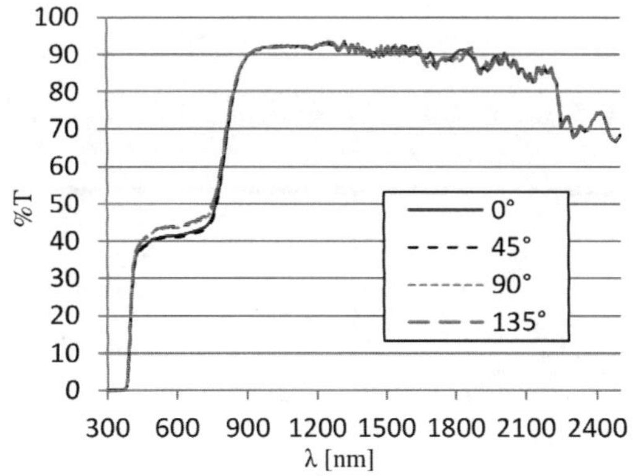

(a)

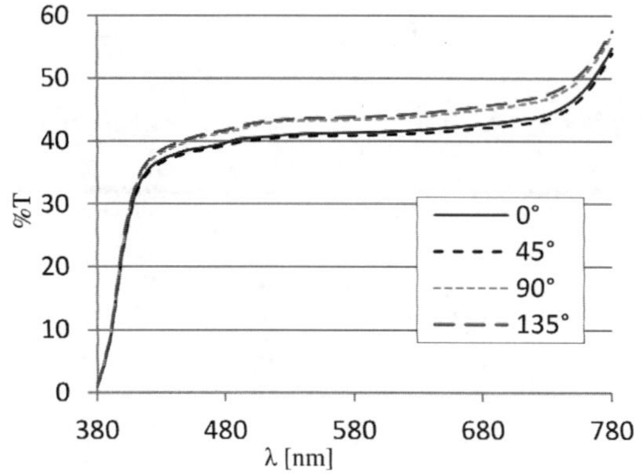

(b)

Figure 8. $\tau(\lambda)$, single Polaroid film, various orientations θ: full spectrum (a) and Vis range detail (b)

To reduce this effect, the subsequent transmittance and reflectance measurements on the coupled films are averaged on four orientations, as in equation (11), also applicable to $\rho(\lambda)$.

$$\tau(\lambda) = \bar{\tau}(\lambda) = \frac{\sum_{i=1}^{4} \tau_{\theta_i}(\lambda)}{4} \quad \text{for } \theta_i = 0°, 45°, 90°, 135° \tag{11}$$

3.1.2. Double film, clear configuration (φ=0°). The total optical proprieties of the double-film specimen have been tested. Compared to the single film, $\tau(\lambda)$ is reduced almost uniformly by 6-10%, whereas $\rho(\lambda)$ is almost constant around 10%. The calculated absorptivity, $\alpha(\lambda)$, shows clearly the filter behavior; since reflectance is uniformely low, UV radiation is almost totally absorbed, just like the non-transmitted part of visible radiation (figure 9).

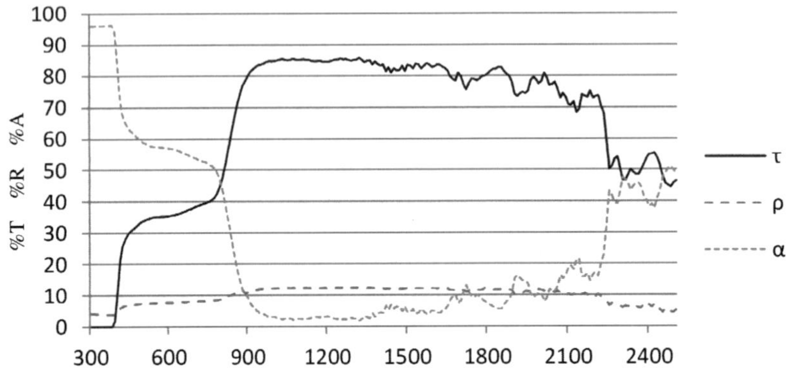

Figure 9. Transmittance, reflectance and absorptivity vs. wavelength for a double Polaroid film, φ = 0°

3.1.3. Double film, variable configuration. This test shows, as graphed in figure 10, the good agreement with Malus' theory for the Polaroid filters in the Vis region; $\tau(\lambda)$ is reduced by half for φ = 45°, and completely blocked for φ = 90°. The UV and NIR behavior remains almost unchanged, as well as spectral reflectance (not shown); in particular, even in the dark configuration, the filter is almost transparent within most of the NIR range.

.

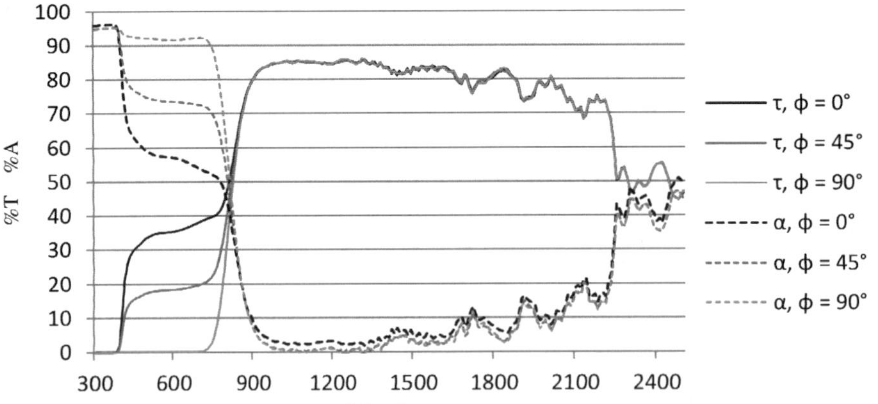

Figure 10. Transmittance and absorptivity vs. wavelength for a double Polaroid film, various φ

Journal of Physics: Conference Series **547** (2014) 012013 doi:10.1088/1742-6596/547/1/012013

3.2. Wire Grid polarizer

3.2.1. *Single film.* Just like the Polaroid film, the single WG film shows little differences in the Vis region for $\tau(\lambda)$, while varying its orientation θ. In particular, artifacts of 1.5% for $\theta = 45°$ and 1.7% for $\theta = 135°$ are visible in Figure 11 at 860 nm, where detector is swiched, due to polarized light dependence of PMT readings [11]. Transmittance values are almost constant and just below 50%, and this is a hint of the main feature of the WG film, that is the ability to polarize NIR radiation. UV radiation is totally blocked, but in a different way with respect to the Polaroid filter; i.e. radiation is absorbed and reflected in equal parts, whereas it is completely absorbed by the Polaroid filter.

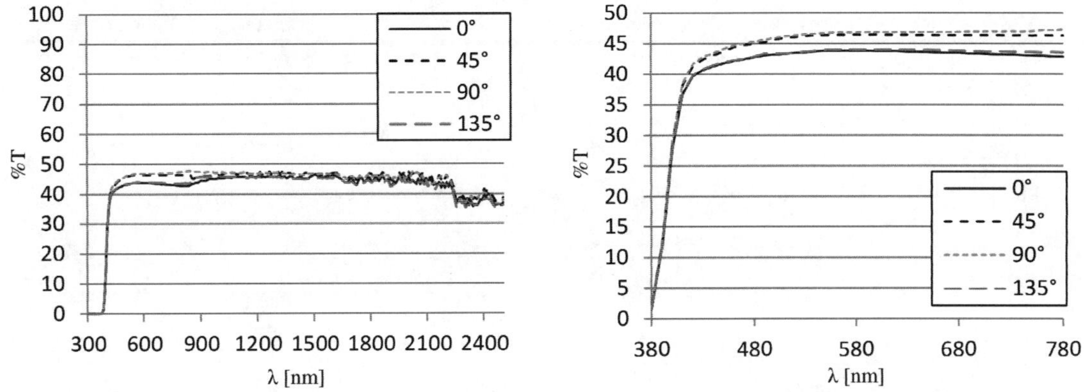

(a) (b)

Figure 11. $\tau(\lambda)$, single WG film, various orientations θ: full spectrum (a) and Vis range detail (b)

3.2.2. *Double film, clear configuration (φ=0°).* The double-film specimen shows, in the $\varphi = 0°$ (clear) configuration, a slight and almost uniform transmittance reduction of 5% as compared with the single film. As showed in figure 12, reflectance is around 50% in the whole range, while absorptivity values show different behaviors from UV (50%), to Vis (15%) and to NIR (5% from 850 nm to 2200 nm, 20% to 2500 nm).

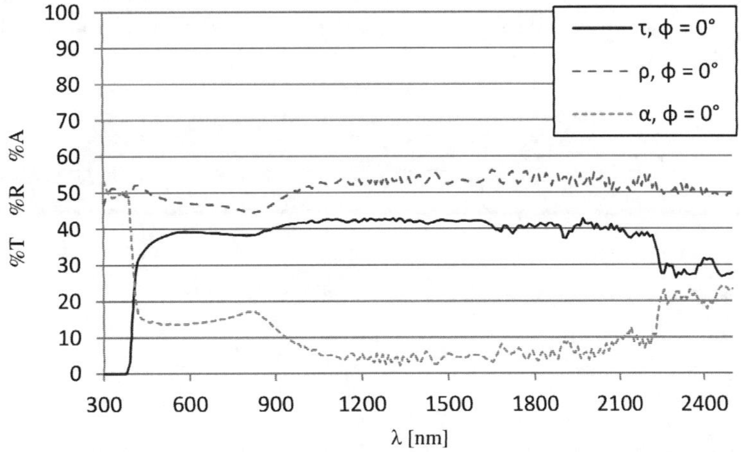

Figure 12. Transmittance, reflectance and absorptivity vs. wavelength for a double WG film, $\varphi = 0°$

3.2.3. Double film, variable configuration. Varying the rotation angle φ shows some interesting behaviors, namely, the NIR radiation polarization effect, the non-linearity between φ and the transmittance attenuation (or reflectance enhancement), mainly caused by inter-reflections, the independence of UV behavior from φ, and the total radiation blockage for $\varphi = 90°$. Moreover, the local reflectance minimum around 800 nm, typical of aluminum, is enhanced by closing the filter, as it can be seen in figure 13(b).

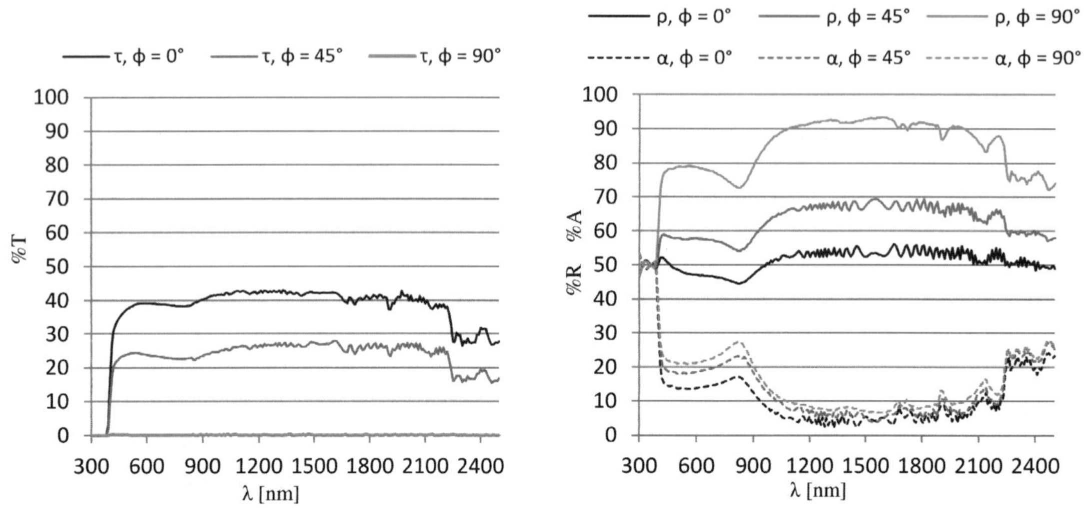

(a) (b)

Figure 13. Transmittance (a), reflectance and absorptivity (b) vs. wavelength for a double WG film, various φ

3.3. Global parameters

The calculated global solar radiation glazing factors are reported in table 1. Clear ($\varphi = 0°$), gray ($\varphi = 45°$) and dark ($\varphi = 90°$) configurations are considered for visible and solar transmittance parameters, clear and dark for UV transmittance, visible and solar transmittance, and solar absorption. These parameters summarize what can be seen in figures 8-13, i.e., UV radiation is completely blocked by both Polaroid and WG samples, while the WG specimen shows a better overall visible transmission and the capability of NIR transmission control.

Table 1. Calculated solar radiation glazing factors

	τ_{uv}	τ_{uv}	τ_v			ρ_v		τ_e			ρ_e		α_e	
φ[°]	0	90	0	45	90	0	90	0	45	90	0	90	0	90
P_1	0.1	-	44.0	-	-	6.6		57.6	-	-	6.9	-	35.5	-
WG_1	0.1	-	47.0	-	-	46.0	-	42.3	-	-	45.9	-	11.9	-
P_2	0.0	0.0	36.1	18.3	0.0	7.5	8.2	49.4	40.3	30.7	8.8	9.7	41.8	59.6
WG_2	0.0	0.0	40.6	24.5	0.1	46.8	85.2	36.7	22.4	0.1	47.0	78.9	16.2	21.0

P and WG Subscripts indicate the number of films

4. Conclusions

The Polaroid crystalline polarizer shows excellent UV blocking proprieties and a good agreement in the Vis range with Malus theory, so that it can be used as a test reference. However, the poor Vis transmittance in the double film clear state ($\tau_{v,\ \varphi=0°} = 36.1$), and the high NIR transmittance, as well, in

every configuration do not allow its use for effective smart windows. Wire Grid polarizer shows a much more interesting behavior, due to the ability of polarizing NIR radiation, as well as to both low solar absorption ($\alpha_{e\ ,\ \varphi\ =\ 90°} = 21.0$) and high reflectance ($\rho_{e\ ,\ \varphi\ =\ 90°} = 78.9$) especially in the dark state. In particular, low values of α_e are required in hot climates to avoid the thermal shocks in glasses. More advances are expected with the diffusion of nanolitographic techniques that will allow, at a certain degree, the control over the spectral characteristics of Wire Grid filters.
The measurements have been made using the UNI EN 410 as reference, but the norm needs to be improved to take into account this kind of materials and, in general, smart glasses. In particular, polarization effects must be taken into account for more reliable spectrophotometric measurements, especially in the visible wavelength range. Further work is needed on the spectral characterization of polarizing technologies, especially in layered glass-integrated configurations, and with variable angle of incidence.

References

[1] Jelle B P 2013 *Solar En. Mat. & Solar cells* **116** 291-323
[2] Lee E S, Fernandes L L, Goudey C H, Jonsson C J, Curcija D C, Pang X, DiBartolomeo D and Hoffmann S 2013 A Pilot Demonstration of Electrochromic and Thermochromic Windows in the Denver Federal Center, Building 41, Denver, Colorado *U.S. General Services Administration report* available online
[3] Sbar N L, Podbelski L, Yang H M and Pease B 2012 *Int. J. of Sust. Built Env.***1** 125-39
[4] Jonsson A and Roos A 2010 *Sol. En.* **84** 1-9
[5] Asdrubali F, Buratti C and Baldinelli G 2000 Misure spettrofotometriche su materiali trasparenti per chiusure perimetrali in edilizia *55th Congresso Nazionale ATI (Bari-Matera)*
[6] UNI EN 410:2011 Glass in Building – determination of luminous and solar characteristics of glazing
[7] Rottman G 2006 *Space Sci. Rev.* **125** 39-51
[8] Taylor J L 2009 Eliminating the inherent polarization artifact of UV/Vis/NIR spectrophotometers *Perkin Elmer UV/Vis/NIR spectroscopy resource page* online at http://pe.taylorjl.net/PE_Blog/
[9] 2009 Transmission measurements using integrating spheres for the Lambda 950/850/650 UV/Vis/NIR and UV/Vis spectrophotometers *Perkin Elmer application note* available online
[10] Hamamatsu Photonics K. K. Editorial Committee 2007 *Photomultiplier tubes – basics and applications* Word Technical Writing, Inc. (Hamamatsu Photonics K. K. Electron Tube Division) chapter 4 pp 78–80 available online
[11] Taylor J L 2009 The grating/detector change in a UV/Vis/NIR instrument: polarization in specular reflectance measurements *Perkin Elmer UV/Vis/NIR spectroscopy resource page* online at http://pe.taylorjl.net/PE_Blog/

Journal of Physics: Conference Series **547** (2014) 012014 doi:10.1088/1742-6596/547/1/012014

Computation of linear transmittance of thermal bridges in precast concrete sandwich panels

Davide Luscietti[1], **Paola Gervasio**[2], **Adriano M Lezzi**[1]

[1] Dipartimento di Ingegneria Meccanica ed Industriale,
Università degli Studi di Brescia, via Branze 38, 25123 Brescia
[2] Dipartimento di Ingegneria Civile, Architettura, Territorio, Ambiente e di Matematica,
Università degli Studi di Brescia, via Branze 43, 25123 Brescia

E-mail: adriano.lezzi@unibs.it

Abstract. Precast concrete lightened sandwich panels are widely used building elements. They are made by two concrete wythes separated by a layer of lightweight material: the central layer is inhomogeneous due to the presence of concrete ribs which tie the external wythe and act as thermal bridges.

Computation of thermal transmittance of sandwich panels is clearly described in European Standards, but in many cases it requires numerical simulations to determine the linear transmittance ψ associated with lightweight material-concrete interfaces in the inhomogeneous layer. Although simple, these simulations represent a critical issue for many panel manufacturers and they would much rather prefer correlations to compute ψ.

In this work we present a correlation based on an artificial neural network (ANN) to estimate linear trasmittance values for current Italian sandwich panel production. Five input parameters are considered: rib width, lightweight material conductivity, and thickness of the three panel layers. To obtain the data which are necessary to train and test the ANN, a fast and accurate Spectral Element Method is used to solve Laplace equation in the neighborhood of a rib. 5460 ψ values are collected which ensure an accurate network response.

1. Introduction

Precast concrete sandwich wall panels are building elements that allow fast and economical constructions. Here we are mostly interested in panels used for one- or two-store buildings such as factories, warehouses, and malls: these panels are designed to span from the foundation to the roof and can be used as load bearing walls, or exterior claddings, or both.

Panels are produced in factories which can be far from the site of the building under construction: weight is a crucial point with regard to handling, transportation, and installation issues. The easiest way to meet both the requirements of structure robustness and weight containment, is to have a frame of concrete and fill the empty zones with lightweight materials like expanded polystyrene. In this paper this kind of panel will be referred to as LSP, precast concrete Lightened Sandwich wall Panel.

In a LSP two reinforced or prestressed concrete wythes are separated by a heterogeneous layer made by lightweight slabs and concrete ribs: so there are panel regions made of solid concrete which act as thermal bridges. Goal of this paper is to present an accurate correlation to evaluate the effect of such thermal bridges.

LSPs are not thermal efficient because the use of insulating slabs is aimed to reduce weight, reduction of the average thermal transmittance is just a by-product effect. There are other types of sandwich panels much more thermal efficient because of the presence of a continuous layer of homogeneous insulations between the concrete wythes: they are not considered here since the computation of their thermal transmittance is straightforward.

Buildings designers in order to fulfill the requirements of the European directives on the energy performance of buildings (Directives 2002/91/EC and 2010/31/EU) as implemented by the EU Member States, need reliable estimates of the thermal transmittance of building elements, including the effects of thermal bridges. In principle, computation of the thermal transmittance U, or U-value, of precast concrete sandwich panels is not a critical issue: European Standards [1–3] provide accurate methods to do that. Nonetheless, for many panel manufacturers it does represent a practical issue.

The U-value of LSPs has to be calculated according to Standard EN ISO 14683 which requires the evaluation of the linear transmittance associated with the conductivity discontinuity at the lightweight material-concrete interfaces in the central heterogeneous layer. Linear transmittance values can be obtained upon numerical calculations performed as described in Standard EN ISO 10211. The effort required by the computation of linear transmittances for a LSP is rather small, since simulations based on 2D model provide results of adequate accuracy. However, most panel manufacturers are SMEs (Small-Medium Enterprises) and their technical staff does not have either the know-how, nor the time to numerically compute the linear transmittances necessary to determine the U-value of their product range: a linear transmittance catalogue or correlations easily implemented in a spreadsheet or in an in-house code would be of relevant practical utility for them.

Other recent studies [4, 5] pursue a goal similar to ours that is to find correlations to evaluate thermal properties of building components which satisfy the following requirements: they must reproduce with good approximation the results of standardized methods; in addition, they must be suited for routine use in industry. In [4], Tenpierik *et al.* propose a closed analytical model for calculating linear transmittances in vacuum insulation panels (VIP) much more accurate than models based on electric circuits, but less laborious than numerical simulations. In [5], Buratti *et al.* develop an artificial neural network (ANN) for estimating U-value of wooden framed windows. Based on 278 experimental and numerical data obtained in accordance with European Standards, the ANN returns an estimate of window transmittance as a function of 10 simple parameters like window typology, wood kind, frame thickness, etc.

In previous works we have presented a database of about one thousand linear transmittance values of thermal bridges in LSPs, obtained using ANSYS Fluent, and proposed a set of simple correlations to evaluate linear thermal transmittance in terms of five independent variables [6, 7]. Although we were rather satisfied with the practical results obtained, we did see two main drawbacks of our work. First, the database build-up had been time-consuming. The bottle-neck being the preprocessing stage that was not automated and required both creation of a new grid and problem set-up for any new combination of geometric parameters. Second, the data analysis to obtain correlations for the linear transmittance was based on a heuristic rather than a systematic approach. We were not sure that we could have extended successfully our correlations to include dependence of the linear transmittance on a sixth parameter (concrete conductivity, for example).

To overcome these limits a new approach has been taken. On one hand, we use a fast, high accurate conformal quadrilateral Spectral Element Method (SEM) to compute the linear transmittance [8, 9]. The method is implemented in an in-house MATLAB code with automated configuration set-up: thousands of different cases are solved within a single run. The solver is used to evaluate linear transmittance varying systematically five parameters within the range of values typical of current panel production.

On the other hand, the large dataset obtained numerically is used to build an artificial neural network for accurate prediction of linear thermal transmittance values of LSPs. This last tool can be supplied to LSPs manufacturers for fast computation of panel U-value without any need of further numerical simulations.

In literature there are only a few other studies on thermal bridges in precast concrete panels and none of them is concerned with accurate evaluation of linear transmittance. Pessiki and Lee have investigated the thermal performance of a modified type of LSP: precast concrete three-wythe sandwich wall panel, that is three concrete wythes separated by two heterogeneous layers with concrete ribs staggered in location to increase the overall thermal resistance [10, 11]. The goal of Pessiki and Lee is to compare the U-value of a small set of two- and three-wythe panel configurations: to do that there is no point in evaluating linear trasmittances associated with the staggered ribs, therefore U is directly calculated upon computation of the heat flux through the entire panel, using a FEM commercial code. Studies [12, 13] consider point transmittance of metal connectors used to tie the concrete wythes in insulated sandwich wall panels and propose either a method based on electric circuit [12] or a correlation [13] to evaluate it.

2. Evaluation of the U-value of a lightened sandwich panel

The heat flow rate q through a wall panel is usually written as

$$q = H\,\Delta T = A\,U\,\Delta T \tag{1}$$

where H denotes the heat transfer coefficient through the wall panel, ΔT is the temperature difference between the internal and the external environments separated by the panel, A is the panel area, and U is the panel average thermal transmittance. In Eq. (1) the four panel edges are considered adiabatic.

H can be calculated upon numerical solution of the conduction equation for the entire panel or – much more efficiently – by means of the following expression,

$$H = \sum_i A_i U_i + \sum_j l_j \psi_j + \sum_k n_k \chi_k \tag{2}$$

In Eq. (2) A_i and U_i are area and thermal transmittance of the i-th *section* of the panel; l_j and ψ_j are length and linear transmittance of the j-th linear thermal bridge; n_k and χ_k are number and point transmittance of the k-th point thermal bridge. Following Standard EN ISO 6946 [1, Clause 6.2.2], here *section* denotes a panel part made of thermally homogeneous layers. As clearly shown in figure 1, a LSP is made of two *sections*: the solid concrete part, corresponding to ribs (*section a*); and the three-layer lightened part, made of the concrete wythes and the lightened layer that separates them (*section b*).

The thermal transmittance of the two *sections*, U_a and U_b, is easily calculated in terms of surface resistances, R_{se} and R_{si}, and of thermal resistances of the homogeneous layers. Therefore, the first sum in Eq. (2), $\sum_i A_i U_i$ represents the heat coefficient through the panel as if the *sections* were thermally insulated one from the other, and the temperature field were 1D within each *section*.

The other two sums, $\sum_j l_j \psi_j$ and $\sum_k n_k \chi_k$, represent the corrections associated to linear and point thermal bridges, that is to the regions where the temperature field is 2D and 3D.

In LSPs, the temperature field is 2D close to the interfaces between concrete ribs and lightweight slabs (see thick lines in figure 1), whereas it is 3D in a neighborhood of the intersections of ribs (see open circles in figure 1). Each point thermal bridge can be considered as the intersection of two or more linear thermal bridges: according to Standard EN ISO 14683 [2, Clause 4.2] the influence of such a kind of point thermal bridge can be neglected and the third

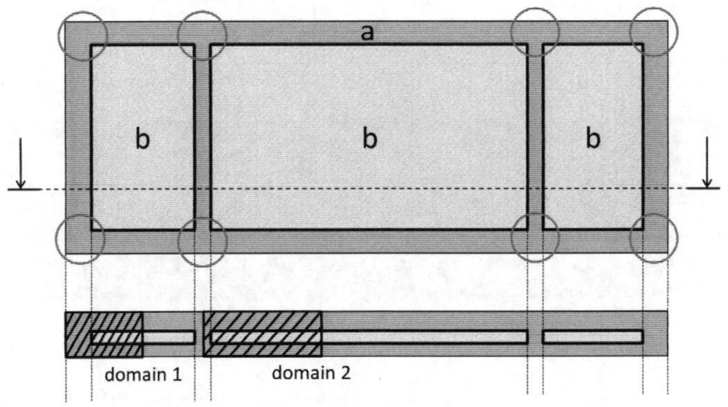

Figure 1. Plan and section view of a lightened sandwich panels: *a*, solid concrete *section*, *b*, lightened *section*. Thick lines and open circles indicate linear and point thermal bridges, respectively. Domains 1 and 2 are examples of 2D models used in numerical simulations.

term in Eq. (2) can be dropped. Therefore, for all practical purposes, an accurate evaluation of the U-value of an LSP is given by

$$U = \frac{A_a U_a + A_b U_b + \sum_j l_j \psi_j}{A} \tag{3}$$

where $A = A_a + A_b$. The linear thermal transmittances ψ_j must be evaluated upon solving the conduction equation on proper 2D domains, in accordance with Standard EN ISO 10211.

3. Linear transmittance calculation: problem description

In figure 2 it is sketched the 2D model used to determine linear thermal transmittance values: it represents a panel section around the interface between a concrete rib and a lightweight slab, where the temperature field is 2D. The rectangle A′B′B″A″ belongs to the solid concrete *section a* of the panel, whereas the other part B′C′C″B″ belongs to the three-layer lightened *section b*.

The geometrical model coincides with a rectangular domain Ω of width $L = L_a + L_b$ and height (panel thickness) $d = d_1 + d_2 + d_3$. d_1 and d_3 denote the thickness of the external and internal wythe, respectively, whereas d_2 is the ligthweight material thickness.

According to Italian Standard UNI 10351 that supplies thermal conductivity values of building materials, the effective thermal conductivity of the outer layer of an external wall must be increased to take into account the stronger effect of aging and moisture content [14, Note 3, p. 16]. For this reason, the external wythe concrete is considered a material different from the internal wythe and rib concrete. Their thermal conductivities will be denoted λ_{ce} and λ_{ci}, respectively. Thermal conductivity of lightweight slabs is λ_{lw}.

On the domain boundaries the following conditions must be satisfied. On the external and internal surfaces A′C′ and A″C″ Robin (mixed) condition is imposed. The heat transfer from surface A′C′ to the external environment at temperature T_e is characterized by a heat transfer coefficient h_e and a surface resistance $R_{se} = 1/h_e$. Internal environment temperature T_i, heat transfer coefficient h_i, and surface resistance $R_{si} = 1/h_i$ characterize heat transfer to the internal surface A″C″.

A′A″ and C′C″ are cut-off planes as defined in Standard EN ISO 10211 [3, Clause 5.2], which coincide with adiabatic surfaces. A cut-off plane is either a symmetry plane or a plane placed in a region so far from any thermal bridge that the temperature field can be considered 1D.

Boundary A′A″ represents either one of the adiabatic lateral surfaces of the panel (as in domain 1 in figure 1) or the symmetry plane of an internal rib (as in domain 2 in figure 1): therefore L_a is either the width of a bounding rib or the half-width of an internal rib.

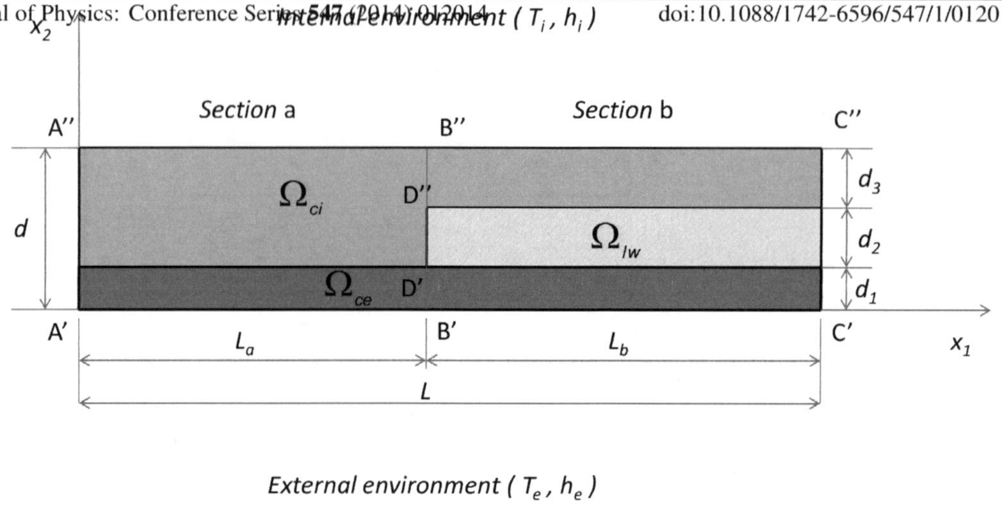

Figure 2. The computational domain Ω.

Boundary C′C″ either belongs to the symmetry plane of a lightweight slab (as in domain 1 in figure 1), or it falls far enough from the closest thermal bridge (as in domain 2 in figure 1). With reference to figure 2 the interface D′D″ between the concrete and the lightweight material represents the thermal bridge: according to Standard EN ISO 10211 the distance L_b between the cut-off plane C′C″ and the thermal bridge must be equal to $\max(1\,\text{m}, 3d)$, where d is the total thickness of the panel.

Denoting q_l the heat flux per unit length through the 2D domain Ω, from Eqs. (1), (2), and (3) it holds,

$$q_l = (L_a U_a + L_b U_b + \psi)\,\Delta T, \tag{4}$$

therefore, upon numerical approximation of q_l (see Sec. 4), the linear thermal transmittance associated with the conductivity discontinuity at the interface D′D″ in figure 2, can be computed by the following equation

$$\psi = \frac{q_l}{\Delta T} - (L_a U_a + L_b U_b) \tag{5}$$

4. Approximation of the heat flux by Spectral Element Methods

The conduction problem described in Sec. 3 is recasted in the following mathematical setting in order to solve it through Spectral Element Methods.

With reference to figure 2, Ω is the union of three non-overlapping regions Ω_{ce}, Ω_{ci}, and Ω_{lw}, corresponding to the different materials forming the panel. We introduce a cartesian system of coordinates with origin in the left bottom vertex of Ω: $\Omega = [0, L] \times [0, d]$.

In order to approximate the heat flux q_l we compute the temperature $T = T(\mathbf{x})$ inside the panel. It solves a homogeneous elliptic equation with discontinuous thermal conductivity λ such that $\lambda|_{\Omega_{ce}} = \lambda_{ce}$, $\lambda|_{\Omega_{ci}} = \lambda_{ci}$, and $\lambda|_{\Omega_{lw}} = \lambda_{lw}$, i.e., T is the solution of

$$\begin{cases} -\nabla \cdot (\lambda \nabla T) = 0 & \text{in } \Omega, \\ \lambda \dfrac{\partial T}{\partial n} + h_e T = h_e T_e & \text{on } \partial\Omega_e, \\ \lambda \dfrac{\partial T}{\partial n} + h_i T = h_i T_i & \text{on } \partial\Omega_i, \\ \lambda \dfrac{\partial T}{\partial n} = 0 & \text{on } \partial\Omega_a. \end{cases} \tag{6}$$

For simplicity the bottom side of Ω is denoted by $\partial\Omega_e$ (external surface), the top side by $\partial\Omega_i$ (internal surface), whereas $\partial\Omega_a$ denotes the union of the adiabatic vertical sides.

Problem (6) can be reformulated in a weak sense, and it can be proved that it admits a unique weak solution $T \in H^1(\Omega)$, being $H^1(\Omega) = \{v \in L^2(\Omega) : \nabla v \in [L^2(\Omega)]^2\}$ the Sobolev space of order one (see, e.g. [15]). Notice that the weak form of (6) is well posed even if λ is a discontinuous bounded function.

To approximate the temperature field inside the panel we consider conformal quadrilateral Spectral Element Methods (SEM) (see, e.g. [8, 9]). SEM, sometimes known as hp–Finite Element Methods (hp–FEM), are high accurate methods (more largely accurate than classical FEM or Finite Volume Methods) designed to discretize partial differential equations. They do their best when the differential problem is set on cartesian geometries, exactly as in the problem we are dealing with, and when the data are regular.

To this aim the computational domain Ω is partitioned in N_e non-overlapping rectangles Q_k (also named elements) of size h (tipically h denotes the diagonal), such that two adjacent elements share a common vertex or a complete edge, and we denote by $\mathcal{Q}_h = \bigcup_{k=1}^{N_e} Q_k$ such a partition. We accept that the elements Q_k can have different size h_k, in such a case we set $h = \max_k h_k$. Finally, let us denote by $\mathbb{Q}_p$ the space of polynomials defined on $\mathbb{R}^2$ that are of degree less than or equal to p (≥ 1) with respect to each variable. Given a partition $\mathcal{Q}_h$ of Ω and fixed the polynomial degree p, we look for an approximation T_h of T that is globally continuous on $\overline{\Omega}$ and locally (that is in each element Q_k) a polynomial of degree p. By introducing the finite dimensional space

$$X_h = \{v_h \in C^0(\overline{\Omega}) : v_h|_{Q_k} \in \mathbb{Q}_p, \ \forall Q_k \in \mathcal{Q}_h\} \subset H^1(\Omega),$$

the weak discrete counterpart of (6) reads:

$$\begin{aligned} &\text{find } T_h \in X_h : \int_\Omega \lambda \nabla T_h \cdot \nabla v_h \, d\Omega + \int_{\partial\Omega_e} h_e T_h v_h \, d\partial\Omega + \int_{\partial\Omega_i} h_i T_h v_h \, d\partial\Omega = \\ &\int_{\partial\Omega_e} h_e T_e v_h \, d\partial\Omega + \int_{\partial\Omega_i} h_i T_i v_h \, d\partial\Omega \qquad \forall v_h \in X_h. \end{aligned} \tag{7}$$

It can be proved that the discrete solution T_h converges in a suitable norm to the weak solution T of (6) when the mesh size h tends to zero and/or the polynomial degree p grows to infinity.

As for classical Lagrangian FEM, the degrees of freedom of the discrete problem are the values of the function T_h at suitable points $\{\mathbf{x}_j\}_{j=1}^{N_h}$ in Ω, and, when SEM are adopted, these points are the Legendre-Gauss-Lobatto (LGL) nodes, that are defined on $[-1, 1]^2$ and mapped on each element Q_k of the partition ([16]). If p is the polynomial degree chosen in each element Q_k, there are $(p+1)^2$ LGL points in Q_k matching, on the edges of Q_k, with those of the adjacent elements.

The choice of LGL nodes guarantees the high accuracy of the method, but also a low computational effort since LGL nodes are used to both represent and integrate the approximating function, thus avoiding time-consuming interpolation processes typical of FEM.

We remark that solving problem (7) guarantees that the discrete solution is continuous on the whole domain Ω and no additional constraints have to be imposed where the thermal conductivity jumps, since we are dealing with a unique global well-posed problem.

Once the discrete temperature T_h is available, the outgoing heat flux per unit length from horizontal edges $\partial\Omega_e$ and $\partial\Omega_i$ can be computed by the following formulas:

$$\phi_{i,h} = \int_{\partial\Omega_i} \lambda \frac{\partial T_h(\mathbf{x})}{\partial n} \, d\partial\Omega = \lambda_{ci} \int_0^L \left. \frac{\partial T_h(\mathbf{x})}{\partial x_2} \right|_{x_2=d} dx_1, \tag{8}$$

$$\phi_{e,h} = \int_{\partial\Omega_e} \lambda \frac{\partial T_h(\mathbf{x})}{\partial n} \, d\partial\Omega = -\lambda_{ce} \int_0^L \left. \frac{\partial T_h(\mathbf{x})}{\partial x_2} \right|_{x_2=0} dx_1. \tag{9}$$

Journal of Physics: Conference Series **547** (2014) 012014 doi:10.1088/1742-6596/547/1/012014

The computation of derivatives in (8) and (9) is simple and affordable, and the integrals are evaluated by Legendre-Gauss-Lobatto quadrature formulas [16].

The moduli of $\phi_{i,h}$ and $\phi_{e,h}$ are the discrete approximation of the heat flux per unit length through Ω, q_l, that is required to calculate the linear transmittance ψ by Eq. (5). In particular, for $T_i > T_e$ we obtain $-\phi_{i,h} \rightarrow q_l$ and $+\phi_{e,h} \rightarrow q_l$ when $h \rightarrow 0$ and/or $p \rightarrow \infty$, i.e. when the discretization errors tend to zero.

For a more in-depth description of SEM or $hp-$FEM methods and in their use to approximate partial differential equations, we refer, e.g., to [9, 15, 17].

5. Numerical results

The SEM described in Sec. 4 has been implemented in MATLAB together with an automated procedure to generate partition $\mathcal{Q}_h$ depending on the geometrical parameters $(L_a, L_b, d_1, d_2, d_3)$.

In order to get a good trade-off between accuracy and computational time one has to choose properly the partition $\mathcal{Q}_h$ and the polynomial degree p valid for all elements Q_k.

Accuracy of SEM depends on both size h of elements Q_k and polynomial degree p. Error tends to zero as $h^p p^{-s}$, being $s \geq 1$ the regularity order of the solution. Error decreases rather quickly for increasing p, but only algebraically w.r.t. h, as h tends to zero. During the initial simulation trials, we have found that for $p \geq 12$ computational time required to solve a single case was too large, so we decided to set $p = 8$, a value that guaranteed the requested accuracy on the partition chosen with computational time ten times shorter: as a matter of fact in most applications of SEM p is equal to 6 [16].

With regard to partition $\mathcal{Q}_h$, elements Q_k of non-uniform size have been taken in order to improve resolution where temperature variations are larger. The element width (along direction x_1) increases moving from the concrete-lightweight material interface at $x_1 = L_a$, towards the adiabatic walls. This is justified by the fact that far from the interface $x_1 = L_a$ the temperature field tends to become independent of x_1. As an example of partition along x_1 consider the case $L_a = L_b = 1$ m. The element width sequence on both sides of the interface is: $[0.5, 0.5, 1, 1, 1.5, 1.5, 2, 2, 5, 5, 20, 20, 20, 20]$ cm.

The element thickness (along direction x_2) is set uniform and equal to 2 cm, except for the rows of elements close to the material interfaces at $x_2 = d_1$ and $x_2 = d_1 + d_2$. On both sides of lines $x_2 = d_1$ and $x_2 = d_1 + d_2$ we place two rows of 0.5 cm thick elements, followed by one row of 1 cm thick elements: in this way interfaces between different materials always coincide with the common boundary of adjacent elements. In principle, the discrete temperature field T_h converges to T even if the conductivity discontinuities fall inside elements of the partition, in practice, however, for large h and small p, one obtains more accurate results when discontinuities fall on element boundaries.

In our simulation the total number of elements Q_k ranges from 18×14 for the smallest computational domain ($L = 1.025$ m, $d = 0.12$ m), to 28×20 for the largest one ($L = 2$ m, $d = 0.24$ m).

To assess accuracy requirements we referred to criteria A.2.d and A.2.e of Standard EN ISO 10211 and to comparison with data in the database presented in [6, 7]. In all our computations the relative difference between the two fluxes $|\phi_{i,h}|$ and $\phi_{e,h}$ is less than 10^{-4} as required by Clause A.2.e of [3] – actually it is always less then 10^{-9}.

With regard to the number of nodes to be used in the computational model, Clause A.2.d of [3] requires that upon doubling the number of nodes, the computed heat flux through the building element must not vary more than 1%. The total number of LGL nodes in Ω increases of a factor slightly larger than 2 when the polynomial degree changes from $p = 8$ to $p = 12$. In a few test cases we have checked how q_l and ψ vary and we have verified that the requirement was satisfied: the relative change of q_l is of order 10^{-4} or smaller, whereas the relative change of ψ is of order 10^{-3} or smaller. For these test cases the value of ψ is available in the database presented in [6, 7]. The relative difference with respect to the value obtained with SEM is close to 8×10^{-3}.

The linear transmittance ψ as defined by Eq. (5) in Sec. 3 depends on several parameters: geometrical, L_a, L_b, d_1, d_2, d_3; and thermophysical, λ_{ce}, λ_{ci}, λ_{lw}, R_{se}, R_{si}.

In this study L_b, λ_{ce}, λ_{ci}, R_{se}, and R_{si} are fixed, and ψ is computed for varying values of the other five variables:

$$\psi = f(L_a, d_1, d_2, d_3, \lambda_{lw}) \tag{10}$$

The surface resistances are set equal to the conventional values prescribed by Standard EN ISO 6946 [1, Clause 5.2]: $R_{se} = 0.04$ m^{2}K/W and $R_{se} = 0.13$ m^{2}K/W. Concrete conductivity is set equal to the values used by most Italian panel manufacturers to estimate thermal performances of their products [6, 7]. They correspond to a concrete density of 2400 kg/m^3: $\lambda_{ci} = 1.909$ W/m K for external walls, and $\lambda_{ce} = 2.075$ W/m K for the outer layer of external walls [14, p. 5].

In figure 3 the dependence of ψ on the *section* width L_b is shown – for $L_a = 1$ m, data are taken from [6, 7]. When $L_b = 0$, the domain Ω is made only of homogeneous layers, temperature field is 1D and ψ vanishes. As L_b increases ψ tends asimptotically to a constant value: the growth is fast, for L_b larger than 25 cm the value of ψ is hardly distinguishable from the asymptotic value. Since in real panels the lightweight slab width is rarely smaller than 50 cm, the asymptotic value only is needed for all practical purposes: following EN ISO 10211 in all computations we set $L_b = 1$ m.

The dependence of ψ on solid concrete *section* width L_a is similar to the one on L_b (see figure 3): ψ attains its asympotic value for L_a larger than 20 cm, approximately. Since in current LSP production concrete rib width ranges between 10 and 25 cm, that is in the range of significant variation of ψ, numerical computations have been performed for $L_a = 2.5$, 5, 7.5, 10, 12.5, 15, 20, 25, 30, 40, 60, 80, 100 cm: this choice includes the most frequent values of rib width or half-width.

Thermal conductivity of expanded polystyrene used in most LSPs ranges between 0.032 and 0.056 W/m K, but smaller or larger values are not uncommon. ψ has been calculated for $\lambda_{lw} = 0.02$, 0.03, 0.04, 0.05, 0.06 W/m K. ψ decreases monotonically with λ_{lw}.

With regard to the thickness of the two wythes, d_1 and d_3, and to that of the lightened layer, d_2, each one of them has been varied from 4 up to 16 cm by 2 cm steps. However, the following constrain on panel total thickness has been imposed: $d \leq 24$ cm. Only 84 combinations of values of d_1, d_2 and d_3 satisfy condition above. They span a layer thickness parameter space slightly larger than current production of LSPs.

5460 accurate numerical estimates of ψ have been obtained upon varying L_a, λ_{lw}, d_1, d_2, d_3 as specified above. This large dataset has been used, in turn, to build an artificial neural network (ANN) for computation of ψ.

6. Artificial neural network for ψ computation

In recent years several studies have appeared in the heat transfer literature describing the use of ANNs to correlate experimental and/or numerical data [5, 18–20]. The work by Buratti *et al.* on U-value of wooden windows has already been quoted in Sec. 1 [5].

In [18, 19] ANNs are successfully tested to predict Nusselt number and pressure drop for either compact or fin-and-tube heat exchangers. For ANN training and testing 170 (experimental) and 277 (experimental and numerical) data were used in [18] and [19], respectively.

In [20] Karadağ and Akgöbek applied an ANN to a floor-heated room to correlate Nusselt and Rayleigh numbers to wall, ceiling and floor temperatures and room size. 235 data obtained through CFD simulations were used for training and testing.

Fitting of function (10) has been obtained through applications provided by the Neural Network Toolbox in MATLAB (NNT). In particular, we have chosen a standard architecture consisting in a two-layer feed-forward network with tan-sigmoid transfer function in the hidden

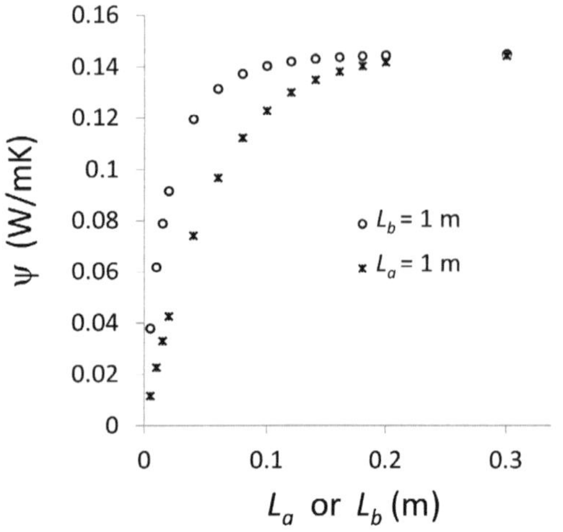

Figure 3. ψ vs. concrete rib width L_a (for $L_b = 1$ m) and lightweight material width L_b (for $L_a = 1$ m) [6, 7].

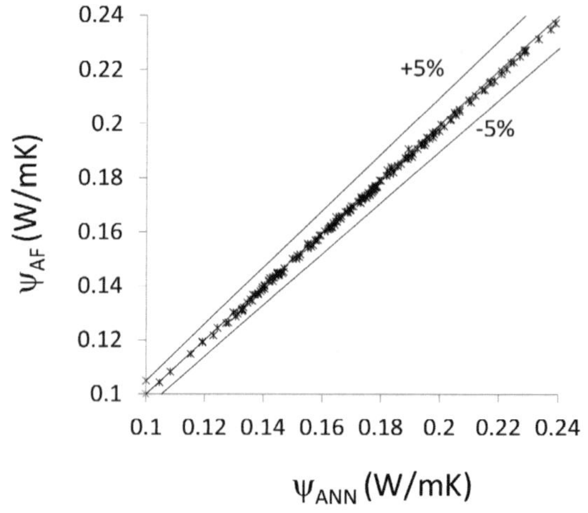

Figure 4. Comparison between the ANN response, ψ_{ANN}, and data from [6], ψ_{AF}.

layer and linear transfer function in the output layer. As noted in [21, p. 5-11] this kind of network allows approximation of any function for a sufficient number of neurons in the hidden layer: after a few trials, this number was set equal to 10.

The default Levenberg-Marquardt backpropagation algorithm has been adopted for training. It divides the 5460 data in three sets, one for *training*, one for *validation* and one for *testing*. It is possible to decide the percentage of data to store for the three steps: here we have used the default values, 60% for training, 20% for validation, and 20% for testing. The training data are used by the network to estimate the weights and biases associated with each neuron layer, the validation data are used to measure the network capability of generalization to new data and avoid overfitting, and the testing data provides a measure of the performance of the network.

To assess the accuracy of the network, a regression analysis between the network response and the original data can be performed. The resulting correlation coefficient, R-value, is $R = 0.99930$ for the test data set (1092 data). More than 97% of predicted data fall within 0.5% from computed results: the maximum relative difference is smaller than 1.3%. These results show that the fit is very good.

Finally, values obtained with the ANN have been compared with data presented in [6, 7]. In the database built up using ANSYS Fluent simulations, 236 records have values of $(L_a, d_1, d_2, d_3, \lambda_{lw})$ falling in the range of validity of the ANN. For these configurations the ANN response, ψ_{ANN}, has been evaluated and compared with ψ_{AF} from the database. Results are shown in figure 4: the relative difference is within 1% for 91% of data, as expected from comparison between results obtained with SEM and ANSYS Fluent (see Sec. 5).

7. Conclusions

This paper deals with the problem of finding an accurate correlation for computing the linear thermal transmittance associated with the interface between a concrete rib and a lightweight slab in precast concrete lightened sandwich panels. If such a correlation is available, evaluation of the U-value of the panel is straightforward and easily implemented in a spreadsheet.

To reach this goal a two-step approach has been followed. A conformal quadrilateral Spectral Element Method was developed in MATLAB to obtain accurate and fast numerical computation of the linear transmittance. The numerical procedure is in accordance with Standard EN ISO 10211. All the preprocessing operations (domain partition generation, node mapping on the partition elements, problem set-up) are automated, thus in a single run thousands of ψ values are collected.

A data set of 5460 configurations has been obtained as a function of five parameters: rib width, lightweight material conductivity, thickness of the two concrete wythes and of the internal lightened layer. The parameters span a range of values slightly larger than those of current Italian panel production.

The data set has been used for training and testing an ANN developed within the Neural Network Toolbox environment. The ANN is a two-layer feed-forward network with tan-sigmoid and linear transfer functions which is a standard choice for function fitting: 10 neurons are placed in the hidden layer. The network response is rather good. A comparison with an existing database of ψ values built by numerical simulations with a finite volume code shows relative differences smaller than 2%.

The ANN can be distributed to panel manufacturers as a standalone executable, although the issue of its dissemination has not been fully addressed yet.

Finally, the results presented here can be quickly extended to include other parameters such as concrete conductivity, or to change the variability range of the input parameters to fit other countries production. Beside, the procedure that has been discussed could be successfully applied to other similar problems like the determination of correlations for point transmittances associated with metal fasteners used to connect the external wythes in insulated sandwich panels.

References

[1] EN ISO 6946 2007 *Building components and building elements - Thermal resistance and thermal transmittance - Calculation method*
[2] EN ISO 14683 2007 *Thermal bridges in building construction - Linear thermal transmittance - Simplified methods and default values*
[3] EN ISO 10211 2007 *Thermal bridges in building construction - Heat flows and surface temperatures - Detailed calculations*
[4] Tenpierik M, van Der Spoel W and Cauberg H 2008 *J. Building Phys.* **31** 361–87
[5] Buratti C, Barelli L and Moretti E 2012 *Applied Energy* **98** 425–32
[6] Luscietti D 2012 *Metodo per il calcolo della trasmittanza termica di pareti di tamponamento a strati eterogenei* Ph.D. thesis University of Brescia
[7] Luscietti D and Lezzi A M 2012 *Atti 19° Congresso C.T.E. (Bologna)* (Bologna: C.T.E.) pp 563–72
[8] Maday Y and Patera A T 1989 *Spectral element methods for the incompressible Navier-Stokes equations* in *State-of-the-Art Surveys on Computational Mechanics* ed Noor A K and Oden J T (ASME)
[9] Canuto C, Hussaini M H, Quarteroni A and Zang T A 2007 *Spectral Methods. Evolution to Complex Geometries and Applications to Fluid Dynamics* (Heidelberg: Springer)
[10] Lee B J and Pessiki S 2004 *PCI J.* **49** 88–101
[11] Lee B J and Pessiki S 2006 *Energy Buildings* **38** 1006–14
[12] Lee B J and Pessiki S 2008 *PCI J.* **53** 86–100
[13] Willems W and Hellinger G 2010 *Bauphysik* **32** 275–287
[14] UNI 10351 1994 *Materiali da costruzione. Conduttività termica e permeabilità al vapore.*
[15] Quarteroni A and Valli A 1994 *Numerical Approximation of Partial Differential Equations* (Heidelberg: Springer)
[16] Canuto C, Hussaini M H, Quarteroni A and Zang T A 2006 *Spectral Methods. Fundamentals in Single Domains* (Heidelberg: Springer)
[17] Quarteroni A 2014 *Numerical Models for Differential Problems* (Heidelberg: Springer)
[18] Ermis K 2008 *Int. J. Energy Res.* **32** 581–94
[19] Xie G, Sunden B, Wang Q and Tang L 2009 *Int. J. Heat Mass Transfer* **52** 2484–97
[20] Karadağ R and Akgöbek Ö 2008 *Int. Comm. Heat Mass Transfer* **35** 312–25
[21] Demuth H, Beale M and Hagan M 2009 *Neural Network Toolbox User's Guide* (Natick, MA: The MathWork Inc.)

Numerical Simulation of a Solar Domestic Hot Water System

L Mongibello[1], N Bianco[2], M Di Somma[2], G Graditi[1], V Naso[2]

[1] ENEA – Italian National Agency for New Technologies, Energy and Sustainable Economic Development, Portici RC – Napoli – Italy
[2] Dipartimento di Ingegneria Industriale (DII) – Università degli Studi Federico II – Napoli – Italy

Abstract. An innovative transient numerical model is presented for the simulation of a solar Domestic Hot Water (DHW) system. The solar collectors have been simulated by using a zero-dimensional analytical model. The temperature distributions in the heat transfer fluid and in the water inside the tank have been evaluated by one-dimensional models. The reversion elimination algorithm has been used to include the effects of natural convection among the water layers at different heights in the tank on the thermal stratification. A finite difference implicit scheme has been implemented to solve the energy conservation equation in the coil heat exchanger, and the energy conservation equation in the tank has been solved by using the finite difference Euler implicit scheme. Energy conservation equations for the solar DHW components models have been coupled by means of a home-made implicit algorithm. Results of the simulation performed using as input data the experimental values of the ambient temperature and the solar irradiance in a summer day are presented and discussed.

1. Introduction

The production of Domestic Hot Water (DHW) by means of non-concentrating solar thermal collectors is one of the most efficient and cost effective ways to utilize the solar energy.

In the last decades, several papers have focused on transient mathematical models of solar DHW systems. Rodriguez-Hidalgo et al. [1] developed and validated experimentally a transient model in order to achieve the design criteria of solar DHW systems under daily transient conditions. Dehghan et al. [2] investigated numerically the transient performance of a solar DHW system including a vertical storage tank with a mantle heat exchanger. Kenjo et al. [3] modeled the mantle tank of a solar DHW system by using a zonal approach, and they validated the model with experimental tests.

Several transient models of the components of solar DHW systems have been proposed. Klein et al. [4] studied numerically the effects of the thermal capacitance on the performance of a flat plate solar collector. Rodriguez-Hidalgo et al. [5, 6] investigated the efficiency of a flat plate solar collector by using a transient zero-dimensional mathematical model. Cadafalch [7] analysed different configurations of a flat-plate solar collector using a one-dimensional transient numerical model. Transient mathematical models for solar DHW systems were also proposed by Logie et al. [8] and Chauvet et al. [9].

This paper is focused on the numerical simulation of a solar DHW system consisting of two flat plate collectors connected in series, a vertical cylindrical water tank for the heat storage, and a coil heat exchanger immersed into the water tank. The performance of the system has been simulated using a Matlab home-made code. The temperature distributions in the heat transfer fluid and in the water inside the tank have been evaluated by unsteady one-dimensional models. The reversion elimination algorithm has been used to include the effects of natural convection among the water layers at different heights in the tank on the thermal stratification. In the first part of the paper, the analytical models, the numerical schemes and the algorithms used in the simulation are reported in detail. In the second part, results of the simulation performed using as input data the measured values of the ambient temperature and the solar irradiance in a summer day are presented and discussed.

2. Models of the solar DHW system components

Figure 1 shows the sketch of the investigated system.

2.1. Solar collector model

In the present study reference is made to a couple of commercial flat-plate solar collectors Vitosol 100 w2.5 by Viessmann Werke GmbH and Co KG®. Table 1 reports their main data. The collectors were assumed to be connected in series and installed in Portici (Italy), due south, with a 30° inclination above the horizontal. The wind velocity was assumed to be equal to zero and a solution of water and propylene glycol (30%) was considered as the heat transfer fluid (HTF).

A zero-dimensional analytical model has been developed to evaluate the time-variable temperature of the collectors components and the temperature of the HTF at the exit sections of the collectors. Figure 2 shows a sketch of the collector cross section and the thermal circuit for the lumped-element model that has been used to carry out the energy balances. The conductive thermal resistances of the box aluminum back cover, the glass, the copper absorber plate were neglected, since they are negligible compared with the insulation thermal resistance.

With reference to figure 2, the following correlations hold for each collector.

The energy balance on the collector insulation gives

$$\frac{T_{abs} - T_{ins,coll}}{R_{ins,coll}} - \frac{T_{ins,coll} - T_{box}}{R_{ins,coll}} = c_{ins,coll}\, \rho_{ins,coll}\, A_{ins,coll}\, L_{ins,coll}\, \frac{d(T_{ins,coll})}{dt} \quad (1)$$

where T_{abs} and T_{box} are the temperatures of the absorber plate and the box back cover, respectively; $T_{ins,coll}$ is the lumped temperature of the insulation; $c_{ins,coll}$, $\rho_{ins,coll}$, $L_{ins,coll}$, are the specific heat, the density, the thickness of the insulation, respectively, t is the time; and the insulation thermal resistance, $R_{ins,coll}$, is given by

$$R_{ins,coll} = (L_{ins,coll}/2)(A_{ins,coll}\, k_{ins,coll})^{-1} \quad (2)$$

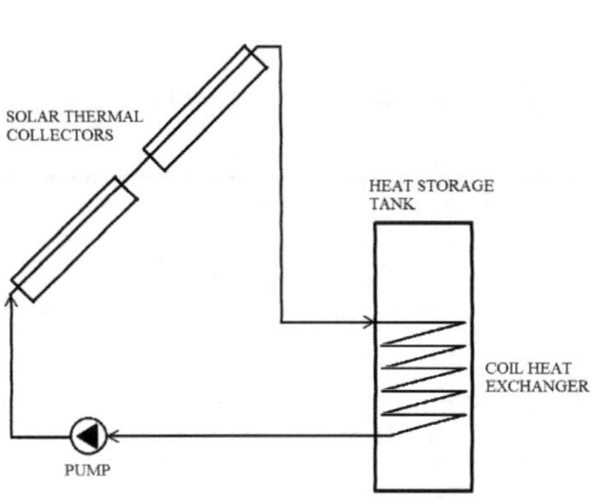

Figure 1. Sketch of the solar DHW system.

Table 1. Collector data.

Collector width: 2.39 m; Collector length: 1.14 m
Absorber plate area: 2.50 m^2
Glass cover thickness: 4.0 10^{-3} m
Inner air layer thickness: $2.6 \cdot 10^{-2}$ m
Absorber sheet thickness (Cu): $2.5 \cdot 10^{-4}$ m
Outer tube diameter (Cu): $1.0 \cdot 10^{-2}$ m
Inner tube diameter (Cu): $8.0 \cdot 10^{-3}$ m
Insulation thickness (mineral fiber): $5.0 \cdot 10^{-2}$ m
Box back cover thickness (Al): $2.0 \cdot 10^{-3}$ m
Glass emittance: 0.85; Glass transmittance: 0.90
Absorber plate emittance: 0.70
Absorber plate absorptance: 0.94

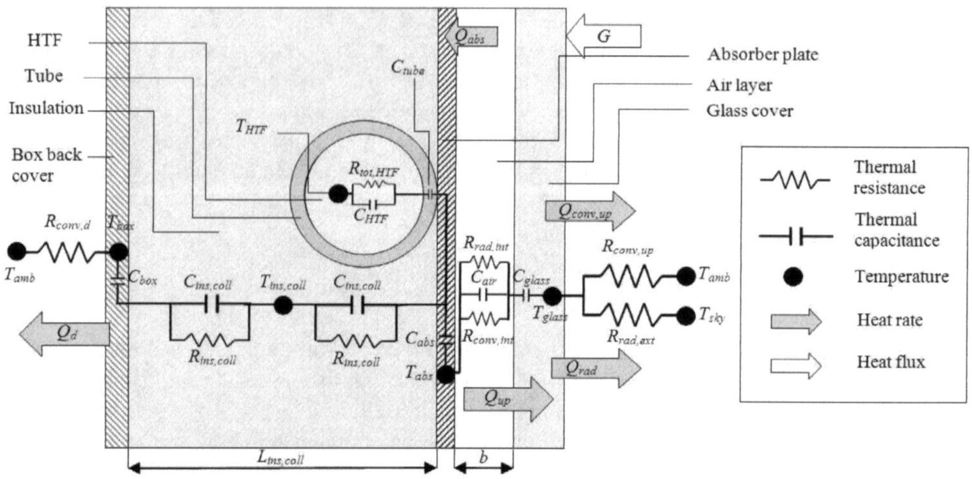

Figure 2. Cross section and thermal circuit of the collector.

with $A_{ins,coll}$ the insulation section area; $k_{ins,coll}$ the insulation thermal conductivity.

The heat rate dissipated through the box back cover, Q_d, is

$$Q_d = (T_{box} - T_{amb})\left(R_{conv,d}\right)^{-1} \tag{3}$$

where T_{amb} is the ambient air temperature and the convective thermal resistance at the box back cover surface, $R_{conv,d}$, is given by

$$R_{conv,d} = 1/\left(A_{box}\,\bar{h}_{conv,d}\right) \tag{4}$$

with A_{box} the area of the box back surface in contact with the air and $\bar{h}_{conv,d}$ the mean convective heat transfer coefficient at the box back surface, evaluated as in [10].

$$R_{int}\,Q_{up} = T_{abs} - T_{glass} \tag{5}$$

where Q_{up} is the heat rate transferred by the absorber plate to the glass, T_{glass} is the temperature of the glass and the thermal resistance of the enclosure between the absorber plate and the glass cover, R_{int}, accounts for the parallel contributions of radiation and convection in the enclosure

$$R_{int} = \left(R_{rad,int}^{-1} + R_{conv,int}^{-1}\right)^{-1} \tag{6}$$

The internal radiative thermal resistance, $R_{rad,int}$, assuming the surfaces to be grey, is given by

$$R_{rad,int} = \left(1/\varepsilon_{glass}\,\varepsilon_{abs}\right)\left[\sigma\,A_{abs}\left(T_{glass}^2 + T_{abs}^2\right)\left(T_{glass} + T_{abs}\right)\right]^{-1} \tag{7}$$

where ε_{glass} and ε_{abs} are the emittancies of the glass and the absorber plate surfaces, respectively; σ is the Stephan-Boltzmann constant; A_{abs} is the absorber plate area.
The internal convective thermal resistance, $R_{conv,int}$, is given by

$$R_{conv,int} = 1/\left(A_{abs}\,\bar{h}_{conv,int}\right) \tag{8}$$

where $\bar{h}_{conv,int}$ is the mean natural convective heat transfer coefficient in the enclosure

$$\bar{h}_{conv,int} = (\overline{Nu}_{air}\,k_{air})\,b^{-1} \tag{9}$$

with $\overline{Nu}_{air}$ the mean Nusselt number evaluated using the Hollands et al. correlation [11]; k_{air} the thermal conductivity of the air; b the enclosure thickness.

The convective heat rate from the glass to the ambient air, $Q_{conv,up}$, is given by

$$Q_{conv,up} = (T_{glass} - T_{amb})/R_{conv,up} \tag{10}$$

where the convective thermal resistance at the glass cover surface, $R_{conv,up}$, is given by

$$R_{conv,up} = 1/(\, A_{abs}\,\bar{h}_{conv,up}) \tag{11}$$

with $\bar{h}_{conv,up}$ the mean convective heat transfer coefficient at the glass surface, evaluated as in [10].

The radiative heat rate from the glass to the ambient, Q_{rad}, is given by

$$R_{rad,ext}\, Q_{rad} = T_{glass} - T_{sky} \tag{12}$$

with the external radiative thermal resistance, $R_{rad,ext}$,

$$R_{rad,ext} = 1/\left[\sigma\, \varepsilon_{glass}\, A_{abs}\, \left(T_{glass}^2 + T_{sky}^2\right)\left(T_{glass} + T_{sky}\right)\right] \tag{13}$$

where T_{sky} is the “sky temperature” evaluated as in [12]

$$T_{sky} = 0.0552\, T_{amb}{}^{1/5} \tag{14}$$

The radiative solar energy rate absorbed by the plate, Q_{abs}, is

$$Q_{abs} = q_{abs}\, A_{abs} \tag{15}$$

where the fraction of the solar irradiance absorbed by the absorber plate, q_{abs}, is

$$q_{abs} = \eta\, G \tag{16}$$

with G the solar irradiance measured on a due south 30° inclined surface and η the glass transmittance (0.90) times the absorber plate absorptance (0.94), that is assumed to be constant.

The energy balance on the heat transfer fluid gives

$$\frac{T_{abs} - T_{HTF,avg}}{R_{tot,HTF}} - \dot{V}_{HTF}\, \rho_{HTF}\, c_{HTF}\left(T_{HTF,out} - T_{HTF,in}\right) = \rho_{HTF}\, c_{HTF}\, V_{HTF}\, \frac{d\left(T_{HTF,avg}\right)}{dt} \tag{17}$$

with $T_{HTF,out}$ and $T_{HTF,in}$ the time varying temperatures of the HTF in the outlet and inlet sections of the collectors; $T_{HTF,avg} = (T_{HTF,out} + T_{HTF,in})/2$; $\dot{V}_{HTF}$, ρ_{HTF}, c_{HTF}, V_{HTF}, the volumetric flow rate, the density, the specific heat and the volume of the HTF in the collectors, respectively.

The thermal resistance between the HTF flowing in the tube and the absorber plate, $R_{tot,HTF}$, is evaluated taking into account only the contribution of the convective resistance and neglecting the far smaller conductive resistances in the tube and the absorber plate walls

$$R_{tot,HTF} = 1/\left(\bar{h}_{tube}\, A_{i,tube}\right) \tag{18}$$

where $A_{i,tube}$ is the area of the inner surface of the tube and $\bar{h}_{tube}$ is the mean convective heat transfer coefficient

$$\bar{h}_{tube} = (\overline{Nu}_{tube}\, k_{HTF})\, D_{i,tube}{}^{-1} \tag{19}$$

with k_{HTF} the thermal conductivity of the HTF; $D_{i,tube}$ the inner diameter of the tube; $\overline{Nu}_{tube}$ the mean Nusselt number evaluated using the correlation for thermally fully developed laminar flow in ducts, the Reynolds number being lower than 2300.

The term $\dot{V}_{HTF}\, \rho_{HTF}\, c_{HTF}\left(T_{HTF,out} - T_{HTF,in}\right)$ in equation (17) represents the heat rate delivered to the coil heat exchanger, Q_{useful}.

Each collector has circular distributor pipes, with a diameter of $2.2 \cdot 10^{-2}$ m, in the HTF entrance and exit sections. Their contribution to the heat transfer from the absorber to the HTF was also considered.

The energy balances on the overall collector and on the glass give, respectively,

$$Q_{abs} = Q_{up} + Q_d + Q_{useful} + \sum_k V_k \rho_k c_k (dT_k/dt) \quad (20)$$

$$Q_{up} = Q_{conv,up} + Q_{rad} + V_{glass} \rho_{glass} c_{glass} (dT_{glass}/dt) \quad (21)$$

with V_k, ρ_k, c_k and T_k the volume, the density, the specific heat and the temperature of each component of the collector (absorber plate, back box cover, tube, air). The thermal capacitance of each component in equation (20) has been modelled for each component of the collector. The reference temperatures for the tube, the insulation and the air layer have been chosen equal to the absorber plate temperature, the arithmetic mean of the absorber and the box back cover temperatures, and the arithmetic mean of the absorber plate and the glass temperatures, respectively. The first order backward discretization of the time derivative has been applied.

At each time-step, the system of equations (1, 3, 5, 10, 12, 15, 17, 20, 21) is solved implicitly. The input temperature of the HTF at the inlet of the first collector is given by the coil heat exchanger simulation, while the HTF temperature at the inlet of the second collector is assumed equal to the HTF temperature at the outlet of the first collector. The model outputs are the time-variable temperatures of all the collectors components and the time-variable HTF temperature at the exit of the collectors tubes.

2.2. Immersed coil heat exchanger model

A transient one-dimensional model has been developed for the evaluation of the temperature field in the HTF flowing through the immersed coil heat exchanger. The data of the enameled crude steel coil heat exchanger that has been considered in the present study are reported in Table 2.

The heat exchanger has been subdivided into I isothermal nodes, having the same volume. The energy balance equation for the ith node is

$$\rho_{HTF,i}\ c_{HTF,i}\ V_{HTF,i} \frac{dT_{HTF,i}}{dt} = \frac{T_{w,tank} - T_{HTF,i}}{R_{coil}} + \rho_{HTF,i}\ c_{HTF,i}\ \dot{V}_{HTF,i} \left(T_{HTF_{in,i}} - T_{HTF_{out,i}} \right) \quad (22)$$

where $\rho_{HTF,i}$ and $c_{HTF,i}$ are the density and the specific heat of the HTF at the ith node, respectively; $V_{HTF,i}$ is the volume of the ith node; $T_{w,tank}$ is the time and space varying temperature of the water inside the tank; $\dot{V}_{HTF,i}$is the volumetric flow rate in the ith node; $T_{HTF_{in,i}}$ and $T_{HTF_{out,i}}$ are the time varying HTF inlet and outlet temperatures at the boundaries of the ith node, respectively;

The thermal resistance between the HTF in the coil heat exchanger and the water in the tank, R_{coil}, is

$$R_{coil} = R_{conv,i_{coil}} + R_{conv,e_{coil}} + R_{cond} \quad (23)$$

The internal convective thermal resistance, $R_{conv,i_{coil}}$, is given by

$$R_{conv,i_{coil}} = 1/\left(A_{i,coil}\ \bar{h}_{conv,i_{coil}} \right) \quad (24)$$

Table 2. Immersed coil heat exchanger data.

Thermal conductivity	40.0 W/m K
Inner diameter	$3.0 \cdot 10^{-2}$ m
Outer diameter	$3.4 \cdot 10^{-2}$ m
Helix diameter	$44.0 \cdot 10^{-2}$ m
Helix height	$75.0 \cdot 10^{-2}$ m
Helix length	16.63 m
Helix outer area	1.76 m^2

where $A_{i,coil}$ is the area of the internal surface of each node of the coil and $\bar{h}_{conv,i_{coil}}$ is the mean internal convective heat transfer coefficient

$$\bar{h}_{conv,i_{coil}} = \left(\overline{Nu}_{i_{coil}} k_{HTF}\right)/D_{i,coil} \tag{25}$$

with $\overline{Nu}_{i_{coil}}$ the mean Nusselt number, evaluated using the Gnielinski's correlation for a coiled tube heat exchanger [13], and $D_{i,coil}$ the inner diameter of the coil.
The external convective thermal resistance, $R_{conv,e_{coil}}$, is given by

$$R_{conv,e_{coil}} = 1/\left(A_{e,coil}\, \bar{h}_{conv,e_{coil}}\right) \tag{26}$$

where $A_{e,coil}$ is the area of the external surface of each node of the coil and $\bar{h}_{conv,e}$ is the mean external convective heat transfer coefficient

$$\bar{h}_{conv,e_{coil}} = \left(\overline{Nu}_{e_{coil}} k_{w,tank}\right)/D_{e_{coil}} \tag{27}$$

with $D_{e_{coil}}$ the outer diameter of the coil, $k_{w,tank}$ the thermal conductivity of the water inside the tank and $\overline{Nu}_{e_{coil}}$ the mean Nusselt number evaluated with the Morgan's correlation for natural convection over horizontal cylinders [14].

The conductive thermal resistance of the coil has been neglected, since it is negligible compared with the internal and external convective thermal resistances.

For each time-step and for each node, equation (22) is solved using the implicit Euler method. The input for the numerical simulation of the HTF inside the coil are the HTF temperature at the inlet section, that has been considered equal to the HTF temperature at the outlet of the second collector, and the temperature distribution of the water inside the tank. The output of the model is the time-variable HTF outlet temperature at each node.

2.3. *Heat storage tank model*

A vertical cylindrical heat storage tank has been considered in the present study. The inner tank diameter is 0.58 m and its height is H_{tank} = 1.5 m, with a 0.40 m^3 volume. The thermal insulation consists of a 50 mm thick polyurethane layer. The inlet section of the immersed coil heat exchanger is located on the tank side wall, 0.75 m above the tank bottom, while the outlet section is at the tank bottom. No water inlet or outlet and no auxiliary heaters have been simulated. Figure 3 shows the temperature array used for the simulation of the storage tank.

The water is subdivided into J isothermal nodes, each representing an equal volume water layer. The energy balance for the jth node can be written as

$$\rho_{w,j}\, c_{w,j}\, V_{w,j} \frac{dT_{w,j}}{dt} = \frac{T_{HTF} - T_{w,j}}{R_{coil}} + Q_{cond} - \frac{T_{w,j} - T_{amb}}{R_{tank}} \tag{28}$$

where $\rho_{w,j}$, $c_{w,j}$, $V_{w,j}$ and $T_{w,j}$ are the density, the specific heat, the volume, and the temperature of each water layer, respectively; T_{HTF} is the time and space varying temperature of the water inside the coil; Q_{cond} is the conductive heat rate between the jth layer and the adjacent water layers, that is the sum of the following two contributions

$$Q_{cond} = \frac{\pi D_{i_{tank}}^2}{4} \frac{k_{w_{tank}}}{\Delta_l} \left(T_{j-1} - T_j\right) + \frac{\pi D_{i_{tank}}^2}{4} \frac{k_{w_{tank}}}{\Delta_l} \left(T_{j+1} - T_j\right) \tag{29}$$

where $D_{i,tank}$ is the inner tank diameter and Δ_l is the distance between two adjacent nodes.
The thermal resistance of each layer of the tank wall, R_{tank}, was evaluated as

$$R_{tank} = R_{conv,i_{tank}} + R_{cond_{tank}} + R_{conv,e_{tank}} \tag{30}$$

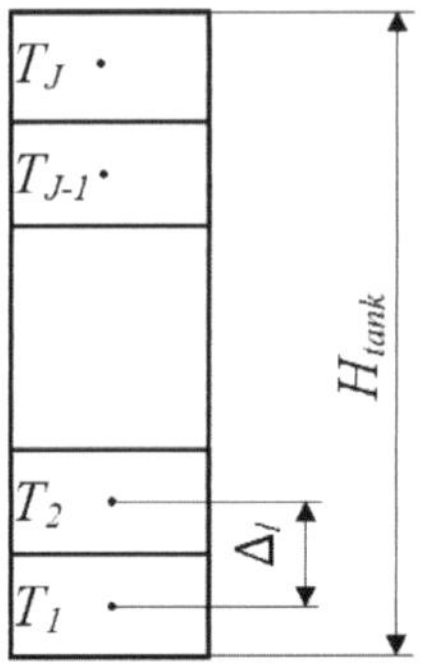

Figure 3. Temperatures array for the heat storage tank simulation.

where the inner convective resistance, $R_{conv,i_{tank}}$, is

$$R_{conv,i_{tank}} = 1/\left(A_{i_{tank}}\, \bar{h}_{conv,i_{tank}}\right) \tag{31}$$

with $A_{i_{tank}}$ the area of the inner lateral surface of each layer and $\bar{h}_{conv,i_{tank}}$ the mean convective coefficient at the inner tank wall, which was evaluated with the correlation for natural convection on a vertical flat plate.
The conductive thermal resistance of the tank wall, $R_{cond_{tank}}$, was evaluated as

$$R_{cond_{tank}} = \ln\left(r_{o_{tank}}/r_{i_{tank}}\right)/\left(2\,\pi\, k_{ins_{tank}}\, \Delta_l\right) \tag{32}$$

where $r_{o_{tank}}$ and $r_{i_{tank}}$ are the outer and the inner radius of the tank, respectively, and $k_{ins,tank}$ is the thermal conductivity of the insulation. The tank thickness was assumed to be equal to the insulation thickness, since the conductive thermal resistance of the tank steel was neglected.
The convective thermal resistance at the external wall of the tank, $R_{conv,e_{tank}}$, is given by

$$R_{conv,e_{tank}} = 1/\left(A_{e_{tank}}\, \bar{h}_{conv,e_{tank}}\right) \tag{33}$$

where $A_{e_{tank}}$ is the external lateral area of each layer of the tank and $\bar{h}_{conv,e_{tank}}$ is the mean convective heat transfer coefficient at the external wall of the tank, that was set equal to 5 W/m^2 K. For each time-step, the system of energy balance equations (29) is solved using the implicit Euler method. The empirical reversion-elimination algorithm [15, 16] was adopted, in order to account for the effects of natural convection between the water layers at different heights in the thermal stratification inside the tank. Indeed, at each time-step, should an unstable gradient arise, namely when a node temperature becomes higher than the temperature of the adjacent above node, the mixing of the fluid in the two layers occurs. The mixing process continues until the above layer is warmer than the below mixed layers. Then, the algorithm goes bottom-down to allow layers to mix with the warmer layers below. The input for the heat storage tank model are the ambient temperature and the HTF temperatures.

2.4. Components coupling

The solar DHW system was simulated numerically by means of a Matlab home-made code. Heat losses through the piping walls have been neglected. The system components were coupled with an iterative approach. At each time step, the first collector is simulated by using as input temperature at its inlet section the HTF temperature at the exit of the coil heat exchanger at the previous time-step. Then, the second collector is simulated using as input temperature at its inlet section the updated value of the HTF temperature at the first collector outlet. The coil heat exchanger is simulated using as input temperature at its inlet section the HTF temperature at the exit section of the second collector, given by the above described simulation of the collectors; the temperature distribution in the tank water is that evaluated in the previous time step. Finally, the tank is simulated using the updated temperatures of the HTF inside the coil heat exchanger. Then the process restarts from the collectors simulation using the updated temperatures and goes on till convergence is attained.

3. Simulations results

The results of a one-day simulation of the solar DHW system are presented in the following,. The input time-varying ambient temperature and solar irradiance, measured at the ENEA Portici RC, as a function of the hours of the day are reported in figure 4. The HTF flow rate was set equal to $6.0{\cdot}10^{-5}$ m^3/s and a 60 s time step was used. The initial temperatures of the collectors components were set

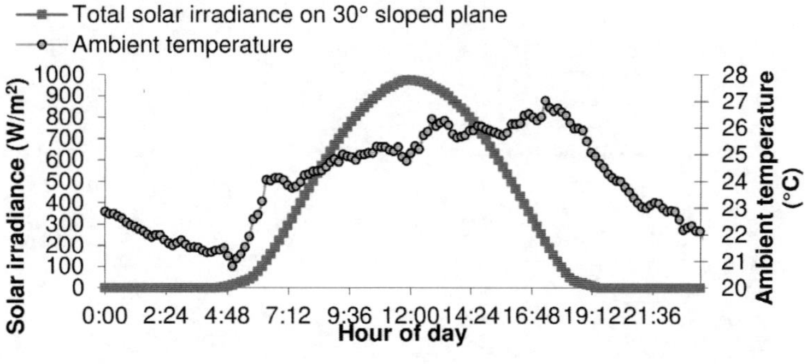

Figure 4. Solar irradiance and ambient temperature *vs.* hour of the day.

equal to the ambient temperature at midnight, while the initial temperatures of the water in the tank and of the HTF in the heat exchanger were chosen equal to 18°C, that is the average temperature of mains water in July at Portici.

The HTF temperature in the heat exchanger inlet and outlet sections as well as the mean water tank temperature as a function of the hour of the day are reported in figure 5. Since the heat losses through the piping system have been neglected, the HTF temperature at the heat exchanger inlet coincides with the HTF temperature in the second collector outlet section and the HTF temperature in the heat exchanger outlet section is the same as the HTF temperature at the first collector inlet section.

As it was to be expected, the HTF and water temperatures undergo a strong increase after the sunrise and, due to the thermal inertia of solar collectors, the maximum value of the HTF temperature in the inlet section of the heat exchanger is attained at about 14.00. It is worth noticing that the temperature of the HTF in the inlet section of the heat exchanger and the mean temperature of the water tank become equal at about 18.00. Afterwards, till midnight, the HTF circulation is switched off to avoid heat transfer from the heat storage tank to the solar collectors, and the HTF temperature in the heat exchanger is assumed equal to the mean water temperature. The slight decrease in the mean water temperature when there is not the HTF circulation is due to the heat losses through the tank walls.

Figure 6 shows the collectors components temperatures, the mean HTF temperature in the tube, and the ambient temperature during the day. Before sunrise, the temperatures of the external components of each collector, namely the box and the glass, are nearly equal to the ambient temperature. The

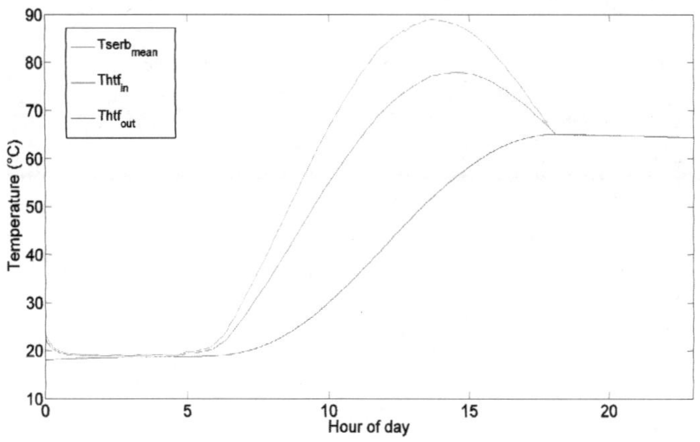

Figure 5. HTF temperature in the heat exchanger inlet and outletsections and mean water tank temperature *vs.* the hour of the day.

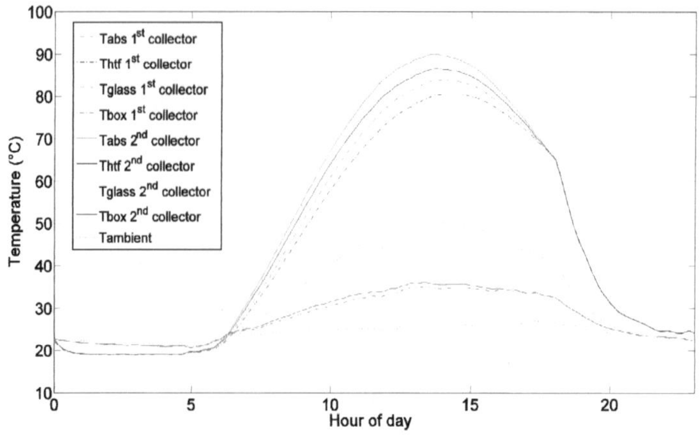

Figure 6. Collectors components temperatures, mean HTF temperatures in the tube, and ambient temperature *vs.* the hour of the day.

temperatures of the absorbers and of the heat transfer fluids are lower than the ambient temperature, since the initial temperature of the tank water is lower than the ambient temperature, and also because of the thermally resistive layers in the collectors. The above layers consist of the insulation and the air gap that separate the tube and the absorber from the box and the glass, respectively. The hour at which the HTF circulation switches off can be easily seen in figure 6, which points also out a strong decrease in the temperature of the collector components due to the energy losses to the ambient.

The useful heat rate and the heat rate dissipated at the top and bottom sides of the collectors are reported in figure 7. It can be noticed that the useful heat rate attains the maximum value at the same time, when the solar irradiance is the maximum. Figure 5 showed that the HTF temperature in the inlet section of the first collector becomes slightly higher than the outlet temperature in the second collector at the switch-off time. This is the reason why the useful heat rate becomes negative at the switch-off time, as shown in figure 7. Due to the larger heat losses from the second collector, the useful heat rate is higher in first collector until 16.30. Afterwards the useful heat rate in the second collector becomes slightly higher, due to the larger thermal energy stored in the absorber plate and pipes.

The collectors efficiency, that is the ratio of the collector useful heat rate to the solar irradiation on

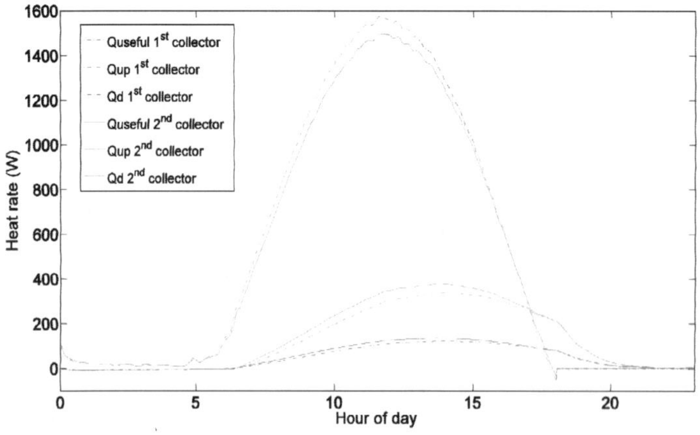

Figure 7. Collectors useful and dissipated heat rates *vs.* the hour of the day.

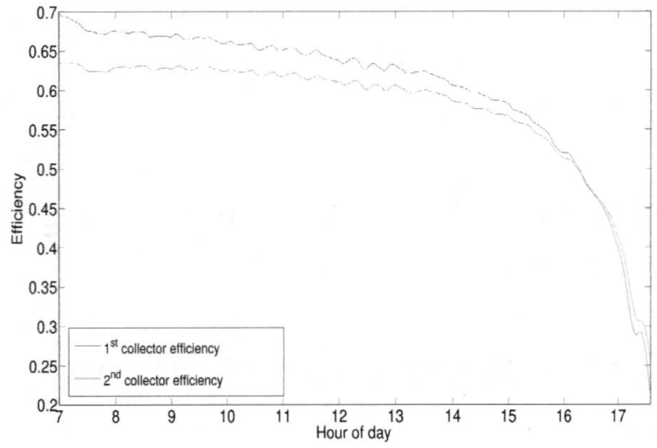

Figure 8. Collectors efficiency.

the collector, as a function of the hour of the day, is reported in figure 8. It can be noticed that for the same reason, again at about 16.30 the efficiency of the second collector becomes higher than that of the first collector.

4. Conclusions

The numerical simulation of a solar DHW system, made up by two flat plate solar collectors, a vertical cylindrical water tank for the heat storage and an immersed coil heat exchanger, has been addressed. Mathematical models and numerical schemes for each system component, described in the paper, were implemented in a Matlab home-made code, allowing the numerical simulation of the system.

Results of the simulation performed using as input data the ambient temperature and the solar irradiance at the ENEA Portici Research Center in a summer day were presented and discussed.

5. References

[1] Rodrígues-Hidalgo M C, Rodríguez-Aumente P A, Lecuona A, Legrand M and Ventas R 2012 Appl. Energy **97** 897-906.
[2] Dehghan A A and Barzegar A 2011 Energy Conversion and Management **52** 468-76.
[3] Kenjo L, Inard C and Caccavelli D 2007 Appl. Thermal Engineering **27** 1986-95.
[4] Klein S A, Duffie J A and Beckman A 2010 J. Engineering Gas Turb.Power **96** 109-13.
[5] Rodrìguez-Hidalgo M C, Rodriguez-Aumente P A, Lecuona A, Gutìerrez-Urueta G L and Ventas R 2011 Appl. Thermal Engineering **31** 2394-404.
[6] Rodrìguez-Hidalgo M C, Rodriguez-Aumente P A, Lecuona A, Gutìerrez-Urueta G L and Ventas R 2011 Appl. Thermal Engineering **31** 2385-393.
[7] Cadafalch J 2009 Solar Energy **83** 2157-64.
[8] Logie W R and Frank E 2013 J. Solar Energy Engineering **135** 1-7.
[9] Chauvet L P and Nevrala D J 1993 Appl. Energy **44** 283-314.
[10] Bejan, A and Kraus A D 2003 *Heat Transfer Handbook*, John Wiley & Sons Inc. Hoboken, NJ
[11] Hollands K G T, Unny T E, Raithby G D and Konicek L 1976 J. Heat Transfer **98** 189–193.
[12] Ashrae HVAC 1999 *Application Handbook*,.
[13] Gnielinski Y 1986 *Heat Transfer 1986*, San Francisco, USA, pp. 2847-54.
[14] Morgan V T 1975 *Adv. Heat Transfer*, vol 11, pp. 199-264.
[15] Mather D W, Hollands K G T and Wright J L 2002 Solar Energy **73** 3-13.
[16] Newton B J, Schmid M, Mitchell J W and Beckman W A 1995 Proc. ASME/JSME/JSME Int. Solar Energy Conf., March 19-24 Maui, Hawaii. ASME, New York, Part 2; 1111.

Journal of Physics: Conference Series **547** (2014) 012016 doi:10.1088/1742-6596/547/1/012016

Quantitative thermography for the estimation of the U-value: state of the art and a case study

Iole Nardi, Stefano Sfarra, Dario Ambrosini

Department of Industrial and Information Engineering and Economics, University of L'Aquila, via G. Gronchi 18, I-67100 L'Aquila, Italy

E-mail: iole.nardi@graduate.univaq.it

Abstract. Energy consumption of buildings could be significantly reduced by improving the efficiency of the envelope. Currently, the estimation of the energy performance of existing buildings requires the knowledge of the overall heat transfer coefficient (U-value) of the walls. U-values can be calculated through a theoretical approach, knowing the thermal conductivity and thickness of each material that constitutes the wall stratigraphy, from project data or coring. Alternatively, U-values can be obtained experimentally, through the ISO recommended heat flow meter measurements. Although generally accepted, the heat flow meter method suffers from some disadvantages. Recently, an alternative approach based on infrared thermography (IRT) has been proposed for in situ measurements. Main advantages of this new approach are non invasivity and the possibility of inspecting relatively large areas in real time. In this paper, after a brief description of the state of the art in the field of U-value measurement by IRT, a case study is described. In particular, the results obtained by IRT on an existing building are compared with U-values given by the standard ISO calculation and heat flow meter measurements; advantages and limitations of the new method are outlined. Some suggestions for a successful exploiting of the IRT approach are also given.

1. Introduction

According to IEA data, residential buildings are responsible of about 32% of total final energy consumption; in terms of primary energy consumption, they represent around 40% in most IEA countries.

Heat transfer through a composite wall, a key parameter for the assessment of the energy sustainability of the whole structure, can be described by a single number, the U-value (in Europe) or its reciprocal, the R-value (in the USA).

The estimation of thermal transmittance (U-value) is held through two methodologies, both standardized.

The first approach [1, 2], is based on the knowledge of wall stratigraphy, gathered from project data, coring or endoscopic surveys through the wall, or by similarity to coeval buildings.

This method, however, is quite uncertain: it is not always possible to know the wall stratigraphy from projects, because they are often missing. Furthermore, for historical buildings, often of cultural interest, it is not possible to carry out the destructive surveys mentioned above to identify the stratigraphy.

Considering that more than 60% of world historical constructions is located in Italy, the optimization of the energy efficiency of this architectural heritage is very important.

Moreover, conductivity values provided in literature often do not match with the real one, because of aging and installation features.
The second approach, normalized by ISO 9869 [3], uses a heat flow meter (HFM), an instrument able to calculate instantaneous and averaged value of transmittance as ratio between heat flux through the wall and the difference between interior and exterior wall surface temperature, after an acquisition time of at least 72 hours (more for high thermal inertia walls).
One of the disadvantages of this method is that it gives only punctual values, so the transmittance determination in another point requires another measurement campaign. Furthermore, in order to have reliable values, a minimum gradient of 10 °C between indoor and outdoor temperature is required.
The measured values can also be influenced by thermal bridges, humidity, mold and partial adhesion of the sensors to the wall (especially if the wall façade is rough).
In conclusion, the experimental values obtained by the heat flow meter method are often very different from the calculated ones. This mismatch is particularly important for historical buildings, for which the differences between standard and measured U-values can vary from 2% to 58%, depending on the composition of the ancient walls [4].
In the last years, induced by the advances in infrared detection technology and infrared imaging applications [5, 6], the use of infrared thermography (IRT) for applications in buildings has been proposed by many authors [7-12].
IRT is widely accepted as a tool for the analysis of the energy efficiency of new and existing buildings in a qualitative way [13, 14] but, recently, several research groups proposed a method based on quantitative thermography for the in situ measurement of wall transmittance.
Albatici and Tonelli [15, 16], for evaluating wall transmittance, use a relation that involves parameters measured by IRT, except for the wind velocity (measured by a hot-wire anemometer in proximity of the wall). They also give an experimental procedure to measure the surface emissivity in the spectrum range related to the thermography. They reported three case studies [16]: differences with U-values determined by HFM are of the order of 30%, while differences with calculated U-values (ISO 6946) can grow up to 80%. Environmental conditions are very important for a reliable measurement: direct solar irradiation must be avoided; wind speed must be lower than 1m/s (best condition is lower than 0.2 m/s) and the difference between inner and outer temperature during the test realization must be of at least 10–15°C.
In the earlier work by Madding [17], a better matching between measured and calculated values (within 13%) is reported. No comparison is given with HFM results as well as no details about the thermo physical parameters involved in the measurements. According to Madding's uncertainty analysis, the most sensitive variables are the emissivity and the difference between the reflected apparent temperature and the wall temperature.
Grinzato *et al.* [18] proposed a more rigorous and complex procedure: a light metallic frame is inserted near the wall and in the field of view of the thermal camera; the frame holds several targets on which particular parameters, useful for accurate measurements of wall and air temperatures, are recorded. Then the entire system is scanned by IRT in an automatic way. Reported experimental results show a difference with U-values determined by HFM of the order of 30-35%.
Fokaides and Kalogirou [19] examined 5 case studies. They refer that U-values by IRT are within 10-20% of notional values (ISO 6946). For one case study, they also made a comparison with HFM value, the difference being well within 10%.
More recently, Dall'O *et al.* [20] presented an extensive investigation on different buildings. They reported that percentage absolute deviation between the notional and the measured U-values for IR thermography has an acceptable level, being within 40-45%, as long as suitable cautions are taken. Their survey also showed that IRT evaluation is more reliable for solid-mass walls, for which the error percentage is lower if compared to HFM method.
In the following, a case study is described. In particular, the results obtained by IRT are compared with U-values given by the standard ISO calculation and heat flow meter measurements; advantages

Journal of Physics: Conference Series **547** (2014) 012016 doi:10.1088/1742-6596/547/1/012016

and limitations of the new method are outlined. Some suggestions for a successful exploiting of the IRT approach are also given.

2. Methodology

There isn't a normative concerning the U-value estimation by means of IRT, since this technique is still a topic of study by several research groups, as explained in section 1.

In this study, transmittance is calculated approximating the overall heat exchanged through the wall with the heat exchanged with the outdoor by radiation, that is [16]:

$$U = \frac{5.67\,\varepsilon_{tot}\left[\left(\frac{T_w}{100}\right)^4 - \left(\frac{T_{out}}{100}\right)^4\right] + 3.8054\,v\,(T_w - T_{out})}{(T_{\text{int}} - T_{out})} \qquad (1)$$

where T_{int} is the indoor environment temperature [K], T_{out} is the outdoor environment temperature [K], T_w is the surface temperature of the element [K], ε_{tot} is the emissivity on the entire spectrum and v is the wind velocity.

All the values involved in Equation (1) can be estimated through IRT.

First of all, mean apparent reflected temperature (defined as "the apparent temperature of other objects that are reflected by the target into the infrared thermography camera" [21]) must be calculated. This can be accomplished, as suggested in [21], by using a high reflective surface, like an aluminum foil stretched after crumpling. The foil must be located near the wall, with the reflective leaf towards the outside: if the IR camera emissivity is set to equal 1, the mean foil temperature is the reflected ambient temperature.

The second parameter to be evaluated is emissivity. In Equation (1) total emissivity, gathered from literature, is involved, since the aim is to calculate the whole radiant energy emanated from the surface. Bearing in mind that wall temperature is determined through a tool (IR camera) that works between specific wavelengths, it is clear that for detecting wall temperature it is necessary to know walls' spectral emissivity, that strongly depends on its roughness and conditions (like humidity and pollution). Wall spectral emissivity can be evaluated using target of known emissivity higher than 0.9 located on the wall. After a suitable time, needed to the target for reaching wall temperature, a thermogram is acquired setting on the IR camera target emissivity.

Target and wall temperature can be calculated using a proprietary software (in our case, ThermaCAM Researcher Pro 2.8) and through the placement of two spots. Theoretically, those temperatures should be the same, since the target, well sticked to the wall, has the wall temperature.

However, target and wall have different emissivities, so two different values of radiative energy arrive to the thermal camera sensor, and they are converted, at the same emissivity (target emissivity, set on thermal camera) , in two temperature values. In fact, target shape can be clearly individuated. Then, the emissivity set on IR camera is varied till the wall temperature equals the one previously detected on the target. This procedure, once again, complies with international standard [21].

Indoor temperature has been measured using thermography, as suggested by Albatici and Tonelli [16], through a partial opening of a window, that can be assimilated to a black body.

Outdoor temperature can be assessed setting emissivity equal to 1 and detecting temperature of a black PVC tube left outside (to reach outdoor temperature), with a tapped side and whose length is 5-6 time the opening diameter (which is comparable to the diameter of the thermal camera lens).

The wind velocity v can be measured near the wall by a hot-wire anemometer. In the following experimental campaigns, we performed the measurements in quiet days, therefore a wind velocity $v \approx 0$ m/s was assumed.

3. Case study

As a case study, an university small building was considered: the "Solar house", placed in Monteluco di Roio (960 m a.s.l.), district of L'Aquila (climatic zone: E; degree days: 2514).

The building (category E.7 according to D.P.R. 412/93) was built in the 1970s; it has a masonry of perforated bricks, coated with an insulating layer of polystyrene after the refurbishment consequent to 2009 earthquake, and painted green. Details are given in Table 1, while the building and its stratigraphy are shown in Figure 1. The design (ISO 6946) U-value is U = 0.31 W $m^{-2}K^{-1}$.

Table 1. Case study stratigraphy.

Layer	Thickness [m]	Conductivity [W $m^{-1}K^{-1}$]	Thermal resistance for unit area [m^2K W^{-1}]
Outside plaster	0.02	0.9	0.022
EPS layer	0.12	0.054	2.22
Bricks	0.25	0.3	0.833
Inside plaster	0.01	0.9	0.011

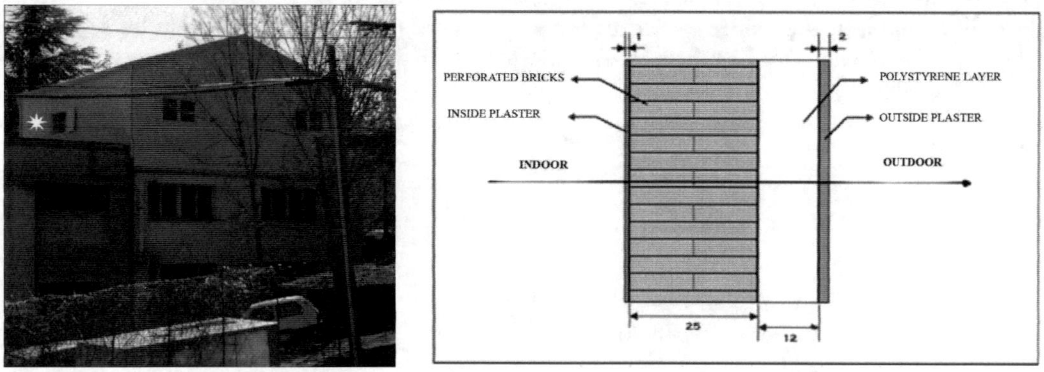

Figure 1. Case study "Solar house"; exterior of the building (with the asterisk marking the wall under test, North – North East façade) and stratigraphy.

4. Experimental results

The acquisition of thermograms has been carried out by an infrared camera model ThermaCAM S65 HS from FLIR System, equipped with a focal plane array uncooled microbolometer. Some technical specifications are listed in Table 2.

Table 2. Main technical specifications of ThermaCAM S65 HS.

IFOV	Spectral range	Image size	Accuracy	Thermal sensitivity
1.1 mrad	7,5 ~ 13 μm	320 x 240 pixels	± 2 ° C or ± 2% of reading	< 0.05 °C with respect to 30 °C blackbody temperature

At first, a thermal inspection of outdoor walls has been performed in order to detect heat losses and the possible presence of thermal bridges or surface humidity.
The main difficulty for in-situ examination is linked to non-stationary boundary conditions. Due to changes of ambient temperatures, not only thermal conductivity but also specific heat and density of materials vary, and this influences in a significant way the heat transfer through buildings envelope. The influence on non-stationary conditions may be, to some extent, reduced by making the inspection during a period carefully selected and characterized by limited ambient parameters variations.
To quantify the thermal field, a series of parameters have been acquired, according to the procedure described in section 2.
The ambient parameters have been monitored for checking and comparison during the measurements with thermo-hygrometric sensors (KIMO HD100 and with HOBO®).
U-values determined by IRT have been compared with U-values measured by HFM, according to ISO 9869; the thermopile heat flow sensor (ESR240, uncertainty 5% on 12 hrs measurement) and the surface thermocouples (EST124, uncertainty 0.15 °C) were by LSI Lastem, while the measurements were recorded by M-LOG Mini data logger (LSI Lastem). Data were processed by INFOflux program.
For the case study, a long measurement campaign was carried out. Taking into account the preliminary thermal inspection, the HFM was positioned far from thermal bridges and non homogeneous regions of the wall. The measured data were analyzed by the average method [3].
Table 3 shows the results obtained by HFM; percentage absolute deviations between the design and the HFM U-values range from 11% to 29%, while the value averaged over the whole period has a deviation of about 14%.

Table 3. U-values obtained by HFM.

Period	U-value [$Wm^{-2}K^{-1}$]
08 Feb – 11 Feb 2013	0.276
11 Feb – 14 Feb 2013	0.392
14 Feb – 17 Feb 2013	0.356
17 Feb – 20 Feb 2013	0.375
20 Feb – 22 Feb 2013	0.400
08 Feb – 22 Feb 2013	0.354

Then U-values have been determined by IRT and thermal images have been processed in MATLAB® environment. Figure 2 shows the results obtained on 22nd Feb 2013, a cloudy day.

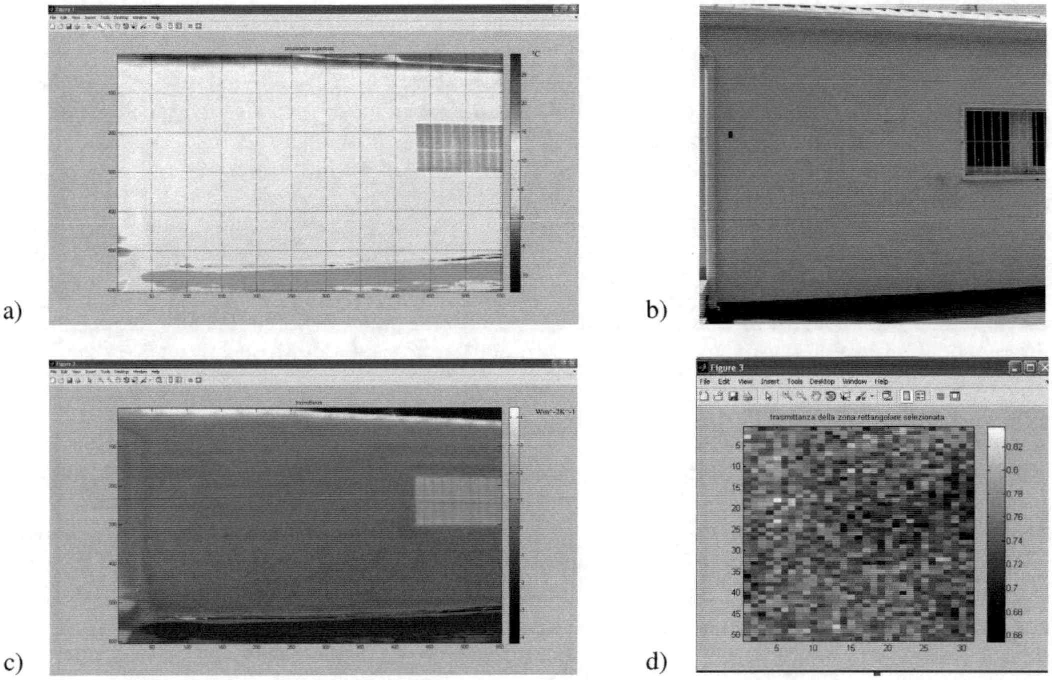

Figure 2. Case study: (a) surface temperatures; (b) visible image; (c) U-values calculated according to Equation 1; (d) U-values in the selected area.

A homogeneous area of about 20 cm × 20 cm, close to the known emissivity label and near the position on which the heat flow sensor was mounted, was selected. U-value has been calculated for each pixel of the selected area (see Figure 2 d)) and these values have been averaged to estimate the wall transmittance; this, to avoid to re-set the emissivity for the calculation of wall temperature underneath the label, and therefore to avoid possible contact resistance.

Measurement was repeated on 27th Feb 2013, a sunny day.

Results are shown in Table 4. Δ_1 and Δ_2 are the percentage deviations with respect to the design U-value (0.31 $Wm^{-2}K^{-1}$) and the average HFM U-value (0.354 $Wm^{-2}K^{-1}$); the deviations range from very good (about 2%) to acceptable (about 37%).

The difference between outdoor and reflected temperature is equal to 3.6 °C (22nd Febr.) and 11.59 °C (27th Febr.).

Table 4. U-values obtained by IRT.

Date	U-value [$Wm^{-2}K^{-1}$]	Δ_1	Δ_2
22 Feb 2013	0.360	16.12%	1.69%
27 Feb 2013	0.222	28.38%	37.28%

5. Conclusions

In this paper, after a brief description of the state of the art in the field of U-value measurement by IRT, a case study is described. The results obtained by IRT on an existing building are compared with U-values given by the standard ISO calculation and heat flow meter measurements.

Measurement campaign involved an extensive investigation by HFM in different weather conditions. Then, measurements by IRT were performed.

Another building, of recent construction and high energy efficiency, has been investigated too. In this case, it was not possible to perform a long HFM measurement campaign, and IRT has been applied only once, being the building a private one. For this reason, the case has not been discussed in this work, although preliminary results show a more accurate transmittance estimation. In fact, percentage deviations of the U-value with respect to the design and the average HFM ones range, respectively, from very good (less than 1%) to good (within 12%).

According to our experience, IRT can give reliable results if a series of conditions (some of them outlined in literature) are fulfilled.

In particular, the surface temperature of the wall should be greater than the outdoor environment temperature (at least 2 °C) and the difference between indoor environment temperature and outdoor environment temperature should be great. The measurement of the spectral emissivity should be accurate and the weather conditions should be stable.

The next step would consist in experimental measurements in a controlled environment to fully characterize the influence of different parameters on the final results.

References

[1] UNI TS 11300: 2008 Prestazioni energetiche degli edifici

[2] EN ISO 6946: 2007. Building components and building elements. Thermal resistance and thermal transmittance Calculation method

[3] ISO 9869:1994 Thermal insulation. Building elements. In-situ measurement of thermal resistance and thermal transmittance

[4] Adhikari RS, Lucchi E and Pracchi V 2012 Experimental Measurements on Thermal Transmittance of the Opaque Vertical Walls in the Historical Buildings *Proc PLEA2012 - 28th Conference, Opportunities, Limits & Needs Towards an environmentally responsible architecture (Lima, Perù)*

[5] Maldague X 2001 *Theory and practice of infrared technology for nondestructive testing* (New York: Wiley)

[6] Bagavathiappan S, Lahiri BB, Saravanan T, Philip J and Jayakumar T 2013 Infrared thermography for condition monitoring – A review *Infrared Phys. Technol.* **60** 35

[7] Balaras CA and Argiriou AA 2002 Infrared technology for building diagnostics *Energy Build* **34** 171

[8] Grinzato E, Bison PG and Marinetti S 2002 Monitoring of ancient buildings by the thermal method *J Cult Herit* **3** 21

[9] Avdelidis NP and Moropoulou A 2004 Applications of infrared thermography for the investigation of historic structures *J Cult Herit* **5** 119

[10] Grinzato E, Cadelano G, Bison PG and Petracca A 2009 Seismic risk evaluation aided by IR thermography *Thermosense XXXI Proc. SPIE* **7299** 72990C

[11] Paoletti D, Ambrosini D, Sfarra S and Bisegna F 2013 Preventive thermographic diagnosis of historical buildings for consolidation *J Cult Herit* **14** 116

[12] Bisegna F, Ambrosini D, Paoletti D, Sfarra S and Gugliermetti F 2014 A qualitative method for

combining thermal imprints to emerging weak points of ancient wall structures by passive infrared thermography – A case study *J Cult Herit* **15** 199

[13] Kauppinen T 2009 The use of thermography in energy performance of buildings-directive (EPBD)-applications *Thermosense XXXI Proc. SPIE* **7299** 729908

[14] EN 13187: 1998 Thermal performance of buildings. Qualitative detection of thermal irregularities in building envelopes. Infrared method

[15] Albatici R and Tonelli AM 2008 On site evaluation of U-value of opaque building elements: a new methodology *Proc PLEA 2008 - 25th Conference on Passive and Low Energy Architecture (Dublin, Ireland)*

[16] Albatici R and Tonelli AM 2010 Infrared thermovision technique for the assessment of thermal transmittance of opaque buildings elements on site *Energ Build* **42** 2177

[17] Madding R 2008 Finding R-values of stud frame constructed houses with IR thermography *Proc InfraMation 2008 (Reno, USA)*

[18] Grinzato E, Bison P, Cadelano G and Peron F 2010 R-value estimation by local thermographic analysis *Thermosense XXXII Proc. SPIE* **7661** 76610H

[19] Fokaides PA and Kalogirou SA 2011 Application of infrared thermography for the determination of the overall heat transfer coefficient (U-value) in building envelopes *Appl Energ* **88** 4358

[20] Dall'O G, Sarto L and Panza A 2013 Infrared screening of residential buildings for energy audit purposes: results of a field test *Energies* **6** 3859 www.mdpi.com/journal/energies

[21] ISO 18434-1:2008. Condition monitoring and diagnostics of machines – Thermography. Part 1: General procedures. Annex A "Field measurements of reflected apparent temperature and emissivity"

Journal of Physics: Conference Series **547** (2014) 012017 doi:10.1088/1742-6596/547/1/012017

Modeling operation mode of pellet boilers for residential heating

D Petrocelli, A M Lezzi

Dipartimento di Ingegneria Meccanica ed Industriale,
Università degli Studi di Brescia, via Branze 38, 25123 Brescia

E-mail: adriano.lezzi@unibs.it

Abstract. In recent years the consumption of wood pellets as energy source for residential heating has increased, not only as fuel for stoves, but also for small-scale residential boilers that produce hot water used for both space heating and domestic hot water.

Reduction of fuel consumption and pollutant emissions (CO, dust, HC) is an obvious target of wood pellet boiler manufacturers, however they are also quite interested in producing low-maintenance appliances. The need of frequent maintenance turns in higher operating costs and inconvenience for the user, and in lower boiler efficiency and higher emissions also.

The aim of this paper is to present a theoretical model able to simulate the dynamic behavior of a pellet boiler. The model takes into account many features of real pellet boilers. Furthermore, with this model, it is possible to pay more attention to the influence of the boiler control strategy. Control strategy evaluation is based not only on pellet consumption and on total emissions, but also on critical operating conditions such as start-up and stop or prolonged operation at substantially reduced power level. Results are obtained for a residential heating system based on a wood pellet boiler coupled with a thermal energy storage.

Results obtained so far show a weak dependence of performance – in terms of fuel consumption and total emissions – on control strategy, however some control strategies present some critical issues regarding maintenance frequency.

1. Introduction

In recent years the diffusion of boilers fueled by wood pellet has increased [1, 2]. On one hand some governments encourage usage of renewable energy sources by means of financial incentives and tax reduction. On the other hand, in some areas, especially areas not reached by the natural gas distribution network, pellet boilers are an alternative to coal, oil and LGP boilers.

End users and legislation have paid increasing attention to energy efficiency and pollutant emissions and that has prompted manufacturers to improve their products. Pellet boilers must comply the European Standard EN 303-5 which defines methods to determine the quality of combustion and thermal efficiency, and establishes also minimum requirements that must be satisfied by the appliances. This Standard classifies boilers in three classes and the best class requires an efficiency of at least $87+\log_{10}(\mathrm{Pn})$, where Pn is the nominal power expressed in kW. Limits on emissions and efficiency may be further restricted by national and regional legislation that regulate the access to the market or distribution of subsidies and tax incentives [3–5].

Results of tests to obtain certification show that over the last few years emissions of pollutants have reached a very low level [6]. These tests are performed according to the procedure specified by EN303-5, which prescribes to measure emissions and efficiency at steady continuous operation, both at the maximum and at the minimum power level (not larger than 30% of maximum power level), excluding transient periods during which emissions can be much higher [7–9].

Conversely, in everyday use, duration of transient conditions amounts to a considerable fraction of total operating time. Thermal load usually varies within a day and from day to day depending on weather and on user's usage. This implies that the boiler does not usually work at maximum power or that frequent changes from maximum to minimum power may occur or that start-ups and shutdowns may alternate several times. Reduction of fuel consumption and pollutant emissions (CO, dust, HC) is an obvious target of wood pellet boiler manufacturers, however they are also quite interested in producing low-maintenance appliances, because the need of frequent maintenance turns in higher operating costs and inconvenience for the user. In addition, factors that increase maintenance frequency are responsible for lower boiler efficiency and higher emissions. Some physical characteristics of the boiler, such as geometry of the combustion chamber and of the heat exchanger, have great influence on performances of course, but also control strategies of pellet boiler can influence consumption, emissions, ash production and fouling deposition.

In literature, efficiency and emissions of pellet boilers during real working conditions have been addressed through both experimental studies and numerical modeling. Measurement of pellet stove and boiler emissions in real conditions can be found in [9–12]. In all these works authors focus their attention also on emissions of stoves and boilers in transient conditions measuring emissions in real cases or realizing a test bench for boilers in which some conditions are simulated [7].

The study of performances in real conditions can be done by computer simulation upon performing a dynamic simulation of the whole heating system. Numerical simulation is useful for optimization of heating systems without the need of long experimental tests, in fact several variables, that influence the operation of the boiler, can be varied. In these simulations models of the heating system and of the house must be used. A few models for simulation of a pellet boiler are available. They calculate fuel consumption, heat transferred to water and emissions. Earlier models consider only steady state conditions, neglecting the typical long transient of these appliances. They compute efficiency and emission as a function of the power of the boiler only. Later on dynamical models that simulate also start-up and stop conditions have been developed [13–16]. The model proposed in [14], for instance, calculates the CO emissions as function of power for normal operation and a constant amount for each start and stop. These models are implemented to be used in TRNSYS or similar programs.

In this paper we present a dynamical model of a pellet boiler in which the control strategy can be modified. In this way it is possible to evaluate its influence on performances. After a description of the model, we present the preliminary results on the behavior of a pellet boiler for different control strategies. For the case presented here, the boiler is connected to the heating system through a thermal energy storage (water tank). Assuming a time dependent thermal load as input data, the time history of thermal energy storage temperature is computed. On the basis of the thermal energy storage temperature and of control strategies, the control system of the boiler determines whether to start-up or to stop the boiler, and controls the current power level. Several quantities are estimated on a weekly base, such as emissions, pellet consumption, number of start-ups and shutdowns, duration of operation at minimum power, etc, with particular regard to those that may influence maintenance frequency.

2. Characteristics of pellet boilers

Pellet boilers are appliances designed to produce hot water from combustion of wood pellet. Wood pellet derives from the compression in grains of sawdust of untreated wood. The main components of wood are mainly carbon, hydrogen, oxygen and water, therefore the products of complete combustion are CO_2, H_2O. The most relevant pollutant substances are CO, dust,

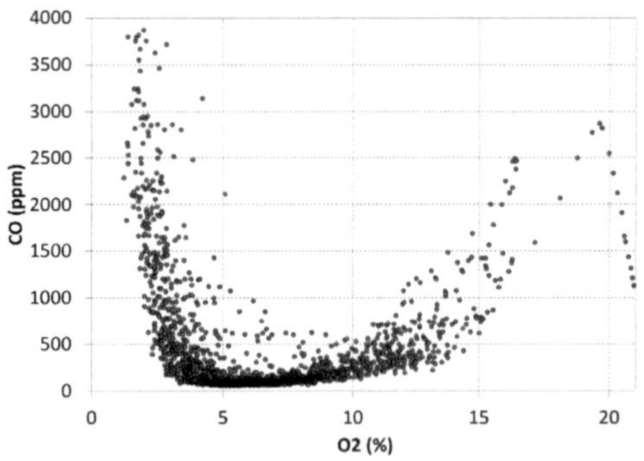

Figure 1. CO concentration in flue gas of a real pellet boiler as function of O_2 concentration for the whole cycle (from start up to switch off). The optimal range of O_2 is between 5% to 7%. [21]

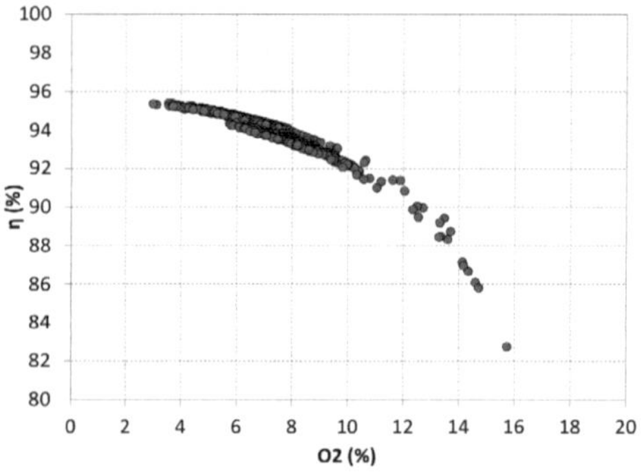

Figure 2. Efficiency of a real pellet boiler as a function of O_2 concentration for steady state regime. [21]

hydrocarbons (HC) and NOx. Their concentrations, in modern pellet boilers, are of the order of 10^{-2} g/m^3 of flue gas. Other substances with concentrations of the order of 10^{-5} g/m^3 of flue gas such as SOx, HCl, PCDD/PCDF Policyclic Aromatic Hydrocarbons (PAHs), may also be present [12, 17, 18], but they are currently neglected by standards and legislation. Also NOx is often omitted. Excess air is the parameter that has the greatest influence on pollutant emission [19, 20]. It has been shown that it exists an interval of excess air (i.e. interval of concentration of oxygen in flue gas) in which combustion takes place in optimal conditions, yielding a low concentrations of pollutant [21], as one can see in figure 1.

All modern pellet boilers are equipped with an electronic control system. On the basis of user's needs and of few measured temperatures, the control system regulates power, start-up and stop procedures, and other functions. In particular, it regulates the pellet mass flow rate and the combustion air mass flow rate.

The typical pellet feeding system of boilers is a screw auger conveyor. Pellet mass flow rate is never exactly constant [21]: consequently the instantaneous burning rate (i.e. the mass of pellet burned per unit time) fluctuates as well as the power supplied to water, the composition and temperature of flue gas, etc. Only quantities averaged over time intervals of a few minutes can be considered roughly constant.

The combustion air mass flow rate is regulated by controlling the speed of rotation of a blower that is usually placed at the flue gas outlet port.

An important component of pellet boiler is the heat exchanger. To reach high efficiency values it is necessary that flue gas transfers as much heat as possible to water, while the heat released to the room is considered as wasted.

Because flue gas of wood pellets is dirtier than that of LGP, for example, ducts inside the boiler must be large enough to avoid obstruction due to fouling, therefore compact flue gas-water heat exchangers cannot be used. This turns in large thermal inertia of pellet boilers, longer response time to a load demand change and the need of a periodic cleaning to remove bottom ashes and fouling deposits on exchange surfaces.

Effectiveness of heat exchanger is a decreasing function of flue gas mass flow rate $\dot{m}_f$ [21] and, as a consequence, of the excess air of combustion n. Since $\dot{m}_f = \dot{m}_p(1+n)$, where $\dot{m}_p$ is the pellet mass flow rate, high values of excess air n cause low effectiveness and should be avoided. At the same time excess air cannot be too low otherwise combustion quality would be poor. Typically, excess air is kept larger than 1.3-1.5, corresponding to an oxygen concentration

Journal of Physics: Conference Series **547** (2014) 012017 doi:10.1088/1742-6596/547/1/012017

in flue gas of about 6% . Figure 2 shows the actual efficiency of the boiler as a function of oxygen concentration calculated from experimental data [21].

A pellet boiler is characterized also by long start-up/shutdown processes. After the start signal of the control unit, it takes some minutes before the flame lights. Afterwards, it takes some other minutes before the rate of combustion reach the maximum.

With regard to the maintenance of boilers, high efficiency of exchange implies a low temperature of flue gas and this can be a problem because if flue gas temperature is too low, some substances, in particular unburnts, can condensate and attach on cold surfaces of the boiler or of the chimney, forming a layer of deposits (often called creosote). Deposits worsen the heat exchange and can obstruct the flow of gas. Deposition of creosote is also dangerous because it is flammable.

There are no law requirements about the exhaust gas temperature from the boiler: at the nominal power it usually ranges between 120 and 160°C [2,22,23]. But, for instance, at the minimum power the exchange area results over-designed and flue gas temperature is easily under 100°C. Furthermore, combustion occurs in non optimal condition because of the high excess air. All that turns into higher concentration of unburnt substances in flue gas and increasing rate of deposit build up on exchange surfaces and chimney walls. Working at the minimum power, as well as at start-up and at shutdown, are operating conditions characterized by low temperature of flue gas and high concentration of dust and HC, thus promoting formation of deposits on heat exchanger surfaces and reducing the time interval between maintenance operations.

3. Pellet boiler model

In the model presented here the boiler is composed by three submodules: the control unit, the combustion chamber, and the flue gas-water heat exchanger.

The control unit determines whether the boiler has to be started up or to be shut off, and regulates the nominal power, on the basis of water temperature T_t measured in the hot water tank. Two threshold values for T_t are introduced: $T_{t,lo}$ and $T_{t,hi}$. In this study they have been set equal to $T_{t,lo} = 60°C$ and $T_{t,hi} = 70°C$, respectively.

The control strategies considered here are the following:

- CS 1 When water temperature is increasing: the boiler works at maximum power until $T_t \leq T_{t,lo}$; it switches to minimum power (as in real case, 30% of the maximum power [24]) between the two thresholds, $T_{t,lo} < T_t \leq T_{t,hi}$; above the higher threshold, $T_{t,hi} < T_t$, the boiler is switched off. As water temperature decreases, the boilers either starts up again or resumes operating at maximum power when $T_t < T_{t,lo}$.
- CS 2 It is a modified version of CS 1: the minimum power is set equal to 70% of the maximum power.
- CS 3 It is a modified version of CS 2: as water temperature is increasing between $T_{t,lo}$ and $T_{t,hi}$, the boiler power is varied linearly between maximum power and minimum power (equal to 70% of maximum power).
- CS 4 When the boiler is on, it works at maximum power only: when water temperature in the hot water tank reaches the higher threshold, $T_{t,hi} < T_t$, the boilers is switched off. As water temperature decreases, the boilers starts up again when it falls below the lower threshold $T_t < T_{t,lo}$.
- CS 5 It is a modified version of CS 4. When water temperature in the hot water tank reaches the higher threshold, $T_{t,hi} < T_t$, the power is reduced to minimum (30% of the maximum power): if temperature remains higher than $T_{t,hi}$ for 5 minutes, the boilers is switched off, otherwise it keeps on operating at minimum power. As water temperature decreases, the boilers either starts up again or resumes operating at maximum power when $T_t < T_{t,lo}$.

Journal of Physics: Conference Series **547** (2014) 012017 doi:10.1088/1742-6596/547/1/012017

CS1, CS4 and CS5 are implemented in boiler produced by some manufacturer. On the basis of water temperature, the control unit computes the mass flow rate of pellet and combustion air. The fluctuations in time of pellet mass flow rate are modeled summing up an increment randomly distributed between ±10% of the nominal pellet flow rate.

Characteristic time delays of pellet boiler are taken into account: it takes 4 minutes from the start signal until the actual ignition of the flame; it takes additional 5 minutes from ignition until pellet flow rate reaches its maximum value.

Finally, as in real boilers, the control unit governs the pump that forces water through the boiler heat exchanger (P1 in figure 3): in particular, the pump is switched off when the outlet water temperature, $T_{b,o}$, falls below 40°C.

The combustion chamber submodule simulates the reactions of combustion. On the basis of imposed mass flow rate of air and pellet we compute the mass flow rate $\dot{m}_f$ and composition of flue gas and the adiabatic flame temperature T_{ad}. The pollutant substances taken into account are CO, PM, HC, NOx. They are calculated as a function of the excess air with these methods:

(i) CO and NOx by use of correlations based on experimental data measured during the operation of a real pellet boiler, including start-up and shutdown, and shown in figure 1 as a function of oxygen concentration in flue gas;

(ii) PM and HC by use of interpolation of data taken from [25, 26], and from [7] for start-up and shutdown.

In the heat exchanger the flue gas produced in the combustion chamber heats the water that returns from the hot water tank. Heat exchanger is characterized by its effectiveness ε which is calculated as a function of flue gas mass flow rate through the equation

$$\varepsilon = A\left(\dot{m}_f\right)^{-B} \tag{1}$$

coefficients A and B are chosen in order to match the experimental results. The outputs from the heat exchanger submodule are the outlet water temperature, $T_{b,o}$, and the flue gas temperature at the discharge, T_f. We use a plug flow model for the heat exchanger, dividing it in 4 parts in which the temperature of water is T_k and the inlet temperature of flue gas is $T_{f,k}$. The energy balance for each part of the exchanger is

$$\frac{C_b}{4}\frac{dT_k}{dt} = \dot{m}_b\left(T_{k-1} - T_k\right) + \frac{\varepsilon\,\dot{m}_f c_f\left(T_{k-1} - T_{f,k-1}\right)}{4} \qquad \text{for k} = 1, 2, 3, 4 \tag{2}$$

where $T_0 = T_{b,i}$, $T_{f,0} = T_{\text{ad}}$, $T_4 = T_{b,o}$, $T_{f,4} = T_f$, C_b is the heat capacity of the boiler that takes into account both the mass of water in the boiler and the mass of the metal parts.

In the model the start-up is divided in two periods: the first in which no combustion occurs, and the second in which the rate of combustion grows linearly from zero to maximum value.

4. Heating system model

In order to study whether pollutant emissions, pellet consumption and maintenance frequency can be significantly modified by the choice of a proper control strategy, we implemented a model of a heating system, for which a dynamical simulation was run on MatLab. For simplicity we describe the case of a pellet boiler used as energy source for space heating, with only one temperature probe monitoring the hot water tank. However the model has been extended to the case of production of hot water used for both space heating and domestic hot water, with two temperature probes. A sketch of the model is shown in figure 3: from right to left one can recognize the boiler (represented by three submodules, control unit CU, combustion chamber CC, and heat exchanger HE), the hot water tank, the heat distribution system, and the house or heated space. Each model element is briefly described in what follows.

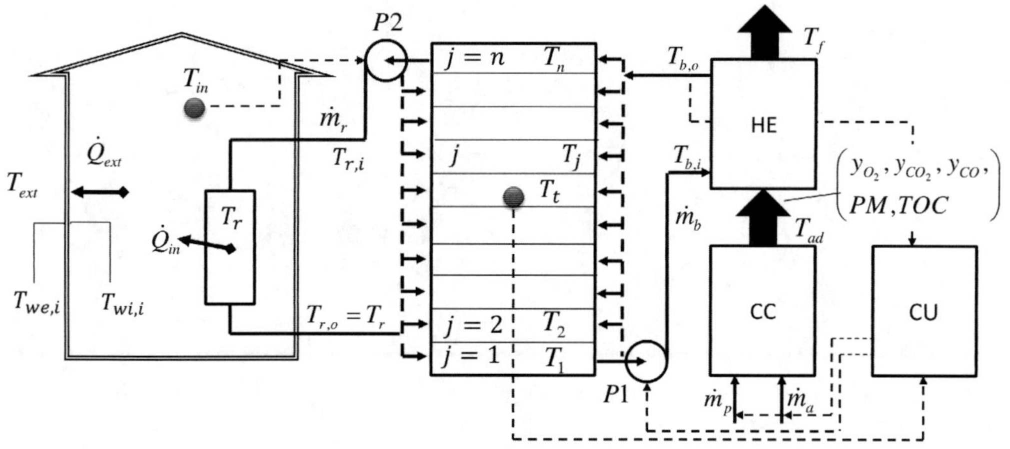

Figure 3. Scheme of the heating system modelled.

4.1. Water tank

In heating systems with a pellet boiler there should be a water tank for the storage of energy. It uncouples the boiler from the heating distribution system. That allows to reduce the duration of transient conditions and, hopefully, total emissions, fouling and maintenance frequency. Here we use a vertical cylindrical stratified water tank that we model as in [27]: temperature depends on time and elevation only. Space dependence is discretized upon dividing the tank in n horizontal disks of equal thickness.

The water outlet towards the boiler is in the disk 1, while the water outlet towards the heating system is in disk n. The position of water inlets from the boiler and from the heating system is chosen based on temperature: water enters in the tank in the disk with the closest temperature. Results of this model were successfully compared with experimental data in [28].

4.2. Heating system

In this example space heating is provided by means of radiators. They are characterized by their heat capacity C_r and by the heat exchange coefficient UA with the air of the room. For the radiators the heat balance is

$$\frac{dE_r}{dt} = C_r \frac{dT_r}{dt} = \dot{m}_r c_r \left(T_{r,i} - T_{r,o}\right) - UA\left(T_r - T_{\rm in}\right) \tag{3}$$

On the basis of the room temperature, $T_{\rm in}$, a thermostat switches on or off the pump (P2 in figure 3) that forces hot water flow $\dot{m}_r$ through the radiators.

4.3. House

The energy balance for the house (heated space) is rather simple,

$$\frac{dE_{\rm in}}{dt} = C_{\rm in} \frac{dT_{\rm in}}{dt} = UA\left(T_r - T_{\rm in}\right) - \sum_i h_{\rm w,i} A_{\rm w,i}\left(T_{\rm in} - T_{\rm wi,i}\right) \tag{4}$$

In equation (4) the house heat capacity $C_{\rm in}$ takes into account heat capacity of air and inner elements like furniture and internal walls. $h_{\rm w,i}A_{\rm w,i}$ represents the heat exchange coefficient between air and the ith wall of the building shell. The building shell was modeled as in [29] with a single-node model. In this model the shell is composed by two layers, hence it is characterized by three temperatures: inner side temperature $T_{wi,i}$, contact temperature between layers and the

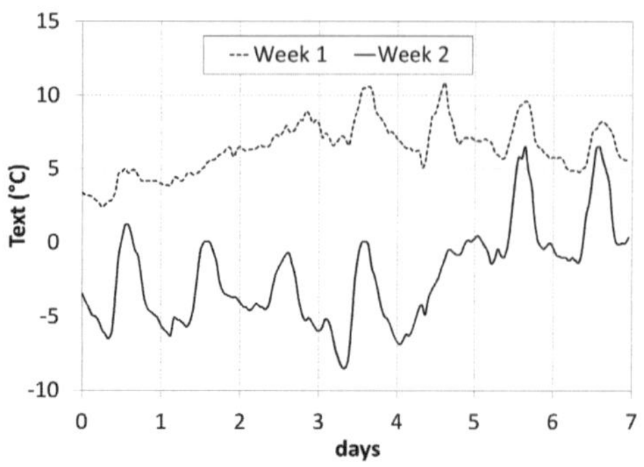

Figure 4. Time history of the temperature of the external environment.

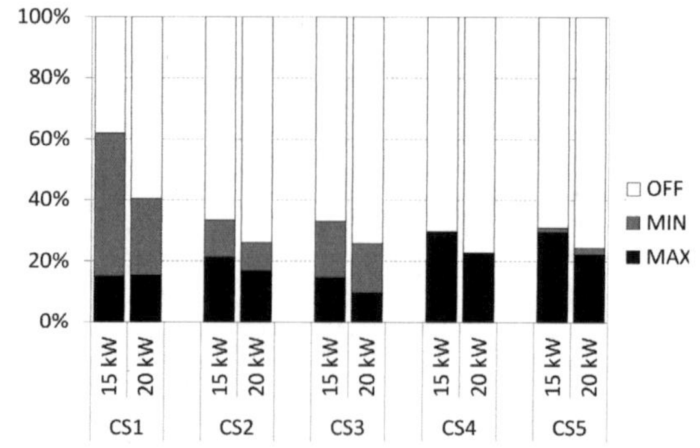

Figure 5. Fraction of time at maximum and minimum power for a 15 kW and a 20 kW boilers.

outer side temperature $T_{we,i}$. In this way it is possible to take into account also the heat capacity of the layers of the shell and their temperatures. The external side of the building shell exchanges heat with external environment only by convection, h_{ext} denoting the convective heat transfer coefficient. Real data are used for the time history of the external environment temperature. In particular, data refer to hourly temperature in Lonato (Bs, Italy) from January 3 to January 9, 2014 (week 1) and from February 3 to February 9, 2012 (week 2) [30].

5. Results

The model briefly sketched in Sec. 3 and 4 has been implemented in MatLab.

With reference to figure 3 the house size is 100 m^2 × 2.7 m filled by air and other objects (heat capacity 7.3 MJ/K); heat capacity of the external walls is 1.25 MJ/m^{3}K, their thermal conductivity 0.2 W/m K; convective heat transfer coefficients are $h_{\text{w,i}} = 8$ W/m^{2}K and $h_{\text{ext}} = 20$ W/m^{2}K. With regard to the heating distribution system, we consider 55 cast iron radiators elements each one has mass equal to 9.27 kg and is filled with 1 kg of water.

The desired room temperature T_{set} is set equal to 18°C from 5:00 to 7:00, 20°C from 7:00 to 23:00, 18°C from 23:00 to 24:00, while the heating system is off from 00:00 to 5:00. The dead-band is set equal to 1.5 K.

Simulations have been performed over a time interval of one week, for:

- two different external temperature time histories, denoted "week 1" and "week 2" (shown in figure 4);
- two boiler nominal powers: 15 kW, 20 kW;
- six water tank volumes: 283 l, 503 l, 754 l, 1005 l, 1571 l, 1963 l;
- five different control strategies (see Sec. 3).

The power of the boiler was chosen after a preliminary calculation of the thermal power demand of the house.

The water mass flow rate circulating through the boiler heat exchanger is such as to have a 15 K difference between outlet and inlet when boiler works at the maximum power ($T_{b,o} - T_{b,i} = 15$ K). As long as $T_{b,o} < 40$°C water recirculates in the boiler.

The water tank is discretized in $n = 40$ disks. Temperature of the 20th disk is taken as the water temperature T_t used by the control unit.

Fuel consumption does not seem to depend on control strategy nor on water tank volume: it has been estimated close to 155 kg/week for week 2 and 94 kg/week for week 1 for boiler of 15 kW. For boiler of 20 kW 159 kg/week for week 2 and 96 kg/week for week 1. Differences for different control strategies and water tank volumes fall within ±1.5%. Figure 5 reports the

Journal of Physics: Conference Series **547** (2014) 012017 doi:10.1088/1742-6596/547/1/012017

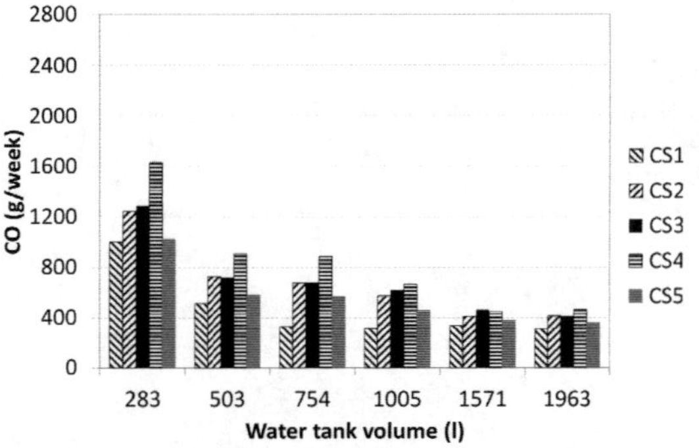

Figure 6. Total CO emissions for a 15 kW boiler during the week 2.

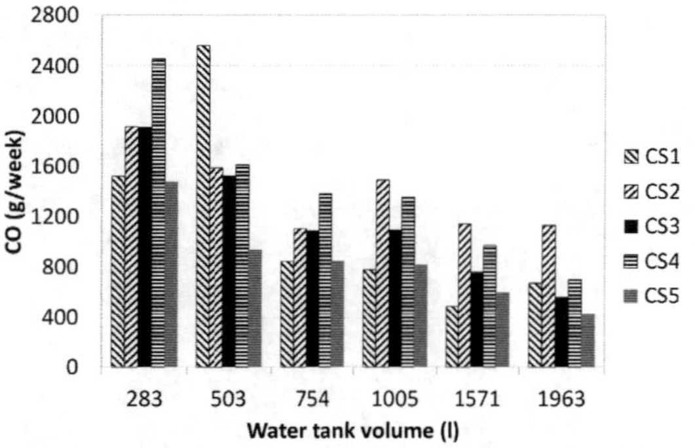

Figure 7. Total CO emissions for a 20 kW boiler during the week 2.

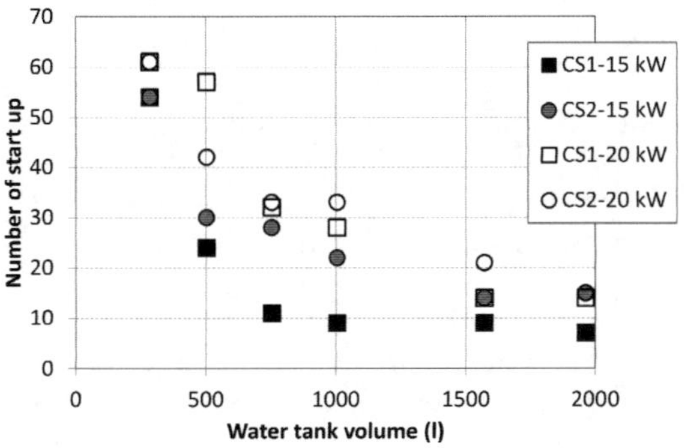

Figure 8. Number of start-ups vs. water tank volume, for strategies CS1 and CS2 for both 15 kW and 20 kW boilers.

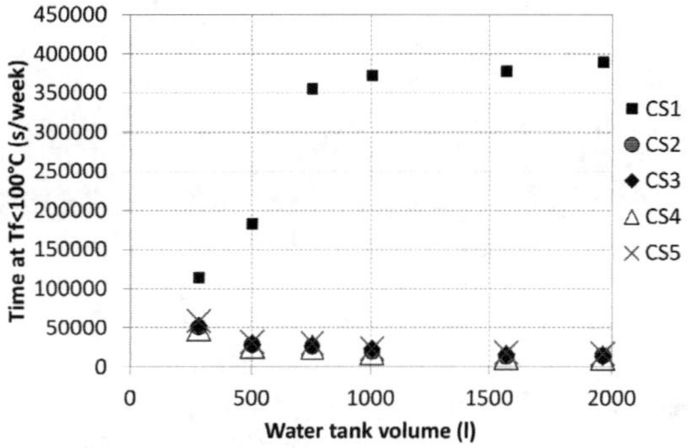

Figure 9. Total time flue gas temperature is below 100°C vs. water tank volume, for different strategies and 15 kW boiler.

fraction of time the boilers of 15 kW and 20 kW are off or are working either at maximum or minimum power. From figures 6 and 7 it is evident that the CO production of boiler of 20 kW is higher than the CO production of boiler of 15 kW. Production rates of CO at the maximum power for both 15 kW and 20 kW boilers are about 28 mg/MJ. The large difference is due to the CO emitted during the start-up and stop phases. In figure 8 the number of start-ups is plotted versus the water tank volume for control strategies CS1 and CS2. Number of start-ups of CS3, CS4 and CS5 are very close to those of CS2. Control strategy CS1 ensures less start-ups/stops than other control strategies.

From these first observations, we can assert that the 20 kW boiler, which is over-designed for this house, presents more critical issues than the 15 kW boiler, therefore we focus our attention on the latter only. Figure 9 reports the total time flue gas temperature is below 100°C. For CS1 flue gas temperature is below 100°C for about 2/3 of the time: this is due to the long time of working at the minimum power and to start-ups and stops. Figures 10, and 11 show total emissions versus water tank volume for different control strategies.

The calculated emission rate are the following: during operation at the nominal power the average emission rate are 28 mg/MJ for CO and 6 mg/MJ for PM. When boiler operates at the 70% of nominal power emission rates are 42 mg/MJ and 11 mg/MJ, while when it operates at 30% they are 75 mg/MJ and 16 mg/MJ. For each start-up, which takes place in the same way for all strategies, the CO and PM production are 20 mg/start-up and 1 mg/start-up, respectively. During the switch off the production of CO and PM is strongly influenced by the power at which the boiler works just before the stop: if the boiler works at the maximum power (CS4) CO and PM production are 15 mg/switch off and 1 mg/switch off, respectively, if it works at the 70% of

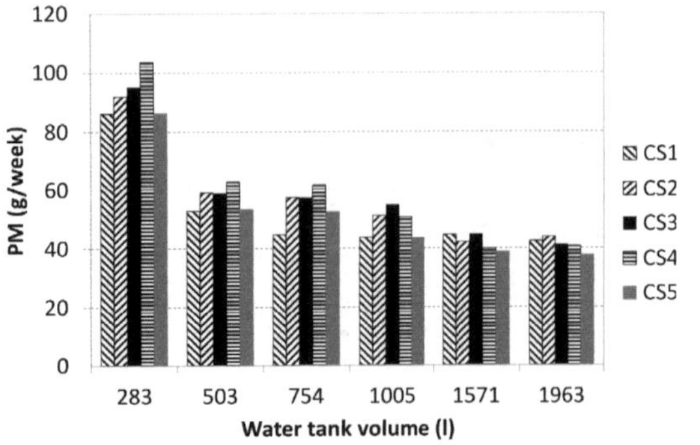

Figure 10. Total PM emissions for a 15 kW boiler during the week 2.

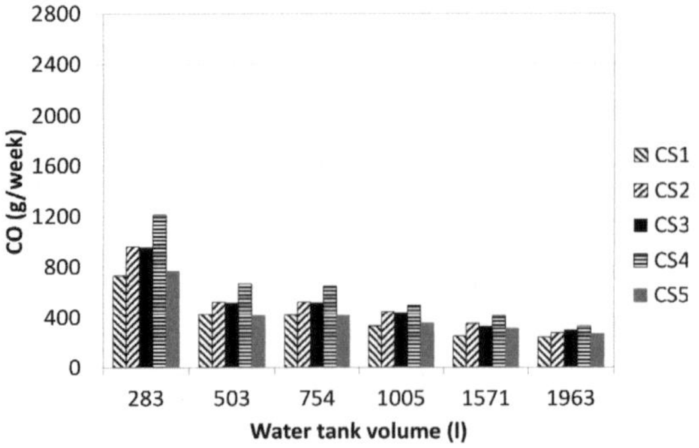

Figure 11. Total CO emissions for a 15 kW boiler during the week 1.

the maximum power (CS2 and CS3) 5 mg/switch off and 0.2 mg/switch off, while if it work at the 30% (CS1 and CS5) 1 mg/switch off and $<$ 0.2 mg/switch off.

Although these results must be considered preliminary and need confirmation, we observe what follows with regard to the five control strategies.

CS1) The number of start-ups is smaller than those for the other control strategies for water tank volume greater than 250 l: that should reduce emissions and fouling deposition. However this effect is counterbalanced by the low temperature of flue gas and by the long time of operation at minimum power (characterized by higher pollutant concentration).

CS2) This control strategy produces more CO and PM emission than CS1 even if number of start-ups is much larger and flue gas temperature is below 100°C for a shorter time.

CS3) There are not clear differences between CS3 and CS2.

CS4) This control strategy yields the highest value of CO emissions. CO and PM emissions per unit of energy produced in normal operation are the lowest, but the high number of start ups/shutdowns and the shutdown from the maximum power give a large contribution on pollutant emissions.

CS5) This control strategy provides low emissions: they are larger than emissions of CS1, but smaller than emissions of other strategies. With respect to CS1 emissions during normal operation are smaller because the time of working at the minimum power is significantly reduced, but total emissions result larger because of the higher number of start-ups/shutdowns especially for tank volumes between 503 l and 1005 l. This control strategy is currently implemented in some pellet boilers to avoid short time interval between a stop and the subsequent start-up.

The volume of the hot water tank appears to be more important than control strategy in reducing both the number of start-ups and emissions. As expected, number of start-ups is a decreasing function of water tank volume for all strategies. Also CO and PM emissions decrease with tank volume and differences between strategies become less important. However it can be observed that PM emissions fall abruptly when tank volume is increased from 250 l to 500 l.

Finally, results for HC have similar trend to those for PM.

In order to find the control strategy that optimizes the operation of the boiler in most conditions as possible, it is necessary to extend this sensitivity study varying other variables, such as thermal load, heating system, temperatures $T_{t,lo}$, $T_{t,hi}$, minimum power, etc, and the simulation can be performed over a longer time interval. For the example presented here, the optimal solution is the 15 kW boiler, working with strategy CS5 and a water tank of 500 liters. Larger water tanks imply higher cost and the differences in term of pollutant emissions and time at low temperature are negligible. Even if there are relations to calculate the rate of deposition of fouling and creosote are not available yet, experience suggests to discard CS1 due to the long time of working with low temperature of flue gas.

6. Conclusions

In this work we studied the performance of a heating system based on a wood pellet boiler coupled with a thermal energy storage. Upon use of a theoretical model implemented in MatLab the heating system performance can be estimated in terms of fuel consumption, pollutant emissions, number of boiler start-ups, time of operation at minimum power. Performances are studied for five different control strategies and for varying volume of the hot water tank.

Only few preliminary results have been obtained insofar. Although they need confirmation and a systematic and careful analysis, nevertheless they do show a weak dependence of performance on control strategy. In particular,

(i) fuel consumption is independent for all practical purposes from both control strategy and water tank volume;

(ii) total CO and PM emissions over a week substantially decrease for increasing volume of the water tank, but for water tank larger than 500 l the difference can be considered negligible.

(iii) control strategies do affect parameters like number of start-ups or fraction of time at minimum power, however the combined effects on total emissions tend to compensate each others: only CS 1 seems to show CO emission somehow lower than other CS, but the use of this strategy can increase the formation creosote and fouling.

References

[1] Miguez J, Moran J, Granada E and Porteiro J 2012 *Renewable and Sustainable Energy Reviews* **16** 3867–3875
[2] Thek G and Obernberger I 2012 *The Pellet Handbook: The Production and Thermal Utilization of Biomass Pellets* (Taylor & Francis) ISBN 9781136539916
[3] BAFA (Germany) 2013
[4] DM 28/12/12 (Italy) 2012
[5] Art 15a B-VG (Austria) 2012
[6] Brunner T, Barnthaler G and Obernberger I 2006 *Proceedings 2nd World Pellets Conference* 87–91
[7] Win K, Kaung M, Persson T and Bales C 2012 *Atmospheric Environment* **59** 320–327
[8] Boman C, Nordin A and Ohman M 2005 *Umea University*
[9] Bafver L, Leckner B, Tullin C and Berntsen M 2011 *Biomass and Bioenergy* **35** 3648–3655
[10] Dias J, Costa M and Azevedo J 2004 *Biomass and Bioenergy* **27** 531–539
[11] Wiinikka H and Gebart R 2004 *Biomass and Bioenergy* **27** 645–652
[12] Johansson L, Leckner B, Gustavsson L, Cooper D, Tullin C and Potter A 2004 *Atmospheric Environment* **38** 4183–4195
[13] Persson T, Fiedler F, Nordlander S, Bales C and Paavilainen J 2009 *Applied Energy* **86** 645–656
[14] Nordlander S and Persson T 2003 *ISES Solar World Congress, june 2003*
[15] Schnetzinger R, Jofre M, Hebenstreit B, Schwarz M and Hoftberger E 2012 *WSED 2012*
[16] Haller M, Haberl R, Droscher A and Frank E 2009 *11th Int. IBPSA Conference*
[17] Kjallstrand J and Olsson M 2004 *Biomass and Bioenergy* **27** 557–561
[18] Olsson M, Kjallstrand J and Petersson G 2003 *Biomass and Bioenergy* **24** 51–57
[19] Eskilsson D, Rönnbäck M, Samuelsson J and Tullin C 2004 *Biomass and Bioenergy* **27** 541–546
[20] Petrocelli D and Lezzi A 2014 *J. Phys.: Conf. Ser.* **501** 012036
[21] Petrocelli D 2013 *Indicazioni sperimentali per la progettazione di stufe a pellet ad alta efficienza* Ph.D. thesis University of Brescia
[22] Obenberger I and Thek G 2006 *Proceedings 2nd World Pellets Conference* 31–41
[23] BLT 2011 *http://blt.josephinium.at*
[24] EN 303-5:2012 2012 *Heating boilers - Part 5: Heating boilers for solid fuels, manually and automatically stoked, nominal heat output of up to 500 kW - Terminology, requirements, testing and marking*
[25] Kelz J, Brunner T and Obernberger I 2012 *Environmental Sciences Europe* 1–18
[26] Brunner T 2009 *Proceedings of the 17th European Biomass Conference & Exhibition, June 2009, Hamburg*
[27] Klein S 1989 *A design procedure for solar heating system* Ph.D. thesis University of Wisconsin
[28] De Cesaro, Oliveski, Rejane, Krenzinger, Arno, Vielmo and Horacio 2003 *Solar Energy* **75** 121–134
[29] De Rosa M, Bianco V, Scarpa F and Tagliafico L A 2014 *Applied Energy* **128** 217–229
[30] Provincia di Brescia 2011 *http://meteo.provincia.brescia.it*

Journal of Physics: Conference Series **547** (2014) 012018 doi:10.1088/1742-6596/547/1/012018

Energy retrofit of an office building by substitution of the generation system: performance evaluation via dynamic simulation versus current technical standards

D Testi[1], E Schito[1], E Menchetti[2] and W Grassi[1]

[1] DESTEC, Department of Energy, Systems, Territory and Constructions Engineering, University of Pisa, Italy
[2] Directorate of Buildings, Real Estate Optimization Unit, University of Pisa, Italy

E-mail: daniele.testi@unipi.it

Abstract. Constructions built in Italy before 1945 (about 30% of the total built stock) feature low energy efficiency. Retrofit actions in this field can lead to valuable energetic and economic savings. In this work, we ran a dynamic simulation of a historical building of the University of Pisa during the heating season. We firstly evaluated the energy requirements of the building and the performance of the existing natural gas boiler, validated with past billings of natural gas. We also verified the energetic savings obtainable by the substitution of the boiler with an air-to-water electrically-driven modulating heat pump, simulated through a cycle-based model, evaluating the main economic metrics. The cycle-based model of the heat pump, validated with manufacturers' data available only at specified temperature and load conditions, can provide more accurate results than the simplified models adopted by current technical standards, thus increasing the effectiveness of energy audits.

1. Introduction

Energy requirements for the residential and tertiary sector account for a significant share of the total energy uses. In Italy, this share is about 30% [1]. In the European Union, according to Directive 2010/31/EU [2], energy costs for heating and cooling of buildings represent about 40% of the total energy uses. For this reason, many programs have been developed in order to encourage policies of energy efficiency and measure progress in this field. Odyssee project, for example, has collected energy efficiency data from all countries in the European Union for the last 2 decades and verified that energy consumption in the building sector has increased by around 1%/year since 1990, mainly in non-residential buildings (characterized by an average specific consumption of about 295 kWh/m^2) [3]. Retrofit actions in this field can lead to significant energy savings.

In Italy, about 30% of constructions were built before 1945 [4]: many of these buildings have low energy efficiency, but are protected against refurbishment, having a significant heritage value, so retrofit actions involving the envelope cannot be applied. In these cases, great energetic and economic savings can be obtained through the substitution of traditional low-efficiency boilers with new high-efficiency generators such as Heat Pumps (HPs).

Current technical standards (see for instance UNI/TS 11300-4 [5]) define the procedure to calculate the performance of HPs during the heating season and evaluate the energetic savings obtained by the installation of these generators. These models are "black boxes", because they require only few manufacturers' data and do not accurately consider the influence of some parameters, viz.: temperature evolution of the external source and of the end-user loop or destination thermal source, conditions of partial load, properties of the working refrigerant, efficiencies of individual components (heat exchangers, compressor or absorption and generation unit), control types, defrosting mode.

A development of a cycle-based model for the simulation of a HP can provide more accurate outputs than "black box" methods: these outputs are necessary for a correct energy audit, hence for a precise estimate of the obtainable energetic and economic savings.

2. Definition of the test case

With the aim of verifying the effective savings that could be obtained by a substitution of an inefficient generator with a HP (simulated through a dynamic model), we chose an existing building of historical value, which can be considered as representative of many Italian PA and office buildings, in terms of characteristics of structure, systems and uses and also in terms of the proposed retrofit measure.

The building used for the test case is known as Palazzo Venera, property of the University of Pisa.

Palazzo Venera is a five-story building in the centre of Pisa (gross volume about 13,000 m^3), facing west on the central street Via Santa Maria and facing east on a courtyard garden. It is bordered on both the southern and northern side by private buildings (heated in the winter season, thus considered as adiabatic surfaces).

We considered only the three central levels (first, second and third floor that are for office use only), excluding the ground floor with classrooms and the under roof space. We divided this volume into six zones, in order to distinguish the east and the west zone of each floor, characterized by the presence of different obstructions and shadings and so by different solar radiation. Table 1 shows the characteristics of each zone in terms of transmittance of opaque walls and windows. We implemented schedules for the presence of people, use of electrical equipment and artificial lighting; they were chosen after inspection and users' interviews. They are different for weekdays and Saturdays; during festivities (in total 14 days of the heating period) and Sundays the building is closed.

There is a fan coil in every office, but not all fan coils are activated during the heating season: so we estimated a number of active fan coils for each zone, as a result of both building inspection and simulated audit. Fan coils are controlled within the proportional band 18°C-20°C.

The distribution system is divided into two loops: the hot water provided by the boiler is distributed in the whole building through a primary loop and the secondary loop of each zone provides energy independently from the other zones.

The heat generator is a traditional natural gas boiler of nominal power 230 kW, with climate control. This boiler is switched on from 7 a.m to 6 p.m., from Monday to Friday, and from 7 a.m. to 1 p.m. on Saturdays, during the heating season of Pisa (1st of November – 15th of April, Heating Degree Days: 1694).

The gas consumption of the whole building has been assigned to the zones in proportion to the building total installed power of fan coil units; as a consequence, 49.9% of the primary energy use is attributed to the 3 simulated floors, corresponding to 102 MWh, according to the last 5 years of natural gas billings.

Table 1. Characteristics of each thermal zone (MTT: Mean Thermal Transmittance, in $W/(m^2K)$, E east, W west).

Thermal zone	External opaque walls		External glazed elements		N° of fan coils
	Area [m^2]	MTT	Area [m^2]	MTT	
1st floor E	157.7	1.59	31.1	4.08	15
1st floor W	193.3	1.17	42.8	4.08	8
2nd floor E	153.9	1.88	25.9	4.08	15
2nd floor W	192.5	1.17	43.6	4.08	8
3rd floor E	112.2	1.78	25.9	4.08	15
3rd floor W	139.7	1.46	33.1	4.08	8

Further details about the characteristics of the building under exam, including its use, and about the assumptions of the simulation can be found in [6].

Journal of Physics: Conference Series **547** (2014) 012018 doi:10.1088/1742-6596/547/1/012018

3. Dynamic simulation of the building

A full dynamic simulation of Palazzo Venera was run on TRNSYS 17, interfaced with Matlab. As shown in Figure 1, a TRNSYS routine and a Matlab script calculated the energy requirements of the building, simulating the heat exchange of the fan coils. A second Matlab script calculated the performance of the generation system (first the existing natural gas boiler, then a HP with a back-up boiler). The simulation time step was 15 minutes.

In the TRNSYS routine, we modeled the characteristics of the envelope of the analyzed area in terms of transmittance of the opaque walls and windows. We also implemented the air exchange due to infiltrations (0.3 volumes per hour) and the profiles of the presence of users in offices, of the uses of electrical equipment and of artificial lighting.

As for the climatic data, we used hourly external temperature, beam and diffuse solar radiation on the horizontal plane, relative air humidity and wind velocity: all these data were provided in 2012 by the Italian Thermotechnical Committee (CTI). These values were interpolated in order to obtain four data points per hour, according to the chosen simulation time step. On the west side, the buildings in front of Palazzo Venera were considered as obstructions for the radiation and the sloped roof as a horizontal shading for the third floor only. On the east side, the south-east wing of the building itself is a vertical shading.

A Matlab routine, directly linked to the TRNSYS simulation, calculated the heat provided in each zone by the fan coils, considered by TRNSYS as a purely convective internal gain. The supply temperature to the six zones was calculated for every time step as a function of the corresponding external temperature.

When fan coils in a zone were active, the routine would calculate the return temperature with the energy balance expressed by Equation 1, valid for each emitter:

$$\dot{m}c\left(T_s - T_r\right) = UA\left(\frac{T_s + T_r}{2} - T_i\right) \quad (1)$$

where $\dot{m}$ is the water mass flow rate of a fan coil unit (250 kg/h), c is the water specific heat, T_s and T_r are, respectively, the water supply and return temperatures, U is the total heat transfer coefficient and A the area of the heat exchange coil of the unit (from manufacturer's datasheets), and T_i is the internal air temperature, calculated by TRNSYS at each time step and for each air node. The U-value of the fan coil unit was found by manufacturers' datasheets.

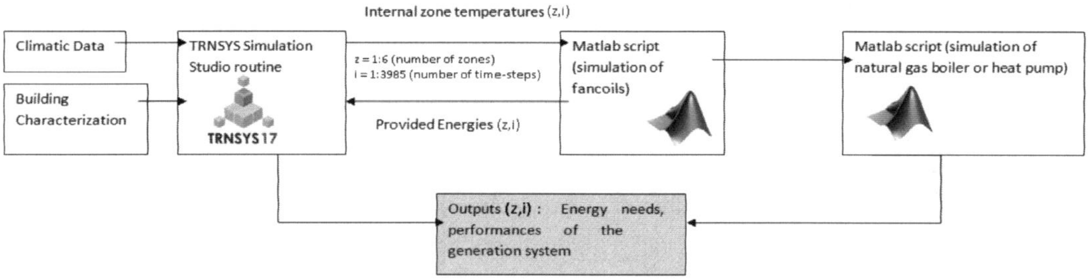

Figure 1. Simulation flow diagram of the routines and software.

Due to the low amount of contained water, fan coils can be considered steady systems, thus no transient terms were included in Equation 1. Once T_r is obtained, the energy Q_{heat} delivered to the zone by all the fan coil units is given by Equation 2:

$$Q_{heat} = n\,\dot{m}\,c\left(T_s - T_r\right)\cdot t \qquad (2)$$

where n is the number of fan coil units in the zone and t is the duration of the time step. The energy gain calculated for each zone and time step was used in the TRNSYS simulation at the subsequent time step due to convergence issues.

As a result of this first part of simulation, TRNSYS provided, for the entire heating season, the evolution of internal temperatures of the examined zones (e.g., air, mean radiant and operative temperatures), share-outs of global energy use among the zones and specific contributions of heat losses (transmission and infiltration) and gains (internal and solar). Figure 2 presents these results through a pie chart, relative to the whole heating season and to all the zones: in the graph legend we reported the value of each contribution and its relative weight.

In the second part of the simulation, we evaluated the primary energy used by the natural gas boiler. We summed the energy needs provided by TRNSYS with the heat losses due to the distribution and the generation system. The two latter terms were calculated using the methodologies described in the Italian standard UNI/TS 11300-2:2008 [7] and the efficiency data provided by the manufacturer of the generator.

The resulting primary energy use for the analyzed zones (accounting for the 49.9% of the energy requirements of the whole building) is equal to 111 MWh.

The 8.8% deviation between this output and the average consumption based on the last five years of billings (102 MWh) confirms appropriate accuracy of the developed method. The seasonal efficiency of the boiler is 94.9%. We also calculated that the total electric energy absorbed by pumps, fans and other auxiliaries is 5.2 MWh. Using the primary energy factor for electric energy (equal to 2.174, according to the Italian Regulatory Authority for Electricity and Gas), we found that the overall primary energy use per gross volume is 14.9 kWh/m^3.

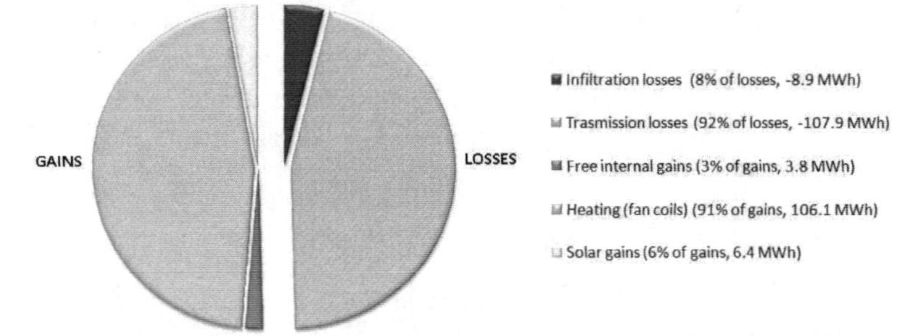

Figure 2. Energy balance of the building envelope in the heating season.

4. Description of the heat pump system cycle-based model

In order to evaluate the effectiveness of a retrofit action consisting in the substitution of the generator, we hypothesized that the energy requirements of the analyzed volume would be satisfied by an air-to-water electrically-driven modulating HP. Based on the manufacturer's data, the nominal thermal power of the HP was 100 kW, while the maximum power was 130 kW. The declared COP at external source temperatures 7°C and mean end-user loop temperature 42.5 °C was 3.48. An auxiliary natural gas boiler, with the same nominal performance of the existing one, provided back-up energy out of the working range of the HP.

The HP seasonal heating performance was obtained by simulating the evolving inverse thermodynamic cycle, calculated at every time step. We modeled all the thermodynamic transformations of the refrigerant (compression, condensation, expansion and evaporation), similarly to Jin [7], Armstrong et al. [9], Zakula [9], Corberan et al. [11] and Carbonell et al. [11]. In literature, there are several models characterized by different procedures for calculating the thermodynamic states and flow rate of the refrigerant during the cycle. For the cycle-based model that we propose, we used the following assumptions:

- the isentropic efficiency and the overall electro-mechanical efficiency of the scroll compressor depend on the compression ratio, similarly to [12];
- the UA-value of the evaporator and of the whole condenser are determined from manufacturers' data (U: total heat transfer coefficient of the coils of the heat exchangers; A: heat transfer area of the coils of the heat exchangers);
- the ratio between U of the condenser and U of the superheater is fixed;
- the heat exchanged in the subcooler of the condenser is a fixed percentage of the total heat exchanged in the condenser;
- the temperature of the refrigerant at the exit of the evaporator is 1 K higher than the evaporation temperature;
- energy losses due to defrosting cycles are considered through a COP penalization, depending on external temperature and relative humidity, as proposed in [13].

The HP under exam uses R134a as refrigerant: this fluid has a high critical temperature (over 100°C), so its use is strongly recommended when the emitters are fan coils. The thermodynamic characteristics of R134a were provided to the Matlab routine by a property database. For every time step, the routine calculated temperatures, pressures, enthalpies and densities of the refrigerant in correspondence of every single HP component. The compressor speed was also determined. If the speed of the scroll were less than the minimum modulation value, the HP would switch off and the energy requirement of that time step would be satisfied by the back-up boiler. In addition, if the speed of the scroll were greater than its maximum value, the scroll would work at maximum power and the residual energy would be provided to the building by the boiler.

As for the condenser and the evaporator, the Matlab routine used the ε-NTU method [15], verifying the energy balance in the heat exchanger for both the refrigerant side and the secondary side (water in the condenser, air in the evaporator). The throttle valve was considered isenthalpic and controlled the superheat of the refrigerant at the evaporator exit. As already mentioned, the compressor was a scroll type connected to an inverter: as every reciprocating compressor, the flow rate in the scroll is proportional to the compressor speed and independent from the provided pressure ratio.

In summary, at each time step, the Matlab routine calculated the heat exchanges, the electric energy requirement and the performance of the HP.

5. Performance results: cycle-based simulation vs. technical standard

5.1 Calculation by cycle-based model

First, we verified that, at the same source temperatures, the calculated COP differed by less than 5% from the COP declared by the manufacturer.

The calculated balance temperature of the building – heating system was 2.1°C, as shown in Figure 3. If the building required a greater heat power than the one provided by the HP (red solid line), the boiler would provide the residual energy. If the building required a lower heat power than the minimum deliverable by the HP (red dashed line), this power would be provided by the boiler only.

The resulting Seasonal Performance Factor (SPF) of the HP was 3.8, with minimum values of 1.9 reached at the lowest external temperatures and maximum values above 6 under partial load conditions and mild temperatures. The number of on/off cycles undergone by the HP unit during the 144 heating days was only 165. Unlike black-box models, the present cycle-based simulation could find performance improvement at partial loads, which are the most frequent operative conditions. Defrost cycles entailed a 2.2% penalization of SPF.

The strong influence of the external temperature on COP is shown in Figure 4, in which the profile of external temperature and of COP during five working days (between January and February) are represented. The *inertia* of the building also influences the profile of COP, as it is shown in the third curve of the figure (Monday).

Figure 5 shows the Primary Energy Ratio (PER) of the whole heating system (HP and back-up) as a function of the load factor (or Capacity Ratio, CR). Minimum values were reached when only the boiler was active (lowest partial loads); high efficiency values were reached at intermediate partial loads and tended to decrease when the ratio of the energy provided by the boiler and the total energy required by the building increased. The strips of data observable in the graph are due to the particular heating control, that is a zone heating control: when a zone reaches the upper dead-band temperature (20°C), all its fan coil units switch off, so there is not a control on each single unit, but a control on sets of fan coil units.

The HP required 24 MWh_{el} to heat the building. The natural gas boiler required 14 MWh of primary energy and its mean seasonal efficiency was 95.1%. The primary energy required for the whole heating system was equal to 77.4 MWh (also considering the electric energy absorbed by pumps, fans and other auxiliaries, equal to 5.2 MWh) leading to a 38% saving of primary energy (about 46 MWh) with respect to the original system. The seasonal PER was 1.6 (69% higher than the efficiency of the existing boiler).

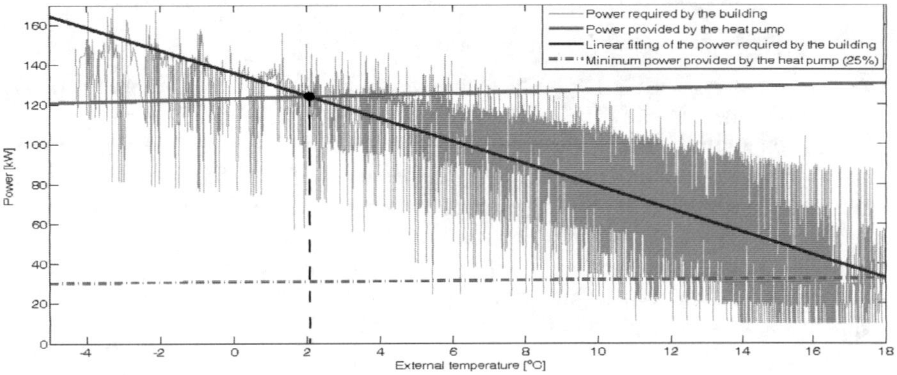

Figure 3. Balance temperature of the system.

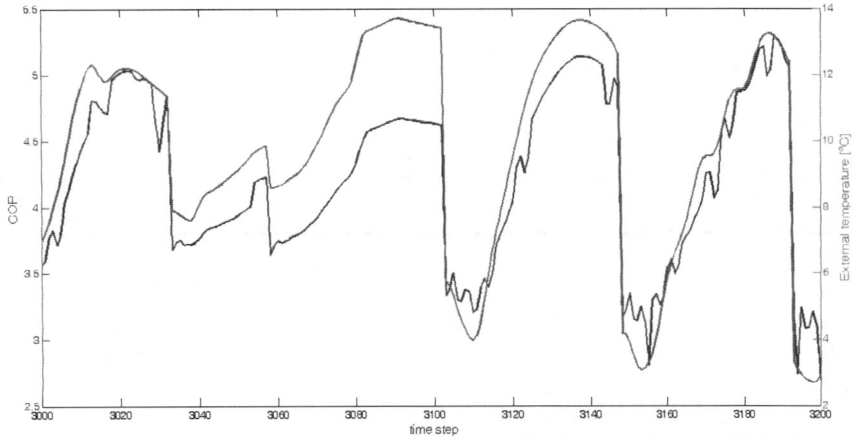

Figure 4. Profiles of external temperature and COP during five working days.

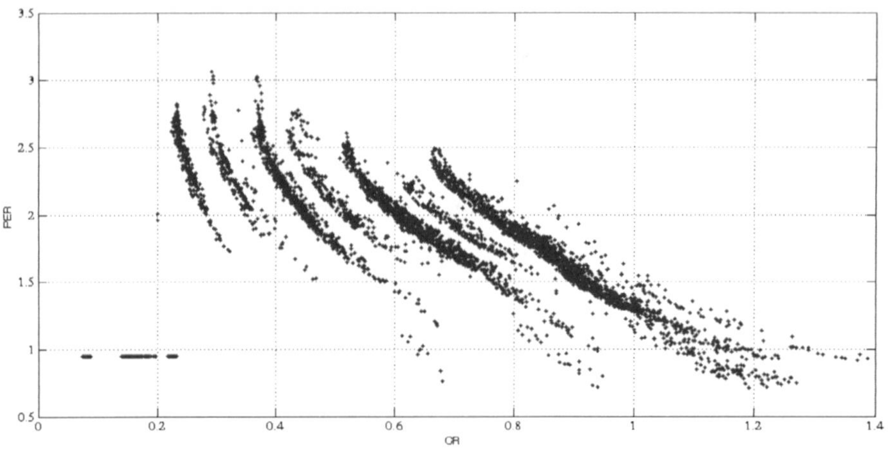

Figure 5. PER vs. CR of the whole heating system (HP and boiler).

5.2 Calculation by current technical standard

An alternative simulation of the HP performance was run on spreadsheets implementing the quasi-steady state procedure described in Italian Standard UNI/TS 11300-4 [5]. According to this standard, the energy provided and the monthly and seasonal performance of a HP can be calculated by interpolation of manufacturers' data of COP at particular temperatures of the secondary fluids in the heat exchangers. We used the same HP of thermal nominal power 100 kW employed for the cycle-based simulation.

The calculation showed that the chosen HP could satisfy the monthly energy requirements of the building, without the auxiliary boiler, but with a SPF of 2.3 (39% lower than one obtained with the cycle-based simulation). The procedure described in technical standard UNI/TS 11300-4 leads to underestimate values of COP because it does not take into account the effects of intermittent heating and climate control, thus leading to lower load factors during the heating season and a higher source temperature of the end-user loop. These two elements negatively affect the HP performance; particularly, low CR values entail great performance penalization, according to the COP reduction factor defined in [15].

6. Economic evaluation of the investment

Once we verified the energetic effectiveness of the substitution of the traditional boiler with a HP, we performed an economic analysis of the retrofit action. We developed a specific routine for the investment analysis, requiring, as inputs, the energetic savings obtained with the retrofit action, costs of the energetic vectors and of the retrofit action, maintenance costs for the original and the retrofit system, inflation and discount rates and possible government incentives.

The cost of the new heating system (HP and auxiliary boiler) was 47 k€. The estimated lifetime of this new system was 15 years. The cost of a new boiler, with the same performance and characteristics of the current one, whose remaining useful life was 2 years, was 9 k€. The maintenance costs increased of 100 € per year with the substitution of the boiler. According to current state laws, the installation of a HP is boosted by the Italian government with a 10-year refund of 65% of the investment.

Two simulations were run, considering different scenarios for the inflation rates. In any case, the economic analysis showed that the substitution of the generator would lead to significant profits. Table 2 summarizes the inputs of the analysis and the resulting indexes for the evaluation of the investment, using the energetic savings obtained with the cycle-based model.

Table 2. Economic analysis: inputs and resulting indexes.

Inputs					
Discount rate	Electric energy cost	Natural gas cost	General inflation rate	Electric energy inflation rate	Natural gas inflation rate
2%	165 €/MWh_{el}	72 €/MWh_{th}	1st scenario: 2%	3%	3%
			2nd scenario: 4%	7%	7%

Results			
Discounted Payback Period (DPP)	Net Present Value (NPV)	Profitability Index (PI)	Internal Rate of Return (IRR)
1st scenario: 7 years	45 k€	0.95	14%
2nd scenario: 6 years	49 k€	1.0	17%

Two similar simulations were run, using the same specific routine for the investment analysis but using the energetic savings obtained with the methodology described in current technical standard. In this cases, the substitution of the boiler with the heat pump was not cost-effective, with no payback of the investment.

7. Conclusions

We presented a procedure for accurately predicting the energetic and economic impact of a paradigmatic retrofit measure in a protected office building: the substitution of a traditional boiler with an air-to-water modulating electric Heat Pump (HP). Acting on the generation system can have a large impact on the overall system efficiency; besides, particularly in historical buildings, it is a feasible and common solution, unlike other ones involving the protected envelope.

By modeling all the subsystems and coupling software like TRNSYS, Matlab and a refrigerants' property database, it was possible to perform a full dynamic simulation of the building and its plant. Similar all-inclusive methods are recently under development (see, for instance, [17]).

The calculated baseline energy consumption of the test case was coherent with the available history of natural gas billings. The main results for the HP system were: 38% of primary energy savings with respect to the existing boiler and around 15% of internal rate of return of the investment. The proposed energy efficiency measure could have wide application, given that the accuracy of the underlying dynamic models allows operating in a frame of reliability.

On the other hand, following the HP performance evaluation procedure defined in current technical standards, the energetic efficiency of the HP system (in terms of primary energy ratio) resulted only 11% higher than the original boiler, with no payback of the investment. This major underestimation by the technical standard in terms of HP performance pushes towards the adoption of dynamic physically-based methods, as the one presented herein, in order not to penalize the HP technology in evaluations for energy audits or in design trade-offs.

Experimental validation of the proposed model by application to existing plants and by continuous monitoring in laboratory facilities is in view.

Possible further developments are:

- optimization of the climate control curve and of the control logic that switches on/off the HP and the back-up generator, in order to maximize the system PER;
- proposal and evaluation of quasi-steady state methods, alternative to the ones suggested by current standards and tailored to energy audits (e.g., taking into account the actual heating hours of the building), to be implemented in dedicated software, like SEAS (Simplified Energy Auditing Software) [17].

Acknowledgements

We acknowledge the precious advices of Ph.D. candidate Mr. Paolo Conti and engineering technologist Mr. Davide Della Vista on the modeling of the generation system and the implementation of

Journal of Physics: Conference Series **547** (2014) 012018 doi:10.1088/1742-6596/547/1/012018

the dynamic simulation routine. We gratefully thank the University staff for the cooperation during surveys and interviews and for providing the necessary data regarding Palazzo Venera.

References

[1] ENEA 2003 *Risparmio Energetico nella Casa. Sviluppo Sostenibile* Booklet G1-034-0
[2] European Parliament 2010 *Directive 2010/31/EU* Official J. European Union
[3] 2012 *Odyssee Project* European Energy Efficiency Database
[4] ISTAT 2001 *14° General Census of Population and Housing* Italian Institute of Statistics
[5] UNI 2012 *UNI/TS 11300-4* Italian Unification Body
[6] Testi D, Schito E, Menchetti E and Grassi W 2014 Dynamic simulation of a heat pump serving a historical building: seasonal performance and economic evaluation *Proc. 49th AiCARR Int. Conf. (Rome)*
[7] UNI 2008 *UNI/TS 11300-2* Italian Unification Body
[8] Jin H 2002 *Parameter Estimation Based Models of Water Source Heat Pumps* Ph.D. Thesis Oklahoma State University
[9] Armstrong P R , Jiang W, Winiarski D, Katipamula S, Norford L K and Willingham R A 2009 *HVAC&R ASHRAE* **15** 366
[10] Zakula T 2010 *Heat Pump Simulation Model and Optimal Variable Speed Control for a Wide Range of Cooling Conditions* Master Thesis Massachusetts Institute of Technology
[11] Corberan J M , Finn D P, Montagud C, Murphy F T and Edwards K C 2011 *Energy and Buildings* **43** 82
[12] Carbonell S D, Cadafalch R J, Pärlisch P and Consul S R 2012 Numerical analysis of heat pumps models: comparative study between equation-fit and refrigerant cycle based models *Proc. Int. Conf. on Solar Heating, Cooling and Buildings, EuroSun 2012 (Rijeka, HR)*
[13] Nardotto D 2009 *Analisi Sperimentale e Modellizzazione di Circuiti Frigoriferi con Compressore ad Inverter Impiegati nella Refrigerazione Commerciale* Ph.D. Thesis University of Padua
[14] Lazzarin R, Busato F, Nardotto D and Noro M 2009 *Studio del Brinamento delle Batterie su Pompe di Calore a Compressione* Report RSE/2009/31 (Rome: ENEA)
[15] Incropera F P and DeWitt D P 2002 *Fundamentals of Heat and Mass Transfer (5th Ed.)* (New York: Wiley)
[16] CEN 2012 *EN 14825* European Committee for Standardization
[17] Buonomano A, Calise F and Palombo A 2012 *Applied Energy* **91** 222
[18] Testi D, Menchetti E, Schito E and Grassi W 2013 An open access tool for building energy audits harmonizing European standards *Proc. Building Simulation Applications Conf., BSA 2013 (Bozen-Bolzano, IT)*

Flow boiling of a new low-GWP refrigerant inside a single square cross section microchannel

S Bortolin and D Del Col
Dipartimento di Ingegneria Industriale, Università degli Studi di Padova
Via Venezia 1, I-35131 Padova, Italy

E-mail: stefano.bortolin@unipd.it

Abstract. In this work new experimental heat transfer coefficients measured during flow boiling of HFO-1234ze(E) in a horizontally-assembled square cross section microchannel having an hydraulic diameter of 1.23 mm are presented. The test runs have been performed at around 30°C saturation temperature, correspondent to 5.8 bar, with mass flux ranging between 300 kg m^{-2} s^{-1} and 500 kg m^{-2} s^{-1}. As a peculiar characteristic of the present technique, the heat transfer coefficient is not measured by imposing the heat flux; instead, the boiling process is governed by controlling the inlet temperature of the heating secondary fluid. On this regard the present data is new and original since the large majority of data in the literature is taken by means of Joule effect heating. The heat transfer coefficients are compared against two predicting models available in the literature. Finally, the local heat transfer coefficients measured during flow boiling of R1234ze(E) inside the square cross section microchannel are compared against the values measured with the same refrigerant in a 0.96 mm diameter circular microchannel, with same surface roughness, with the aim of investigating the effect of channel shape on the heat transfer process.

1. Introduction

During the last years, flow boiling heat transfer inside microchannels has received increasing attention due to the rapid advances in microfabrication technology along with increasing demand for space saving and energy conservation in many applications. Although miniaturization offers many benefits, however, flow boiling in micro-devices itself involves a very complex combination of thermophysical processes which are still far from being entirely understood. Differently from single-phase flow heat transfer, the current knowledge of flow boiling heat transfer in macroscale cannot be extended without modifications to microscale, where bubble confinement plays a more relevant role with the decreasing of the channel size. In two-phase microchannel flow, capillary (surface tension) forces become predominant over gravity force, thus impeding stratification of liquid. As reported by Thome *et al.* [1], bubble flow, slug flow, annular flow and mist flow are the primary flow regimes observed in microchannels. For evaporating flow, the zone with bubbles smaller than channel diameter is very short as bubbles grow to the channel size very quickly. Chen *et al.* [2] investigated flow patterns in four circular tubes with internal diameters of 1.10, 2.01, 2.88 and 4.26 mm. The experiments were conducted in vertical upward two-phase flow using R134a as working fluid at 6, 10 and 14 bar pressure. The flow characteristics in the 2.88 and 4.26 mm tubes are similar to those typically observed in normal size tubes, while the 1.10 mm and the 2.01 mm tubes exhibit small tube characteristics. Dispersed bubble, bubbly, confined bubble, slug, churn and annular are the flow regimes observed by those authors inside the 1.1 mm diameter channel. The confined bubble flow appears in the 1.10 mm tube at all experimental pressure, suggesting that the surface tension becomes the dominant force. Tibiriçá and Ribatsky [3] presented flow pattern and bubble departure characteristics during R134a and R245fa flow boiling in a 0.4 mm circular horizontal channel at mass velocity ranging from 100 to 900 kg $m^{-2}s^{-1}$ and heat flux up to 226 kW m^{-2}. Four main flow patterns were observed (bubble, slug, annular and dryout) and the absence of the stratified flow is remarked.

In spite of the large number of papers published in the flow boiling area, many aspects still need to be better explained in order to provide a full understanding of local two-phase flow boiling characteristics in microchannels. Such knowledge is essential to develop more reliable prediction methods that can be used for designing new high-performance microchannel evaporators. The relative importance of nucleate boiling, thin film evaporation and convective boiling in the individual flow patterns, that are characteristics to a microchannel flow, is still unclear. Some recent studies have put some light on the main heat transfer mechanism. However, additional innovative studies are necessary to investigate local phenomena (Tibiriçá and Ribatski [4]). Some experimental studies show that the heat transfer coefficients obtained during vaporization in microchannels are not a function of vapor quality nor mass velocity (in contrast with the macro-channel trend), but they are a function of heat flux and saturation pressure, concluding that flow boiling in small channels is dominated by nucleate boiling (Bao *et al.* [5], Tran *et al.* [6]). Other experimental studies demonstrate that the heat transfer coefficient also depends on vapor quality and mass velocity (Yan and Lin [7], Lin *et al.* [8], Ong *et al.* [9]). Thome [10] suggests that during slug flow, nucleate boiling is completely suppressed and heat is transfer primarily by conduction through the thin evaporating film surrounding the elongated bubbles. According to the authors, the strong dependency of the boiling heat transfer coefficients on the heat flux is mechanistically explainable by the thin-film evaporation process and the cyclic heat transfer process occurring in elongated bubble flows.

Generally speaking, most of the boiling studies were performed in multiport rectangular channels. In single channels, studies were generally performed inside circular cross section microchannels with halocarbon refrigerants and fluid temperatures typical of the room ambient. Very few works deal with boiling in non-circular single-channels. In the great majority of cases, the fluid evaporation is obtained by applying electric current directly to the channel. With regard to the test fluids, most studies have been performed with HFC refrigerants which have relatively large values of GWP, but recent development efforts have been initiated to investigate halogenated olefins as possible refrigerants, with fluorinated propene isomers, in particular, having emerged as possible solutions. R1234ze(E), which has a 100-years horizon global warming potential less than 1 (IPCC [11]), seems to be suitable to substitute R134a in refrigeration applications.

Tibiriçá *et al.* [12] obtained flow boiling experimental data for the refrigerant R1234ze(E) in two horizontal stainless steel circular channels with diameter of 1.0 and 2.2 mm. The experimental campaign includes mass velocities ranging from 50 to 1500 kg m^{-2} s^{-1}, heat fluxes from 10 to 300 kW m^{-2} and saturation temperature of 25°C, 31°C and 35°C. R1234ze(E) demonstrated similar thermal performance with respect to R134a when running at similar conditions. Flow pattern characterization was performed, indicating that for both channel diameters microscale phenomena have already shown to be important. With regard to the heat transfer, they found that an increase in mass velocity or in vapor quality causes an increase in heat transfer coefficient at constant heat flux.

Del Col *et al.* [13] investigated the flow boiling process of R1234yf and R1234ze(E) inside a horizontal single circular microchannel with 0.96 mm internal diameter at 31°C saturation temperature. During the tests, the heat flux is not imposed, but it is transferred from a secondary fluid (water) to the evaporating refrigerant. The heat transfer coefficient shows to be highly dependent on the heat flux while it decreases with vapor quality when all the others parameters are constant. The mass velocity was found to have a minor effect. From the comparison between the tested refrigerants, the heat transfer coefficients of R1234yf are around 20% higher than those of R1234ze(E) and this may be related to the different reduced pressure of the fluids.

Recently, Vakili-Farahani *et al.* [14] carried out upward flow boiling experiments in a flat aluminum extruded multiport tube, which is composed of 7 parallel rectangular channels (1.1 mm x 2.1 mm) with hydraulic diameter of 1.4 mm. Two refrigerants, R245fa and R1234ze(E) were tested. A new hot water heating technique that accounts for either uniform or nonuniform local heat flux distribution along the channel was developed to obtain and reduce the data. Effects of heat flux, mass flux, vapor quality, and saturation temperature on flow boiling heat transfer in multiport tubes were considered. Once saturated boiling begins, the heat transfer coefficient profile remains either rather flat with vapor quality. Moreover, the heat transfer coefficient increases with increase of saturation temperature, heat flux and mass flux.

With reference to microchannel flow boiling databases in the open literature, it appears that most experimentalists use electrical heating. Very few researchers took data of flow boiling of a refrigerant against a secondary fluid, and this data when available refers to multiport microchannels, instead of a single microchannel. However, there are many applications where the heat flux is a dependent variable, as it happens when it is transferred from a secondary fluid. This is the case of automotive and domestic air conditioning, where the heat is removed from a secondary fluid (air).

Starting form this background, in this work new experimental heat transfer coefficients measured during flow boiling of HFO-1234ze(E) in a square cross section microchannel having an hydraulic diameter of 1.23 mm are presented. As a peculiar characteristic of the present technique, the heat transfer coefficient is not measured by imposing the heat flux; instead, the boiling process is governed by controlling the inlet temperature of the heating secondary fluid. The flow boiling data taken in the present test section is discussed, with particular regard to the effect of heat flux, mass velocity and vapor quality.

2. Experimental apparatus

The test rig used for the experimental tests is depicted in figure 1. It consists of the primary (refrigerant) loop and of two auxiliary loops: the hot water loop to promote boiling and the cold water loop to control the refrigerant condensation after the test section. The subcooled refrigerant from the condenser is sent through a filter drier into an independently controlled gear pump, which is magnetically coupled to a variable speed electric motor. During flow boiling measurements, the fluid is pumped through the Coriolis-effect mass flow meter into the test section as a subcooled liquid. The refrigerant enters the test section, which is made of two counter-flow heat exchangers. The first one (pre-sector) is used to achieve the desired inlet condition and the second one is the actual measuring sector.

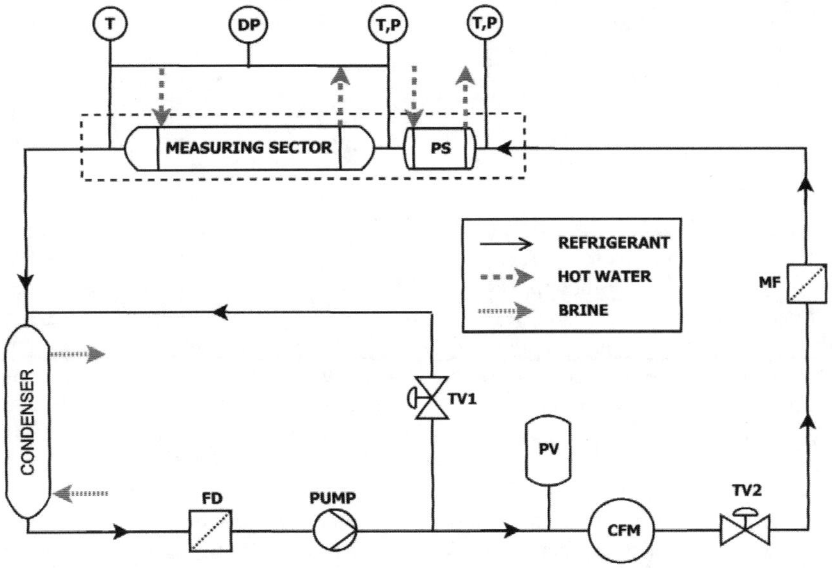

Figure 1. Experimental test rig. (FD=filter drier, PV=pressure vessel, CFM=Coriolis-effect mass flow meter, TV=throttling valve, MF=mechanical filter, PS=pre-sector, P=pressure transducer, T=temperature transducer, DP=differential pressure transducer).

Journal of Physics: Conference Series **547** (2014) 012019 doi:10.1088/1742-6596/547/1/012019

The pressure is gauged through two digital pressure transducers, connected to manometric taps to measure the fluid pressure upstream and downstream of the test tube. Two refrigerated thermal baths are used in the flow boiling tests. The first one provides the water entering the measuring sector and the pre-sector at a desired temperature and the other one provides the coolant for the condenser. When necessary, the water entering the pre-sector and the measuring sector can be maintained at different temperatures through an electric heater installed downstream of the thermal bath. The water flow rates, in the pre-sector and in the measuring sector, are measured by means of two Coriolis-effect mass flow meters and the total temperature difference of water across both sectors is measured with four copper-constantan thermocouples.

A sketch of the microchannel test section is reported in figure 2. The channel is obtained from a copper rod and has a square cross section with 1.18 mm side length. Each corner has a curvature radius equal to 0.15 mm, which leads to a hydraulic diameter equal to 1.23 mm. The internal surface roughness of the copper channel has been measured with the digital surface roughness machine ZEISS-TSK Surfcom 1400A. The measurements have been performed at different positions. The mean roughness *Ra*, as defined by the ISO 4287:1997, ranges between 0.80 μm and 1.32 μm, with a mean value equal to 1.02 μm. Figure 2 also shows the passage for the water. The geometry has been designed with the aim of increasing the external heat transfer area and thus decreasing the external heat transfer resistance. This complex flow passage causes an internal local mixing of the water and thus high heat transfer coefficients by changing continuously the flow direction and disturbing the boundary layer. During the fabrication process of the test section, the water channel in the copper is externally contained by means of a covering made with epoxy resin. The thermocouples embedded in the wall are installed in 0.6 mm diameter cylindrical holes, machined 0.5 mm away from the internal tube surface. The measuring sector is 225 mm long and is equipped with a large number of thermocouples (19 in the water, 45 in the wall). Position of thermocouples in the wall and in the water is shown in figure 2. The measuring sector is thermally separated from the pre-sector and the exit tube through stainless steel capillary tubes (adiabatic segments).

The present test tube, with the described water channel, allows the measurement of local "quasi-mixing cup" temperatures of the water at low flow rates directly in the external annular space, permitting evaluation of the local heat flux from the water temperature gradient. In particular, water thermocouples wires are fixed in the cover made with epoxy resin that surrounds the water channel. In this way the water thermocouples are not in contact with the cooper block, avoiding undesired temperature gradients. Secondly, the present design of the test section allows to embed several thermocouples in the tube without passing them through the water flow and thus preventing also the wall thermocouple wires from being exposed to severe temperature gradients.

3. Data reduction

In the present setup, the heat flux is determined from the temperature profile of the secondary fluid in the measuring sector. The wall temperature is measured by means of thermocouples embedded in the copper tube. Since the wall thermocouples are installed 0.5 mm far from the internal tube surface, a correction is required for the calculation of the heat transfer coefficient due to the distance between the thermocouple and the inner tube surface. This correction is accounted for by calculating the conductive thermal resistance of the copper between inner tube surface and wall thermocouple.

The saturation temperature is directly measured in the adiabatic segments at the inlet and outlet of the measuring sector (figure 1), by means of two T-type thermocouples soldered to the stainless steel adiabatic capillary tubes. The presence of both temperature and pressure sensors at the inlet of the measuring sector allows a double check of the saturation temperatures.

Figure 3 displays the refrigerant, wall and water temperature measurements along the measuring sector during flow boiling of R1234ze(E) at mass velocity $G = 300$ kg m^{-2} s^{-1}. The refrigerant enters the test channel at 30.2°C saturation temperature (corresponding to 5.8 bar saturation pressure) with about 5 K subcooling. At the exit, saturation temperature is decreased by 1.3 K due to pressure drop.

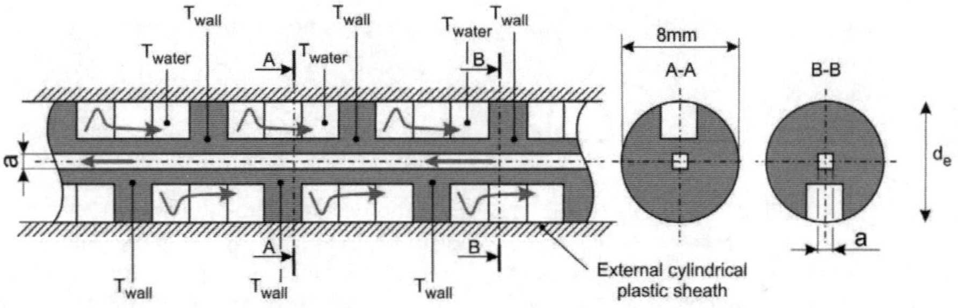

Figure 2. Detailed sketch of the test section: geometry of the water flow passage.

The water enters from the opposite side of the test tube at 37.2°C and its temperature decreases due to the heat transfer to the boiling refrigerant. As the vaporization proceeds, the wall temperature increases uniformly up to the end of the test section. Figure 3 also shows the standard deviation of the temperature readings in the wall: each temperature measurement is the average value of fifty readings taken in fifty seconds and standard deviations are pretty much the same in all the channel. It should be mentioned that, during experimental tests, when dryout takes place, a great increase of the wall temperature standard deviation is found just at the location where the heat transfer starts to decrease due to the thermal crisis. The standard deviation of the fifty temperature readings goes back to the initial values in the superheated vapor zone. This trend was previously detected by Del Col *et al.* [15] and it is related to the dryout of the film adjacent to the heated wall and the temperature fluctuation is the symptom of an oscillating drying process of the film at the wall. It is worth noting the relevance of such a criterion for the determination of the critical quality in the channel when the heat flux is not imposed. During tests with electrical heating, when the critical conditions occur, an abrupt change in the wall temperature is evident. When vaporizing by means of a secondary fluid, still there is a temperature change in the wall, but it is limited by the temperature of the secondary fluid. Only the data points characterized by vaporization prior to onset of dryout are considered in the present paper.

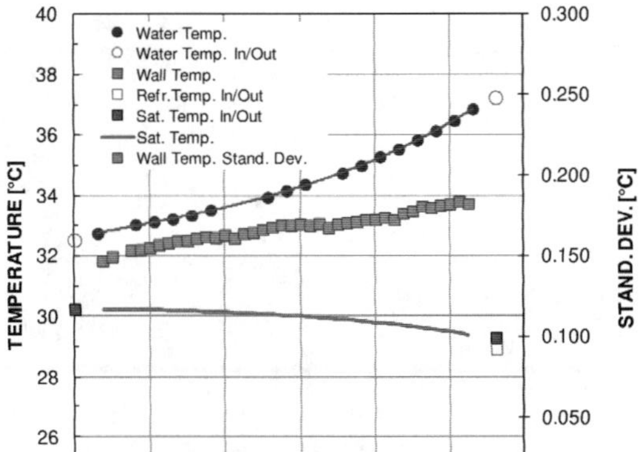

Figure 3. Water, wall and saturation temperature at G = 400 kg m^{-2} s^{-1}. The big square dots represent the standard deviation of the wall temperature recording.

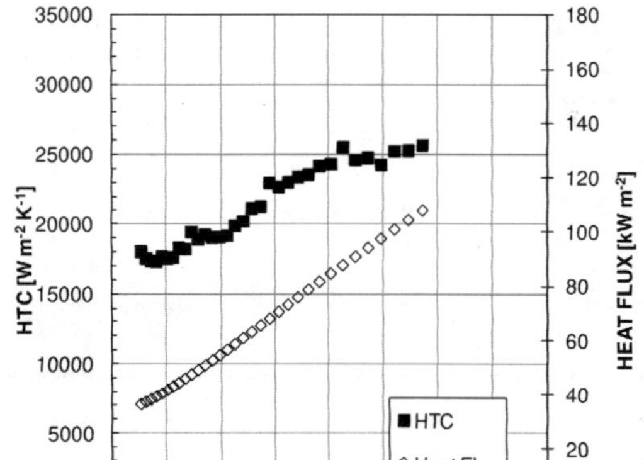

Figure 4. R1234ze(E) flow boiling test at G = 400 kg m^{-2} s^{-1}: heat transfer coefficient and heat flux versus vapor quality.

3.1. Heat flux

Since the heat flux q' is not directly fixed here, it must be obtained indirectly, from the slope of the secondary fluid temperature profile:

$$q' = \dot{m}_w \cdot c_{p,w} \cdot \frac{1}{p}\frac{\mathrm{d}T_w}{\mathrm{d}z} \tag{1}$$

where z is the axial coordinate along the tube, $\mathrm{d}T_w/\mathrm{d}z$ is the derivative of the water temperature along z, $\dot{m}_w$ is the water mass flow rate, $c_{p,w}$ is the water specific heat capacity and p is the channel perimeter. In the calculation, a polynomial function is used to interpolate the water temperature profile along the channel. Two different independent procedures have been adopted for the determination of the fitting polynomial degree of the water temperature and thus the calculation of the heat flux (equation 1). The first criterion, dubbed as physical, is based on the assumption that all the calculated values by the polynomial interpolation of the water temperatures should be below ±0.05 K with respect to the experimental values, and 68% of the predicted values should be inside the ±0.03 K prediction boundary. Beside this, a statistical method based on the choice of the simplest fitting polynomial has been used. This statistic approach uses the R^2 statistic parameter, and adjusts it on the basis of the residual degrees of freedom: this is the reason of the name R^2_{adj} (Rawlings *et al.* [16]). The polynomial degree is the one that can satisfy both criteria.

3.2. Saturation temperature

The fluid temperature is directly measured by means of adiabatic segments only at inlet and outlet of the test tube. Nevertheless the local saturation temperature can be determined from the pressure profile along the channel. When the pressure is measured at inlet and outlet, as in the present case, a linear variation is usually assumed in the channel. However, during vaporization, the vapor quality changes along the microchannel and therefore the pressure gradient varies in the measuring sector; for this reason, a linear interpolation of the saturation temperature between inlet and outlet may lead to errors in the determination of the heat transfer coefficient. Instead, in the present work, the pressure profile inside the microchannel is calculated by implementing the Del Col *et al.* [17] two-phase pressure gradient correlation and then integrating the calculated pressure gradient over the total length: the resulting value is corrected by an empirical coefficient to match the total pressure drop to the value measured by the differential pressure transducer. Once the pressure profile is known, the local saturation temperature can be determined. In figure 3 the saturation temperature profile is plotted with a solid line.

3.3. Heat transfer coefficient and vapor quality

The local heat transfer coefficient α inside the microchannel is obtained as the ratio of heat flux q' to temperature difference:

$$\alpha = \frac{q'}{\left(T_{wall} - T_{sat}\right)} \tag{2}$$

where the saturation temperature T_{sat} is obtained from pressure measurements. The measured saturation temperature at the outlet can be compared with the saturation temperature obtained from the pressure, provided that no local superheating occurs.

The vapor quality x at any axial position z is obtained from:

$$x = \frac{q - \dot{m}_r \cdot (h_{SL} - h_{sub})}{\dot{m}_r \cdot h_{LG}} \tag{3}$$

as a function of the total heat flow rate q up to that location, the specific enthalpy of saturated liquid (h_{SL}), the specific enthalpy of the subcooled liquid (h_{sub}), the latent heat (h_{LG}) and the refrigerant mass flow rate $\dot{m}_{ref}$. The enthalpy of the subcooled refrigerant is determined from the inlet pressure and temperature, while the total heat flow rate transferred to the refrigerant is obtained by integrating the local heat flux q' from 0 (refrigerant inlet) to z.

3.4. Axial conduction
Given the measurement of the wall temperature along the channel, it is possible to calculate the contribution of the axial conduction in the copper wall. The effect of axial conduction on the heat transfer coefficient can be determined from an energy balance at the wall: this allows to determine the actual heat flux to be used in the evaluation of the heat transfer coefficient. The heat transfer coefficient is corrected by the same amount as the heat flux, being fixed the measured values of wall and saturation temperature. The correction for axial conduction does not significantly affect the values of heat flux and heat transfer coefficient.

4. Flow boiling results
Flow boiling tests have been performed with R1234ze(E) at mass velocity ranging between 300 and 500 kg m^{-2} s^{-1}, at a saturation temperature around 30°C in the present 1.23 mm hydraulic diameter square cross section channel. During the test runs, only mass velocities and inlet fluid temperatures can be fixed; thus the heat flux varies along the process and this complicates the comparison of the present tests to the ones obtained with electrical heating. Wall and water temperature profiles during flow boiling process at $G = 400$ kg m^{-2} s^{-1} are illustrated in figure 4. Since refrigerant and water flow in counter-current, the heat flux usually increases along the channel in the refrigerant flow direction. Figure 4 reports heat flux, vapor quality and heat transfer coefficient. The heat flux and the heat transfer coefficient increase with vapor quality. No dryout is reached during the test run. All the data points reported in the present chapter were taken after steady state conditions were reached by at least ten minutes. Furthermore experimental data refer to boiling conditions before the dryout. Test runs are performed with decreasing heat flux, to make sure that all nucleation sites are activated.

4.1. Heat flux
In this work the heat flux is not electrically imposed: the controlled parameters are the inlet temperature and the mass flow rate of the secondary fluid. As a consequence, for each test run, not only the heat transfer coefficient and the vapor quality vary along the channel but also the heat flux is a dependent variable (figure 4). Therefore, with the present technique, it is impossible to get the heat transfer coefficient variation with vapor quality in one test run as reported in literature studies using electric heating. In order to explore the effects of heat flux and vapor quality on the heat transfer coefficient, several tests were made at constant refrigerant mass velocity and varying the inlet water conditions (mass flux and temperature). A further data processing is then required to investigate the effects of vapor quality, mass velocity and heat flux on the heat transfer coefficient.

The effect of heat flux on the heat transfer coefficient can be better understood in figure 5 where the heat transfer data has been filtered at constant quality $x = 0.2$ for three different mass velocities. The data points show a clear dependence of the heat transfer coefficient on the heat flux. The rising heat transfer coefficients with heat flux in the saturated region conform with the results observed by many authors (Lazarek and Black [18], Ong and Thome [19], Del Col *et al.* [20]).

4.2. Mass velocity and vapor quality
The effect of mass velocity on the heat transfer coefficient can be found in figure 5. It can be seen that at a given heat flux, the corresponding heat transfer coefficients are independent of mass velocity. Such behavior has been confirmed by other experiments (Shiferaw *et al.* [21] and Maqbool *et al.* [22]).

By processing data at constant heat flux, it is possible to get information on the influence of vapor quality. Figure 6 shows the experimental trend of heat transfer coefficient at 300 and 400 kg m^{-2} s^{-1} mass velocity. Two different values of heat flux are considered. The heat transfer coefficient is pretty constant with vapor quality for both conditions.

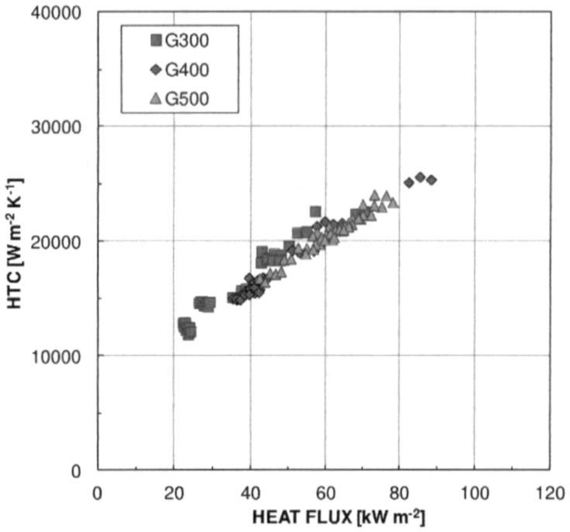

Figure 5. Local heat transfer coefficient versus heat flux at constant vapor quality x = 0.2.

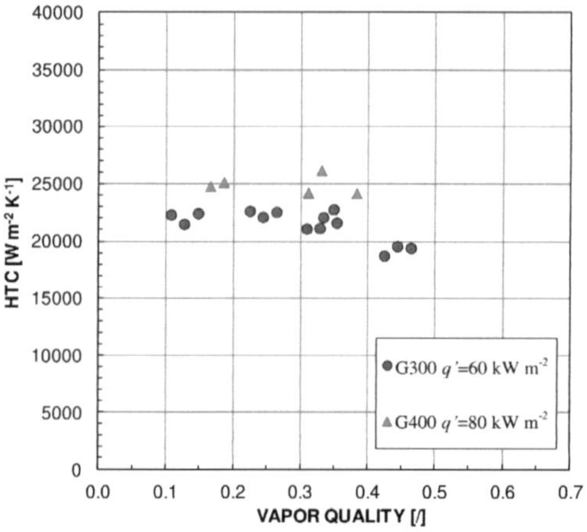

Figure 6. Local heat transfer coefficient versus vapor quality at two different values of heat flux q' and mass flux G [kg m^{-2} s^{-1}].

4.3. Effect of channel shape

In figure 7, the present data is compared with R1234ze(E) flow boiling measurements by Del Col *et al.* [13] in a round channel. The test section was a copper cold drawn tube of internal diameter 0.96 mm, and roughness Ra = 1.3 μm. Del Col *et al.* [13] data is taken at mass velocity ranging between 300 and 500 kg m^{-2} s^{-1}, heat flux from 20 to 100 kW m^{-2}, 0.2 quality and 30°C saturation temperature. The experimental conditions are similar to the ones tested by the present authors. From the graph in figure 7, it is clear that the heat transfer coefficients in the square channel are higher than those obtained in the circular channel. Since flow visualization inside the square channel is not possible, only an attempt to give a physical explanation of such behavior can be made.

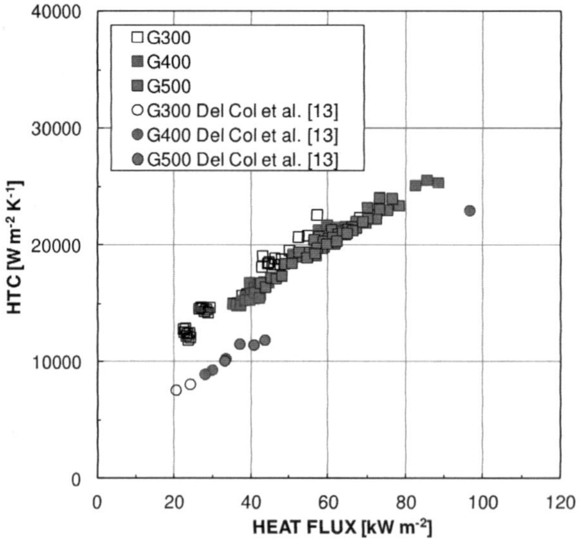

Figure 7. Comparison between the heat transfer coefficients obtained inside the present square channel and the data by Del Col *et al.* [13] inside a 0.96 mm circular channel at 0.2 vapor quality.

In the work by Tibiriçá and Ribatski [3] it is reported that bubbles can detach from the wall with diameters smaller than the tube diameter and the authors observed bubble active sites for all flow patterns, also in the slug flow regime. They found that, according to the heat flux level over the tube surface, the number of nucleation sites increases. For lower heat flux conditions, one nucleation site is possible but increasing the heat flux, multiple nucleation sites become active. Furthermore, nucleation sites can also be active over all vapor quality range, depending on the level of the applied heat flux and the presence of liquid in contact with the wall.

The surface tension in a square channel is known to push le liquid towards the corners of the channel, leading to a thinner liquid film on the flat sides (Bortolin *et al.* [23]). Starting from the visualizations by Tibiriçá and Ribatski [3] it can be argued that the heat transfer coefficient enhancement in the square channel could be due to the presence of several active nucleation sites at the corners of the channel, where the liquid is pulled by surface tension. However, more investigation is needed to confirm such hypothesis. Also, it is worth mentioning that the diameters of the square and the circular channels are not exactly the same.

5. Comparison with the models

More than 650 data points have been collected during the present experimental campaign. Vapor quality varies between 0.05 and 0.7, mass velocity between 300 and 500 kg m^{-2} s^{-1}, while the heat flux ranges from 20 to 140 kW m^{-2}. The present database has been compared against two models available in the open literature: the model by Lazarek and Black [18] and the model by Sun and Mishima [24].

The comparison with Lazarek and Black [18] is shown in figure 8. On average, this model seems to be able to predict the experimental trend versus heat flux, while it strongly underpredicts the experimental results.

In figure 9 the experimental data are compared with predicted heat transfer coefficient values by Sun and Mishima [24] model. Sun and Mishima developed a new correlation based upon the Lazarek and Black [18] correlation and taking the effect of Weber number into account. This model is able to catch the experimental trend with heat flux but the data are underpredicted by 50%.

6. Conclusions

In the present paper the authors describe a different experimental technique to measure the flow boiling heat transfer coefficient in a square cross section microchannel having an hydraulic diameter of 1.23 mm, using a secondary fluid instead of fixing the heat flux by electrical heating.

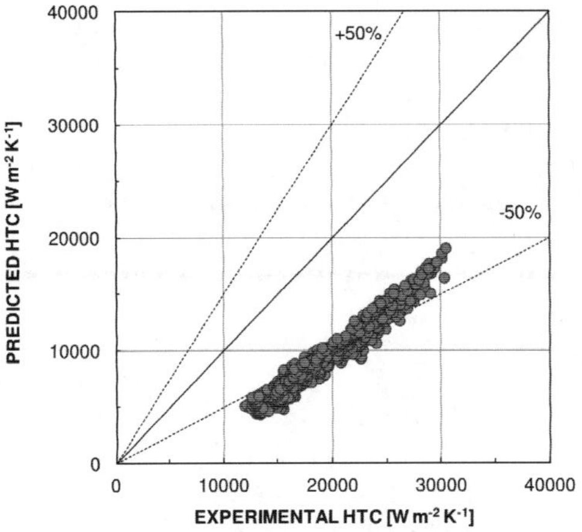

Figure 8. Experimental data compared against the model by Lazarek and Black [18].

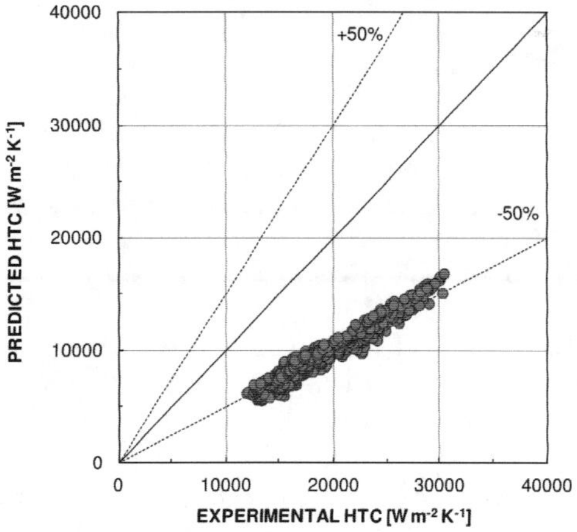

Figure 9. Experimental data compared against the model by Sun and Mishima [24].

The heat transfer coefficient is measured at around 30°C saturation temperature during flow boiling of R1234ze(E). The experimental heat transfer coefficient is highly dependent on the heat flux. The behavior of the heat transfer coefficients, that increase with heat flux in the saturated region, agrees with the results reported in the literature by many authors. Mass velocity and vapor quality were found to have a negligible effect on the heat transfer coefficient.

The present data is compared with R1234ze(E) flow boiling measurements by Del Col *et al.* [13] taken inside a 0.96 mm microchannel displaying a similar surface roughness. At the same experimental conditions (saturation pressure, mass velocity and heat flux), the heat transfer coefficient is higher in the square channel as compared to the circular one. It should be noticed that, the boiling process described in this paper and in the work by Del Col *et al.* [13] is governed by controlling the inlet temperature of the heating secondary fluid.

The present data has been compared with two predicting correlations available in the open literature: the model by Lazarek and Black [18] and the model by Sun and Mishima [24]. Both models are able to catch the heat transfer coefficient trend versus heat flux but none of them provides an accurate estimation of the heat transfer coefficients.

7. References

[1] Thome J R, Bar-Cohen A, Revellin R and Zun I 2013 *Exp. Therm. Fluid Sci* **44** 1-22
[2] Chen L, Tian Y S and Karayiannis T G 2006 *Int. J. Heat and Mass Transfer* **49** 4220-4230
[3] Tibiriçá C B and Ribatski G 2014 *Exp. Therm Fluid Sci.* http://dx.doi.org/10.1016/j.expthermflusci.2014.02.017
[4] Tibiriçá C B and Ribatski G 2013 *Int. J. Refrigeration* **36** 301-324
[5] Bao Z Y, Fletcher D F and Haynes B S 2000 *Int. J. Heat and Mass Transfer* **43** 3347–3358
[6] Tran T M, Wambganss M W and France D M 1996 *Int. J. Multiphase Flow* **22** 485–498
[7] Yan Y Y and Lin T F 1998 *Int. J. Heat and Mass Transfer* **41** 4183–4194
[8] Lin S, Kew P A and Cornwell K 2001 *Int. J. Refrigeration* **24** 51–56
[9] Ong C L and Thome J R 2004 *Exp. Therm Fluid Sci.* **33** 651-663
[10] Thome J R 2004 *Int. J. Heat and Fluid Flow* **25** 128-139
[11] IPCC 2013 *Climate Change 2013: The Physical Science Basis. Contribution of Working Group I to the Fifth Assessment Report of the Intergovernmental Panel on Climate Change* ed Stocker T F, Qin D, Plattner G-K, Tignor M, Allen S K, Boschung J, Nauels A, Xia Y, Bex V and Midgley P M (Cambridge and New York: Cambridge University Press) p 733
[12] Tibiriçá C B, Ribatski G and Thome J R 2012 *J. Heat Transfer* **134** 020906
[13] Del Col D, Bortolato M and Bortolin S Measurements of heat transfer coefficient during flow boiling of two HFOs inside a circular minichannel, *Proc. 8th World Conference on Experimental Heat Transfer, Fluid Mechanics, and Thermodynamics* June 16-20 2013 Lisbon
[14] Vakili-Farahani F, Agostini B and Thome J R 2013 *Int. J. Refrigeration* **36** 335-352
[15] Del Col D and Bortolin S 2012 *Int. J. Thermal Sciences* **57** 25-36
[16] Rawlings J O, Pantula S G and Dickey D A 1998 *Applied Regression Analysis* (New York: Springer)
[17] Del Col D, Bisetto A, Bortolato M, Torresin D and Rossetto L 2013 *Int. J. Heat and Mass Transfer* **67** 326–337
[18] Lazarek G M and Black S H 1982 *Int. J. Heat and Mass Transfer* **25** 945-960
[19] Ong C L and Thome J R 2011 *Exp. Therm. Fluid Sci.* **35** 873-886
[20] Del Col D, Bortolin S and Rossetto L 2013 *Int. J. Heat and Mass Transfer* **67** 1231-1245
[21] Shiferaw D, Karayiannis T G and Kenning D B R 2009 *Int. J. Therm. Sci.* **48** 331-341
[22] Maqbool M M, Palm B and Khodabandeh R 2013 *Exp. Therm. Fluid Sci.* **46** 120-130
[23] Bortolin S, Da Riva E and Del Col D 2014 *Heat Transfer Engineering* **35** 193-203
[24] Sun L and Mishima K 2009 *Int. J. Heat and Mass Transfer* **52** 5323-5329

Application of a hybrid kinetic-continuum solver to the near wall modelling

O Rovenskaya and G Croce
University of Udine, via delle Scienze 208, 33100 Udine, Italy

E-mail: olga.rovenskaya@uniud.it, giulio.croce@uniud.it

Abstract. A hybrid method dynamically coupling the direct numerical solution of the S-model kinetic equation and Navier-Stokes equations is applied to a numerical simulation of the flow through the channel of a finite length due to arbitrarily pressure ratios and for a wide range of Knudsen number. The decomposition of the physical domain into kinetic and hydrodynamic sub-domains is updated at each time step. The solution is advanced in time simultaneously in both kinetic and hydrodynamic domains: the coupling is achieved by matching half fluxes at the interface of the kinetic and Navier-Stokes domains, thus taking care of the conservation of momentum, energy and mass through the interface. Solver efficiency is increased via MPI (Message Passing Interface) parallelization. Accuracy and reliability of the method, for different decomposition criteria, are assessed via comparison with a pure kinetic solution.

1. Introduction

At microscale, the analysis of gaseous flow may require to take into account rarefaction effects. The first area where such effect becomes significant is the near-wall region. This is due to the well known fact that the flow near a solid surface can be divided into a thin boundary layer, of the order of a few mean free path (so thin to be negligible for macro configuration), which is a rarefied regime area and the internal core flow, which is a continuum one. The rarefied domain is naturally described by kinetic equation for the velocity distribution function, which involves a considerable effort in terms of CPU time and memory requirements, due to the discretization in both physical and velocity spaces. On the other hand, the continuum domain is well described by the fluid Navier-Stokes (NS) equations in terms of mere average flow velocity, gas density and temperature. These equations are more efficient, but less accurate in critical rarefied areas. The inaccuracy of NS equations in boundary layer can be partially overcome by introducing slip boundary and temperature jump conditions on the solid surface. However, as indicated in the literature, the slip conditions are valid only for the local Knudsen number $Kn \leq 0.1$, and any attempt to increase their range, resorting to higher order slip boundary conditions, is not trivial and highly geometrical dependent. Therefore, the development of hybrid solvers applying kinetic model in the region with high rarefaction, while keeping continuum model in the rest of the flow allow us to simulate flows with an accuracy close to the full kinetic solution and computational expenses close to the continuum one. Thus, this became an important area of research over the last decade. Potential applications of such solvers range from gas flows in micro systems to aerospace applications, such as high altitude flights. Major challenges in the development of hybrid code are the identification of kinetic and continuum domains and the choice of the coupling technique.

The most popular methods, in the open literature, are based on domain decomposition in physical space: the computational domain is decomposed into kinetic and continuum sub-domains using appropriate criteria [1-7]. Typically, particle methods such as DSMC or Molecular Dynamics are used in regions with strong deviations from equilibrium, and a continuum fluid (Euler or NS, depending on problem features) solver is used elsewhere [4]. Nonetheless, the Direct Numerical Solution (DNS) of kinetic equations is a viable alternative to DSMC and may even be preferable to it, for coupling purposes, since both kinetic and continuum models use similar numerical techniques. Recent efforts to combine the DNS of kinetic equation with a NS solver used a priori domain decomposition in [1, 2] and later evolved in dynamically updated decomposition in [3]. In [3] a wall boundary independent problem (the gas flow through the slit) was considered and the decomposition of the physical domain into kinetic and continuum sub-domains was done by computing gradient-length Knudsen number Kn_{GL}, based on the local Knudsen number and macroparameters gradients proposed in [4, 6].

In the present paper the same hybrid solver is applied to the simulation of the rarefied gas flow through a channel of finite length caused by an arbitrary pressure ratio, where the flow is dominated by the wall boundary effect. The hybrid results are compared with pure kinetic and NS solutions, for different pressure ratios and a wide range of Knudsen number. The results are discussed in terms of both accuracy and computational efficiency.

2. Statement of the problem

The two-dimensional pressure–driven monoatomic gas flow through a planar microchannel of width H and length $l = 10H$ connecting two reservoirs of $L_x \times L_y = 30H \times 15H$ is considered. The gas in reservoirs far from the channel is in equilibrium at constant pressures p_0 and p_e, $(p_0 > p_e)$, and temperature T_0. The temperature of the wall T_w is equal to the temperature in reservoirs T_0. Since the flow is symmetric about $y = 0$ only a half of the domain is simulated. The interface I_c between kinetic and NS sub-domains is dynamically updated during the computation.

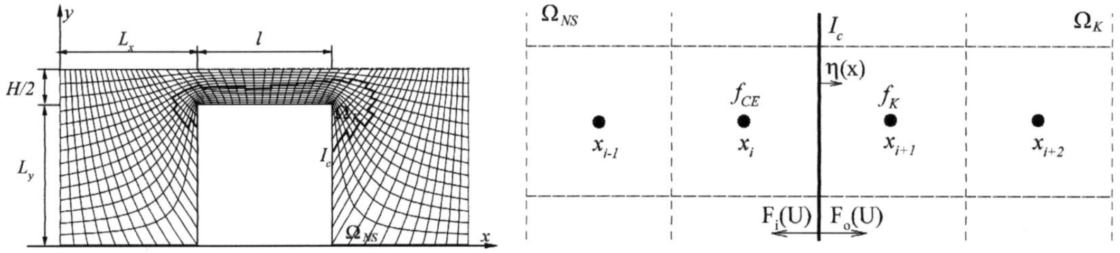

Figure 1. Computational domain and coupling procedure sketch.

The gas flow is determined by the pressure ratio p_0/p_e and the rarefaction parameter δ:

$$\delta = \frac{p_0 H}{\mu_0 v_0} = \frac{8}{5Kn\sqrt{\pi}} \tag{1}$$

$v_0 = (2RT_0)^{0.5}$ is the thermal speed, R is the specific gas constant, μ_0 is the dynamic viscosity at T_0.

The dimensionless flow rate W is introduced as a global characteristic of the flow:

$$W = \frac{\dot{m}}{m_{fm}} \qquad m_{fm} = \frac{p_0}{\sqrt{\pi}v_0} \tag{2}$$

m_{fm} is the analytically deduced in the limit of free molecular regime mass flow rate. The mass flow rate through the channel is constant and computed at channel mid-section as:

$$\dot{m} = 2\int_0^{H/2} \rho(x, y)u(x, y)dy \tag{3}$$

where $\rho(x, y)$ is the gas density, $u(x, y)$ is the gas bulk velocity.

The friction factor f_r is defined as an average value over a channel section of $\Delta L \approx 9$ (i.e., skipping small regions $\Delta l \approx 0.5$ both at the inlet and exit of channel to avoid entrance and exit effects):

$$f_r = 2\frac{\Delta\overline{p}}{\Delta L}\frac{D_h}{\overline{\rho}_{av}\overline{u}_{av}^{\,2}} \tag{4}$$

where D_h is the hydraulic diameter, $D_h = 2H$, overbar means average values over cross section, Δ and av are the difference and the average between inlet and outlet values. The Poiseuille number Po is then written in terms of the local Re as:

$$Po = f_r \cdot Re = 2\frac{\Delta\overline{p}}{\Delta L}\frac{D_h^2}{\overline{\mu}_{av}\overline{u}_{av}} \qquad Re = \frac{\overline{\rho}_{av}\overline{u}_{av}D_h}{\overline{\mu}_{av}} \tag{5}$$

Due to the conservation of mass flow rate, Re will change only slightly along the channel, because of the dynamic viscosity μ variations with temperature.

3. Coupling algorithm and numerical method

The coupling strategy between kinetic and continuum solvers is applied to the S-model of Boltzmann equation [8] and NS equations. The S-model can be written in dimensionless form as:

$$\frac{\partial f}{\partial t} + \xi\frac{\partial f}{\partial \mathbf{x}} = \delta\,\rho\sqrt{T}\,(S - f) \tag{6}$$

where $f = f(t, \mathbf{x}, \xi)$ is the velocity distribution function, i.e. the probability of finding a molecule with velocity $\xi = (\xi_x, \xi_y, \xi_z)$ in the position $\mathbf{x} = (x, y)$ at the time moment t. S is the standard Shakhov function [8], ρ is the gas density, T is the temperature, $\mathbf{c} = \xi - \mathbf{V}$ is the relative speed of a molecule against a background gas with bulk velocity $\mathbf{V} = (u, v)$. The problem was recast in terms of dimensionless variables using the inlet reservoir equilibrium values and the height of the channel H. Assuming the hard-sphere molecular model the dimensionless viscosity coefficient is $\mu = T^{0.5}$. The macroscopic density, momentum and internal energy e per unit mass, respectively, are computed as:

$$\rho = \int f d\xi \qquad \rho(u,v)^T = \int (\xi_x, \xi_y)^T f d\xi \qquad \rho e = \frac{1}{2}\int \mathbf{c}^2 f d\xi \tag{7}$$

Since the flow considered is two-dimensional it is possible to eliminate the third velocity component [1-3]. The S- model equation is discretized in both velocity and physical spaces and solved using the explicit-implicit numerical scheme [1-3]. In particular, the transport term in equation (6) is treated explicitly and approximated by a standard finite volume scheme. Numerical fluxes are determined by the TVD scheme with minmod limiter.

According to the Chapman-Enskog (CE) theory [9] the NS equations are the first order approximation of the Boltzmann equation. Thus, introducing dimensionless CE function f_{CE} as:

$$f_{CE} = M(\mathbf{c})(1 + f_1) = M(\mathbf{c})\Psi(\mathbf{c}) \tag{8}$$

$$f_1 = \frac{1}{pT\delta}\left(\tau_{ij}c_ic_j - q_ic_i\left(1 - \frac{2}{5}\frac{\mathbf{c}^2}{T}\right)\right) \tag{9}$$

$$\tau_{ij} = -\mu\left(\frac{\partial u_i}{\partial x_j} + \frac{\partial u_j}{\partial x_i} - \frac{2}{3}\delta_{ij}\frac{\partial u_k}{\partial x_k}\right), \qquad q_i = -\kappa\frac{\partial T}{\partial x_i} \tag{10}$$

where $M(\mathbf{c})$ is the standard Maxwellian, f_1 is the correction term, τ_{ij} is the shear stress tensor, δ_{ij} is the Kronecker delta, $\kappa = 15/4\mu$ is the thermal conductivity for the hard sphere molecular model. Substituting f_{CE} for f in equation (6), multiplying by the collision vector $\varphi(\xi) = (1, \xi, \xi^2/2)$ and integrating over the whole velocity domain $[-\infty;\infty]$ we recover the conservative form of NS equations:

$$\frac{\partial \mathbf{U}}{\partial t} + \frac{\partial \mathbf{F}(\mathbf{U})}{\partial \mathbf{x}} = 0, \tag{11}$$

$$\mathbf{U} = \int f_{CE}\varphi(\xi)^T d\xi = \left(\rho, \rho\mathbf{V}, \rho(e + \mathbf{V}^2/2)\right)^T \tag{12}$$

$$\mathbf{F}(\mathbf{U}) = \int \xi \, f_{CE} \varphi(\xi)^T d\xi \tag{13}$$

where **U** is the vector of macroscopic values, **F(U)** is the flux.

The Navier-Stokes solver is based on a hybrid finite difference-finite volume method and has formal second order accuracy in space and time [1-3]. Fluxes are defined via neighbouring values averaging, and an artificial dissipation term is added to prevent checker boarding and numerical instabilities. Artificial dissipation terms are given by a blend of second and fourth order differences, scaled by the maximum eigenvalue of jacobian matrix of the convective terms in vector **F**, as suggested in [10]. Second order terms are switched on near discontinuities. Viscous flux vectors are evaluated with second order finite differences centred at mesh midpoint i+1/2. The solution is advanced in time via Crank Nicolson integration scheme. The use of the spatially factored ADI scheme proposed by Beam and Warming leads, at each time step, to the resolution of two series of cheap block tridiagonal algebraic systems, rather than the original pentadiagonal block system arising from flux discretization.

3.1. Breakdown parameter

The choice of the breakdown criterion defining the size of Ω_K region is important, since a wrong domain decomposition could even lead to a non-positive velocity distribution function. One of the options proposed in the literature is the gradient-length Knudsen number $Kn_{GL}(x, y)$ [4, 6]:

$$Kn_{GL}(x, y) = \max(Kn_{GL\rho}, Kn_{GL\,V}, Kn_{GL\,T}), \tag{14}$$

$$Kn_{GL\,\Phi}(x, y) = Kn \frac{|\nabla\Phi(x, y)|}{\Phi(x, y)}, \tag{15}$$

where $\Phi(x, y)$ represents the parameter of interest: the local density ρ, the magnitude $|\mathbf{V}|$ of local bulk velocity and the local temperature T. The kinetic solution is performed where $Kn_{GL}(x, y) > \varepsilon = 0.05$. It should be noticed that here Kn_{GL}, based on the gradient of the velocity magnitude $|\mathbf{V}|$, is different from what proposed in [4, 6]. The new definition takes into account share stresses which are essential for the considered problem. Moreover, the velocity magnitude gradient in $Kn_{GL\,|\mathbf{V}|}$ is normalized via the maximum between $|\mathbf{V}|$ and a minimum value $v_{\min}$, here chosen as 5% of the theoretical isentropic exit velocity, in order to avoid singularities where the velocity approaches to zero.

Another breakdown parameter is based on the CE distribution function [5, 6]. When f_1 (equation (8)) is small $\Psi(\mathbf{c})$ slightly deviates from 1 and NS equations are valid; when $\Psi(\mathbf{c})$ is sufficiently far from unity then NS equations may fail, and a kinetic approach is required. The direct evaluation of f_1 is difficult and numerically expensive because it is a function of not only flow field gradients but also the random velocity **c** and so either an average or maximum value of f_1 must be evaluated over the full distribution function. Thus, in [5, 6] an approximate breakdown parameter based on the normalized shear stress τ^*_{ij} and heat flux q^*_i was proposed:

$$B_{q\tau} = \max(|\tau^*_{ij}|, |q^*_i|) \tag{16}$$

$$\tau^*_{ij} = \tau_{ij} / p\delta, \; q^*_i = q_i / p\sqrt{T}\delta \tag{17}$$

The kinetic solution is activated when $B_{q\tau} > 0.1$.

Tiwari [7] has proposed a similar criterion based on the norm of additional term in CE distribution function $\|f_1\|$ which can be written in dimensionless form as:

$$\| f_1 \| = \frac{1}{p\delta} \sqrt{2|\tau_{ij}|^2 + \frac{1}{5}\frac{|q_i|^2}{T}} \tag{18}$$

For the continuum approximation to hold this parameter should be much less than unity.

3.2. Coupling procedure

To solve a kinetic equation in a physical cell x_{i+1} we need the distribution function of incoming particles from the neighbouring cell x_i (see Fig. 1). On the coupling interface I_c this information has to be provided by the continuum solution: therefore, we impose f_{CE} distribution function:

$$f\,|_{I_c}(\mathbf{x}) = f_{CE}(x_i) \qquad \text{if } \xi\cdot\boldsymbol{\eta}(\mathbf{x}) > 0 \tag{19}$$

where $\boldsymbol{\eta}(\mathbf{x})$ is the inward normal vector to the boundary of Ω_K. Macroscopic values ρ, **V**, T appearing in $f_{CE}(x_i)$ are computed at the grid point $x_i \in \Omega_{NS}$ and the evaluation of parameters gradients involves also values in the neighbor to x_i points: x_{i-1} and x_{i+1}. The interface I_c position is updated at each time step and depends on the chosen breakdown criterion and threshold value ε; thus, some nodes considered as continuum at previous time step may become kinetic ones at current time step. In this case, the kinetic distribution function f in "new" kinetic nodes will be initialized as $f_{CE}(\mathbf{x})$ computed using current macroscopic values and their gradients.

To solve NS equations at the coupling interface I_c the following boundary condition is imposed:

$$\mathbf{F}(\mathbf{U})\cdot\boldsymbol{\eta}(\mathbf{x})\,|_{I_c} = \mathbf{F}_i(\mathbf{U})\cdot\boldsymbol{\eta}(\mathbf{x}) + \mathbf{F}_o(\mathbf{U})\cdot\boldsymbol{\eta}(\mathbf{x}) \tag{20}$$

$\mathbf{F}_i(\mathbf{U})\cdot\boldsymbol{\eta}(\mathbf{x})$ and $\mathbf{F}_o(\mathbf{U})\cdot\boldsymbol{\eta}(\mathbf{x})$ are half fluxes associated with incoming (from Ω_K to Ω_{NS}) and outcoming (from Ω_{NS} to Ω_K) molecules:

$$\mathbf{F}_i(\mathbf{U})\cdot\boldsymbol{\eta}(\mathbf{x}) = \int_{\xi\cdot\boldsymbol{\eta}(\mathbf{x})<0} \xi\cdot\boldsymbol{\eta}(\mathbf{x}) f(t,\xi,\mathbf{x})\varphi(\xi)^T d\xi \tag{21}$$

$$\mathbf{F}_o(\mathbf{U})\cdot\boldsymbol{\eta}(\mathbf{x}) = \int_{\xi\cdot\boldsymbol{\eta}(\mathbf{x})>0} \xi\cdot\boldsymbol{\eta}(\mathbf{x}) f_{CE}(t,\xi,\mathbf{x})\varphi(\xi)^T d\xi \tag{22}$$

where $f(t, \mathbf{x}, \xi) \equiv f_K$ is the current solution of kinetic equation for molecules exiting from Ω_K.

If a boundary node at the inlet/outlet falls in continuum domain the conditions in left and right reservoirs specify the inlet total temperature and pressure at the inlet, or the static pressure at the outlet. If an inlet/outlet boundary node falls in the kinetic domain a standard Maxwell distribution function, based on inlet/outlet total temperature and pressure, is assumed for incoming particles.

At solid wall in kinetic domain Ω_K Maxwell diffuse reflecting boundary condition with the full accommodation is applied. In Ω_{NS} domain the Maxwell first order slip boundary condition and a Dirichlet temperature boundary condition with Smoluchowski temperature jump are imposed.

The hybrid code is parallelized in order to improve its efficiency using MPI message passing protocol. The solution of kinetic equation (5) is local in velocity node, and therefore completely parallelizable in the velocity space. For the solution of NS equation a set of block tridiagonal algebraic systems is solved sequentially: the solution of each system may thus be considered as independent and can easily be parallelized. The software code was written in C++ (for the kinetic and coupling coding) and Fortran (for the NS related procedures) with the use of MPI. Computations were carried out on Multi Core system consisting of 2 quad core processors, Intel(R) Xeon(R) E5520 CPU, 2.27 (2.93) GHz, 8 MB Cache, for a maximum of 8 concurrent cores.

4. Results and discussion

4.1. Parameters of modeling

The size of uniform two-dimensional velocity grid should be selected large enough to capture all of the important features of the problem: here, the velocity space boundary satisfies the condition $v_{\max} \geq \max(|u|, |v|) + 3.5T_{\max}^{0.5}$. For most of the computations the number of grid points for each velocity component is 24 and velocity space is bounded by $v_{\max} \approx 5$. For high gradient flow $p_0/p_e = 10$ $v_{\max}$ increases up to 7.2, while number of points up to 40. The optimal number of grid points in velocity space was chosen by checking that doubling it produces a change in mass flow rate lower than 1-1.5%. In the physical space, a non uniform structured single block curvilinear grid of 360 nodes in the streamwise direction, with minimum grid spacing 0.02H and 40 nodes in the transverse direction (with minimum grid spacing $0.008H$), is used. Grid independence test has been done using a coarse grid with 240 × 40 nodes. The mass flow difference between finer and coarse meshes is less than 1%. The time step is unique for both solvers and it should satisfy the stricter stability (or accuracy) constraint $\Delta t = \min(\Delta t_K, \Delta t_{NS})$. The solution is considered to be converged when the criterion $\|\mathbf{U}^{n+1}-\mathbf{U}^n\|_{L2} < \Delta$ is fulfilled with L_2 norm and $\Delta = 10^{-7}$.

4.2. Breakdown criteria comparison

In [3], for the flow through a slit, the breakdown criterion $Kn_{GL} > 0.05$ guaranteed a difference between hybrid and kinetic solutions within 1%. Here, we evaluate the effectiveness of modified Kn_{GL} (using gradient of velocity magnitude $|\mathbf{V}|$) and the two more complex criteria $B_{q\tau}$ and $\|f_1\|$ applied to the boundary value problem, via comparison with pure kinetic Boltzmann computations [11].

Table 1. Mass flow rate W_h and Po number

	Criterion	N_K	W_h	W_K, [11]	Po_h	Po_K
	$Kn_{GL}(\varepsilon = 0.1)$	5640	0.1840		52.3	
$p_0/p_e = 2$	$B_{q\tau}(\varepsilon = 0.015)$	5640	0.1844	0.184	52.2	51.3
$\delta = 10$	$\|f_1\|(\varepsilon = 0.05)$	5700	0.1843		52.16	
	$Kn_{GL}(\varepsilon = 0.1)$	3030	0.515		92.47	
$p_0/p_e = 2$	$B_{q\tau}(\varepsilon = 0.015)$	6800	0.515	0.51	92.4	91.6
$\delta = 50$	$\|f_1\|(\varepsilon = 0.05)$	5300	0.511		92.38	
	$Kn_{GL}(\varepsilon = 0.1)$	7050	0.277		47.26	
$p_0/p_e = 10$	$B_{q\tau}(\varepsilon = 0.015)$	8160	0.280	0.280	46.95	46.2
$\delta = 10$	$\|f_1\|(\varepsilon = 0.05)$	7050	0.280		47.16	
	$Kn_{GL}(\varepsilon = 0.1)$	4270	0.604		103	
$p_0/p_e = 10$	$B_{q\tau}(\varepsilon = 0.015)$	7770	0.605	0.604	102.8	102.5
$\delta = 50$	$\|f_1\|(\varepsilon = 0.05)$	8268	0.6045		102.8	

In table 1 mass flow rates W_h and global Po_h using different criteria at different pressure ratios and rarefaction levels are presented. At $p_0/p_e = 2$ and $\delta = 10$ the value of ε, for each criterion, was chosen such as to provide hybrid solution close to kinetic one while using the same number of kinetic points N_K. This 'optimal' value is retained for further computations with the same criterion, i.e.: $\varepsilon = 0.1$ for Kn_{GL}, $\varepsilon = 0.015$ for $B_{q\tau}$ and $\varepsilon = 0.05$ for $\|f_1\|$. For both pressure ratios the maximum deviation of mass flow rate W_h from kinetic value is around 1% and 2% for Po number. At the same time the difference between hybrid solutions computed with different criteria is less than 1%. It should be noticed that for low rarefaction level ($\delta = 50$) the use of criterion $\|f_1\|$ and $B_{q\tau}$ generates larger number of the kinetic points than the use of $Kn_{GL}$with little or no gain in accuracy. As can be seen in figure 2 (top), showing contour line of local Knudsen number, all criteria create almost identical kinetic regions at $p_0/p_e = 2$ and $\delta = 10$. On the other hand, for high rarefaction $\delta = 50$ at the same pressure ratio (figure 2, bottom) criteria $\|f_1\|$ and $B_{q\tau}$ create significantly larger kinetic region: 6800 points for $\|f_1\|$, 5300 for $B_{q\tau}$, 3030 for Kn_{GL}. Thus, for further hybrid computations breakdown parameter Kn_{GL} with $\varepsilon = 0.1$ will be used.

4.3. Hybrid results discussion

Assuming the kinetic solution as the reference one, we may define the relative error induced by the use of hybrid and pure NS solvers as:

$$\Delta W = 1 - \frac{W}{W_S}, \qquad \Delta Po = 1 - \frac{Po}{Po_S}, \tag{23}$$

for mass flow rate W and Poiseuille number Po, respectively.

These relative differences ΔW_{NS}, ΔPo_{NS}, ΔW_h and ΔPo_h at pressure ratios p_0/p_e of 10, 3, 2 and 1.1 and rarefaction parameter δ ranging from 10 to 100 are shown in figure 3. In the slip regime ($\delta \geq 50$) mass flow rates and Po obtained by hybrid, kinetic and NS solvers are close to each other: maximum difference between results is less than 2%. Starting from $\delta = 20$ the difference becomes noticeable, around 4-6%. When rarefaction increases the difference growth up to 10-14% at $\delta = 10$.

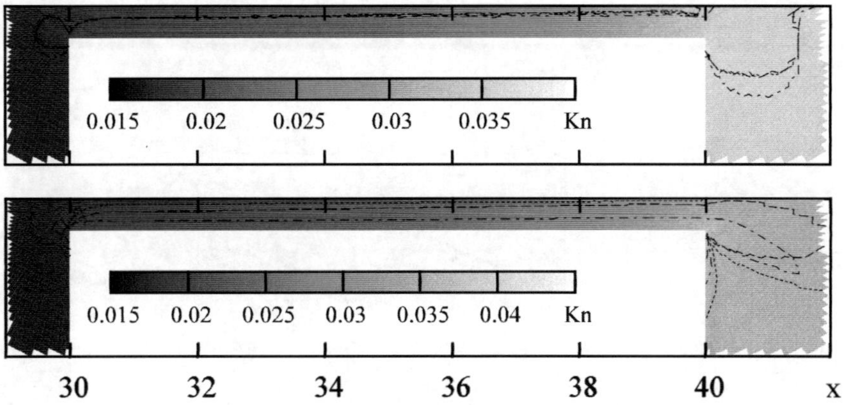

Figure 2. Domain decomposition: Kn_{GL} (dashed dotted), $B_{q\tau}$ (dashed), $\|f_1\|$ (dotted): $p_0/p_e = 2$, $\delta = 10$ (top); $\delta = 50$ (bottom).

When gradients of macroparameters become higher the local Knudsen number increases: thus, at the same rarefaction parameter δ, we can have higher local Knudsen for higher pressure ratio; e.g., for rarefaction $\delta = 10$ and for pressure ratio $p_0/p_e = 2$ the difference between NS and kinetic results is around 10%; for $p_0/p_e = 3$ the difference is around 12% and 14% for $p_0/p_e = 10$. On the other hand, the hybrid code successfully reproduces kinetic values for both W and Po under any condition (difference is within 1% for W and 2% for Po).

Profiles of density and temperature, normalized by inlet values, and velocity along the symmetry axis $y = 0$ are shown in figure 4 for selected cases, chosen in order to demonstrate noticeable difference between NS and S-model solutions.

Density (or pressure) variations are qualitatively similar in all cases. Before and after the channel density tends to upstream and downstream conditions, while decreasing along the channel.

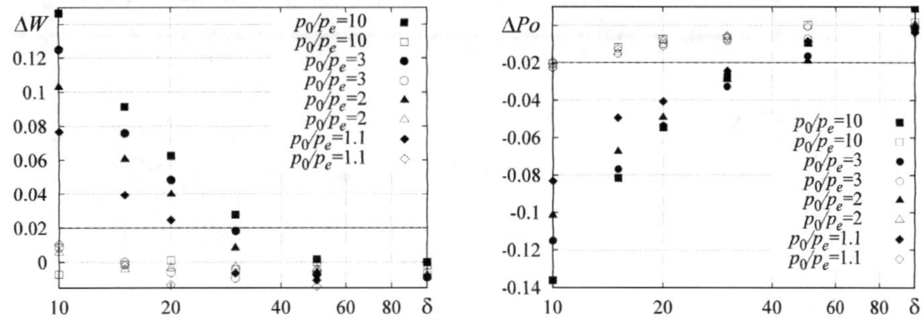

Figure 3. ΔW_{NS}, ΔPo_{NS} (solid symbols) and ΔW_h, ΔPo_h (empty symbols) via δ at different p_0/p_e

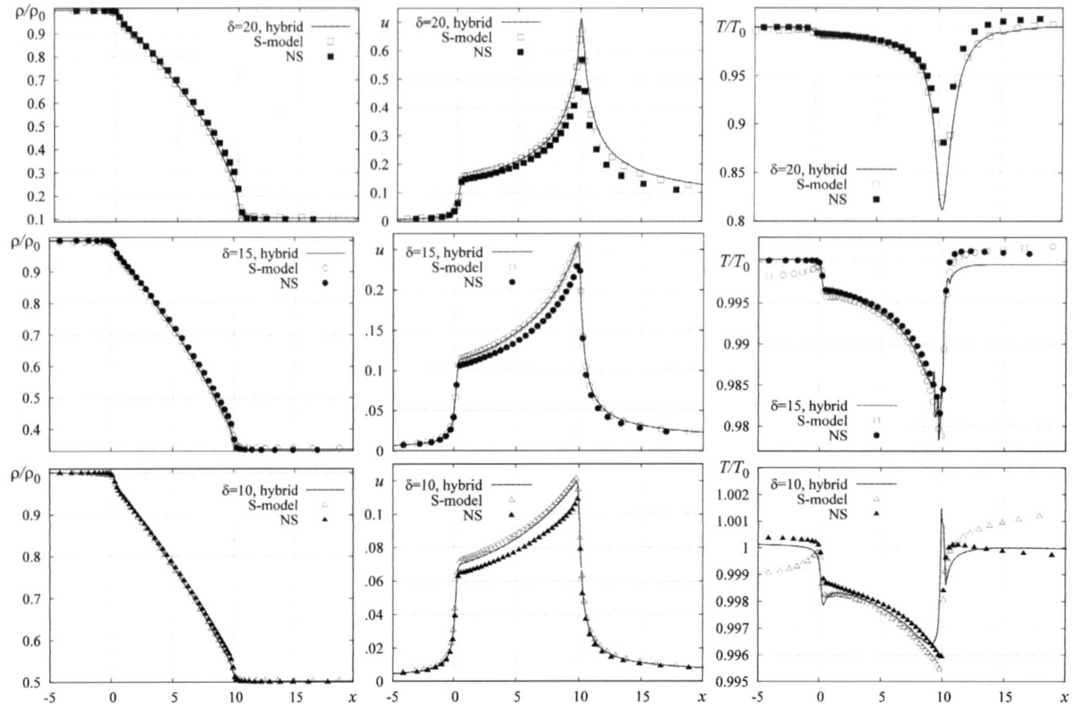

Figure 4. Density ρ/ρ_0, velocity u and temperature T/T_0 along symmetry line. Top: $p_0/p_e = 10$, $\delta = 20$; middle: $p_0/p_e = 3$, $\delta = 15$; bottom: $p_0/p_e = 2$, $\delta = 10$.

At high pressure ratio the density profile is nonlinear. The axial velocity, close to zero in the large reservoirs, grows along the channel. Accordingly, the temperature decreases through the channel, while in the low velocity reservoirs it approaches to the reference temperature. The maximum variations of macroparameters appear for the highest pressure ratio $p_0/p_e = 10$ (figure 4, top). The hybrid solutions are close enough to the kinetic ones for all values of rarefaction and pressure ratios.

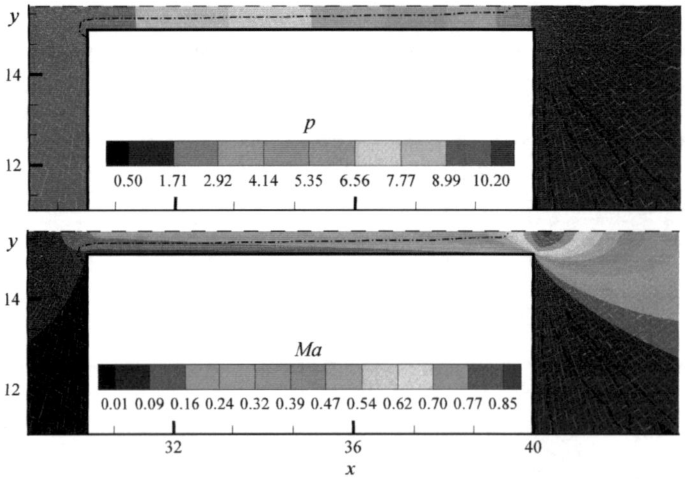

Figure 5. Dimensionless pressure (top) and Mach number (bottom) contours at $\delta = 20$ and $p_0/p_e = 10$; dashed dotted line shows interface I_c at convergence.

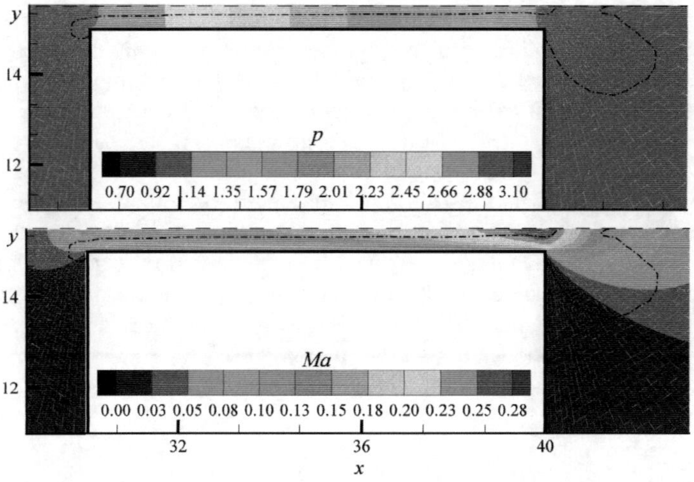

Figure 6. Dimensionless pressure (top) and Mach number (bottom) contours at $\delta = 15$ and $p_0/p_e = 3$; dashed dotted line shows interface I_c at convergence.

Contour lines of pressure and Mach number in the channel for different rarefaction are presented in figures 5-7, together with the extension of the kinetic region at convergence. For low rarefaction $\delta = 20$ (figure 5), but high pressure ratio 10, the local Kn near the coupling boundary I_c (dashed dotted line) varies from 0.045 at the inlet of the channel to 0.14. At the exit maxim local Kn is 0.5. For rarefaction $\delta = 15$ and $p_0/p_e = 3$, figure 6, the local Knudsen number near the coupling boundary varies from 0.06 to 0.16, while the maximum local Kn occurs at the exit is equal to 0.2.

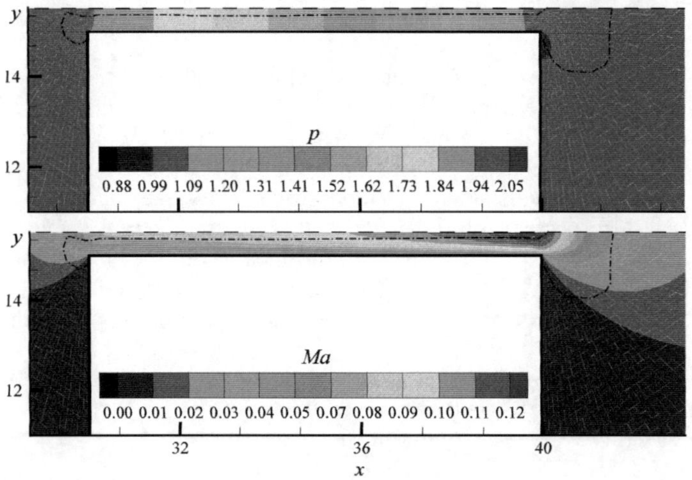

Figure 7. Dimensionless pressure (top) and Mach number (bottom) contours at $\delta = 10$ and $p_0/p_e = 2$; dashed dotted line shows interface I_c at convergence.

For the smaller rarefaction parameter value, $\delta = 10$, and $p_0/p_e = 2$, shown in figure 7, the local Knudsen near the boundary is higher, varying from 0.092 to 0.17, and its maximum value is 0.19 near the exit. The coupling between kinetic and NS solutions shows a smooth transition along the contour lines crossing the domain interface I_c for both Ma and pressure contours. Moreover, as can be seen in figure 4, NS equations do not give appropriate solution especially for velocity and temperature.

CPU time per time step of hybrid code t_h is the sum of the time required for the solution of the S-model equation in N_K kinetic points (N_K is estimated from converged hybrid solution), of NS equations and the coupling procedures. Since the sum of CPU times relative to NS solution and coupling computation is quite small in comparison to the kinetic solution requirements, time t_h is essentially dictated by the number of N_K kinetic points should be solved during computation. Figure 8 shows the hybrid code speedup compared to the pure kinetic S-model code as a function of rarefaction parameter. The hybrid code offers a CPU time speedup ranging from 2, at a rarefaction level where NS solution would be completely inadequate, up to 5 in the slip flow regime. For example, pure kinetic computation at $p_0/p_e = 10$ and $\delta = 20$ on physical mesh of 360×40 and velocity mesh of 40×40 requires 2.51 s, while hybrid time step t_h is 1.3 s (number of kinetic points $N_K = 7050$) and the NS step is $t_{NS} = 0.14$ s. The total CPU advantage over the whole computation, with the dynamic coupling, is however appreciably greater, since at the beginning of the computation the kinetic area is typically much smaller than at converged state. It can be seen in figure 8 that with decrease in δ the speedup decreases, since the number of kinetic points in Ω_K increases.

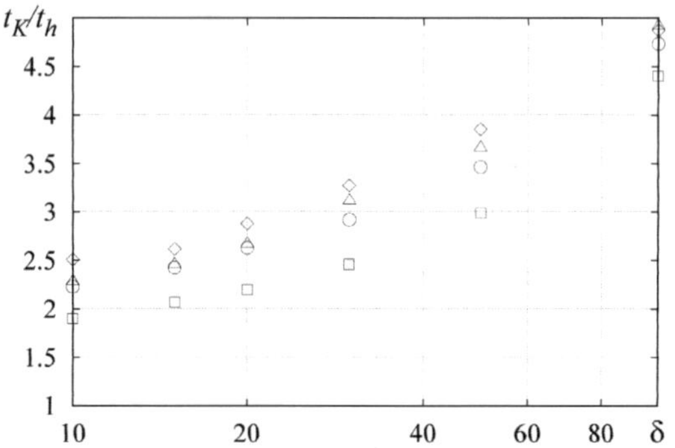

Figure 8. Speedup for hybrid simulations: □- $p_0/p_e = 10$; ○ - $p_0/p_e = 3$;Δ - $p_0/p_e = 2$; ◊- $p_0/p_e = 1.1$.

5. Conclusion

A hybrid algorithm based on the direct numerical solution of the kinetic S-model equation coupled with a Navier-Stokes model was applied to the numerical investigation of a gas flow through the channel. It was shown that hybrid code gives results close to full kinetic solutions for flow regimes where the Navier-Stokes solution clearly fails. It was found that the breakdown criterion based on the modified gradient length Knudsen number is more suitable for the boundary value problem. The obtained speed up, with respect to full S -model solutions, is between 2 and 5, although strongly depends on the size of the kinetic region. Moreover, the accuracy of the NS solution provided with slip boundary conditions is reasonable up to $Kn \leq 0.05$ (deviation from kinetic solution does not exceed 5%). Thus, for higher Kn the use of first order slip boundary condition is not recommended.

References

[1] Rovenskaya O, Croce G 2013 *Heat Transfer Eng.* **34** 192
[2] Rovenskaya O, Croce G 2013 *J. Heat Transfer* **135** (10) 101005
[3] Rovenskaya O, Croce G 2014 *Vacuum* **109** 266
[4] Schwartzentruber T E, Boyd I D 2006 *J. of Comput. Phys.* **215** 402
[5] Garcia A L, Alder BJ 1998 *J. of Comput. Phys.* **140** 66
[6] Boyd ID 2003 *AIP Conf. Proc.* **663** 899
[7] Tiwari S, Klar A, Hardt S 2009 *J. of Comput. Phys.* **228** 7109
[8] Shakhov EM 1968 *Fluid Dyn.* **3** 95
[9] Chapman S, Cowling T G 1990 *The mathematical theory of nonuniform gases* (Cambridge: Cambridge Univ. Press).
[10] Pulliam T H 1986 *AIAA Journal* **24** 1931
[11] Rovenskaya OI 2013 *Comput.& Fluids* **81** 45

Journal of Physics: Conference Series **547** (2014) 012021 doi:10.1088/1742-6596/547/1/012021

On the effective thermal conductivity of metal foams

Paola Ranut and Enrico Nobile

Department of Engineering and Architecture, University of Trieste, Piazzale Europa 1, 34127 Trieste (Italy)

E-mail: pranut@units.it

Abstract. Knowing the effective thermal conductivity is essential in order to design a metal foam heat transfer device. Beside the experimental characterization tests, this quantity can be deduced from empirical correlations and theoretical models. Moreover, CFD (Computational Fluid Dynamics) and numerical modeling in general, at the pore scale, are becoming a promising alternative, especially when coupled with a realistic description of the foam structure, which can be recovered from X-ray computed microtomography (μ-CT). In this work, a review of the most relevant correlations and models published in the literature, usable for the estimation of the effective thermal conductivity of metal foams, will be outlined. In addition, a validation of the models with the experimental values available in the literature will be presented, for both air and water as working fluids. Furthermore, the results of a strategy based on μ-CT - CFD coupling at the pore level will be illustrated.

1. Introduction

In the last decades, research in flow and heat transfer in porous media has received increased attention, in view of the broad range of applications that they find in science and engineering. Porous media are employed, for instance, in packed and fluidized beds, catalytic converters, filtration equipments, light-weight structural panels, heat and energy absorption devices. In recent years, high-porosity metal foam materials have caught the attention in view of their low density and peculiar transport properties, which make them attractive for enhancing the thermal performances of heat transfer devices, while allowing the use of smaller and lighter equipments. Most common metal foams are made of aluminum and copper alloys. Usually, their porosity ε, i.e., the volume fraction occupied by the voids, falls between 80% to 97%, while the pore density of uncompressed open-cell foams ranges from 5 to 100 PPI (pores per inch). The cross section of the solid ligaments varies according to the porosity, from an inner concave triangle at $\varepsilon = 0.97$ to a circle at $\varepsilon = 0.85$. In relation to heat transfer, the attractiveness of metal foams derives from their specific geometric structure, as the high surface area to volume ratio, the high conductivity of the solid matrix and the enhanced flow mixing induced by the tortuous flow paths, which is sometimes referred to as thermal dispersion.

In order to effectively utilize metal foams in heat transfer applications, accurate knowledge of their thermal transport properties are needed. From a practical point of view, it is common to assimilate the foam to a homogeneous medium having an effective thermal conductivity k_{eff} which depends, to varying degrees, on conduction in both the solid and fluid phase, convection and radiation [1]. Conduction in the solid phase is the principal heat transfer modality, even in the case of less conducting solid matrices. As Calmidi and Mahajan [2] observed, the role

of thermal dispersion induced by hydrodynamic phenomena can be considerable for water-foam systems but it is almost negligible in the case of air-foam systems in view of the relatively high conductivity of the solid phase. Furthermore, at high temperatures, the effective thermal conductivity may depend even on radiation. The simplest heat transfer mechanism in a porous medium involves only conduction and occurs when the fluid inside is stagnant. In this case, the energy transport is controlled by the so called effective stagnant thermal conductivity, or effective thermal conductivity hereafter.
Interest in heat transfer in porous media dates back to the 19th century. Since Maxwell's work in 1865, there have been many efforts to estimate the effective thermal conductivity but, despite the constant advances, the problem is still open. Among the large amount of works about porous media, several papers provide relations specifically derived for metal foams, in which convection and thermal radiation are neglected. According to the methodology followed, these works can be classified as (1) asymptotic solutions, (2) empirical correlations and (3) unit cell approaches. In addition, the literature offers several contributions in which the effective thermal conductivity is estimated numerically, both on simplified models [3] and realistic geometries [1, 4, 5].
The paper is organized in the following way. In Sec. 2, a review of the most significant empirical and analytical models for the estimation of the effective thermal conductivity of metal foams, summarized in Table 1, is described. In Sec. 3, the results of three characterization tests based on a μ-CT - CFD strategy, involving numerical simulations at the pore level, are illustrated.
It is worth pointing out that not all the models delineated in Sec. 2 were developed directly for metallic foams. The literature includes further models which have not been considered here since they should be optimized before being applied to metal foams. For the model illustrated, the predicting capabilities have been tested against a wide set of experimental values taken from the literature, which comprises data for both air and water [6–9]. The RMS values reported in Table 1 corresponds to the mean square error between the available experimental data and the values calculated with the model.

2. Available models for the effective thermal conductivity

2.1. Asymptotic solutions

This group incorporates the theoretical correlations corresponding to two limiting cases for the effective conductivity, called upper and lower bounds.
The Parallel model $k_{||}$, Eq. (1), is an upper bound which combines the thermal conductivity of the solid, k_s, and the fluid, k_f, in an arithmetic mean, through the volume fractions of the two phases. The lower bound Series model, $k_{\perp}$, Eq. (2), is obtained by arranging k_s and k_f in series. Two other limiting solutions exist, the Maxwell-Eucken correlations, Eq. (3) and Eq. (4), originally formulated for a solid medium having a random distribution of small spheres
As it can be appreciated from Table 1, these asymptotic solutions are not suited for describing the real trend of the experimental data.

2.2. Empirical correlations

This category comprises the relations obtained by fitting experimental data. Their ability to accurately describe the behaviour of metal foams is often demanded to the fitting parameters, whose values can be calibrated over the experimental measurements available.
Calmidi and Mahajan [6] formulated an alternative of the parallel model, as described in Eq. (5). The values of the constants A and n were calibrated on their experimental data on aluminum foams, having porosities higher than 0.90. Equation (5) fits well the available experimental data [6–9] with a RMS of 0.309% for air and 0.180% for water.
Bhattacharya et al. [10] combined the Series and the Parallel models in Eq. (6), for which the value of the constant A was estimated after a fitting with the data of Ref. [6] and with new data

on reticulated vitreous carbon foams. Equation (6) gives a RMS of 0.360% and 0.353% when applied to air and water, respectively.
Also Singh and Kasana [11] expressed the effective thermal conductivity as a combination of $k_{||}$ and $k_{\perp}$, Eq. (7). The parameter F, corresponding to the fraction of solid oriented in the direction of the heat flow, comprises an empirical constant, whose value was calibrated over the data of Ref. [10]. This model produces a RMS of 0.329% and 0.038% in the case of air and water, respectively.
Krupiczka, reported from Ref. [12], formulated an empirical correlation, Eq. (8), for packed beds of spherical particles. Unfortunately, as can be appreciated from Table 1, this expression can not be applied satisfactorily to metal foams.
In addition, even if it seems not to work for metal foams, it is worth mentioning the work of Wang et al. [13], who expressed the effective thermal conductivity of complex materials by combining some basic structural models (Series, Parallel, Maxwell upper and lower bound, and the so called Effective Medium Theory) through combinatory rules based on structure volume fractions.

2.3. Unit cell approaches

Fall in this group the theoretical models formulated on an idealized and simplified foam geometry. In this case it is customary to model only a single unit cell, which is assumed to repeat itself throughout the medium in view of periodicity.
Calmidi and Mahajan [6] used a two-dimensional array of hexagonal cells, with square lumps of metal at the intersection of the ligaments. The geometry is constituted of three layers in series, within which the solid and fluid phase are arranged in parallel. The resulting equation for k_{eff}, Eq. (9), fits well the experimental data with a RMS of 0.330% and 0.048% for air and water, respectively.
Boomsma and Poulikakos [14] followed a strategy similar to [6], by using a 3D tetrakaidecahedron cell, made of four distinct layers in series. The ligaments had cylindrical cross section and cubic nodes are considered at their intersection. The model presents a parameter, to be estimated from the experimental data. However, Dai et al. [15] detected some errors in their development and presentation: with their corrections, the predictive capability of the model worsen significantly. Even if omitted in [16], the optimal value of the fitting parameter for the revised model leads to an unfeasible geometry. For this reason, the model is not reported in Table 1.
Jagjiwanram and Singh [17] considered a cubic unit cell having, at each vertex, one eighth of an ellipsoidal inclusion. The model, Eq. (10), presents a porosity-correction term F, which depends on two constants, C_1 and C_2, calibrated over the experimental data of Ref. [10]. A careful analysis of [17] shows that there is a typing error in reporting the value of C_2 for water: the proposed value of 0.1535 must be replaced with 0.5134. In this way the predicted values agree with those reported by the authors. The RMS of the fitting is, for air, 2.445% and, for water, 0.159%.
A cubic unit cell with cylindric struts and square lumps of metal at their intersection was used by Edouard [18]. The model can account for two different foams structures, namely fat and slim, as shown in Eq. (11), depending on whether there is accumulation of material at the struts' connections. The value of $1-\varepsilon$ was derived for a regular pentagonal dodecahedral cell having triangular struts, reported in Ref. [19]. The predicting capability of the model is quite scarce: in the case of the fat model, the RMS is 1.847% in the case of air and 2.616% for water.
A different approach was followed by Fourie and Du Plessis [20], under the assumption of thermal nonequilibrium: empirical correlations were developed after having computed numerically the temperature field in a 3D unit cell, constituted by three mutually perpendicular solid square prisms. The proposed correlations account both of the effective (k_{ss}, k_{ff}) and the coupled (k_{fs}, k_{sf}) thermal conductivities of the medium. Coupled conductivities are considered since,

in the case of thermal nonequilibrium, conduction within a phase is function of the local mean temperature gradients of both phases. Their correlations presents a RMS of 1.495% when applied to air, and 0.657% in the case of water.
In addition to the aforementioned models, specifically formulated for metal foams, in the literature there are other works which consider other types of porous media. As an example, Schuetz and Glicksmann developed an analytical model for high-porosity polymeric foams ($\varepsilon > 0.95$), reported in Ref. [21], in which the only structural parameters are the porosity and the fraction of solid phase in the struts, f_s. As it can be seen from Eq. (13), the model resembles the Parallel model, Eq. (1), with the exception of the multiplicative factor. Although quite simple, the predictive capability of this model for metal foams is rather good: the RMS of the fitting is 0.257% and 0.414% for air and water, respectively.
Dul'nev [22] proposed a model, Eq. (14), for disperse solid systems, based on a cubic unit cell formed by 12 solid prisms of square section. This model fits the experimental data with a RMS of 0.795% in the case of air, and 0.340% in the case of water.
Ahern et al. [21] developed a correlation, based on the Maxwell relation for mixtures of materials having different electrical conductivities, Eq. (15), as function of the fraction of solid in the faces (called windows by the authors), f_w. If applied to the available data set, the RMS of the fitting is 0.258% for air and 0.345% for water.

A graphical comparison of the most effectual models for the effective thermal conductivity is illustrated in Fig. 1 and Fig. 2.

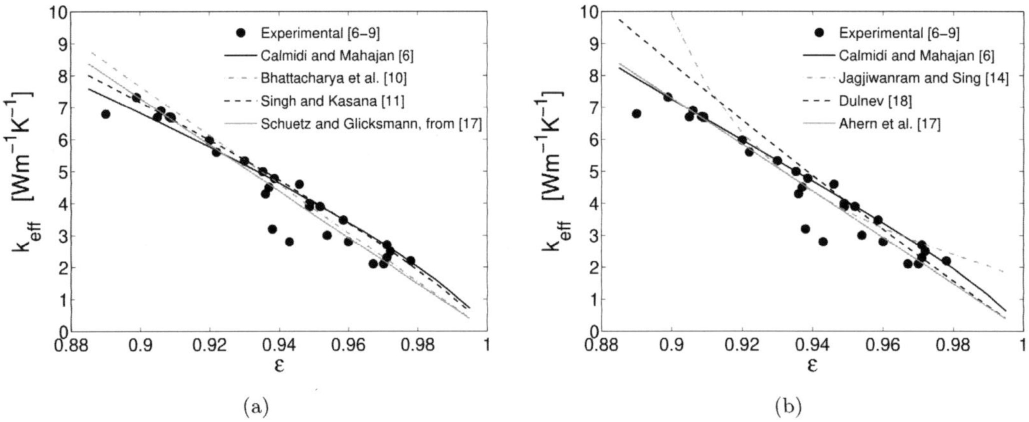

Figure 1: Validation of the models for the prediction of the effective thermal conductivity in the case of air.

Table 1: Available models for effective thermal conductivity. (A) stands for air while (W) for water.

Authors	**Model**		**RMS** (%)
	Asymptotic solutions		
Parallel model	$k_{\|\|} = \varepsilon k_f + (1-\varepsilon)k_s$	(1)	15.23 (A) 67.64 (W)
Series model	$k_\perp = \left(\dfrac{\varepsilon}{k_f} + \dfrac{1-\varepsilon}{k_s}\right)^{-1}$	(2)	22.36 (A) 19.70 (W)

Table 1: continues in the next page.

Table 1: continues from the previous page

Authors	**Model**		**RMS** (%)
Maxwell upper bound	$k_{\text{eff}} = k_s \dfrac{2k_s + k_f - 2(k_s - k_f)\varepsilon}{2k_s + k_f + (k_s - k_f)\varepsilon}$	(3)	26.79 (A) 14.33 (W)
Maxwell lower bound	$k_{\text{eff}} = k_f \dfrac{3k_s - 2(k_s - k_f)\varepsilon}{3k_f + (k_s - k_f)\varepsilon}$	(4)	22.33 (A) 19.04 (W)
	Empirical correlations		
Calmidi and Mahajan [6]	$k_{\text{eff}} = \varepsilon k_f + A(1-\varepsilon)^n k_s$ $A = 0.181, n = 0.763$	(5)	0.309 (A) 0.180 (W)
Bhattacharya et al. [10]	$k_{\text{eff}} = A k_{\|\|} + (1-A)k_{\perp}$ $A = 0.35$	(6)	0.360 (A) 0.239 (W)
Singh and Kasana [11]	$k_{\text{eff}} = k_{\|\|}^F k_{\perp}^{1-F}$ $F = C\left[0.3031 + 0.0623 \ln\left(\varepsilon k_s/k_f\right)\right]$ $0 \le F \le 1$, $C = 0.9683$ (A), $C = 1.0647$ (W)	(7a) (7b)	0.329 (A) 0.038 (W)
Krupiczka, from Ref. [12]	$k_{\text{eff}} = k_f \left(\dfrac{k_s}{k_f}\right)^{0.28 - 0.757\log\varepsilon - 0.057\log\left(k_s/k_f\right)}$ Developed for packed beds of spherical particles	(8)	21.13 (A) 16.601 (W)
	Analytical models		
Calmidi and Mahajan [6]	$k_{\text{eff}} = \left(\dfrac{2}{\sqrt{3}}\left(\dfrac{r\frac{b}{L}}{k_f + \left(1+\frac{b}{L}\right)\frac{k_s-k_f}{3}} + \dfrac{(1-r)\frac{b}{L}}{k_f + \frac{2}{3}\frac{b}{L}\left(k_s-k_f\right)} + \dfrac{\frac{\sqrt{3}}{2} - \frac{b}{L}}{k_f + \frac{4r}{3\sqrt{3}}\frac{b}{L}\left(k_s-k_f\right)}\right)\right)^{-1}$ $\dfrac{b}{L} = \dfrac{-r + \sqrt{r^2 + \frac{2}{\sqrt{3}}(1-\varepsilon)\left(2 - r\left(1+\frac{4}{\sqrt{3}}\right)\right)}}{\frac{2}{3}\left(2 - r\left(1+\frac{4}{\sqrt{3}}\right)\right)}$ $r = 0.09$, $r = t/b < 0.0336$, b: half thickness of the lump, L: length of the strut, t: half thickness of the strut	(9a) (9b)	0.330 (A) 0.048 (W)
Jagjiwanram and Singh [17]	$k_{\text{eff}} = \dfrac{k_f\left[(k_s-k_f)\sqrt{\pi/6}\sqrt{F} + k_f\right]}{k_f + (1-\sqrt{6/\pi}\sqrt{F})(k_s-k_f)\sqrt{\pi/6}\sqrt{F}}$ $\sqrt{F} = C_1\sqrt{1-\varepsilon}\exp(k_f/k_s) + C_2$ $C_1 = 0.034, C_2 = 0.7111$ (A) $C_1 = 0.5217, C_2 = 0.5135$ (W)	(10a) (10b)	2.45 (A) 0.159 (W)

Table 1: continues in the next page.

Table 1: continues from the previous page

Authors	Model	RMS (%)
Edouard [18]	$\frac{1}{k_{\text{eff}}} = \frac{2d}{(d^2(4y^2-2\pi y)+\pi d)k_s + (d^2(2\pi y - 4y^2) - \pi d + 1)k_f} + \frac{2d(y-1)}{4y^2d^2k_s + (1-4y^2d^2)k_f} + \frac{1-2dy}{\pi d^2 k_s + (1-\pi d^2)k_f}$ (11a) $1-\varepsilon = \underbrace{3\pi d^2(1-2d) + 8d^3}_{A} + \underbrace{\delta d^3(8\delta^2 + 24\delta + 24 - 6\pi)}_{B}$ (11b) If $\delta = 0: \quad A = k^2\frac{\sqrt{15}}{\phi^4} - k^3\frac{\sqrt{10}}{3\phi^4}$ (11c) If $\delta \neq 0: \begin{cases} A = k^2\frac{\sqrt{15}}{\phi^4} - k^3\frac{\sqrt{10}}{3\phi^4} \\ B = \frac{40}{\sqrt{5}\phi^4}\frac{\sin^2(\frac{\pi}{5})\phi^2\left(1-\frac{k}{2}\sqrt{\frac{2}{3}}\right)}{32\sqrt{3}(3-\phi)}\sqrt{\frac{1}{4}-\frac{\sin^2(\frac{\pi}{5})\phi^2}{9-3\phi}} \\ \quad + \frac{12k}{\sqrt{5}\phi^2}\left(\frac{5\phi}{4\sqrt{3-\phi}} - \frac{\pi}{4}\frac{\phi^2}{3-\phi}\right)\left(1-\frac{k}{2}\sqrt{\frac{2}{3}}\right)^2 \end{cases}$ (11d) $d = b/L$, $y = x/b = 1+\delta$, $\delta \geq 0$, $\phi = 1.618$ b: radius of the cylindric strut, L: lenght of the cylindric strut, x: length of the square lump, k: ratio between strut side and the side of a perfect pentagon	For $\delta \neq 0$: 1.847 (A) 2.616 (W)
Fourie and Du Plessis [20]	$k_{\text{eff}} = (1-\varepsilon)(k_{ss} + k_{fs}) + \varepsilon(k_{sf} + k_{ff})$ (12a) $k_{ss} = 0.559(1.64-\varepsilon)k_s + 0.562k_f\left(\frac{\varepsilon}{1-\varepsilon}\right)^{0.235}$ (12b) $k_{sf} = 1.18k_f\varepsilon^2 - 2.51k_f\varepsilon + 1.32k_f$ (12c) $k_{ff} = 0.645(\varepsilon - 0.961)(k_f^2/k_s) + 1.82k_f\varepsilon^2 - 3.29k_f\varepsilon + 2.43k_f$ (12d) $k_{fs} = (k_f^2/k_s)(-14.2\varepsilon^2 + 22.6\varepsilon - 9.62) + 0.328k_f\left(\frac{\varepsilon}{1-\varepsilon}\right)^{0.539}$ (12e) Under thermal nonequilibrium conditions	1.495 (A) 0.657 (W)
Schuetz and Glicksmann, from Ref. [21]	$k_{\text{eff}} = \varepsilon k_f + (1-\varepsilon)\frac{2-f_s}{3}k_s$ (13) Derived for polimeric foams. For metal foams, $f_s = 1$	0.257 (A) 0.406 (W)
Dulnev [22]	$k_{\text{eff}} = k_s t^2 + k_f(1-t)^2 + \frac{2t(1-t)k_f k_s}{k_s(1-t) + k_f t}$ (14a) $t = \frac{1}{2} + \cos\left(\frac{1}{3}\arccos(2\varepsilon - 1) + \frac{4\pi}{3}\right)$ (14b) t: dimensionless thickness of the solid ligament	0.795 (A) 0.340 (W)
Ahern et al. [21]	$k_{\text{eff}} = k_f + (k_s - k_f)(1-\varepsilon)\beta$ (15a) $\beta = f_w\beta_w + (1-f_w)\beta_s$ (15b) $\beta_s = \frac{1}{3}\left(1 + \frac{4k_f}{k_f + k_s}\right), \quad \beta_w = \frac{2}{3}\left(1 + \frac{k_f}{2k_s}\right)$ (15c) $f_w = 1 - f_s$: fraction of solid in the windows. For metal foams, $f_w = 0$	0.258 (A) 0.345 (W)

Table 1: end

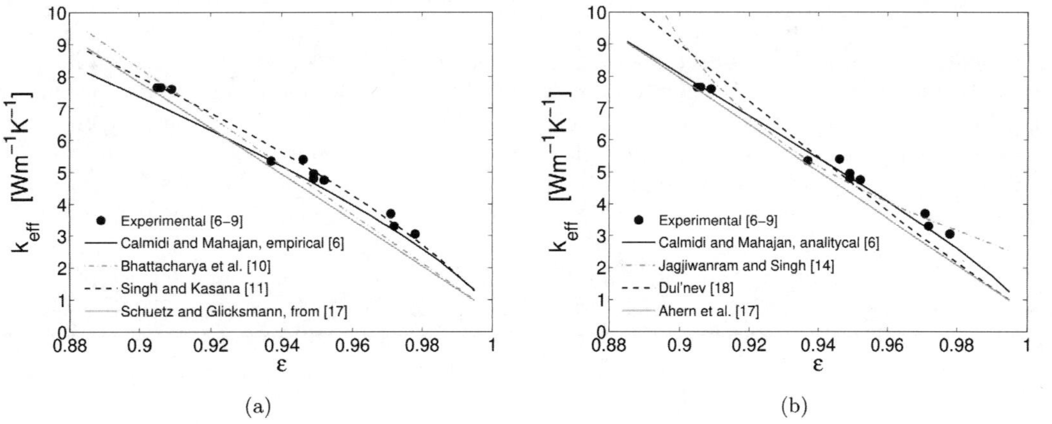

Figure 2: Validation of the models for the prediction of the effective thermal conductivity in the case of water.

3. Microtomography-CFD based approach

Besides the models illustrated so far, the literature offers several examples in which the effective thermal conductivity is estimated, numerically, by means of CFD. In order to perform a numerical simulation, the foam structure can be idealized [3, 23, 24] or described in a realistic way [1, 4, 25, 26]. In the second case, the foam geometry can be recovered, with high fidelity, after a X-ray microtomography (μ-CT) scan, as described in [1]. The result of a μ-CT is a series of 2D X-ray images, recorded during a 360° rotation of the sample around its vertical axis, which can be afterwards combined together in order to obtain a 3D volume. The elemental digital units constituting this reconstructed volume are called voxels.

In [1], three aluminum foams samples of different pore densities (10-20-30 PPI) were analyzed by means of the combined methodology X-ray μ-CT - CFD. The tomographic scan, of 9.1 µm-resolution, was performed at the *TomoLab station* [27] of the Elettra Synchrotron Light Laboratory in Trieste. It should be emphasized that the chosen resolution is significantly higher, at least twice, than most of the studies published in the literature [4, 25, 26, 28, 29].

The domains used for the CFD simulations are a fraction of the volume reconstructed from the μ-CT; the dimensions of the computational domains are chosen according to the PPI of the foam: 548^3 voxels for the 30 PPI sample, 848^3 for the 20 PPI sample and 1000^3 for the 10 PPI one. The commercial software ANSYS ICEM CFD was employed for the generation of geometrically exact, feature-preserving grids: in view of the complexity of the geometry, a tetrahedral grid was employed.

The CFD simulations were performed with ANSYS CFX 13.0. In this case, the problem was only conductive since the interstitial fluid (air or water) was considered as motionless. The energy equation was solved both in the fluid and in the solid region. An arbitrary temperature difference was applied along one direction, while the other lateral faces were kept adiabatic. The simulations were performed along the three space directions, in order to test the degree of isotropy of the medium.

The effective thermal conductivity was computed as:

$$k_{\text{eff}} = \frac{-\int \mathbf{J}\cdot d\mathbf{A}}{\dfrac{\partial T}{\partial x_i} A} = \frac{-\left(\int_s \mathbf{J}\cdot d\mathbf{A}_s + \int_s \mathbf{J}\cdot d\mathbf{A}_f\right)}{\dfrac{\partial T}{\partial x_i}\left(A_s + A_f\right)} \tag{16}$$

where **J** and $d\mathbf{A}$ are the heat flux and the outward pointing area vector, respectively, and the subscripts s and f refer to the solid and the fluid.

The simulations demonstrated that the degree of isotropy of the medium is related to the pore density of the sample. In fact, while for the 30 PPI foam the thermal conductivities evaluated along the three spatial directions are quite close, for the other two samples the differences are more significant. There are two possible explanations for this behaviour. The first is the peculiar and different geometrical characteristics of the samples. In fact, while the cells of the 30 PPI sample are almost visually identical along the three spatial coordinates, for the other two samples they are rather stretched along one specific direction. The highest value of k_{eff} is registered precisely along the direction of elongation of the cells. In addition, as already asserted in [4], this anisotropy might be affected by the size of the domains employed. However, 9.1 μm resolution limited the maximum size of the scanned samples and precluded the employment of significantly larger domains in the CFD simulations. Therefore, in view of these considerations, and following the example of [4], in Table 2 only the mean values of k_{eff}, averaged along the three spatial coordinates, are summarized. A graphical comparison with the experimental and numerical values published in the literature is reported in Figure 3.

As it can be seen, the values of k_{eff} evaluated in this study agree well with those available in the literature. The values of k_{eff} estimated with water are slightly larger than those with air, thus proving that, for aluminum foams, the thermal conductivity of the fluid influences, but not in a significant way, the effective thermal conductivity of the medium.

Table 2: Prediction of the effective thermal conductivity values [$\text{Wm}^{-1}\text{K}^{-1}$] with air or water as the interstitial fluid.

PPI	Porosity	$k_{\text{eff,air}}$	$k_{\text{eff,water}}$
10	0.944	3.49	4.13
20	0.927	4.97	5.76
30	0.929	5.31	5.99

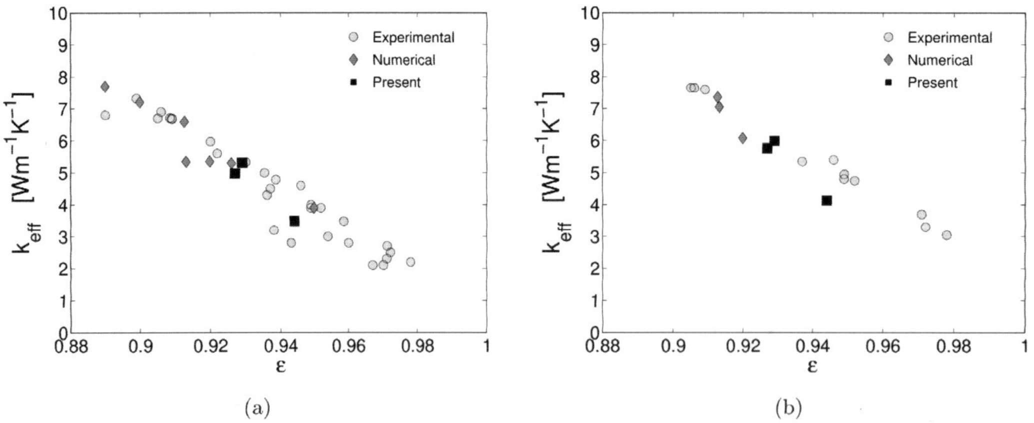

Figure 3: Comparison of the values of the effective thermal conductivity calculated numerically by the present authors with the experimental and numerical data available in the literature. (a) Air and (b) Water.

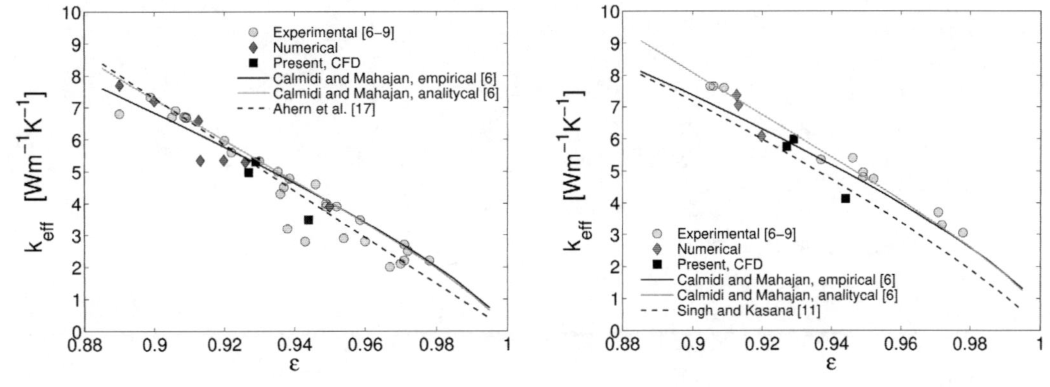

(a) (b)

Figure 4: Comparison of the values of the effective thermal conductivity calculated numerically by the present authors with the experimental and numerical data available in the literature and some of the best models described in Sec. 2. (a) Air and (b) Water.

4. Comments and concluding remarks

In this paper, the most significant models published in the literature for the estimation of the effective thermal conductivity k_{eff} of open-cell metal foams have been tested against a wide set of experimental data. From the results obtained, the following conclusions can be drawn:

- The analysis has shown that the use of a more complex theoretical approach does not always lead to better results than those estimated with an empirical model. In fact, if one considers the empirical and the analytical models of Calmidi and Mahajan [6], it can be observed that their predictions are quite similar (RMS = 0.309% vs RMS = 0.330% in the case of air, and 0.180% vs 0.048% in the case of water). On the contrary, the derivation of a theoretical model is significantly more laborious than the definition of an empirical model.
- In Fig. 4 the values of k_{eff} estimated with a combined μ-CT - CFD strategy are compared against the models, introduced in Sec. 2, which provided the best results for open-cell metal foams. As it can be seen, although the best models are generally satisfactory in predicting k_{eff}, for some applications the accuracy may not be adequate: in these cases, therefore, the proposed μ-CT - CFD strategy seems to be the best choice. However, it is considerably time and resource consuming and, therefore, its application is justified only when higher accuracy is required or, together with experimental measurements, for the validation and/or calibration of simpler models.
- In view of their greater simplicity, empirical models are sufficient, for most applications, to obtain satisfactory estimates of the effective thermal conductivity of metal foams, without resorting to complex geometrical and mathematical descriptions, leaving the μ-CT - CFD methods to specific cases where higher accuracy is required.

Acknowledgment

This work was financially supported by Italian government, MIUR grant PRIN-2009KSSKL3.

NOMENCLATURE

A Area, see Eq. (16)
$d\mathbf{A}$ Outward pointing area vector, see Eq. (16)

J Heat flux vector, see Eq. (16)
k_f Thermal conductivity of the fluid phase
k_{eff} Effective thermal conductivity of the foam
k_s Thermal conductivity of the solid phase
$k_{||}$ Thermal conductivity given by k_f and k_s arranged in parallel
$k_{\perp}$ Thermal conductivity given by k_f and k_s arranged in series
T Temperature, see Eq. (16)
x Generic space coordinate, see Eq. (16)
ε Porosity of the foam (fraction volume of the fluid phase)

References

[1] Ranut P, Nobile E and Mancini L 2014 *Appl. Therm. Eng.* **69** 230–240
[2] Calmidi V and Mahajan R 2000 *ASME J. Heat Transf.* **122** 557–565
[3] Krishnan S, Murthy J and Garimella S 2006 *ASME J. Heat Transf.* **128** 793–799
[4] Bodla K, Murthy J and Garimella S 2010 *Numer. Heat. Tr. A-APPL* **58** 527–544
[5] Bianchi E, Heidig T, Visconti C, Groppi G, Freund H and Tronconi E 2012 *Chem. Eng. J.* **198-199** 512–528
[6] Calmidi V and Mahajan R 1999 *ASME J. Heat Transf.* **121** 466–471
[7] Bhattacharya A and Mahajan R 2006 *J. Electron. Packaging* **128** 259–266
[8] Paek J, Kang B, Kim S and Hyun J 2000 *Int. J. Thermophys.* **21** 453–464
[9] Takegoshi E, Hirasawa Y, Matsuo J, and Okui K 1992 *Trans. Jpn. Soc. Mech. Eng.* **58(547B)** 879–884 in Japanese
[10] Bhattacharya A, Calmidi V and Mahajan R 2002 *Int. J. Heat Mass Tran.* **45** 1017–1031
[11] Singh R and Kasana H 2004 *Appl. Therm. Eng.* **24** 1841–1849
[12] Kaviany M 1999 *Principles of Heat Transfer in Porous Media* 2nd ed (Springer)
[13] Wang J, Carson J, North M and Cleland D 2006 *Int. J. Heat Mass Transfer* **49** 3075–3083
[14] Boomsma K and Poulikakos D 2001 *Int. J. Heat Mass Tran.* **44(4)** 1827–836
[15] Dai Z, Nawaz K, Park Y, Bock J and Jacobi A 2010 *Int. Commun. Heat Mass* **37** 575–580
[16] Boomsma K and Poulikakos D 2011 *Int. J. Heat Mass Tran.* **54(1)** 746–748
[17] Jagjiwanram and Singh R 2004 *Bull. Mater. Sci.* **27(4)** 373–381
[18] Edouard D 2011 *AIChE Journal* **57(6)** 1646–1651
[19] Truong Huu T, Lacroix M, Pham Huu C, Schweich D and Edouard D 2009 *Chem. Eng. Sci.* **64** 5131–5142
[20] Fourie J and Du Plessis J 2005 *AIChE Journal* **50(3)** 547–556
[21] Ahern A, Verbist G, Waire D, Phelan R and Fleurent H 2005 *Colloids and Surfaces A: Psycochem. Eng. Aspects* **263** 275–279
[22] Dul'nev G 1965 *J. Eng. Phys. Thermophys.* **9** 399–404
[23] Boomsma K, Poulikakos D and Ventikos Y 2003 *Int. J. Heat Fluid Flow* **24** 825–834
[24] Kopanidis A, Theodorakakos A, Gavaises E and Bouris D 2010 *Int. J. Heat Mass Tran.* **53** 2539–2550
[25] Petrasch J, Meier F, Friess H and Steinfeld A 2008 *Int. J. Heat Fluid Flow* **29** 315–326
[26] Gerbaux O, Buyens F, Mourzenko V, Memponteil A, Vabre A, Thovert J F and Adler P 2010 *J. Colloid Interface Sci.* **35** 155–165
[27] `http://www.elettra.trieste.it/lightsources/labs-and-services/tomolab/tomolab.html`

[28] Laschet G, Sauerhering J, Reutter O, Fend T and Scheele J 2009 *Comp. Mat. Sc.* **45** 597–603
[29] Magnico P 2009 *Chem. Eng. Sc.* **64** 3564–3575

Journal of Physics: Conference Series **547** (2014) 012022 doi:10.1088/1742-6596/547/1/012022

The thermal entrance region in a porous medium saturated by a nanofluid: analysis of the Brinkman's model

Eugenia Rossi di Schio

Alma Mater Studiorum - Università di Bologna.
Department of Industrial Engineering. Viale Risorgimento 2, 40136 Bologna, Italy

E-mail: eugenia.rossidischio@unibo.it

Abstract. The Darcy-Graetz problem for a channel filled by a nanofluid saturated porous medium is studied. The flow is assumed to be fully developed and described through Brinkman's model. For the model of the nanofluid, both thermophoresis and Brownian diffusion are taken into account. After an adiabatic preparation region, a boundary temperature linearly varying with the longitudinal coordinate is prescribed. A study of the thermal behaviour of the nanofluid is performed by solving numerically the fully–elliptic coupled equations, with reference both to the thermal entrance region and to the fully developed region. With reference to the fully developed region the solution has been obtained analytically, while for the thermal entrance region it has been obtained numerically, by a Galerkin finite element method implemented through the software package Comsol Multiphysics (© Comsol, Inc.). The analysis shows that, for physically interesting values of the Péclet number, the concentration field depends very weakly on the temperature distribution, for any given value assumed by the Darcy number. Indeed, since the effects of thermophoresis and Brownian diffusion are negligible, the homogeneous model could be employed effectively.

1. Introduction

As is well known, a nanofluid, i.e. a suspension of nanoparticles in a base fluid, is an innovative technique that uses ultra fine solid particles in the fluid to improve the heat transfer occurring in industrial and technology applications like power manifacturing, transportation, electronics, etc [1–4]. A comprehensive survey of convective transport in nanofluids was made by Buongiorno [5] and Kakaç and Pramuanjaroenkij [6].

Recently, the topic of heat transfer in nanofluids has been deeply investigated [7–15]. Two different approaches can be distinguished in order to model the nanofluid behaviour: one may consider the distribution of the nanoparticles in the base fluid as homogeneous or may consider the distribution of nanoparticles in the base fluid as non-homogeneous. These two different approaches lead to different sets of governing equations: the homogeneous model is totally equivalent to the model employed for clear fluids except for a suitable rescaling of the governing parameters due to the presence of nanoparticles inside the base fluid [11]. Among all possible homogeneous models, an example is that defined by Tiwari and Das [12]. A non-homogeneous model, on the other hand, adds one or more equations to the set of governing equations in order to describe the nanoparticle distribution and/or the nanoparticle velocity. Among the

Journal of Physics: Conference Series **547** (2014) 012022 doi:10.1088/1742-6596/547/1/012022

possible non-homogeneous models, the most widely employed is the model by Buongiorno [5], a model suitable for the heat transfer analysis and for the investigation of the nonhomogeneous distributions of nanoparticles. In this paper, the set of local balance equations appropriate for the study of the flow of nanofluids is presented, with an emphasis on the physical effects peculiar for nanofluids: thermophoresis and Brownian diffusion.
According to the author's knowledge, despite the growing amount of studies dealing with nanofluid saturated porous media, very few papers analyze the forced convection. In particular, the problem is treated by means of Buongiorno's model by Nield and Kuznetsov [14], and by means of the alternative approach of considering the porous medium as composed by an array of microchannels by Hung [15].
The Graetz-like problems, namely the problems dealing with the thermal entrance region in a channel or a duct, have been widely investigated. The aim of the present contribution is to extend the analysis presented in [13] for a clear nanofluid to the special case of a nanofluid saturating a porous medium.
In the present contribution, we aim to investigate the steady laminar forced convection in a parallel-plane channel, filledy by a nanofluid saturated porous medium.

2. Mathematical model

We consider a channel bounded by two impermeable plates having distance $2L$, and infinite longitudinal length. Due to the symmetry, the problem is two-dimensional: let $\overline{x}$ be the longitudinal coordinate, while $\overline{y}$ is the transverse one. The flow is assumed to be fully developed, according to Brinkman law. We refer to the nanofluid model introduced by Buongiorno [5]. The thermal entrance region is studied assuming an adiabatic preparation of the duct [13] upstream of the thermal entrance region, where a boundary temperature varying with the longitudinal coordinate is prescribed. For the nanoparticle concentration, an inlet condition given by a uniform distribution is assumed while the boundary walls are considered as impermeable. Downstream of the adiabatic preparation region, a boundary temperature given by a linear function of the longitudinal coordinate is prescribed. The aim of this paper is to solve the fully elliptic coupled equations, for the temperature and concentration fields. First, the temperature and concentration fields are determined analytically. Then, the numerical solution for the thermal entrance region is obtained by a Galerkin finite element method implemented through the software package Comsol Multiphysics.
In the hydrodynamically developed region, the assumptions imply that the velocity is parallel to the axial direction $\overline{x}$, namely

$$U = \overline{\Lambda} L\, U_m \frac{\cosh(\overline{\Lambda}L) - \cosh(\overline{\Lambda}\overline{y})}{\overline{\Lambda}L\cosh(\overline{\Lambda}L) - \sinh(\overline{\Lambda}L)}, \tag{1}$$

where U_m is the man value of the velocity distribution and

$$\overline{\Lambda} = \sqrt{\frac{\mu}{K\,\mu'}}\ , \tag{2}$$

with K permeability of the porous medium, μ viscosity and μ' effective viscosity of the porous medium. By invoking that local thermal equilibrium holds, the diffusion equation for the nanoparticles and the local energy balance equation for the fluid are given by

$$U\frac{\partial\tilde{\phi}}{\partial\overline{x}} = \varepsilon\overline{\boldsymbol{\nabla}}\cdot\left(D_B\overline{\boldsymbol{\nabla}}\tilde{\phi} + D_T\frac{\overline{\boldsymbol{\nabla}}T}{\overline{T}}\right), \tag{3}$$

$$U\frac{\partial\overline{T}}{\partial\overline{x}} = \alpha_m\overline{\nabla}^2\overline{T} + \tau\left(D_B\overline{\boldsymbol{\nabla}}\tilde{\phi}\cdot\overline{\boldsymbol{\nabla}}T + D_T\frac{\overline{\boldsymbol{\nabla}}T\cdot\overline{\boldsymbol{\nabla}}T}{\overline{T}}\right), \tag{4}$$

where

$$\alpha_m = \frac{k_m}{(\varrho c)_f}, \qquad \tau = \frac{\varepsilon(\varrho c)_{np}}{(\varrho c)_f}. \tag{5}$$

In Eqs. (3) and (4) ε is the porosity of the porous medium, $\tilde{\phi}$ is the nanoparticle concentration, $\overline{T}$ is the temperature, D_B and D_T are the Brownian diffusion and the thermophoresis coefficients, defined as

$$D_B = \frac{k_B\, T_i}{3\,\pi\,\mu\, d_P}, \qquad D_T = \beta\,\nu\,\phi_0, \tag{6}$$

and β is a proportionality factor defined by McNab and Meisen [16].
The thermal boundary conditions can be described as follows: at the inlet section the fluid temperature is uniform and denoted as T_i; then, after an adiabatic preparation, the wall temperature assumes the value T_w at $\overline{x} = 0$ and starts changing linearly in the longitudinal direction, namely

$$\overline{T}(\overline{x}, \pm L) = T_w + (T_w - T_i)\frac{\overline{x}}{L}. \tag{7}$$

Let us introduce the following dimensionless quantities:

$$\begin{aligned} &x = \frac{\overline{x}}{L}\ ,\ y = \frac{\overline{y}}{L}\ ,\ u = \frac{U}{U_m}\ ,\ T = \frac{\overline{T}}{T_i}\ ,\ N_{BT} = \frac{D_B\,\phi_0}{D_T}\ , \\ &Le = \frac{\alpha_m}{\varepsilon D_B}\ ,\ Pe = \frac{U_0\,L}{\alpha_m}\ ,\ G = \frac{\phi_0(\varrho c)_{np}}{(\varrho c)_f}\ ,\ \gamma = \frac{T_w}{T_i}\ ,\ \Lambda = \overline{\Lambda} L\ , \end{aligned} \tag{8}$$

where Le is the Lewis number and Λ is Darcy number. As is well known, Darcy number usually assumes very small values compared with unity in highly porous media. According to the bridging role of Brinkman model, the limit $\Lambda \to 0$ yields to Darcy law, while the limit $\Lambda \to \infty$ yields to the clear fluid behavior, thus allowing a qualitative comparison with the results presented in [13].
Moreover, let us rescale the volumetric fraction of nanoparticle with respect to the inlet value, i.e.

$$\phi = \frac{\tilde{\phi}}{\phi_0}. \tag{9}$$

By employing the dimensionless quantities, Eqs. (1), (3) and (4) can be rewritten as

$$u = \Lambda\frac{\cosh(\Lambda) - \cosh(\Lambda\, y)}{\Lambda\cosh(\Lambda) - \sinh(\Lambda)}, \tag{10}$$

$$u\frac{\partial\phi}{\partial x} = \frac{1}{Le\,Pe}\left[\nabla^2\phi + \frac{1}{N_{BT}}\left(\frac{\nabla^2 T}{T} - \frac{\boldsymbol{\nabla} T\cdot\boldsymbol{\nabla} T}{T^2}\right)\right], \tag{11}$$

$$u\frac{\partial T}{\partial x} = \frac{\nabla^2 T}{Pe} + \frac{G}{Le\,Pe}\left(\boldsymbol{\nabla}\phi\cdot\boldsymbol{\nabla} T + \frac{1}{N_{BT}}\frac{\boldsymbol{\nabla} T\cdot\boldsymbol{\nabla} T}{\overline{T}}\right), \tag{12}$$

to be solved together with the dimensionless boundary conditions,

$$T(x, \pm 1) = \gamma + x(\gamma - 1). \tag{13}$$

In order to evaluate the order of magnitude of the dimensionless parameters, we refer to typical values of the thermophysical properties for water and copper flowing in a porous medium with porosity 0.5. Of course, the parametric analysis allows a different choice of the values. One may now choose the typical size of the nanoparticle $d_p \approx 10^{-9}$m and an inlet volumetric fraction of the nanoparticle $\phi_0 \approx 10^{-2}$. According to this choice, the values of N_{BT}, Le and G become

$$N_{BT} = 0.46 \ , \quad Le = 1.7 \cdot 10^4 \ , \quad G = 8.61 \cdot 10^{-3}. \tag{14}$$

Eq. (14) shows that the fraction G/Le that appears in Eq. (12) is of order of magnitude 10^{-7}, and this justifies neglecting the last term on the right hand side of Eq. (12). Equation (12) thus becomes

$$u\frac{\partial T}{\partial x} = \frac{\nabla^2 T}{Pe}. \tag{15}$$

Eq. (15), together with the boundary conditions, shows that the dimensionless temperature depends only on the parameters Pe and Λ and on the boundary condition, i.e. Eq. (13), and namely on the parameter γ. On the contrary, Eq. (11) shows that the distribution of the volumetric fraction of the nanoparticles is influenced by the temperature field as well.
One can introduce the following dimensionless temperature,

$$\tilde{T} = \frac{\overline{T} - T_i}{T_w - T_i} = \frac{T - 1}{\gamma - 1}. \tag{16}$$

By employing Eq. (16), Eq. (15) becomes

$$u\frac{\partial \tilde{T}}{\partial x} = \frac{\nabla^2 \tilde{T}}{Pe}. \tag{17}$$

to be solved together with the inlet condition, $\tilde{T} = 0$, and with the boundary conditions,

$$\begin{aligned} &\frac{\partial \tilde{T}}{\partial y} = 0, \quad \text{for } x < 0, \ y = \pm 1; \\ &\tilde{T} = 1 + x \quad \text{for } x \geq 0, \ y = \pm 1. \end{aligned} \tag{18}$$

Eqs. (17) and (18) show that the dimensionless temperature $\tilde{T}$ does not depend on the parameter γ.

2.1. *Fully developed region*

Let us first investigate the region far from the entrance region, where the solution of Eq. (15), together with the boundary conditions (13), can be obtained analytically. The analytical solution obtained for the fully developed region will be employed to determine the dimensionless functions to be prescribed on the outlet section, in order to investigate the thermal entrance region.
One may assume that, far from the entrance region, the dimensionless temperature distribution is given by the sum of a linear function of the longitudinal coordinate and a function of the transverse coordinate, namely

$$T(x, y) = \gamma + x(\gamma - 1) + \Theta(y). \tag{19}$$

By substituting Eq. (19) into Eq. (15), and by employing Eq. (10), one has

$$\frac{\mathrm{d}^2\Theta}{\mathrm{d}y^2} = (\gamma - 1) Pe\, \Lambda \frac{\cosh(\Lambda) - \cosh(\Lambda\, y)}{\Lambda \cosh(\Lambda) - \sinh(\Lambda)}, \tag{20}$$

to be solved together with the boundary condition $\Theta(y) = 0$ for $y = \pm 1$. Indeed, one obtains

$$\Theta(y) = \frac{(\gamma - 1)Pe}{2\Lambda} \frac{\left[2 + \Lambda^2 \left(y^2 - 1\right)\right] \cosh(\Lambda) - 2\cosh(\Lambda y)}{\Lambda \cosh(\Lambda) - \sinh(\Lambda)}. \tag{21}$$

With reference to the concentration, in the fully developed region, the temperature gradient in the longitudinal direction remains constant while the dimensionless temperature increases with the longitudinal coordinate, thus allowing us to infer that the last terms on the righthand side of Eq. (11) can be dropped. Indeed, far from the thermal entrance region, the governing equation for the concentration becomes

$$u\frac{\partial \phi}{\partial x} = \frac{\nabla^2 \phi}{Le\,Pe}. \tag{22}$$

Similarly, one can infer that the boundary condition for the nanoparticle concentration becomes

$$\left.\frac{\partial \phi}{\partial y}\right|_{y=\pm 1} = 0. \tag{23}$$

Indeed, one may assume that, far from the entrance region, the nanoparticle concentration distribution is given by the sum of a linear function of the longitudinal coordinate and a function of the transverse coordinate, namely

$$\phi(x, y) = A\,x + \Phi(y). \tag{24}$$

By substituting Eq. (24) into Eq. (22), one has

$$Le\,Pe\,A\,\Lambda\frac{\cosh(\Lambda) - \cosh(\Lambda\,y)}{\Lambda\cosh(\Lambda) - \sinh(\Lambda)} = \frac{\mathrm{d}^2\Phi}{\mathrm{d}y^2}, \tag{25}$$

to be solved together with the boundary condition (23). By integrating Eq.(25) with respect to y in the interval $[-1, 1]$, and by employing Eq.(23), one obtains

$$A = 0 \ , \quad \Phi = const. \tag{26}$$

Equations (21), (24) and (26) yield the dimensionless temperature and concentration distributions to be prescribed on the outlet section in the numerical solution for the thermal entrance region.

2.2. Thermal entrance region

The governing equations (11) and (15), together with the dimensionless boundary conditions (13) and (23), have been numerically solved by employing Galerkin's finite element method, implemented through the software package Comsol Multiphysics. The numerical simulations have been performed by considering a half-channel, i.e. $0 \leq y \leq 1$, and by replacing the boundary conditions in $y = -1$ with the symmetry conditions in $y = 0$, namely

$$\frac{\partial T}{\partial y} = \frac{\partial \phi}{\partial y} = 0. \tag{27}$$

The numerical simulations have been performed by taking into account Eq. (14) and for different values assumed by the parameters Pe, γ and Λ. In particular, the values $Pe = 0.1$ and $Pe = 0.01$, $\gamma = 1.01$ and $\gamma = 1.001$, $\Lambda = 0.1$ and $\Lambda = 10$ have been considered. If reference is made to copper particles suspended in water flowing in a channel with $L = 1\,mm$, then $Pe = 0.01$ corresponds to a mean velocity proportional to 1 mm/s. Moreover, on account of the definition of the dimensionless parameters, the value $\gamma = 1.01$ corresponds, if an inlet temperature of 300 K is assumed, to a boundary temperature of 303 K prescribed at the thermal entrance section, $x = 0$, and then to a boundary temperature increase of 10 K per millimeter of axial length of the channel.

First, the solution has been checked to be independent of the dimensionless length of the

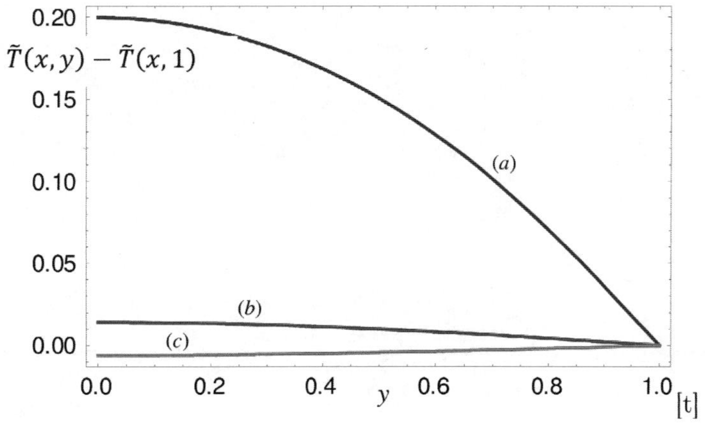

Figure 1. Difference between the dimensionless temperature and the boundary temperature distribution in a half channel for $\Lambda = 0.1$ and $Pe = 0.01$, evaluated at three different axial positions: $x = 0.5$ (a), $x = 2$ (b) and $x = 5$ (c).

computational domain, as well as of the mesh employed. First, different lengths of the thermal entrance region have been compared in the range $10 \leq x_{out} \leq 40$, where x_{out} denotes the dimensionless axial position of the outlet section. In order to perform this comparison, an unstructured mesh of triangular elements has been generated by fixing the maximum element size $m.e.s. = 0.05$. According to this choice, one has, for instance, a mesh consisting of 20100 elements when $x_{out} = 10$; obviously, for increasing lengths of the computational domain, the number of elements increases. Both the cases $Pe = 0.1$ and $Pe = 0.01$ have been tested, assuming $\Lambda = 0.1$. Then, in order to test the independence from the maximum element size, reference has been made to $\gamma = 1.01$ and to the computational domain $-10 \leq x \leq 10$, where $x = 0$ corresponds to the thermal entrance section and $x = -10$ corresponds to the inlet section. The value $\Lambda = 0.1$ has been considered, and both the cases $Pe = 0.1$ and $Pe = 0.01$ have been tested, and different values of the maximum element size have been prescribed, in order to obtain different unstructured meshes characterized by a uniform distribution of triangular elements. The numerical solution has been proved to be independent of the mesh, especially with reference to the axial positions far from the thermal entrance region.

3. Discussion of the results

In this section, the main features of the numerical solution for the thermal entrance region will be discussed. In the present analysis, as explained in the previous section, the values of the parameters given by Eq. (14) have been assumed and a dimensionless axial position of the outlet section $x_{out} = 10$ has been used to define the computational domain. Moreover, an unstructured mesh of triangular elements with maximum element size 0.04 is employed. With reference to the parameters Pe, γ, and Λ different values have been taken into account in order to better investigate the features of the mathematical model.

In Figs. 1 and 2 the difference between the dimensionless temperature distribution $\tilde{T}$ and its value at the boundary is reported, in the thermal entrance region, versus the transverse coordinate y for three different axial positions x and for $Pe = 0.01$. Fig. 1 refers to $\Lambda = 0.1$, while Fig. 2 refers to $\Lambda = 10$. The figures show that far from the thermal entrance region, the temperature distribution tends to become independent of the transverse coordinate y and that this behaviour is reached very soon. Moreover, a comparison between the two figures shows that the solution is

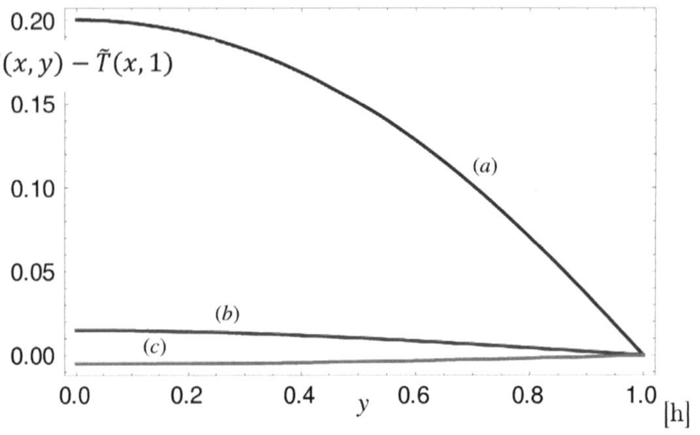

Figure 2. Difference between the dimensionless temperature and the boundary temperature distribution in a half channel for $\Lambda = 10$ and $Pe = 0.01$, evaluated at three different axial positions: $x = 0.5$ (a), $x = 2$ (b) and $x = 5$ (c).

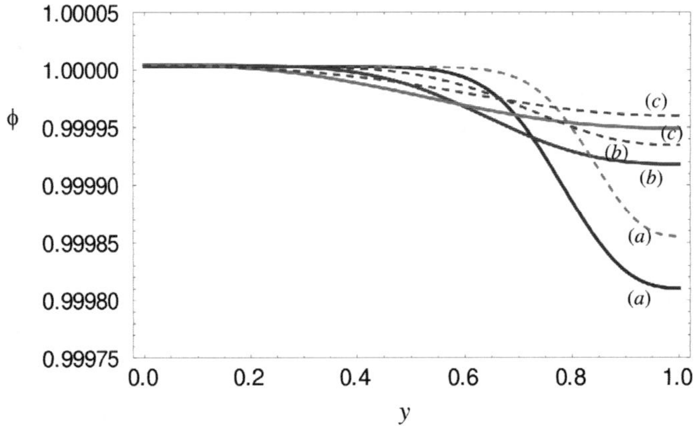

Figure 3. Nanoparticle concentration distribution in a half channel for $Pe = 0.01$ and $\gamma = 1.01$, evaluated at three different axial positions: $x = 0.5$ (a), $x = 2$ (b) and $x = 5$ (c). The solid lines refer to $\Lambda = 10$, while the dashed ones refer to $\Lambda = 0.1$.

not appreciably affected by the velocity distribution, i.e. by the parameter Λ. To underline this feature the unphysical value $\Lambda = 10$ has been assumed in Fig. 2.

In Figs. 3 and 4 the volumetric nanoparticle concentration is reported versus the transverse coordinate y for different axial positions x and for $Pe = 0.01$. Figure 3 refers to $\gamma = 1.01$, while Figure 4 refers to $\gamma = 1.001$. The figures show that the volumetric concentration of nanoparticle varies with the transverse coordinate only in a narrow region close to the boundary of the channel. Moreover, the variability becomes smaller for increasing values of the parameter Λ. However, if one looks at the range of values of the variability of ϕ, one discovers that this range is, indeed, extremely small. The results presented show that the volumetric concentration of nanoparticle is not much affected by the velocity distribution, since it depende very weakly on the parameter Λ, and is not much affected by the temperature distribution, since it depends very weakly on

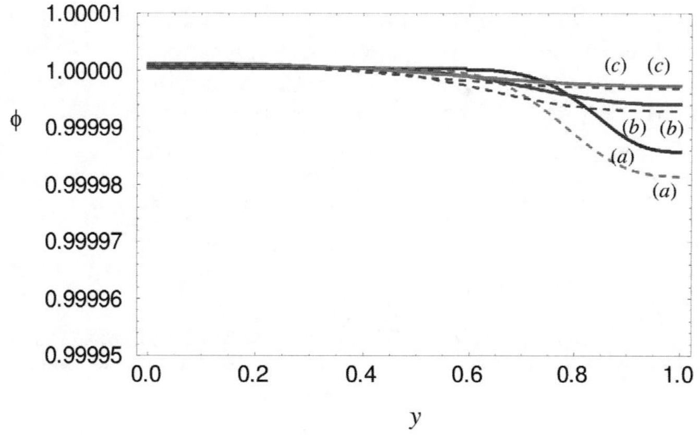

Figure 4. Nanoparticle concentration distribution in a half channel for $Pe = 0.01$ and $\gamma = 1.001$, evaluated at three different axial positions: $x = 0.5$ (a), $x = 2$ (b) and $x = 5$ (c). The solid lines refer to $\Lambda = 10$, while the dashed ones refer to $\Lambda = 0.1$.

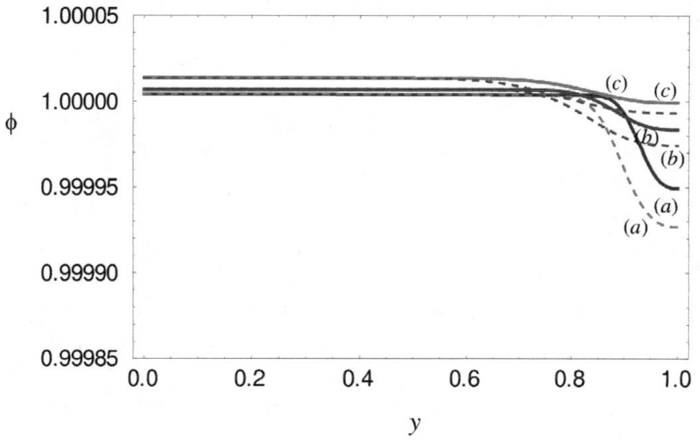

Figure 5. Nanoparticle concentration distribution in a half channel for $Pe = 0.1$ and $\gamma = 1.01$, evaluated at three different axial positions: $x = 0.5$ (a), $x = 2$ (b) and $x = 5$ (c). The solid lines refer to $\Lambda = 10$, while the dashed ones refer to $\Lambda = 0.1$.

the parameters γ and Pe. Moreover, as discussed at the beginning of this section, the value $\gamma = 1.01$ (Fig. 3) is a barely interesting case, since it implies that, if an inlet temperature of 300 K is assumed, the boundary temperature increases 10 K per millimeter along the channel. But the range of values of the variability of ϕ is even smaller for $\gamma = 1.001$, as shown in Fig. 3. In Figs. 3 and 4, the dashed lines refer to the limit of a clear nanofluid, and the results repeat that presented in [13]. Only a qualitative comparison with the results regarding the clear fluid and presented in [13] can be done, since the definition of the dimensionless parameters slightly differs.

Finally, it should be pointed out that Figs. 3 and 4 refer to $Pe = 0.01$. For higher values of the Péclet number the distribution of the nanoparticle concentration depends more on the dimensionless temperature field, but always within a negligicle range of variability. For instance,

results of the nanoparticle concentration referring to $Pe = 0.1$ are reported in Fig. 5, and higher values of Pe are less interesting under a physical point of view.
To sum up, the resolution of the fully-elliptic coupled equations, which arise from the model proposed by Buongiorno [5], requires a computational effort and yields to the determination of a field, the nanoparticle concentration distribution, which may be considered as uniform in many cases of practical interest. Recently, many authors have proposed to consider the nanofluid not through the complete model, but as a normal fluid with properly defined thermophysical properties, i.e. the so-called homogeneous nanofluid model (see, for instance, [11] and [12]). The results arising from the present paper suggest to neglect the nanoparticle concentration gradient, and to consider the nanofluid saturating the porous medium as a "normal fluid", since the peculiar effects introduced in the model by Buongiorno [5] do not affect sensibly the numerical solution.

4. Conclusions

In the present paper, the steady laminar forced convection in a parallel-plane channel involving a nanofluid saturating a porous medium is studied. The flow is assumed to be fully developed, and described through the Brinkman's model. Downstream of an adiabatic preparation of the channel, a boundary temperature given by a linear function of the longitudinal coordinate is prescribed. A study of the thermal behaviour of the nanofluid is performed by solving numerically the fully elliptic governing equations. More precisely, the mathematical model proposed by Buongiorno [5] is employed, in order to evaluate the effects of thermophoresis and Brownian diffusion. In particular, the model yields to a pair of partial differential equations: one for the temperature field and one for the nanoparticle concentration distribution. The solution for the fully developed region is determined analytically, while the solution for the thermal entrance region is evaluated numerically by a Galerkin finite element method implemented through the software package Comsol Multiphysics (© Comsol, Inc.). The numerical solution is fairly independent of the dimensionless length of the channel, as well as of the mesh refinement.
The present analysis leads to the conclusion that thermophoresis and Brownian diffusion display negligible effects in realistic cases. In fact, the volumetric concentration of nanoparticle is a weak function of the temperature distribution, since it depends very weakly on the parameters γ and Pe. Moreover, both dimensionless temperature and nanoparticle concentration are weak function of the velocity distribution, since they depend very weakly on the parameter Λ as well. Indeed, the resolution of the fully-elliptic coupled equations, which arise from the model proposed by Buongiorno, requires a computational effort and yields to the determination of the nanoparticle concentration gradient, that may be neglected in many cases of practical interest.
A foreseen future development of the present analysis is to consider different boundary conditions, such as periodic boundary condition, since for the case of clear fluid the effect of Brownian diffusion and thermophoresis is not negligible for that boundary conditions.

References

[1] Choi S 1995 Enhancing thermal conductivity of fluids with nanoparticle, in D.A. Siginer, H.P. Wang, H.P. (Eds.), *Developments and Applications of Non-Newtonian Flows* ASME FED **231**/MD **66** ASME, New York 99
[2] Masuda H, Ebata A, Teramae K and Hishinuma N 1993 *Netsu Bussei* **4** 227
[3] Lee S, Choi SUS, Li S and Eastman JA 1999 *J. Heat Transfer* **121** 280
[4] Eastman JA, Choi SUS, Li S, Yu W and Thompson LJ 2001 *Appl. Phys. Lett.* **78** 718
[5] Buongiorno J 2006 *J. Heat Transfer* **128** 240
[6] Kakaç S amd Pramuanjaroenkij A 2009 *Int. J. Heat Mass Transfer* **52** 3187
[7] Wang XQ and Mujumdar AS 2007 *Int. J. Therm. Sci.* **461**
[8] Daungthongsuk W and Wongwises S 2007 *Renew. Sust. Energ. Rev.* **11** 797
[9] Rao Y 2010 *Particuology* **8** 549
[10] Nield DA and Kuznetsov AV 2013 *J. Heat Transfer* **135** 061201

[11] Magyari E 2011 *Acta Mech.* **222** 381
[12] Tiwari R and Das M 2007 *Int. J. Heat Mass Transfer* **50** 2002
[13] Rossi di Schio E, Celli M and Barletta A 2014 *J. Heat Transfer* **136** 022401
[14] Nield DA and Kuznetsov AV 2014 *Int. J. Heat Mass Transfer* **70** 430
[15] Hung YM 2011 *Heat Transfer Eng.* **31**1184
[16] McNab GS and Meisen A 1973 *J. Colloid Interface Sci.* **44** 339

Journal of Physics: Conference Series **547** (2014) 012023 doi:10.1088/1742-6596/547/1/012023

Numerical Investigation of Viscous Dissipation in Elliptic Microducts

P Vocale[1], G Puccetti[2], GL Morini[2], M Spiga[1]

[1]Department of Industrial Engineering, University of Parma, Parco Area delle Scienze 181/A, 43124 Parma, Italy.

[2] DIN – Alma Mater Studiorum Università di Bologna, Laboratorio di Microfluidica, Via del Lazzaretto 15/5, Bologna 40131, Italy.

E-mail: pamela.vocale@unipr.it

Abstract. In this work a numerical analysis of heat transfer in elliptical microchannels heated at constant and uniform heat flux is presented. A gaseous flow has been considered, in laminar steady state condition, in hydrodynamically and thermally fully developed forced convection, accounting for the rarefaction effects. The velocity and temperature distributions have been determined in the elliptic cross section, for different values of aspect ratio, Knudsen number and Brinkman number, solving the Navier-Stokes and energy equations within the Comsol Multiphysics® environment. The numerical procedure has been validated resorting to data available in literature for slip flow in elliptic cross sections with Br =0 and for slip flow in circular ducts with Br >0. The comparison between numerical results and data available in literature shows a perfect agreement. The velocity and temperature distributions thus found have been used to calculate the average Nusselt number in the cross section. The numerical results for Nusselt number are presented in terms of rarefaction degree (Knudsen number), of viscous dissipation (Brinkman number), and of the aspect ratio. The results point out that the thermal fluid behavior is significantly affected by the viscous dissipation for low rarefaction degrees and for aspect ratios of the elliptic cross-section higher than 0.2.

1. Introduction

The tendency of increasing power rating in electronic systems requires advanced technology in thermal management, and the introduction of microchannels into power cooling has significantly increased the need of more refined models to determine the thermal performance.

Experimental results on single-phase flows in microchannels have evidenced that the conventional models, can no longer be considered able to predict pressure drop and convective heat transfer coefficients. Researchers justify this conclusion by invoking new micro-effects (such as viscous forces, rarefaction, compressibility, axial heat conduction, conjugate heat transfer, wall roughness and so on), which Herwig and Hausner [1] called "scaling effects with respect to a standard macro-analysis". In the analysis of microchannels, scaling effects are of paramount importance to describe the behavior of the fluid.

With regards to the viscous forces, it is well-know that their influence on the heat transfer is significant for highly viscous fluids or for flows characterized by high velocity. However it has to be pointed out that the role of the viscous dissipation becomes significant also when the temperature difference between the fluid and the wall is small and when the channel diameter is reduced (as it occurs in microscale) [2].

In the last years many researchers have investigated the combined effect of rarefaction and viscous dissipation on the fluid behavior in microducts featuring different cross-sections, as highlighted by Colin in a recent review [3]. In particular, the thermal problem in uniform heat flux (CHF) boundary conditions has been tackled by many researchers. By applying the integral transform technique, Tunc and Bayazitoglu [4] analytically investigated the influence of Knudsen number and Brinkman number on the convective heat transfer in microducts. They found that an increase in the Knudsen number leads to a decrease in the Nusselt number, due to the temperature jump. By including the contribution of the viscous dissipation, they observed that the Nusselt number decreases as the Brinkman number increases. It has to be pointed out that for CHF boundary conditions a positive value of the Brinkman number means that the fluid is being heated.

The same problem has been theoretically investigated by Aydın and Avcı [5]. They found that for low values of the Brinkman number the Nusselt number decreases as the Knudsen number increases, while Nu presents a maximum for high values of Br. Their results show also that the influence of the Brinkman number becomes less relevant as the rarefaction effects become more pronounced because of the presence of the slip velocity which reduces the velocity gradient. These results have been confirmed by the analysis carried out by Jeong and Jeong [6] and by Çetin et al. [7] who have investigated the effect of rarefaction, viscous dissipation and axial conduction in microtubes by extending the classical Graetz problem. The same trend has been also observed by Sun et al. [8] who have performed a numerical investigation on slip flow in microducts accounting for rarefaction and viscous dissipation effects. The influence of viscous dissipation and rarefaction in micropipes has been also investigated by Hooman [9] who has extended the analysis to the Second Law presenting correlations for the dimensionless entropy generation number and for the Bejan number depending on the Brinkman and Knudsen numbers.

Closed form solutions for temperature and Nusselt number have been derived by Sadeghi and Saidi [10] in a microannulus. They have also investigated the influence of viscous dissipation accounting for the rarefaction effects in microchannels bounded by parallel plate as well. The role of viscous forces in the latter geometry has been also analytically analyzed by Jeong and Jeong [11], Aydın and Avcı [12] and more recently by Zhang et al. [13].

Moreover the combined effect of viscous dissipation and rarefaction has been numerically investigated in other cross-section geometries, even considering uniform heat flux (CHF) boundary conditions. van Rij [14,15] and Sun and Jaluria [16] have analyzed convective heat transfer in rectangular microducts. Their results show that viscous heating significantly affects the fluid thermal behavior, leading to a decrease in the Nusselt number, when the fluid is being heated. The amount of this reduction depends on the rarefaction degree, the aspect ratio and on the value of the Brinkman number. The same conclusions have been stated by Kuddusi [17] who have performed a numerical investigation of slip flow in trapezoidal microchannels accounting for the viscous heating.

Despite these important contributions, the influence of the viscous and rarefaction effect in elliptic microducts has not yet been investigated in literature, at least to the best knowledge of the authors.

In this work convective heat transfer for a laminar and fully developed gaseous flow, through elliptical microchannels, is investigated by adopting a numerical approach. The momentum conservation and energy equations are solved within the Comsol Multiphysics® environment considering a constant and uniform heat flux and accounting for the rarefaction effects. The numerical procedure is validated by comparing the numerical results with the data available in literature for both elliptic and circular microducts. The numerical data for the Nusselt number are presented as a function of the main parameters that influence the fluid behavior, highlighting the role of the viscous forces, of the rarefaction degree and of the aspect ratio of the considered cross-section.

The main goal of this work is to demonstrate that the problem of heat transfer enhancement in microdevices cannot be solved simply by indefinitely reducing the microchannel dimensions, because the viscous dissipation effects shall offset the gains of high heat transfer coefficients associated with a reduction in the channel size.

2. Mathematical model

Experimental investigations highlight that, within the slip flow regime, at a distance of approximately one mean free path from the wall the continuum hypothesis is still valid, hence the problem is commonly tackled using the Navier-Stokes and energy equations, but modifying the boundary conditions to account for the rarefaction effects in the channels. To represent these effects, Knudsen number, defined as the ratio of the molecular mean free path to the hydraulic diameter of the cross section, is invoked. According to Karniadakis et al. [18] a slip flow model can be used for $0.001<Kn<0.1$.

By considering a Newtonian gas flowing in an elliptical microchannel (Fig. 1) and assuming that the physical properties are constant and the flow is laminar and fully developed, the Navier-Stokes equation can be written as follows:

$$\mu\left(\frac{\partial^2 u}{\partial \xi^2}+\frac{\partial^2 u}{\partial \psi^2}\right)=\frac{\partial p}{\partial \zeta} \tag{1}$$

being μ, u and $\partial p/\partial\zeta$, the fluid dynamic viscosity, the local velocity and the axial component of the pressure gradient, respectively. Due to low value of the Mach number, in the present analysis the compressibility effects are not considered [19].

The temperature field can be found by the solution of the energy equation:

$$\rho c_p u\frac{\partial T}{\partial \zeta}=\lambda\left(\frac{\partial^2 T}{\partial \xi^2}+\frac{\partial^2 T}{\partial \psi^2}\right)+\Phi \tag{2}$$

being ρ and λ the fluid density and thermal conductivity, respectively, and c_p the specific heat at constant pressure. The last term in Equation (2) indicates the viscous dissipation function which is defined by:

$$\Phi=\mu\left[\left(\frac{\partial u}{\partial \xi}\right)^2+\left(\frac{\partial u}{\partial \psi}\right)^2\right] \tag{3}$$

In fully developed flow, $dT_b/d\zeta=\partial T/\partial\zeta$, therefore the axial component of the temperature gradient can be evaluated by an overall energy balance for an elemental control volume:

$$\rho c_p WA\frac{\partial T}{\partial \zeta}=qL_P+\int_A \Phi dA \tag{4}$$

where W is the average fluid velocity, A indicates the cross-sectional area, q the wall heat flux and L_P the heated perimeter of the cross-section.

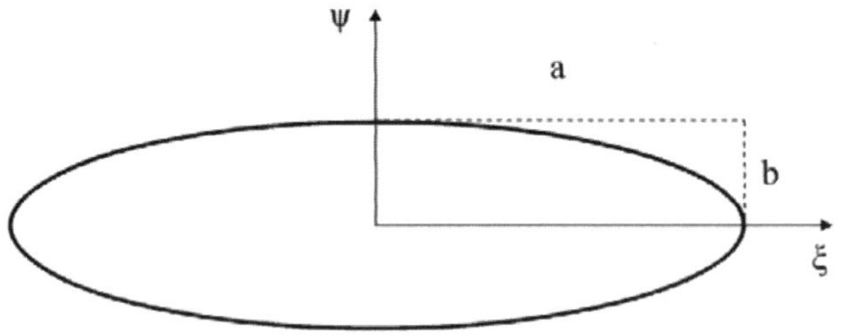

Figure 1. Sketch of the considered cross-section.

To obtain a general solution for the considered model, the following dimensionless coordinates are introduced:

$$x = \frac{\xi}{D_h} \qquad y = \frac{\psi}{D_h} \qquad z = \frac{\zeta}{D_h} \tag{5}$$

together with the dimensionless functions listed below:

$$P = -\frac{D_h^2}{\mu W}\frac{\partial p}{\partial \zeta} \qquad U = \frac{u}{W} \qquad \theta = \frac{\lambda T}{q\, D_h} \tag{6}$$

By considering the above non-dimensional quantities, the Equations (1-2) can be rewritten as follow:

$$\frac{\partial^2 U}{\partial x^2} + \frac{\partial^2 U}{\partial y^2} = P \tag{7}$$

$$\frac{\partial^2 \theta}{\partial x^2} + \frac{\partial^2 \theta}{\partial y^2} = \frac{U}{A^*}\left[L_P^* + Br \int_{A^*} \Phi^* \, dA^* \right] - Br\Phi^* \tag{8}$$

in which A^* denotes the dimensionless cross-sectional area ($A^*=A/D_h^2$), L_P^* is the dimensionless heated perimeter of the cross-section ($L_P^*=L_P/D_h$), and Br indicates the Brinkman number defined by:

$$Br = \frac{\mu W^2}{q D_h} \tag{9}$$

To take into account the rarefaction effects, for the dimensionless velocity at the wall the usual first-order slip boundary condition is considered, according to [20]:

$$U - U_w = \frac{2-\sigma_v}{\sigma_v} Kn \left(\frac{\partial U}{\partial n} \right)_w \tag{10}$$

The energy equation is solved by considering a constant and uniform heat flux at the wall (i.e. H2 boundary conditions), hence the thermal boundary condition in the non-dimensional form reads as:

$$\left(\frac{\partial \theta}{\partial n} \right)_w = -1 \tag{11}$$

By considering the temperature jump at the wall, the non-dimensional wall temperature can be evaluated as follows:

$$\theta_w(x_w, y_w) = \theta(x_w, y_w) + \frac{2-\sigma_T}{\sigma_T}\frac{2k}{k+1}\frac{Kn}{\mathrm{Pr}} \tag{12}$$

The knowledge of the velocity and temperature fields enables to calculate the main physical parameters, such as the dimensionless bulk temperature and the average Nusselt number:

$$\theta_b = \frac{1}{A^*}\int_{A^*} U\theta \, dA^* \tag{13}$$

$$Nu = \frac{h\, D_h}{\lambda} = \frac{1}{(\overline{\theta_w} - \theta_b)} \tag{14}$$

in which $\overline{\theta_w}$ indicates the non-dimensional average wall temperature.

3. Results and discussion

The Navier-Stokes and energy equations in the non-dimensional form have been solved within the COMSOL Multiphysics® environment. To guarantee accurate solutions for the governing equations the convergence criterion has been set 1e-06.

The numerical runs have been performed considering air as working fluid (Pr=0.7, k=1.4) and assuming accommodation coefficients $\sigma_v = \sigma_T = 1$, as usual in literature [3].

To obtain mesh-independent solutions, several numerical simulations have been carried out on different grid sizes by monitoring the value of the Nusselt number. For each analysed grid resolution the relative error in Nu has been calculated by comparing the value of Nusselt number for the tested mesh size with a reference value of Nu obtained adopting the finest mesh (i.e. for each value of the aspect ratio considered in the present study a reference grid characterized by at least 400000 elements has been chosen):

$$\varepsilon_r = \frac{Nu - Nu_{ref}}{Nu_{ref}} \tag{15}$$

In Figure 2 the relative error (ε_r) in Nu as a function of the number of the grid elements (N) for $\beta = 0.1$ and for different values of the Knudsen and Brinkman numbers, is presented. It is evident that for a number of elements N higher than 2000 the value of the relative error tends to 0 which means that the numerical solution becomes independent on the grid resolution. The same trend has been also observed for the other values of the aspect ratio, therefore in all the numerical runs a grid consisting of at least 2000 elements has been adopted.

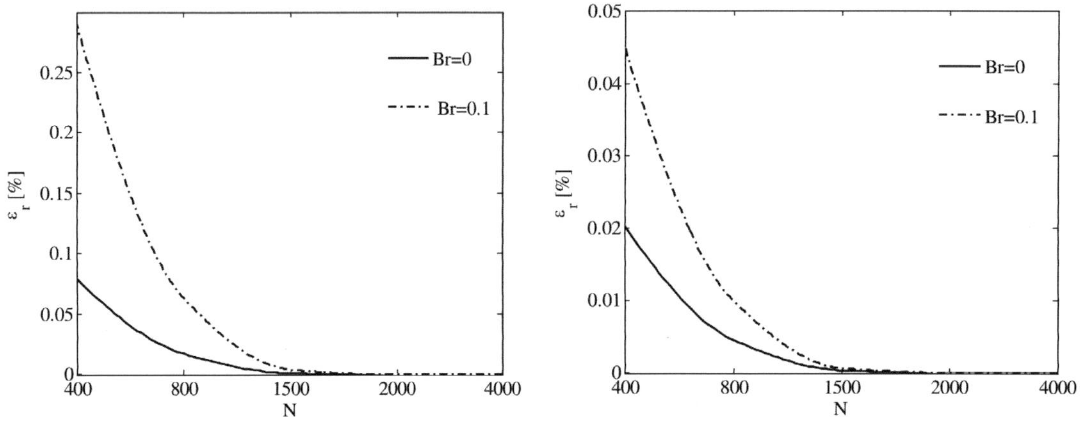

(a) (b)

Figure 2. Relative error in Nusselt number as a function of the grid elements for different values of the Brinkman number: (a) Continuum flow (Kn=0); (b) Slip flow (Kn=0.1).

The validation of the numerical procedure has been performed by comparing the numerical results with the data available in literature. For circular microducts (β=1) an excellent agreement with the analytical values of the Nusselt number obtained by applying the solutions proposed by Aydın and Avcı [5] and by Jeong and Jeong [6] has been found for all values of the Knudsen and Brinkman numbers analysed, as shown in Table 1. The maximum discrepancy between the numerical Nu and the analytical ones is less than 1e-05.

To validate the numerical solution for the elliptical microchannels as well, a comparison between the present numerical results of the Nusselt number for Br =0 and other numerical values available in literature [2, 21] has been performed; as shown in Table 2; a perfect agreement has been found even in this case.

Both benchmarks confirm that the numerical procedure adopted in the present analysis can be considered robust and accurate.

To investigate the influence of the viscous dissipation and rarefaction degree in the considered cross-section, several numerical runs have been performed by considering different values of the main parameters that influence the fluid behavior.

Table 1. Comparison between numerical Nusselt numbers for circular ducts (β=1) and analytical ones.

	Br									
	0		0.005		0.01		0.05		0.1	
Kn	[5] [6]	Present work	[5] [6]	Present work	[5] [6]	Present work	[5] [6]	Present work	[5] [6]	Present work
0	4.364	4.364	4.271	4.271	4.181	4.181	3.582	3.582	3.038	3.038
0.02	4.071	4.071	4.017	4.017	3.964	3.964	3.589	3.589	3.209	3.209
0.04	3.749	3.749	3.717	3.717	3.685	3.685	3.449	3.449	3.194	3.194
0.06	3.439	3.439	3.419	3.419	3.399	3.399	3.248	3.248	3.078	3.078
0.08	3.156	3.156	3.143	3.143	3.130	3.130	3.031	3.031	2.916	2.916
0.10	2.904	2.904	2.895	2.895	2.887	2.887	2.820	2.820	2.741	2.741

Table 2. Comparison between numerical Nusselt numbers for elliptic cross-section and data available in literature for *Br*=0.

	Kn								
	0			0.04		0.06		0.1	
β	[2]	[21]	Present work	[21]	Present work	[21]	Present work	[21]	Present work
0.100	0.6562	0.656	0.656	0.710	0.710	0.720	0.720	0.724	0.724
0.125	0.9433	0.943	0.943	0.999	0.999	1.003	1.003	0.988	0.988
0.200	1.820	1.820	1.820	1.811	1.811	1.768	1.768	1.655	1.655
0.250	2.333	2.333	2.333	2.246	2.246	2.161	2.161	1.973	1.973
0.333	3.006	3.006	3.006	2.778	2.778	2.627	2.627	2.331	2.331
0.500	3.802	3.802	3.802	3.362	3.362	3.121	3.121	2.678	2.678
0.667	-	4.171	4.171	3.618	3.618	3.332	3.332	2.832	2.832
0.750	-	4.266	4.266	3.683	3.683	3.385	3.385	2.868	2.868
0.833	-	4.325	4.325	3.723	3.723	3.417	3.417	2.889	2.889

With regards to the aspect ratio of the elliptic cross-section, the investigation has been carried out assuming the most common values of the aspect ratio, namely $\beta = 0.10$, $\beta = 0.125$, $\beta = 0.20$, $\beta = 0.25$, $\beta = 0.333$, $\beta = 0.50$, $\beta = 0.667$, $\beta = 0.75$, $\beta = 0.833$ and $\beta = 1$. It has to be pointed out that to keep the same value of the hydraulic diameter, for each value of aspect ratio both the semi-minor axis (*b*) and the semi-major axis (*a*) have been varied. The role of the viscous forces in the convective heat transfer has been investigated by varying the Brinkman number in the range 0÷0.1 while the rarefaction effects have been taken into account by changing the Knudsen number from 0 (continuum flow) up to 0.1.

In Figure 3 the dimensionless fluid temperature as a function of the semi-minor axis of the ellipse (y) is depicted, for a shallow elliptical channel (β=0.1), for different values of the Brinkman number, namely *Br* =0.005, *Br* =0.01, *Br* =0.05 and *Br* =0.1, and for both continuum flow (*Kn* =0) and slip flow (*Kn* =0.1). It has to be pointed out that in H2 boundary conditions the temperature distribution is calculated apart from an arbitrary constant; the here presented results have been obtained by considering that the dimensionless temperature at the centre of the cross-section is zero. By observing Figure 3a, it can be noticed that the thermal fluid behavior is significantly affected by the viscous heating, when the rarefaction effects are disregarded (*Kn* =0); as the Brinkman number increases from 0.005 to 0.1, the maximum of the non-dimensional fluid temperature, reached at the wall, increases by about 80%. On the contrary the influence of the viscous heating on the fluid behavior becomes less relevant in slip flow regime. It has been observed that as the rarefaction effects become more pronounced the influence of the Brinkman number is reduced. This effect is due to the presence of the slip velocity at the wall which leads to a reduction of the velocity gradient. This trend is evident in Figure 3b in which the non-dimensional temperature profiles as a function of the shorter axis of the ellipse for Kn =0.1, are presented. An increase in the Brinkman number from 0.005 to 0.1 leads an increase in the wall dimensionless temperature of about 19%.

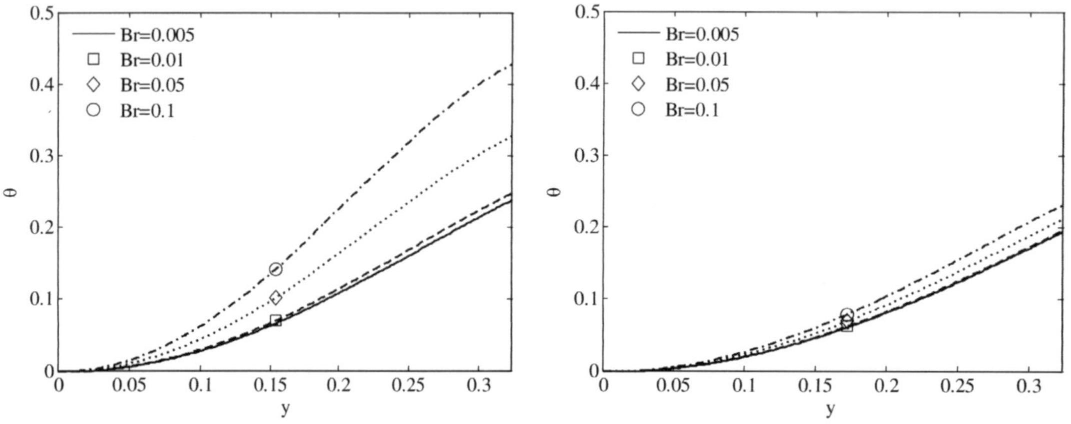

(a) (b)

Figure 3. Non-dimensional temperature profile for β=0.1: (a) Continuum flow (Kn=0); (b) Slip flow (Kn=0.1).

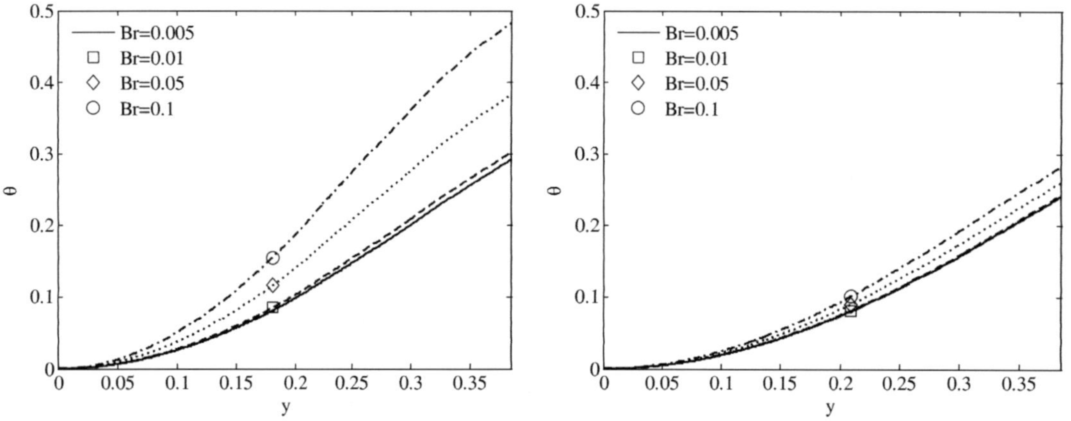

(a) (b)

Figure 4. Non-dimensional temperature profile for β=0.5: (a) Continuum flow (Kn=0); (b) Slip flow (Kn =0.1).

The same trends have been observed for each value of the aspect ratio considered in the present analysis. Figure 4 shows the non-dimensional temperature profiles for β=0.5, keeping the same values of the Brinkman and Knudsen numbers. However by comparing the graphs presented in Figure 3a with those in Figure 4a, it can be noticed that the influence of the Brinkman number becomes less pronounced as the aspect ratio increases. In fact by increasing Br from 0.005 up to 0.1 the dimensionless wall temperature increases by about 65%. On the contrary, within the slip flow regime the influence of the aspect ratio becomes less relevant (i.e. for Br changing from 0.005 to 0.1, the wall temperature increases by about 17%).

The influence of the aspect ratio is more evident by observing the graphs shown in Figures 5 and 6. In Figure 5 the Nusselt number as a function of both Knudsen and Brinkman numbers is depicted for several values of the aspect ratio. With regards to the effect of the viscous heating, numerical results have pointed out that viscous dissipation degrades the heat transfer: the Nusselt number decreases as the Brinkman number increases for all values of the Knudsen number and of aspect ratio investigated in the present study. Similar trends have been also observed by other researchers in different geometry cross-sections (see for instance Tunc and Bayazitoglu [4], Aydın and Avcı [5] and Jeong and Jeong [6] for circular ducts, van Rij et al.[14] for rectangular microducts and Kuddusi [17] for trapezoidal microchannels).

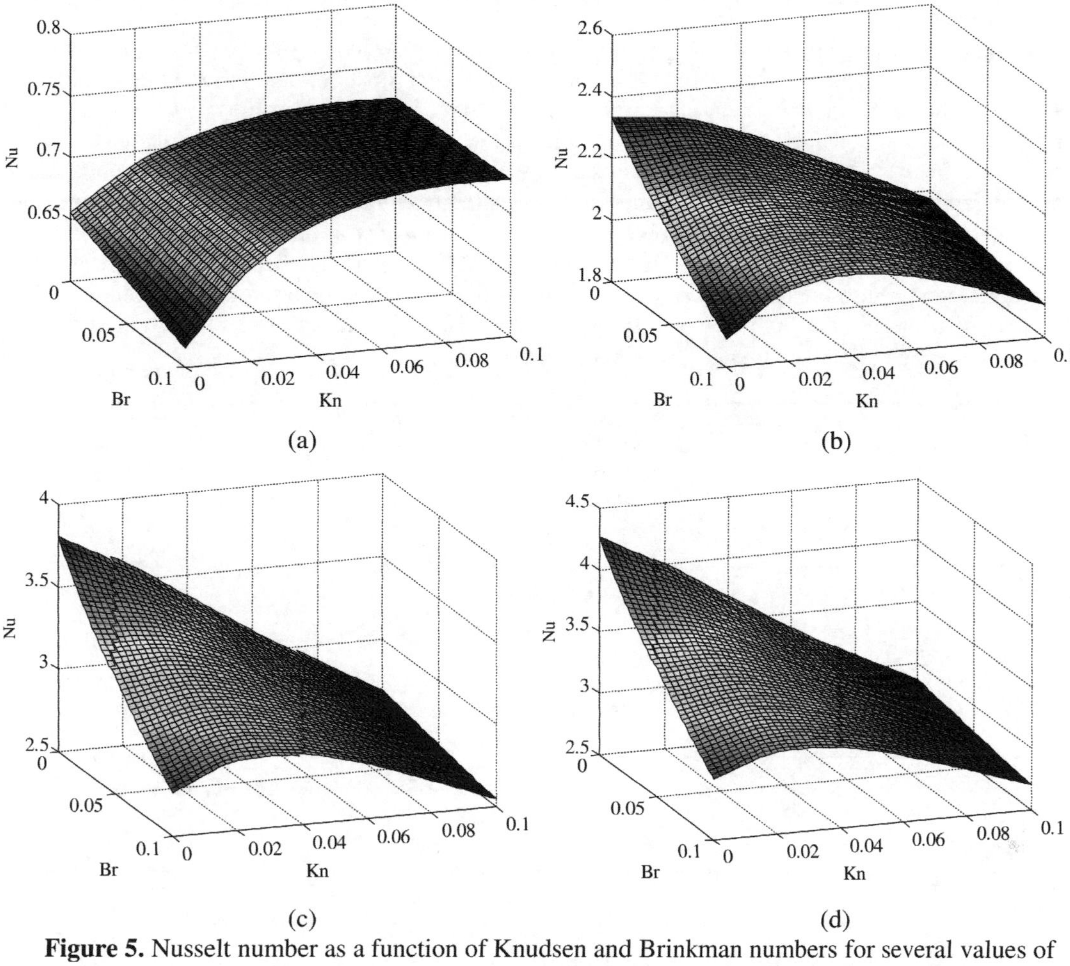

Figure 5. Nusselt number as a function of Knudsen and Brinkman numbers for several values of aspect ratio: (a) β=0.1; (b) β=0.25; (c) β=0.5; (d) β=0.75.

On the other hand the rarefaction effects may increase or decrease the Nusselt number depending on the value of the aspect ratio and of the Brinkman number. In fact for $\beta \geq 0.25$ and for low values of the Brinkman number (i.e. Br ranges between 0.01 and 0.04, depending on the value of the aspect ratio), Nusselt number decreases as the Knudsen number increases, while for $\beta \geq 0.25$ and for high values of the Brinkman number, Nusselt number presents a maximum for Kn ranging between 0.02 and 0.04 (depending on the values of β and Br).

For shallow microchannels the Nusselt number presents a maximum for all values of the Brinkman number considered in the present work; the value of the Knudsen number in which the maximum Nu occurs depends on the value of β and Br, as it can be noticed by observing Figures 5a and 5b. However, it is important to highlight that for very shallow elliptical microchannels the Nusselt number becomes lower than 1, which means that the convective heat transfer is less efficient than pure heat conduction.

Journal of Physics: Conference Series **547** (2014) 012023 doi:10.1088/1742-6596/547/1/012023

To highlight the influence of the viscous dissipation on the heat transfer coefficient, a comparison between the Nusselt number obtained for $Br > 0$ and the corresponding values for $Br = 0$ computed by Vocale et al. [21], has been performed.

The value assumed by the normalized Nusselt number (i.e. $Nu_N=Nu_{Br>0,Kn>0}/Nu_{Br=0,Kn=0}$) as a function of the aspect ratio β is shown in Figure 6 by taking several values of the Brinkman and Knudsen numbers.

The analysis of the trends of the normalized Nusselt number (Nu_N) leads to the same conclusions as regards the combined effect of the rarefaction and the viscous dissipation, highlighting the influence of the aspect ratio. In fact by observing Figure 6 it can be noticed that for all values of the Knudsen and Brinkman numbers considered in this work, the influence of the viscous dissipation increases as the aspect ratio increases. Values of the normalized Nusselt number larger than 1 can be obtained in the slip flow regime for shallow elliptical channels (β <0.2); the maximum increase of the Nusselt number is obtained for β=0.1 when the rarefaction effects are more evident (Kn=0.1) and this increase is of the order of 10% independently from the value assumed by the Brinkman number. On the contrary, when the elliptic cross section tends to the circular geometry the normalized Nusselt number is always less than 1, which means that both rarefaction and viscous dissipation make worse the convective heat transfer, reducing the Nusselt number by about 30% for β=1 (i.e. circular microducts).

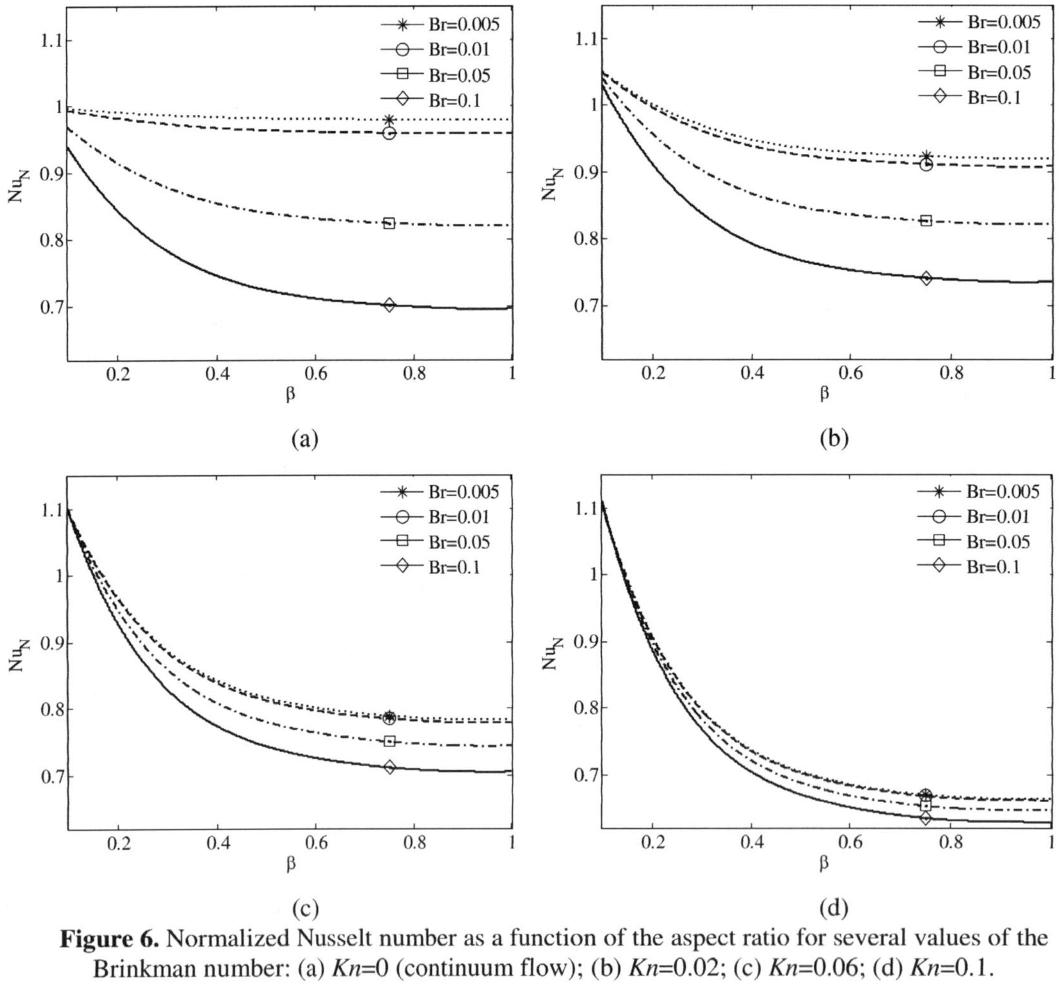

Figure 6. Normalized Nusselt number as a function of the aspect ratio for several values of the Brinkman number: (a) Kn=0 (continuum flow); (b) Kn=0.02; (c) Kn=0.06; (d) Kn=0.1.

4. Conclusions

In this work the role of the viscous dissipation on the convective heat transfer for a laminar and fully developed gaseous flow, in elliptic microducts, has been numerically investigated. The governing equations, which have been solved within the Comsol Multiphysics® environment, have taken also into account the presence of rarefaction effects. The thermal problem has been tackled by considering a constant and uniform heat flux at the wall.

The numerical results, validated against the data available in literature for circular and elliptical cross-sections, show that the viscous forces play a leading role in the convective heat transfer. In particular for high values of the aspect ratio and rarefaction degree, neglecting the viscous dissipation effect may lead to an overestimation of the Nusselt number up to about 59%.

It has been demonstrated that for shallow elliptical microchannels with an aspect ratio β lower than 0.2 the presence of rarefaction effects is beneficial for convective heat transfer and an increase of the Knudsen number determines an increase of the Nusselt number up to 10%. However, it is important to highlight that shallow elliptic channels having an aspect ratio lower than 1/8 are characterized by Nusselt numbers lower than 1 in fully developed laminar regime which means that, even in the case in which rarefaction effects and viscous dissipation effects are negligible, forced convection is ineffective with respect pure heat conduction.

These results confirm that reducing the channel size is not enough to enhance the convective heat transfer coefficient; in fact the viscous dissipation effects shall offset the gains of high heat transfer coefficients associated with a reduction in the channel size.

Acknowledgments

This work has been carried out thanks to the financial support of the PRIN2009TSYPM7_001 project “Single-phase and two-phase heat transfer for microtechnologies. Heat transfer and fluid flow in microscale”.

References

[1] Herwig H and Hausner O 2003 *Int. J. Heat Mass Tran.* **46** 935-937
[2] Shah RK and London AL 1978 *Laminar flow forced convection in ducts* ed Academic Press (New York) chapter III
[3] Colin S 2012 *ASME J. Heat Transf.* **134** 020908
[4] Tunc G and Bayazitoglu Y 2001 *Int. J. Heat Mass Tran.* **44** 2395–2403
[5] Aydın O and Avcı M 2006 *Int. J. Heat Mass Tran.* **49** 1723-1730
[6] Jeong HE and Jeong JT 2006 *J.Mech. Sci Technol.* **20** 158-166
[7] Çetin B, Yazicioglu AG and Kakac S 2009 *Int. J. Therm. Sci.* **48** 1673-1678
[8] Sun W, Kakac S and Yazicioglu AG 2007 *Int. J. Therm. Sci.* **46** 1084-1094
[9] Hooman K 2007 *Int. J. Heat Mass Tran.* **34** 945-957
[10] Sadeghi A and Saidi MH 2010 *J Heat Transf.* **132** 072401
[11] Jeong HE and Jeong JT 2006 *Int. J. Heat Mass Tran.* **49** 2151-2157
[12] Aydın O and Avcı M 2007 *Int. J. Therm. Sci.* **46** 30-37
[13] Zhang T, Jia L, Yang L and Jaluria Y 2010 *Int. J. Heat Mass Tran.* **53** 4927-4934
[14] van Rij J, Ameel T and Harman T 2009 *Int. J. Therm. Sci.* **48** 271-281
[15] van Rij J, Ameel T and Harman T 2009 *Int. J. Heat Mass Tran.* **52** 2792-2801
[16] Sun Z and Jaluria Y 2012 *Int. J. Heat Mass Tran.* **55** 3488-3497
[17] Kuddusi L 2011 *Int. J. Heat Mass Tran.* **54** 52-64
[18] Karniadakis G, Beskok A and Aluru N 2005 *Microflows and Nanoflows – Fundamentals and Simulation* ed Springer (New York) chapter 2
[19] Morini GL, Spiga M and Tartarini P 2004 *Superlattice Microst.* **35** 587-599
[20] Spiga M and Vocale P 2012 *Adv.Mech. Eng.* **2012** 481280
[21] Vocale P, Morini GL and Spiga M 2014 *Int. J. Heat Mass Tran.* **71** 376-385

Journal of Physics: Conference Series **547** (2014) 012024 doi:10.1088/1742-6596/547/1/012024

Experimental investigation of counter current air-water flow in a large diameter vertical pipe with inners

Giorgio Besagni, Gaël Guédon, Fabio Inzoli
Politecnico di Milano, Department of Energy, Via Lambruschini 4a, 20156, Milano

E-mail: giorgio.besagni@polimi.it

Abstract. Counter current two phase flow is encountered in a wide variety of industrial applications. This paper describes the experimental results obtained in a circular column of 240 mm diameter with two inner pipes. The counter current flow studied concerns an upward flow of air and a downward flow of water at ambient temperature and pressure. The following range of operating conditions was analysed: superficial air velocities up to 23 cm/s and superficial water velocities up to - 11 cm/s, corresponding to a global air volume fractions (gas holdup) up to 29%. The experimental investigation concerned (i) flow visualization, (ii) local data from a double fibre optical probe and (iii) gas holdup measurements. The images obtained from an optical camera were used to observe the general flow patterns. The data obtained from the double fibre optical probe were used to study the local flow characteristics. In particular, the local void fractions, the bubble velocities, the bubble mean diameters and the bubble diameter distributions and are presented and discussed. The bed expansion technique was used to obtain the gas holdup measurements for every operating condition. The gas holdup measurements are discussed, compared with existing correlations and used for investigating the flow regime transitions. Finally, the gas holdup and the local void fraction measurements data are compared and used for understanding the local hydrodynamics.

1. Introduction

Two-phase flow in vertical pipes is encountered in several plants in the chemical and in the nuclear fields. The chemical industries utilize vertical pipes in bubble columns and in fluidized beds for heat and mass transfer operations. In the nuclear industry, a typical example of vertical pipe is the natural circulating flow in the chimney region of Boiling Water Reactor (BWR). The correct design and operation of these devices can be managed with the proper prediction of the flow patterns and flow properties at the operating conditions of interest. These information can be obtained by using experimental investigations. In the broader framework of two phase flow in vertical pipes, we focus on counter current air-water flow in a large diameter vertical pipe.

Two-phase vertical counter-current flow has been investigated experimentally by several researchers using intrusive and non-intrusive techniques. Yamaguchi and Yasaburo [1] investigated cross-sectional void fractions of co-current and counter current air–water flow in vertical pipes with inner diameters of 40 and 80 mm. The cross-sectional void fraction was measured by a quick-closing valve method. These experiments were carried out in bubbly and slug flow regimes. Hasan et al. [2] investigated cross-sectional void fractions of air–water co-current and counter current flow in a vertical pipe with an inner diameter of 127 mm. The cross-sectional void fraction was calculated using a pressure drop technique. These experiments were carried out in bubbly and slug flow regimes. Aritomi et al. [3-5] investigated the behaviours of counter current flow in a rectangular channel. Local void fraction was measured by an ultrasonic technique. These experiments were carried out in bubbly flow regime. Fuangworawong et al. [6] investigated local void fractions of air–water counter current flow in a vertical pipe with an inner diameter of 50 mm. The local void fractions were calculated using a wire mesh tomography technique. These experiments were carried out in bubbly flow regime. Ghosh et al. [7, 8] investigated cross-sectional void fractions of air–water counter current flow in a vertical pipe with an inner diameter of 26.4 mm. The cross-sectional void fractions were calculated using a wire conductivity probe technique. These experiments were carried out in slug, churn and annular regimes, including the flooding condition.

All the previous studies were focused on small diameter pipes. The present investigation concerns a pipe with inner diameter of 240 mm, which is a large diameter pipe under the ambient operating conditions. The size range of a pipe is determined by the non-dimensional hydraulic diameter D^*_H [9]:

$$D_H^* = \frac{D_H}{\sqrt{\sigma / g\left(\rho_l - \rho_g\right)}} \qquad (1)$$

where D_H is the hydraulic diameter, σ is the surface tension, g is the force due to gravity and ρ_l-ρ_g is the density difference between the phases. Pipes with non-dimensional diameters greater than a critical (or transitional) value $D^*_{H,cr}$ are considered to be large diameter pipes. The critical value is reported to be $D^*_{H,cr}$ = 52 [10]. The above critical diameter corresponds to $D_H \approx$ 130 mm for air and water at atmospheric conditions and $D_H \approx$ 63 mm for BWR operating conditions. The critical value $D^*_{H,cr}$ is given in terms of the Taylor wavelength. When the pipe diameter is larger than such a value, the stabilizing effect of the channel wall on the interface of the Taylor bubbles become lower and Taylor bubble can no longer be sustained due to the Raleigh–Taylor instability. The interaction of these two instabilities results in large bubbles being broken into many smaller cap-shaped bubbles. Hence, large diameter pipes can be defined as flow channels in which stable slug bubbles cannot be sustained [9]. The above-mentioned studies have a non-dimensional diameter of D^*_H = 14.69 – 29.38 [1], 46.64 [2], 18.18 [3-5], 18.38 [6] and 9.69 [7.8]. All of these values are below the critical value of 52. Indeed in these studies the slug flow is reported and/or can exist. This is particularly interesting for [2], which has a non-dimensional diameter D^*_H = 46.64. This pipe can be considered as an intermediate pipe. Indeed, the values is a little bit below the critical value (indeed, the slug flow appears). Generally speaking, flow in large pipes has several significant differences from flow in small pipes because of the many changes to the hydrodynamics: change in bubble drag, presence of additional turbulence [11] and strong secondary recirculation in the flow. The three-dimensional circulatory behaviour is due to the liquid flowing around the cap bubbles rather than being forced out of the way, as is the case with slug bubbles. These changes result in very different physical mechanisms behind the transport of gas and liquid and causes significant changes to the void fraction and velocity profiles flow in smaller pipes. Indeed, turbulence, along with the lift, turbulent dispersion and the wall lubrication force, plays an important role in determining the void profiles [12] and can also result in enhanced bubble breakup. It should be noted that these differences may become less significant for higher liquid velocities where the turbulent intensity is already very high [13]. For all these reasons, the models typically used to predict the behaviour of two-phase flow in small diameter pipes, in which slug flow exists, may not be scaled for use in understanding flow in large diameter pipes [14]. Furthermore, flow regime maps and criteria to determine flow regime transitions for large pipes differs from small diameter ones [15, 16].

Even with the several progresses made in the field of two-phase flow, there is still little knowledge of counter-current flow in large diameter pipes. In addition, studies involving also the presence of inner pipes are rare [17] and are focused on a small diameter pipe. Moreover, we focus on a layout never studied in the literature. For the above mentioned reasons, experimental data and correlations must be specifically developed for the analysis of this system. The present study is aimed at providing insights into such flow configuration through an experimental investigation. The following range of operating conditions was analysed: superficial air velocities up to 23 cm/s and superficial water velocities up to -11 cm/s, corresponding to global air volume fractions (holdup) up to 29%. The experimental investigations concerned (i) flow visualization, (ii) local data from a double optical probe and (iii)

Journal of Physics: Conference Series **547** (2014) 012024 doi:10.1088/1742-6596/547/1/012024

holdup measurements. Images obtained from an optical camera were used to observe the general flow pattern. Data obtained from the double optical probe were used to study local flow characteristics. In particular, the local void fractions, the bubble velocities, the bubble mean diameters and the bubble diameter distributions and are presented and discussed. The bed expansion technique was used to obtain the gas holdup measurements for every operating condition. The gas holdup measurements are discussed, compared with existing correlations and used for investigating flow regime transition. Finally, the gas holdup and the local void fraction measurements are compared and used for understanding the local hydrodynamics.

2. Experimental setup

The experimental facility (figure 1) consists of a non-pressurised vertical pipe made of Plexiglas with d_{column} = 240 mm and H_{column} = 5.3 m. An internal pipe, made of PVC, with an external diameter of 60 mm, is centrally positioned, while another internal pipe, also made of PVC, with an external diameter of 75 mm, is positioned asymmetrically. A pump, controlled by a by-pass valve, provides water recirculation. A rotameter (3) measures the water flow rate. A pressure redactor controls the pressure upstream the rotameters (1) and (2), used to measure the air flow rate. The air distributor, which is positioned asymmetrically, is made of a stainless steel tube with an external diameter of 70 mm, an height of 340 mm and has holes positioned along the circumference. Various hole diameters ranging from d_{holes} = 1 mm to d_{holes} = 5 mm were investigated and it was found that the hole diameter has no influence on the bubble Sauter mean diameter (measured in the flow developed region). In this study, a distributor with d_{holes} = 3.5 mm is used. Due to the nature of the observed flow phenomena and their sensitivity to surface tension forces, clean filtered deionized water was used. Furthermore, the system was flushed beforehand for long periods of time to remove contaminants. During the experiments, the air was maintained at atmospheric pressure at the outlet of the test section, and the air and water temperatures were maintained constant. At the start of each test, the water height is set to z = 3.245 m. In this study, the values of gas density and superficial gas velocity are based upon the operating conditions existing at the column mid-point, such as suggested by [18]. The mid-point column pressure, used to compute gas density, was assumed to be equal to the column outlet pressure plus one-half the total experimental hydrostatic pressure head.

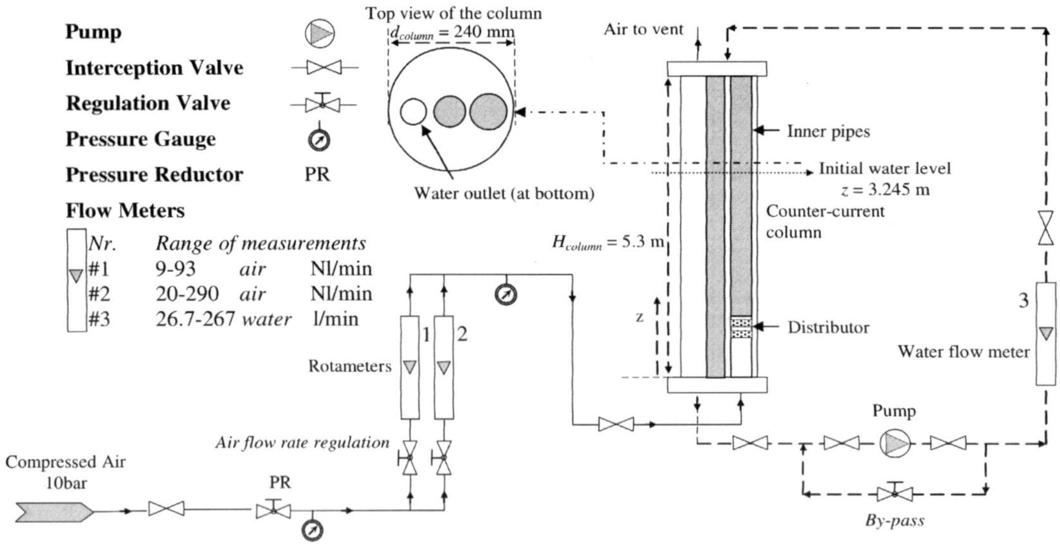

Figure 1. Experimental facility.

3. Measurement techniques

The local bubble properties were measured using a double fibre optical probe. The gas holdup measurements were determined using the bed expansion technique. Photos were taken at selected conditions. All the measurements were taken after the system reached the steady state.

3.1. Local measurements

Local flow properties have been measured by means of a double fibre optical probe system [19] manufactured by RBI (France). The optical probe is inserted, via an access port (figure 2), into the flow at a height of z = 2.3 m from the bottom (z = 1.9 m from the air distributor). Optical probes are intrusive phase detection tools that distinguish the gas and liquid phases by measuring the intensity of a laser light that is reflected at the probe tip when submerged in either phase. The laser is reflected and/or refracted at varying intensities depending on the probe tip geometry and the refractive indexes of the probe tip ($n_{tip} \approx 1.6$), gas ($n_{air} \approx 1$), and liquid ($n_{water} \approx 1.33$) phases. The probe signal is measured via an optoelectronic module which emits the laser to the probe tip and converts the reflected optical signal into a digital signal. The resulting signal distinguishes between the gas and the liquid phases. The response time of the cell is 0.5–1 μs, thus local phase interfaces are detected instantaneously. The time averaged local flow properties measured are the local void fraction, bubble vertical velocity, bubble size, interfacial area concentration and bubble chord length distributions. All the acquisitions are performed using a sampling period equal to $\Delta t_{sampling}$ = 1000 s. This measurement period is large enough to produce reliable time-averaged values. However, there are some potential sources of errors when characterizing bubbles using optical probes. Among the various effects we may state: (i) the improper dewetting at the probe tip (the blinding effect), (ii) the alteration of bubble trajectory prior to or during the piercing process (the drifting effect) and (iii) bubble deformation and/or deceleration at the probe tip (the crawling effect) [19, 20].

Figure 2. The double fibre optical probe and its position within the pipe cross-section.

3.2. Holdup measurements

Measurements of the bed expansion allowed the evaluation of the global void fraction ε_{air} (gas holdup). The procedure involves measuring the change of liquid height when air is introduced in the column. The global void fraction is then obtained using:

$$\varepsilon_{air} = \left(H_D - H_0\right)/H_D \tag{2}$$

where H_D and H_0 are the heights of the free-surface after and before aeration, respectively. The change in liquid height is measured after a steady averaged liquid level is reached. The "holdup" and the "global void fraction" have the same meaning and refer to the same definition, the equation (2).

3.3. Photography

Photos were taken using a Canon α200 camera (1/1000s, ISO400). The back light method is employed in the experiments and the light source is provided by a 500 W halogen lamp. Visualization sections consist in squared boxes (filled with water) around the vertical pipe for correcting the distorted image. The camera was aligned horizontally to the visualization sections.

Journal of Physics: Conference Series **547** (2014) 012024 doi:10.1088/1742-6596/547/1/012024

4. Results

4.1. Flow regimes description and discussion

4.1.1. Flow regime description
Ranging from low to high superficial gas velocity we have observed the homogeneous-bubbly and the heterogeneous-churn turbulent flow regime, with a transition zone between the two regimes. We have not observed a stable slug flow, the annular flow and the flooding condition. Qualitatively speaking, for low air superficial velocities, bubbly flow is observed (figure 3). In this regime, bubbles are uniformly distributed in the cross section of the pipe, traveling vertically with minor transverse and axial oscillations. Bubbles have not a regular spherical shape. Around the inner pipes we have observed bubble recirculation and the periodic appearance of cap-bubbles. The cap-bubbles seem to originate because of the bubble coalescence around the inner pipes. Due to the presence of these cap-bubbles this regime can be better defined as a pseudo-homogeneous regime. Increasing air flow rate, we have observed a transition zone (figure 4) toward the heterogeneous regime (figure 5). Here, we have noticed an increased coalescence rate. Finally, a fully developed churn turbulent regime is reached (figure 6). In this condition we have observed periodic very large bubble which occupy the major part of the cross section of the pipe. Those bubbles stake the form of spherical caps with a very mobile and flexible interface. Increasing the water flow rate, we have observed an increase in bubble number, a reduced bubble rising velocity and highly recirculating phenomena.

Figure 3. J_{air} = 0.87 cm/s– J_{water} = 0 cm/s.

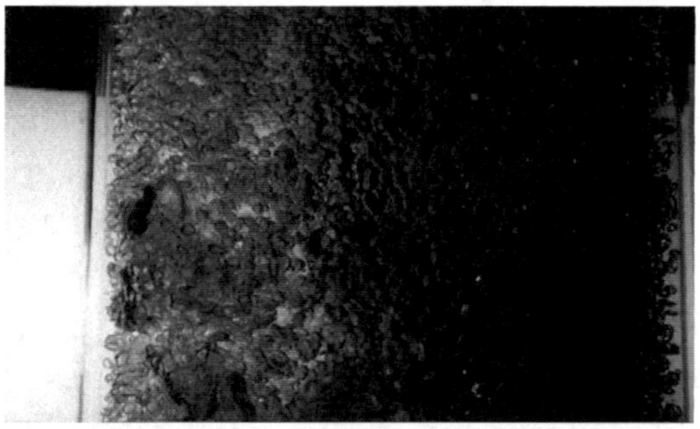

Figure 4. J_{air} = 4.09 cm/s – J_{water} = 0 cm/s.

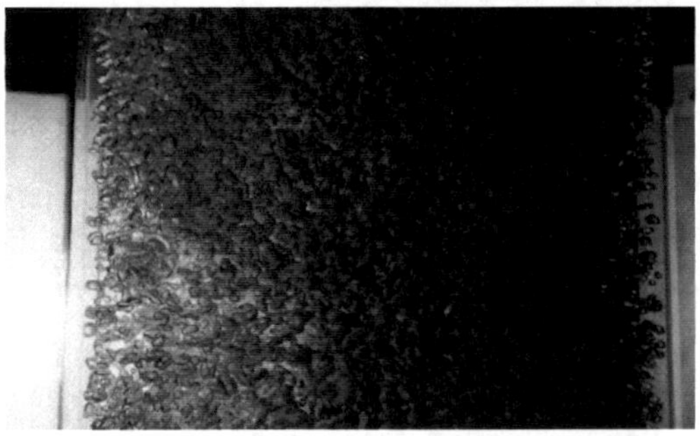

Figure 5. J_{air} = 11.94 cm/s – J_{water} = 0 cm/s.

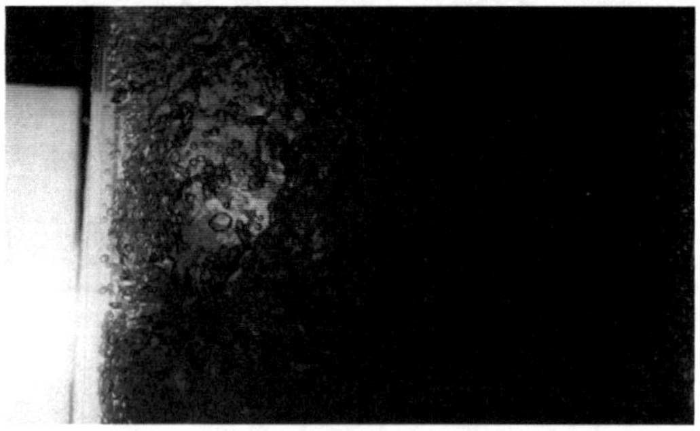

Figure 6. J_{air} = 19.91 cm/s – J_{water} = 0 cm/s.

4.1.2. Flow regime discussion
Flow regimes can be discussed with three main considerations: (i) counter current flow, (ii) large diameter pipe and (iii) the presence of inner pipes. Firstly, the two-phase flow dynamics in the counter-current flow configuration is very similar to the bubble column configuration. Large periodic eddies accompanied with flow recirculation are observed all along the vertical development of the

pipe. Secondly, flow phenomena can be detailed with reference to large diameter pipe theory. In the present case, the dimensionless diameter is $D^*_H = 88.13$, if the inner pipes are not considered, and $D^*_H = 47.37$, if the inner pipes are considered. The former value if above the critical value of 52 [10] and the latter is below. Hence, we are studying an intermediate pipe that originates from a large diameter pipe. This explains the absence of a fully developed stable slug regime and the existence of periodic very large bubbles which occupies a large part of the pipe. Stable Taylor bubble has not been detected. Taylor bubbles were noticed only when the air flowmeters were opened and air was introduced in the pipe with stagnant water. After this occurrence stable Taylor bubbles could no more exist because of the turbulence in the pipe and the above-discussed instabilities. Moreover, in a large diameter pipe, high turbulence is expected and, in the present case, the turbulence intensity is such that bubble coalescence and break-up constantly occur even at low gas flow rate. The high flow mixing is the reason why the flow field rapidly becomes developed: distance of about 5÷7 d_{inner} is required for the flow to develop downstream the air distributor. Accordingly, the double optical probe measurements are taken at $z = 1.9$ m $\approx 8\ d_{inner}$ from the air distributor. Thirdly, the presence of inner pipes may influence the flow characteristics. At first, cap-bubbles are noticed to originate along the inner pipes. At second they may provide additional recirculating flow phenomena, affecting the turbulence intensity and the size and frequency of the turbulent eddies. The presence of additional turbulence is verified in the present experiments: an analysis of different distributor hole diameters and configurations did not influenced the flow dynamics and bubble sauter mean diameters.

4.2. Gas holdup measurements and flow regime transitions

4.2.1. Gas holdup

The gas holdup data has been measured for superficial air velocities up to 23 cm/s and superficial water velocities up to - 11 cm/s (figure 7, left). At low air superficial, the relation between the gas holdup and the superficial gas velocity is linear, followed by a change in tendency at a certain transition superficial gas velocity. The linear trend corresponds to bubbly flow and the change in tendency is due to flow regime transition toward the bubbly-churn transition zone. Indeed, in dispersed flow and assuming proper gas distribution, gas holdup increases linearly as a function of the gas flow rate, where more bubbles of similar size occupy more volume in the column. Increasing the superficial water velocity, the holdup increase because bubble rise velocity decrease. Above the transition velocity, bubble coalescence increases the average rise velocity (figure 9) and reduces gas residence time in the column, hence decreasing the gas holdup versus gas velocity slope. From flow visualisations, a progressive change in flow regime is observed in correspondence to this phenomenon, i.e., the flow turbulence starts to increase significantly and large deformed bubbles starts to appear. The transition toward the heterogeneous regime is completed at about $\varepsilon_{air} \approx 16\%$. Above $\varepsilon_{air} \approx$ 16-17 %, another phenomenon is identified when the counter-current flow configuration is considered: the water superficial velocity has no influence on the gas holdup. The non-influence of water velocity on the gas holdup is a matter of further investigation. Here, we propose two possible explanations for this phenomena. The first explanation concerns the role of the turbulence. The two-phase flow dynamics is mainly dominated by turbulence in large diameter pipes. Moreover, it is known from the literature that, for high liquid velocities, when the turbulence intensity is already high, the discrepancies in the fluid dynamics become less significant [13]. The second explanation concerns the distribution of the phases in the column. At such conditions large bubbles, that occupies the major part of the cross section of the pipes, appears and the liquid is mainly concentrated along the wall pipe. These hypotheses can be validated only performing investigation without the presence of the inner pipes. The reported void fraction trend can be justified by considering the contribution of small and large bubbles [21]. In the churn-turbulent flow regime, the contribution of small bubbles to overall holdup is constant, whereas the large bubble holdup increases with increasing superficial velocity. On the contrary, in bubbly flow, small bubble holdup is not constant but changes significantly as the superficial velocity is changed.

4.2.2. Flow regime transition

The flow regime transition description proposed in the previous paragraph is here supported using a more quantitative analysis. In particular, the flow regime transition from bubbly flow to the transition zone can be predicted using the method proposed by Krishna et al. [22] by plotting the swarm rise velocity (defined as the ratio of superficial gas velocity to gas holdup) versus corresponding superficial gas velocity (figure 7, right). The swarm velocity was found to be constant in the homogeneous regime, but it starts to increase as the system enters the heterogeneous regime at a certain transition superficial velocity. The transition velocity decreases while increasing the superficial liquid velocity. The appearance of the first large bubble is responsible for such sudden increase in swarm velocity and is an indication of flow regime transition. Further investigations concerning flow regime transition are matter of future works.

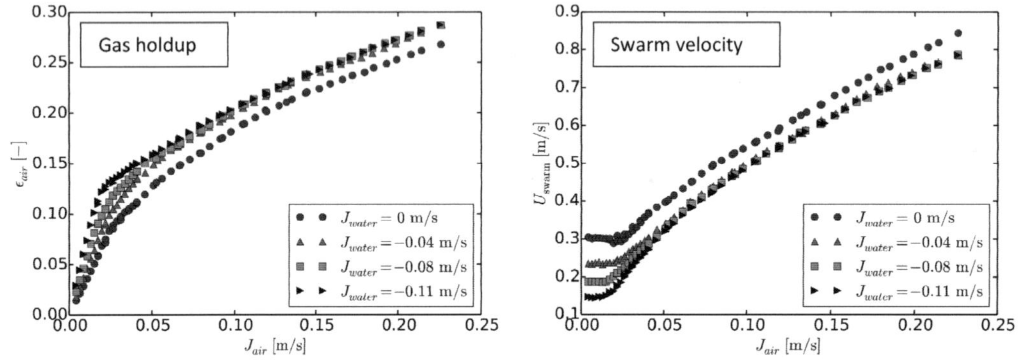

Figure 7. Gas holdup measurements (left) and swarm velocity (right).

4.2.3. Comparison with correlation

The data taken in the bubble column configuration (J_{water} = 0 cm/s) are compared (figure 8) with four correlations for the holdup prediction in bubble column reactors: Reilly et al. [23], Joshi and Sharma [24], Hughmark [25] the Kawase and Moo-Young [26]. For both low and higher air superficial velocities, the correlation of Hughmark gives the closest predictions. For low air superficial velocities, the correlation of Joshi and Sharma and the correlation of Kawase and Moo-Young well predict the experimental data. The correlation of Reilly et al. gives a similar trend to the experimental data; however a constant shift of volume fraction is noted. The inner pipes may influence the trend of the holdup, thus further investigations without the presence of the inner pipes is required.

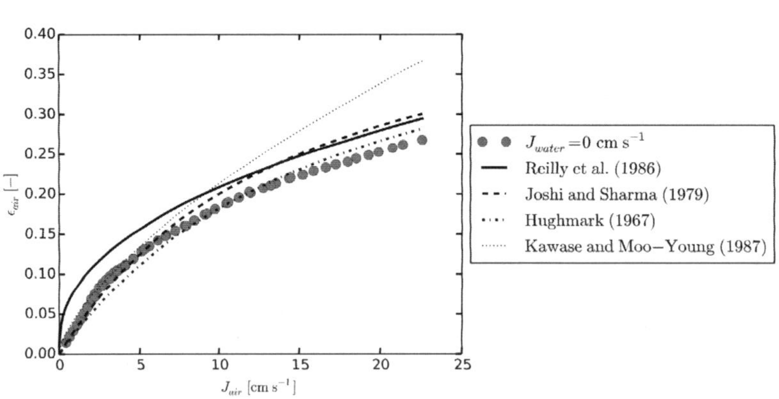

Figure 8. Holdup measurement (J_{water} = 0 cm/s): comparison with literature correlations.

Journal of Physics: Conference Series **547** (2014) 012024 doi:10.1088/1742-6596/547/1/012024

4.3. Optical probe measurements

Three to four measurements were performed at low air superficial velocities while two measurements were taken at moderate and high superficial air velocities. These measurements were performed for the four water superficial velocities investigated in the holdup analysis (figure 9). Four of the six point analysed of the bubble column configuration (J_{water} = 0 cm/s) are the same of flow visualisation previously described (figures 3 to 6). In the following, some comments are outlined.

Concerning the local void fraction (figure 9a), there are three considerations. Firstly, it increases with superficial air velocity, such as expected. Secondly, it increases with superficial water velocity. This is because bubbles move in a non-stagnant surrounding liquid, which is forced to move downward. In a time averaged point of view, this means that the same amount of mass flow for unit of volume needs more time to flow out from a control volume. Therefore void fraction increases. Thirdly, comparing local and global void fraction (figure 10). some consideration on void fraction profiles can be outlined: (i) a negative difference value (gas holdup < local void fraction) may indicate a centre peak void fraction profile; (ii) a zero difference value (gas holdup ≈ local void fraction) may indicate a flat void fraction profile; (iii) a positive difference value (gas holdup > local void fraction) may indicate a wall peak void fraction profile. The global-local void fraction comparison is a matter of further studies.

Concerning the bubble vertical velocity (figure 9b), an increase in air superficial velocity leads to an increase in bubble vertical velocity as one would expect. The increase of the bubble vertical velocity after the transitional superficial gas velocity is due to the increased coalescence rate. The increase in superficial water velocity, which we recall flows downward, also leads to increase in bubble vertical velocity. This indicates that the velocity in the recirculation cells may become higher. Furthermore, liquid may be more concentrated near walls and an increase of liquid velocity may results in the

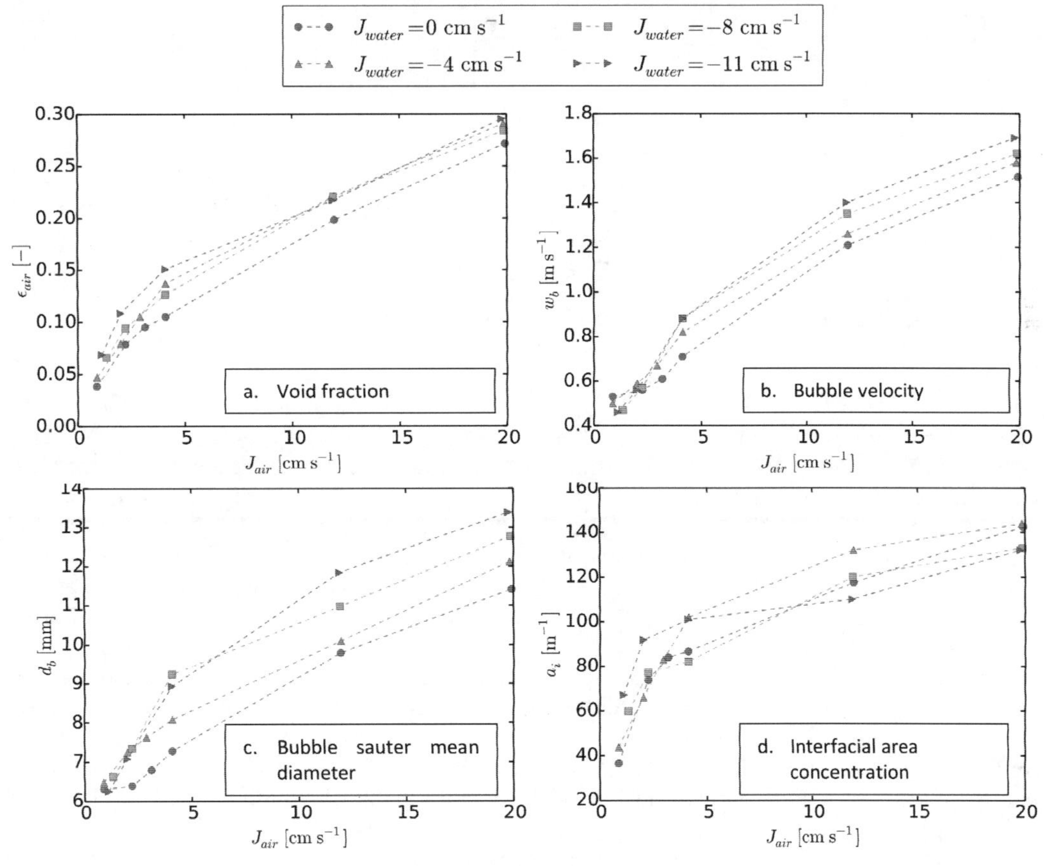

Figure 9. Optical probe measurements.

increase of the thickness of the liquid film near walls. This would result in the increase of the velocity of the gas phase in the core of the section.

Concerning the bubble Sauter mean diameter (figure 9c), it increases with air and water flow rate. The authors think that this trend is caused by the increased turbulence. An increased turbulence would cause an increase in the coalescence rate due to the higher number of collisions. The reasons for this trend can be linked to the coalescence phenomena [27]. It should not escape notice that the sauter mean diameter refers to the hypothesis of spherical shape. In the churn turbulent flow regime it is better to refer to bubble size distribution rather to a bubble Sauter mean diameter. The bubble diameter distribution for the churn turbulent regime is presented in figure 11. It should not escape notice that above distribution is obtained from a chord length distribution and the assumption of spherical bubbles. Above distribution should be corrected taking into account the shape of the bubbles. However, from the above figure some considerations can be outlined. It seems that in the fully developed churn turbulent flow, the peaks of the distributions increases decreasing water flow rate. This suggests that higher liquid flow rate induces a more uniform bubble size distribution.

Concerning the interfacial area concentration (figure 9d), it shows an increasing trend with the increase of water superficial velocity at low air superficial velocity and an opposite trend at higher air superficial velocity. This means that, as the bubble diameter increases, the interfacial area concentration decreases after some point.

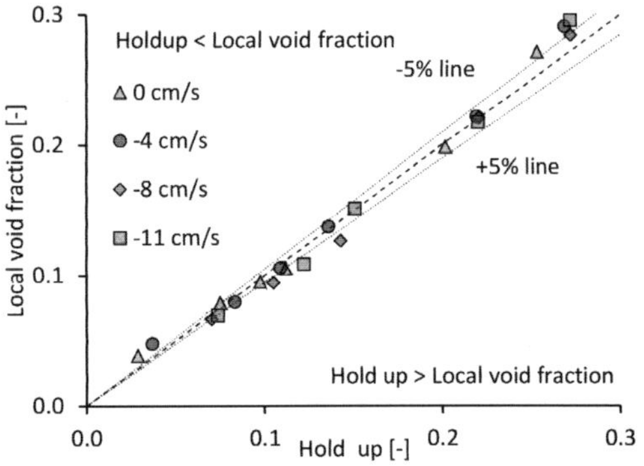

Figure 10. Local and global void fraction comparison. -- lines represent the ±5% difference.

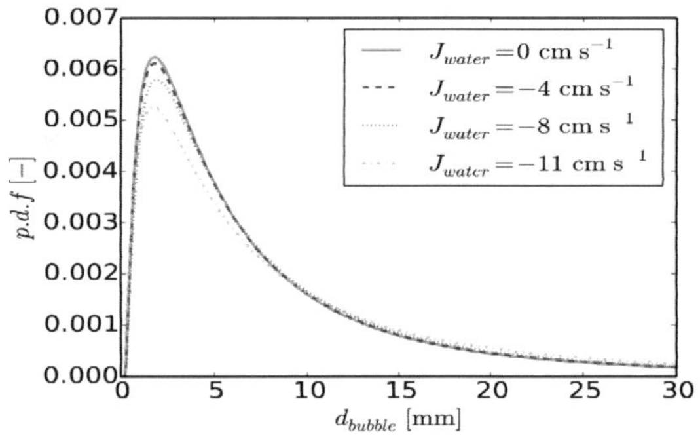

Figure 11. Bubble diameter distributions for the churn turbulent regime.

5. Conclusions

This paper investigates the air-water counter-current flow in a large diameter vertical pipe at ambient temperature and pressure. The inner pipe is 240 mm diameter and has two inner pipes. This layout has never studied in the literature The following range of operating conditions was analysed: superficial air velocities up to 23 cm/s and superficial water velocities up to - 11 cm/s, corresponding to global air volume fractions up to 29%. The experimental investigations concerned (i) flow visualization, (ii) local data from a double fibre optical probe and (iii) holdup data. A homogeneous-bubbly flow regime and a heterogeneous-churn turbulent flow regime have been reported with a transition zone between the two regimes. In the homogeneous regime bubble recirculation occurs around the inner pipes along with the periodic appearance of cap-bubbles. The cap-bubbles seems to originate along the inner pipes because of the coalescence phenomena. Due to the presence of these cap-bubbles this regime can be better defined as a pseudo-homogeneous regime. In the heterogeneous regime periodic very large bubbles which occupy the majority of the cross section of the pipe are observed. The stable slug flow and the annular flow and the flooding condition were not observed. The absence of slug flow is motivated by theoretical consideration on large diameter pipe. The flow regime transition has been analysed using holdup measurements. Flow regime transition between bubbly and transition regime starts at a certain transitional gas velocity, which decrease while increasing the superficial liquid velocity. In the bubbly flow regime gas holdup increases linearly as a function of the gas flow rate. Moreover, it increases with water flow rate. The transition toward the heterogeneous regime is completed at about $\varepsilon_{air} \approx 16\%$. Above $\varepsilon_{air} \approx 17$ %, the water superficial velocity has no influence on the holdup. Possible explanations for the phenomena are given and discussed. At the present state, the role of inner pipes on flow behaviour is not clear and is a matter of further study. A comparison between holdup measurements and literature correlations has been performed. Local flow behaviour has been analysed considering the local void fraction, the bubble velocity, the bubble diameter and the bubble diameter distribution. Local void fraction, bubble mean diameter and bubbles velocity increase with air and water flow rate, which have been critically analysed and commented. Analysing the bubble diameter distribution in the churn turbulent flow regime, it seems that higher liquid flow rate induces a more uniform bubble size distribution. At certain operating conditions global and local void fraction values are very similar and this may suggest a flat void fraction profile. At other operating conditions the difference may suggest a wall peak or a centre peak void fraction profile. Further studies may concern the influence of the inner pipes of the flow behaviour, a more detailed analysis of the flow regime transition and the study of the local flow behaviour. Finally, it would be interesting to analyse the images for providing data concerning the bubble shape.

References

[1] Yamaguchi K and Y. Yamazaki Y 1982, *J. Nucl. Sci. Technol.***19** 985-96
[2] Hasan A R, Kabir C S and Srinivasan S 1994 *Chem. Eng. Sci.* **49** 2567-74
[3] Aritomi M, Zhou S, Nakajima M, Takeda Y, Mori M and Yoshioka Y 1996 *J. Nucl. Sci. Technol.***33** 915-23
[4] Aritomi M, Zhou S, Nakajima M, Takeda Y and Mori M, 1997 *J. Nucl. Sci. Technol.***34** 783-91
[5] Zhou S, Suzuki Y, Aritomi M, Matsuzaki M, Takeda Y and Mori M 1998 *J. Nucl. Sci. Technol.***35** 335-43
[6] Fuangworawong N, Kikura H, Aritomi M and Komeno T 2007 *Chem. Eng. J.***130** 111-8
[7] Ghosh S, Pratihar D K, Maiti B and Das P K 2012 *Chem. Eng. Sci.* **84** 417-36
[8] Ghosh S, Pratihar D K, Maiti B and Das P K 2013 *Int. J. Multiphas. Flow* **52** 102-20
[9] Kataoka I and Ishii M 1987 *Int. J. Heat Mass Tran.***30** 1927-39
[10] Brooks C S, Paranjape S S, Ozar B, Hibiki T and Ishii M 2012 *Int. J. Heat Fluid Fl.* **37** 196-208
[11] Serizawa A and Kataoka I 1990 *Nucl. Eng. Des.* **122** 1-16
[12] Ohnuki A and Akimoto H, 1998 *Proc. Third International Conference on Multiphase Flow, ICMF '98* (Lyon, France)
[13] Elghobashi S and Abou-Arab T W 1983 *Physics of Fluids* **26** 931-8
[14] Hibiki T and Ishii M 2003 *Int. J. Heat Mass Tran.***46** 1773-90
[15] Ohnuki A and Akimoto H 2000 *Int. J. Multiphas. Flow* **26** 367-86
[16] Schlegel J P, Sawant P, Paranjape S, Ozar B, Hibiki T and Ishii M 2009 *Nucl. Eng. Des.* **239** 2864-74
[17] Al-Oufi F M, Cumming I W and Rielly C D 2010 *Can. J. Chem. Eng.* **88** 482-90
[18] Reilly I, Scott D, Debruijn T and MacIntyre D 1994 *Can. J. Chem. Eng.* **72** 3-12
[19] Barrau E, Rivière N, Poupot C and Cartellier A 1999 *Int. J. Multiphas. Flow* **25** 229-56
[20] Vejražka J, Večeř M, Orvalho S, Sechet P, Ruzicka M C and Cartellier A 2010 *Int. J. Multiphas. Flow* **36** 533-48
[21] Hyndman C L, Larachi F and Guy C 1997 *Chem. Eng. Sci.* **52** 63-77
[22] Krishna R, Wilkinson P M and Van Dierendonck L L 1991 *Chem. Eng. Sci.* **46** 2491-6
[23] Reilly I, Scott D, De Bruijn T, Jain A and J. Piskorz 1986 *Can. J. Chem. Eng.* **64** 705-17
[24] Joshi J and Sharma M 1979 *Chem. Eng. Resw. Des.***57** 244-51
[25] Hughmark G A, 1967 *Ind. Eng. Chem. Proc. DD.* **6** 218-20
[26] Kawase Y and Moo-Young M 1987 *Ind. Eng. Chem. Res.* **26** 933-7
[27] Oguz H N and Prosperetti A 1993 *J. Fluid. Mech.* **257** 111-45

Journal of Physics: Conference Series **547** (2014) 012025 doi:10.1088/1742-6596/547/1/012025

A detailed characterization of viscous oil-water flows downward sudden contractions in horizontal pipes

Luigi P.M. Colombo, Manfredo Guilizzoni, Giorgio Sotgia
Politecnico di Milano, Dipartimento di Energia, Via Lambruschini 4, 20156 Milano

luigi.colombo@polimi.it

Abstract. Two-phase flows of viscous oil and water through singularities such as sudden area contractions/expansions have been taken into limited consideration in the relevant scientific literature. Nevertheless, they play a role of primary importance in industrial systems, for instance, but not only, in the exploitation of oil wells and pipelines. The proposed work is based on the comparison of photographic images of the flow patterns taken from three points of view, i.e. upper, lower and frontal, thanks to a couple of mirrors ±45° inclined with respect to the horizontal plane. Oil-water flow regimes have been observed both upward and downward of five horizontal test sections with diameter ratios d/D = 40/50, 30/50, 30/40, respectively. The observed structures of the oil-water interface, especially for core-annular flows, has suggested also detecting flow patterns in a 30 mm straight pipe for sake of comparison. Actually, the shape of the oil-core interface appears significantly influenced by the sharp-edged area change as well as by the expected momentum variation.

1. Introduction

Geometrical singularities like sudden contractions and expansions, elbows, valves, T-junctions are often present in industrial duct systems. Two-phase flows of liquid (water) – gas (air, vapour, steam) through such singularities have been widely investigated both by experimental and theoretical approaches. In particular, flow patterns and pressure drop have been studied for sudden contractions [1-4], expansions [5-8], valves and orifices [9-12]. Some studies have been also dedicated to the flow distribution and stability in systems with junctions, parallel channels and manifold collectors [13-16].

On the contrary, there are few works on liquid-liquid flows especially if one liquid phase shows medium-high viscosity such as heavy hydrocarbons, which play a non-negligible industrial interest. Actually, though very low viscosity oils, e.g. kerosene [17] can be assumed to behave almost like water, in contrast, heavy oils in mixture with water (and/or gas) show different flow patterns such as the typical core annular flow (CAF) and a variety of wavy stratified flows (WSF).

The main question is: how much does a geometrical singularity influence the flow characteristics in the downward pipe, i.e. flow patterns, pressure gradients and holdup, resulting in the concentrated pressure drop? In fact, researches on water-air flows through sudden contractions of several area ratios and diameters [3] have shown that the method of pressure gradients, usually adopted to measure the concentrated pressure drop across the singularity for single phase flows, cannot be trivially extended to two-phase flows because it may lead to absurd trends of the concentrated pressure drop. The reason is found in the flow pattern change downstream the contraction due to the influence of the upstream one. Furthermore, the results for gas-liquid flows are hardly extended to liquid-liquid flows even for straight pipes, confirming the G. Hewitt dilemma: "From gas-liquid to liquid-liquid flow: a difficult journey?"

The experience gained by the authors suggests that it is worth paying attention to the effects of singularities on the structure of the oil-water interface indicating the perturbation of the flow regime, which in turn affects the concentrated pressure drop.

In particular, the proposed work is based on the comparison of photographic images of the flow patterns taken from three points of view, i.e. upper, lower and frontal, thanks to a couple of mirrors ±45° inclined with respect to the horizontal plane. Oil-water flow regimes have been observed both upward and downward of five horizontal test sections with diameter ratios d/D = 40/50, 30/50, 30/40, respectively. The observed structures of the oil-water interface, especially for core-annular flows, has suggested also detecting flow patterns in a 30 mm straight pipe for sake of comparison. Actually, the shape of the oil-core interface appears significantly influenced by the sharp edged area change as well as by the expected momentum variation.

2. Test facility

The liquid-liquid flow facility available in the Multiphase Thermo-Fluid Dynamics Laboratory at the Department of Energy, Politecnico di Milano is sketched in figure 1.

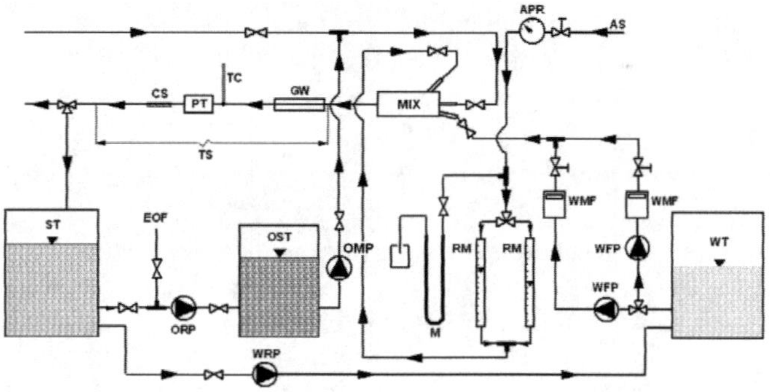

Figure 1. Schematic representation of the oil-water loop. APR air pressure regulation, AS air supplying line, CS capacitance sensor, EOF external oil feeding, GW glass window, M manometer, MIX phase inlet mixer, OMP oil metering pump, ORP oil recovering pump, OST oil supply tank (0.5 m^3), PT pressure transducer, RM rotameter, ST phase collector/separator tank (1.0 m^3), TC thermocouple (K type), TS test section, WFP water feeding pump, WMF water magnetic flow meter, WRP water recovering pump, WT water supply tank (5 m^3).

Oil (Milpar 220, μ_o = 890 kg/m^3, μ_o = 0.9 Pa·s at 20 °C, σ_o = 0.035 N/m) and water (tap water, μ_w = 1.026×10^{-3} Pa·s at 20 °C, σ_w = 0.070 N/m) are pumped separately from their storage tanks. The water flow rate is measured by a magnetic flowmeter (accuracy ±0.5% of the reading), while a calibrated metering pump is used for the oil. The two liquids pass through a coaxial mixer, where oil flows parallel to the pipe axis while water is injected through an annulus into the oil stream, then the mixture enters the test section. Plexiglas® pipes are used to allow flow visualization.

The test section consists of a 12 m long horizontal duct made of two circular pipes: a 7 m long upstream pipe, D = 50 mm or 40 mm inner diameter, and a 5 m long downstream pipe of lower inner diameter, namely d = 30 mm. According to the definition of contraction area ratio

$$\varsigma = \left(\frac{d}{D}\right)^2 \tag{1}$$

three configurations with $\varsigma = 0.64$ (D = 50 mm, d = 40 mm), $\varsigma = 0.56$ (D = 40 mm, d = 30 mm) and $\varsigma = 0.36$ (D = 50 mm, d = 30 mm) have been investigated.

Flow patterns have been detected by photographic and video recordings of all the upstream and downstream pipes and of a straight tube 30 mm i.d. for comparison.

3. Experimental procedure

Tests were run by introducing in the test section the water starting from the maximum value of the superficial velocity $J_{w,max}$. Then, oil was added at the selected superficial velocity J_o. At each run J_w was decreased until its minimum value was reached. The value of J_o was then changed and the sequence was repeated. Table 1 reports the ranges of the oil and water superficial velocities.

D [mm]	J_o [m/s]	J_w [m/s]
50	0.27 – 0.54	0.42 – 0. 85
40	0.56 – 0.84	0.66 – 1.33
30	0.75 – 1.49	1.18 – 2.36

Table 1. Superficial velocities of oil and water in the investigated ducts.

Another parameter useful to characterize the operating conditions is the water input volume fraction, defined as

$$\varepsilon_w = \frac{J_w}{J_w + J_o} \tag{2}$$

These operating conditions correspond to flow regimes with water wetting the wall, both annular and dispersed, in order to avoid stratified flows with oil adjoining the upper wall that are characterized by poor repeatability. Such flow regimes are not suitable for pumping due to their increased pressure drop. The stability of the core annular flow has been investigated by the authors in [18].

Referring to figure 2 the photographs were taken by means of a photo camera Nikon D90 with an AF-S Nikkor 60 mm macro lens, mounted on a suitable tripod. Exposure time was 1/4000 s and the necessary lighting was provided by two 800 W halogen lamps. Two mirrors lying parallel to the pipe and ±45° inclined with respect to the horizontal plane provide top and bottom views in addition to the frontal one. The optical path followed by the light in the frontal view is only 5-6% lower than the others; moreover, the images were used for a qualitative analysis; therefore, no correction was performed to equalize the dimensions of the images relative to the different views. This simultaneous triple view, not reported in the literature where only frontal and top views are available [19], allows a complete observation of the oil-water interface.

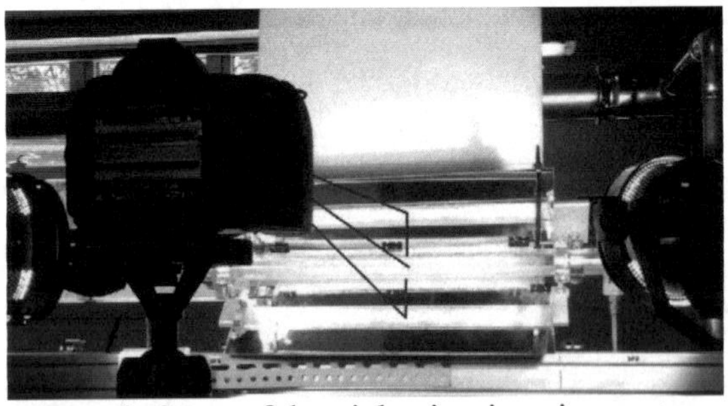

Figure 2. Setup of the triple view imaging system.

4. Results and discussion

The investigation has aimed at detecting differences in the flow structure downstream the sudden area contraction due to both the sharp-edged change in diameter and the increase in momentum flux.

In the conditions reported in table 1, the flow regime is mainly core annular, more or less eccentric, and gradually evolving to disperse patterns of oil drops as the water superficial velocity is increased.

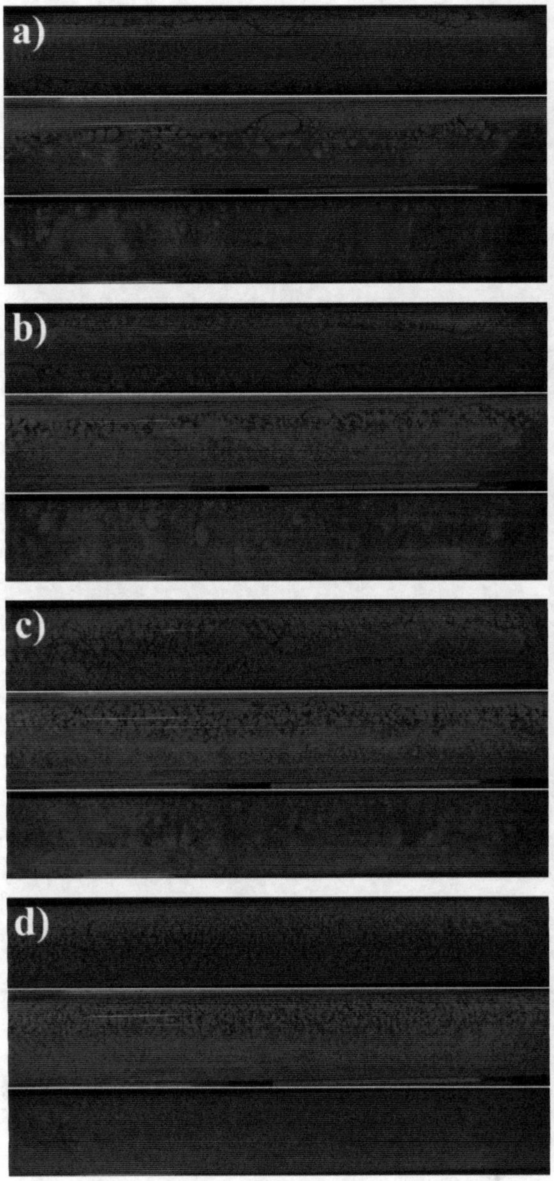

Figure 3-1. Flow patterns for the D50 duct, J_o = 0.27 m/s, J_w = 0.42 (a), 0.57 (b), 0.71 (c), 0.85 (d) m/s

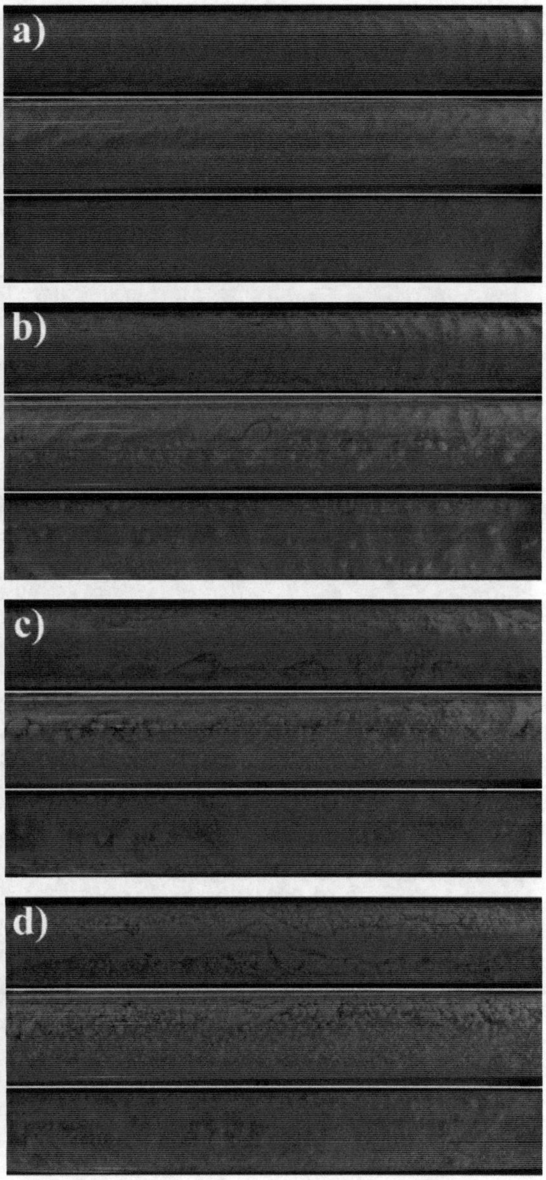

Figure 3-2. Flow patterns for the D50 duct, J_o = 0.54 m/s, J_w = 0.42 (a), 0.57 (b), 0.71 (c), 0.85 (d) m/s

As a principal observation, eccentricity of the downstream flow patterns is reduced compared to the upstream ones for both core annular and dispersed regimes, whereas transitions between different regimes have not been observed. Moreover, the developing length downward the contraction appears to be about 10 diameters. Actually, pressure drop measurements taken in a parallel test campaign have shown a constant pressure gradient along the downstream pipe with regression coefficients very close to unity. Along the upstream pipe 50 mm i.d., the flow patterns at the minimum and maximum value of J_o (see table 1) are depicted in figures 3-1 (a), (b), (c), (d) and 3-2 (a), (b), (c), (d), respectively, in order to show the flow evolution. In particular, at the lowest ε_w and the highest J_o, a strongly eccentric oil core flow takes place, as it is seen in figures 3-1 (a) and 3-2 (a).

The presence of oil drops in the lower side of the duct increases with ε_w, figure 3-2 (b), with the flow evolving toward dispersion, figure 3-2 (d). Therefore, in our opinion, this kind of regime cannot be labelled as "core annular" but "eccentric core flow" because the oil occupies less than the upper half of the pipe and presents its lower interface almost flat, figure 3-2 (a).

At the highest ε_w, regimes of oil dispersed drops in water have been observed showing more or less pronounced eccentricity according to J_o, figures 3-1 (d) and 3-2 (a).

In figures 3-1 (c) and 3-2 (d) the regimes of transition from the eccentric core with oil drop to the dispersed flow are depicted. Note that in the figure 3-2 (d) the dispersed flow is not completely established. In the intermediate regimes – figures 3-1 (b), (c) and 3-2 (b), (c) – oil cores of lower size with several oil drops also present more irregular interfaces.

Along the upstream pipe of 40 mm i.d., the flow patterns at the minimum and maximum value of J_o, table 1, are shown in figures 3-3 (a), (b), (c) and 3-4 (a), (b), (c), respectively. The figures with the label (b) correspond to the transition from core to dispersed flow.

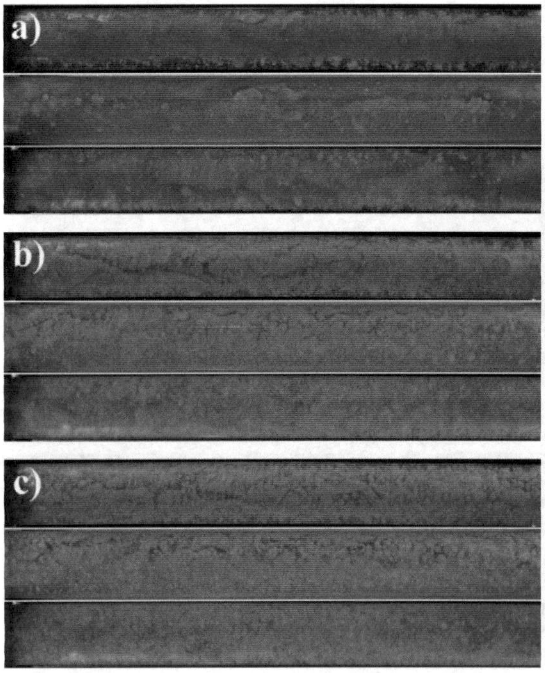

Figure 3-3. Flow patterns for the D40 duct, J_o = 0.42 m/s, J_w = 0.66 (a), 1.22 (b), 1.33 (c) m/s

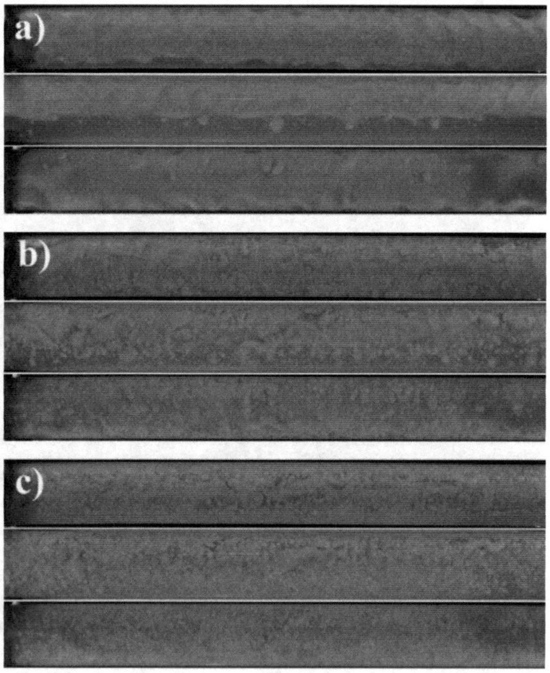

Figure 3-4. Flow patterns for the D40 duct, J_o = 0.84 m/s, J_w = 0.66 (a), 1.11 (b), 1.33 (c) m/s

In figure 3-3 (a) the oil core appears eccentric, though with larger holdup than for the 50 mm pipe even by increasing ε_w, figure 3-3 (b). In contrast, the dispersed flow at the maximum ε_w appears practically axisymmetric, figure 3-3 (c).

The mixture at the highest J_o shows a nearly centered core for the lowest ε_w, figure 3-4 (a), eccentric core increasing ε_w, figure 3-4 (b), towards a complete dispersion, figure 3-4 (c).

Flow patterns in the 30 mm i.d. pipe, downstream the 50 mm i.d. one, appear strongly influenced by the singularity, not only by the increased momentum flux, but also by the sharp-edged change in diameter. The flow patterns at the minimum and maximum value of J_o (see table 1) are depicted in figures 3-5 (a), (b), (c), (d) and 3-6 (a), (b), (c), (d), respectively, in order to show the flow evolution.

In particular, it is seen that increasing ε_w at the maximum J_o eccentricity disappears (figures 4.6 all). On the contrary, at the minimum J_o eccentricity is still shown though at a lower extent in comparison with the 50 mm upstream tube. Dispersed flow patterns, however, appear in any case axisymmetric. Moreover, the oil-water interface for any regime with oil core, with or without drops, is extremely uneven, likely due to the effect of the sharp change in the cross-section diameter.

Figure 3-5. Flow patterns for the D30 duct downstream the D50 duct, J_o = 0.75 m/s, J_w = 1.18 (a), 1.57 (b), 1.97 (c), 2.36 (d) m/s

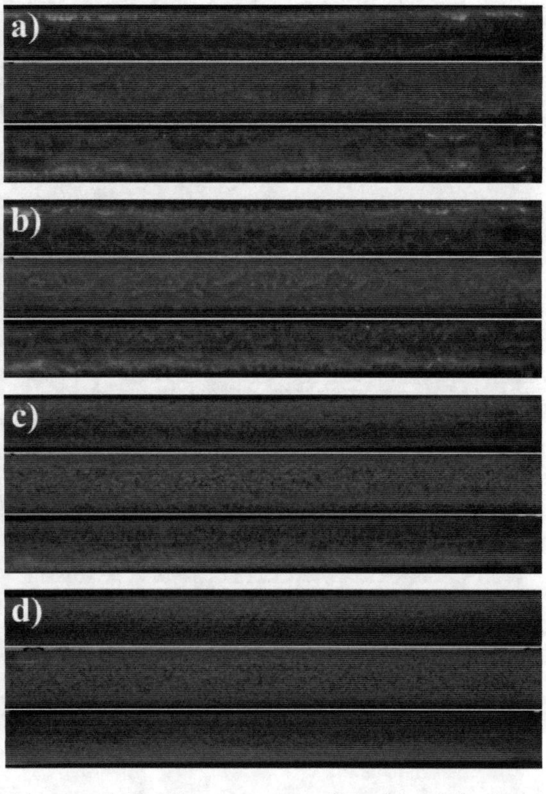

Figure 3-6. Flow patterns for the D30 duct downstream the D50 duct, J_o = 1.49 m/s, J_w = 1.18 (a), 1.57 (b), 1.97 (c), 2.36 (d) m/s

Flow patterns in the pipe of 30 mm i.d., downstream the 40 mm one, are reported in figures 3-7 (a), (b), (c) and 3-8 (a), (b), (c) for the minimum and maximum J_o, respectively.

Core flow patterns still appear strongly influenced by the geometrical singularity, in spite of both the lower variation in the momentum flux and the reduced upstream eccentricity. Consequently, the sharp-edged change in diameter appears the most important factor affecting flow patterns as it is also inferred by the comparison between figures 3-5 (a) and 3-7 (b).

These observations are strengthened in considering the flow evolution along the constant 30 mm diameter pipe is shown in the figures 3-9 (a), (b), (c), (d) and 3-10 (a), (b), (c), (d) for the minimum and maximum J_o, respectively. It is clearly seen that the core interface appears more regular than in the pipes downstream the contraction, as described above, whereas it is not possible to observe significant differences in the dispersed regimes.

The changes in the flow patterns produced by the sudden contraction have also a quantitative impact. In fact, the couples (ε_w, J_o) corresponding to the transition between core and dispersed pattern are plotted in figure 4 for both the downstream and the upstream flows, as well as for the flow in the straight pipe. J_o has been normalized ($J_{o,n} = J_o/J_{o,max}$) to make the comparison easier. It is evident that the contraction promotes the transition that takes place at lower J_w.

Figure 3-7. Flow patterns for the D30 duct downstream the D40 duct, J_o = 0.75 m/s, J_w = 1.18 (a), 1.57 (b), 2.36 (c) m/s

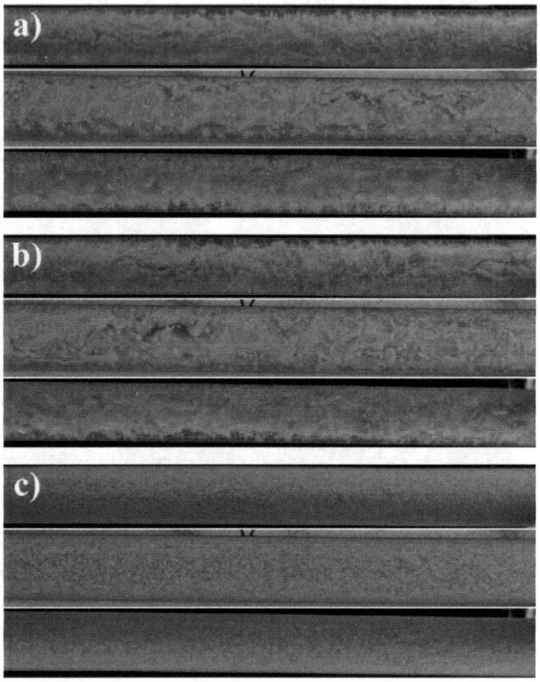

Figure 3-8. Flow patterns for the D30 duct downstream the D40 duct, J_o = 1.49 m/s, J_w = 1.18 (a), 1.38 (b), 2.36 (c) m/s

5. Conclusions

Detecting the flow patterns from a threefold view, i.e. frontal, top and bottom, turns out to be a very useful tool to characterize the structure of the oil-water interface without lack of information. In particular, the top view allows verifying whether oil is sticking the wall, and detecting the transition toward dispersed flow as the elongated oil drops, resulting from breaking the core, disappear. The more usual front view adopted to classify the flow regimes highlights the eccentricity in horizontal streams for both core and dispersed patterns. The bottom view, never reported in the literature to the best knowledge of the authors, is useful to study the shape of the lower interface in core flows and, in particular, the formation and the motion of oil drops.

Journal of Physics: Conference Series **547** (2014) 012025 doi:10.1088/1742-6596/547/1/012025

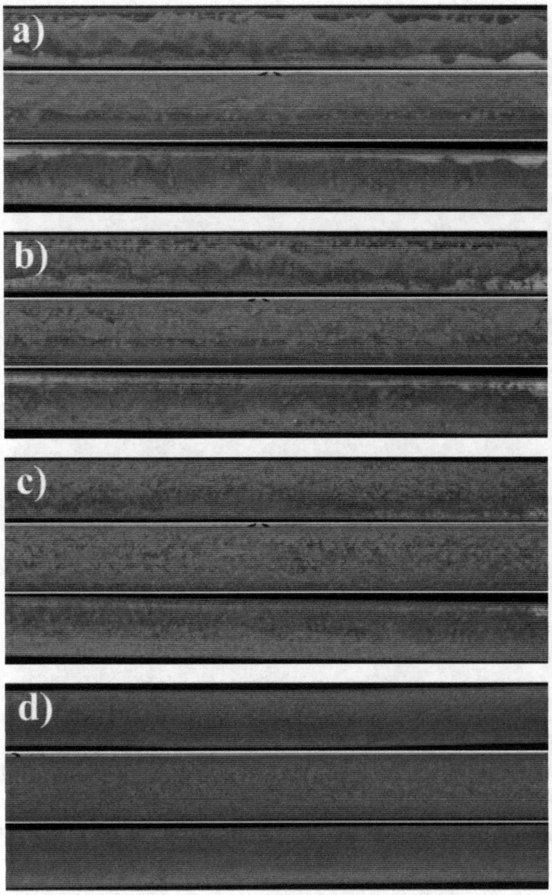

Figure 3-9. Flow patterns for the D30 duct, J_o = 0.75 m/s, J_w = 1.18 (a), 1.57 (b), 1.97 (c), 2.36 (d) m/s

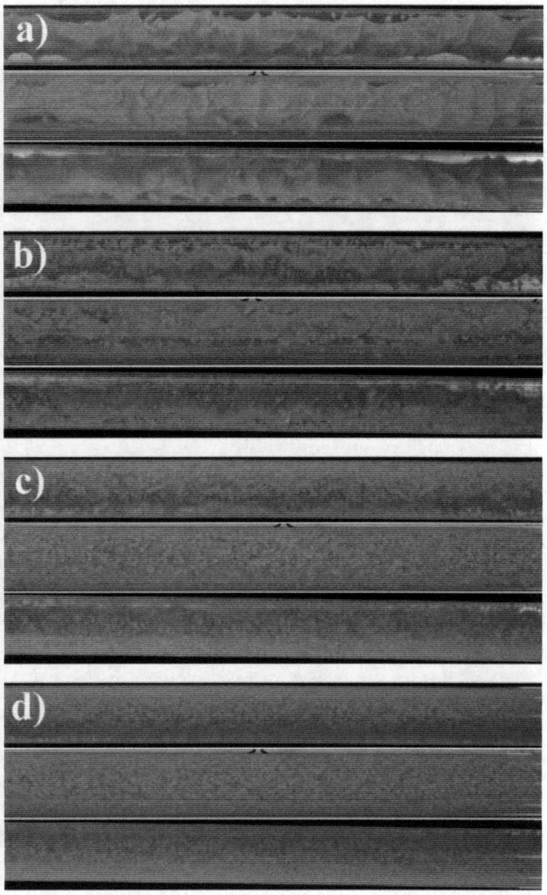

Figure 3-10. Flow patterns for the D30 duct, J_o = 1.49 m/s, J_w = 1.18 (a), 1.57 (b), 1.97 (c), 2.36 (d) m/s

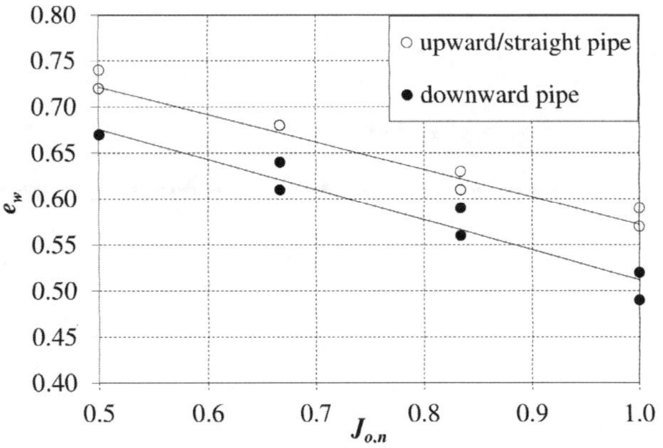

Figure 4. Water input volume fraction versus normalized oil superficial velocity. Comparison between downward and upward/straight pipe flows.

With reference to the specific subject of this work, the complete inspection of the oil-water interface is important to put in evidence the influence of sharp-edged contractions compared with the same flow conditions in constant diameter pipes.

In fact the qualitative changes in the flow patterns are connected with important quantitative findings that need further investigation: in particular, the transition between core bubbly and dispersed flow occurs at a lower value of the water superficial velocity at constant oil superficial velocity.

References

[1] Jannsen E 1966 *Proc. Int. Heat Transfer Conf.* (New York) 13-23
[2] Schmidt J and Friedel L 1997 *Int. J. Multiphase Flow* **23** (2) 238-299
[3] Guglielmini G, Muzzio A and Sotgia G 1997 *Proc. 4th World Conference on Experimental Heat Transfer Fluid mechanics and Thermodynamics* (Bruxelles) **2** 1023-1036
[4] Fossa M and Guglielmini G 1998 *Int. J. Heat and Mass Transfer* **41** (23) 3807-3815
[5] Schmidt J and Friedel L 1996 *Chem. Eng. Comm.* 175-190
[6] Arosio S and Guilizzoni M 2004 *22nd UIT Heat Transfer Conference and 42nd European Two Phase Flow Group Meeting* (Genova)
[7] Ahmed W H, Ching C Y and Shoukri M 2007 *Int. J. Multiphase Flow* **33** 575-594
[8] Arosio S and Guilizzoni M 2008 *Journal of Visualization* **11** (2) 143-152
[9] Simpson C, Rooney D H and Grattan E 1983 *Int. Conf. on the Physical Modelling of Multiphase Flow* (Coventry) 25-40
[10] Morris S D 1985 *European Two Phase Flow Group Meeting* (Southampton)
[11] Kojasoy, Kwame-Mensah P and Chang C T 1997 *Exp. Thermal Fluid Sci.* **15** 347-358
[12] Fossa M and Guglielmini G 2002 *Exp. Thermal Fluid Sci.* **26** 513-523
[13] Azzopardi B J and Whalley P B 1982 *Int. J. Multiphase Flow* **8** 491-507
[14] Wang S F, Mosdorf R and Shoji M 2003 *Int. J. Heat and Mass Transfer* **46** 1519-1528
[15] Pustylnik L, Barnea D and Taitel Y 2006 *AIChE Journal* **52** 3345-3352
[16] Minzer U, Barnea D and Taitel Y 2006 *Chem. Eng. Science* **61** 7249-7259
[17] Angeli P and Hewitt G F 2000 *Int. J. Multiphase Flow* **26** 1117-1140
[18] Colombo L P M, Guilizzoni M and Sotgia G 2012 *Exp. in Fluids* **53** 1617-1625
[19] Acikgöz F, Franca F and Lahey R T 1992 *Int. J. Multiphase Flow* **18**(3) 327-226

R1234yf vs. R134a Flow Boiling Heat Transfer Inside a 3.4 mm ID Microfin Tube

A Diani, S Mancin, L Rossetto
Università di Padova, Dipartimento di Ingegneria Industriale,
Via Venezia 1, 35131 – Padova, ITALY

E-mail: luisa.rossetto@unipd.it

Abstract. The refrigerant charge minimization as well as the use of eco-friendly fluids can be considered two of the most important targets for these applications to cope with the new environmental challenges. This paper compares the R1234yf and R134a flow boiling heat transfer and pressure drop measurements inside a small microfin tube with internal diameter at the fin tip of 3.4 mm. This study is carried out in an experimental facility built at the Dipartimento di Ingegneria Industriale of the University of Padova especially designed to study both single and two phase heat transfer processes. The microfin tube is brazed inside a copper plate and electrically heated from the bottom. Several T-type thermocouples are inserted in the wall to measure the temperature distribution during the phase change process. In particular, the experimental measurements were carried out at constant saturation temperature of 30 °C, by varying the refrigerant mass velocity between 190 kg m^{-2} s^{-1} and 940 kg m^{-2} s^{-1}, the vapour quality from 0.2 to 0.99, at different imposed heat fluxes. The two refrigerants are compared considering the values of the two-phase heat transfer coefficient and pressure drop.

1. Introduction

The use of synthetic refrigerants with a non-negligible Global Warming Potential or, on the contrary, of natural but flammable or toxic natural fluids calls for the charge minimization of the refrigerating and air conditioning equipment. The refrigerant charge minimization as well as the use of eco-friendly fluids can therefore be considered two of the most important targets for these applications to cope with the new environmental challenges. Traditional microfin tubes are also widely used in air and water heat exchangers for heat pump and refrigerating applications during condensation or evaporation. The possible downsizing of microfin tubes can lead to more efficient and compact heat exchangers and thus to a reduction of the refrigerant charge of the systems. Furthermore, over the last several years, much research and development effort has been focused on potential refrigerants possessing low GWPs. Among the fluorinated propene isomers which have normal boiling point temperature data published in the public domain, several have low GWPs and normal boiling temperatures relatively close to R134a; among them, R1234yf has as a normal boiling temperature approximately 3.4°C lower than that of R134a, with a GWP<1. Recently, Domanski et al. [1] and McLinden et al. [2] have performed a detailed thermodynamic analysis of refrigerants, studying the performance limits of the vapour compression cycle and testing the possible low-GWP refrigerant candidates suitable for use in common types of refrigeration and air conditioning equipment. In particular, Domanski et al. [1] suggested that the critical temperature of a refrigerant can be considered the most dominant parameter influencing the tradeoff between the Coefficient of Performance (COP) and the volumetric capacity (Q_{vol}). McLinden et al. [2] carried out a screening of the possible refrigerants, eliminating those toxic or unstable and focusing on those that presented critical temperatures between 300 K and 400 K; they analysed 62 refrigerants. Two of those were R1234ze(E) and R1234yf, which have recently been matter of research and investigation of the scientific community because they present values of low-GWP and normal boiling temperature close to that of R134a, the traditional fluid commonly used in refrigeration and air conditioning equipment.

The R1234yf has as a normal boiling temperature approximately 3.15 K lower than that of R134a, whereas that of R1234ze(E) is 7.3 K lower than that of R134a; these new refrigerants have a GWP<1. These two fluids are candidates to substitute the traditional R134a in several applications: from the automotive air conditioning to the high temperature heat pump.

Fukuda et al. [3] thermodynamically, experimentally, and numerically analyzed the feasibility of R1234ze(E) and R1234ze(Z) for high temperature heat pumps, demonstrating that these new low-GWP fluids are capable of being a potential refrigerant in high-temperature heat pump systems for industrial purposes, rather than typical air conditioners or refrigeration systems. However, up today only few works experimentally investigated the heat transfer capabilities of these refrigerants during single and two-phase flow inside conventional and enhanced tubes.

Grauso et al. [4] studied the heat transfer and pressure drop during evaporation of R1234ze(E) in a circular smooth tube of 3 mm OD; the authors measured the two-phase heat transfer coefficient at different mass velocity from 146 to 520 kg m^{-2} s^{-1}, and vapour quality, saturation temperature between -2.9 °C and 12.1 °C (p_{red} = 0.05-0.09), imposing two different heat fluxes: 5 and 20 kW m^{-2}, provided by electrical heating.

Hossain et al. [5] experimentally investigated the evaporation of R1234ze(E), R32, R410A, and a mixture of R1234ze(E) and R32 inside a horizontal smooth tube. The experiments were carried out under the conditions of mass flux varying from 150 to 445 kg m^{-2} s^{-1}, the saturation temperature were 5 °C and 10 °C (R1234ze(E) p_{red} = 0.07 – 0.09; R32 p_{red} = 0.19; R410A p_{red} = 0.22; R1234ze(E)/R32 mixture p_{red} =0.15 – 0.18), over vapour quality from 0 to 1. Refrigerant flowed inside a smooth copper tube with an inner diameter of 4.35 mm and an outer diameter of 6.35 mm, whereas heating water flowed through an annular space in counter current. With this technique, authors achieved heat fluxes up to 70 kW m^{-2}. The authors found that the heat transfer coefficients exploited by pure R1234ze(E) are lower than those measured for R1234ze(E)/R32 mixture, R410A, and R32.

Del Col et al. [6] investigated the R1234yf flow boiling heat transfer in a 1 mm diameter circular microchannel. During tests, the heat was provided to the boiling fluid by using a secondary fluid. Flow boiling tests were carried out at 31 °C of saturation temperature (p_{red} = 0.24), mass fluxes ranging between 200 and 600 kg m^{-2} s^{-1}, and heat fluxes from 10 to 130 kW m^{-2}. Authors compared the results with those measured for R134a, and no significant differences between the flow boiling performance of R1234yf and R134a were reported.

Lu et al. [7] investigated the influences of heat flux and mass flux on the two-phase convective boiling heat transfer performance for refrigerants R1234yf and R134a in a 3.9 mm smooth diameter tube. The horizontal test section is a double-pipe heat exchanger: water flows countercurrently in the test annulus, while refrigerant is evaporated inside the test tube. Tests were performed with a saturation temperature of 10 °C (p_{red} = 0.13), with a mass velocity ranging from 200 to 500 kg m^{-2} s^{-1}, and heat flux from 5 to 20 kW m^{-2}. A noticeable deterioration of the heat transfer coefficient for R1234yf was encountered whereas the pressure drops for R134a are about 5–15% higher than those of R1234yf.

Only one work relative to flow boiling heat transfer of new low-GWP refrigerants inside microfin tubes can be found in the open literature; Kondou et al. [8] experimentally investigated the flow boiling of R32, R1234ze(E) and R32/R1234ze(E) non-azeotropic mixtures in a horizontal microfin tube of 5.2 mm inner diameter at 10 °C of saturation temperature (R32 p_{red} = 0.19, R1234ze(E) p_{red} = 0.09, R32/R1234ze(E) 0.2/0.8 p_{red} = 0.100; R32/R1234ze(E) 0.5/0.5 p_{red} = 0.179), with heat fluxes of 10 and 15 kW m^{-2} with a water heating, and mass velocities from 150 to 400 kg m^{-2} s^{-1}. The heat transfer coefficients of R1234ze(E) are lower than those of R32 but they are greater than those obtained for R32/R1234ze(E) mixtures.

However, a few of other works studied the flow boiling heat transfer inside microfin tubes with an internal diameter lower than 5 mm, among those: Mancin et al. [9-11], Gao et al. [12-13], Dang et al. [14], and Wu et al. [15].

Recently, Mancin et al. [9-11] have experimentally investigated the flow boiling heat transfer of R134a inside a mini microfin tube, with an internal diameter at the fin tip of 3.4 mm, electrically heated. The collected data permitted to study the effects of refrigerant mass velocity, vapour quality, and heat flux on the liquid-vapour phase change process at constant saturation temperature of 30 °C ($p_{red} = 0.19$).

Gao et al. [12-13] conducted experiments on flow boiling of CO_2 and oil mixtures in electrically heated horizontal smooth and microfin tubes. The microfin was a copper tube with an inner diameter of 3.04 mm. Experiments were carried out at mass velocities from 190 to 1300 kg m^{-2} s^{-1}, at a saturation temperature of 10 °C ($p_{red} = 0.61$), heat fluxes from 5 to 30 kW m^{-2}, and an oil circulation ratio from <0.01 to 0.72 wt%.

Dang et al. [14] investigated the flow boiling of CO_2 inside a horizontal internally-grooved tube, with an internal diameter of 2.0 mm at a saturation temperature of 15 °C ($p_{red} = 0.67$), the heat flux ranged between 4.5 to 18 kW m^{-2}, and the mass velocity from 360 to 720 kg m^{-2} s^{-1}. The test section was placed horizontally, and the copper test tube was heated by directly supplying DC current to it. The heat transfer coefficient for the grooved tube was 1.9 - 2.3 times higher than that for the smooth tube, and the dryout quality was much higher, ranging between 0.90 and 0.95, while the pressure drops are 1.5 – 2.7 higher.

Wu et al. [15] performed experiments during flow boiling of R22 and R410A inside one smooth tube and five microfin tubes with the same outer diameter of 5 mm, the mass velocity was varied from 100 to 620 kg m^{-2} s^{-1}, the heat flux from 5 to 31 kW m^{-2}, at a saturation temperature of around 6 °C (R22 $p_{red} = 0.12$; R410A $p_{red} = 0.20$). The test section was a 2 m long straight, horizontal tube-in-tube heat exchanger, where the refrigerant flows in counter current with heating water. They also developed a new general semi-empirical model based on present data and recent data from literature. This model is applicable for intermittent and annular flow patterns.

This paper presents the experimental results of R1234yf flow boiling heat transfer inside a 3.4 mm ID microfin tube electrically heated; the effects of mass flow rate, vapour quality, and heat flux at constant saturation temperature of 30°C (R1234yf $p_{red} = 0.23$; R134a $p_{red} = 0.19$) are presented and analysed. Furthermore, the experimental measurements for R1234yf are compared with those obtained for R134a in the same tube at the same operating test conditions [9-10].

2. Experimental setup and data reduction

The experimental setup is located at the Heat Transfer in Micro-geometries Lab (HTMg-Lab) at the Dipartimento di Ingegneria Industriale of the University of Padova. As shown in Fig. 1, the experimental facility consists of three loops: refrigerant, cooling water and hot water loops. The rig was designed for heat transfer and pressure drop measurements and flow visualization during either vaporization or condensation of pure refrigerants and refrigerants mixtures inside structured micro-geometries. The facility has a maximum working pressure of 3 MPa, while refrigerant mass fluxes can be varied up to G=400 kg m^{-2} s^{-1} in a section of 50 mm^2.

In the first loop the refrigerant is pumped through the circuit by means of a magnetically coupled gear pump, it is vaporized and superheated in a brazed plate heat exchanger fed with hot water. Superheated vapour then partially condenses in a pre-condenser fed with cold water to achieve the set quality at the inlet of the test section. The refrigerant enters the test section at a known mass velocity and vapour quality and then it is vaporized by means of a calibrated Ni-Cr wire resistance.

Journal of Physics: Conference Series **547** (2014) 012026 doi:10.1088/1742-6596/547/1/012026

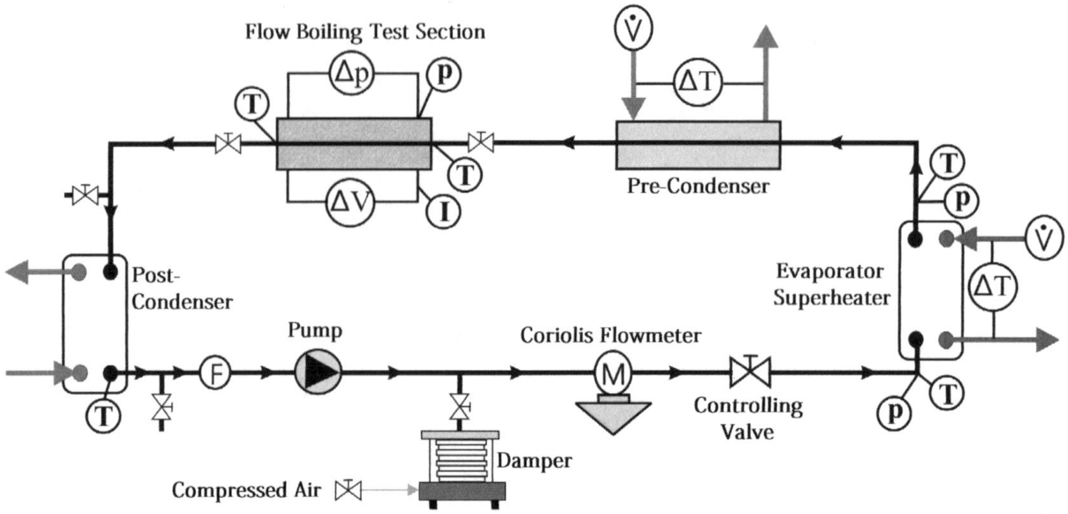

Figure 1. Schematic of the experimental setup.

The fluid leaves the test section and enters in a post-condenser, a brazed plate heat exchanger, where it is fully condensed and subcooled. The subcooled liquid passes through a drier filter and then is sent back to the boiler by a pump. A damper connected to the compressed air line operates as pressure regulator to control the saturation condition in the refrigerant loop. As shown in Fig. 1, the refrigerant pressure and temperature are measured in several locations throughout the circuit to know the refrigerant properties at the inlet and outlet of each heat exchanger. The refrigerant mass flow rate can be independently controlled by the gear pump and it is measured by means of a Coriolis effect flowmeter. The inlet vapour quality to the test section is determined by the heat extracted in the precondenser, which can be controlled by varying water temperature and flow rate. The cold water loop consists of a chiller with thermostatic control connected to the precondenser. The hot water circuit consists of a pump, an electrical heater and a controlling valve; it permits to set both the water flow rate and the inlet water temperature. Water flow rates in the precondenser and boiler sections are measured by means of magnetic type flow meters, while the water temperature differences are measured using 4-junction T-type thermopiles. Table 1 lists the values of accuracy of the instruments implemented in the experimental facility.

Table 1. Accuracy of the implemented instruments.

Transducer	Accuracy
T-type thermocouples	± 0.05 K
T-type thermopiles	± 0.03 K
Electric power	± 0.13% of the reading
Coriolis mass flowmeter (refrigerant loop)	± 0.10% of the reading
Magnetic volumetric flowmeters	± 0.25% of the reading
Differential pressure transducer (test section)	± 25 Pa
Absolute pressure transducers	± 1950 Pa

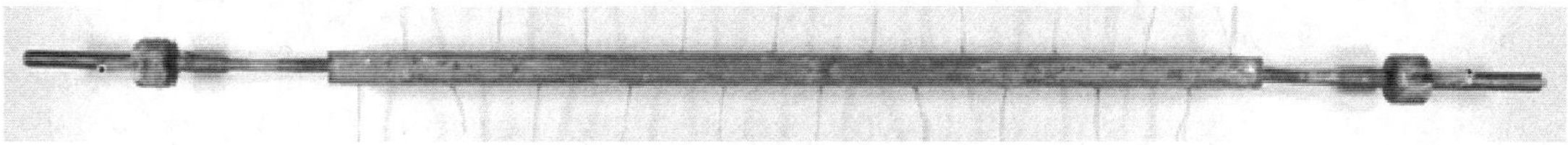

Figure 2. Photo of the test section.

As shown in Fig. 2, the tested microfin tube is brazed inside a groove milled in a 20 mm thick, 10 mm wide, and 300 mm long copper plate. The pressure taps are located around 50 mm upstream and downstream of the heated tube; a smooth connection was designed and manufactured in order to prevent any possible abrupt pressure loss. The test tube has a heated length of 300 mm whereas the total length for the pressure drop measurement is 410 mm. The microfin tube is heated from the bottom by means of a calibrated Ni-Cr wire resistance inserted in a 2 mm deep groove milled on the bottom face of the copper plate. The heat is supplied by means of stabilized DC power supplier, which is able to supply up to 900 W. The instrumented test section is located inside an aluminum housing filled up with a 30 mm thick layer of rock wool to limit as much as possible the heat loss.
According to the nomenclature described in Fig. 3, the microfin tube has an OD of 4.0 mm, an ID at the fin's tip of D=3.4 mm, it has 40 fins with a fin's height of h=0.12 mm, the helix angle β=18°. Twenty 1 mm ID holes were drilled along the centreline of the copper plate just 1 mm under the tested microfin tube. Twenty calibrated T-type thermocouples are inserted in those 5 mm deep holes to measure the wall temperature distribution during the heat transfer process. Figure 3 also reports a photo of the cross sectional area of the mini microfin tube tested where several fins are clearly visible.
Preliminary heat transfer measurements permitted to estimate the heat loss (q_{loss}) due to conduction through the test section as a function of the mean wall temperature. The tests were run under vacuum conditions by supplying the power needed to maintain the mean wall temperature at a fixed value. The measurements were carried out by varying the wall temperature from around 30 °C to 60 °C, at different ambient temperature from 21 °C to 24 °C. In this range, there is not any appreciable effect of the ambient temperature on the actual heat loss. The relationship between the heat loss and the wall temperature is linear; in this way, the actual value of heat supplied to the sample can be evaluated. The heat loss was never higher than 4% of the electrical power when the ambient temperature is kept constant within 21 °C and 24 °C.

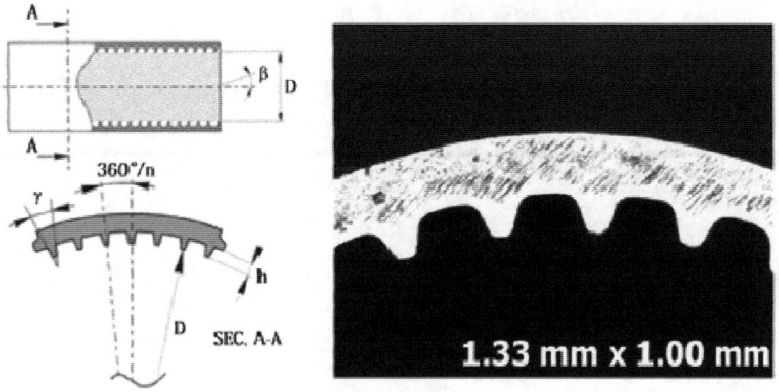

Figure 3. Drawing of a typical microfin tube and photo of the cross section of the tested tube.

As described before, the subcooled liquid is pumped to the boiler where it is vaporized and superheated; the refrigerant temperature and pressure are measured at both inlet and outlet of the heat exchanger. Preliminary tests were run to verify the heat balance at both pre-condenser and evaporator; the misbalance was always less than 2%. The vapour quality at the inlet of test section depends on the refrigerant conditions at the inlet of the precondenser and on the heat flow rate exchanged in the tube-in-tube heat exchanger and it can be obtained from a thermal balance on the cooling water side as given by:

$$q_{pc} = \dot{m}_{w,pc} \cdot c_{p,w} \cdot \left(t_{w,pc,out} - t_{w,pc,in}\right) = \dot{m}_{ref}\left(h_{vs} - h_{TS,in}\right) \quad (1)$$

where $\dot{m}_{w,pc}$ is the water mass flow rate at the precondenser, $c_{p,w}$ the water specific heat at constant pressure, $t_{w,pc,out}$ and $t_{w,pc,in}$ the water temperatures at the outlet and inlet of the precondenser, respectively. Considering the right-hand side of eq. (1), $\dot{m}_{ref}$ is the refrigerant mass flow rate, while h_{vs} is the enthalpy of the superheated gas at the inlet of the precondenser, and $h_{TS,in}$ the enthalpy of the refrigerant at the inlet of the test section. The vapour quality at the inlet of the test section (x_{in}) can be calculated from the heat balance, as:

$$x_{in} = \frac{h_{TS,in} - h_L}{h_V - h_L} \quad (2)$$

where h_L and h_V are the specific enthalpies of the saturated liquid and vapour, respectively, evaluated at the saturation pressure of the refrigerant measured at the inlet of the test section. All the thermophysical properties of the refrigerant were estimated using Refprop ver. 9.1 [16] database. The electrical power supplied to the sample is indirectly measured by means of a calibrated reference resistance (shunt) and by the measurement of the effective EDP (Electrical Difference Potential) of the resistance wire inserted in the copper heater. The current can be calculated from the Ohm's law. From preliminary measurements, the heat loss through the test section can be estimated by:

$$q_{loss}[\mathrm{W}] = 0.1121 \cdot \bar{t}_{wall}[^\circ\mathrm{C}] - 2.4042 \quad (3)$$

where $\bar{t}_{wall}$ is the mean wall temperature; thus, the actual heat flow rate supplied to the tested microfin tube is given by:

$$q_{TS} = P_{EL} - q_{loss} = \Delta V \cdot I - q_{loss} \quad (4)$$

The two-phase heat transfer coefficient *HTC*, referred to the area of the smooth tube with the same inner diameter of that at the fin tip, A_D, can now be defined as:

$$HTC = \frac{q_{TS}}{A_D \cdot (\bar{t}_{wall} - \bar{t}_{sat})} \quad (5)$$

where $\bar{t}_{wall}$ and $\bar{t}_{sat}$ are the average values of the wall and saturation temperatures, respectively. Their values are given by:

$$\bar{t}_{wall} = \frac{1}{20}\sum_{i=1}^{20} t_{wall,i} \qquad \bar{t}_{sat} = \frac{t_{sat,in}(p_{sat,in}) + t_{sat,out}(p_{sat,out})}{2} \quad (6)$$

From the error propagation analysis, it was estimated that the mean uncertainty on the two-phase heat transfer coefficient is ±2.5%, whereas the vapour quality has an uncertainty of ±0.035.
The frictional pressure gradient is obtained from the measured value of the total pressure gradient by subtracting the momentum pressure gradient and neglecting the gravitational term, as given by:

$$\left(-\frac{\mathrm{d}p}{\mathrm{d}z}\right)_f = \left(-\frac{\mathrm{d}p}{\mathrm{d}z}\right)_{tot} - \left(-\frac{\mathrm{d}p}{\mathrm{d}z}\right)_M \tag{7}$$

The model proposed by Rouhani and Axelsson [17] was used to estimate the void fraction values to account for the momentum pressure gradient.

3. R1234yf vs. R134a

This section compares the experimental results obtained during flow boiling heat transfer of R1234yf and R134a at 30 °C of saturation temperature at the inlet of the mini microfin tube. Figures 4 and 5 report the data collected for R1234yf and R134a, respectively, at HF= 10 kW m^{-2}; the heat flux is referred to the outer heat transfer area of the microfin tube having a diameter equal to 4 mm. At a glance, the two fluids show different behaviours; starting from the R1234yf (Figure 4), it clearly appears that the heat transfer coefficient is only weakly affected by the refrigerant mass velocity while it increases as the vapour quality increases. In fact, at low vapour quality the heat transfer coefficients are similar at all the investigated mass velocities. It is interesting to highlight that the highest heat transfer coefficients are exhibited by G=375 kg m^{-2} s^{-1} and, at high vapour quality, the heat transfer coefficient appears to slightly decrease as the mass velocity increases. It seems that the nucleate boiling controls the phase-change process limiting the two-phase forced convection especially at the high mass velocity. At G=190 kg m^{-2} s^{-1} the dryout occurred at a critical vapour quality around x_{cr}=0.97; at higher mass velocity the dryout was not observed. The R134a (Figure 5) exploits a different behaviour: considering the lowest mass velocity, G=190 kg m^{-2} s^{-1}, the heat transfer coefficient remains almost constant at around 6100 W m^{-2} K^{-1} up to x_{mean}=0.5, this might mean that the phase change process is controlled by the nucleate boiling; then, it increases with vapour quality. When increasing the mass velocity, the plateau at low vapour quality, where the heat transfer coefficient remains constant disappears, and the heat transfer coefficient increases almost linearly with the vapour quality meaning that the two phase forced convection is affecting in the phase change process. It is worthy to point out that, at x_{mean}>0.65, the values of the heat transfer coefficient measured at G=375 kg m^{-2} s^{-1} are greater than those obtained at higher mass velocities; this can be linked to a particular effect, due to the presence of the helical micro-fins that might be enhanced at this operating test condition. Furthermore, at this heat flux, no experimental evidence of the onset of dryout was observed at any mass velocity.

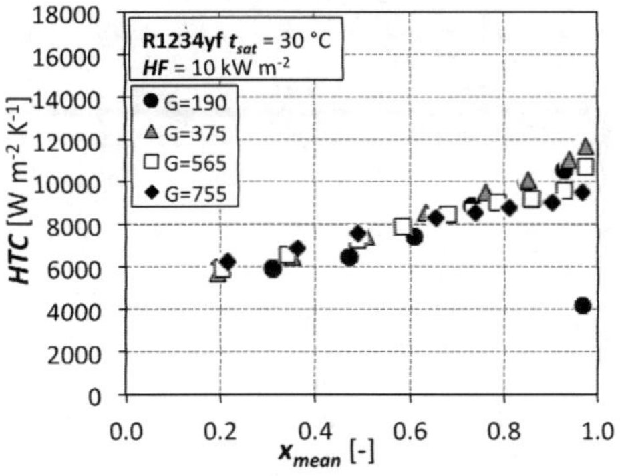

Figure 4. R1234yf flow boiling heat transfer coefficient at 10 kW m^{-2}.

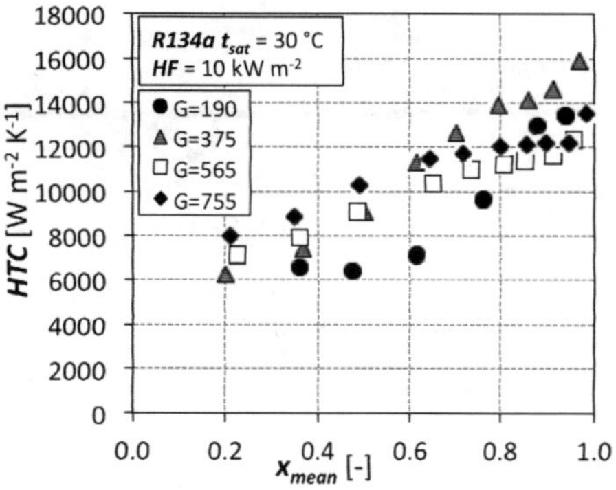

Figure 5. R134a flow boiling heat transfer coefficient at 10 kW m^{-2}.

Journal of Physics: Conference Series **547** (2014) 012026 doi:10.1088/1742-6596/547/1/012026

Table 2. Major thermophysical properties of R1234yf and R134a at 30 °C of saturation temperature. Data from Refprop 9.1 [16]

Property	R134a	R1234yf
p_{sat} [bar]	7.70	7.84
p_{red} [-]	0.19	0.23
ρ_L [kg m^{-3}]	1188	1073
ρ_V [kg m^{-3}]	37.5	43.7
h_{LV} [kJ kg^{-1}]	173	141
λ_L [W m^{-1} K^{-1}]	0.079	0.062
μ_L [µPa s]	183	145
σ [mN m^{-1}]	7.38	5.56
$(dp/dT)_{sat}$ [MPa K^{-1}]	0.0221	0.0212

The results described before can be discussed considering the properties of the two refrigerants at 30 °C of saturation temperature, as listed in Table 2. The reduced pressure and vapour density of the R1234yf are 22% and 17%, respectively, higher than those of R134a while its thermal conductivity is some 22% lower than that of the more traditional fluid. As a result, it might be expected that R1234yf shows greater attitude to the nucleate boiling by virtue of the higher reduced pressure. On the other hand, the two-phase forced convection would be better exploited by the R134a, which has lower vapour density that, for a given mass velocity, means higher vapour velocity. The lower thermal conductivity of the R1234yf as compared to that of R134a might also explain its lower values of heat transfer coefficient.

The results reported in Figures 6 and 7 confirmed the considerations drawn before; in particular, Figure 6 compares the heat transfer coefficients for R134a and R1234yf as a function of the mean vapour quality at HF=25 kW m^{-2} and G=375 kg m^{-2} s^{-1}. At low vapour quality, the two fluids show similar values of the heat transfer coefficients. As the vaporization proceeds, the amount of liquid decreases and the two-phase forced convection becomes more and more important, thus the attitude of R134a to two-phase forced convection leads to higher heat transfer coefficients as compared to those measured for the R1234yf. The onset of the dryout occurred at almost the same value of vapour quality, around x_{cr}=0.89.

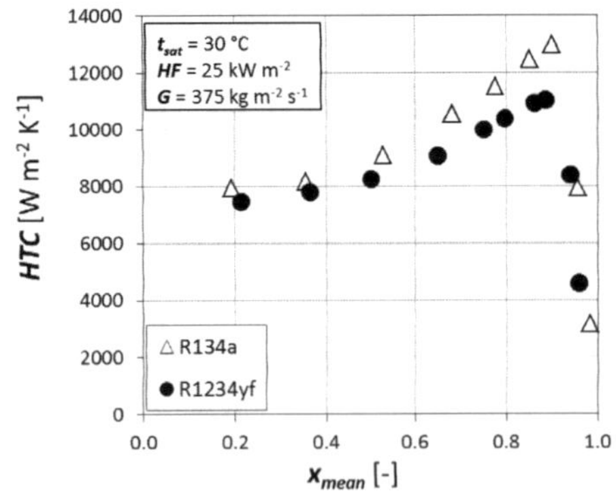

Figure 6. R1234yf vs. R134a at 25 kW m^{-2} and G=375 kg m^{-2} s^{-1}

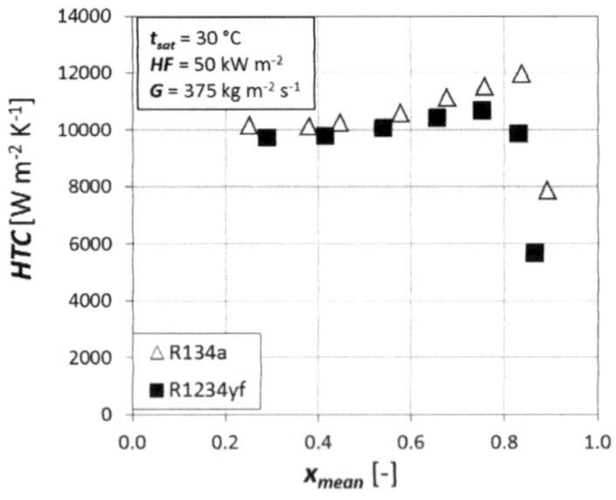

Figure 7. R1234yf vs. R134a at 50 kW m^{-2} and G=375 kg m^{-2} s^{-1}

Figure 7 compares the behaviour of the two refrigerants at the same mass velocity G=375 kg m^{-2} s^{-1} but at higher heat flux, HF=50 kW m^{-2}. The higher heat flux flattens the values of the heat transfer coefficients, which are almost constant up to around x_{mean}=0.5 and then they slightly increase with vapour quality. The two refrigerants show similar values of the heat transfer coefficient at low vapour qualities. For higher vapour qualities, R134a shows slightly higher values: in this case, the difference between the two fluids is lower than that at 25 kW m^{-2}, since at 50 kW m^{-2} the effect of nucleate boiling is stronger, and thus the better thermal performance of R1234yf in nucleate boiling seems to mitigate its worse performance in two-phase forced convection at high vapour qualities. The onset of the dryout occurred at lower vapour qualities for R1234yf (x_{cr}=0.81) as compared to R134a (x_{cr}=0.85), in any case earlier than those exhibited at lower heat flux.

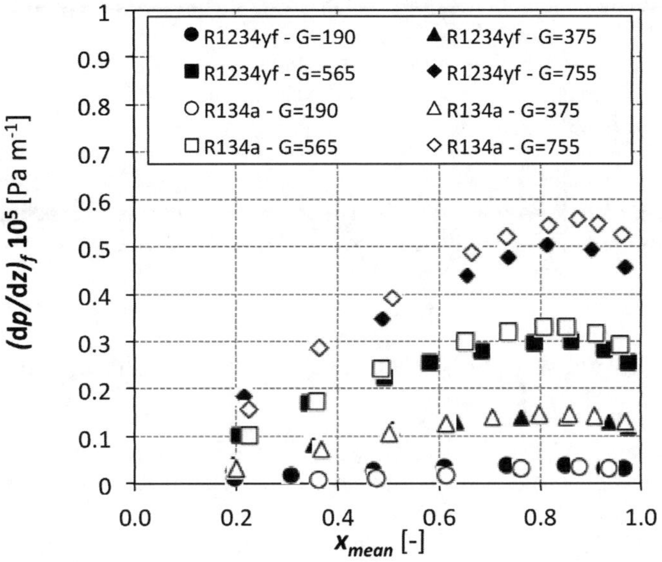

Figure 8. R1234yf vs. R134a frictional pressure gradient at HF=10 kW m^{-2}.

Figure 8 compares the frictional pressure gradients measured during flow boiling heat transfer of R1234yf and R134a inside the tested microfin tubes at HF=10 kW m^{-2}. It clearly appears that, for given vapour quality and refrigerant mass velocity, the frictional pressure gradients shown by R1234yf are slightly lower, at least similar, to those measured for R134a. In particular, at low mass velocities, the pressure drops of the two fluids are almost the same whereas when G>375 kg m^{-2} s^{-1}, the values exhibited by the R1234yf are lower than those obtained for R134a. This behaviour might be explained considering that the dynamic viscosity of the liquid of the R1234yf is some 20% lower than that of R134a while the vapour density is around 20% higher leading to a somewhat lower pressure drop.
The proposed comparison merely considers the heat transfer and fluid flow behaviours of the two refrigerants; however, for a given cooling capacity, keeping constant the evaporating and condensing temperatures, the R1234yf refrigerant mass flow rate needed is some 20% higher than that of R134a because the latent heat of the new refrigerant is as much lower. This means that for a comprehensive and detailed comparison of the two refrigerants, the different mass flow rate must be taken into account to obtain meaningful results using the data reported in this work.

4. Conclusions

This paper compares the flow boiling heat transfer and fluid flow behaviours of the R1234yf and R134a inside a mini microfin tube having an internal diameter at the fin tip of 3.4 mm. The experimental measurements were collected at 30 °C of saturation temperature, by varying the refrigerant mass velocity, the vapour quality, and the heat flux.

The two refrigerants show different results as a function of the operating test conditions due to their slightly different thermophysical properties which control their attitude to enhance either the nucleate boiling or the two-phase forced convection phase change mechanisms.

The presented results might also be useful to compare the performance of these two refrigerants in real air conditioning and refrigerating applications, which has to be optimized as a function of the selected fluid.

Finally, this paper highlights the interesting heat transfer features of this new mini microfin tube, which exhibits high heat transfer enhancement and large charge minimization capability as compared to the traditional microfin tubes.

5. Acknowledgment

The support of Wieland-Werke AG and Dr. Christoph Walther on this research activity is gratefully acknowledged. The support of University of Padova is gratefully acknowledged.

References

[1] Domanski P A, Brown J S, Heo J, Wojtusiak J and Mc Linden M O 2013 *Int. J. Refrig.* **38** 71-79
[2] Mc Linden M O, Kazanov A F, Brown J S and Domanski P A 2013 *Int. J. Refrig.* **38** 80-92
[3] Fukuda S, Kondou C, Takata N and Koyama S 2014 *Int. J. Refrig.* **40** 161–173
[4] Grauso S, Mastrullo R, Mauro A W, Thome J R and Vanoli G P 2013 *Int. J. Refrig.* **36** 478-491
[5] Hossain Md A, Onaka Y, Afroz H M M and Miyara A 2013 *Int. J. Refrig.* **36** 465-477
[6] Del Col D, Bortolin S and Rossetto L 2013 *Int. J. Heat Mass Transf.* **67** 1231-1245
[7] Lu M C, Tong J R and Wang C C 2013 *Int. J. Heat Mass Transf.* **65** 545–551
[8] Kondou C, BaBa D, Mishima F and Koyama S 2013 *Int. J. Refrig.* **36** 2366-2378
[9] Mancin S, Diani A, Zilio C and Rossetto L 2013 Experimental Measurements of R134a Flow Boiling Inside a 3.4 mm ID Microfin Tube *Proc. 13th UK Heat Transfer Conference, UKHTC2013 (London)*
[10] Mancin S, Diani A and Rossetto L 2013 Experimental Measurements of R134a Flow Boiling Inside a 3.4 mm ID Microfin Tube *Submitted and accepted to Heat Transf. Eng.*
[11] Mancin S, Diani A and Rossetto L 2014 *Energy Procedia* **45** 608–615
[12] Gao L, Honda T and Koyama S 2007 *HVAC&R Research* **13** 415-425
[13] Gao L, Ono T and Honda T 2007 Pressure drops of CO_2 and oil mixtures in horizontal smooth and micro-fin tube evaporators *Proc. International Congress Refrig. (Beijing)*, ICR07-B1-1262
[14] Dang C, Haraguchi N and Hihara E 2010 *Int. J. Refrig.* **33** 655-663
[15] Wu Z, Wu Y, Sunden B and Li W 2013 *Exp. Therm. Fluid Sci.* **44** 298-408
[16] Lemmon E W, Huber M L and McLinden M O 2013 NIST Standard Reference Database 23: Reference Fluid Thermodynamic and Transport Properties-REFPROP, Version 9.1 *National Institute of Standards and Technology, Standard Reference Data Program (Gaithersburg)*
[17] Rouhani S Z and Axelsson E 1970 *Int. J. Heat Mass Transf.* **13** 383-393

Low Cost True Monofiber Optical Probe for Local Void Fraction Measurements in Minichannels

G. Guidi[1], P. Di Marco[1], S. Filippeschi[1], M. Mameli[2]

[1]Università di Pisa, DESTEC, Largo Lucio Lazzarino 2, 56122 Pisa, Italy
[2]Università di Bergamo, Viale Marconi 5, 24044 Dalmine (BG), Italy
[*]Corresponding author: e-mail s.filippeschi@ing.unipi.it

Abstract. Two phase flow inside minichannels is one of the most investigated research topic at present. The measurement of the flow rate parameters is fundamental to characterize the flow pattern and its evolution over time. This paper shows that an optical technique, well-known for large diameter pipes, can be applied to mini channels with a laminar mass flow rate. In particular, a Y-junction mono-fiber optic system with a chamfered tip probe has been built and tested. This method is applied to the local void fraction measurement in a copper capillary pipe with internal diameter of 2 mm and external diameter of 3.00 mm. Different probes have been developed and tested. The accuracy of the method depends on the size, the shape of the tip and on the tip distance from the pipe centre. Different distances and liquid flow rate have been tested. The two-phase flow pattern is also visualized and recorded by a high speed camera (FASTEC Troubleshooter 16000 fps) and post processed with an image analysis technique. A good agreement between the optical and the video signal has been observed.

1 Introduction

Two-phase flow prediction in minichannels is a basic very active research field at present, because of the increasing number of mini equipments in ground and space applications which involves capillary pipes. The measurement techniques of parameters as pressure, velocity, temperature, mass flow rate are well assessed at the moment even in capillary pipes. On the other hand the measurement of void fraction inside two-phase flow for minichannels is currently an unsolved problem.

The simplest measurement technique of void fraction is the quick-closing valve method, still used at present. This method traps liquid in a tube to measure mass and liquid volume [1]. Until now there are several studies concerning void fraction measurements and involving different configurations and setups. They are largely applied to large pipes [2]-[4].

Maurus et al. [5] visually observed the void fraction in a subcooled flow of boiling water through a horizontal rectangular channel. A high-speed camera's top and side views of vapor shape helped to estimate the void fraction. Bowers and Hrnjak [6] visually observed the void fraction measurement of R-134a stratified flow through three different tube diameters, 7.2, 8.7, and 15.3 mm. They stated that void fraction depended on tube size diameter, while mass velocity had little effect on it. Winkler et

al. [7] used image analysis to measure the void fraction of R-134a's condensing process while flowing through a minichannel ranging from 2.00 to 4.91 mm. Their results show that neither mass flux nor channel size had a significant effect on the void fraction of a wavy flow pattern.
The optical system based on the images acquisition has, however, a big limitation because it cannot be applied to opaque pipes. Among the void fraction measurement techniques, the employment of optical fiber probes seems to be one of the most interesting one, because of they can be applied to opaque pipes of different materials (metallic, plastic, glass), they are appropriate for a large number of working fluids and they can be used with a very high sampling rate [8]-[9]. This technique is based on the difference, in terms of refraction index, between the vapour and liquid phases where the tip of the probe is periodically immersed.
Unfortunately the fiber optic probe is an intrusive measurement because the tip is immersed in the fluid and the sensitive surface, depending on the size of the probe, could be not negligible respect the cross sectional area of the pipe. In the past it is usually applied to large ducts measurements.
In addition optical fiber probe measures the local void fraction evolution over the time of the zone where the tip is immersed, so that a big number of probes, in different points of the cross section, must be used if the flow pattern must be characterised.
However, in the case of two-phase flow pattern characterisation for mini-channels the number of probes can be drastically reduced down to one. The highly confined flow rate shows different flow patterns respect to the large channels case: patterns as slug/plug, annular and semi annular are the most frequently observed [8][11]. In annular, semi-annular and slug/plug patterns the position of the probe tip inside the cross section and its distance from the pipe centre assumes lower importance in the flow pattern characterisation.
This paper deals with the limitations and the accuracy of the application of a fiber optic technique to capillary pipes to detect the two-phase flow pattern over time. Capillary pipes are characterised by very low diameters, given by Bond number lower than 2. This configuration is typical of Pulsating Heat Pipes devices and other automotive heat exchangers.
Different probes have been built and tested in a water/air loop. The void fraction experimentally detected by the optical probe inside a copper pipe with an internal diameter of 2 mm, has been compared with that measured with an image acquisition system, showing good agreement. Some preliminary qualitative studies on the influence of the probe tip position with respect to pipe centre have been presented.

2 Fiber Optic void fraction measurements

The fiber optic void fraction measurement is based on the physical principle that a light beam ray passing through a transparent surface can be refracted with different angles according to medium in contact with the surface. The basic relationship between governing refraction at the interface between two media is given by the Snell's law:

$$n_1 \sin(\theta_1) = n_2 \sin(\theta_2) \quad (1)$$

where n_1 and n_2 are the refraction indexes of the two media and θ_1 and θ_2 are and the incidence angles of the incoming and outcoming light beam, respectively. It is interesting to note that as θ_1 decreases, θ_2 tends to 0 and the intensity of the light beam, that crosses the interface, decreases to 0, therefore the light beam is reflected back. The value of θ_1 , named θ_{TIR} and corresponding to the total reflection, is given by the limit of θ_2 that tends to 0, so that $\cos\theta_2 \approx 1$

$$\theta_{TIR} = \cos^{-1}\left(\frac{n_2}{n_1}\right). \quad (2)$$

Eq(2) shows that if $\theta_1 \leq \theta_{TIR}$ the beam inside the fiber core is totally reflected, and it is completely contained inside it. This phenomenon allows optical fibers to correctly work, because the refraction index of the core is higher than the refraction index of the external cladding and, therefore, the beam isconfined inside the core channel. In this method the material of the probe immersed in the fluid assumes great importance in the feasibility of the measurement.

One of the most applied probe in large pipes in the past is the *U-shaped probe.* This type of probe tip is characterized by a single optical fiber bended with a relative small curvature radius. The U-turn of the fiber is then inserted at the measurement point. This configuration is the simplest because it needs no fiber junctions or splitter, as there is only one fiber where the beam can pass. However this technique presents several disadvantages: the fiber optic must be sufficient flexible to be curved, and the probe assumes large dimensions. It cannot be applied to capillary pipes. Another probe tip is named "Joined-tip". It is characterized by a welded probe composed of two optical fibers polished and fused or joined together by means of a suitable glue or resin. Thismanufacturing process is shown in [12] . The tip is fixed by means of a suitable epoxy resin. This probe also has large dimensions, and it cannot be applied to capillary pipes.

The true monofiber optical probe is characterized by a single fiber section of variable length. The light beams, both the one coming from the laser source and going to the detector, pass through the same core channel and are then splitted by means of a Y- Junction.

In the past this technique has been used by Carrica et al. [13] to characterise the void fraction in a pool boiling process. Carrica et al. state that the tip of the true monofiber probe must be opportunely chamfered to obtain a back signal as the light beam impacts on the fiber external surface. An angle of 45° is advised. Figure 1 shows the different behaviour of the chamfered tip in case of total reflection or refraction. The beam (not necessarily parallel to the fiber axis) incident on the tip surface (interface between the two media) is partially reflected and partially refracted. The quantity of reflected light is higher when the tip is in contact with gas phase than with liquid phase.

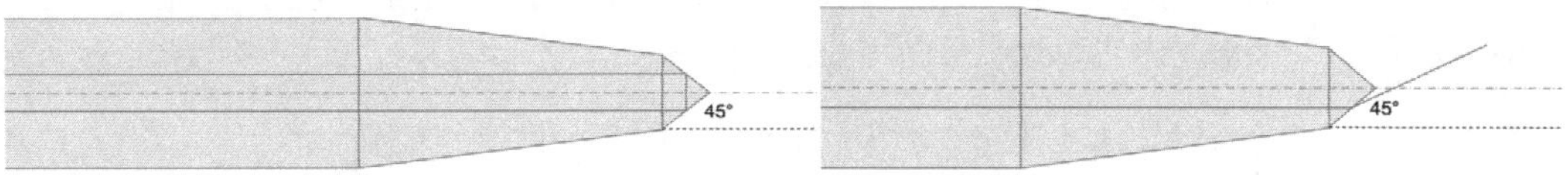

(a) (b)

Figure 1: Probe tip ideal working principle (a) gas phase (b) liquid phase

3 Experimental Activity

A Y-junction optical system has been built and tested inside a capillary pipe (D=2 mm). The experimental facility consists of: a measurement loop, the optical fiber void fraction system and the visualization section. A scheme of the setup is shown in shown in Figure 2.

The measurement loop consists of a reservoir, a capillary pipe loop, a pump and a syringe pump.

The materials: of the pipes are silicon, copper in the fiber optic section and glass in the visualization section.

The pump is peristaltic (ISMATECH® MCP-Z Standard Gear Pump ISM 405) with water flow rate of 1-7020 l/min.

The two phase flow pattern is created by a syringe pump pushing air into the loop with a mass flow rate up to 5 cm^3/s. The temperature of the water inside the tank is measured by a PT100 probe.

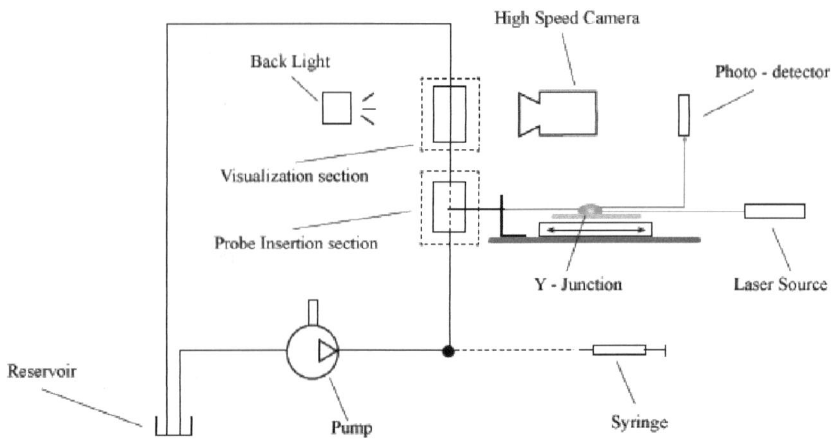

Figure 2: Sketch of the experimental setup

1.1. The optical fiber measurement system

The optical fiber system consists of: light source, fiber connections, Y-junction, tip probe, photo detector and acquisition system. The light source is a He-Ne Class 3B Laser working at 632,9 μm with a nominal power P_0 < 5mW. The light beam coming from laser source is collimated in air into the fiber connection, that has been fixed on a stainless steel tube, placed with a ferrule clamp (FCM/M) on an optical desk.
The fiber optics chosen to connect the laser source to the Y-junction and the junction with the photo detector are multimode pure slice fiber manufactured by THORSLAB® and listed in Table 1. They are characterised by high flexibility and high Numerical Aperture (N.A.).

TABLE 1: Optical fibers employed for the probe

Code	Range μm	Cladding diameter μm	Core diameter μm	N.A.
BFH48-400	300 – 1200	430	400	0,48
FT200UMT	300 – 1200	225	200	0,39

The Y-junction is the most critical part of the optical system. It conveys the light beams, coming from the source, into the monofiber probe and splits the light coming back from the probe to the photodetector. It operates as a circulator. The incorrect alignment of the fiber causes high light losses, therefore and the signal going back to the photodetector could be insufficient or very noisy. The basic scheme was proposed by Carrica et al. [13] and is shown in Figure 3.
The two fibers coming from the light source and going back to photodetector are collimated with the aid of a 20x microscope. The fibers are then placed and glued in a stainless steel micropipe. Then the two tips are cut and polished with polish paper. The same polishing procedure is used to manufacture the larger diameter fiber used for the chamfered probe. The surface finish is checked by means of a microscope , and a first signal check is performed in order to enhance fiber alignment. A bi-component resin, used for connectorizing fibers, is placed on the junction in order to fix the components and reduce losses.

Journal of Physics: Conference Series **547** (2014) 012027 doi:10.1088/1742-6596/547/1/012027

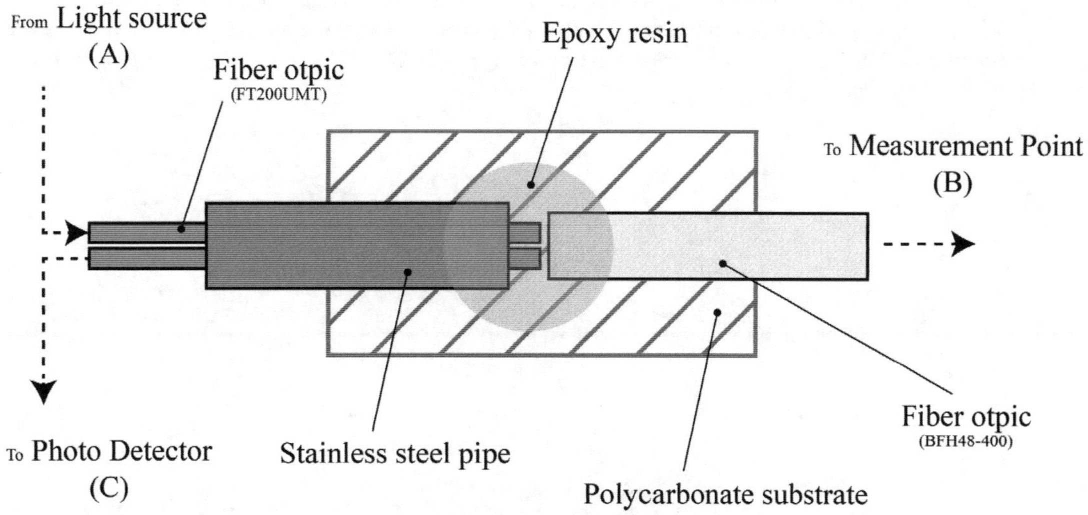

Figure 3: Y-Junction scheme

The probe is made of a 400 μm fiber and the tip is - chamfered at an angle of 45°. The fiber has not been stripped, and there was the need to remove part of the coating. The polishing procedure is the same as the junction. At first the tip has been cut to length and polished with 1000K paper, then finished with 2000K paper. During this procedure, it has been constantly checked by means of a microscope. Due to the fact the tip must fit into a minichannel, the chamfering has been extended to the section before the tip, obtaining a conical shape. An enlarged (20X) photo of the tip probe is shown in Figure 4a.

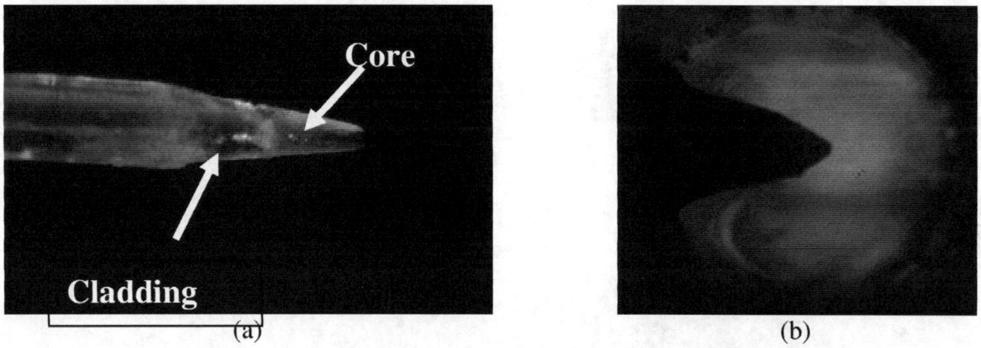

Figure 4: (a) Enlarged photo of the true monofiber tip probe (b) Enlarged photo of tip probe inside the pipe

The tip probe has been inserted inside the void fraction measurement cell, that consists of a copper T junction between the copper pipe (OD=3 mm, ID = 2mm) and a stainless steel pipe (OD=1 mm, ID=900 μm). The Y junction has been placed over a micro-regulated chassis that allows the regulation of the position of the tip inside the pipe with an accuracy of ±0.05 mm.
The enlarged photo of the probe with the tip positioned at the pipe centre, see Figure 4b, shows that the probe is highly intrusive, and a big portion of the cross sectional area is occupied by the probe. The ratio between the surface occupied by the probe and the entire cross sectional area, versus the position of its tip along the pipe radius, is shown in Figure 5.

The signal coming from the photo detector has been acquired by an acquisition board Arduino ATmega328 . It has 6 analogue inputs, and features the Atmega16U2 (Atmega8U2 up to version R2) programmed as a USB-to-serial converter. The clock speed is 16 MHz and the acquisition sampling rate is 40 Hz.

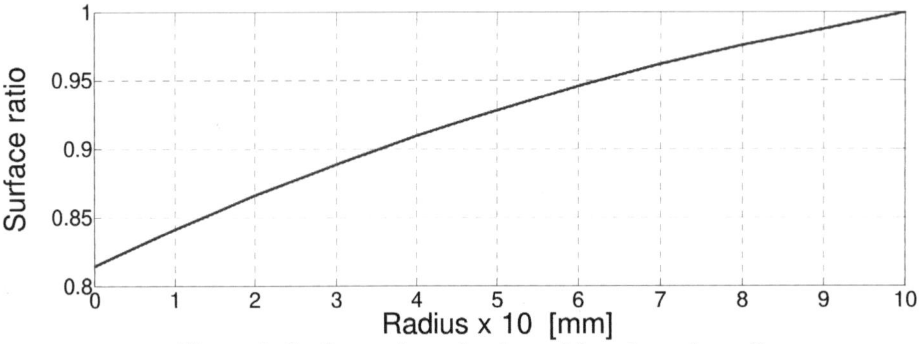

Figure 5: Surface ratio vs the tip position along the radius

1.2. Video acquisition system

The void fraction has been detected also by acquiring a video with a high speed videocamera (FASTEC Troubleshooter HR, up to 16000 fps @ 1280x32), and post processing the collected images. The video section consists of a glass pipe (ID=2 mm) and a video camera, externally triggered to synchronize the video capture with the signal coming from the photo-detector. The back light behind the glass pipe is a white LED Source 5,5 W DC12V.It has a good light diffusion, a constant background colour and reduced reflection noise effects. The video has been taken at 125 fps with a resolution of 1280x32 pixels. The images have been post-processed as shown in Figure 6.

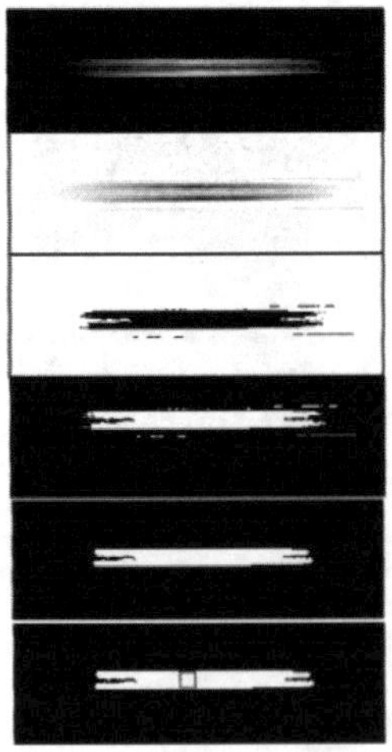

Figure 6: Images post-processing sequence

4 Results and discussion

4.1 Optical probe and video capture void fraction measurement

In order to c to understand the reliability of the fiber probe when applied to minichannels, a test has been carried out at constant water mass flow, pushing a measured volume of air, at the same temperature, into the loop, in order to obtain a two phase flow pattern. In accordance with Ong et al. [10] the water mass flow rate is 2.5 ml/s and the volume of air pushed inside the loop has been calculated to obtain a slug flow. Three different vapour slugs have been visualized.
Figure 7 shows the comparison between the optical signal with a tip probe placed at the pipe centre, with that resulting from image acquisition.

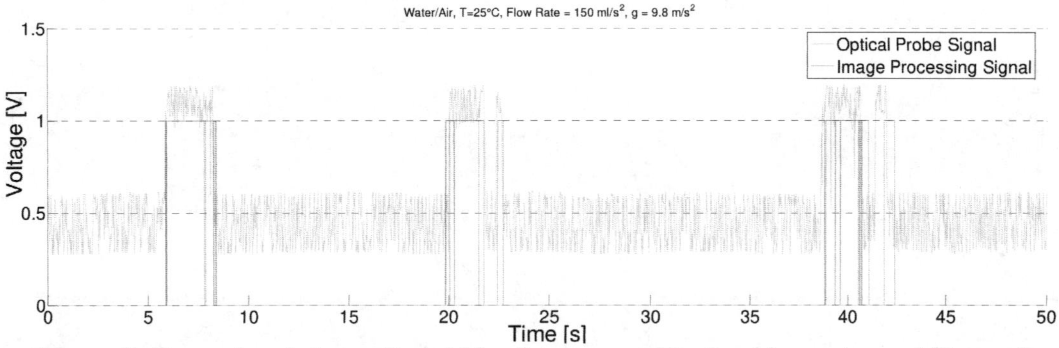

Figure 7: Comparison between the void fraction measured by the video capture and fiber probe

According to the Snell's law and taking into account that the refraction index n_l of the silica core is higher than the refraction index of liquid/gas, it is possible to define the critical total reflection angles of the two phases. By considering water (n_{water} = 1.33) and air (n_{air} = 1) and silica core (n_{silica} = 1,46) the critical angles are 0.42 rad (24°) for water and 0.72 rad (41.2°) for air.
For this reason the quantity of light which is reflected back to the light detector is greater when the probe tip is in contact with air. Figure 7 shows with great evidence three slugs passing through the probe. The local void fraction measured by the optical probe and that obtained by image processing show a good agreement and errors lower than 12%. A good repeatability has been observed.

4.2 Preliminary test on the influence of the tip position

A preliminary experimental activity has been carried out in order to understand to what extent the optical probe can be considered intrusive in flow pattern detection. Three different tests at different liquid flow rates (2.8, 5 and 8.3 ml/s) and two radial positions of the tip probe (r=0 and 5 mm) have been tested. The first position (r=0 mm) corresponds to the tip placed at the pipe centre with 82% of the cross sectional area left free by the probe, the second position is with the tip placed at r=5 mm (free surface of 92%). A visualization of the different flow patterns at different regimes is shown in Figure 8.
The images in figure qualitatively shows that the optical probe modifies the two-phase pattern distribution mostly at low flow rates. Figure (a) shows the end of a slug of vapour for a flow rate of 2.8 ml/s. It possible to see as the probe breaks the slug backward meniscus, and several following microbubbles can observed. This problem is evident at high mass flow rates and even when the tip is placed at r=5 mm, even if 92% of the surface is free in this case
However it is possible to note that at high flow rates and at r=0.5 mm both the advancing (Figures 8c, g, i) and the receeding one menisci, (Figures 8c, g, i, h) tend to be less disturbed than in the case with probe at r=0 mm.

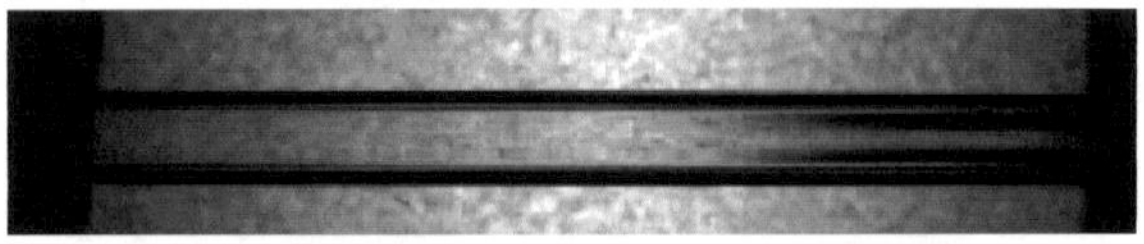

(a) ml=2.8 ml/s – tip position r=0 mm

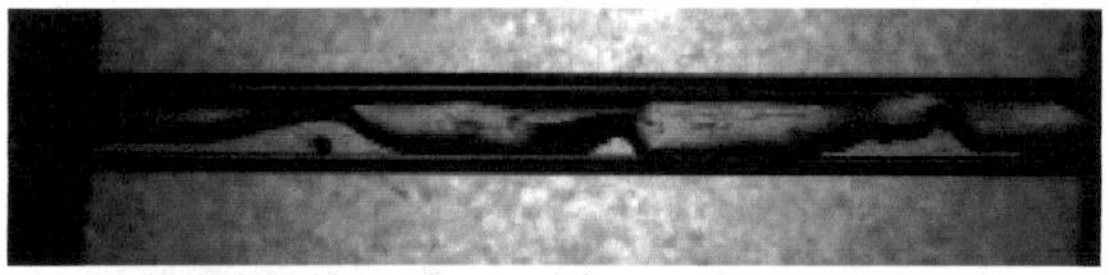

(b) ml=2.8 ml/s – tip position r=0 mm

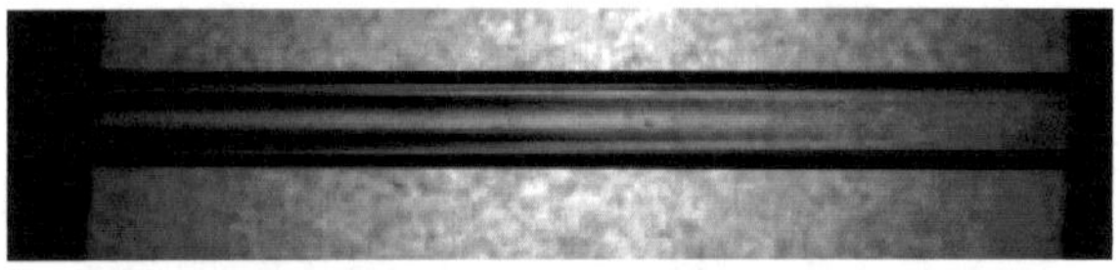

(c) ml=8.3 ml/s – tip position r=0 mm

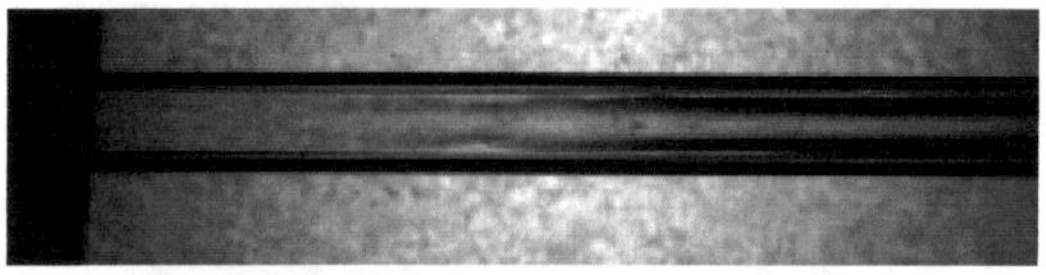

(d) ml=8.3 ml/s – tip position r=0 mm

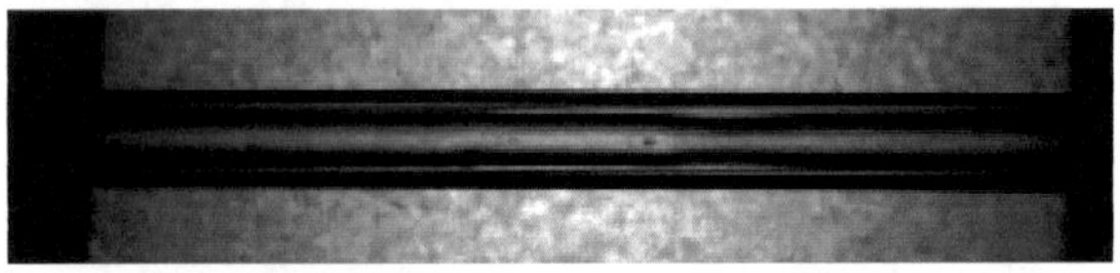

(e) ml=8.3 ml/s – tip position r=0 mm

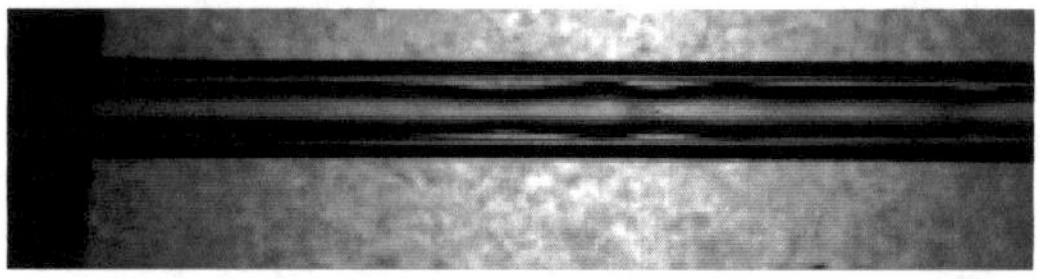

(f) 2.8 ml/s – tip position r=0.5 mm

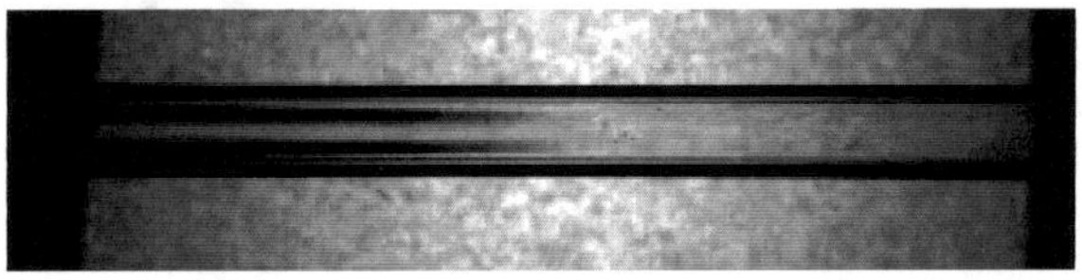

(g) ml=2.8 ml/s – tip position r=0.5 mm

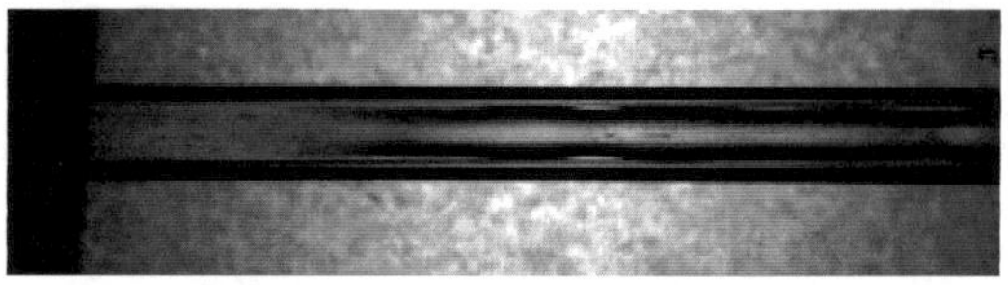

(h) ml=2.8 ml/s – tip position r=0.5 mm

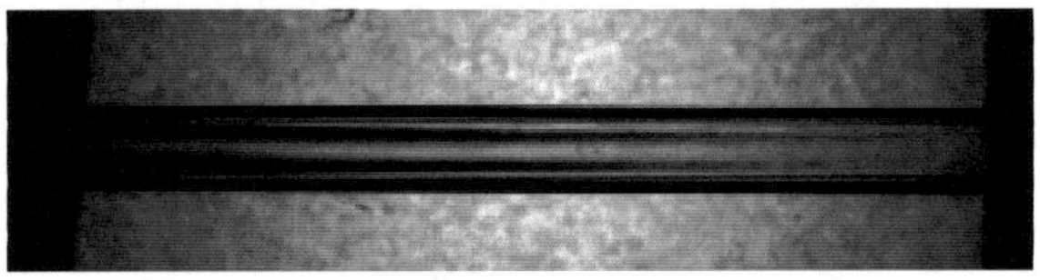

(i) ml=8.3 ml/s – tip position r=0.5 mm

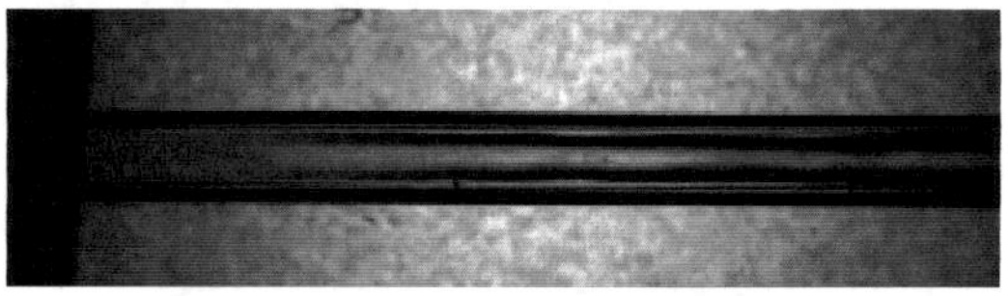

(j) ml=8.3 ml/s – tip position r=0.5 mm

Figure 8: Different flow pattern at different liquid flow rate and tip radial positions

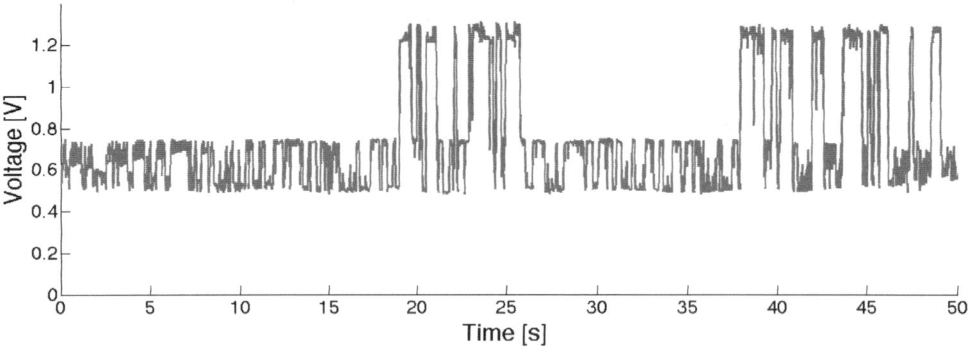

Figure 9: Photo-detector signal over time for the probe at r=0 mm and ml=2.8 ml/s.

Another interesting qualitative tendency, observable as the probe is receeded toward the inner surface, regards the pattern of the slug. For low flow rates (Figure 8b) the flow pattern is higly disturbed by the fiber probe, but this disturbance becomes very low as the probe is placed at r=5 mm (Figure 8f, j) .
Figure 9 and 10 show the voltage measured by the photo detector vs. time as two-phase flow is passing through the probe, at r=0 mm and flow rate of 2.8 ml/s, and at r=0.5 mm and flow rate of 8.3 ml/s, respectively.

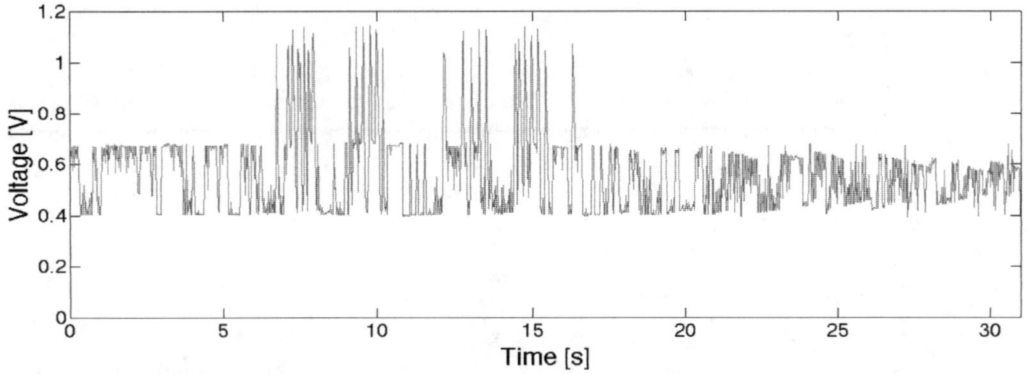

Figure 10: Photo-detector signal over time for the probe at r=5 mm and ml=8.3 ml/s.

It is interesting to note as the probe is able to detect the flow pattern in both cases. Figure 10 shows a greater number of peaks, because the the slugs are shorter than the in the other case. The comparison between the probe signal and the void fraction calculated by the image processing has however confirmed a good agremeent also in these two cases.

5 Conclusions

This paper deals with the limitations and the accuracy of the application of a fiber optic technique to detect the flow pattern over time in capillary pipes. Among the void fraction measurement techniques, the fiber optic probes can be applied to opaque pipes of different materials (metallic, plastic, glass), they can be used with a large number of working fluids with a very high sampling rate. Different probes have been built and tested in a water/air loop. The void fraction detected by the optical probe has been compared with that measured with an image acquisition system, and a good agreement has been observed (errors lower than 12%). Unfortunately the fiber optic probe is an intrusive measurement: some preliminary qualitative studies on the influence of the probe tip position with respect to pipe centre have been presented. Experimental tests have confirmed that this probe highly disturbs the flow pattern by breaking the advancing and receeding menisci in slug flow. As expected, this disturbance increases when the probe is more deeply inserted along the tube diameter. Further studies will be aimed to optimize the radial position of the probe, in order to detect the void fraction making as low as possible the disturbances and modifications of the flow pattern.

Nomenclature

Bo	Bond Number	
D	Pipe diameter	(m)
g	Gravity acceleration	(m/s^2)
ml	Liquid flow rate	(m^3/s]
n	Refraction index	(-)

ρ_v	Vapour density	(kg/m^3)
ρ_l	Liquid density	(kg/m^3)
θ	Angle	(rad)

References

[1] M. Ishii, "One-Dimensional Drift-Flux Model and Constitutive Equations for Relative Motion Between Phases in Various Two-Phase Flow Regimes," Argonne National Lab Report, ANL 77-47 , October 1977.

[2] G.B. Wallis, One-Dimensional Two-Phase Flow, McGraw Hill, 2nd Edition, New York, 1979.

[3] Srisombaa R., Mahianb O., Dalkilicc A.S., Wongwisesa S., Measurement of the void fraction of R-134a flowing through a horizontal tube International Communications in Heat and Mass TransferVolume 56, August 2014, Pages 8–14.

[4] S.M. Bhagwat, A.J. Ghajar A flow pattern independent drift flux model based void fraction correlation for a wide range of gas–liquid two phase flow, International Journal of Multiphase Flow 59, (2014) 186–205

[5] R. Maurus, V. Ilchenko, T. Sattelmayer, Study of the bubble characteristics and the local void fraction in subcooled flow boiling using digital imaging and analysing techniques, Exp. Thermal Fluid Sci., 26 (2002), pp. 147–155.

[6] C.D. Bowers, S.P. Hrnjak, Determination of void fraction in separated two-phase flows using optical techniques International Refrigeration and Air Conditioning Conference, 2010, Paper 1083 (2010).

[7] J. Winkler, J. Killion, S. Garimella, Void fractions for condensing refrigerant flow in small channels. Part II: void fraction measurement and modeling, Int. J. Refrig., 35 (2012), pp. 246–262

[8] B. Stankovic An experimental study on the local void fraction measurement in large diameter vertical pipes using optical probes Open Access Dissertations and Theses (1997), Paper 3244

[9] N. Abuaf, O.C. Jones Jr Optical probe for local void fraction and interface velocity measurements Review of Scientific Instruments 49, 1090 (1978); doi: 10.1063/1.1135524

[10] L. Ong, J.R. Thome: Macro-to-microchannel transition in two-phase flow: Part 1 – Two-phase flow patterns and film thickness measurements, Experimental Thermal and Fluid Science 35 (2011) 37–47

[11] C.L. Ong, J.R. Thome: Macro-to-microchannel transition in two-phase flow: Part 2 – Two-phase flow patterns and film thickness measurements, Experimental Thermal and Fluid Science 35 (2011) 873–886

[12] C.M. Davis, E.F.Carome, et al.: Fiberoptic Sensors: technology handbook, Dynamic System Inc., (2000) Data Book III, Worverine Inc., Cap. 17

[13] P. Carrica, V. Masson, P. Di Marco Application of a low-cost true-monofiber local phase detection probe to void fraction measurements in pool boiling, Proc. of 4th World Conference on Experimental Heat Transfer, Fluid Mechanics and Thermodynamics, ExHFT 4, 1997, 293-300, Brussel, June 2-6

Micro computed tomography and CFD simulation of drop deposition on gas diffusion layers

M Guilizzoni[1], M Santini[2], M Lorenzi[3], V Knisel, S Fest-Santini[2]

[1] Dipartimento di Energia, Politecnico di Milano, via Lambruschini 4, 20156 Milano
[2] Dipartimento di Ingegneria, Università di Bergamo, viale Marconi 5, 24044 Dalmine (BG)
[3] School of Engineering and Mathematical Sciences, City University London, Northampton Square, London EC1V 0HB, UK

E-mail: maurizio.santini@unibg.it

Abstract. Fuel cells are electrochemical power generation system which may achieve high energy efficiencies with environmentally friendly emissions. Among the different types, Proton Exchange Membrane fuel cells (PEMFC) seem at present one of the most promising choices. A very important component of a PEMFC is the gas diffusion layer (GDL), which has the primary role of managing water in the cell, allowing reactant gases transport to the catalyst layer while keeping the membrane correctly hydrated and preventing electrode flooding. Therefore, GDLs have to be porous and very hydrophobic. Carbon clothes or carbon papers coated with a hydrophobizing agent – typically a fluoropolymer – are used. Given the complex chemistry and morphology of the GDLs, wettability analyses on them present some critical issues when using the conventional contact angle measurement techniques. In this paper, the deposition of a drop on a GDL (produced using polytetrafluoroethylene-co-perfluoroalcoxy vinyl ether as the fluorinated polymer) was investigated by means of micro computed tomography (microCT) and numerical simulation. The microCT facility operational at the University of Bergamo was used to acquire a 3D tomography of a water drop deposed on a sample GDL. The reconstructed drop dataset allows thorough understanding of the real drop shape, of its contact area and contact line. The GDL dataset was used to create a realistic mesh for the numerical simulation of the drop deposition, which was performed using the OpenFOAM® *interFOAM* solver.

1. Introduction

Fuel cells are electrochemical power generation systems where chemical energy is directly converted into electrical energy, without passing through combustion [1,2]. Their high energy efficiencies and environmentally friendly emissions make them very promising, even if a series of issues still has to be solved before their large scale application might be feasible. They also constitute a highly interesting topic from the theoretical point of view, as their design involves chemical, electrical, mechanical, thermal and fluid dynamic aspects at both the macro- and the microscale. Therefore, they constitute a field of very active research. Among the different available types of fuel cell, Proton Exchange Membrane fuel cells (PEM-FC) work at low temperatures with hydrogen as fuel, offering high power densities and flexibility to load changes. A fundamental component in a PEMFC (and also in other types, e.g. microbial fuel cells [3]) is the Gas Diffusion Layers (GDL). One of the major tasks for a GDL is to manage water within the cell [4], acting as a barrier to water transport – thus maintaining the electrolyte membrane correctly hydrated and keeping the ionic conductivity high – and letting water pass only through some of the pores, keeping the most of the pores free for the reactants fluxes. To carry out such task, it has to be porous and highly hydrophobic. GDLs are therefore made of carbon fibers hydrophobized by a fluoropolymer dispersed on them. The fibers may be arranged either as a proper carbon fabric with a real warp-and-weft structure (carbon cloth) or in a woven-non-woven texture (carbon papers) obtained by high temperature sintering. The most used hydrophobizing agent is PTFE, but other fluoropolymers are under study too, e.g. perfluoroalcoxy (PFA), fluorinated ethylene propylene (FEP), fluorinated polyurethane based on perfluoropolyether (PFPE) blocks [5].

The high importance of the GDL hydrophobicity for the fuel cell performance is matched by the complexity of its evaluation. GDLs are absolutely far from the ideal surfaces on which the Young equation [6] and the conventional contact angle measurement techniques can be easily applied. They are chemically inhomogeneous, not flat, with a multi-scale roughness (from the fiber roughness to the warp-and-weft texture) which is only pseudo-isotropic. Moreover, they have to be extremely hydrophobic, up to the limit of superhydrophobicity (contact angle $\geq 150°$, contact angle hysteresis $\leq 10°$). All these conditions pose very significant difficulties [7] also to the state-of-the-art techniques for wettability analysis using sessile drops (which is the most commonly used method), e.g. Axisymmetric Drop Shape Analysis (ADSA) [8]. The major issues are that on such surfaces:

- the contact angle may be, and in general is, different from point to point along the triple line due to the surface morphology and to the triple line pinning on the surface edges [9];
- the apparent contact angle, i.e. the angle that the drop contour forms with an ideal horizontal line cutting the base surface at the contact points, is the only one accessible to conventional measurements (based on optical side-views acquisitions). The apparent contact angle may be related to the real contact angle by means of different models, among which the most credited are still the Wenzel model [10] – for homogeneous surfaces in which the drop liquid completely fills the grooves in the surface texture – and the Cassie-Baxter model [11], to be used for chemically heterogeneous surfaces or when the drop does not fill the grooves. The effectiveness of such models for surfaces characterized by multivariate wetting scenarios is still debated in the literature, so that detailed experiments about chemically and morphologically complex surfaces are needed for their validation;
- when conventional optics are used, the measurement may be affected by lens distortion, small depth of field with focus problems (particularly for the identification of the baseline), low or excess lighting or non-parallel light direction, low resolution of the image, diffraction near the contact points (particularly for very high contact angles). Moreover, side views of the drop-surface system may not be equivalent to real cross sections of the same;

Thus, there is a need to develop new techniques to improve the contact angle measurement and more in general the wettability analysis (triple line, wetted area, adhesion) of complex surfaces. GDLs are excellent test surfaces, as they join complexity and usefulness for practical applications.

In this work the use of micro computed tomography (microCT) is proposed to acquire a real three-dimensional representation of a sessile drop on a GDL surface, with the aim of extracting from the tomographic volume both direct experimental information about the drop shape and the surface wettability, and a realistic mesh for the numerical simulation of a drop deposition onto such surfaces.

2. Micro computed tomography

Among several emerging tomographic imaging techniques (derived using different physical phenomena, e.g. gamma rays, radio-frequency waves, electrical resistance, electrons, magnetic particles) currently available for the non-invasive visualization of the inside of an opaque material, the microCT is the one with highest spatial resolution and good penetrating capability. Thus, it is the most suited non-destructive technique (NDT) for the morphological reconstruction of multiphase interfaces at microscale.

A prototype microCT set-up (Fig. 1) is operational at the University of Bergamo composed by a X-ray microfocus open type emission tube (160 kVp @ 200 microA) allowing the emission as Bremsstrahlung (i.e. polychromatic and incoherent) with energies ranges suitable for micrometric spatial resolution tomography in a millimetric size sample. Projections are acquired by digital detectors (CCD flat panels with optimized scintillators are used to capture X-ray intensity). A rotary air-bearing direct drive stage provides superior angular positioning. The object itself is placed on the stage and rotated by angular steps according to Nyquist-Shannon theorem. Volumes are reconstructed with filtered back-projection standard algorithm, using commercial software VGStudio MAX, generating a 3D digital image of the object.

Figure 1. microCT set-up at the University of Bergamo.

3. Numerical simulations

Numerical simulations were performed using the *interFoam* solver of the OpenFOAM® 2.2.0 open source CFD toolbox [12]. OpenFOAM® was selected due to its free and open source nature and to the many favorable reviews and successful cases of application described in the literature [e.g. 13-17]. *interFoam* is a isothermal finite volume solver implementing a modified version of the Volume of Fluid (VOF) [18] method to simulate two-phase flow using a single set of equations. Both phases are assumed as incompressible and surface tension is included by means of the continuum surface force (CSF) model [19]. Volume tracking and interface reconstruction, with no explicit interface tracking, is performed. With respect to the original VOF formulation, the *interFoam* model includes an additional term in the volume fraction equation (Eqn. 1), to obtain interface compression down to just 2-4 cells. The full set of the governing equations solved by *interFoam* is thoroughly described in [13,14,17]. In summary, it is constituted by the incompressible continuity and Navier-Stokes equation system, plus an equation for the transport of the volume fraction (also called color function) α:

$$\frac{\partial \alpha}{\partial t} + \nabla \cdot (\mathbf{u}\alpha) + \nabla \cdot [\mathbf{u_r}\,\alpha(1-\alpha)] = 0 \qquad (1)$$

In the last equation, u_r is the so-called "compression velocity" that performs interface compression. In strict terms, such an additional term is a mass source, but both literature results [17] and verification by the authors proved that mass variation is completely negligible.

The drop water and the surrounding air were both assumed as Newtonian fluids, and the flow as laminar. For each phase the values of all the relevant thermophysical properties were taken at ambient temperature (20 °C). The drop diameter and the initial position of the drop with respect to the solid substrate were fixed using the *setFields* OpenFOAM® utility, assuming an initially spherical and still drop.

Previous tests by the authors [20] confirmed that the discretization schemes (both for space and time), solution algorithms and solver settings indicated in [17] are the most suitable for the investigated cases. Automatic time step adaptation was used to keep the Courant–Friedrichs–Lewy (CFL) number under the limit of 0.3 which is requested by the peculiar interface compression scheme used in *interFoam* for 3D simulations [13]. Some volume fraction sub-cycles were also performed to further improve the accuracy. The side effect of these choices is obviously that for the very fine meshes required by the GDL structure such a low value of the CFL number implies very small time steps and consequently very long simulation times.

Both complete 3D simulations and 2D simulations on single slices of the investigated volume were performed. For the 3D simulations, the domain was modeled as a "box" having sides of a few millimeters and whose bottom is constituted by the real GDL surface. The latter was extracted from the tomographic volume as an isosurface at the proper intensity value and exported as a stereolithography (STL) file. A structured mesh was initially created filling a box with regular hexaedra and then the OpenFOAM® *snappyHexMesh* utility was used to modify such basic mesh. For 2D simulations, a completely analogous procedure was followed, with the only difference that an extrusion of the selected volume slice was used as the bottom of the domain.

snappyHexMesh is able to perform different steps in mesh processing:

- select, from the base mesh, only the cells which are on one side of the surface described by the STL file; this creates a mesh with jagged boundaries;
- snap the vertices of the selected cells to their nearest points on the surface, so that the resulting mesh boundaries follows the original limiting surface;
- add boundary layer cells, if desired to better capture boundary layer phenomena.

For this work, only the first two steps were performed. In Fig. 2 examples of the resulting mesh after the first step and after the second are presented, both for 3D (top row) and 2D (bottom row) domains. The zooms from the 2D meshes are shown to better evidence the effect of the snapping step. Regrettably, the GDL surface is extremely complex, so that the snapping step creates a mesh that may include a certain number of extremely small and skewed cells, particularly in 3D. Great care has therefore to be devoted to obtaining a correct mesh, and in any case the CFL number limit requires time steps so small that 3D simulations become unfeasible on common personal computers. Therefore, the simulations described in this work were run on the snapped mesh in 2D, while using the jagged meshes in 3D. The use of jagged meshes worsens the well-known issues about contact angle and surface tension modeling – common to all non-smoothed interface-tracking and interface-reconstruction algorithms – that result in the development of spurious velocities near the interface. On small domains, the latter may heavily affect the reliability of the simulation. This is a quite common situation for analysis concerning complex surfaces (e.g. for porous media) and suitable countermeasures are under development to overcome these issues [21]. They require significant modifications of the solvers, which were not taken on here. Moreover, the simulation time in 3D was very long (some weeks for each simulation), with the consequence that rigorous mesh independence studies could not be carried out on 3D simulations. Therefore, they were performed on 2D simulations, while only preliminary checks about mesh independence were performed in 3D. Table 1 reports the domain and mesh characteristics for the performed simulations, while Fig. 3 shows the mesh dependence of the drop profile during the first part of the deposition in 2D simulations. It can be noticed how the drop profile and main drop dimensions are mesh independent. Concerning 3D simulations, no significant difference in the results were found with the finer mesh, apart from minor improvements in the resolution of the results and in the reduction of the spurious velocities.

The described issues concerning the mesh should be kept in mind in the evaluation of the simulation results, which at present must be considered mainly as a preliminary "feasibility study" about this topic, to trace the route for more in-depth studies.

Concerning boundary conditions:
- the top and the side boundaries of the domain were set as open boundaries at fixed total pressure, with volume fraction equal to 1 for the air phase (the drop must not reach such boundaries during the interesting part of the simulation);
- the bottom (modeling the GDL surface) was set as a no-slip wall at fixed flux pressure, with a dynamic contact angle condition for the volume fraction. The latter b.c. is defined in OpenFOAM® on the basis of the maximum advancing and minimum receding contact angles, plus a scaling velocity which allows the calculation of the effective contact angle in dependence of the velocity of the interface in contact with the boundary.

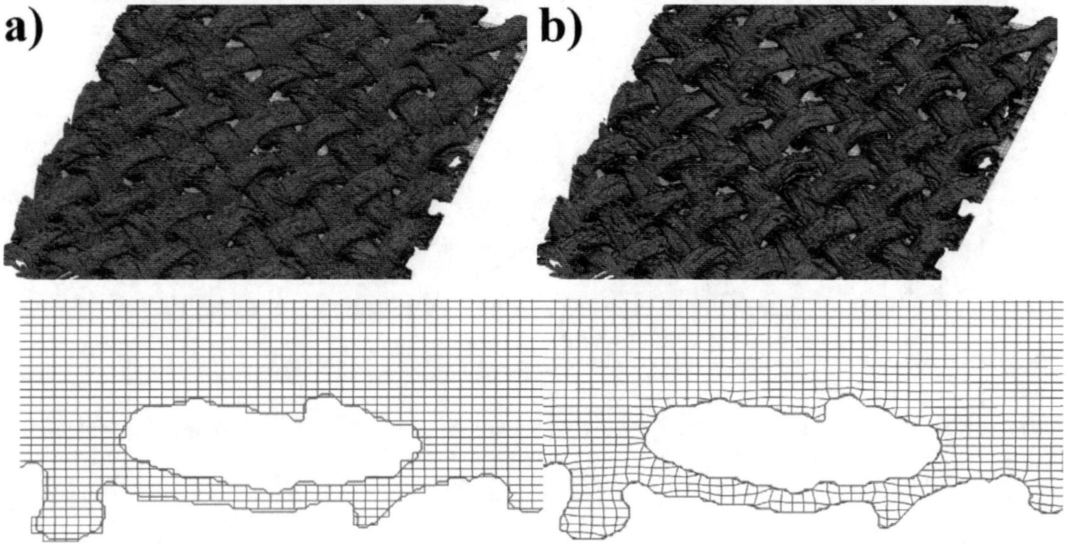

Figure 2. Mesh after running *snappyHexMesh* for the GDL-PFA cases: a) 3D case, cell selection alone, b) 3D case, cell selection and snapping to the limiting surface. On the bottom row, a zoom from the 2D intermediate mesh (see Tab. 1) is shown, to better evidence the effect of the snapping step (the limiting surface is drawn in red). (For interpretation of the references to color in this figure caption, the reader is referred to the Web version of this article.)

Table 1. Domain and mesh details for the performed simulations.

	Domain dimensions [m]	Number of cells [kcell]
GDL-PFA 2D coarse	0.0064 x 0.0042	37.17
GDL-PFA 2D intermediate	0.0064 x 0.0042	148.9
GDL-PFA 2D fine	0.0064 x 0.0042	595.6
GDL-PTFE 3D coarse	0.0040 x 0.0040 x 0.0020	1701
GDL-PTFE 3D fine	0.0030 x 0.0030 x 0.0020	4100
GDL-PFA 3D	0.0040 x 0.0040 x 0.0040	3916

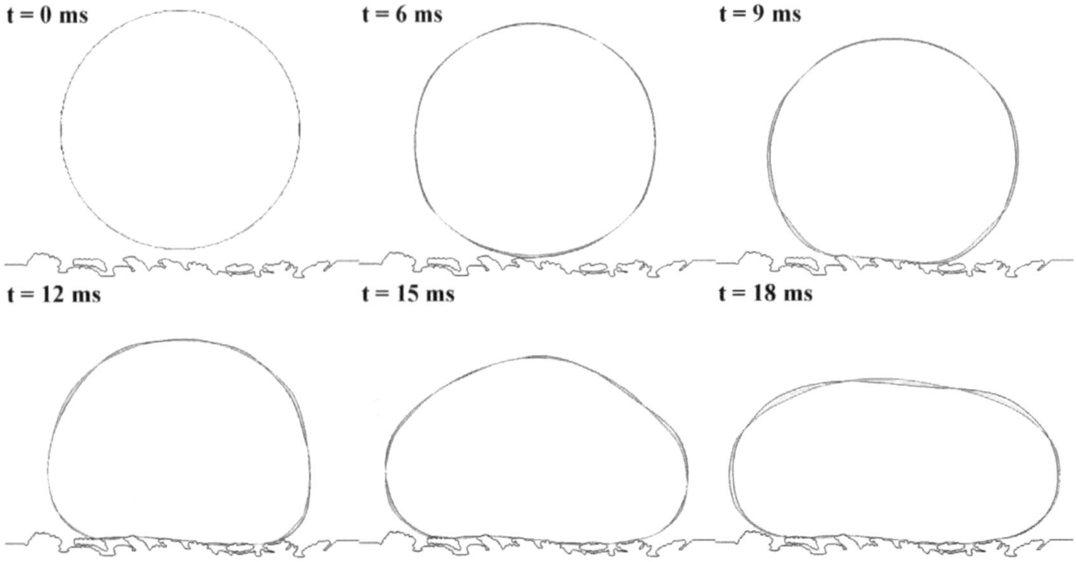

Figure 3. Mesh dependence of the drop profile during the first part of the deposition in 2D simulations: cases GDL-PFA 2D coarse (blue), GDL-PFA 2D intermediate (green), GDL-PFA 2D fine (red). (For interpretation of the references to color in this figure caption, the reader is referred to the Web version of this article.)

4. Results

As test surfaces, two GDLs – prepared by the Materials for Energy and Environment group of the Chemistry, Materials and Chemical Engineering Department of Politecnico di Milano – were used. The carbon matrix is the same for both, but one was hydrophobized using PTFE and the other using PFA. The voxel resolutions of the tomographic scans are 8 μm and 5.12 μm for the GDL-PTFE and GDL-PFA surfaces, respectively. Figures 4 a) and b) show the isosurface extracted from the tomographic volume for the two GDLs. As already described, such isosurfaces were saved as STL files and used to prepare the mesh for the numerical simulations.

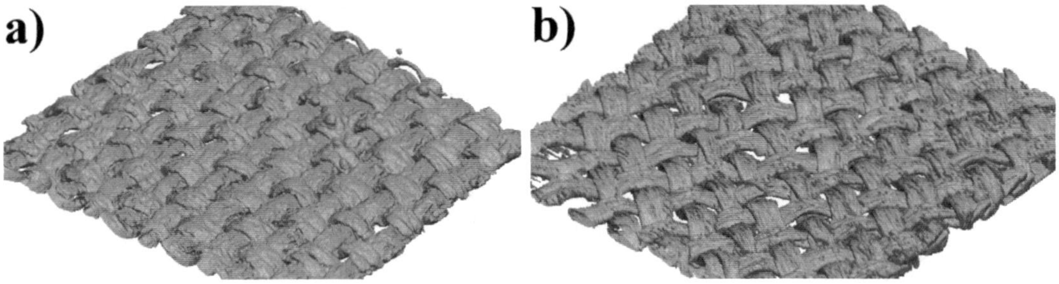

Figure 4. GDL-PTFE (a) and the GDL-PFA (b) surfaces extracted as intensity isosurfaces from the tomographic volume.

The GDL-PFA was also acquired after a drop was deposed on it. Figures 5 a) and b) show such situation both by means of a conventional side view taken with a SLR camera with macro lens and as the isosurfaces for drop and GDL after tomographic acquisition. It is evident how the tomographic reconstruction offers a much higher level of detail, with a micrometric resolution which is unreachable by the conventional optical investigation techniques and in a real three-dimensional reconstructed volume, which cannot be obtained with conventional electron microscopy.

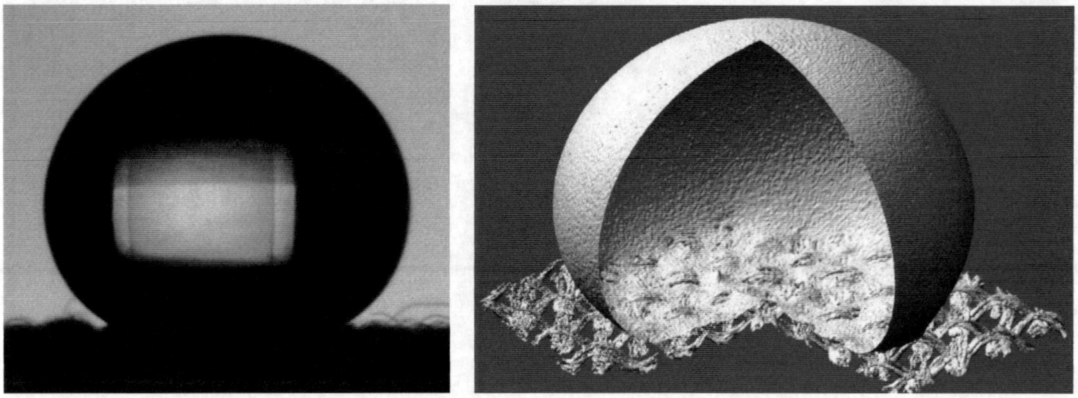

Figure 5. Optical side view of a sessile drop deposed on a GDL-PFA surface (left) and drop and GDL surfaces extracted from the reconstructed tomographic volume (right).

The tomographic output can be further processed to get accurate measurements of the drop volume and surface area, of the wetted area and of the real shape of the contact line [22,23]. It can also be sliced both horizontally, to get information about the water permeation within the GDL fibers (Fig. 6), and vertically, to evidence the complexity of the wetting, e.g. with pinning on the fibers (Fig. 7). If polar vertical slices are extracted and the drop is assumed to be macroscopically axisymmetric, axisymmetric drop shape analysis (ADSA) can also be applied to get information about the contact angle on a slice-wise basis [22]. Using such an approach, the average contact angle on the GDL-PFA was calculated and compared with the measurement by conventional ADSA on sessile drop pictures. Good agreement was obtained between the two approaches, with an average value of the "as placed" static contact angle $\theta = 151°$ (std. dev. 0.6°) for the conventional measurements and $\theta = 154°$ (std. dev. 2.6°) for the microCT analysis. The higher standard deviation for the microCT case reflects the variability of the contact angle along the triple line, which can be easily missed when using the conventional approach. Dynamic contact angle measurements by the conventional approach evidenced a very low hysteresis.

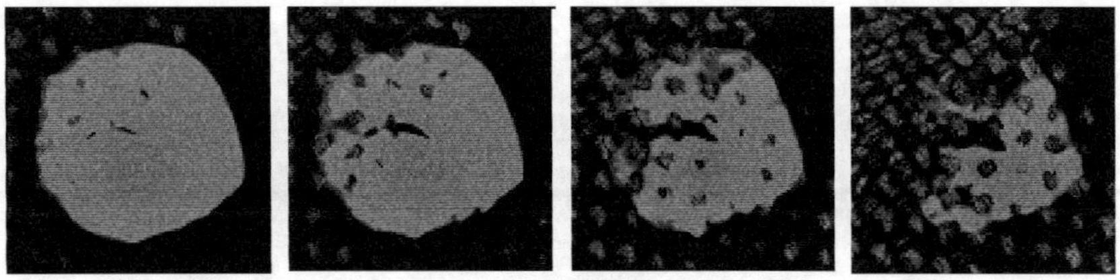

Figure 6. Horizontal slices of the tomographic volume for the GDL-PFA in the region where the drop wets the surface. Water volume included in each slice is 0.0518 μl, 0.0452 μl, 0.0315 μl, 0.0174 μl respectively.

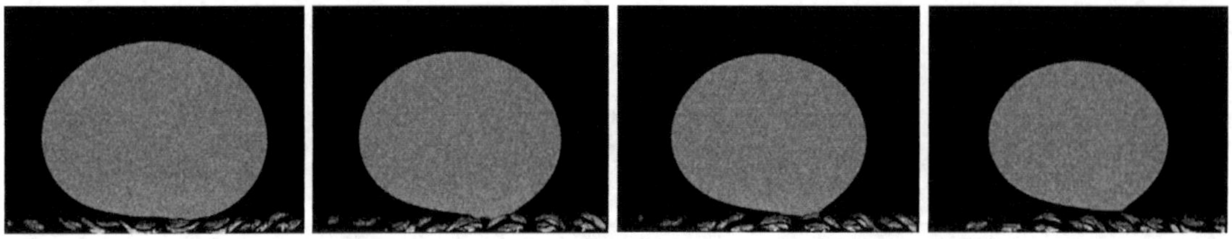

Figure 7. Vertical slices of the tomographic volume for the GDL-PFA, evidencing the complexity of the wetting with pinning on the fibers.

Therefore, a value of the static contact angle equal to 150° and a hysteresis of 10° were used as input for the simulation of the drop deposition on the GDL-PFA surface. On the contrary, a contact angle equal to 110° with and hysteresis of 20° – reflecting literature values for smooth PTFE – was arbitrarily set for the simulation on the GDL-PTFE surface, to simulate a less hydrophobic scenario. It is worth noting that the real wettability of GDL-PTFE is extremely similar to the one on the GDL-PFA, due to the dominating effect of morphology. Figure 8 shows some images of the drop shape during the deposition on the two surfaces. Figure 9 shows the time evolution of the apparent spreading ratio during the simulated deposition on the GDL-PTFE (case 3D fine). The apparent spreading ratio is defined here as the ratio between the wetted area vertically projected onto a horizontal plane and the equatorial section area of the drop before impact.

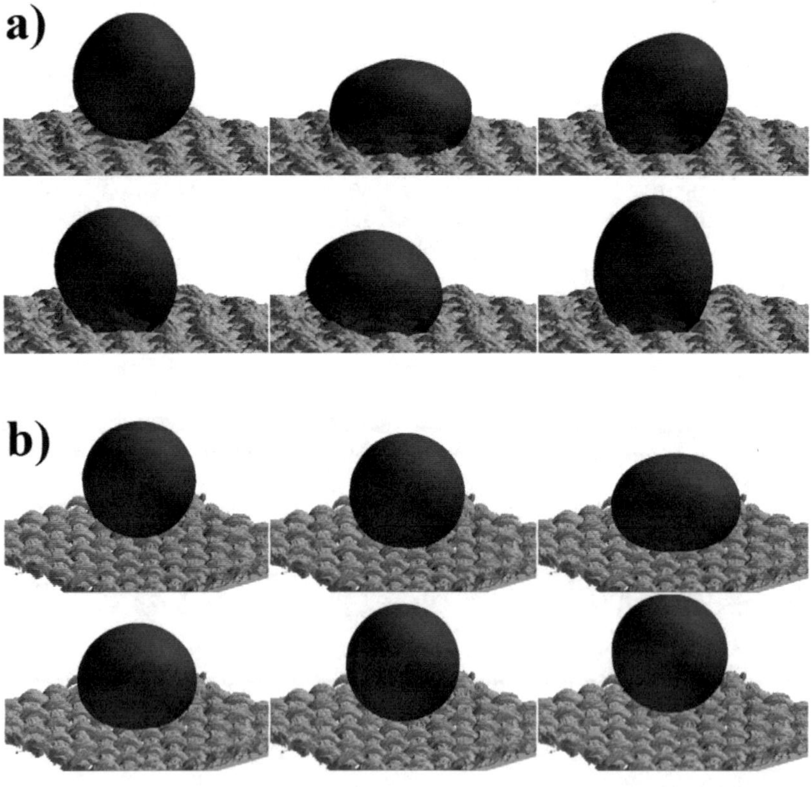

Figure 8. Drop shape during the first instants of the simulated deposition on the two investigated surfaces: GDL-PTFE (a) and GDL-PFA (b). The time step between the images is 2 ms.

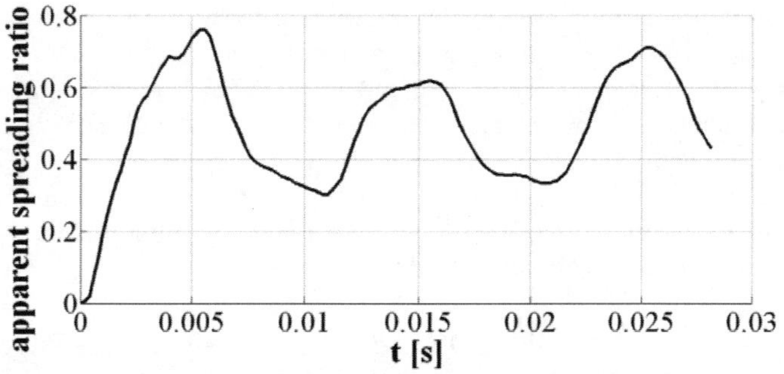

Figure 9. Apparent spreading ratio during simulated drop deposition on the GDL-PTFE.

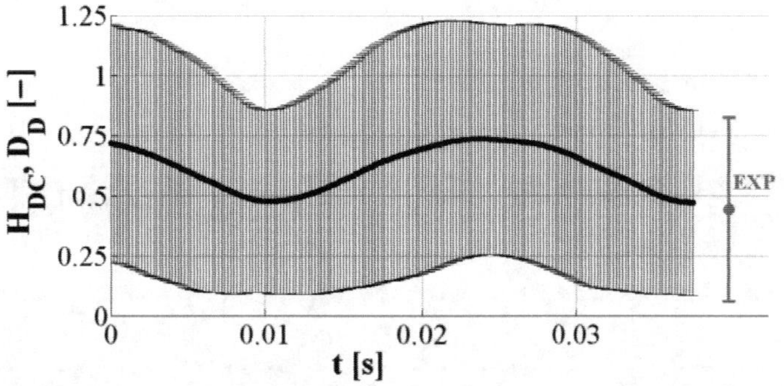

Figure 10. Normalized height (H_{DC}) of the drop center of mass and vertical drop diameter (D_D, shown by means of the error bars) during simulated drop deposition on the GDL-PFA. The equilibrium values for the corresponding experimental drop are represented in red on the right of the chart. (For interpretation of the references to color in this figure caption, the reader is referred to the Web version of this article.)

The drop on such less hydrophobic surface has a much wider wetted area and it remains pinned to the surface after the contact, while on the more hydrophobic surface the drop bounces back, despite the extremely low "impact" velocity. Figure 10 shows the time evolution of the normalized (with respect to the drop diameter before impact) height of the drop center of mass and drop vertical diameter during simulation on GDL-PFA (case 3D). The first was calculated from the reference plane identified by the lowest point of the GDL surface. The drop normalized diameter is shown in the figure by means of the error bars. The equilibrium values of such quantities for the experimental drop on the GDL-PFA substrate are also shown in red in the chart for comparison. The bouncing and deformation of the drop after impact is clearly evidenced and it is qualitatively consistent with direct observation during real drop deposition, even if the bouncing height in the simulations appears (no rigorous quantitative measurement was performed during the experimental part) excessive.

Concerning the final drop shape, the spurious velocities induce distortions and oscillations of the drop surface for a time which is much longer than the experimentally observed one. As a consequence, the condition of static equilibrium for the drops on the surfaces could only be reached in 2D simulations, which are obviously much faster, and thanks to a "trick": once the deposition dynamics of interests were captured, the viscosities of the phases were increased 1000 times to reduce the spurious velocities and dampen the oscillations (as in [24]). In fact, viscosity for Newtonian fluids has no influence on the drop equilibrium shape, which is only related to the capillary constant and curvature radii at the drop apex. Thus the ADSA technique to evaluate the drop shape and contact angle from vertical slices of the simulated drops could be applied. The contact angle in this case is a trivial result:

Journal of Physics: Conference Series **547** (2014) 012028 doi:10.1088/1742-6596/547/1/012028

being one of the boundary condition of the simulation, it is obviously reproduced. On the contrary, analysis of the drop shape is a good validation of the simulation performances. Vertical slices of the simulated volume (Fig. 11, right part) show that on the more hydrophobic surface the drop is in a Cassie-Baxter wetting state, with air pockets entrapped between the drop and the GDL. This is consistent with the experimental evidence from microtomographic scans (Fig. 11, left part). Table 2 summarizes the results of ADSA analysis on the tomographic slices and on the results of the 2D simulation for GDL-PFA, while Fig. 12 provide an example of the fitted theoretical drop contours on the slices (experimental and numerical) shown in Fig. 11.

Table 2. ADSA results for experimental and simulated drops.

	Experimental drop	**Simulated drop (2D)**
Estimated drop volume (real value directly obtained by microtomography: 40.945 μl)	40.960 μl	41.429 μl
Eötvös number	0.919	0.946
Height over maximum diameter ratio	0.709	0.714
Curvature radius at the drop apex	2.60 mm	2.64 mm

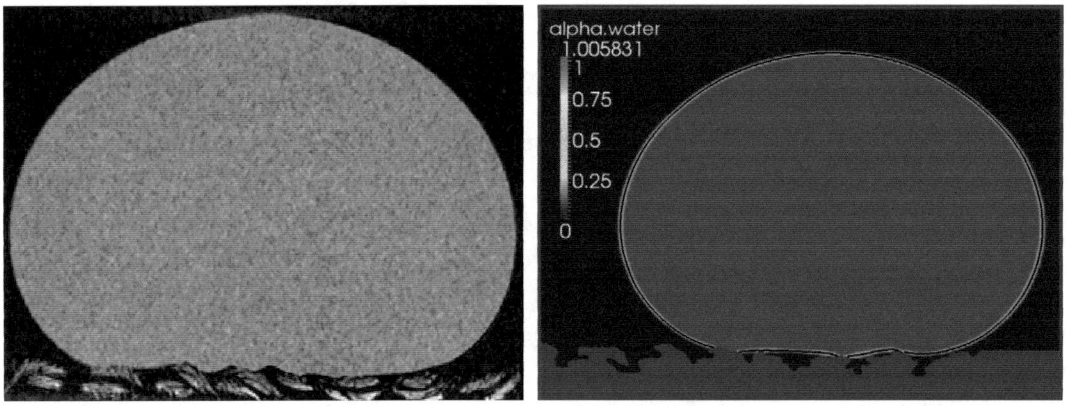

Figure 11. Vertical slices of the domain in correspondence of the drop apex, evidencing the Cassie-Baxter wetting state, for the tomographic (left) and simulated (right, color scale proportional to the water volume fraction) cases on GDL-PFA.

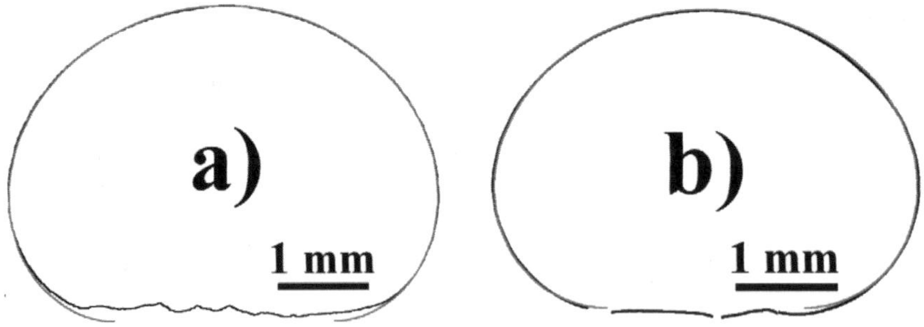

Figure 12. Theoretical drop profile from ADSA analysis (red) superposed to the drop contour (black) extracted from a tomographic slice (a) and 2D simulation (b). Contours have been enlarged for better visualization and the scale factor is slightly different between the two images. (For interpretation of the references to color in this figure caption, the reader is referred to the Web version of this article.)

Journal of Physics: Conference Series **547** (2014) 012028 doi:10.1088/1742-6596/547/1/012028

It is evident how both drop profiles can be satisfactorily fitted with the numerical integration of the Laplace-Young equation, apart from the contact region where the drop pinning on the GDL fibers alters the drop shape with respect to the theoretical one (which is developed for a perfectly smooth surface). The curvature radius at the apex, the height over maximum diameter ratio, the Eötvös number are well reproduced by the simulation.

5. Conclusions

Computer-aided microtomography (microCT) was used to investigate wettability for two gas diffusion layers for fuel cells, both of cloth type but hydrophobized using different fluoropolimers (PTFE and PFA). Such surfaces pose high difficulties for conventional wettability studies. A sessile drop deposed on the GDL-PFA surface was acquired by the microCT and information about the real contact area, contact line and contact angle could be obtained from the tomographic volume. The microCT scan also allows direct visualization of the contact region, confirming that on the highly hydrophobic and morphologically complex GDL structure the drop is in a Cassie-Baxter wetting state. The two GDL surfaces extracted from the tomographic volume were then used to create a realistic mesh for the numerical simulation of drop deposition on such surfaces, using the VOF technique. The complexity of the resulting mesh is a serious issue for the reliability of the simulations, particularly due to spurious velocities and dynamic contact angle modeling. Nevertheless, simulation results were consistent both qualitatively and partially quantitatively with experimental observations.

Acknowledgments

The financial support from the Regione Lombardia (Italy) within the call “Cooperazione Scientifica e Tecnologica Internazionale nelle aree tematiche agroalimentare, energia-ambiente, salute e manifatturiero avanzato” Project ID-MAN11 and the research fund grant to Maurizio Santini “Fondi di Ricerca di Ateneo” and also the prize “5 per 1000” of the University of Bergamo (Italy) are gratefully acknowledged. The research Grant of Stephanie Fest-Santini was financed within “Progetto ITALY® – Azione Giovani in Ricerca anno 2013” of the University of Bergamo (Italy).

References

[1] Barbir F 2005 *PEM fuel cells: theory and practice* (Amsterdam: Elsevier Academic Press)

[2] Wee J, 2007 Applications of proton exchange membrane fuel cell systems *Renew Sustain Energy Rev* **11** 1720–38

[3] Santoro C, Guilizzoni M, Correa Baena J P, Pasaogullari U, Casalegno A, Li B, Babanova S., Artyushkova K and Atanassov P 2014 The effects of carbon electrode surface properties on bacteria attachment and start up time of microbial fuel cells *Carbon* **67** 128–39

[4] Dai W, Wang H, Yuan X-Z, Martin J J, Yang D, Qiao J and Ma J 2009 A review on water balance in the membrane electrode assembly of proton exchange membrane fuel cells *Int J Hydrogen Energy* **34** 9461–78

[5] Latorrata S, Balzarotti R, Gallo Stampino P, Cristiani C, Dotelli G and Guilizzoni M, 2014, Design of properties and performances of innovative gas diffusion media for polymer electrolyte membrane fuel cells, *Progress in Organic Coatings*, in press

[6] Young T 1805 An essay on the cohesion of fluids *Philos Trans R Soc Lond* **95** 65–87

[7] Extrand C W, Moon S I 2010 Contact Angles of Liquid Drops on Super Hydrophobic Surfaces: Understanding the Role of Flattening of Drops by Gravity *Langmuir* **26(22)** 17090–99

Journal of Physics: Conference Series **547** (2014) 012028 doi:10.1088/1742-6596/547/1/012028

[8] del Río O I, Neumann A W 1997 Axisymmetric Drop Shape Analysis: Computational Methods for the Measurement of Interfacial Properties from the Shape and Dimensions of Pendant and Sessile Drops *J. Colloid and Interface Sci* **196(2)** 136–47
[9] Tadmor R 2004 Line Energy and the Relation between Advancing, Receding, and Young Contact Angles *Langmuir* **20** 7659–64
[10] Wenzel R N 1936 Resistance of solid surfaces to wetting by water *Ind. Eng. Chem.* **28** 988–94
[11] Cassie A B D, Baxter S 1944 Wettability of porous surfaces *Trans. Faraday Soc.* **40** 546–51
[12] OpenFOAM®, The open source CFD toolbox, 2013. <http://www.openfoam.com/> (accessed October 2013)
[13] Gopala V R, van Wachem B G M 2008 Volume of fluid methods for immiscible-fluid and free-surface flows *Chemical Engineering Journal* **141** 204–21
[14] Deshpande S, Anumolu L and Trujillo M F 2012 Evaluating the performance of the two-phase flow solver interFoam *Computational Science & Discovery* **5** 014016 1–37
[15] Berberovic E, van Hinsberg N P, Jakirlic S, Roisman I V and Tropea C 2009 Drop impact onto a liquid layer of finite thickness: Dynamics of the cavity evolution *Physical Review E* **79** 036306
[16] Trujillo M F, Lewis S R 2012 Thermal boundary layer analysis corresponding to droplet train impingement *Physics of Fluids* **24** 112102 1–23
[17] Deshpande S, Trujillo M F, Wu X and Chahine G 2012 Computational and experimental characterization of a liquid jet plunging into a quiescent pool at shallow inclination *Int. J. Heat and Fluid Flow* **34** 1–14
[18] Hirt C W, Nichols B D 1981 Volume of Fluid (VOF) Method for the Dynamics of Free Boundaries *J. Computational Physics* **39(1)** 201–225
[19] Brackbill J U, Khote D B and Zemach C 1992 A Continuum Method for Modeling Surface Tension *J. Computational Physics* **100** 335–354
[20] Fest-Santini S, Guilizzoni M, Santini M and Cossali G E 2012 Water drop impact into a deep pool: influence of the liquid pool temperature, Proceedings of the Droplet Impact Phenomena & Spray Investigations (DIPSI) Workshop 2012, Dalmine (Italy), http://aisberg.unibg.it/handle/10446/27138
[21] Raeini A Q, Blunt M J and Bijeljic B 2012 Modelling two-phase flow in porous media at the pore scale using the volume-of-fluid method *J. Computational Physics* **231** 5653–68
[22] Santini M, Guilizzoni M and Fest-Santini S 2013 X-ray computed microtomography for drop shape analysis and contact angle measurement *J. Colloid and Interface Sci.* **409** 204–210
[23] Santini M, Guilizzoni M 2014 3D X-ray micro computed tomography on multiphase drop interfaces: from biomimetic to functional applications *Colloid and Interface Sci. Comm.* **1** 14–17
[24] Guilizzoni M 2011 Drop shape visualization and contact angle measurement on curved surfaces *J. Colloid and Interface Sci.* **364** 230–236

Journal of Physics: Conference Series **547** (2014) 012029 doi:10.1088/1742-6596/547/1/012029

Visual analysis of flow boiling at different gravity levels in 4.0 mm tube

C M Valencia-Castillo[1], G P Celata[2], L Saraceno[2] and G Zummo[2]
[1]University of Guanajuato, Salamanca, México
[2]ENEA, via Anguillarese 301, Rome, Italy

E-mail: giuseppe.zummo@enea.it

Abstract. The aim of the present paper is to describe the results of flow boiling heat transfer at low gravity and compare them with those obtained at earth gravity, evaluating possible differences. The experimental campaigns at low gravity have been performed during the parabolic flight campaign of October-November 2013. The paper will show the analysis of differences between the heat transfer coefficients and vapour bubble parameters at normal and at zero gravity. The results of 4.0 mm tube are presented and discussed. With respect to terrestrial gravity, heat transfer is systematically lower at microgravity in the range of the experimental conditions. Heat transfer differences for the two gravity conditions are related to the different bubble size in each of them. The size of a bubble in flow boiling is affected by the gravity level, being larger at low gravity, unless inertial forces are largely predominant over buoyancy and other forces acting on the bubble itself when detaching from a heated wall. Vapour bubble parameters (bubble diameter, bubble length, width, and nose velocity) have been measured.

1. Introduction

Flow boiling heat transfer (FBHT) is encountered in many engineering fields (energy conversion, environmental applications, food, chemical and other process industries, etc.) and in particular, in space applications. In the coming years, expectations for space-based systems such as communication satellites and manned space-platforms or missions will grow rapidly. Due to the increasing size and capabilities of these systems, their power requirements will also greatly increase. More sophisticated thermal management systems capable of dealing with greater loads will have to be designed. FBHT may offer the solution for increasing heat transfer rates under future, challenging space conditions, at least under certain conditions and/or in certain areas of the system. High performance boiling heat transfer systems, which take advantage of the large latent heat transportation, are therefore important to reduce the size and weight of space platforms and satellites. Nonetheless, FBHT knowledge is also very important for the safe operation of existing single-phase liquid systems that may enter this mode of operation in case of accidental increase of the heat generation rate. An accurate understanding of the off-design situations is therefore essential for properly managing accidental situations.

Among the available platforms for microgravity experiments, such as parabolic flight (gravity level 10^{-2}g, duration 20 s, repeated for about 30 times in a day), sounding rocket (gravity level 10^{-5}g, duration 2-15 min), orbital flight (gravity level 10^{-2}-10^{-5}g, duration unlimited), drop tower (gravity level 10^{-5}g, duration 2-10 s), because of the complexity of an experimental loop for flow boiling heat transfer tests the most widely used microgravity platform is the parabolic flight.

The few data available shows small coherence, maybe due to severe restrictions in the test apparatus specification, strict prescription of experimental conditions and not enough chance to repeat experiments for repeatability, short lasting of 0-g conditions, etc. A complete overview (including pool boiling) can be found in Ohta et al. [1], Ohta [2], Di Marco [3], Kim [4], Ohta [5], and Baldassarri and Marengo[6].

Although the available experimental data are limited, some general conclusion may already be drawn: i) results on heat transfer are contradictory, spanning from increase to decrease with respect to terrestrial gravity, and include no effect of gravity level; ii) the effect of gravity level on heat transfer strongly depends on the flow pattern; therefore, its knowledge is crucial and systematic visualisation tests are required; iii) forced convection flow (inertial effects) plays a fundamental role in microgravity flow boiling heat transfer as, added up to the buoyancy force, it may overcome it; the thresholds beyond which inertial effects are dominant over buoyancy effects have to be carefully determined; iv) a systematic study of flow boiling heat transfer is necessary in order to create a consistent data set for design purposes and to better establish the flow boiling heat transfer knowledge at microgravity.

The aim of the present paper is to describe the results of flow boiling heat transfer at low gravity and compare them with those obtained at earth gravity, evaluating possible differences. The experimental campaigns at low gravity have been performed during the 59th ESA parabolic flight campaign of October-November 2013.

2. Experimental loop

The activity is carried out on the experimental facility called MICROBO (MICROgravity BOiling), specifically designed and built to be boarded on the Airbus A300 ZERO-Gravity managed by Novespace for parabolic flight experiments of ESA, French and German Space Agencies. The loop, schematised in Figure 1, consists of a gear pump (Gmax=500 mL/min), a filter, a Coriolis flow-meter, two electric pre-heaters, the test section (where boiling phenomena occurs), two condensers, a bellows, and a tank for the storage of the process fluid (FC-72).

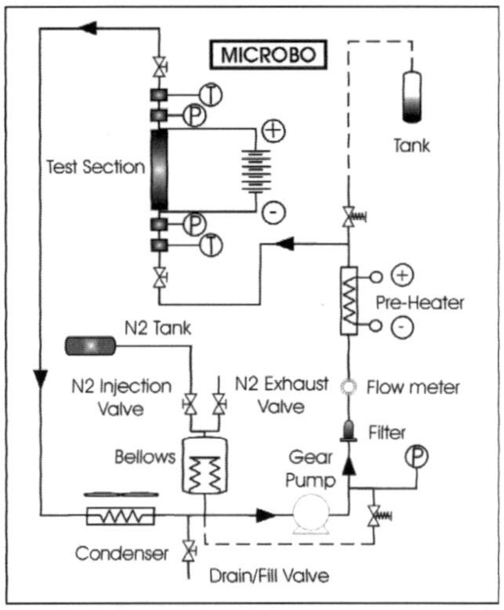

Figure 1. Schematic of the experimental loop.

The experiments were performed on a vertical tubular test section made of a Pyrex tube with a transparent metal layer deposited on the outer surface (Figure 2). Inner tube diameter is 4.0 mm. The metal layer is made of ITO (Indium Tin Oxide) and its thickness is of the order of 100-150 nanometers. The heated length is 155 mm. The test section is heated by Joule effect by an electric supplier. This configuration allows obtaining constant heat flux at the inner surface of the test section.

The metal layer is transparent in order to allow tube heating as well as boiling phenomena visualization. Flow pattern visualisation is performed with a high speed digital video camera and images are analysed with an image processing software. This test section and the visualisation system allow the observation of flow patterns in diabatic flow conditions directly along the boiling channel.
Wall temperatures are measured by a set of 10 thermocouples attached to the outer tube wall. The picture in Figure 2 shows the transparent test section with the wall thermocouples. The inlet section is at the left of the image. Working fluid is FC-72, perfluorohexane C6F14, a fluorinert liquid manufactured by 3M, used in electronic cooling. To reduce the heat losses, the test section is confined in a special box made of ERTACETAL® closed with two transparent plates made of polycarbonate
The test matrix for the experimental campaign presented in the paper are summarised in Table 1.
The experiments were conducted with a constant mass flux G of 173 kg/(m^2·s) and a heat flux q" which stretches from 7.6 kW/m^2 in single phase convection regime up to 65 kW/m^2 in flow boiling regime. The maximum heat flux causes the dryout of the liquid film. In this heat flux range the analysis of video images gives detailed observations of flow patterns and measurements of bubble parameters as dimensions, frequency, and velocity. At the same time, heat transfer is measured through the wall thermocouples and correlated with the flow pattern observations and bubble parameter measurements.

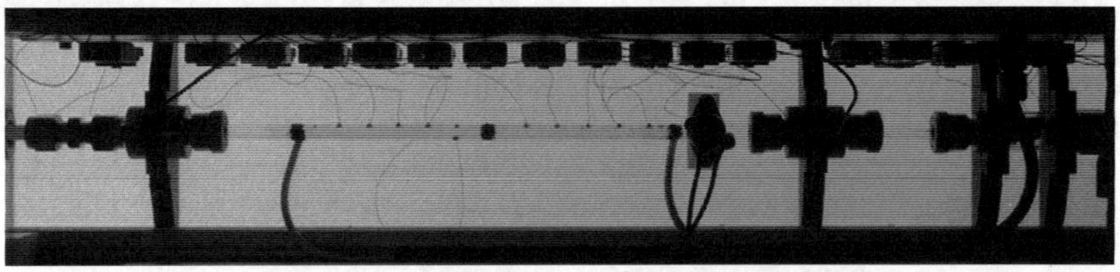

Figure 2. Test section in the container box.

Table 1. Test matrix

D_i [mm]	4
L [mm]	155
ΔT_{sub} [K]	24
P_{in} [bar]	1.8
G [kg/m^2-s]	173
Re_{in}[-]	1472
q" [kW/m^2]	7.6 - 65.9
x_{out} [-]	-0.22 - 0.48

3. Vapour bubble parameters

High frequency video has been registered trough the transparent wall of the tube, at the position of thermocouples 7, 8, 9 and 10, the closest thermocouples to the tube outlet. Bubble video clips are recorded at frequency of 600 Hz. Video analyses are performed with the aim to measure the bubble geometry (width and length), nose velocity, frequency, and distance between two consecutive bubbles (liquid slug length in slug flow).

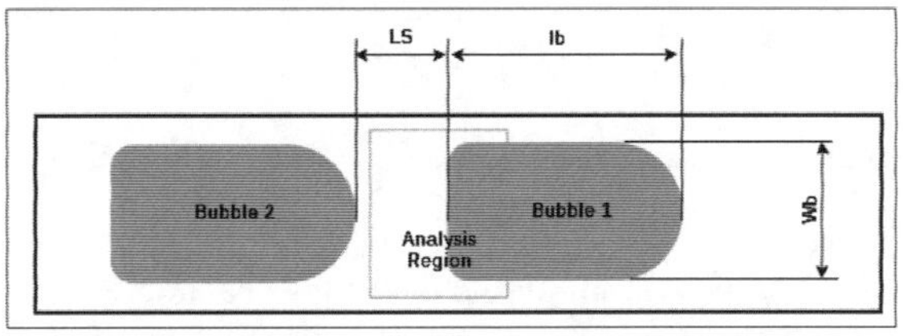

Figure 3. Geometrical parameters of bubbles.

The measured parameters are shown in Figure 3. For each specified parameter, five bubbles have been analysed and their measurements are averaged. Measurements and analysis of video images are performed by the operator with the aid of video analysis software.
For each thermocouple, all bubble measurements are carried out in an analysis zone centred on the thermocouple position for an extension of $D_i/2$ upwardly and $-D_i/2$ downwardly.
When a bubble enter the analysis zone, then the operator starts the bubble measurements during the period of bubble transit in the analysis zone. Nose velocity is the velocity measured in the direction of the channel axis. Bubble length, the distance between the nose and the tail, is measured when the geometrical centre of bubble is inside the analysis zone.

4. Results and discussion

4.1. Flow pattern

Flow patterns were determined as a result of video and images analysis operated by a human operator. The results are shown in Figure 4 for normal gravity and Figure 5 for low gravity tests. The graphs in these figures show local thermodynamic quality (calculated at the thermocouple position along the channel axis) on the horizontal axis, and mass flux on the vertical axis. Flow images recorded with the high speed video camera have been analysed frame by frame in order to study the flow patterns at different heat fluxes and positions along the flow channel. The vapour-liquid structures observed in the images are classified according to the flow pattern classifications in upward two-phase flow, given by McQuillan and Whalley [7]. These flow patterns are shown in Figure 6. The only exception is that in the present article, "plug flow" is called "slug flow". In order to give an extended description of vapour-liquid structures, a number of additional flow patterns have been introduced between these fundamental flow patterns at normal gravity conditions. At low gravity the evolution of flow patterns is simpler and does not need additional flow patterns. As it is well known, this method of analysis is affected by a degree of subjectivity of the operator in the identification of the flow patterns and this causes a certain level of uncertainty in the flow pattern transitions.
The two groups of pictures in Figure 7, show a complete view of the evolution of two-phase flow, from ONB up to annular flow and CHF, for increasing values of heat fluxes (from left to right). The right portion of Figure 7, shows the images recorded at normal gravity, while the left portion shows the images recorded at low gravity.

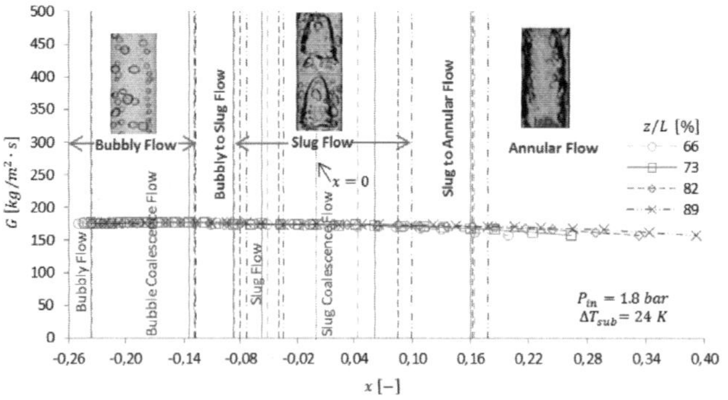

Figure 4. Flow patterns, and their boundaries for 1-g tests

Flow pattern evolution is progressive, i.e., is in evolution as seeing through the tube, being the bubbles larger as they flows toward the outlet. Taking into the account the axial position 10 (z/L=89 %), i.e., the closest position with respect to the outlet, the corresponding flow patterns are: a) and b) bubble

Journal of Physics: Conference Series **547** (2014) 012029 doi:10.1088/1742-6596/547/1/012029

coalescence flow, c) bubbly to slug flow, d) and e) slug coalescence flow, f) slug to annular flow, g) annular flow, and h) annular flow (up to critical heat flux, CHF).
Figure 7 Right, shows the pictures of all flow pattern characteristics for tests at microgravity.
In bubbly flow, the vapour phase is distributed as discrete bubbles in a continuous liquid phase and the bubble diameters are smaller than the diameter of the tube as shown in Figure 7 Right a,b,c.

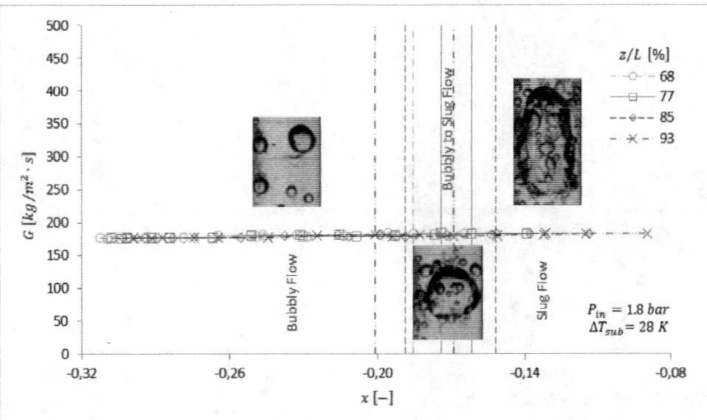

Figure 5. Flow patterns, and their boundaries for 0-g tests

Bubbles at low gravity are larger than those at normal gravity and characterised by lower velocity (due to the absence of buoyancy) and almost linear trajectories and absence of deformation (spherical shape) in comparison to those at 1-g. This leads to a more ordered flow with a reduced relative velocity among the vapour bubbles. In these conditions (ordered flow, linear trajectories, spherical shape of bubble) it is reasonable to expect a reduction of the turbulence level due to the absence of bubble-induced turbulence.

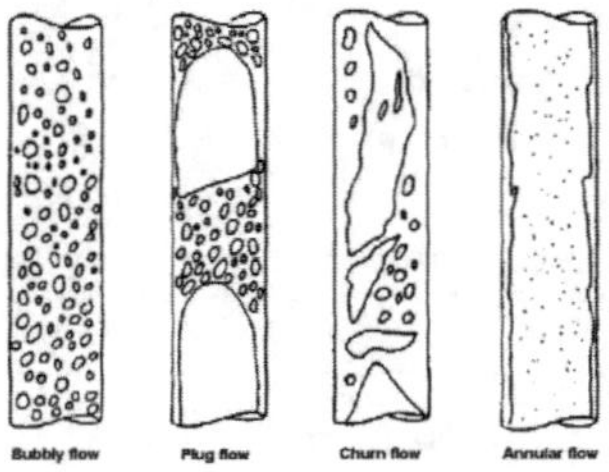

Figure 6. Flow pattern schematization in two-phase flow according to McQuillan and Whalley [7]

An interesting issue for discussion concerns the larger diameter of bubbles at zero-g. It is assumed that two mechanisms control the bubble diameter: detachment diameter and coalescence rate. Detachment diameter is expected larger due to the absence of buoyancy in the balance of forces acting on a bubble during the nucleation phase. In order to have the bubble detachment, the diameter of the bubble must be large enough to generate the drag force sufficient to detach the bubble from the nucleation site.
The other mechanism responsible for the increase of bubble diameter at low gravity could be the higher rate of coalescence. The coalescence of two bubbles is strongly dependant on their approaching velocity (Ribeiro and Mewes [8]). For low relative velocities, coalescence of bubbles occurs; for approaching velocities higher than a given value (critical velocity), bubbles do not coalesce but bounce.
At low gravity conditions, bubbles in bubbly flow were observed to be larger, with spherical shape (without deformation), following linear trajectories and lower velocities. These conditions seem ideal for promoting coalescence of bubbles. In fact, linear trajectories reduce the lateral motion of bubbles.

Therefore, for the same flow conditions low gravity bubbly flow shows higher coalescence rate and larger bubble diameters in comparison to the flow in similar conditions at normal gravity conditions.
In bubbly/slug flow, bubbly flow and slug flow are both present as shown in Figure 7 Right, d,e. This is a transition from pure bubbly to pure slug flow.

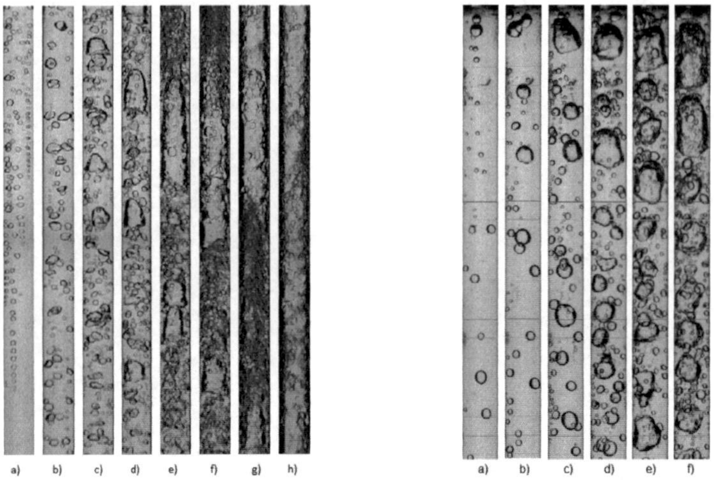

Figure 7. Left: Flow patterns visualisation ***at 1-g*** (z/L from 66 to 89 %) with increasing the heat flux: a) 8.0, b) 16.7, c) 23.5, d) 29.5, e) 39.8, f) 44.9, g) 52.6 and h) 65.9 kW/m^2
Right: Flow patterns visualisation ***at 0-g*** (z/L from 68 to 93 %) with increasing the heat flux: a) 5.1, b) 9.6, c) 15.1, d) 20.6, e) 24.9 and f) 28.8 kW/m^2

In slug flow, vapour bubbles become elongated and reach approximately the same diameter as the tube as shown in Figure 7 Right, f. The nose of the bubble is characterised by a hemispherical shape and the vapour in the bubbles is separated from the tube wall by a thin film of liquid. The liquid flow is contained mostly in the liquid slugs, which separate successive vapour bubbles. The length of the vapour bubbles can vary considerably and is increased for increasing heat fluxes. At low gravity conditions the liquid film is significantly thinner than that at normal gravity. The tail of a Taylor bubble at low gravity is different from that at normal gravity (Figures 7 Left, e,f and Figure 7 Right, f). The shape of the bubble tail is more regular and the motion in the liquid slug is less chaotic. The length of Taylor bubbles at low gravity is significantly longer than those for the same conditions at 1-g.

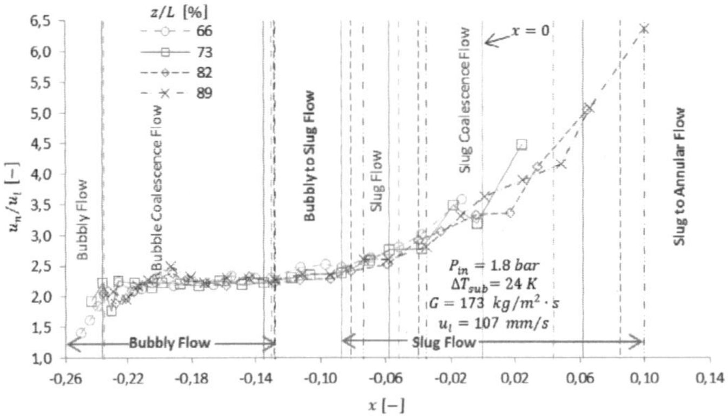

Figure 8. Bubble nose velocity versus local quality at 1-g

4.2. Bubble parameters

Graphs of figures 8 and 9 show the results of bubble nose velocity and length measurements versus local quality at normal gravity. The figures show also the boundaries of the observed flow patterns. Bubble nose is represented (Figure 8) as the ratio of the bubble nose velocity on the liquid velocity (average liquid velocity calculated at the inlet of the test section). The bubble length is represented in Figure 9 as the ratio between the bubble length and the tube diameter (4 mm for the present case). In the bubbly and bubbly coalescence regions, bubbles are almost spherical and the reported length corresponds the bubble diameter.

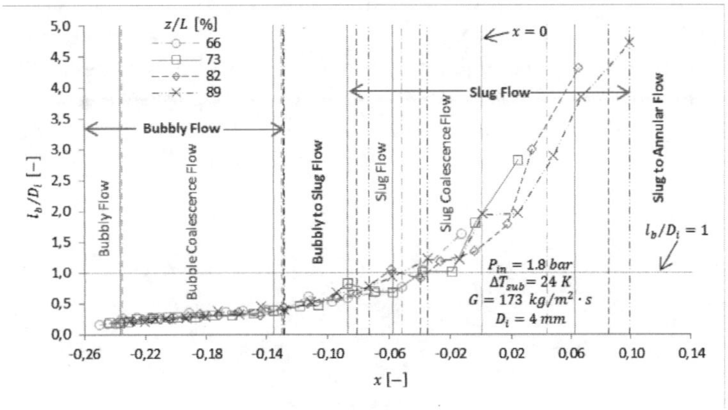

Figure 9. Bubble length versus local quality at 1-g

The figures show the parameters measured at the different thermocouple positions along the tube (Tw7, Tw8, Tw9, and Tw10), characterised by different values of local quality. Figure 8 shows the effect of local quality on vapour bubble nose velocity. Although, the graph shows few results in the bubbly flow region, it exhibits a notable velocity increase for increasing values of thermodynamic quality. In the regions of bubble coalescence flow and bubbly to slug flow, bubble nose velocity is independent of local quality and has an almost flat slope with a value of 2.31.

In the slug flow region, nose velocity shows a dependence on local quality, and in the slug coalescence flow region, this dependency becomes significant. In the latter region, two associated effects cause the increase of the nose velocity slope: the acceleration of Taylor bubbles due to coalescence and liquid film evaporation. The region of slug to annular flow does not show any results because as the Taylor bubbles start the axial coalescence, they increase their length and the flow becomes a semi-annular flow; in these conditions it is almost impossible to distinguish the tail and the nose of two consecutive bubbles due to the reduced distances, high velocities, and more chaotic flow. The maximum measured bubble nose velocity on liquid velocity is 6.36, equivalent to bubble nose velocity of 682 mm/s.

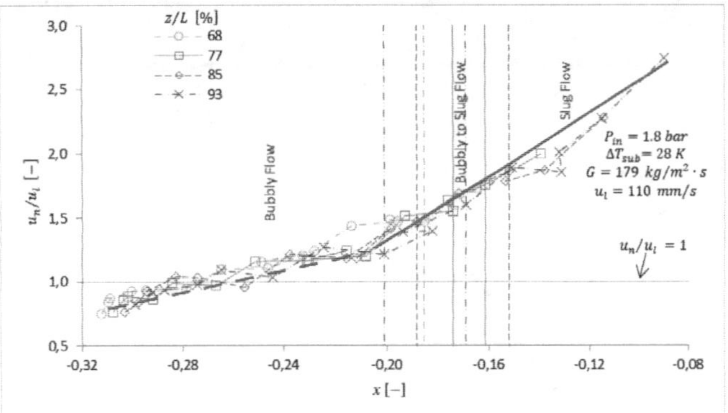

Figure 10. Bubble velocity versus local quality at 0-g

The graph of figure 9 shows the effect of local quality on vapour bubble length at low gravity. In bubbly and bubble coalescence flow, bubble lengths show a linear dependence on local quality, with a small slope. In the bubbly to slug flow and slug flow regions the trend is not linear any longer, and the slope increases rapidly for increasing values of local quality (as shown in the slug coalescence region). Measurements in this region are affected by the same considerations made for the bubble nose velocity. Reduced distances between consecutive bubbles, high velocities, and more chaotic flow, result in more difficulties in parameter measurements from images. The longest measured bubble is 18.84 mm (l_b/D_i = 4.71).

The graph of figure 10 shows the results of vapour bubble nose velocity at microgravity. Nose velocity of bubbles increases linearly with local quality and shows two different regions characterised by different slopes. The first region is the region where bubbly flow occurs (local quality less than -0.20). In this region, the flow pattern evolves from sparse small bubbles, dispersed in the liquid phase to large bubbles (Figures 7 Left, a, b, c). The measured bubble nose velocity ratio is in the range from about 0.75 up to 1.25. In this flow pattern regime, it was expected to find a near zero slip velocity, i.e. u_n/u_l = 1 (due to the absence of buoyancy).

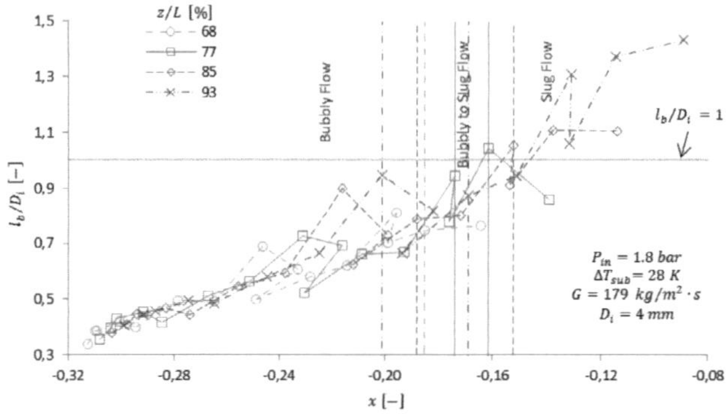

Figure 11. Bubble length versus local quality at 0-g

The explanation for this discrepancy can be caused by the fact that the velocities reported in the graph are the average value of velocities of all bubbles at certain condition of heat flux. This average value does not take into the account the position of the bubble in the tube (bubbles near the wall have a smaller velocity than the liquid, while bubbles in the centre have same mean liquid velocity). Further analysis will be performed on this point in the future.

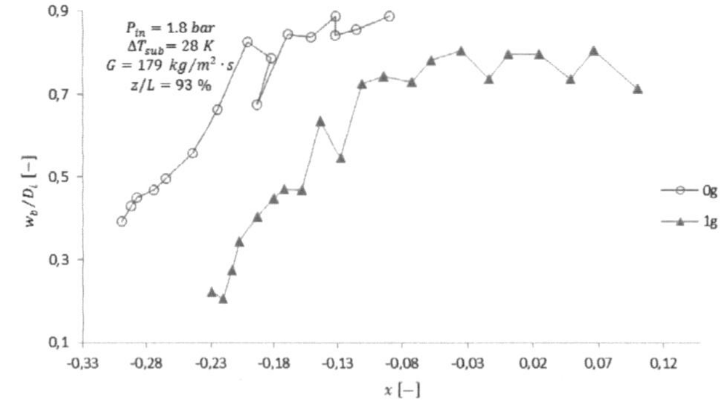

Figure 12. Comparison of bubble width at 1-g and 0-g

The second region (local quality larger than -0.20) is characterised by the transition bubbly to slug flow and by slug flow. Bubble nose velocities are in the range 1.25 – 2.7. In this region the higher velocities could be caused by the acceleration of the vapour phase and axial coalescence of Taylor bubbles.

Figure 11 shows the effect of local quality on vapour bubble length (actually, the ratio l_b/D_i). Bubble lenght in bubbly flow region shows a linear dependence on local quality. Bubbly to slug flow and slug flow regions are characterised by data scatter caused by the effect of coalescence of bubbles. In fact, coalescence of two consecutive Taylor bubbles increases the bubble length with a large amount from one frame to the successive.

Figure 12 shows the comparison of bubble width at two different gravity levels. The effect of gravity is significant: the width of bubbles at low gravity is larger than that at earth gravity. It can be explained by both, the larger bubble diameter at detachment and higher rate of coalescence at low gravity condition as discussed in the previous section. The liquid film thickness for slug region (where the width does not increase any more) at normal gravity condition is significantly larger than in microgravity. At normal gravity the thickness is about 400 μm, double than that at low gravity.

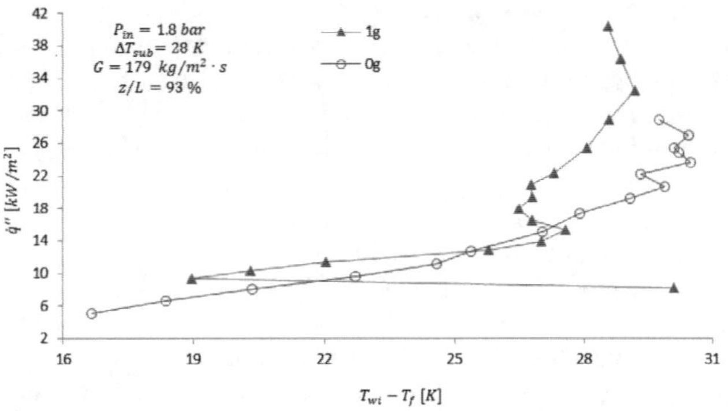

Figure 13. Boiling curves for axial position of Tw10 (z/L=93%) at 1-g and 0-g

4.3. Heat transfer

Figure 13 shows the boiling curves for the two gravity conditions. The low gravity boiling curve shows the results of decreasing heat fluxes starting from the region of slug flow. The curve at normal gravity condition is built with increasing heat fluxes and shows the onset of nucleation. The behaviour of the two curves is quite similar in the nucleate boiling region. Here, the two curves show small differences of curve slope. The transition from bubbly to slug flow is evidenced by a large change of slope. This change in heat transfer mechanism occurs at lower superheating for normal gravity data.

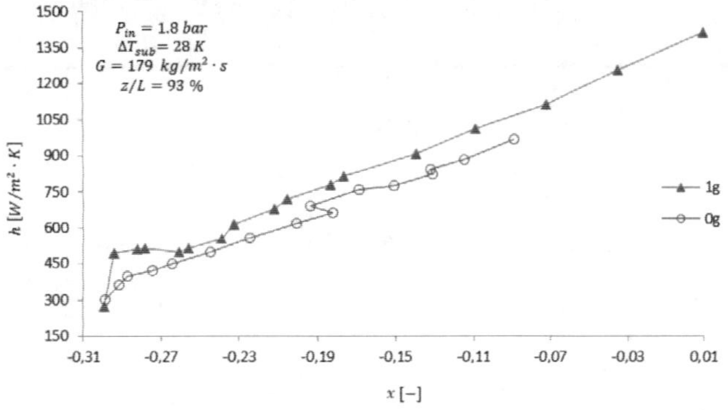

Figure 14. Heat transfer coefficient for axial position of Tw10 (z/L=93%) at 1-g and 0-g

Figure 14 shows the local heat transfer coefficients measured at Tw10 next to the tube outlet. Both gravity heat transfer coefficients are strong functions of heat flux. Low gravity heat transfer is systematically lower than that at normal gravity for all values of local quality and for all the observed flow patterns (bubbly flow and slug flow).
The differences of heat transfer in bubbly flow at the two gravity levels, could be explained with the lower level of turbulence expected at low gravity, as described in the previous section.

5. Conclusions

The paper presents the results of a research dedicated to the study the effect of gravity level on flow patterns and heat transfer in flow boiling inside a small tube of 4.0 mm in diameter. The working fluid is FC-72. The test section and the visualisation system allow observing flow patterns and bubble behaviours in diabatic flow conditions. The observed flow patterns at earth gravity vary from bubbly flow up to the intermittent flow, annular flow and thermal crisis. At low gravity flow patterns are simpler than in earth gravity and vary from bubbly flow, slug flow to annular flow. Experiments provided detailed observation of flow patterns and measurement of vapour bubble dimensions and velocities at different flow patterns. At the same time, heat transfer is measured with the wall thermocouples and correlated with the flow pattern observations and bubble parameter measurements.
At microgravity conditions, bubbles are found larger than those at normal gravity with lower velocity (due to the absence of buoyancy), almost linear trajectories and absence of deformation in comparison to those at 1-g. Larger bubble diameters are controlled by two mechanisms: detachment diameter and coalescence. Liquid film thickness of Taylor bubbles at normal gravity condition is found to be almost double (1.65) than that at low gravity condition.
Heat transfer coefficients and boiling curve are presented and discussed in details. Heat transfer data are correlated with the observed flow patterns and heat transfer mechanisms are highlighted in the different regimes. Heat transfer at low gravity condition is lower than that at normal gravity condition in the tested range. The reduction of heat transfer in nucleate boiling can be explained with the reduction of the turbulence level in bubbly flow at low gravity condition.

6. References

[1] Ohta H, Baba A, and Gabriel K 2002 *Ann. N.Y. Academy of Sciences*, 974, 410-427
[2] Ohta H 2002 *Ann. N.Y. Academy of Sciences* 974 463-480
[3] Di Marco P. 2003 *J. Japanese Society of Microgravity Applications* 20 252-263
[4] J.H. Kim 2003 *J. Japanese Society of Microgravity Applications* 20 264-271
[5] Ohta H *J. Japanese Society of Microgravity Applications* 20 272-285
[6] Baldassarri C, Marengo M 2012 *Progress in Energy and Combustion Science*
[7] McQuillan K W and Whalley P B 1985 *Int. J. Multiphase Flow* 11 161–175
[8] Ribeiro Jr C P and Mewes D 2006 *Chemical Engineering Science* 61 5704 – 5716

Acknowledgments

Authors wish to thank the European Space Agency for the financial assistance and for the opportunity to make microgravity experiments during parabolic flights campaigns realized at Bordeaux, France. A special thanks is due to people of Novespace for their matchless collaboration during the parabolic flights.
Authors wish to thank the Mexican CONACYT for its financial support.

Numerical evaluation of gas core length in free surface vortices

L Cristofano, M Nobili and G Caruso

"Sapienza" University of Rome – Dep. of Astronautic, Electrical and Energy Engineering (DIAEE) – Corso Vittorio Emanuele II, 244 – 00186 Rome, IT

Email: luca.cristofano@uniroma1.it, matteo.nobili@uniroma1.it, gianfranco.caruso@uniroma1.it

Abstract. The formation and evolution of free surface vortices represent an important topic in many hydraulic intakes, since strong whirlpools introduce swirl flow at the intake, and could cause entrainment of floating matters and gas. In particular, gas entrainment phenomena are an important safety issue for Sodium cooled Fast Reactors, because the introduction of gas bubbles within the core causes dangerous reactivity fluctuation. In this paper, a numerical evaluation of the gas core length in free surface vortices is presented, according to two different approaches. In the first one, a prediction method, developed by the Japanese researcher Sakai and his team, has been applied. This method is based on the Burgers vortex model, and it is able to estimate the gas core length of a free surface vortex starting from two parameters calculated with single-phase CFD simulations. The two parameters are the circulation and the downward velocity gradient. The other approach consists in performing a two-phase CFD simulation of a free surface vortex, in order to numerically reproduce the gas-liquid interface deformation.. Mapped convergent mesh is used to reduce numerical error and a VOF (Volume Of Fluid) method was selected to track the gas-liquid interface. Two different turbulence models have been tested and analyzed. Experimental measurements of free surface vortices gas core length have been executed, using optical methods, and numerical results have been compared with experimental measurements. The computational domain and the boundary conditions of the CFD simulations were set consistently with the experimental test conditions.

1. Introduction

The deformation of gas-liquid interface (e.g. the free surface vortex formation) has been studied in several scientific and industrial fields during the last fifty years. Nevertheless, the interface transient behaviors, combined with the swirling motion in the vortex core, make the free surface vortex extremely difficult to understand. Strong whirlpool can cause gas entrainment (GE) phenomena, which consist in the entrapment of gas bubbles under the gas-liquid interface. If the downward flow is strong enough, the entrapped bubbles will be transported within the liquid and can affect equipment as well as plant operations. In sodium-cooled fast reactors (SFRs), entrained bubbles within the reactor core cause a positive reactivity insertion, which jeopardizes the reactor safety operation. Consequently, in the last years, also nuclear industry faces the problem of studying and predicting free surface vortex formation and evolution. Several experimental studies on GE occurrence have been carried out, and empirical correlations have been obtained [1]-[3]; however, experimental tests are always expensive, and the obtained results can be strongly affected by geometry and/or experimental conditions. For these reasons, different researchers started to study GE phenomena performing numerical simulations with CFD codes. The definition of a reliable method, for the numerical prediction of free surface vortex formation and evolution, could be very useful since it allows to easily investigate several computational domains, boundary conditions and operating fluids (and their physical properties). The main drawback of this approach is the needed for accurate modeling of both the gas liquid interface curvature and the vortical flow near the deformed free surface.

Different researchers have tried to exploit CFD codes capabilities for the simulation of GE phenomena, testing different numerical models and schemes. Satpathy et al. [4] applied the VOF model to a 2D idealized slab of a FBR hot pool in order to obtain the threshold sodium velocity, at free surface, to avoid gas entrainment; afterwards, they applied this threshold value to the results from a 3D single-phase simulation, in a 90° symmetric sector of the hot-pool, to verify if GE onset conditions are reached. Chen et al. [5] performed simulations with standard k-ε model and RNG k-ε model (both coupled with VOF); the comparison of numerical results showed that standard k-ε model is not suitable for surface vortex simulations since it does not allow to predict turbulent kinetic energy, turbulent dissipation rate and turbulent viscosity distributions in vortex zone. In two different papers Kuljarni et al [6] and Durve et al. [7] numerically investigated shear type entrainment in surface aeration systems, applying Large Eddy Simulation (LES), coupled with VOF, and showed that shear type GE can be suitably simulated with this model. Merzari et al. [8] analyzed the benchmark case of Moriya [9] using LES and Detached Eddy Simulation (DES) turbulent models for CFD simulations (VOF has been used for two-phase flow calculations). The conclusions were that single phase simulations can be used for identifying the velocity field far from the free surface, while in the vortex region LES simulations, with an appropriate grid, have to be preferred. Škerlavaj et al [11] tested the Monji et al. [10] benchmark case with single-phase CFD simulations, concluding that Scale Adaptive Simulation with curvature correction (SAS-CC) is the best RANS turbulence model since it is in good agreement with laminar and LES simulations. A GE prediction method, based on CFD simulations and on Burgers' analytical model [12], has been proposed by Sakai et al. [13] and it is used to define design criteria for free surface GE in a fast breeder reactor. Recently, Ito et al. [14] developed a high-precision numerical simulation algorithm, for gas-liquid two-phase flows, able to simulate GE phenomena in large-sized Japanese Sodium-cooled Fast Reactor (JSFR). They showed that the algorithm is able to foresee the evolution of the vortex gas core also in complex geometries (e.g. the JSFR tank).

In this paper, the vortex formation transient has been analyzed and several numerical methods, aimed at evaluating the free surface vortex gas core length (L_{gc}), have been tested. A single-phase flow transient simulation has been performed and the Sakai's CFD-based prediction methodology has been applied. With the aim of numerically reproducing the gas liquid interface deformation, two-phase flow transient simulations have been carried out with the same computational domain and boundary conditions. Numerical results obtained with the different methodologies have been compared with experimental measurements.

2. Experimental setup

Onset of GE due to free surface vortex has been experimentally investigated in a test facility (Gas Entrainment Test Section-GETS) that does not provide any imposed rotation to the flow and the results have been presented in previous papers [1]-[2]. In the present study the GETS tank was modified introducing two baffles, which give a tangential inlet to the fluid on both side (see figure 1 and figure 2-a).

The reference vortex formation transient considered in the present paper and reproduced in the modified GETS facility is characterized by cold water as operating fluid, mass flow rate of 0.1 kg/s, water level at 0.05 m and a drain hole diameter of 0.026 m. The initial conditions are: stagnant water in the tank and no inlet mass flow rate; the pump is turned on at beginning of the analysis. Digital images of the occurring vortex (figure 5) have been acquired every 5 s during the transient in order to measure the vortex gas core length (L_{gc}).

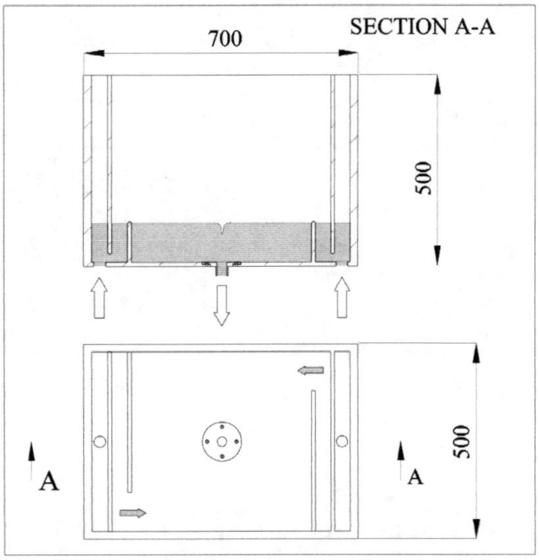

Figure 1. Vertical section and plant of the modified GETS tank.

3. CFD modeling

Three different simulations of the vortex formation transient have been simulated with ANSYS FLUENT® v.15, starting from stagnant conditions in the tank.

A first evaluation of the L_{gc} has been carried out by a single-phase simulation applying the Sakai's prediction method [12] (described in subsection 3.4). The circulation (Γ) and the downward velocity gradient (α) are evaluated from the CFD results and then they are combined with the Burgers' analytical model.

Two other numerical simulations have been carried out with incompressible two-phase flow and constant surface tension. Both of them use the VOF model to describe the interface evolution and a geometric Piecewise Linear Interface Construction (PLIC) method for interpolating the interface. Each of them uses a different turbulence model with the aim of identifying the influence of turbulence models on vortex gas core length evaluation.

Water at 20° C has been used as operating fluid.

The same time step (0.001 s) has been set for all CFD simulations and the Courant number has been kept around 0.5. Using the VOF explicit scheme, time step size near the interface was internally calculated on the basis of a user-defined maximum Courant number: .

$$Courant\ number = \frac{\Delta t}{\Delta x_{cell} / v_{fluid}} \tag{1}$$

In order to limit the computed sub time step to a maximum value equal to a quarter of the minimum transit time, a value of 0.25 has been selected as limit for Courant number near the interface (see [15]).

3.1. Computational domain and boundary conditions

The computational model adopted for single-phase flow simulation is shown in figure 2; it represents the GETS tank geometry and it is composed by a parallelepiped, joined to a cylinder in the bottom center, and connected to two smaller volumes (for tangential inlets) at two opposite sides. The larger volume has a 460 x 460 mm squared base and a height equal to the water level (i.e. 50 mm in the present case); the two smaller inlet zones are 100 mm wide and 20 mm high. The outlet connection, which has an inner diameter of 26 mm, is 30 mm long to reduce possible boundary effects on the CFD results.

Tank walls have been treated as no-slip boundary condition and a mass flow rate equal to 0.1 kg/s has been imposed to inlet and outlet boundaries. Top surface of the tank has been treated as a free-slip wall in single-phase flow simulations.

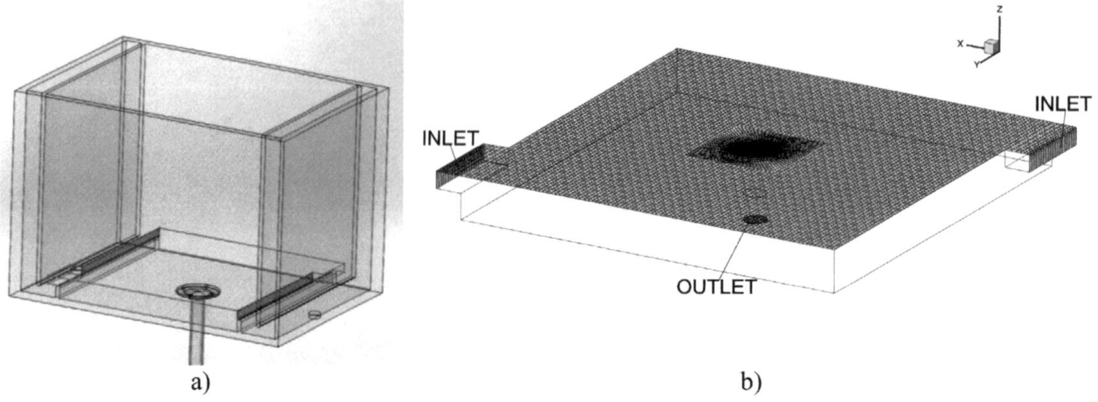

Figure 2. a) 3D view of the GETS tank; b) Computational domain.

The two-phase-flow computational model is slightly different, since it is 20 mm higher in order to consider a volume of the air above the free surface. The upper boundary of the simulation, which is no longer the gas-liquid interface, has been set to a wall with no-slip; moreover, a pressure inlet condition has been tested for this boundary, but errors in interface reconstruction have been detected.

3.2. Turbulence models and discretization scheme

As proposed by Sakai et al. [12], no turbulent model has been used in single-phase flow calculation in order to reduce the dependence of numerical results on the selected model.

The two-phase flow simulations have been tested by using a Scale Adaptive Simulation (SAS) model. The SAS model has been selected since it results the most suitable way to simulate free surface vortices, as per Škerlavaj et al. [11]. The SAS model is an improved URANS (Unsteady RANS) formulation based on the von Karman length-scale into the turbulence scale equation; this enables SAS model to dynamically adjust to the already resolved scales and allows the development of a turbulent spectrum in the detached regions. Therefore, the SAS model provides standard RANS capabilities for stable flow regions and LES-like behavior in unsteady regions.

PISO algorithm was used for pressure velocity coupling, while momentum equation was discretized using QUICK scheme. Pressure term was discretized using PRESTO scheme and volume fraction equation, in two-phase flow simulations, was discretized using the Geo- reconstruct scheme. Convergence criteria (Scaled Residuals) were set at 10^{-5} for all the equations, after a sensitivity analysis reported below. All simulations were carried out on i7 CPUs @2.93 GHz and 8 Gb RAM.

3.3. Grids

Structured mesh has been used in the present calculations; two different grids (Mesh A and Mesh B) were tested, at the beginning of the present analysis, in order to identify the most suitable one for surface vortex simulations. The two grids differ only near the drain region (see figure 3). The horizontal mesh size is between 5 mm in the outer region and 0.5 mm in the center; the axial mesh size is 2 mm except near the free surface where a refined mesh has been used (up to 1 mm).

A convergence criteria sensitivity analysis has been also carried out on two scaled residuals levels; the standard convergence criterion (10^{-3} for scaled residuals) results to be not suitable and a convergence criterion of 10^{-5} for scaled residuals has been selected. In table **1** the results of the sensitive analysis, for grid type and convergence criteria, are reported. Mesh B with scaled residuals limit equal to 10^{-5}, which give the best results, is used as reference configuration in the present study.

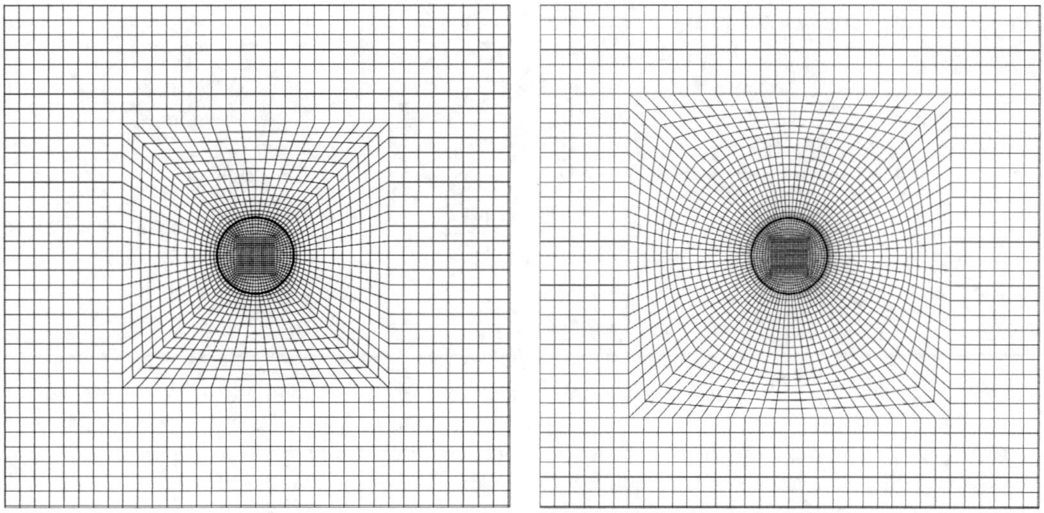

a) b)

Figure 3. Mesh types: a) Mesh A, b) Mesh B.

Table 1. Tangential velocity peak at $h = 35$ mm.

Mesh type	Convergence criteria (Scaled Residuals)	Peak velocity CFD/experimental
A	10^{-3}	0.23
A	10^{-5}	0.53
B	10^{-3}	0.19
B	10^{-5}	1.18

3.4. Sakai's prediction method

Sakai et al. [13] proposed a gas core length evaluation methodology combined with CFD calculations. This method foresees a transient numerical simulation of vortical flows without interfacial deformations to reduce the computational cost (the interfaces are modeled as free-slip walls in the transient numerical simulation), then, the Burgers analytical model is applied to the CFD results to determine the strengths of the vortical flow.

Burgers' model was derived as a strict solution of Navier-Stokes equations and the velocity components are:

$$\begin{cases} u_r = -\dfrac{1}{2}\alpha r \\ u_\vartheta = \dfrac{\Gamma_\infty}{2\pi r}\left[1 - e^{-\frac{r^2}{r_0^2}}\right] \\ u_z = \alpha(z - h_\infty) \end{cases} \tag{2}$$

where r, ϑ e z represent radial, tangential and axial direction respectively, α is the downward velocity gradient, r_0 is the vortex radius, h_∞ is the submergence depth in the tank and Γ_∞ is the circulation of the vortical flow.

In the theoretical model, the relation between α and r_0 is:

$$r_0 = 2\sqrt{\frac{\nu}{\alpha}} \tag{3}$$

Applying a momentum balance in radial and axial directions and neglecting the advection terms in the Navier-Stokes equation, Sakai proposed to calculate the gas core length L_{gc} as:

$$L_{gc} = \frac{\log 2\alpha}{4g\nu}\left(\frac{\Gamma_\infty}{2\pi}\right)^2 \tag{4}$$

while the vortex circulation (Γ_∞) is calculated as:

$$\Gamma_\infty = \int_C \vec{u}\,\vec{ds} \tag{5}$$

In the first step, Γ_∞ is calculated assuming the isoline of the second invariant of the velocity deformation tensor with the value of zero as the initial integrating edge C. Then, the radius of the outer edge C is expanded gradually up to a value twice larger than the initial radius, and the maximum value of the so calculated circulations is selected as the circulation Γ_∞ of the vortical flow.

The downward velocity gradient α is calculated on the initial outer edge C as:

$$\alpha = \frac{\int_C \vec{u}\cdot\vec{n_C}\,ds}{A} \tag{6}$$

where $\vec{n_C}$ is the unit vector normal to C, ds is the local length of the outer edge and A is the area of the inner region. Introducing these two parameters into (4), the L_{gc} is calculated on the Burgers' theory.

4. Results and discussion

The gas core length L_{gc} evaluated through CFD simulations have been compared with experimental measurements; figure 4 shows the time evolution of the numerical and experimental results in the first 200 seconds of the transient. To calculate L_{gc} values of two-phase simulations, the height of the lower point of the iso-surface with a value of water volume fraction equal to 0.5 has been subtracted to the water level in the tank (0.05 m).

Experimentally, from 0 to about 30 seconds L_{gc} is equal to zero but the vortex begins to form. Then, the L_{gc} appears and it starts to increase almost linearly up to about 140 seconds. During the last part of the transient, L_{gc} presents an unsteady behavior with evident fluctuations; however, it seems that the average value of L_{gc} stabilizes around 30 mm. Images are acquired by a high-speed camera with 1024 × 1024 pixel resolution; each pixel corresponds to $8\cdot10^{-5}$ m. The absolute error of experimental data is the same for all L_{gc} values and it is equal to about 0.0015 m. The relative errors are, therefore, between 3% and 40% (greater uncertainties refer to L_{gc} values less than 0.005 m).

During the first seconds of the transient, when the vortex gas core has not yet formed, CFD predictions and experimental data are in good agreement; later, after 40 seconds, the three numerical methods foresee different trends.

It is evident from figure 4 that Sakai's method fails completely as it predicts an increase of L_{gc} up to too large values; for this reason, the single-phase simulation has been stopped after 100 seconds. Probably, the reason of this failure is that, neglecting the free surface deformation (modeling the interface as a flat free slip wall), the resulted velocity fields, and therefore circulation Γ and downward velocity gradient α, are not accurately calculated. In particular, the imposed flat interface affects the velocity component in z-direction near the free surface (and therefore the parameter α); in fact, since the velocity component perpendicular to a *wall-type* boundary by definition must be zero at the boundary, and since no interface deformation is enabled (because of the flat free slip wall condition), the axial velocity assumes small values over the whole free surface. In a real vortex core, instead, the direction perpendicular to the interface no longer coincides with *z*-direction when interface is deformed and axial velocity assumes significant values at the free surface.

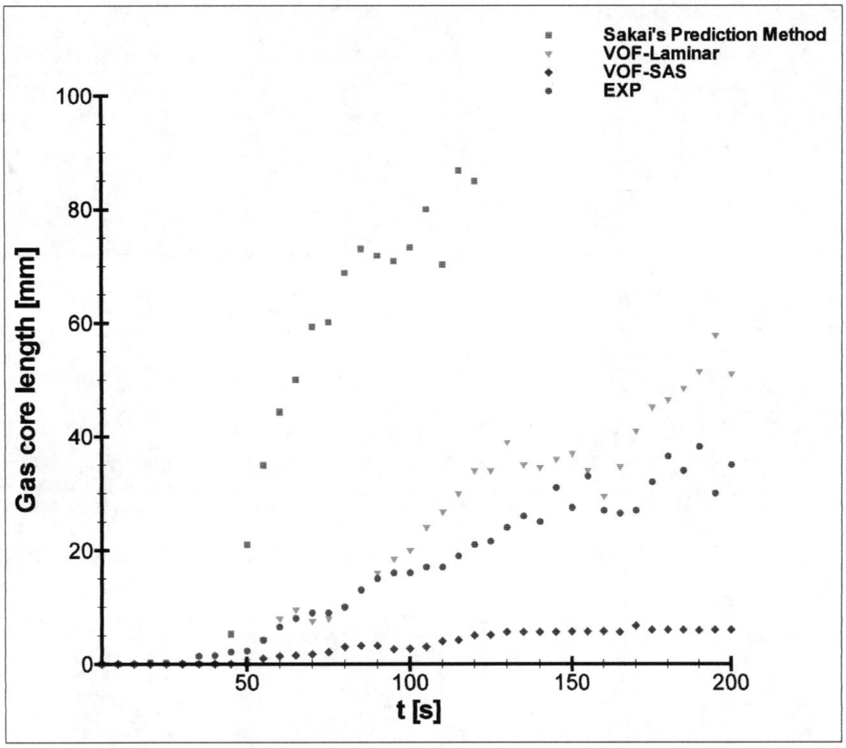

Figure 4. Experimental and numerical calculated L_{gc} during the transient.

Concerning L_{gc} values derived from the two-phase flow simulation with Laminar model agree very well with experimental data in the first 90 seconds of the transient; then, larger values respect to experimental L_{gc} are predicted. At the end of this numerical transient, values of about 50 mm (equal to the water level in the tank) are obtained. This discrepancy can probably be explained by an inaccurate modelling of turbulent phenomena by the Laminar model. In all kind of vortices, tangential velocity increases as the radius decreases (for the conservation of angular moment); in this zone turbulence' effects, like an increased turbulent diffusion, can be significant. Laminar model cannot reproduce this effect, and it results in a lower diffusion of vorticity in the outer zone, and a higher vortex intensity.

Two-phase flow simulation with SAS model, instead, underestimates the L_{gc} during the whole transient (except for the first 30 seconds where L_{gc} is null); it seems that a stationary condition is reached at the end of the transient with a value of L_{gc} equal to about 5 mm. Probably it can be caused by an excessive simplification of the inlet boundary conditions. The GETS tank, in fact, is characterized by a particular shape of the two inlet zones that reduce the free surface perturbations caused by the vertical inlet flow (see figure 1). This shape forces the flow to change several times its direction before reaching the tangential inlet, and provides an initial vorticity to the inlet fluid. In CFD simulation, instead, the flow is uniform and it is directed orthogonally to the inlet boundary, with zero inlet vorticity, resulting probably in lower L_{gc} values.

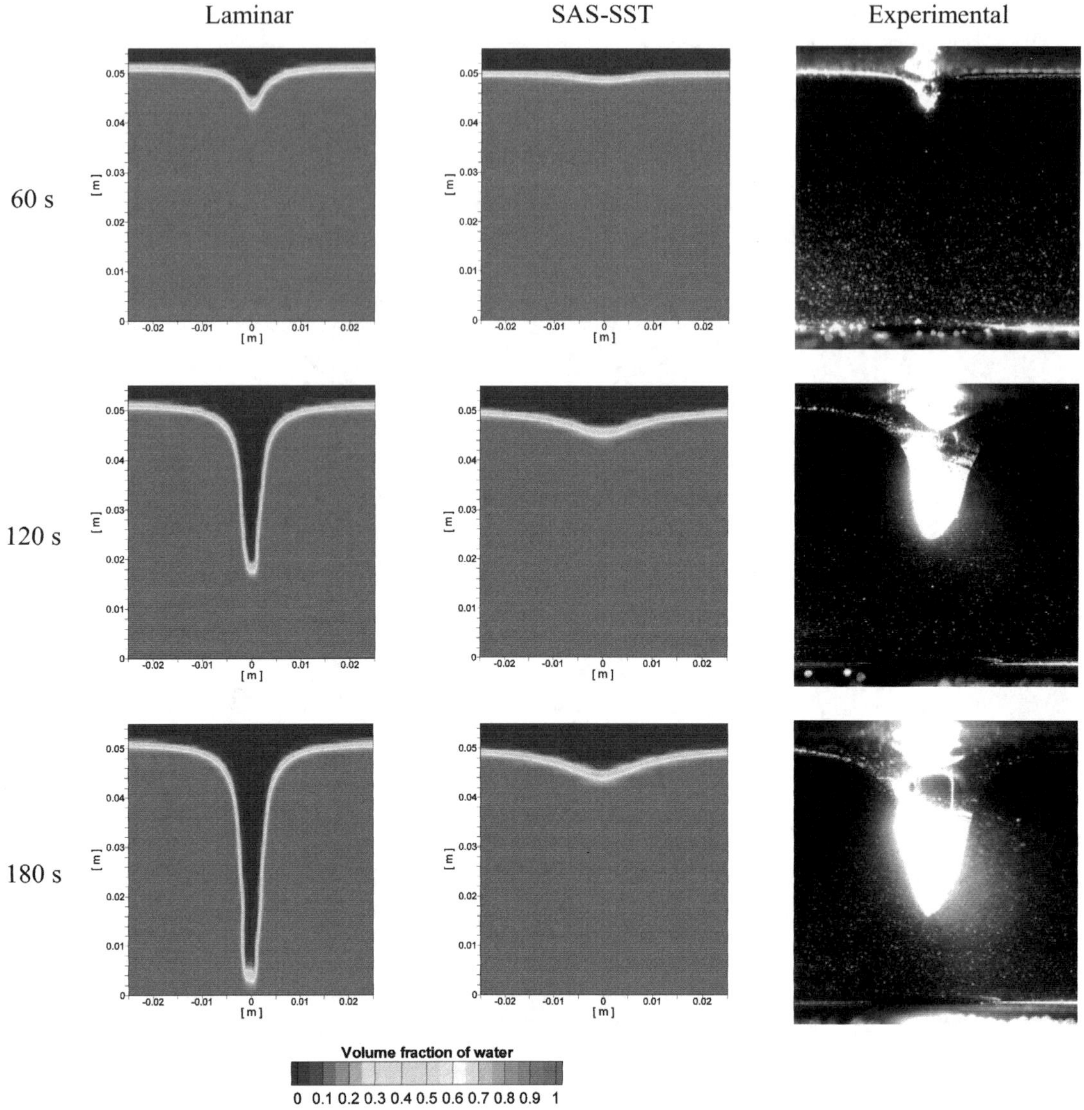

Figure 5. Comparison of experimental and numerical vortex L_{gc} for different phases of the transient.

In figure 5 the volume fraction maps (for the two-phase flow simulations) and the corresponding digital images from the experiment are shown, related to three different times: 60 s, 120 s and 180 s. It is clearly visible that the VOF simulation, with Laminar model, overestimates the L_{gc} during the second half of the transient; moreover it can be noted that the vortex gas core is thinner than the experimental case. As said previously, this is probably due to the inaccurate modeling of turbulent phenomena by Laminar model that results in a lower diffusion of vorticity from the vortex center to the outer zone. VOF simulation with SAS model seems to predict a larger gas core but it does not allow to obtain an acceptable evaluation of the vortex depth.

Figure 6 shows a details of the volume fraction map at 120 s for VOF simulation with Laminar model; it can be seen that the color-shift at the interface is confined in a very thin zone when the interface is almost planar, while it is a bit larger when the curvature of the free surface is significant. However, the color-shift region is everywhere confined within few millimeters.

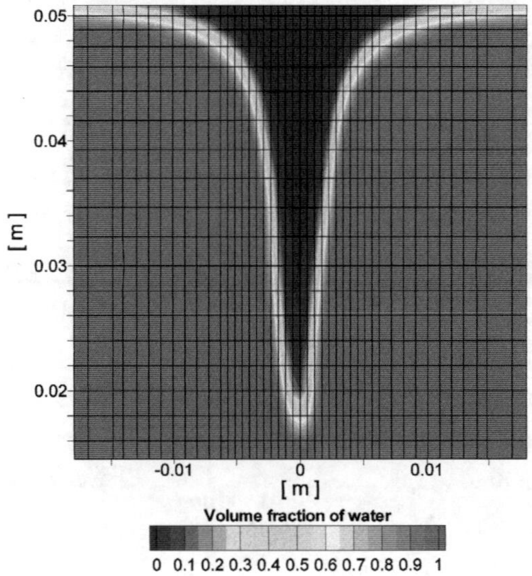

Figure 6. Detail of the volume fraction map at 120 s for VOF simulation with Laminar model with computational mesh.

5. Conclusions

A free surface vortex formation transient has been analyzed experimentally and numerical simulations to evaluate the gas core length L_{gc} have been carried out.

The experimental results show that the vortex gas core formation starts at about 35 seconds, then increases its length almost linearly; after about 140 seconds the L_{gc} has an unsteady behavior with an average value of about 30 mm.

Numerical results, obtained with different methods, predict well the first part of the transient, but then show different trends. Sakai's prediction method provides L_{gc} values completely in disagreement with experimental data; probably because of an incorrect evaluation of the velocity field (and hence of circulation Γ and downward velocity gradient α) near the free surface. Concerning two-phase flow simulations, Laminar model, coupled with VOF method, provides good results during the first 100 seconds, later it overestimates the gas core length L_{gc}. Moreover, the predicted vortex is thinner than the experimental one. SAS model underestimates the vortex depth probably because the inlet boundary conditions are not correctly imposed in term of vorticity.

Future studies on Laminar and SAS models are needed to establish which turbulent model is the most suitable to numerically simulate free surface vortices. Moreover, in order to improve these models, it is definitely required to provide more accurate boundary conditions in term of inlet vorticity; therefore it will be necessary to avoid any simplification of the computational domain, reproducing the exact geometry of GETS tank. Further experimental data, available in the next few months, will be very useful for a better assessment of the turbulence models capability.

Furthermore, it could be interesting to evaluate the capabilities of Large Eddy Simulation (LES) models in the analysis of free surface vortices. Although LES models need larger computational costs, they provide an improved modeling of turbulent phenomena that could lead to more accurate results.

List of symbols

A	area of the inner region, m^2
C	initial outer edge of vortex
g	gravitational acceleration, m s^{-2}
h_∞	submerged depth, m
k	turbulent kinetic energy, m^2 s^{-2}
L_{gc}	gas core length, m
n_C	unit vector normal to C
r	radial coordinate, m
r_0	vortex core radius, m
u	velocity, m s^{-1}
u_r	radial velocity, m s^{-1}
u_z	axial velocity, m s^{-1}
u_ϑ	tangential velocity, m s^{-1}
z	axial coordinate, m
α	coefficient related to downward velocity, s^{-1}
Γ	circulation, m^2 s^{-1}
Γ_∞	circulation of the vertical flow, m^2 s^{-1}
Δt	time step, s
Δx_{cell}	cell dimension, m
ε	turbulent dissipation, m^2 s^{-3}
ν	kinematic viscosity, m^2 s^{-1}

LES	Large Eddy Simulation
DES	Detached Eddy Simulation
SAS	Scale Adaptive Simulation
RANS	Reynolds Averaged Navier-Stokes
JSFR	Japanese Sodium-cooled Fast Reactor
GETS	Gas Entrainment Test Section
VOF	Volume of Fluid
GE	Gas Entrainment
SFR	Sodium-cooled Fast Reactor
FBR	Fast Breeder Reactor
RNG	Re-Normalisation Group
URANS	Unsteady Reynolds Navier Stokes
PRESTO	Pressure Staggering Option
QUICK	Quadratic Upstream Interpolation for Convective Kinematic

References

[1] Cristofano L, Nobili M, Caruso G 2014 *Exp. Therm. Fluid Sci.* **52** 221-229
[2] Caruso G, Cristofano L, Nobili M, Vitale Di Maio D 2014 *J. Phys. Conf. Ser.* **501**
[3] Baum M R and Cook M E 1975 Gas entrainment at the free surface of a liquid: entrainment inception at a vortex with an unstable gas core, *Nucl. Eng. Des.* **32-2** 239-245
[4] Satpathy K, Velusamy K, Chellapandi P 2011 *Energy Procedia* **7** 333-339
[5] Chen Y, Wu C, Wang B, Du M 2012 *Procedia Engineering* **28** 55-60
[6] Kulkarni A L, Patwardhan A W 2013 *Chem. Eng. Res. Des.* 025 (in press) http://dx.doi.org/10.1016/j.cherd.2013.10
[7] Durve A P, Patwardhan A W 2012 *Chem. Eng. Sci.* **73** 140-150
[8] Merzari E, Ninokata H, Wang S, Baglietto E 2009 *Nucl Technol* **165** 313-320
[9] Moriya S. 1993 *Report U93004, CRIEPI Abiko Research Laboratory* (in Japanese)
[10] Monji H, Shinozaki T, Kamide H and Sakai T 2009 *J. Eng. Gas Turb. Power* **132(1)**
[11] Škerlavaj A, Lipej A, Ravnik J, Škerget L 2010 *IOP Conf. Ser.: Earth Environ. Sci.* **12**
[12] Burgers J M 1948 *Adv. In Appl. Mech* 171-99 (New York:Academic Press)
[13] Sakai T, Eguchi Y, Monji H, Ito K, Ohshima H 2008 *Heat Transfer Eng.* **29** 731-739
[14] Ito K, Kunugi T, Ohshima H, Kawamura T 2009 *J. Nucl. Sci. Technol.* **46** 366-373
[15] ANSYS Fluent User's Guide, Release 15.

Mixed convection in inclined lid driven cavity by Lattice Boltzmann Method and heat flux boundary condition

A D'Orazio[a], A Karimipour[b], A H Nezhad[c], E Shirani[d,e]

a Dipartimento di IngegneriaAstronautica, Elettrica ed Energetica, Sapienza Università' di Roma, Via Eudossiana 18, Roma 00184, Italy

b Dipartimento Department of Mechanical Engineering, Najafabad Branch, Islamic Azad University, Isfahan, Iran, Postal Code: 8196848531

c Department of Mechanical Engineering, University of Sistan and Baluchestan, Daneshgah Street, Zahedan, Iran. Postal Code: 98135-987

d Department of Mechanical Engineering, Isfahan University of Technology, Imam Khomeini Street, Daneshgah Street, Isfahan, Iran. Postal Code: 84156-83111

d Foolad Institute of Technology, Shohada Blv, FooladShahr, Isfahan, Iran. Postal Code: 8491663763

E-mail:annunziata.dorazio@uniroma1.it

Abstract. Laminar mixed convective heat transfer in two-dimensional rectangular inclined driven cavity is studied numerically by means of a double population thermal Lattice Boltzmann method. Through the top moving lid the heat flux enters the cavity whereas it leaves the system through the bottom wall; side walls are adiabatic. The counter-slip internal energy density boundary condition, able to simulate an imposed non zero heat flux at the wall, is applied, in order to demonstrate that it can be effectively used to simulate heat transfer phenomena also in case of moving walls. Results are analyzed over a range of the Richardson numbers and tilting angles of the enclosure, encompassing the dominating forced convection, mixed convection, and dominating natural convection flow regimes. As expected, heat transfer rate increases as increases the inclination angle, but this effect is significant for higher Richardson numbers, when buoyancy forces dominate the problem; for horizontal cavity, average Nusselt number decreases with the increase of Richardson number because of the stratified field configuration.

Journal of Physics: Conference Series **547** (2014) 012031 doi:10.1088/1742-6596/547/1/012031

1. Introduction

The lattice Boltzmann Equation (LBE) is a minimal form of the Boltzmann kinetic equation, which is the evolution equation for a continuous one-body distribution function, wherein all details of molecular motion are removed except those that are strictly needed to represent the hydrodynamic behaviour at the macroscopic scale; it has gained much attention for its ability to simulate fluid flows, and for its potential advantages over conventional numerical solution of the Navier–Stokes equations. The kinetic nature of Lattice Boltzmann Method (LBM) introduces key advantages, including easy implementation of boundary conditions and fully parallel algorithms. In addition, the convection operator is linear, no Poisson equation for the pressure must be resolved and the translation of the microscopic distribution function into the macroscopic quantities consists of simple arithmetic calculations [1]. LBM have met with significant success for the numerical simulation of a large variety of fluid flows, including real-world engineering applications and physical phenomena of various complexities as multiphase flows, complex geometries and interfacial flows [2-4]. The application to fluid flow coupled with non negligible heat transfer turned out to be much more difficult. The LBE thermal models fall into three categories: the multi-speed approach, the passive scalar approach and the doubled populations approach. In this latter, successful, strategy [5] thermal energy density and heat flux are expressed as kinetic moments of a thermal distribution function, so that no kinetic moment beyond the first order is ever required, thus providing numerical stability, also in case of significant temperature gradient [6]; in addition, with respect to the previous approaches, viscous heat dissipation and compression work done by the pressure were naturally incorporated and the boundary conditions are easily implemented because both populations live in the same lattice, where additional speeds are not necessary.

Mixed convection in an inclined cavity has not been investigated by LBM, to the authors' knowledge, except for a recent previous work by Karimipour et al. [7], in which laminar mixed convection in a two-dimensional rectangular inclined cavity with moving top lid is investigated numerically for Richardson number ranging from 0.1 to 10 and inclination angle ranging from 0° to 90° in case of imposed temperature at top and bottom walls. Mixed convection in an inclined cavity has not been investigated by LBM in case of imposed non zero heat flux. This type of boundary condition, representing very usual situations in physical world, is not simple to model in lattice Boltzmann schemes. In effect, the only boundary condition able to simulate imposed temperature and imposed heat flux at a boundary has been presented by D'Orazio et al. in [8] where the boundary were at rest. In this work, laminar mixed convection in a two-dimensional rectangular inclined cavity with moving top lid is investigated numerically. Top lid motion results in fluid motion inside. Inclination of the cavity causes horizontal and vertical components of velocity to be affected by buoyancy force, since the heat flux enters the cavity through the top moving lid and it leaves the system through the bottom wall, side walls being adiabatic. To include this effect, collision term of Boltzmann equation is modified and the counter-slip internal energy density boundary condition by [6, 8, 9], able to simulate an imposed heat flux at the wall, is applied. The effects of the variations of Richardson number and inclination angle on the thermal and flow behaviour of the fluid inside the cavity are investigated. The results are presented as stream function contours and isotherms; Nusselt number behavior as a function of Richardson number and inclination angle is discussed. Simulations are performed for $Pr = 0.7$ and $Re = 200$ and for three value of the Richardson number $Ri = 0.1$, $Ri = 1$, $Ri = 10$; the effects of the inclination angle $\gamma = 0°, 30°, 60°, 90°$ on fluid flow and heat transfer are studied in case of positive and negative moving lid velocity.

2. Thermal and hydrodynamic Lattice Boltzmann Method

The lattice Boltzmann equation with a single relaxation time from the BGK model can be expressed as

$$\tilde{f}_i(\vec{x}+\vec{c}_i dt, t+dt) - \tilde{f}_i(\vec{x},t) = -\frac{dt}{\tau_f + 0.5dt}[\tilde{f}_i - \tilde{f}_i^e]$$

$$\tilde{g}_i(\vec{x}+\vec{c}_i dt, t+dt) - \tilde{g}_i(\vec{x},t) = -\frac{dt}{\tau_g + 0.5dt}[\tilde{g}_i - \tilde{g}_i^e] - \frac{\tau_g dt}{\tau_g + 0.5dt} f_i Z_i$$

where the populations $\tilde{f}_i$ carry mass and momentum and the populations $\tilde{g}_i$ carry internal energy and heat flux. The discrete distribution functions $\tilde{f}_i$ and $\tilde{g}_i$ are introduced as in [5]:

$$\tilde{f}_i = f_i + \frac{dt}{2\tau_f}(f_i - f_i^e)$$

$$\tilde{g}_i = g_i + \frac{dt}{2\tau_g}(g_i - g_i^e) + \frac{dt}{2} f_i Z_i$$

$$Z_i = (\vec{c}_i - \vec{u})D_i\vec{u}\ , \quad D_i = \partial_t + \vec{c}_i \cdot \nabla$$

where fi and gi are the discrete populations which evolve when a standard first order integration strategy is adopted, the term

$$Z_i = (\vec{c}_i - \vec{u})D_i\vec{u}$$

represents the effects of viscous heating,

$$D_i = \partial_t + \vec{c}_i \cdot \nabla$$

is the material derivative along direction c_i), τ_f and τ_g are relaxations times and f_i^e and g_i^e are the equilibrium distribution functions. Throughout of this work, two-dimensional square lattice with the nine speeds, as shown in figure 1, is used.

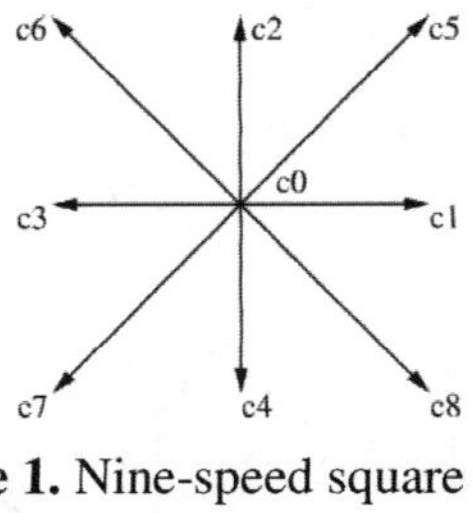

Figure 1. Nine-speed square lattice.

The discrete particle lattice speeds are:

$$c_i = \left(cos\frac{i-1}{2}\pi, sin\frac{i-1}{2}\pi\right)c, \qquad i = 1,2,3,4$$

$$c_i = \sqrt{2}\left[cos\left(\frac{i-5}{2}\pi + \frac{\pi}{4}\right), sin\left(\frac{i-5}{2}\pi + \frac{\pi}{4}\right)\right]c, \qquad i = 5,6,7,8$$

$$c_0 = (0,0)$$

where $c^2 = 3RT$ and T is the temperature. The equilibrium density distributions are chosen as follows:

$$f_i^e = \omega_i\rho\left[1 + \frac{3\vec{c}_i \cdot \vec{u}}{c^2} + \frac{9(\vec{c}_i \cdot \vec{u})^2}{2c^4} - \frac{3(u^2+v^2)}{2c^2}\right]$$

$$g_0^e = -\omega_0 \left[\frac{3\rho e(u^2+v^2)}{2c^2} \right]$$

$$g_{1,2,3,4}^e = \omega_1 \rho e \left[1.5 + 1.5 \frac{\vec{c}_i \cdot \vec{u}}{c^2} + 4.5 \frac{(\vec{c}_i \cdot \vec{u})^2}{c^4} - 1.5 \frac{(u^2+v^2)}{c^2} \right]$$

$$g_{5,6,7,8}^e = \omega_2 \rho e \left[3 + 6 \frac{\vec{c}_i \cdot \vec{u}}{c^2} + 4.5 \frac{(\vec{c}_i \cdot \vec{u})^2}{c^4} - 1.5 \frac{(u^2+v^2)}{c^2} \right]$$

where $\vec{u} = (u, v)$ and $\rho e = \rho RT$ (in two-dimensional geometry). The weights of the different populations are

$$\omega_0 = \frac{4}{9} \qquad \omega_i = \frac{1}{9} \;\; i = 1,2,3,4 \qquad \omega_i = \frac{1}{36} \;\; i = 5,6,7,8$$

The terms enclosed by the square bracket, multiplied by the corresponding weights ω_i, will be called *corresponding form for equilibrium.*

Finally, using $\mathrm{f}_\mathrm{i}^\mathrm{e}$ and $\mathrm{g}_\mathrm{i}^\mathrm{e}$, hydrodynamic and thermal variables are calculated as follows:

$$\rho = \sum_i \tilde{f}_i \qquad \rho e = \sum_i \tilde{g}_i - \frac{dt}{2} \sum_i f_i Z_i$$

$$\rho \vec{u} = \sum_i \vec{c}_i \tilde{f}_i \qquad \vec{q} = \left[\sum_i \vec{c}_i \tilde{g}_i - \rho e \vec{u} - \frac{dt}{2} \sum_i \vec{c}_i f_i Z_i \right] \frac{\tau_g}{\tau_g + 0.5dt}$$

The kinematic viscosity and the thermal diffusivity in the two-dimensional geometry are given by:

$$\nu = \tau_f R\overline{T}, \qquad \chi = 2\tau_g R\overline{T}$$

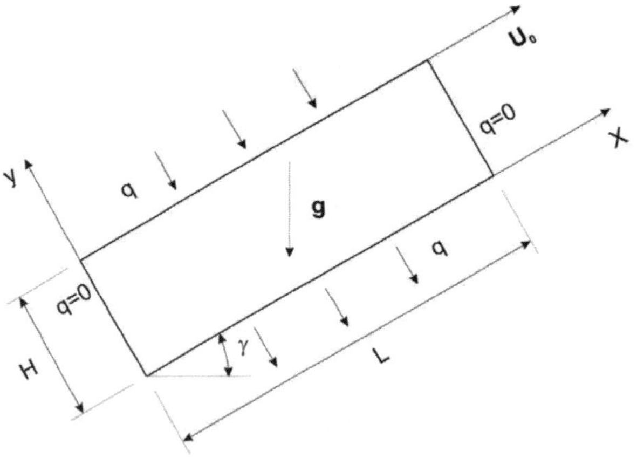

Figure 2. Geometry and coordinates axis in the inclined cavity.

In this problem shear stress applied by moving lid on the fluid layers results in fluid motion, thus creating suitable temperature gradient that enhances buoyancy forces. Therefore, mixed convection is produced in the fluid confined in the cavity. With the Boussinesq approximation, all the fluid properties are considered as constant, except in the body force term in the Navier–Stokes equations, where the fluid density is assumed $\rho = \bar{\rho}[1 - \beta(T - \bar{T})]$, in which β is volumetric expansion coefficient, $\bar{\rho}$ and $\bar{T}$ are the average fluid density and temperature. In order to simulate the mixed convection of nearly incompressible flows, buoyancy force per unit mass is defined as $G = \beta g(T - \bar{T})$ and this is used to drive the flow. By considering inclination angle, coordinate axis and gravity acceleration direction, as shown in figure 2, all of the aforementioned relations are maintained and used except those that are modified below:

$$\tilde{f}_i(\vec{x} + \vec{c}_i dt, t + dt) - \tilde{f}_i(\vec{x}, t) =$$

$$= -\frac{dt}{\tau_f + 0.5dt}[\tilde{f}_i - f_i^e] + \left[\frac{dt\tau_f}{\tau_f + 0.5dt}\frac{3G(c_{ix} - u)}{c^2} f_i^e\right] sin\gamma + \left[\frac{dt\tau_f}{\tau_f + 0.5dt}\frac{3G(c_{iy} - v)}{c^2} f_i^e\right] cos\gamma$$

with:

$$f_i =$$

$$= \frac{\tau_f \tilde{f}_i + 0.5dt f_i^e}{\tau_f + 0.5dt} + \left[\frac{0.5dt\tau_f}{\tau_f + 0.5dt}\frac{3G(c_{ix} - u) f_i^e}{c^2}\right] sin\gamma + \left[\frac{0.5dt\tau_f}{\tau_f + 0.5dt}\frac{3G(c_{iy} - v) f_i^e}{c^2}\right] cos\gamma$$

In this case hydrodynamic macroscopic variables are calculated as follows:

$$\rho = \sum_i \tilde{f}_i, \quad u = \frac{1}{\rho}\sum_i \tilde{f}_i c_{ix} + \frac{dt}{2} G sin\gamma, \quad v = \frac{1}{\rho}\sum_i \tilde{f}_i c_{iy} + \frac{dt}{2} G cos\gamma$$

where $\vec{c}_i = (c_{ix}, c_{iy})$ denote discrete particle speeds.

3. Boundary conditions

With regard to hydrodynamic boundary condition, no slip boundary condition is applied; It is obtained by means of the non-equilibrium bounce back rule as applied to the population perpendicular to the boundary. With regard to thermal boundary conditions, the top cavity lid and bottom wall are heated and cooled respectively by an uniform and constant heat flux entering and leaving respectively, and the sidewalls are insulated. These boundary conditions are obtained by means a thermal counter-slip approach as proposed by [6, 8, 9], in which the incoming unknown thermal populations are assumed to be equilibrium distribution functions with a counter slip thermal energy density e', which is determined so that suitable constraints are verified. For the top wall of the cavity, named as "north wall", in which entering heat flux is constant and equal to q_N, the unknown $\tilde{g}_4$, $\tilde{g}_7$ and $\tilde{g}_8$ are chosen as follows.

$$\tilde{g}_i = \rho(e_N + e') \times [corresponding\ form\ for\ equilibrium] \qquad i = 4, 7, 8$$

By definition:

$$\sum_i c_{iy}\,\tilde{g}_i = \frac{dt}{2}\sum_i c_{iy} f_i Z_i + \rho e_N V_N + \frac{\tau_g + 0.5dt}{\tau_g} q_y$$

which yields to:

$$\rho e_N + \rho e' = \left[K - \frac{dt}{2}\sum_i \frac{c_{iy}}{c} f_i Z_i - \rho e_N \frac{V_N}{c} - \frac{\tau_g + 0.5dt}{\tau_g}\frac{q_y}{c}\right] \Big/ \left[\frac{1}{3} - \frac{1}{2}\frac{V_N}{c} + \frac{1}{2}\frac{V_N^2}{c^2}\right]$$

where V_N is the component normal to the wall of flow velocity at the wall, K is the sum of the three known populations ($\tilde{g}_2$, $\tilde{g}_5$ and $\tilde{g}_6$) e_N denotes the current value of thermal energy density at the north wall. At the insulated walls, the constraint on the heat flux is obtained by imposing $q_x = 0$ on the corresponding equation, for example for the west wall, so that we have [8]

$$\sum_i c_{ix}\tilde{g}_i = 0.5\, dt \sum_i c_{ix} f_i Z_i + \rho e_W U_W$$

where e_W denotes the thermal energy density current value (coming from the run) at the West wall.

The unknown populations $\tilde{g}_1$, $\tilde{g}_5$ and $\tilde{g}_8$ are chosen as

$$\tilde{g}_i = \rho(e_W + e') \times [corresponding\ form\ for\ equilibrium] \qquad i = 1,5,8$$

and become

$$\tilde{g}_i = \frac{\left[K + \frac{dt}{2}\sum_i \frac{c_i}{c} f_i Z_i + \rho e_W \frac{U_W}{c}\right]}{\left[\frac{1}{3} + 0.5\frac{U_W}{c} + 0.5\frac{U_W^2}{c^2}\right]} \times [corresponding\ form\ for\ equilibrium] \qquad i = 1,5,8$$

where U_W is the component normal to the wall of the flow velocity at the wall and K is the sum of the known populations ($\tilde{g}_3$, $\tilde{g}_6$ and $\tilde{g}_7$ in this case). The corners nodes are treated similarly and the counter-slip procedure can be applied to the five unknown incoming populations at the corner. The same relations are used for the bottom wall (named as "south wall"), in which leaving heat flux is constant and equal to q_S and the unknown populations are $\tilde{g}_2$, $\tilde{g}_5$ and $\tilde{g}_6$. With regard to the initial conditions, the velocities of all nodes inside the cavity are taken as zero initially. The initial density is set to a value of 2.7. The initial equilibrium distribution functions are evaluated correspondingly. The initial distribution functions are taken as the corresponding equilibrium values.

4. Results and discussion

Laminar mixed convection of a fluid inside a rectangular cavity with moving top lid and aspect ratio $AR = L/H = 3$, in which L and H are shown in Figure 2, is studied numerically using Lattice Boltzmann Method previously described. The bottom wall is cooled and the top lid is heated, and side walls are assumed insulated. Top lid moves with constant velocity U_0. Characteristic dimensionless number in the analysis of mixed convection problems is Richardson number defined as $Ri = Ra/PrRe^2$. As stated before, lattice Boltzmann method is used for near-incompressible flows and therefore Mach number is assumed as $Ma \ll 1$. More specifically, the characteristic velocity of the flow for both natural regime, $U^* = \sqrt{\beta g H \Delta T}$, and forced regime $U^* = \nu Re/H$ must be small compared with the fluid speed of sound; in present study the velocity U_0 was selected as $U_0 = 0.1$. The Prandtl and Reynolds numbers are assumed as $Pr = 0.7$ and $Re = 200$ respectively and the effects of the variations of the Richardson number Ri and angle γ on the flow and heat transfer are studied. Inclination angle varies from zero degree, horizontal cavity, to 90°, vertical cavity. To avoid ambiguity, the cavity's walls are referred to according to the coordinates shown in Figure 2. In the following the macroscopic variables of fluid flow are made dimensionless as follows, respectively for dimensionless coordinates, velocity components, temperature, time

$$Y = \frac{y}{H}, \quad X = \frac{x}{H}, \quad U = \frac{u}{U_0}, \quad V = \frac{v}{U_0}, \quad \Theta = \frac{T - T_c}{T^{**}}, \quad \tau = \frac{tU_0}{H}, \quad T^{**} = \frac{\chi \nu Ra^{\frac{4}{5}}}{\beta g H^3}$$

Therefore, local and average Nusselt numbers along the lid and bottom wall are calculated using following relations

$$Nu_X = \frac{qH}{k\Delta T} = -\left(\frac{\partial\Theta}{\partial Y}\right)_{wall}, \qquad Nu_m = \frac{1}{AR}\int_0^{AR} Nu_X dX$$

The top lid is the hot moving wall at Y = 1, the bottom wall is the cold wall at Y=0 and two insulated side walls are at X = 0 and X = 3.

In order to obtain grid independent solution, a grid refinement study is performed for a horizontal cavity ($\gamma = 0$). Grid independence of the results has been established in term of average Nusselt number on the lid and dimensionless values of x-velocity U, y-velocity V, and temperature Θ at X = 1.5 and Y = 0.5 (cavity center) for three different grid size, namely 300×100, 450×150 and 600×200 lattice nodes; due to small difference between the results of the last two grid sizes, the 450×150 grid is chosen as a suitable one in this work.

To validate the computer code, the comparison with the analytical solution given by Kimura and Bejan [10] and the value obtained by D'Orazio et al. [6] is examined for the Rayleigh number equal to 10^5 and 10^6 and the Prandtl number equal to 2.0. With regard to the average Nusselt number Nu_m, results show good agreement with those of Bejan et al. [10] and of D'Orazio et al. [8], with a maximum error equal to 18.6%.

In order to show the effect of inclination angles on the flow field and heat transfer, in Figures 3, 4 and 5 and 6, 7 and 8 streamlines and isotherms are reported at different inclination angles $\gamma = 0°$, 30°, 60° and 90°, for the cases for $Ri = 0.1$, 1 and 10, respectively for positive and negative velocity (in this case the inclination angle ranges from 30° to 90°, being the case $\gamma = 0$ not depending on the moving lid direction).

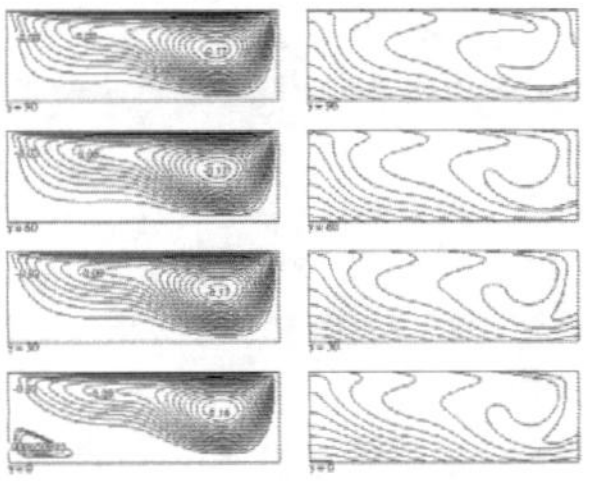

Figure 3. Streamlines and isotherms for $Ri = 0.1$, at $\gamma = 0°$, 30°, 60° and 90°, $U_0 > 0$.

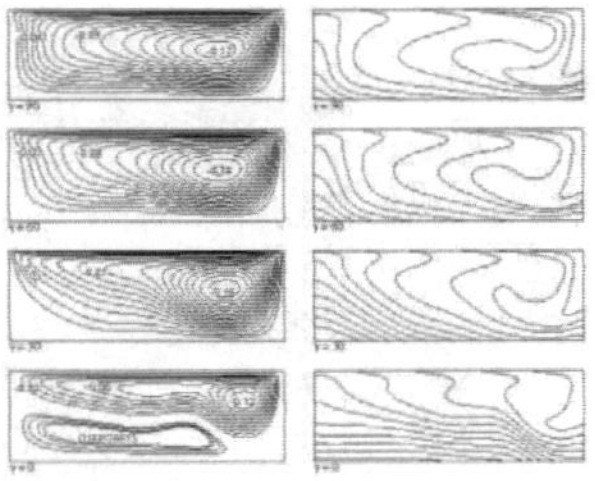

Figure 4. Streamlines and isotherms for $Ri = 1$, at $\gamma = 0°$, 30°, 60° and 90°, $U_0 > 0$.

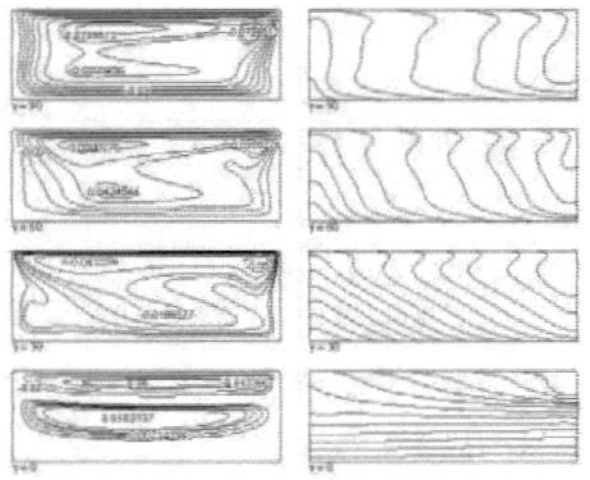

Figure 5. Streamlines and isotherms for $Ri = 10$, at $\gamma = 0°$, 30°, 60° and 90°, $U_0 > 0$.

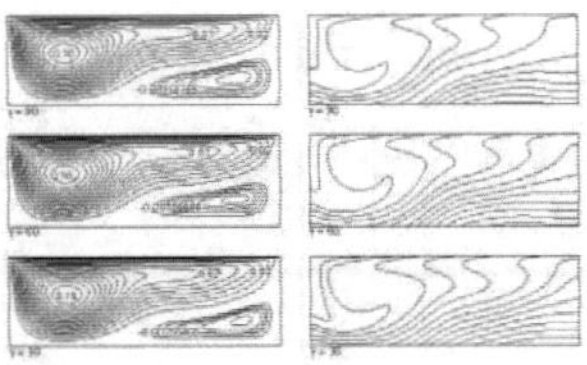

Figure 6. Streamlines and isotherms for $Ri = 0.1$, at $\gamma = 0°$, 30°, 60° and 90°, $U_0 < 0$.

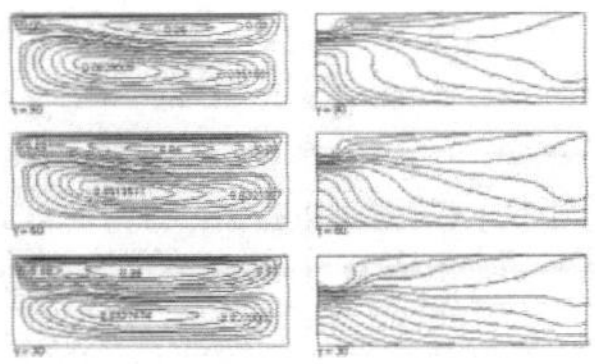

Figure 7. Streamlines and isotherms for $Ri = 1$, at $\gamma = 0°$, 30°, 60° and 90°, $U_0 < 0$.

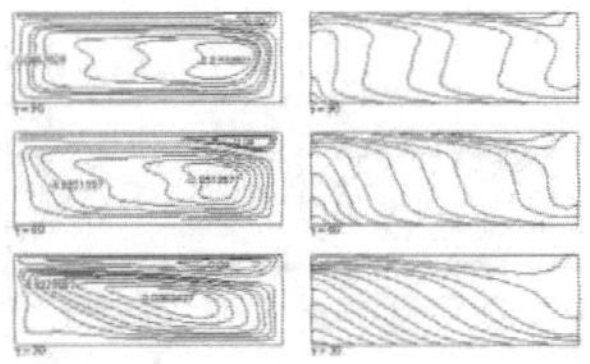

Figure 8. Streamlines and isotherms for $Ri = 10$, at $\gamma = 0°$, 30°, 60° and 90°, $U_0 < 0$.

The motion of the cavity lid, with positive lid velocity, causes the fluid motion in the cavity and produces a strong clockwise rotational cell in it. This motion transfers hot fluid to the lower parts and enhances favourable pressure gradient along the vertical direction, leading to the generating buoyancy motions and transferring hot fluid to the upper parts. Therefore, the combination of free and forced convection, called "mixed convection", is made. In Figure 3, the Richardson number is $Ri = 0.1$, with forced convection dominant with respect to natural convection, and it implies that by increasing γ, the rotational power of the cell in the center of the cavity increases slightly and it does not affect significantly the other moving and thermal behaviour of the fluid.

When buoyancy forces dominate the problem, as for $Ri = 10$ reported in Figure 5, inclination angle has significant effect on the flow field and heat transfer. For $\gamma = 0$, a strong cell in the upper half and a weak cell in the lower half of the cavity, are generated due to the forcing of moving wall. In addition, for $\gamma = 0$, isotherms in the lower half of the cavity are straight lines and perpendicular to the side walls, indicating that conduction heat transfer is dominant in this region and that, without the forced convection contribution, no motion occurs, due the stratified fluid. As the inclination angle increases, the two cells merge, so that for $\gamma = 90$ a large rotational cell covers the whole space of the cavity, with the central part practically motionless.

When the buoyancy effect is predominant ($Ri = 10$) in case of negative velocity, the effect of the direction of the moving wall, in term of modified flow field, can be detected. While for $U_0 < 0$ the two rotational cells lose their individuality already for low inclination angles, for $U_0 < 0$ they remain distinct because the hot moving wall has an opposite effect with respect to the buoyancy.

When $Ri = 1$, it becomes evident as the contribution of the moving wall with negative velocity and the buoyancy due to gravity acceleration have opposite effects on the fluid field; it gives rise to the formation of two rotational cell into the cavity.

For $Ri = 0.1$, when forced convection is the dominant effect on the flow field, streamlines and isotherms for negative velocity, have the same behaviour shown in case of positive wall velocity, although as in a mirror, but the result of the two cited opposing effects is the persistence of the two rotational cells.

In Figure 9, the average Nusselt number on the lid surface is reported as a function of inclination angle for different Richardson number in both case of positive and negative moving wall velocity. It has to be noted that for $\gamma = 90$, Nu_m decreases when Ri increases, since it implies to go to a stratified field configuration. With regard to $\gamma \neq 0$, it is observed that for positive lid velocity when $Ri = 0.1$, Nu_m increases slightly with the increase of γ. At $Ri \geq 1$, Nu_m is increased more intensively. In fact at $Ri = 10$, it is increased by a factor of 5 when γ varies from 0 to 90, indicating that in this case free convection effect is enhanced and its contribution can be added to the forced convection effect. When $\gamma = 0$ (horizontal cavity) Nu_m is maximum at $Ri = 0.1$ (significant forced convection), but for inclined cavity, the maximum Nu_m occurs at $Ri = 10$ (significant natural convection). For negative velocity of moving wall, the opposite effect of buoyancy and forced convection contribution can be observed since the Nusselt number results ever lower than the value corresponding to the case of positive velocity. For low inclination angle, the contribution of natural convection is less important than the hindering effect of the negative moving wall velocity, whereas for increasing γ angle the importance of natural convection becomes predominant.

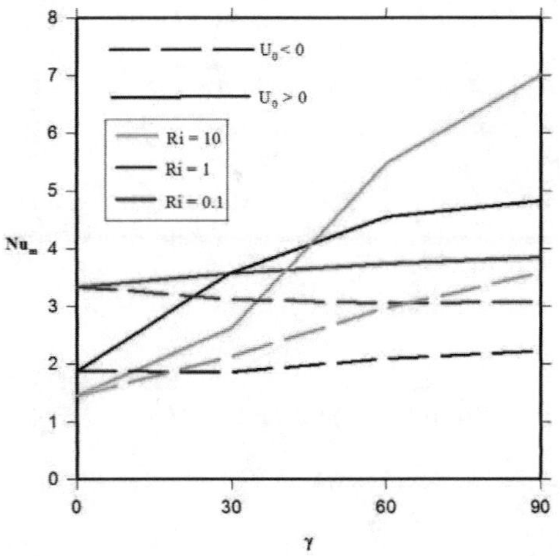

Figure 9. Average Nusselt number $\mathrm{Nu_m}$ on the hot surface as a function of inclination angle γ for different Richardson number and in case of $U_0 > 0$ and $U_0 < 0$

5. Conclusions

A thermal lattice Boltzmann BGK model with a dedicated boundary condition was used to study numerically laminar two-dimensional mixed convection heat transfer inside an inclined rectangular cavity when he heat transfer rate is imposed at the boundaries. Since the inclination of the cavity enhances the buoyancy force, which affects the velocity components of the flow, the forcing term simulating the buoyancy effect in the lattice Boltzmann equation were modified. The results show that, as expected, heat transfer rate increases as increases the inclination angle, but this effect is significant for the higher Richardson numbers, when buoyancy forces dominate the problem; for horizontal cavity, average Nusselt number decreases with the increase of the Richardson number because of the stratified field configuration. The effects of forced convection and natural convection can be considered as cooperating for positive velocity of the moving lid, while on the contrary they can be considered as opposite for negative velocity. This study shows that lattice Boltzmann method together with the counter-slip thermal energy density boundary condition can be effectively used to simulate heat transfer phenomena also in case of moving walls. The method can be successfully applied to simulate a wide class of cooling process where a given thermal power must be removed.

References

[1] Succi S 2001 *The lattice Boltzmann equation for fluid dynamics and beyond* (Oxford: Oxford University Press)

[2] Wang M, He J, Yu J and Pan N 2007 Lattice Boltzmann modeling of the effective thermal conductivity for fibrous materials *Int. J. Thermal Sc.* **46** 848–855

[3] Zhao C Y, Dai L N, Tang G H, Qu Z G and Li Z Y 2010 Numerical study of natural convection in porous media (metals) using Lattice Boltzmann Method (LBM) *Int. J. of Heat and Fluid Flow* **31** 925–934

[4] Wang M and Kang Q 2010 Modeling electrokinetic flows in microchannels using coupled lattice Boltzmann methods *J. of Comp. Phys.* **229** 728-744

Journal of Physics: Conference Series **547** (2014) 012031 doi:10.1088/1742-6596/547/1/012031

[5] He X, Chen S and Doolen G 1998 A novel thermal model for the lattice Boltzmann method in incompressible limit *J. of Comp. Phys.* **146** 282-300

[6] D'Orazio A, Succi S and Arrighetti C 2003 Lattice Boltzmann simulation of open flows with heat transfer *phys. of fluids* **15** 2778-2780

[7] Karimipour A, Nezhad A H, D'Orazio A and Shirani E 2013 The effects of inclination angle and Prandtl number on the mixed convection in the inclined lid driven cavity using lattice Boltzmann method *Theor. Appl. Mech.* **51** 447-462

[8] D'Orazio A, Corcione M and Celata G P 2004 Application to natural convection enclosed flows of a lattice Boltzmann BGK model coupled with a general purpose thermal boundary condition *Int. J. of Thermal Sc.* **43** 575–586

[9] D'Orazio A and Succi S 2004 Simulating two-dimensional thermal channel flows by means of a lattice Boltzmann method *Fut. Gen. Comp. Sys.* **20** 935-944

[10] Kimura S and Bejan A 1984 The boundary layer natural convection regime in a rectangular cavity with uniform heat flux from the side *ASME J. Heat Transfer* **106** 98-103

RELAP5-3D thermal hydraulic analysis of the target cooling system in the SPES experimental facility

M Giardina[1], F Castiglia[1], P Buffa[1], G Palermo[1]
G Prete[2]

[1] Department of Energy, Information Engineering and Mathematical Models (DEIM), University of Palermo, Viale delle Scienze-Parco d'Orléans, 90128 Palermo, Italy

[2] National Institute of Nuclear Physics – National Laboratory of Legnaro, Italy

Abstract. The SPES (Selective Production of Exotic Species) experimental facility, under construction at the Italian National Institute of Nuclear Physics (INFN) Laboratories of Legnaro, Italy, is a second generation Isotope Separation On Line (ISOL) plant for advanced nuclear physic studies. The UCx target-ion source system works at temperature of about 2273 K, producing a high level of radiation (10^5 Sv/h), for this reason a careful risk analysis for the target chamber is among the major safety issues. In this paper, the obtained results of thermo-fluid-dynamics simulations of accidental transients in the SPES target cooling system are reported. The analysis, performed by using the RELAP5-3D 2.4.2 qualified thermal-hydraulic system code, proves good safety performance of this system during different accidental conditions.

Keywords: SPES, thermo-fluid-dynamics simulations, safety analysis, target cooling system, RELAP5.

1. Introduction

SPES (Selective Production of Exotic Species) project is part of the INFN (italian national institute of nuclear physics) road map for the nuclear physics development in Italy, supported by the whole italian nuclear physics community and mainly by LNL (laboratori nazionali di legnaro) and LNS (laboratori nazionali del sud) [1-2].

This project is focused on the development of a radioactive ion beam (RIB) facility to produce intense neutron-rich RIBs according to the Isotope Separation On Line (ISOL) technique [3] with an UCx (Uranium Carbide) direct target.

The facility is devoted to basic research in nuclear physics and astrophysics as well as to interdisciplinary applications, ranging from the production of radionuclides of medical interest to the generation of neutrons for material studies, nuclear technologies and medicine.

The INFN capability to play a role in this research field is supported by the consolidated know-how in accelerators and detectors construction. The most significant facilities are PIAVE superconducting RFQ (electric field provided by a radiofrequency quadrupole), ALPI (acceleratore lineare per ioni) superconductive linac, able to re-accelerate exotic ions at 8÷13 MeV/u, and the EXCYT project, an ISOL RIB facility for light ions.

The major difference between SPES facility and the existing ISOL plants is the target design. This system includes as primary accelerator a cyclotron able to supply at least 40 MeV 0.2 mA proton beam on an UCx direct target to produce a fission rate of about 10^{13} fission/s [4-6]. Very high estimated level of radiation and nuclear activations (about 10^5 Sv/h) are produced in the target area, so special infrastructures, radiation management and control systems are designed.

The target-ion source system must operate stably at temperature of about 2273 K for long periods of time to determine good characteristic of intensity, beam quality, and number of radioactive beams that can be provided for experimental use. High temperature and vacuum are two essential requirements to enhance the radioactive isotopes mobility, extraction and ionization. To assure the operating temperature, the production target and the surface ion source are located inside a cylindrical chamber cooled by an "ad hoc" water cooling system. Obviously, cooling system faults can produce malfunctions in the target-ion source devices, with possible consequent loss of target chamber structural integrity.

The paper reports the analysis of thermo-fluid-dynamics accidental transients of the target cooling system performed by RELAP5-3D 2.4.2 advanced qualified thermal-hydraulic system code. In particular, loss of water coolant and over-power accidental conditions have been examined by using a detailed cooling system nodalization.

The results prove the good safety performance of the water coolant system against the hazards examined in this paper.

2. Target/ion source description

The UCx target is a critical safety item of the facility due to the large level of radioactivity induced by the fission reactions. It follows the design already developed in several laboratories where radioactive beams have operated recently or for long time as EXCYT or ISOLDE (CERN) and HRIBF (ORNL), however the proposed target represent an innovation in term of capability to sustain the primary beam power.

In the context of the facility SPES, the target chamber is used to contain and support the target-ion source complex in an high vacuum environment, providing electrical current to heat by Joule effect both the production target and the ion source in order to operate stably at high temperature of about 2273 K for long periods of time. A water cooling system in the cover of the target compensates the high power deposited by the proton beam channel.

The design is carefully oriented to optimise the radiative cooling taking advantage of the high operating temperature of the target system. In fact a lot of studies have been performed to evaluate the thermo-mechanical behaviour with the ANSYS code [7-11] and, in addition, experimental tests of the target principle were performed at the HRIBF facility (ORNL-USA). The main result is that, in the adopted configuration, the target does not melt and it is rather necessary to supply external power to reach the operating temperature.

The inner container of the target disks is a cylinder with a double entrance window and a beam dump for low energy protons. All these elements are made up of graphite. The window is made by two layers of graphite with a thickness of 0.5 mm. Two window layers are used to separate the reaction chamber from the proton driver beam pipe, this choice is suggested by thermal calculation and in order to improve the safety of the system. The melting point of the graphite is 3773 K well above the temperature of operation of the target.

Number 7 UCx co-axial disks (diameter and thickness of 40 and 1.3 mm respectively) are impinged by 40-50 MeV proton beam (characterized by a current intensity of 0.2 mA), generating approximately 10^{13} fissions per second. Thus, a total beam power of about 8 kW has to be managed [12]. The gap between the disks allows an efficient cooling of the system by thermal radiation. The total amount of the U fissile material is only 28 g and the radioactive isotopes produced by the fissions are driven through the transfer line.

Section view of the production target, transfer line and surface ion-source is showed in figure 1. The target configuration is described in detail in [7, 8].

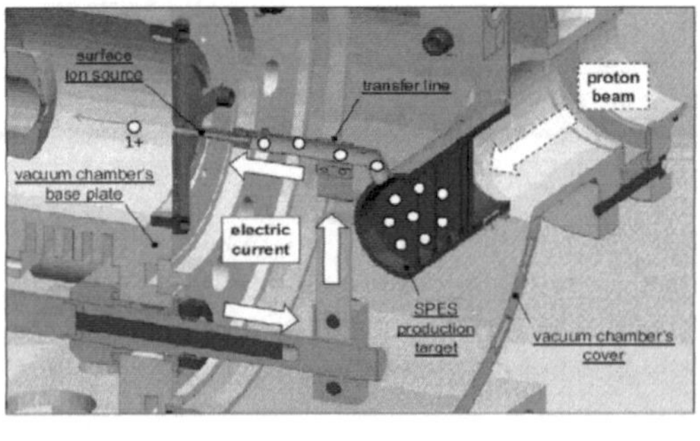

Figure 1. Production target, transfer line and surface ion-source (section view).

The surface ion source and the transfer line are kept at high temperature thanks to a dedicated and independent 10 kW power supply, able to deliver a maximum electrical current and a maximum potential difference of 1000 A and 10 V, respectively. The electrical current flows first through Tantalum (Ta) transfer line and then through the tungsten (W) tubular ionizing cavity, also called hot-cavity, and dissipates by the joule effect the power which enables the system to reach the desired temperature level. The Ta transfer line consists of a tube with an external and an internal diameter of 9 and 8 mm respectively, and a length of approximately 38 mm, connected to the W hot-cavity.

An ion source electrical clamp and two target electrical clamps are used to provide the high electrical currents needed to heat the ion source and the target, respectively. It is worth to mentioning that the electrical clamps should be cooled down to 1357.15 K being the melting point of the Copper (Cu).

Some sub-assemblies of the la target chamber are reported in figure 2. The main plate (figure 2a) supports the cover, the electrical clamps and all secondary components of the target chamber (for example the shutter of the RIB channel). The water cooling channels in the electrical clamps (figure 2a) and in the cover (figure 2b) allow to remove the power produced by the target-ion source system and to operate stably at temperature of 2273 K.

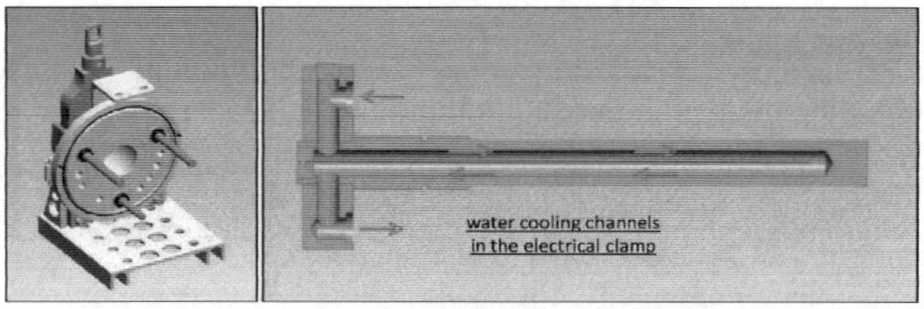

(a)

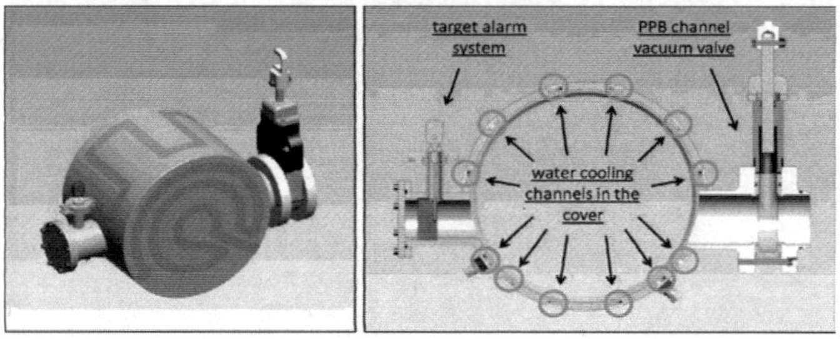

(b)

Figure 2. Sub-assemblies of the target chamber: (a) main plate and cooling channels in the electrical clamp, (b) target cover and cooling channels in the cover.

3. The cooling system

The cooling system consists of two loops (figure 3): the primary loop (Loop A) uses demineralized water as coolant of the target system, the secondary one (Loop B) uses soft water to cool the primary loop. The platform, where the target-ion source system is located, has to be maintained for experimental purposes at 250 kV voltage. The cooling system removes a power of 27 kW in the target chamber and 9 kW in the instrument and measurement devices [2].

A redundant primary coolant feed system allows demineralized water circulation by magnetic drive centrifugal pumps (PS1 and PS2 in figure 3). A distributed control system (DCS) allows to start the auxiliary pump in case of primary pump malfunctions and switches off the beam (target-ion source system trip) if auxiliary pump failures occurs. Each coolant feed pipeline is equipped with check valves to avoid flow inversions.

A pressure gauge installed downstream of the redundant pipeline, called PT1 in figure 3, measures the pressure and proceeds via DCS with pump regulation according to the operating condition of water mass flow rate (0.83 kg/s). The primary coolant is pumped into the corrugated heat exchanger where heat is transferred from the primary coolant (demineralized water) to the secondary coolant (soft water). The inlet and outlet temperatures of the primary coolant in the heat exchanger are 303 and 292 K, respectively;

control and safety systems allow a tolerance of + 0.2 K. The soft water enters in the secondary side of the heat exchanger at temperature of 280 K and exits at temperature of 285 K. A temperature gauge, TT1, allows, via temperature controller system RT, the secondary coolant flow control by using a mixing three-way valve, VM, (figure 3); this regulation allows to keep the primary coolant temperature at 292 K (+ 0.2 K).

At the exit of the heat exchanger, the primary coolant flows into parallel mixed beds (beds ion exchange) to be purified. This operation enables to reduce primary coolant specific conductivity before entering within the target platform.

At the exit of the mixed beds, the primary coolant flows inside a non-conductive material helical pipe (50 m length and 32 mm outer diameter) to minimize parasitic current flows production, then it removes the heat produced in the target chamber, reaching outlet temperature of 303 K. A mass/flow density meter, placed at the inlet of the platform, is used to operate via DCS the target-ion source system trip in presence of vapor or mass flow rate reduction. Two non-conductive material spiral pipes at the platform exit lead the coolant into a return pipeline (50 m length and 25 mm outer diameter), where an expansion tank of 100 liter volume is located.

The facility is also equipped with several automated safety and control systems (i.e. visual emergency alarm system, detection system for target temperature and pressure measurements, radiation monitoring system, interlock systems to stop the beam, mass flow and density meter).

Moreover the control room typically houses three computer working stations: one for accelerators controlling, one for the target, one for controlling and monitoring the radiation protection and the safety systems, cooling system included.

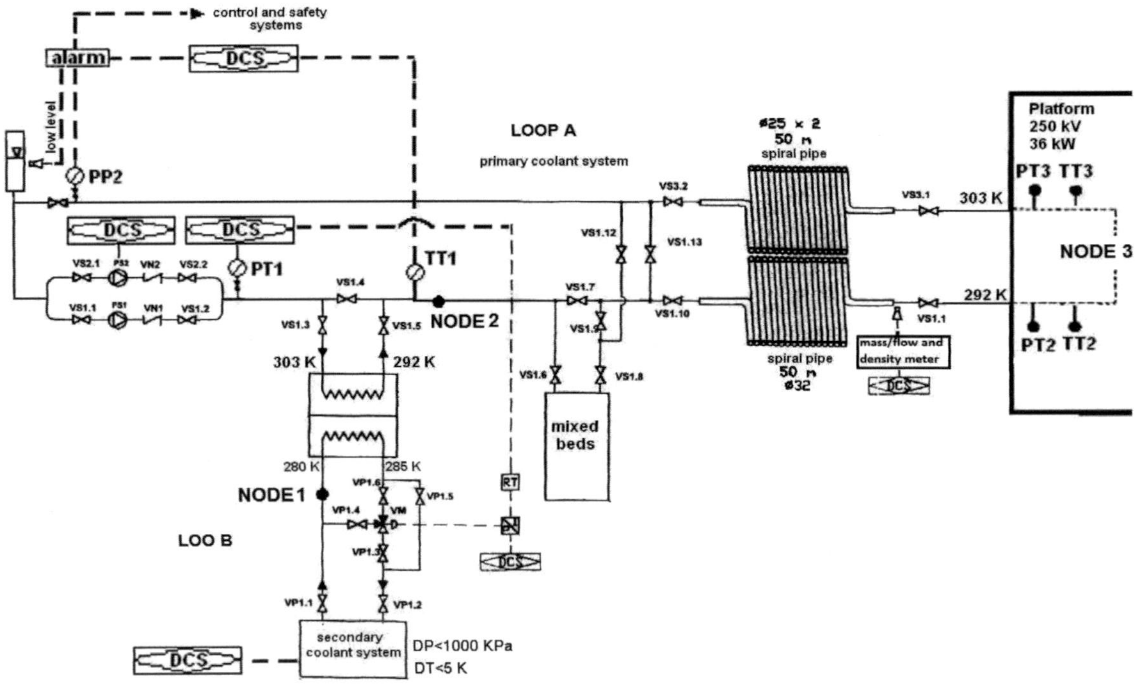

Figure 3. Target cooling system design in the SPES facility.

4. Analysis and results

RELAP5 code has been used, preliminary, to perform the tuning of nodalization behaviours in steady state conditions respect to thermal-hydraulic physical parameters such as pressures drops, water temperatures in the primary and secondary loops, mass flow rates, temperature of target, temperature of instruments and measurement systems.

The variation of each parameter is set according to reasonable variations achieved during the plant nominal operation. The steady-state simulation has shown results congruent to physical expectations.

Subsequently, different accidental scenarios have been hypothesized:

- Loss Of Coolant Accidents (LOCAs), under difference cases for break scenarios;
- Overpower transients, under difference conditions for uncontrolled power production.

Moreover, occurrence of failures of the safety and control devices are assumed to be worse conditions.

4.1 RELAP5 nodalization

The RELAP5 nodalization used for the accidental scenarios simulation is shown in figure 4; moreover the three dimensional representation of the plant model described in the RELAP5 input file, which runs by the RELAP5 Graphical User Interface, is shown in figure 5.

In the primary coolant loop, identified with loop A (figure 3), the expansion tank is represented by pipe (095) of 100 liter, connected to the pipe (097) by using a motor valve (096), which is open during both steady-state and accidental conditions. The branch (130) allows to connect the return primary coolant pipeline to the redundant pipeline, where the pumps are located. The components (171) and (172) schematize the two pumps and impose the nominal operating mass flow rates. The input data needed for the homologous curves which describe the pump behavior under single and two-phase flow conditions are in general known only for a few small-scale pumps, so the pump schematization has been extrapolated from existing data, obtained for different designs and scaling factors, used in the RELAP5 code.

The redundant primary coolant feed pipeline is schematized by using pipes from (141) to (221), and the auxiliary feed pipeline is schematized by using pipes from (142) to (222).

The inlet and outlet water temperatures in the loop B (figure 3) are imposed by using time-dependent volumes (010) and (050), respectively (figure 4). These volumes are connected to heat structures to simulate the heat exchange between primary and secondary coolants flowing into the corrugated heat exchanger.

The time-dependent junction (021) imposes the secondary mass flow rate according to nominal operating conditions.

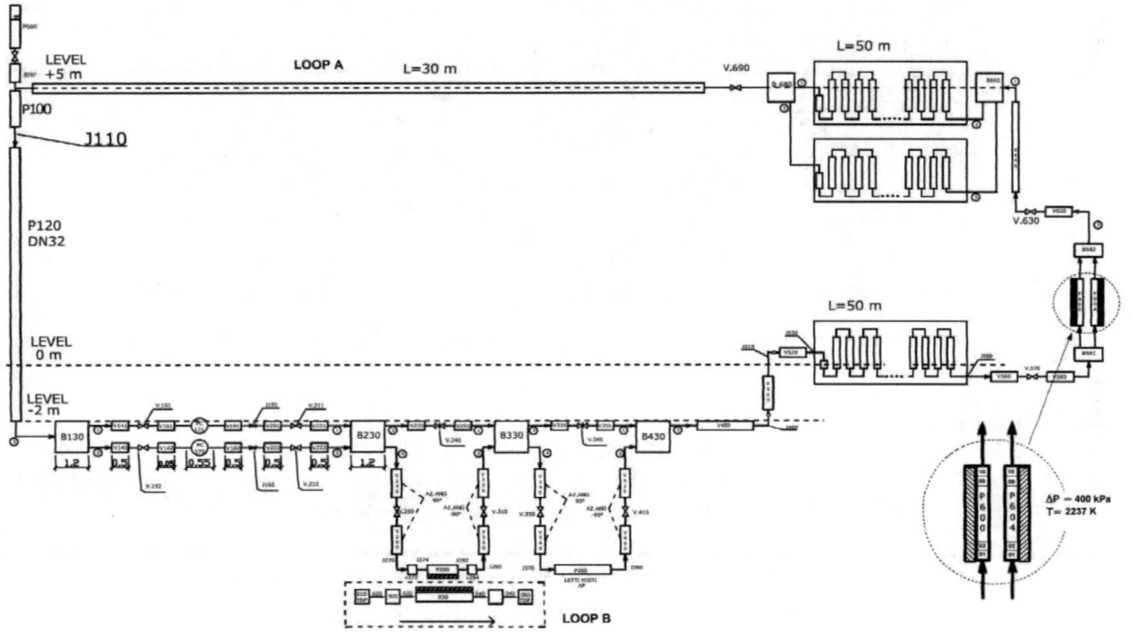

Figure 4. RELAP5 nodalization of the target cooling system.

The pipe (380) represents the mixed beds and the pipes (540), (660) and (670) schematize the helical pipes located at the inlet and outlet of the platform. Ninety-nine axial nodes are used to model the horizontal helical shape of the tubes.

The primary coolant pipelines in the target volume and measurement systems, which allow to remove the heat power, are schematized by using pipes (600) and (604) respectively.

Two heat structure nodes are connected to the pipes (600) and (604) and are used to account for heat exchange between the primary coolant and the target chamber.

Journal of Physics: Conference Series **547** (2014) 012032 doi:10.1088/1742-6596/547/1/012032

Appropriate radial peaking factors are defined to distribute the total power within the target and the measurement systems. Four axial nodes were used for both the hydrodynamic components (600) and (604) and ten corresponding sub-structures.

Finally, dedicated integral control variables are modelled to evaluate the heat power exchanged between the primary coolant and the target chamber (cntrlvar 108 for the target system and cntrlvar 109 for the measurement systems), as well between primary coolant system and secondary coolant one (cntrlvar 110).

The technique used to model the break is the one proposed in the RELAP5 user manuals suggested for different break sizes [13]. For small and medium break sizes, a trip valve component (331) is used. This component is connected to the branch (430) of the cold primary pipeline and stays closed during the steady-state phase and opens at the time of the break initiation.

The abrupt area change option was enabled to account for the additional pressure losses due to the sudden contraction and expansion of the flow at the break. The default chocked flow model is enabled to take into account chocked condition at the break during the phases of the accident. For all the break scenarios, the discharge volume is simulated with a time-dependent volume.

The elevation change of volumes and sub-volumes are set according to plant specifications.

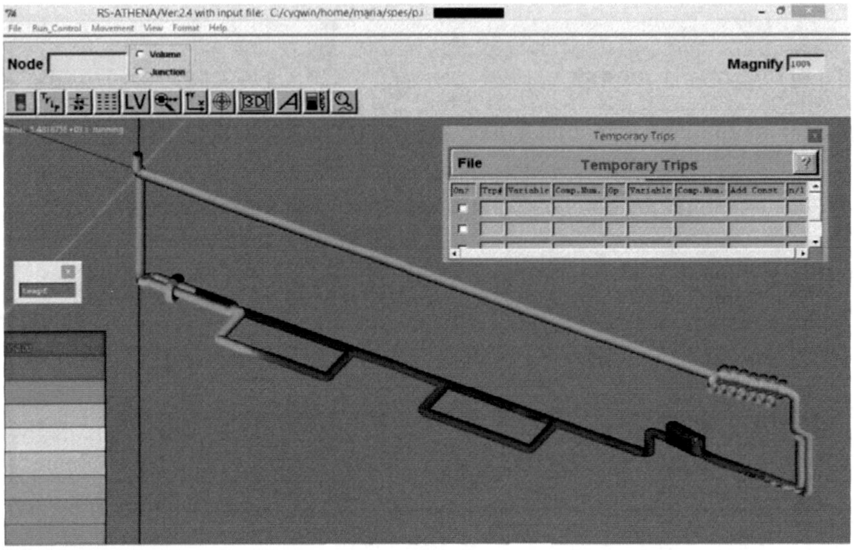

Figure 5. Three dimensional representation of the plant model described in the RELAP5 input, this view is done by using the RELAP5-3D Graphical User Interface.

5. LOCA transients

The accidental event occurs at t =10100 s, during nominal operating condition, when the trip valve (331) opens. Different breaks are tested, as reported in table 1, where the ratio of break area to pipeline area, A/A_0, and the ratio of break mass flow to pipeline mass flow, G/G_0, are shown.

Table 1. Break area used to simulate Loss Of Coolant Accidents.

Break area/pipeline area A/A_0 [%]	Break mass flow rate/pipeline mass flow G/G_0 [%]
1	18
2	36
3	54
4	72
5	85

Figures 6 through 12 show the results obtained by LOCA transients characterized by break of 1% and 5%, which are compared to each other. In particular, figure 6 shows the break mass flow for both the accidental events; as expected the 5% break involves more significant loss of water coolant: approximately, the break mass flow rate is about 85% of the nominal mass flow rate (table 1). The primary coolant system pressure quickly decreases (more for the break of 5%), as shown in figure 7, where the pressure in the inlet pipeline of the target system is represented. Consequently, the pump increases the mass flow rate (figure 8), till vapour production in the pump suction volume (figure 9) causes pump degradation and drastic mass flow rate reduction. The pump degradation occurs much sooner for the break of 5% (after about 130 s from the beginning of the accident).

The water temperatures in the inlet and outlet coolant pipelines placed in the target system and measurement systems, respectively, show a little increase when the break occurs, as shown in figures 10 and 11. In fact, as above said, the pump regulation increases the mass flow rate, so the heat exchange between the primary coolant system and the target chamber is guaranteed. However, when the degradation pump occurs and the regulation stops, the water temperatures quickly increase (figures 10 and 11) and the pressure further decreases (figure 7).

As consequence, the heat structure temperatures, which schematize the target system, increase and exceed about 1500 K at about t=10800 s for break of 1%, and about t=10350 s for break of 5 % (figure 12), moreover the heat structure temperatures, which schematize the measurement systems, increase and exceed about 1000 K at about t=10820 s for break of 1%, and about t=10360 s for break of 5 % (figure 13).

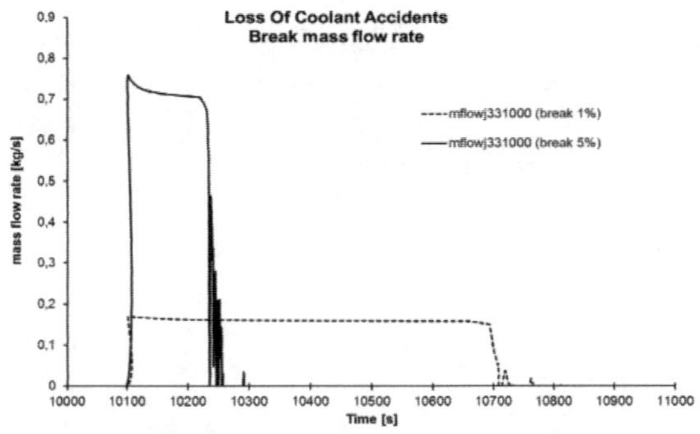

Figure 6. Break mass flow rate during LOCA accidents, for break of 1% and 5%.

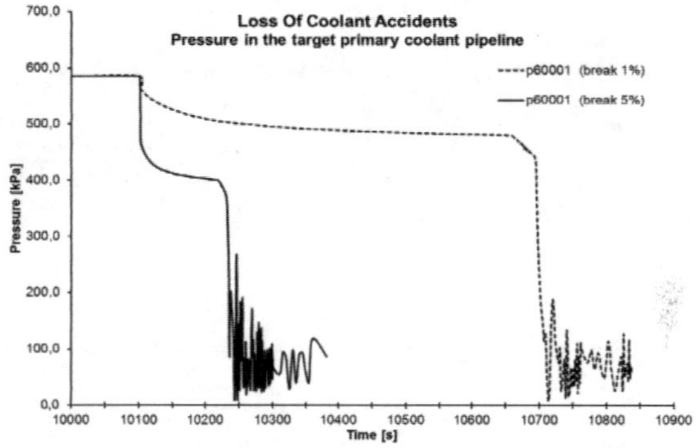

Figure 7. Pressure in the inlet pipeline of the target system during LOCA accidents, break of 1% and 5%.

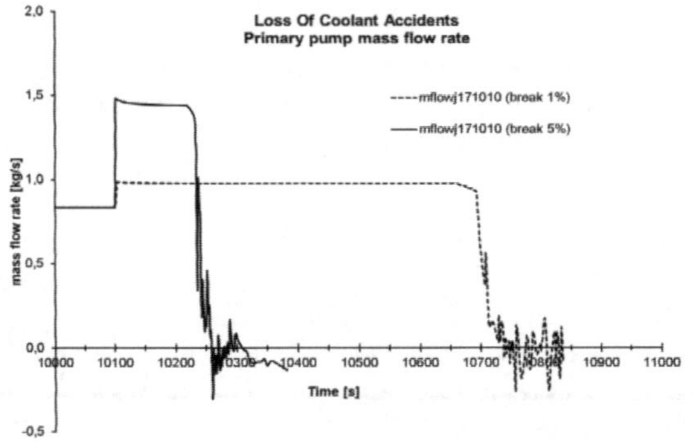

Figure 8. Pump mass flow rate during LOCA accidents, for break of 1% and 5%.

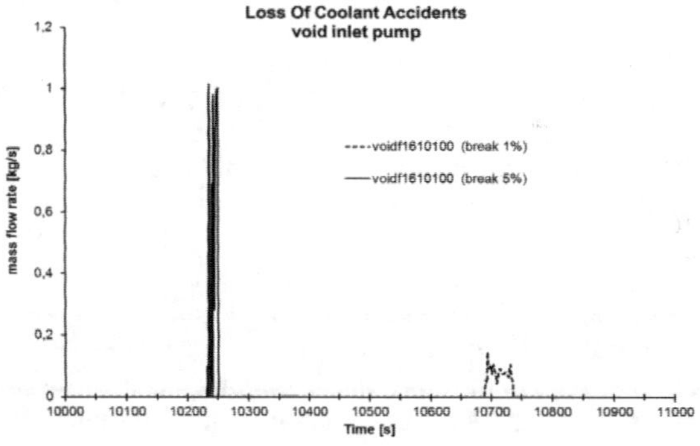

Figure 9. Void fraction in the pump suction volume during LOCA accidents, for break of 1% and 5%.

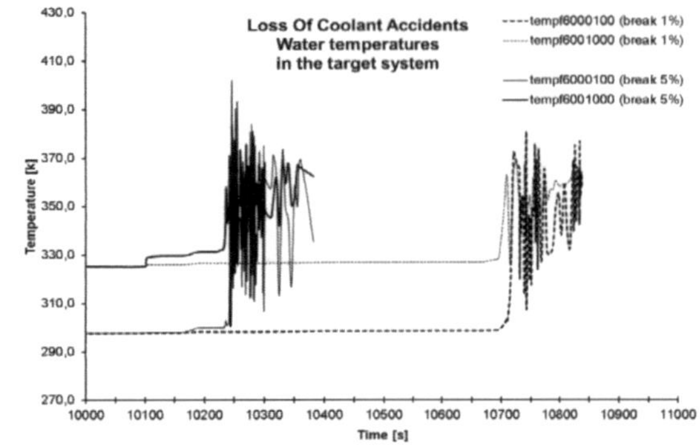

Figure 10. Water coolant temperatures in the inlet and outlet pipelines of the target system during LOCA accidents, for break of 1% and of 5%.

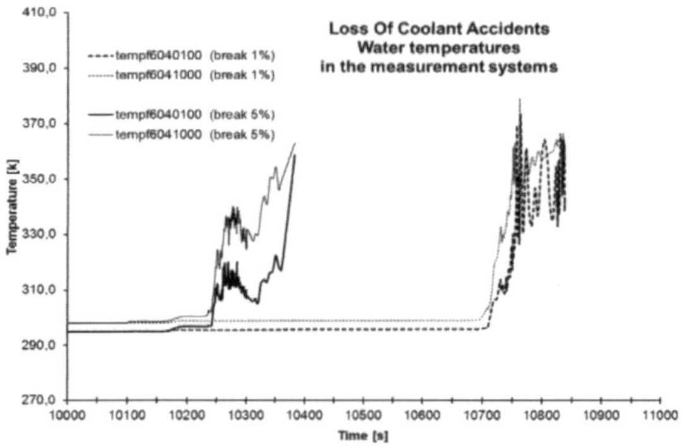

Figure 11. Water coolant temperatures in the inlet and outlet pipelines of the measurement systems during LOCA accidents, for break of 1% and 5%.

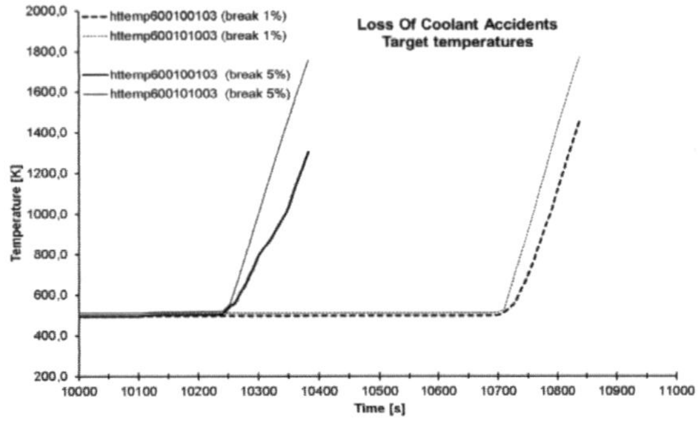

Figure 12. Temperatures of the heat structures used to simulate the target system during LOCA accidents, for break of 1% and 5%.

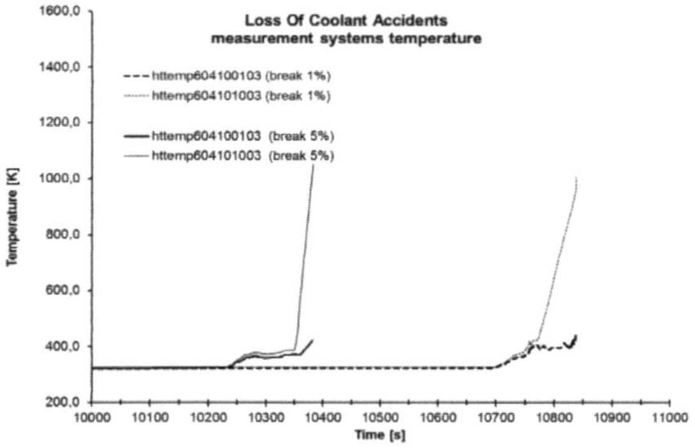

Figure 13. Temperatures of the heat structures used to simulate the measurement systems during LOCA, for break of 1% and 5%.

6. Overpower transients

The beam-target system subjected to uncontrolled primary beam power or electric current increase leads to overpower transients with immediately mismatch between produced and removed heat power and consequent overheating of sub-assemblies.

Failures of the temperature control safety systems, located in the platform of the target chamber, are considered to be worse conditions. Different accidental situations, namely 20%, 40%, 50% and 100% overpower starting from nominal conditions, have been considered. In the following the results obtained by 100% overpower simulation are reported.

The initiating event start at t =10100 s. A quickly power increase up to about 72 kW takes place, as shown in figure 14 where the total power produced in the beam-ion source target is reported. The inlet and outlet water temperatures in the target and measurement systems increase as showed in figure 15.

However, the corresponding heat structure temperatures do not exceed the acceptance criteria in terms of temperature limits (i.e. well below of the copper melting point) allowed for the target chamber, as shown in figures 16 and 17 [5].

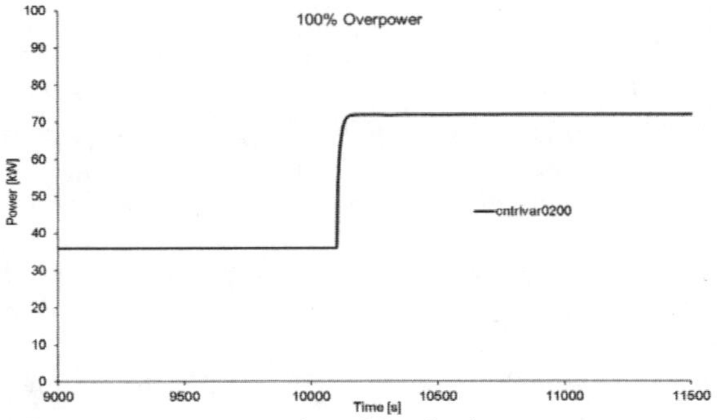

Figure 14. Power produced in the ion-source target system.

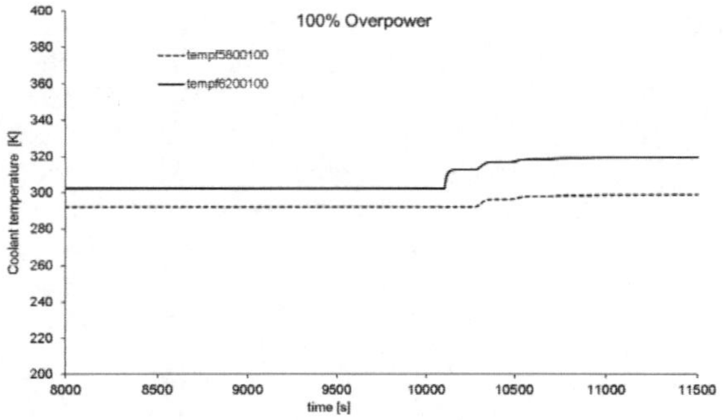

Figure 15. Inlet and outlet primary coolant temperatures in the target and measurement systems.

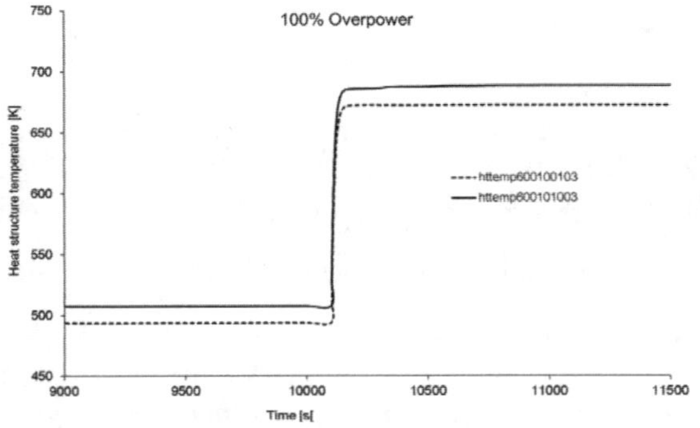

Figure 16. First and last heat structure temperatures used to simulate the target system.

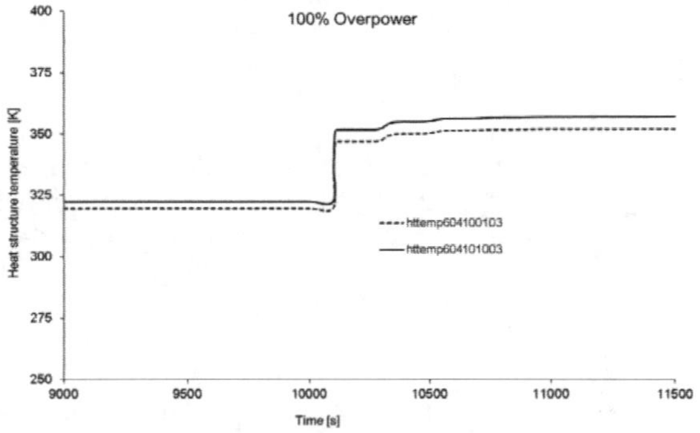

Figure 17. First and last heat structure temperatures used to simulate the measurement systems.

7. Conclusion

The main goal of the SPES experimental facility is to provide an accelerator system to perform forefront research in nuclear physics by studying nuclei far from stability. In particular, the main purpose of the project is the production of neutron-rich radioactive nuclei with mass in the range 80-160. The emphasis to neutron-rich isotopes is justified by the fact that this vast region has been little explored, at exceptions of some decay and in-beam spectroscopy following fission.

The production method is based on the ISOL technique with direct target: a primary proton beam impinges directly on a UCx target inducing the U fission with a maximum rate of 10^{13} f/s. The fission products are extracted, ionized, selected and reaccelerated to produce the secondary radioactive beam. The UCx target is the critical item of the facility from the point of view of the safety due to the large level of radioactivity induced by the fission reactions. As the facility will handle radioactive species, special care is devoted to the radiation protection safety and several systems are added to prevent radiation hazards.

To assure the operational temperature range, the production target and the surface ion source are located inside a cylindrical chamber cooled by a water cooling system.

In this paper safety analysis addressed to verify some design decisions of the water cooling system has been performed.

Different accidental scenarios have been hypothesized and examined by using the RELAP5-3D 2.4.2 advanced qualified thermal-hydraulic system code:

- Loss Of Coolant Accidents (LOCAs), under difference cases for break scenarios;
- Overpower transients, under difference conditions for uncontrolled power production.

In order to simulate worse condition, control and safety system failures have been considered.

The cooling system has shown a good performance during the examined accidental transients, confirming the good safety standards. In fact, the acceptance criteria are not exceeded in case of 100% overpower accident, moreover in case of LOCA accidents the temperature limits in the target chamber go beyond acceptance criteria, but the time interval from beginning of the initiating accidental event is sufficiently to allow the intervention of the operator.

References

[1] Prete G and Covello A, 2008, SPES Technical Design Report, *INFN-LNL* 223, http://www.lnl.infn.it/spes /TDR2008/tech_design08_index.htm.

[2] De Angelis G, Andrighetto A, Bassato G, Biasetto L, Calabretta L, Comunian M, Galatà A, Giacchini M, Gramegna F, Lombardi A, Manzolaro M, Prete G, Sarchiapone L, Scarpa D and Zafiropoulos D, 2011 Future Perspectives of the Legnaro National Laboratories: The SPES project, *Journal of Physics*: Conference Series 267 012003.

[3] Lombardi A, Andrighetto A, Bisoffi G, Comunian M, Favaron P, Gramegna F, Maggiore M, Piazza L A C, Prete G and Zafiropoulos D, 2013, SPES project: a neutron rich ISOL facility for re-accelerated RIBS Proceedings of Cyclotrons, Vancouver, BC, Canada.

[4] Andrighetto A, Cevolani S and Petrovich C, 2005, Fission Fragments Production from Uranium Carbide Disc Targets, *Eur. Phys. J.*, **A25**, 41-47.

[5] Andrighetto A, Antonucci C M, Cevolani S, Petrovich C and Santana Leitner M, 2006, Multifoil UCx target for the SPES project -An update, *Eur. Phys. J.* **A30**, 591-601.

[6] Andrighetto A, Antonucci C, Barbui M, et al., November 2007, The SPES direct UCx target, *The European Physical Journal Special Topics,* , Volume **150**, Issue 1, pp 273-274

[7] Andrighetto A, Antonucci C M, Cevolani S and Petrovich C, 2006, ENEA contribution to the design of the thin target for the SPES project, FIS-P815-020.

[8] Andrighetto A, Antonucci C M, Cevolani S, Petrovich C and Santana Leitner M, 2006, Multifoil UCx target for the SPES project - An update, *Eur. Phys. J.* **A30**, 591.

[9] Cevolani S, 2006, Valutazione approssimata dell'effetto della frequenza di pulsazione del fascio sulla termica del target sottile per SPES., FIS-P815-022.

[10] Manzolaro M, Meneghetti G and Andrighetto A, 2010, Thermal-electric numerical simulation of a surface ion source for the production of radioactive ion beams, *Nuclear Instruments and Methods in Physics Research* A, **623** , 1061–1069.

[11] Manzolaro M, 2003, Study, design and test of the target-ion source system for the INFN SPES facility, *PhD Thesis, XXIII PhD Cycle, Department of Mechanical Engineering, University of Padua.*

[12] Petrovich C, 2006, The SPES target – Power deposition and preliminary radiation protection analyses, FPN-P815-007 ENEA.

[13] June 2005, RELAP5-3D Code manual volume **I**: code structure, system models, and solution methods, Idaho National Laboratory INEEL-EXT-98-00834, Revision 2.4.

Journal of Physics: Conference Series **547** (2014) 012033 doi:10.1088/1742-6596/547/1/012033

Numerical simulation of turbulent forced convection in liquid metals

S Vodret, D Vitale Di Maio and G Caruso

"Sapienza" University of Rome – Dep. of Astronautic, Electrical and Energy Engineering (DIAEE) – Corso Vittorio Emanuele II, 244 – 00186 Rome, IT

Email: gianfranco.caruso@uniroma1.it

Abstract. In the frame of the future generation of nuclear reactors, liquid metals are foreseen to be used as a primary coolant. Liquid metals are characterized by a very low Prandtl number due to their very high heat diffusivity. As such, they do not meet the so-called Reynolds analogy which assumes a complete similarity between the momentum and the thermal boundary layers via the use of the turbulent Prandtl number. Particularly, in the case of industrial fluid-dynamic calculations where a resolved computation near walls could be extremely time consuming and could need very large computational resources, the use of the classical wall function approach could lead to an inaccurate description of the temperature profile close to the wall. The first aim of the present study is to investigate the ability of a well-established commercial code (ANSYS FLUENT v.14) to deal with this issue, validating a suitable expression for the turbulent Prandtl number. Moreover, a thermal wall-function developed at Université Catholique de Louvain has been implemented in FLUENT and validated, overcoming the limits of the solver to define it directly. Both the resolved and unresolved approaches have been carried out for a channel flow case and assessed against available direct numerical and large eddy simulations. A comparison between the numerically evaluated Nusselt number and the main correlations available in the literature has been also carried out. Finally, an application of the proposed methodology to a typical sub-channel case has been performed, comparing the results with literature correlations for tube banks.

1. Introduction

Differently than water, liquefied metals are characterized by a very low range of Prandtl numbers (of the order of 10^{-2} to 10^{-3} due to their high heat conductivity). This means that molecular transfer of momentum is significantly less intense than the molecular thermal diffusivity and so, for turbulent flows, the heat transfer in the channel could be essentially molecular.

Existing engineering computer codes used to simulate turbulent heat transfer adopt the turbulent Prandtl number model based on the Reynolds analogy, that states a coupling between velocity and temperature fields. Although this has already been validated for fluids with a Prandtl number of the order of unit or higher (such as water or air), the applicability for low Prandtl number fluids has to be reassessed.

Since experiments are scarce, expensive and difficult to perform, detailed data regarding turbulent heat transport at different Reynolds and Prandtl numbers are currently available in literature thanks to advanced simulation approaches such as direct numerical simulations (DNS) and large eddy simulations (LES). These data are then used to suggest new laws (wall functions) to be used in Reynolds Averaged Navier-Stokes (RANS) simulations, which provide for most cases the required accuracy and the lowest cost. An exhaustive list of issues to face within liquid metals Computational Fluid Dynamics (CFD) simulations may be found in literature [1].

A first issue of these simulations is the computation of heat diffusivity through the value of kinematic eddy viscosity ν_t and a turbulent Prandtl number. Since turbulent transport of momentum and heat or mass is due to the same mechanism (eddy mixing), the Pr_t value is classically supposed to be approximately 1. From experimental data with water and air as working fluids, indeed, Pr_t has an average value of 0.85 and ranges from 0.7 to 0.9 depending on the Prandtl number of the fluid. This is not valid for liquid metals and different expressions have been proposed for this parameter [2, 3, 4].

The first aim of the paper is to assess ANSYS FLUENT capability to perform wall resolved RANS simulations of heat transport in liquid metals, by using suitable expressions available in the literature for the turbulent Prandtl number. Moreover, the unresolved approach has been analyzed. When the wall function approach is chosen in RANS simulations a proper wall function must be selected to link correctly the value of a certain quantity at the first grid point to the one assumed at the wall (i.e. the temperature or the velocity). While for the velocity profile the linear and log laws are assessed in several cases, the same is not possible to say dealing with the temperature profile in low Pr flows, since no logarithmic region is visible. An original methodology to overcome this limit has been developed and validated using a new law for the thermal wall-function proposed by Bricteux [5, 6].

2. Turbulent Prandtl number correlations

The enhanced heat transport in liquid metals, due to their high thermal conductivity, requires new correlations to link turbulent viscosity to turbulent heat diffusivity, by means of the turbulent Prandtl number ($Pr_t = \nu_t / \alpha_t$). The needing of assessment is evidenced by the numerous correlations proposed in the past through experimental campaigns or analytical studies. Despite this second approach allows a deeper understanding of the physics behind, the first one is more accurate and it is thus preferred. Models of Reynolds [2], Kays [3], Weigand [4, 7], Aoki [8], Dwyer [9] and Cheng and Tak [10] are only some of the suggested correlations and an overview of the other ones can be found in [11].

Among these correlations, two of them were chosen in this study for their assessment: the model of Reynolds [2] and the model of Kays [3], respectively:

$$\mathrm{Pr}_t = (1+100\,\mathrm{Pe}^{-0.5})\left(\frac{1}{1+120\,\mathrm{Re}^{-0.5}}\right) \tag{1}$$

$$\mathrm{Pr}_t = 0.85 + \frac{0.7}{\mathrm{Pr}}\frac{\nu}{\nu_t} \tag{2}$$

3. Wall resolved approach: RANS of turbulent heat transfer

Many channel flow data are available in literature thanks to numerous DNS and LES performed by several authors, such as Kawamura [12] and Abe [13], who performed DNS at Re_τ up to 1020 with Prandtl numbers ranging from 0.025 to 0.71, or Bricteux et al. [5], who made DNS at $Re_\tau = 180$ and 590 case with Pr = 0.01-0.025 together with LES up to $Re_\tau = 2000$ with the same range of Pr number. The works of Moser [14] and Tiselj [15] are also available. A 2D turbulent RANS analysis is performed at turbulent Reynolds number ($=u_\tau\, \delta/\nu$) of 395 with constant wall heat flux and Pr = 0.025 to be compared with the results of the DNS performed by Kawamura [16].

3.1. The channel flow geometry and mesh

The channel flow geometry and mesh are reported in figure 1. The domain consists of two infinite plates separated by a distance 2δ and it is reproduced through a 2D section on the x-y plane. It is assumed that the fluid is flowing along the x axis. The geometry model is characterized by dimensions $L_x \times L_y = 2\pi\delta \times 2\delta$ (where δ is the half channel width) and it is nodalized with $N_x \times N_y = 64 \times 48$ divisions. The mesh is uniform among the streamwise direction while the "bias" tool was used to place more cells next to the wall, ensuring that $y_l^+ < 1$ as required for wall resolved simulations.

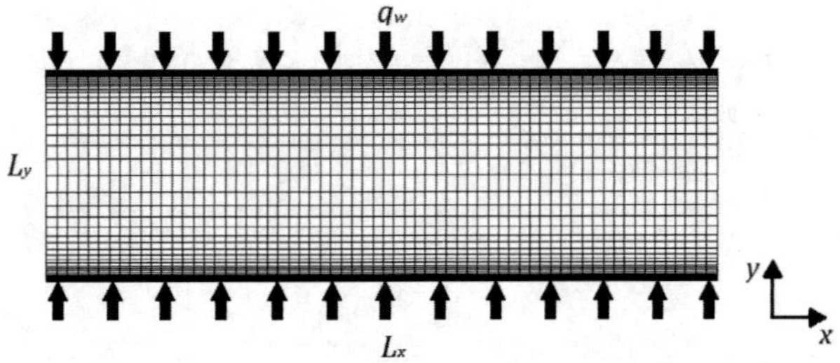

Figure 1. Mesh used for the channel flow at $Re_\tau = 395$.

3.2. Equations and case setup

The RANS equations [17] are modified taking account the hypothesis of steady state and incompressibility. Since the steadiness is simulated through periodic boundary conditions in the streamwise direction the resolution of the energy equation does not converge to a steady state result since the cyclic boundary condition at the channel's inlet and outlet results in a continuous increase of the fluid's absolute temperature in the flow direction. The transformed form of the energy equation is thus employed to tackle this issue [17].

The realizable k-ε and the SST k-ω models are used for the specification of ν_t to solve RANS equations: two transport equations for these quantities are solved assuming μ_t isotropic. Further formulistic details of these methods are available in [17].

Field variables (stored at cell centers) must be interpolated to the faces of the control volumes to solve the discretized transport equations of momentum, turbulent kinetic energy, rate of dissipation of turbulent kinetic energy and temperature. To do this, the second order upwind spatial discretization was used, which ensures a second order accuracy. Gradients of solution variables are also required in order to evaluate diffusive fluxes, velocity derivatives, and for higher-order discretization schemes. A least squares cell-based gradient interpolation method was used, which is recommended for regular meshes, as it is accurate and minimizes false diffusion. Cell-face pressures have to be computed when using the segregated solver. The linear interpolation scheme was used in this case.

3.3. Comparison with DNS results

Obtained profiles are reported in figure 2, together with the reference DNS. As one can see, for the realizable k-ε model there is a good agreement between the simulations, that match up to $y^+ \cong 60$, while the velocity is slightly underestimated in the center of the channel, where a wake is formed and the log law is not more valid. As it concerns the SST k-ω model, despite the perfect matching is up to $y^+ \cong 10$, the overall behavior is in good agreement with the reference DNS.

As expected, the linear sub-layer extends up to $y^+ \cong 5$, while for $10 < y^+ < 40$, where both viscosity and turbulence play an important role to the total momentum transfer, there is the buffer layer, which forms the transition between the viscous wall region and the logarithmic wall region of the boundary layer. At least, the log-law layer starts from $y^+ > 50$. The temperature field has been solved through the classic, Reynolds and Kays correlations for Pr_t (setting a value of 0.85 for the first one). The reported dimensionless temperature profile is built using $\theta^+ = (T_w\text{-}T)/T_\tau$, where T_τ is the friction temperature, $= q_w/(\rho\ c_p\ u_\tau)$. The profiles of the Reynolds' and Kays' turbulent Prandtl numbers for the realizable k-ε model are reported in figure 3, together with the computed Pr_t using data from Kawamura [16]. The three temperature profiles are reported in figure 4 together with the results of the DNS by Kawamura [16] and the linear law for the laminar thermal sub-layer. The dashed curve represents the linear relation $\theta^+ = Pr\ y^+$ valid in the immediate vicinity of the wall.

As one can see, the classic approach fails to predict the temperature profile for this very low Prandtl number, underestimating the temperature difference with a relative error of 29.62 %. Despite the good agreement of the turbulent Prandtl number of the Kays correlation with the DNS' one (see figure 3), the θ^+ profile is slightly underestimated in the center of the channel. This might be because of the enhanced wall function of the realizable k-ε model used, which gives a too low value of the wall temperature. Obtained profile of θ^+ using the SST k-ω model is reported in figure 5.

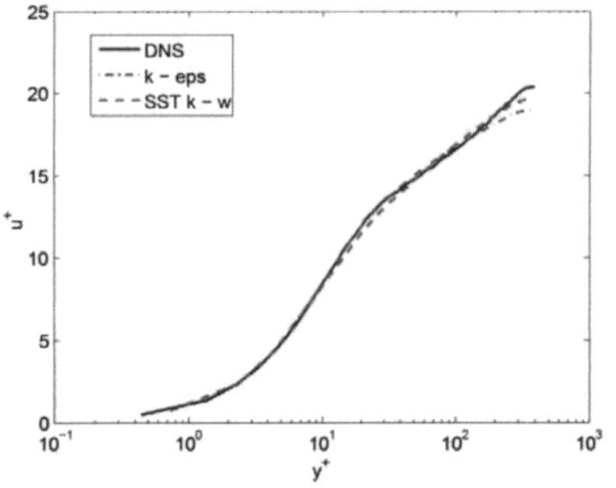

Figure 2. Dimensionless velocity profile for $Re_\tau = 395$.

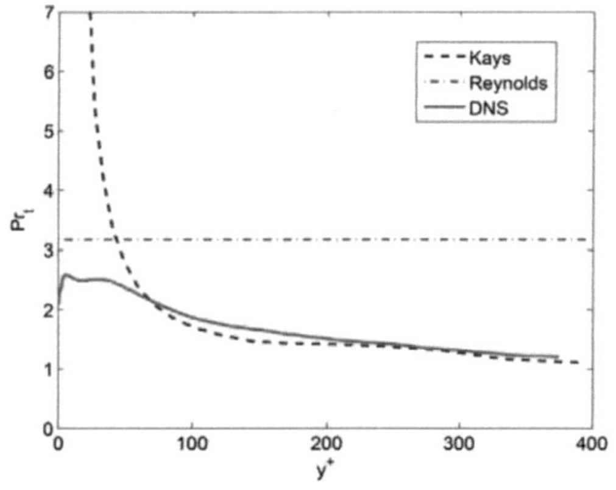

Figure 3. Values of Pr_t used for the realizable k-ε simulations.

In this case the Kays correlation gives a more accurate prediction of the temperature profile, while the Reynolds' one gives a slightly overestimation of θ^+.

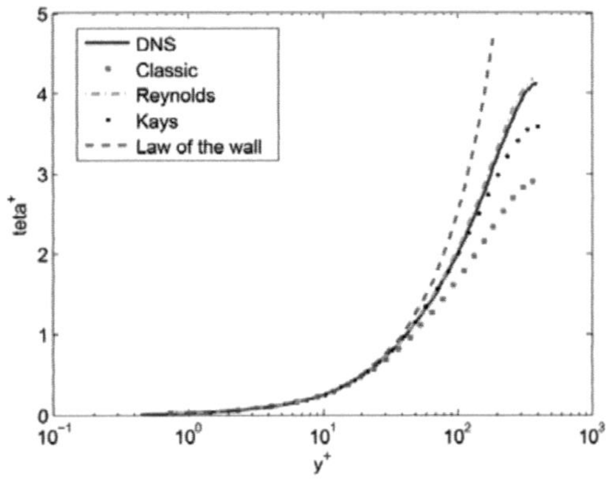

Figure 4. Mean temperature profile at Re_τ=395 and Pr=0.025 - realizable k-ε model.

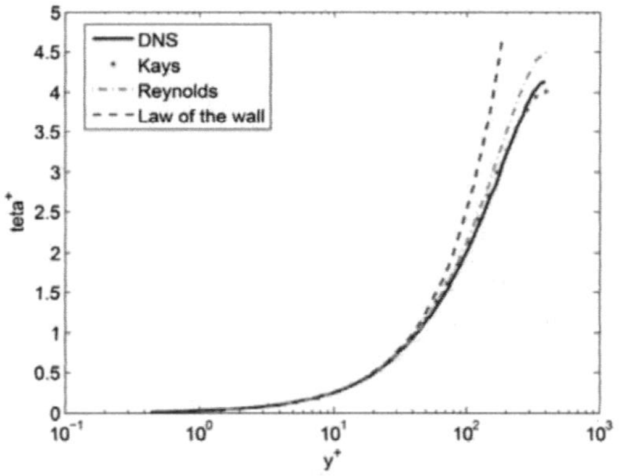

Figure 5. Mean temperature profile at Re_τ=395 and Pr=0.025 - SST k-ω model.

In table 1 the obtained Nusselt numbers are reported, computed using the following formula:

$$\mathrm{Nu} = \frac{q_w 2\delta}{k(T_w - T_b)} \tag{3}$$

Table 1. Computed Nusselt number for Pr=0.025.

Case	Nu	Case	Nu
DNS	6.52	Classic approach	8.19
Reynolds correlation	6.06	Kays correlation	6.52

The classic approach clearly overestimates the heat transfer coefficient, while the other results obtained with the Reynolds and Kays correlations are in good agreement with the reference data.

3.4. RANS calculations for Re_τ = 590 and 2000; comparison with LES results

Similar calculations for RANS at Re_τ = 590 and 2000 (using LES performed by Bricteux et al. [6, 18] as reference) have been performed and just the velocity and temperature profiles are reported below. The mesh was modified each time in order to maintain always the wall resolved approach, ensuring $y_1^+ \leq 1$. Comparing these results with the same obtained for Re_τ=395, a better agreement with literature data is detectable as Re_τ increases, especially for velocity profiles. This is due the better behavior at higher Re_τ of the turbulence models. As it concerns the temperature profiles, if at Re_τ=395 both Reynolds and Kays' correlations gave similar results, here the second one always gives better results and thus might be chosen for future simulations.

4. Wall function approach

So far simulations on a very fine grid have been performed, ensuring that the first node along the y axis was positioned in such a way that $y_1^+ \leq 1$. This leads to very heavy simulations, which are not feasible in industrial field because this method devotes a substantial fraction of the computational time to the near wall region, where velocity and temperature profiles are steep due to viscous effects.

Another way to perform RANS simulations is the so called "wall function approach". This method allows to use an uniform mesh, putting the first point of the grid among the y axis at y_1^+ much greater than one and using semi empirical formulas called "wall functions" to link the viscosity affected region between the wall and the fully-turbulent region. In this section, the wall function proposed by Bricteux et al. [5], applicable for low Pr numbers, is assessed.

Any solver offers several wall functions selectable, giving the user the possibility to build its own. In this case, FLUENT replaces with customized wall functions the standard ones, based on the work of Launder and Spalding [19] using U^* and y^*, defined below:

$$U^* = \frac{\rho U_P C_\mu^{0.25} k_P^{0.5}}{\tau_w} \quad \text{and} \quad y^* = \frac{\rho C_\mu^{0.25} k_P^{0.5} y_P}{\mu} \tag{4}$$

These functions are based on the universal linear and log laws for velocity previously showed and, being built for conventional fluids, the law for temperature is dependent from the velocity one according to the Reynolds analogy. Since no log-law is visible in the temperature profile dealing with liquid metals, a new wall function has to be used. The law proposed by Bricteux et al. [5] is:

$$\theta^+ = \frac{\Pr_t}{\kappa} \lg\left(1 + \frac{\kappa}{\Pr_t} \Pr y^+\right) \tag{5}$$

This law was compared to DNS and LES data for a range of Péclet numbers (Pr=0.025-0.01 and Re_τ=180-2000) and it fits well the data in the range y^+=70-300 where the logarithmic law for the velocity profile is applicable.

Since ANSYS FLUENT does not allow the user to define a custom wall function for the temperature field, a strategy to overcome this limit is proposed in this study and consists in a gradual solution of the flow field, as explained below.

First, the velocity field is computed using the enhanced wall treatment. Once the solution is converged, a user defined wall function for the velocity is hooked up in the solver, thus the momentum and turbulence equations are freezed, while the energy model is activated. When the calculation is run again, the user defined wall function will not affect the velocity field, but will be used by the solver to compute the temperature at the wall. Obviously this procedure is applicable only with the hypothesis of incompressibility of the fluid and considering constant its thermal properties.

The UDF employed is coded reasoning on the standard wall functions (since these wall functions are replaced by custom's one).

Since this approach has been tested using a y_1^+=100, only the Re_τ=2000 case has been solved, as a too coarse mesh would have been necessary to solve the other two cases. The obtained temperature profile is reported in figure 6, solving the energy equation using the Kays correlation for Pr_t since it was more rewarding than the Reynolds' one in the wall resolved approach.

From the results it is evident that the thermal wall function predicts accurately the wall temperature profile, providing results as similar to the reference ones as those obtained using the wall resolved approach, despite the coarse and uniform mesh which drastically reduces the computational time.

5. Forced convection heat transfer in liquid metals

The turbulent heat transfer in circular and flat ducts is one of the most investigated configuration of the liquid metal heat transfer with lead, Lead Bismuth Eutectic (LBE), mercury or sodium.

For forced convection flows of liquid metals, dimensional analysis shows that $Nu = h\,D/k = f(Pe)$, where Pe is the Peclet number ($Pe = \rho c_p D(u_{av})/k = Pr\cdot Re$) [20]. A very large number of correlations, both theoretical and empirical, has been developed in the last sixty years for the constant wall temperature and constant heat flux cases, evidencing the complexity and difficulties related to the study of this phenomenon. Three models are considered in this study.

The first one was proposed by Lyon [21] and it is suitable for pure metals with a perfect wettability of the walls:

$$Nu_{Lyon} = 7 + 0.025Pe^{0.8} \tag{6}$$

The second correlation was obtained by Lubarsky and Kaufman [22] and it should be more representative of real facilities whereby liquid metals are seldom pure:

$$Nu_{Lub} = 0.625Pe^{0.4} \tag{7}$$

This correlation was recently reviewed by Cheng and Tak [10], leading to:

$$Nu_{Cheng} = A + 0.018Pe^{0.8} \tag{8}$$

Where

$$A = \begin{cases} 4.5 & for\, Pe < 1000 \\ 3.6 & for\, Pe > 2000 \end{cases} \tag{9}$$

Global heat transfer performances of the considered cases are analyzed in this section through the obtained values of the Nusselt number. A comparison of the numerically calculated Nu with the reference and the computed through the most famous literature expressions ones is made. As reference Nu parameters, the one obtained through the DNS of Kawamura [16] for the $Re_\tau=395$ case and the two obtained through the LES of Bricteux et al. [6, 18] for the $Re_\tau=590$ and 2000 are used. The Lyon, Lubarsky and Kaufmann [22] and Cheng and Tak [10] correlations were used. All the considered parameters are reported in table 2 and in figure 7.

Table 2. Nusselt numbers and main parameters involved.

Re_τ [-]	T_w [K]	T_b [K]	k [W/m K]	U_b [m/s]	Re [-]	Pe [-]	Nu_{CFD} [-]	Nu_{ref} [-]	Nu_{Lyon} [-]	Nu_{Lub} [-]	Nu_{Cheng} [-]
395	303.027	300	101.26	17.59	13897	347.43	6.52	6.52	9.69	6.49	6.44
590	303.789	300	67.79	18.94	22347	558.67	7.78	7.89	10.94	7.85	7.34
2000	306.773	300	20.00	22.67	90686	2267.14	14.76	14.39	19.09	13.74	12.30

As expected, the Nu numbers from CFD calculations Nu_{CFD} and Nu_{ref} (in good agreement with each other) are slightly higher than Lubarsky and Kaufman and Cheng ones, since CFD does not consider the occurrence of impurities and the wettability problem, leading to higher Nu values. Lyon's correlation is however too optimistic, since it always overestimates the values of Nu (from 48.62 % to 32.66% for higher Re number). Despite it considers the same thermo-fluid dynamic conditions of CFD simulations, the disagreement might be because of the few set of experimental data used to extrapolate Lyon's formulation.

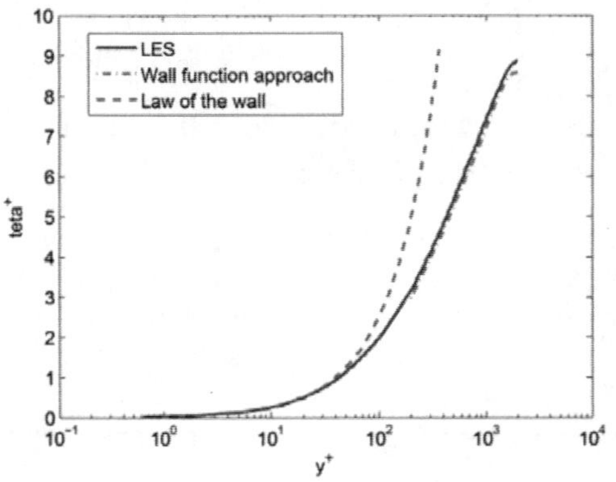

Figure 6. Dimensionless T profile for Re_τ=2000.

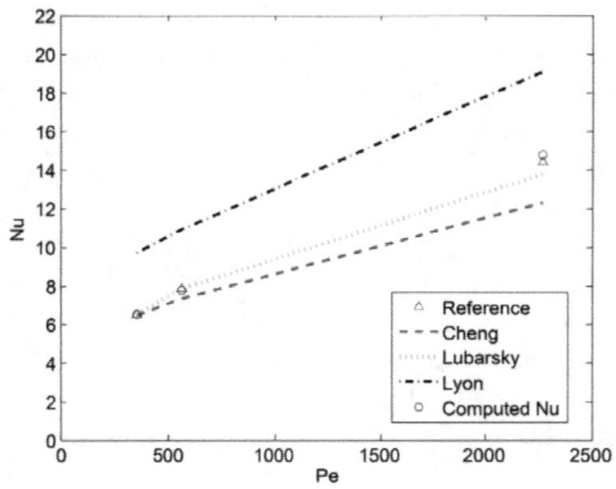

Figure 7. Comparison of Nusselt numbers.

6. The subchannel case

The evaluation of the heat transfer in a real subchannel with constant and cosine heat flux is carried out, on the basis of the previous study of the channel flow. Both the wall resolved and wall function approaches have been used to compute the average Nusselt number in the facility, making use of both the enhanced and user defined wall function for the latter case. The models of Mikityuk [23] and Ushakov [24] were used to assess the obtained results.

The simulations performed are based on the ALFRED (Advanced Lead Fast Reactor European Demonstrator) triangular core lattice geometry. This lead cooled reactor is the third considered within the LEADER (Lead - cooled European Advanced DEmonstration Reactor) project, and is thought to be realized in the short term, relying on presently available technology. This scaled down reactor is thus designed using solutions as much as possible close to the adopted reference conceptual design but considering the essential need to proceed to construction in a short time frame. Many design details of this reactor can be found in [25], here the main ones are reported in table 3. The domain used in the following simulations consists of 1/6 of a subchannel 0.7 m long (composed by the asset zone of 0.1 m and the active length of 0.6 m).

Table 3. ALFRED core parameters.

Parameter (Core)	Value	Unit	Parameter (Fuel pin)	Value	Unit
Thermal power	300	MWth	Total number of FA/Pins per FA	171 / 127	-
Coolant mass flow rate	25984	kg/s	Relative rugosity, ε	$5.17\cdot10^{-4}$	-
Coolant average velocity, $\bar{u}$	1.4	m/s	Cladding outer radius, r	$5.25\cdot10^{-3}$	m
Coolant inlet temperature, T_{in}	401.5	°C	Pitch to diameter ratio, p/D	1.32	-
Coolant outlet temperature, T_{out}	480	°C	Asset length/ Active height	0.1/0.6	m

6.1. The subchannel mesh and setup

The estimation of the y_1^+ is crucial to create the mesh to employ. Since both the wall resolved and the wall function approach have been tested, two meshes are built.

As usual, for the first approach the condition to respect is $y_1^+ < 1$, thus:

$$y_1^+ = \frac{y u_\tau \rho}{\mu} < 1 \tag{10}$$

Taking into account the values of the flow at the half of the active length, y_1=2.67·10^{-3} mm.

In order to gain a regular and hexahedral mesh the mapped face meshing and the multi-zone method have been used. To put more cells near the rod, the inflation method has been employed, setting 2.67·10^{-3} mm as first layer height for 10 layers and a growth equal to 1.43.

The mesh, composed by 240 divisions among the z axis, 19 divisions in the radial direction and 5 in the polar one (22800 elements), is reported in figure 8. As it concerns the wall function approach, the selected value for y_l^+ is ≅ 200. This will imply that y_l=0.5339 mm.

This choice leads to a mesh with only 2 divisions in the radial direction (while the divisions among the z axis and the polar direction are unvaried), leading to a mesh composed by 4338 nodes and 2400 elements (figure 9).

The two approaches have been solved for constant and cosine heat flux using the realizable k-ε model and solving the momentum and energy equations simultaneously (being lead modeled with temperature dependent quantities). The strategy described previously is then inapplicable. As it concerns the wall function approach, both the enhanced wall function and the new wall function have been tested. The considered expression for the cosine heat flux is:

$$q'' = 1.209 \cdot q''_{const} \cdot \cos\left[\pi(z-0.4)/0.9\right] \quad (11)$$

where q''_{const} = 697961 W/m^2 and the extrapolated length is 150% of the active length.

The SIMPLE algorithm is used for the pressure-velocity coupling, and for the spatial discretization the least square cell based method is used for gradient, the standard one for pressure and the second order upwind for every other quantity.

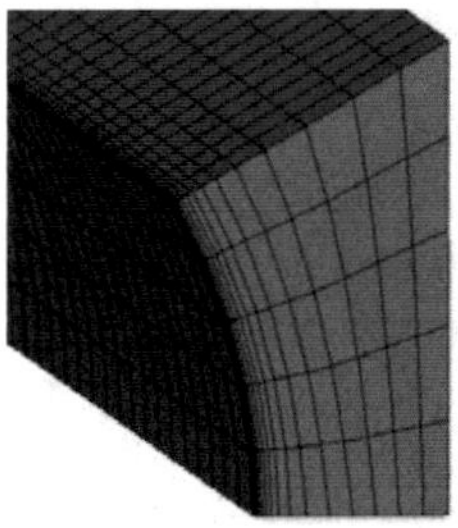

Figure 8. Isometric view of the wall resolved case mesh.

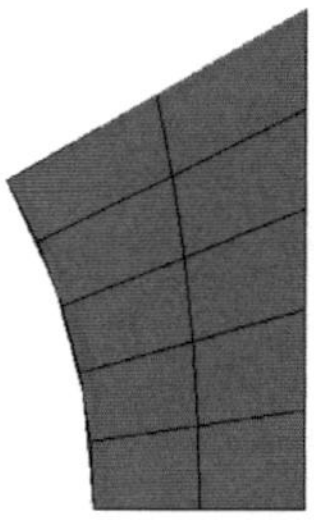

Figure 9. x-y plane section view of the wall function case mesh.

6.2. Results and discussion

The obtained Nusselt numbers for the constant heat flux and the cosine one for the wall resolved (WR) approach, the enhanced wall function (EWF) and the new wall function (NWF) ones are reported in figure 10 and in figure 11 and summarized in table 4 and in table 5.

Firstly, for the constant heat flux case two Nu numbers are obtained. The first one, Nu_{tot}, is averaged among the whole subchannel while the second one is averaged from z~0.2 m. The distinction is made seeing as in this case the "inlet effect" might be too sharp, since the fluid finds immediately a high heat flux, leading to unphysical values of Nu. Starting the average from z~0.2 m removes this effect. An analysis of a periodic channel with constant heat flux could elucidate this duality, but ANSYS FLUENT v.14 does not allow periodic analysis when the fluid properties are temperature dependent. This is not valid for the cosine heat flux case, since the fluid finds a gradually growing heat flux that avoids the above mentioned behavior.

Looking at these results, one can easily notice that the wall resolved and the new wall function approaches give similar results for Nu, but in this second case the number of iterations and the computational time are reduced. Since the two models of Mikityuk and Ushakov give an average value for Nu, they can be used as reference for any situation. As usual, the CFD obtained Nu number is higher than the experimentally obtainable with the models because of the non-perfect wettability of the wall and because of liquid metals impurities. Giving the Ushakov's model a higher value for Nu, this is the closer one to the CFD results. Finally, the enhanced wall treatment clearly underestimates the value of Nu, highlighting again how classic approaches are not suitable for liquid metals.

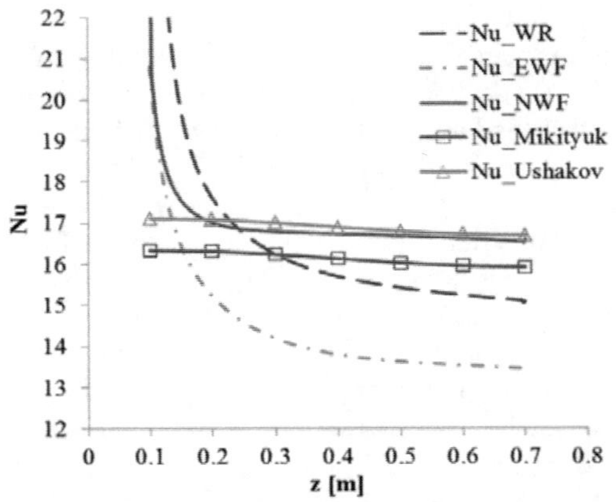

Figure 10. Nu axial profile in the subchannel for constant heat flux.

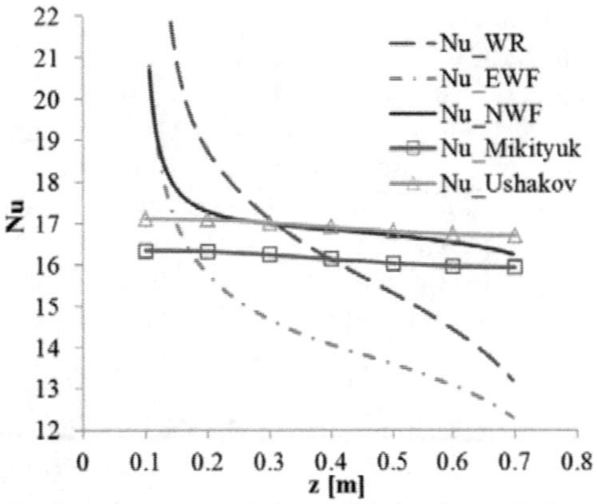

Figure 11. Nu axial profile in the subchannel for cosine heat flux.

Table 4. Average Nu values with mesh statistics and computational time.

Method	Constant heat flux Nu_{tot}	$Nu_{0.2}$	Cosine heat flux Nu_{tot}	Nodes	Elements	Time [s]	Iterations
Wall resolved	17.33	15.76	16.95	28920	22800	390	696
Enhanced wall function	14.52	13.88	14.41	4338	2400	25	307
New wall function	17.03	16.72	16.97	4338	2400	65	293
Mikityuk [23] / Ushakov [24]		16.13 / 16.91					

Table 5. Relative errors on Nu.

Method	Constant heat flux						Cosine heat flux		
	Nu_{tot}	ε_{Mik}	ε_{Ush}	$Nu_{0.2}$	ε_{Mik}	ε_{Ush}	Nu_{tot}	ε_{Mik}	ε_{Ush}
Wall resolved	17.33	7.4%	2.5%	15.76	-2.3%	-6.8%	16.95	5.1%	0.2%
Enhanced wall function	14.52	-10%	-14.1%	13.88	-13.9%	-17.9%	14.41	-10.7%	-14.8%
New wall function	17.03	5.6%	0.7%	16.72	3.7%	-1.1%	16.97	5.2%	0.4%

7. Main issues for the applicability in real cases

Flows involved in real industrial applications are characterized by the variation of density and the other thermo-physical properties with temperature. This implies that the continuity, momentum and energy equations must be solved simultaneously and being every flow parameter variable in space, any user defined function applied (in this case, the thermal wall function) must be as general as possible to account this variability properly.

Whether ANSYS FLUENT is used to deal with these cases, some issues may arise due some limits that are entailed in the code. Firstly, all the flow's equations must be solved at the same time and then the step by step strategy proposed for the presented channel flow case is not applicable. Secondly, the value of u_τ is not accessible through any cell macros in ANSYS FLUENT. A way to overcome this limit might be to preliminary compute this quantity's value in some sections of the domain of a real configuration (i.e. a subchannel) and then to impose a constant value in a thermal UDF.

Thus, a solution strategy to solve real cases through ANSYS FLUENT is the simultaneous solving of all the flow equations using directly the thermal wall function with a constant value of u_τ, but this approach has two limitations: a possible inaccuracy of the velocity field due the improper wall function used for the velocity and the imposition of a constant value for u_τ in the whole domain instead of the local one. Studies addressed to analyze the effect of these limits using this CFD software have then to be carried to validate the wall function approach in liquid metals for industrial applications.

8. Conclusions

The inadequacy of the Reynolds analogy with the classic turbulent Prandtl number concept for liquid metals has been confirmed and other expressions, such as the Reynolds' and Kays' ones have been assessed through wall resolved RANS calculations. Simulations have been carried in a channel flow for several Re_τ cases at Pr=0.025. Thanks to its dependence from local values, the Kays' expression is more accurate and seems more eligible for calculations in complex geometries. Afterwards a wall function approach has been investigated, testing a wall function law for the temperature field, developed by Bricteux et al. [5]. Both this law and the step by step methodology proposed in this study for its implementation in the solver gave rewarding results using ANSYS FLUENT comparing to DNS and LES analyses available in the literature. The comparison of the obtained Nusselt numbers with the reference ones and those coming from other available models, highlighted how the Lubarsky and Kaufman correlation seems to be the most accurate for the considered cases.

Moreover, a real case study has been carried out and from the comparison of the Nusselt numbers obtained it has been highlighted how the wall resolved and the new wall function approaches give results in good agreement with the reference ones, but with a drastically reduced computational time in the second case. Finally the limits of the adopted strategy to implement the new thermal wall function in real cases are discussed, showing a possible way to overcome them.

References

[1] Grötzbach G 2013 *Nucl. Eng. Des.* **264** 41-55
[2] Reynolds A 1975 *Int. J. Heat Mass. Tran.* **18** 1055-69
[3] Kays W 1994 *J. Heat Trans-T. ASME* **2** (116) 284-95
[4] Weigand B, Ferguson J and Crawford M 1997 *Int. J. Heat Mass. Tran.* **17** (40) 4191-6
[5] Bricteux L, Duponcheel M, Winckelmans G, Tiselj I and Bartosiewicz Y 2012 *Nucl. Eng. Des.* **246** 91-7
[6] Bricteux L, Duponcheel M, Manconi M and Bartosiewicz Y 2012 in *Proc. 6th European Thermal Sciences Conference (Eurotherm 2012) (Poitiers, France, 4-7 September 2012)*
[7] Jisha M and Rieke H 1979 *Int. J. Heat Mass. Tran.* **22** (11) 1547-55
[8] Aoki S 1963 *Bull. Tokyo Inst. Tech.* **54** 63-73
[9] Dwyer O 1966 *Atom. Energy Rev.* **I** (4) 3-92
[10] Cheng X and Tak N 2006 *Nucl. Eng. Des.* **236** 385-93
[11] Hasan B 2007 vol X no 1 (Nahrain University, College of Eng. Journal (NUCEJ)) pp 53-64
[12] Kawamura H, Abe H and Matsuo Y 1999 *Int. J. Heat Fluid Fl.* **III** (20) 196-207
[13] Abe H, Kawamura H and Matsuo Y 2004 *Int. J. Heat Fluid Fl.* **III** (25) 404-19
[14] Moser R, Kim J and Monsour N 1999 *Phys. Fluids* **IV** (11) 943-5
[15] Tiselj I, Berar A and Cizelj L 2011 "DNS database of temperature fluctuations in channel flows for liquid metals," Technical report, THINS project, "Josef Stefan" Institute.
[16] Kawamura H http://murasun.me.noda.tus.ac.jp/turbulence/poi/poi.html#poi395 [Online]
[17] ANSYS 2009 "ANSYS FLUENT 12.0 Theory Guide".
[18] Bricteux L, Duponcheel M, Manconi M, Winckelmans G and Bartosiewicz Y 2014 *Int. J. Heat Mass. Tran.* **75** 470-82
[19] Launder B and Spalding D 1974 *Comput. Method. Appl. M.* **3** 269-89
[20] Lienhardt J H I and Lienhardt J H V 2008 *A heat transfer textbook* (Cambridge, MA, USA, Phlogiston Press)
[21] Lyon R 1952 *Liquid Metals Handbook* (Whashington: A.E.C. and Dept of the Navy)
[22] Lubarsky B and Kaufman S J 1955 NACA Thecnical note 3366
[23] Mikityuk K 2009 *Nucl. Eng. Des.* **239** 680-7
[24] Ushakov P, Zhukov A and Matyukhin M 1977 *High Temperature* **5** (15) 868-73
[25] Ponciroli R, Bigoni A, Cammi A, Lorenzi S and Luzzi L 2014 *Prog. Nucl. Energ.* **71** 15-29

Journal of Physics: Conference Series **547** (2014) 012034 doi:10.1088/1742-6596/547/1/012034

Helical coil thermal hydraulic model

M Caramello, C Bertani, M De Salve and B Panella
Energy Department, Politecnico di Torino, Corso Duca degli Abruzzi, 24, 10129 Torino, Italy

E-Mail: bruno.panella@polito.it

Abstract. A model has been developed in Matlab environment for the thermal hydraulic analysis of helical coil and shell steam generators. The model considers the internal flow inside one helix and its associated control volume of water on the external side, both characterized by their inlet thermodynamic conditions and the characteristic geometry data. The model evaluates the behaviour of the thermal-hydraulic parameters of the two fluids, such as temperature, pressure, heat transfer coefficients, flow quality, void fraction and heat flux. The evaluation of the heat transfer coefficients as well as the pressure drops has been performed by means of the most validated literature correlations. The model has been applied to one of the steam generators of the IRIS modular reactor and a comparison has been performed with the RELAP5/Mod.3.3 code applied to an inclined straight pipe that has the same length and the same elevation change between inlet and outlet of the real helix. The predictions of the developed model and RELAP5/Mod.3.3 code are in fairly good agreement before the dryout region, while the dryout front inside the helical pipes is predicted at a lower distance from inlet by the model.

Introduction

Helical coil pipes have been widely used in the past in many industrial applications such as thermal process plants or power plants for the steam generation. In nuclear industry, helical coil steam generators (SG) and intermediate heat exchangers (IHX) have been installed in the fast reactors like Monju (Japan) [1] and Super-Phénix (France). These components are still considered for the future reactors projects such as the small modular reactors (SMR), like IRIS [2], and the fourth generation reactors like the Russian project BREST.

Coiled geometries are of particular interest because they allow to obtain high values of the inner heat transfer coefficients and guarantee an efficient power removal with a very high degree of compactness. Secondary motions in the velocity field are due to the centrifugal forces induced by the pipe curvature: these forces produce a shift of the fluid towards the outer part of the pipe, reducing the boundary layer and increasing the heat transfer. Centrifugal forces induce also a displacement of the maximum values of temperature and velocity from the centre of the axis to the outer side.

Helical coil fluid dynamics has been studied by several authors as regards both the laminar to turbulent transition [3] and the heat transfer and pressure drops in single-phase flow. Many reviews related to the fluid dynamics and heat transfer of curved tubes are available in literature, also recently (Naphon and Wongwises [4] and Vashisth [5]). Most of the work already done considers single-phase heat transfer with constant wall temperature or fixed heat flux boundary conditions, concentrating the efforts on the inner side of the helix. In order to characterize efficiently the heat transfer in steam generators is also important to develop models of heat transfer between fluids, characterized by their

Journal of Physics: Conference Series **547** (2014) 012034 doi:10.1088/1742-6596/547/1/012034

own temperature profile, with variable heat flux and wall temperature distribution in space. Escha et al. [6] studied a helical coil steam generator for a gas cooled reactor with the system code TRACE and the results obtained with different heat transfer coefficient correlations were compared with experimental data. In the present study a model has been implemented in Matlab environment with the aim to evaluate temperature and pressure profiles, heat transfer, pressure drops and two-phase flow dynamic considering an helical coil and shell steam generator. The semiempirical correlations for the evaluation of the heat transfer coefficients and the friction factors have been taken from literature. The model has been applied to the steam generator configuration of the SMR IRIS and the profiles of temperature, pressure and heat transfer coefficients have been compared with the ones obtained with the commercial code RELAP5/Mod.3.3.

Flow dynamics

The correct characterization and modelling of compact steam generators requires the knowledge of the flow field of the fluids, both on the inner and on the outer side. The flow field inside helical coil pipes is a complex result of the balance between inertia, gravity and centrifugal forces. The centrifugal forces are strongly dependent on the geometry of the pipe: considering a constant value of the pipe diameter, its influence is inversely proportional to the radius of the helix. The presence of secondary motions within the pipe cross section affects both velocity and thermal boundary layers: this explains the observed enhancement of the heat transfer coefficient and increase of the friction pressure drops. The helical pipes in the SG are disposed in a parallel configuration at increasing radii as in figure 1, and this involves a different importance of the centrifugal force with respect to inertia and gravity moving from the inner tube bundle to the outer ones.

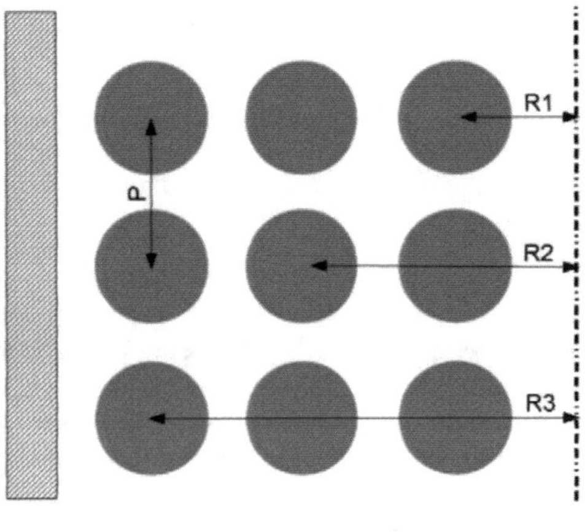

Figure 1. Helical bundle configuration.

When the fluid consists of a mixture of liquid and steam, the presence of centrifugal forces affects the flow pattern and promotes a separation of the phases due to the density difference, causing the heavier phase to move toward the external part of the pipe. From the point of view of the external flow the velocity field is disturbed by the presence of the helical coils and this promotes the formation of vortexes and radial mixing phenomena which increase as the liquid downflows. These effects are strongly dependent on the geometrical characteristics of the pipe bundle. In particular, the compactness of the whole bundle is one of the most important parameter for the velocity field in the external flow. The configuration of heat transfer from the external side lead to the decision to use the correlation of Zhukauskas for the external heat transfer coefficient [7] because it is able to predict the heat transfer in external flow on tube banks.

The thermal hydraulic model

In the reference geometry the fluid inside the helical coils has an upward flow while the fluid in the shell side flows downward in the SG. A countercurrent flow can be considered on the basis of the prevalent flow of the fluids, but in case of low helix curvature the cross-flow configuration can be assumed. The input data required by the model are the inlet conditions of the fluids in terms of pressure, flow rate and temperature and the geometrical characteristics of the helical coil.

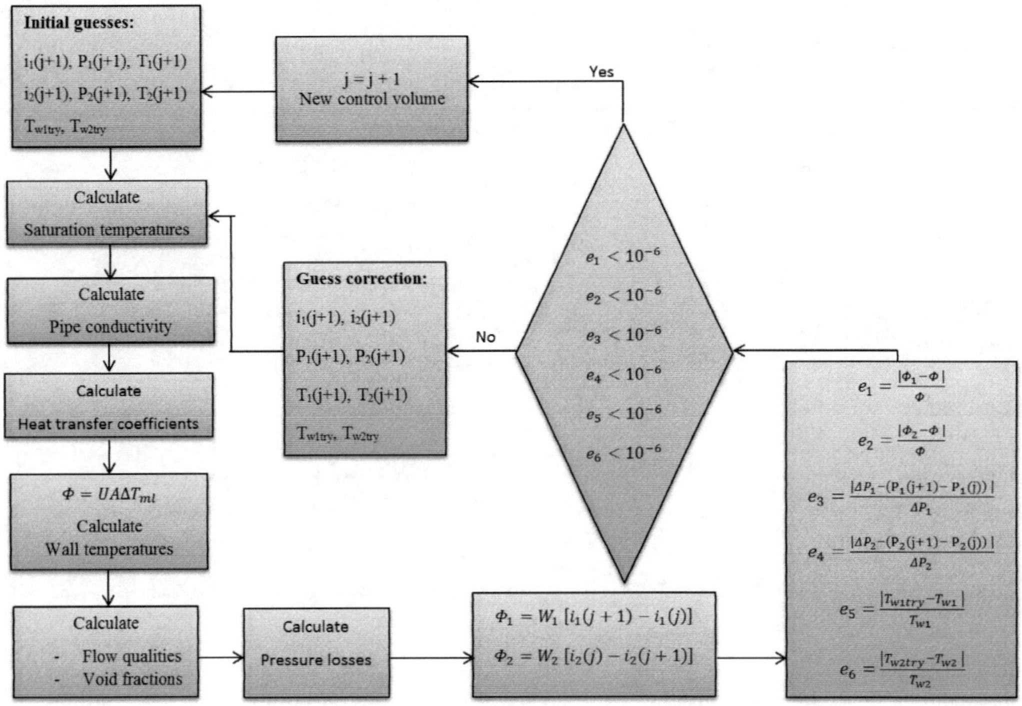

Figure 2. Control volumes flow chart.

The thermal hydraulic model is applied to four characteristic heat transfer regions from the point of view of the fluid inside the helix, namely the liquid single-phase region, the subcooled boiling region, the saturated boiling region and the steam single-phase region. Each region is divided in several control volumes containing the volume of the helical side fluid, the pipe volume itself and the equivalent annular control volume of the shell side fluid. The calculation convergence is guaranteed by a power balance and error flags of the relative error between attempted and calculated temperatures and pressures values. The flow chart describing the steps for the control volumes calculations is shown in figure 2. The parameters calculated in each control volume are heat transfer coefficients, acceleration, friction and elevation pressure drops, as well as flow quality and void fraction. At the end of the calculations the model is able to show the spatial behaviour of the characteristic parameters of the flow, such as temperature, pressure, flow quality, void fraction, heat flux and heat transfer coefficients. The semiempirical correlations and models that have been adopted to evaluate the heat transfer coefficients, the friction pressure drops and the void fraction are taken from the most validated literature correlations taking also into account the effects of centrifugal forces inside the helical coils and the mixing induced by their presence in the flow path of the shell side fluid [7-11]. The adopted correlations are reported in table 1.

Journal of Physics: Conference Series **547** (2014) 012034 doi:10.1088/1742-6596/547/1/012034

Table 1. Empirical correlations adopted in the present model.

Author	Application	Correlation
Mori and Nakayama [8]	Helix side liquid single phase region	$Nu = \frac{1}{41}\left(\frac{Pr}{Pr^{0.6}-0.062}\right) Re^{\frac{5}{6}} \left(\frac{d_i}{D_{hx}}\right)^{\frac{1}{12}} \left(1 + \frac{0.061}{\left\{Re\left(\frac{d_i}{D_{hx}}\right)^{2.5}\right\}^{\frac{1}{6}}}\right)$
Mori and Nakayama [8]	Helix side steam single phase region	$Nu = \frac{1}{26.2}\left(\frac{Pr}{Pr^{0.6}-0.074}\right) Re^{\frac{4}{5}} \left(\frac{d_i}{D_{hx}}\right)^{\frac{1}{10}} \left(1 + \frac{0.098}{\left\{Re\left(\frac{d_i}{D_{hx}}\right)^{2}\right\}^{\frac{1}{5}}}\right)$
Owhadi [9]	Helix side subcooled and saturated boiling	$h = S\, h_b + F\, h_c$ $F = 2.35\,(\chi_{tt}^{-1} + 0.213)^{0.736}$ $F = 1\ in\ subcooled\ boiling$ $S = \begin{cases} (1 + 0.12\, Re_{tp})^{-1.14} & Re_{tp} < 32.5 \\ (1 + 0.42 Re_{tp}^{0.78})^{-1} & 32.5 < Re_{tp} < 70 \\ 0.0797 & Re_{tp} \geq 70 \end{cases}$ $Re_{tp} = \frac{G\,(1-x) D_h}{\mu_f} F^{1.25}\, 10^{-4}$ $h_b = 0.00122 \left(\frac{k_f^{0.79} C_{pf}^{0.45} \rho_f^{0.49}}{\sigma^{0.5}\, \mu_f^{0.29} \lambda^{0.24} \rho_g^{0.24}}\right) \Delta T_{sat}^{0.24} \Delta P_{sat}^{0.75}$ $h_c = \frac{k_f}{41D}\left(\frac{Pr}{Pr^{0.6}-0.062}\right) Re^{\frac{5}{6}} \left(\frac{d}{D}\right)^{\frac{1}{12}} \left(1 + \frac{0.061}{\left\{Re\left(\frac{d}{D}\right)^{2.5}\right\}^{\frac{1}{6}}}\right)$
CISE [10]	Helix side void fraction	$S = 1 + E_1 \left(\frac{y}{1+yE_2} - yE_2\right)^{0.5}$ $y = \frac{\beta}{1-\beta}$ $\beta = \frac{\rho_l x}{\rho_l x + (1-x)\rho_g}$ $E_1 = 1.578\, Re^{-0.19} \left(\frac{\rho_l}{\rho_g}\right)^{0.22}$ $E_2 = 0.0273\, We\, Re^{-0.51} \left(\frac{\rho_l}{\rho_g}\right)^{-0.08}$
Zhukauskas [7]	Shell side heat transfer coefficient	$Nu = 0.033\, Re_{D,max}^{0.8} Pr^{0.36} \left(\frac{Pr}{Pr_s}\right)^{\frac{1}{4}}$
Ito [11]	Helix side friction factor	$Re_{crit} = 20000 * \left(\frac{d_i}{D_{hx}}\right)^{0.32}$ $f = \begin{cases} \frac{344\left(\frac{D}{d}\right)^{-0.5}}{\left\{1.56+\log_{10}\left[Re\left(\frac{D}{d}\right)^{-0.5}\right]\right\}^{5.73}} & Re < Re_{crit} \\ 0.076\, Re^{-0.25} + 0.00725 \left(\frac{D_h}{d_i}\right)^{-0.5} & Re > Re_{crit} \end{cases}$
Chen [12]	Helix side two phase multiplier	$\Phi_{l0}^2 = 2.06 \left(\frac{d_i}{D_h}\right)^{0.05} Re_{tp}^{-0.025} A$ $A = \left[1 + \alpha\left(\frac{\rho_g}{\rho_f} - 1\right)\right]^{0.8} \left[1 + x\left(\frac{\rho_f}{\rho_g} - 1\right)\right]^{1.8} \left[1 + \alpha\left(\frac{\mu_g}{\mu_f} - 1\right)\right]^{0.2}$
Blasius [13]	Shell side friction factor	$f = 0.079 * Re^{-\frac{1}{4}}$

Journal of Physics: Conference Series **547** (2014) 012034 doi:10.1088/1742-6596/547/1/012034

The RELAP5/Mod.3.3 nodalization

RELAP5/Mod.3.3 is a commercial thermal hydraulic system code able to analyse accidents in nuclear power plants [14]. The code is based on a nonhomogeneous and nonequilibrium model for the two-phase system that is solved in a fast, partially implicit numerical scheme to permit economical calculation of system transients [14]. Thanks to its capability to model the fluid dynamics and the heat transfer in various conditions it has been chosen as a comparison of the present model. It is currently not able to predict the heat transfer inside helical pipes because of a lack of correlations for the heat transfer coefficient. As suggested by the Idaho National Laboratory (INL) [15] the helical coil geometry is simulated in the code with an inclined pipe with the same length as the real helix and an equivalent inclination created to obtain the same elevation between inlet and outlet. The equivalent channel for the external flow is simulated by an annulus. The annulus and the inclined pipe are thermally linked with a heat structure (HS) made of T91. The annulus, the inclined pipe and the heat structure have been nodalized with 99 volumes. Time dependent volumes (TMDP VOL) and time dependent junctions (TJ) are used to impose pressure and flow rates at the beginning and at the end of the heated channels. Single junctions (SJ) initialized with temperature, pressure and flow rate initial conditions are used at the outlet of the annulus and of the inclined pipe. The layout of the two hydrodynamic systems is presented in figure 3.

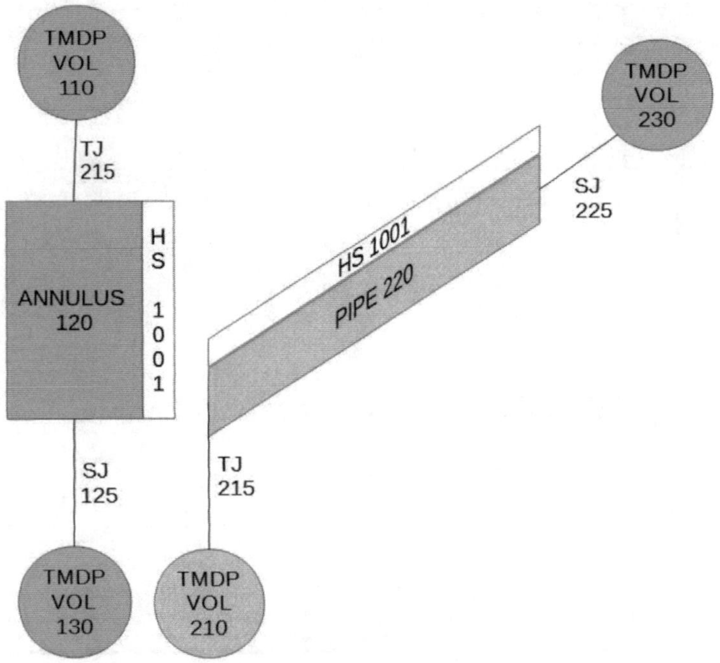

Figure 3. Layout of the hydrodynamic components.

Thermal hydraulic simulation of the IRIS steam generator

The proposed model and the RELAP5/Mod.3.3 code have been used to simulate the thermal hydraulic behaviour of one of the steam generators of the IRIS SMR and the results of the simulations are compared with respect to the nominal values available in literature [2,16]. IRIS is a light water SMR, designed within an international collaboration. The reactor is equipped with 8 once-through helical coil steam generators installed in the annular region between the vessel and the riser above the reactor core. The simulations have been performed by using an average coil in terms of length, pitch and helical radius whose data are shown in table 2.

Table 2. IRIS SG nominal data [2,16].

Parameter	Value
Power [kW]	190.5
Primary inlet temperature [°C]	328.4
Feedwater temperature [°C]	223.9
Primary pressure [bar]	155
Internal diameter [mm]	13.24
Primary flow rate [kg/s]	0.893
Helix length [m]	32.0
Primary outlet temperature [°C]	292.0
Steam temperature [°C]	317.0
Steam pressure [bar]	58
External diameter [mm]	17.46
Feedwater flow rate [kg/s]	0.0958

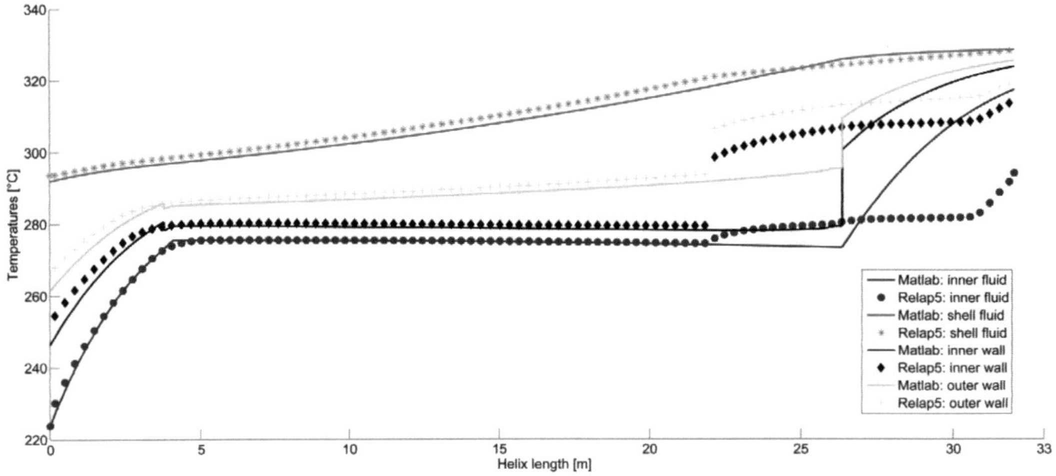

Figure 4. Temperature profiles.

Figure 4 shows the temperature profiles of the fluids as well as the temperature profile of the internal and external walls of the pipe. The main differences between the models are the two-phase nonequilibrium considered by RELAP5/Mod.3.3 and different heat transfer coefficients correlations, which are specific for helical coiled pipes in the proposed model. Nevertheless, the present model shows a good agreement with the code prediction of the temperature profile on the shell side with a maximum relative difference lower than 1%. The wall temperatures predicted by the present model are slightly lower than the code ones in the liquid single-phase and low quality regions. The dryout front is predicted at two different distances from inlet, respectively 22 meters for the code and 27 meters for the proposed model. The dryout occurrence in helical coil pipes is still an open issue and few studies have been performed so far. The heat transfer degradation is found to be asymmetric and strongly dependent on the weight of centrifugal forces [17], and the critical heat fluxes in the helical coil are much higher than in a vertical straight tube [18]. The temperatures of the internal fluid have a different behaviour in the high quality region and this is due to two differences between the models, which are the absence of transition boiling in the proposed model and the treatment of nonequilibrium conditions in the RELAP5/Mod.3.3 code. The outlet temperatures have a discrepancy of 20°C which is consistent with the different power exchanged in the two cases. Figure 5 shows the behaviour of the thermal resistances of the inner and outer heat transfer and the thermal resistance induced by the conduction term in the overall heat transfer. The highest thermal resistance is predicted on the outer side of the pipe in the liquid single-phase and in the low quality regions. After the dryout region the highest thermal resistance is presented by the heat transfer inside the helical pipe. There is a fairly good agreement between the two predictions except in the high quality region inside the helical pipe because of a different prediction of the dryout front.

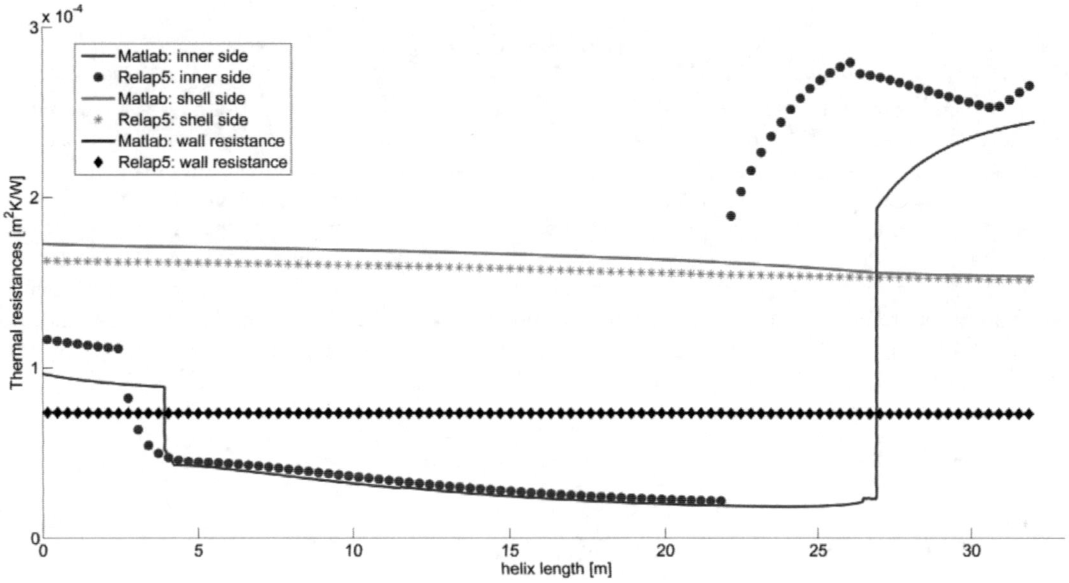

Figure 5. Thermal resistances.

Figure 6 shows the pressure profile as a function of the helix length. The present model, which is implemented with semiempirical correlations specific for the helical geometry, predicts an higher pressure loss compared with the RELAP5/Mod.3.3 code evaluation. The obtained result is consistent with the results available in literature where the frictional pressure drop for coiled geometries is higher than the one of a straight pipe both in single-phase and two-phase flow [11]. The relative difference between the predicted pressure drops is of the order of 25%.

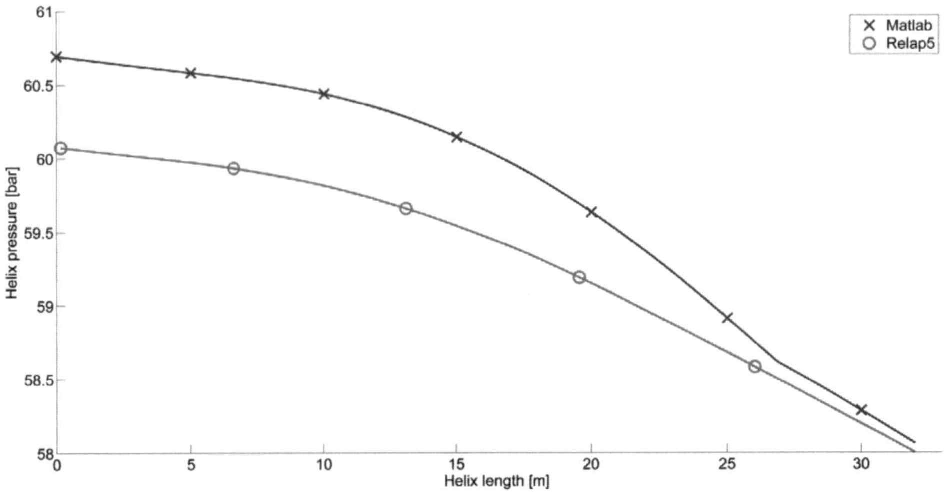

Figure 6. Internal pressure.

Figures 7 and 8 show the behaviour of the heat transfer coefficients for the fluids. The heat transfer coefficient predicted by the present model inside the helical coil is generally higher than the one calculated by RELAP5/Mod.3.3 because the correlations used in the model take into account the heat transfer enhancement due to centrifugal forces. The subcooled region is predicted at a lower distance by the code, and this explains why the heat transfer coefficients result higher than the present model prediction before the fifth meter. The high quality region difference of the temperature profiles shown in figure 5 is also present in the heat transfer coefficient prediction: the region of heat transfer degradation is predicted at an higher distance from the inlet by the present model. As far as the heat transfer coefficient on the shell side is concerned, the proposed model predicts lower values even if the discrepancy is low, with a relative difference lower than 6%.

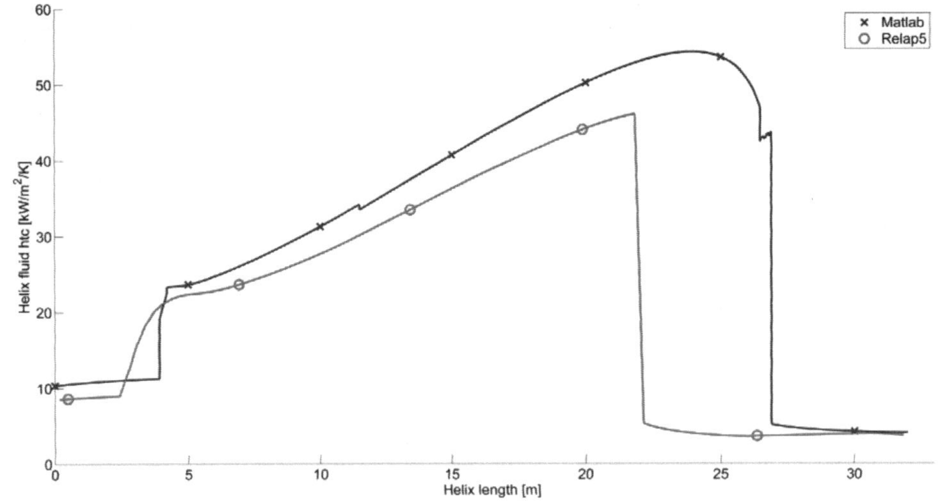

Figure 7. Helix side heat transfer coefficient.

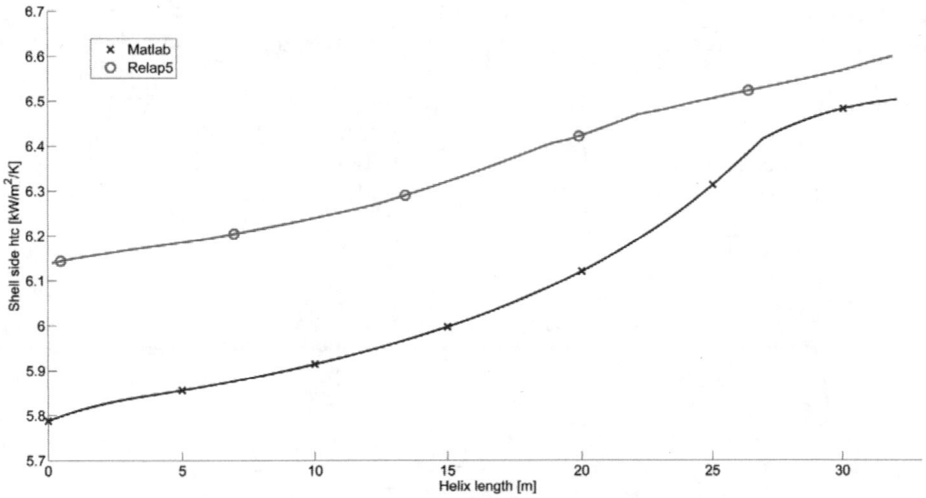

Figure 8. Shell side heat transfer coefficient.

Table 3 presents the comparison between the IRIS steam generator nominal data and the results of the two predictions. The proposed model demonstrates a good capability to reproduce the nominal data.

Table 3. Comparison between nominal and calculated values.

	Nominal data [2,16]	Present model	RELAP5/Mod.3.3
Power [kW]	190.5	189.6	182.03
Liquid single-phase linear power [kW/m]	\	5.97	5.91
Two-phase linear power [kW/m]	\	6.59	5.74
Steam temperature [°C]	317.0	315.5	297.1
Coolant outlet temperature [°C]	292.0	292.1	293.8
Steam pressure [bar]	58	58.06	58

Conclusions and future work

A model for the thermal hydraulic characterization of an helical coil steam generator has been proposed and a comparison with the commercial code RELAP5/Mod.3.3 has been done with respect to the nominal values of the IRIS SMR steam generator. The analysis shows a good agreement between the results except in the high quality region, where the dryout is predicted at two different distances from the inlet: in particular, the proposed model predicts the heat transfer degradation at an higher distance. The present model predicts higher values of in-coil heat transfer coefficient as well as higher pressure drops because of specific correlations for coiled geometries which take into account the effect of centrifugal forces. From the point of view of the shell side heat transfer coefficient the prediction of the two models is comparable. The comparison with the nominal data of the IRIS SMR steam generator show that the proposed model is able to predict the steam generator behaviour with a relative error for the transferred power lower than 1%. Future works to improve the model will be done to analyse the sensitivity of the steam generator performances with respect to the geometry and the fluid dynamics. Also, on the basis of the literature on helical coils, new heat transfer regimes like transition boiling and film boiling will be implemented, as well as by performing predictions for non-equilibrium conditions.

Acknowledgements
The present research has been supported by ENEA and by the ministry of economic development.

Nomenclature

D [m]	Diameter
d [m]	Diameter
P [m]	Pitch
R [m]	Radius
N	Number of
i [J/kg]	Enthalpy
P [bar]	Pressure
T [°C]	Temperature
c_p *[J/kg/K]*	Heat capacity at constant pressure
k [W/m/K]	Thermal conductivity
μ *[Pa s]*	Dynamic viscosity
ρ *[kg/m³]*	Density
σ *[N/m]*	Surface Tension
λ *[J/kg]*	Enthalpy of vaporization
χ_{tt}	Martinelli parameter
S	Suppression factor
F	Reynolds factor
Φ *[W]*	Heat flux
U [W/m²K]	Overall heat transfer coefficient
W [kg/s]	Flow rate

Subscripts

try	Attempt value
SG	Steam generator
IHX	Intermediate heat exchanger
c	Coil
e	External
II	Shell side
I	Inner side
p	Pipes
f	Fluid
sat	Saturation
sub	Subcooled
tp	Two phase
w	Wall

Dimensionless numbers

Re	Reynolds number
Nu	Nusselt number
Pr	Prandtl number

References

[1] Matsuura M, Hatori M and Ikeda M, 2007 *Nuclear engineering and design* **237** 1419-28
[2] Cinotti L, Bruzzone M, Meda N, Corsini G, Conway L E, Lombardi C and Ricotti M, 2002 *Proc. Int. Conf. on Nuclear Engineering* (Arlington: VA USA/ICONE-10)
[3] Cioncolini A and Santini L, 2006 *Experimental Thermal and Fluid Science* **30** 653-61
[4] Naphon P and Wongwises S, 2006 *Renewable Sustain Energy Rev* **10** 463–90
[5] Vashisth S, Kumar V and Nigam K D P, 2008 *Industrial & Engineering Chemistry Research* **47** 3291–337
[6] Escha M, Hurtado A, Knoche D and Tietsch W, 2012 *Nuclear Engineering and Design* **251** 374–80
[7] A. Bejan 2013 *Heat Transfer Handbook* (John Wiley & Sons Inc.)
[8] Mori Y and Nakayama W, 1967 *Int. J. Heat Mass Transfer* **10** 37
[9] Owhadi A, Bell K J and Crain Jr B, 1968 *Int. J. Heat Mass Trans.* **11**, 1779–93
[10] P. B. Whalley 1987 *Boiling, Condensation, and Gas-Liquid Flow* (Clarendon Press - Oxford)
[11] Castiglia F, Giardina M, Morana G, De Salve M and Panella B, 2012 *Nuclear Engineering and Design* **250** 585–91
[12] Guo L, Feng Z and Chen X, 2001 *International Journal of Heat and Mass Transfer* **44** 2601–10
[13] Aroonrat K, Jumpholkul C, Leelaprachakul R, Dalkilic A S, Mahian O and Wongwises S, 2013 *Int. Comm. Heat and Mass transf* **42** 62-8
[14] INL, *RELAP5/MOD3.3* 2001 *Code manual Vol 1* (Information Systems Laboratories Inc.)
[15] Hoffer N, Sabharwall P and Anderson N, 2001 INL/EXT-10-19621
[16] Cioncolini A, Cammi A, Lombardi C, Luzzi L and Ricotti M, 2003 *Nuclear Mathematical and Computational Sciences: A Century in Review, A Century Anew* (Gatlinburg: Tennessee)
[17] Chung W S, Sa Y C and Lee J S, 2002 *KSME International Journal Vol. 16* **11** 1540-9
[18] Styrikovich M A, Polonsky V S and Reshetov V V, 1984 *Int. J. Heat Transfer* **27** 1245-50.

Journal of Physics: Conference Series **547** (2014) 012035 doi:10.1088/1742-6596/547/1/012035

Analysis of a Station Black-Out transient in SMR by using the TRACE and RELAP5 code

F De Rosa[1], C Lombardo[1], F Mascari[1], M Polidori[1]
P Chiovaro[2], S D'Amico[2], I Moscato[2], G Vella[2]

[1] ENEA - UTFISSM/SICSIS
Via Martiri di Monte Sole 4, 40129 Bologna, Italy

[2] UNIVERSITY OF PALERMO
Viale delle Scienze, 90128 Palermo, Italy

E-mail: calogera.lombardo@enea.it

Abstract. The present paper deals with the investigation of the evolution and consequences of a Station Black-Out (SBO) initiating event transient in the SPES3 facility [1]. This facility is an integral simulator of a small modular reactor being built at the SIET laboratories, in the framework of the R&D program on nuclear fission funded by the Italian Ministry of Economic Development and led by ENEA. The SBO transient will be simulated by using the RELAP5 and TRACE nodalizations of the SPES3 facility. Moreover, the analysis will contribute to study the differences on the code predictions considering the different modelling approach with one and/or three-dimensional components and to compare the capability of these codes to describe the SPES3 facility behaviour.

1. Introduction

In the last few years, several new advanced integral Small Modular Reactor (SMR) designs have been considered by the International Community. The integral nature of these reactors requires the analyses and the characterization of the phenomena characteristic of these designs such as, the primary to secondary heat transfer in integrated Steam Generator (SG), the natural circulation in passive systems and Reactor Pressure Vessel (RPV)/containment coupling both in Design and Beyond Design Basis Accidents conditions (DBA and BDBA). The SPES3 facility, being built at the SIET laboratories in the framework of the R&D program on nuclear fission funded by the Italian Ministry of Economic Development and led by ENEA, is a simulator of an integral type SMR simulation. This facility can be used to create a database for the characterization of the SMR phenomenology and to validate and assess best estimate thermal hydraulic system codes, a keypoint in the design and safety analyses of Nuclear Power Plants (NPP). Complete nodalizations of SPES3 were developed for TRACE and RELAP5 codes in order to investigate the code responses by the simulation of the same accidental transients. As a consequence of the accident occurred at Fukushima Daiichi NPP in Japan, the guarantee of reliable safety functions against any initiating event, e.g. earthquake and flooding, has become fundamental to evaluate the consequences of loss of electrical power, including the complete SBO. The SBO scenario involves a loss of offsite power, failure of the redundant emergency diesel generators, failure of alternate current (AC) power restoration and the eventual degradation of the Reactor Coolant Pump (RCP) seals resulting in a long-term loss of coolant.

The objective of the present paper is to investigate the evolution and the consequences of a transient due to a SBO initiating event in the SPES3 facility. Moreover, the analysis will contribute to study the differences on the code predictions considering the different modelling approach with one and/or three-dimensional components and to compare the capability of these codes to describe the SPES3 facility behaviour.

2. The SPES-3 Facility

The scheme of the SPES3 facility is shown in Fig. 1. SPES3 is an integral test facility, based on the IRIS reactor design [2], reproducing the primary, secondary, containment and safety systems. It allows to test the plant response to postulated accidents and to investigate general interest thermal-hydraulic phenomena like primary to containment dynamic coupling.

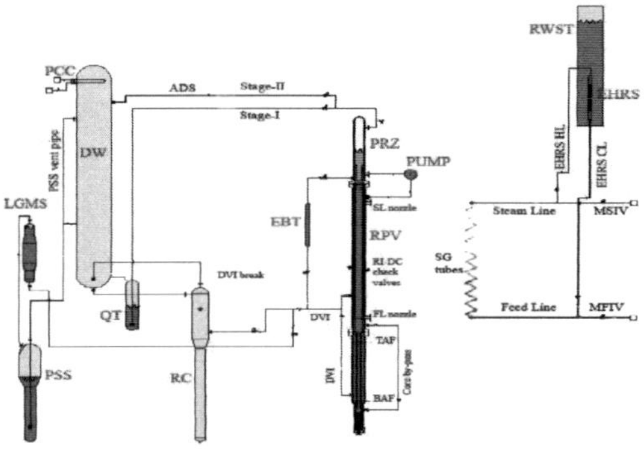

Figure 1. SPES3 facility scheme.

The reactor vessel includes the internals, consisting of the electrically heated core simulator, the riser with Control Rod drive Mechanisms (CRDM), the Pressurizer (PRZ), the pump suction plenum, the helical coil SG, the downcomer (DC) and the lower plenum. Two Emergency Boration Tanks (EBT) are simulated and connected to the Direct Vessel Injection lines (DVI), devoted to direct injection of emergency fluid into the vessel. Each secondary loop is simulated up to the main isolation valves and includes the feed line, the SG, the Steam Line (SL), and the Emergency Heat Removal System (EHRS) with a vertical tube heat exchanger immersed in a Refuelling Water Storage Tank (RWST).

The containment compartments are simulated by tanks, connected to each other by pipes and to the RPV by break lines. They include Dry-Well (DW) and Reactor Cavity (RC), representing the dry zone surrounding the RPV, respectively above and below the mid-deck plane; Pressure Suppression Systems (PSS) representing the wet zone around the lower part of the RPV, suitable to dump pressure in case of containment pressurization; Long-term Gravity Make-up Systems (LGMS) representing the cold water reservoir to be poured into the RPV when depressurized.

The two stages of the Automatic Depressurization System (ADS) are simulated, connected to the PRZ top, with stage I discharging into the Quench Tank (QT) and stage II directly connecting RPV and DW at high plant elevation.

The facility allows to test both LOCAs and secondary side breaks (DBA and BDBA) as well as to perform separate effect tests on particular components such as SG-EHRS thermally coupled to RWST.

The passive EHRS consists of four independent subsystems. Each subsystem has an heat exchanger connected to a separate SG feed steam line and immersed in the RWST, located outside the containment structure.

The RWST water provides the heat sink for the EHRS heat exchangers operating in natural circulation. The secondary fluid in transient conditions comes from the SG outlet and arrives, for natural circulation, in the heat exchanger where it is condensed by transferring energy to the RWST water, then it returns back to the SG.

3. TRACE nodalization

A TRACE model of the SPES3 facility has been developed for design support, pre-test calculations and when the facility will be built, for post-test calculations. The SPES3 TRACE nodalization models the primary and the secondary system, the containment structures and the safety systems simulated in the SPES3 facility. The RPV nodalization models the lower DC , the Lower Plenum (LP), the core, the core bypass, the riser, the PRZ, the pump suction plenum, the outlet pump and connecting piping, the SG primary side and the riser to DC check valves. The three helical coil SGs are modelled with three different equivalent groups of pipes thermally coupled with three different SG primary side regions. The secondary system-A (B, C) consists of the FL-A (B, C), the SG-A (B, C) tubes and the SL-A (B, C). The SG-A (B) consists of a single group of pipes equivalent to the 14 helical coils. The SG-C consists of a two single group of pipes each equivalent to 14 helical coils. The SPES3 TRACE nodalization models the three different EHRS loops A, B and C and the related RWST. The RWSTs are modelled with the 3D component "vessel"; Fig. 2 shows primary and secondary systems. The heat exchanger-A (B) is immersed in the RWST-A/B. The CL-A (B) exits from the heat exchanger-A (B) bottom and connects to the FL-A (B). In particular, the EHRS-A (B) consists of the HL-A (B), the heat exchanger-A (B) and the CL-A (B). The HL-A (B) detaches from the SL-A (B) and enters the heat exchanger-A (B) top.

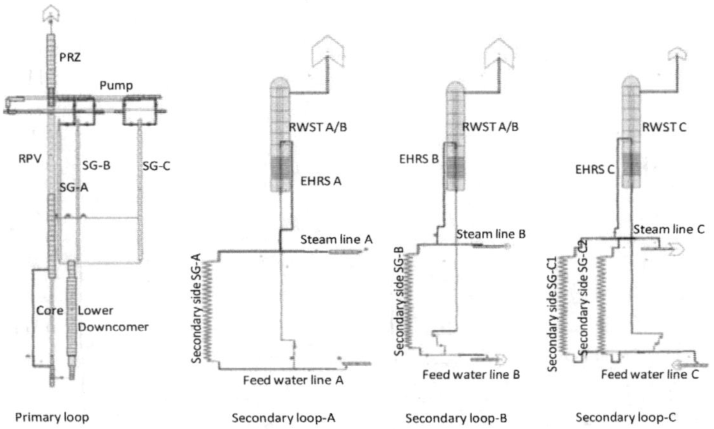

Figure 2. SPES3 TRACE nodalization.

4. RELAP5 noding

The SPES3 facility nodalization [4] contains: the Primary System, the Secondary Systems, the Emergency Heat Removal System and the Containment System. The Primary System RELAP5 noding includes the following main regions: Lower downcomer, Lower plenum Core, Core by-pass, Lower riser, RCCA zone, Upper riser, CRDM zone, Pressurizer Pump Suction plenum, Outer Pump and connecting piping, Primary side of SG-A, -B and -C, Riser-SG primary side connection check valves. Particular attention was focused on the EHRS-RWST modeling (see the Fig.3). The choice to model the RWST with two slices connected by transversal junctions is based on lessons learned by an experimental campaign at SIET and post-test calculations of an similar in-pool immersed heat exchanger [5].

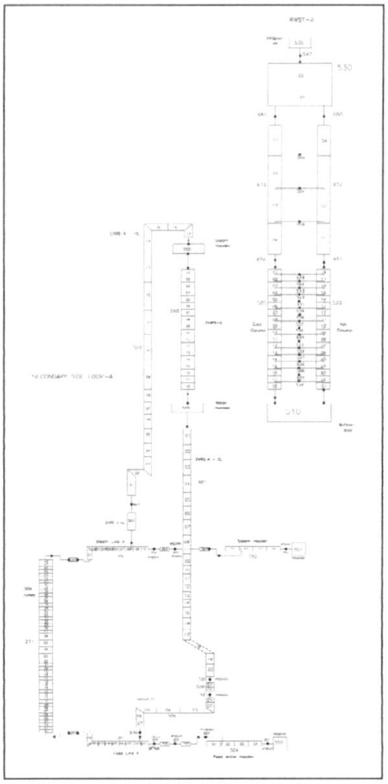

Figure 3. SPES3 facility scheme.

The EHRS-C consists of the HL-C, the heat exchanger-C and the CL-C. The heat exchanger-C is immersed in the RWST-C. The different SPES3 containment compartments and the connecting pipes are modelled. The two EBT tanks are modelled separately and each one is connected to the related DVI line. In particular, the DW, the RC and the PSS are modelled with the 3D "vessel" component. The SPES3 ADS lines are modelled separately [8], [9].

5. SPES3 SBO Transient

The main phases of the transient can be summarised in:

1. The loss of off-site power is assumed to be contemporary to the earthquake and to trigger the RCP and MFW pump coast-down. Natural circulation establishes through the pump-bypass and RI (Riser) –DC (Downcomer) check valves;

2. the low FW signal triggers the reactor scram;

3. the start-up FW signal (postulated for SMRs and not simulated in SPES3) is assumed to fail and this gives permission to isolate the SGs and trigger the EHRS-A and B;

4. the low collapsed level signal in the SG, secondary side, triggers the SG isolation and actuates the EHRS-A and B; RPV level decreases for system cool-down. Pump by-passes uncover and natural circulation occurs only through the RI-DC check valves. The RPV depressurizes;

5. the low core inlet coolant temperature signal actuates the EBTs that inject water until RPV and EBT level equalize;

6. in the long term, the plant continues to be cooled by the EHRSs that removes the core decay heat to the RWSTs.

6. SBO Simulation

The RELAP5 and TRACE simulations have been carried out for 150000 s but the pictures show data for the first 50000 s because this period of time is sufficient to investigate the principal phenomena and differences between the two calculations. The steady-state conditions starting point for the transient are summarized in Tab.1.

Tab. 1 - RELAP5 VS TRACE Steady conditions.

SPES3	Unit	RELAP5	TRACE
Pressurizer pressure	MPa	15.509	15.512
Core power	MW	6.5	6.5
Core inlet fluid temperature	K	563.72	564.52
Core outlet fluid temperature	K	601.64	602.54
Core mass flow	kg/s	29.56	29.32
SG outlet pressure	MPa	6.01	6.11
SG feed water temperature	K	496.95	496.92
SG-A steam outlet temperature	K	548.90	550.00
SG-B steam outlet temperature	K	548.90	550.00
SG-C1 steam outlet temperature	K	549.09	550.20
SG-C2 steam outlet temperature	K	549.09	550.21
SG-A mass flow	kg/s	0.81	0.81
SG-B mass flow	kg/s	0.81	0.81
SG-C1 mass flow	kg/s	0.81	0.81
SG-C2 mass flow	kg/s	0.81	0.81
RWST fluid temperature	K	293.51	293.45

The list of the main events occurring during the transient with timing is reported in Tab. 2 and the main quantities, evidencing the details of the above described phases, are compared in Fig. 6 to Fig. 14. The earthquake occurs at time 0 and the loss of off-site electric power is contemporary. The pumps of the system are switched-off and the coast-down starts for the RCP and the FW pumps. The reactor scram occurs at 1.24 s, on Low FW signal, set at 75% of nominal feed-water mass flow. The Start-up feed-water, which should intervene at the reactor scram, is assumed to fail and this gives permission for secondary loop isolation.

Tab. 2 RELAP5 VS TRACE List of main events.

Phases and events [s]	**RELAP5**	**TRACE**
Earthquake and Loss of off-site power	0	0
RCP coast-down starts	0	0
MFW pump coast-down starts	0	0
Low FW signal	1.24	1.24
SCRAM Signal and Start-up FW actuation fails	1.24	1.24
Natural circulation begins (RI-DC)	20	6
Circulation begins pump by-pass	20, 22	5
Low SGs level signal	43	54
MFIV-A,B,C closure start	43	54
MSIV-A-B-C closure start	43	54
EHRS-A and B opening start	43	54
EHRS-A peak mass flow	44	60
EHRS-B peak mass flow	44	60
RWST-A/B begins to heat-up	90	55
Secondary loop-A pressure peak	143	128
Secondary loop-B pressure peak	143	129
EHRS-A power peak	200	411
EHRS-B power peak	200	502
Low PRZ pressure	1160	748
Low core inlet coolant temperature	2960	3601
EBT-A and B valve opening start	2960	3601
Natural circulation interrupted at SGs top	3680	8900
EBT injection to RPV stop	5310	4344,4908
Core in saturation conditions	37200	16689

The pump head decrease lets the RI-DC and the pump by-pass check valves open and allows natural circulation between riser and SG annuli at lower and high elevation in the RPV. In the secondary side the MFW pump is assumed to stop in 4.9 s. The loss of main feedwater causes the reduction of mass inventory in the SGs, secondary side. The SG low level signal occurs when the collapsed level reaches 0.25 m (see Fig. 4). This signal starts the secondary loop isolation and actuates the EHRS-A and B. Natural circulation starts in SG-A and B, thanks to EHRS-A and B intervention, while SG-C is supposed to fail. For about an hour the SG mass flowrates calculated by the RELAP5 simulation are 13% higher than that obtained from the TRACE simulation, consequently, the power removed by the EHRSs in RELAP5 is 30% greater than in TRACE, Fig. 5, Fig. 6. This is the main reason why within the primary, Fig. 7, and secondary circuits the mass flow rate calculated by RELAP5 is greater than that predicted from the TRACE. Due to the RPV level decrease for system cooling, the pressurizer is empty before the transient reaches the 30 minutes, Fig. 8. Power is rejected to the RWST-A/B that begins to heat-up in one minute and reaches saturation at about 2 hours and 25 minutes, Fig. 9; the RWST A/B mass decreases for water evaporation.

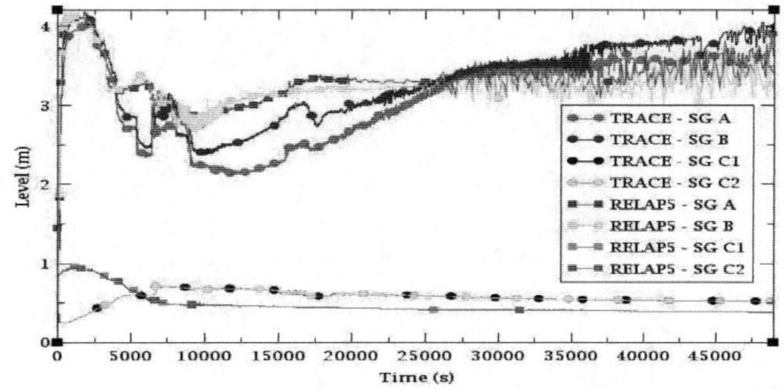

Figure 4. SGs collapsed level.

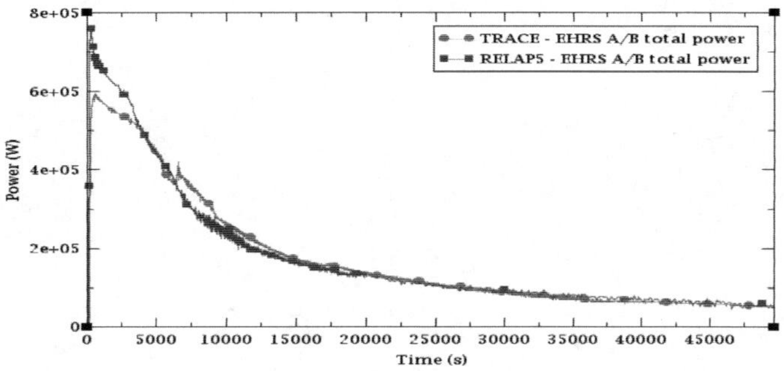

Figure 5. EHRS power.

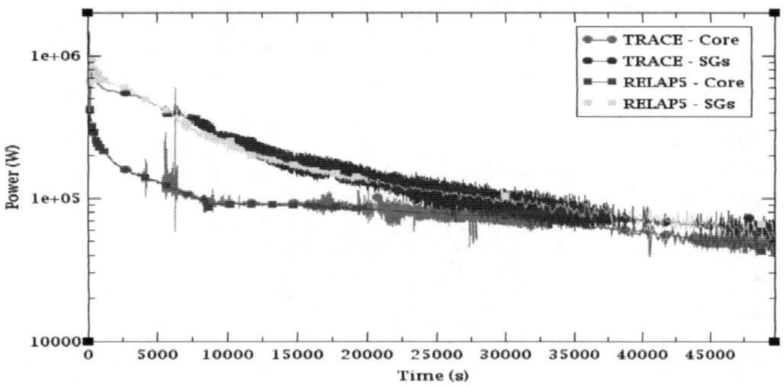

Figure 6. SGs and core power exchanged.

The primary side pressure decreases for the system cool-down, Fig. 10. The inlet and outlet core temperatures together with the saturation temperature corresponding to core outlet pressure are shown in Fig.11 for RELAP5 transient and in Fig.12 for TRACE transient. It is possible to observe that in RELAP5 the fluid temperatures in the core are always lower than in TRACE. For this reason the set-point of 533.15 K for the low core inlet coolant temperature is reached at about 50 minutes in RELAP5 and 10 minutes later in TRACE analysis. The signal triggers the EBT actuation by opening the EBT valves in 15 s. Cold water is injected into the RPV, through the DVI lines, and hot water and steam replace the EBT water through the EBT top line, connected to the RPV. The RPV mass inventory slightly increases thanks to the EBT injection and then it remains stable, Fig.13.

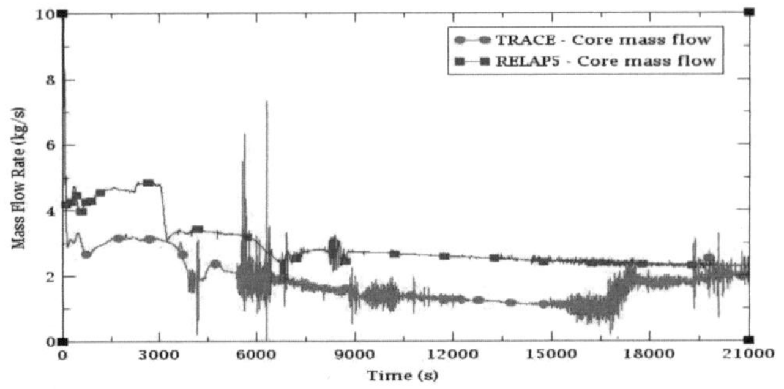

Figure 7. Core mass flow.

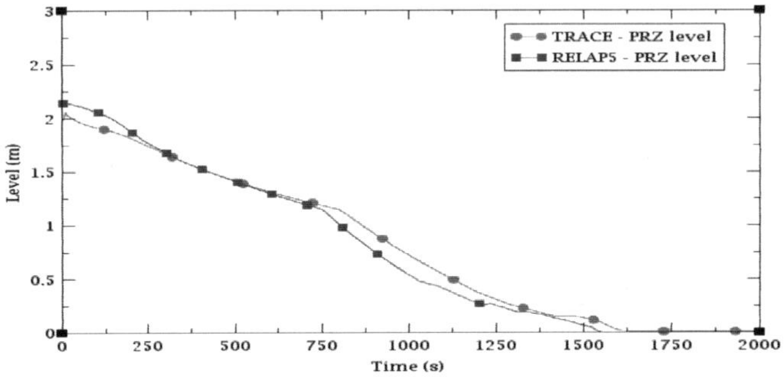

Figure 8. PRZ level.

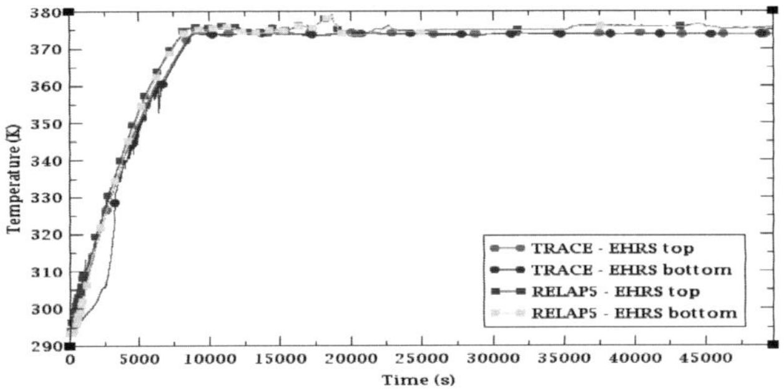

Figure 9. RWST temperatures.

In RELAP5, at about 37000 s, the core outlet temperature is in saturation condition, anyway the core is always under single-phase water, whereas in TRACE the core outlet temperature is in saturation at about 16700 s and at the top of active fuel begins to boil, Tab. 2. However, in both cases no temperature excursion occurs on the heated rods cladding, Fig. 14. After 12 hours, the core decay power is about 1% than nominal value and is reached the long term condition: the EHRS-A and B remove the core decay power and slowly cool-down the system, the mass flow rate is stable around 2 kg/s and the pressure at about 0.3 MPa. The reason for the differences between TRACE and RELAP5 analysis is due to the different models (e.g. RWST noding) and approaches used in the development of the nodalizations for the two codes.

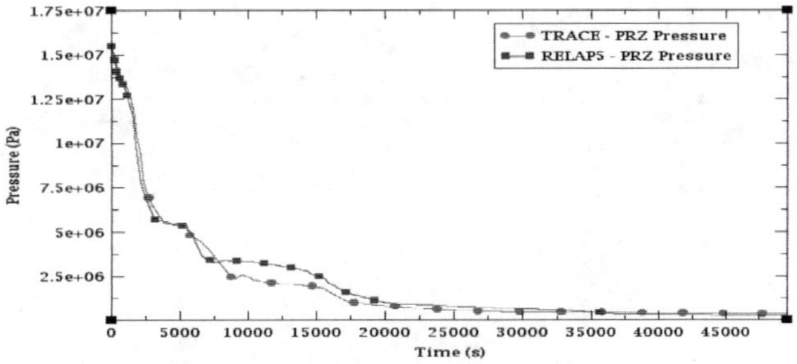

Figure 10. Primary pressure.

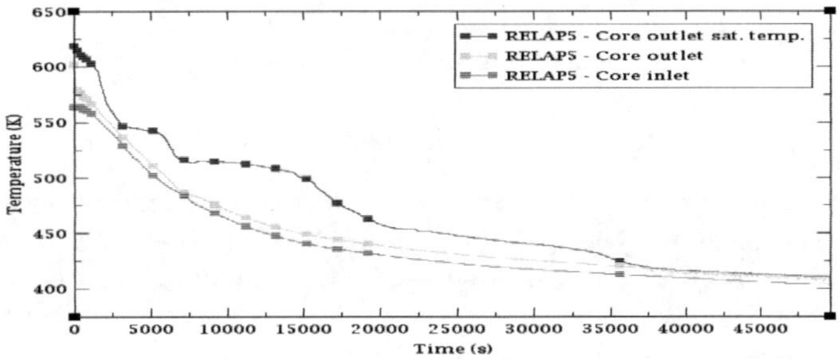

Figure 11. Core inlet, outlet and saturation temperature in RELAP5.

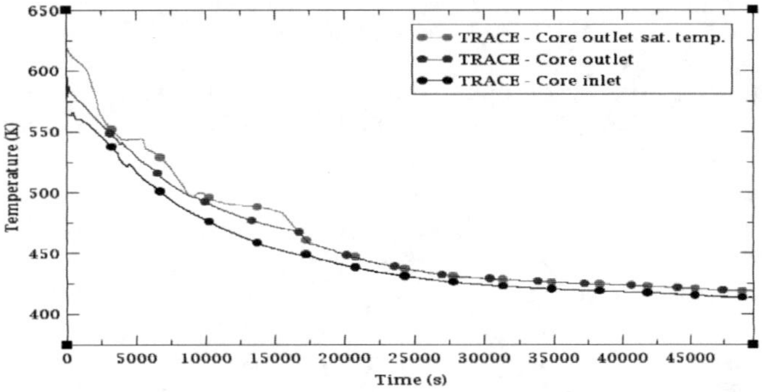

Figure 12. Core inlet, outlet and saturation temperature in TRACE.

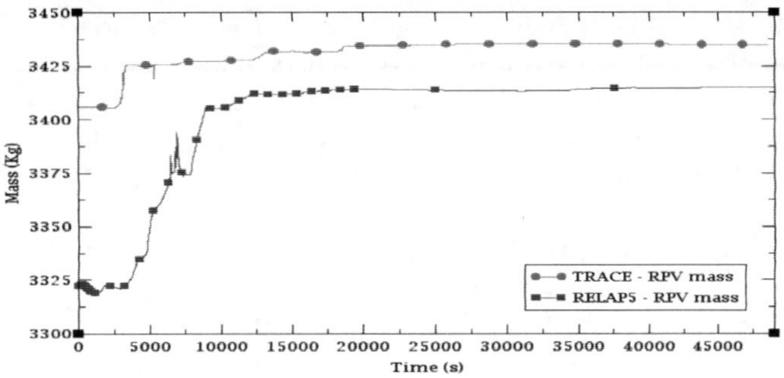

Figure 13. RPV mass inventory.

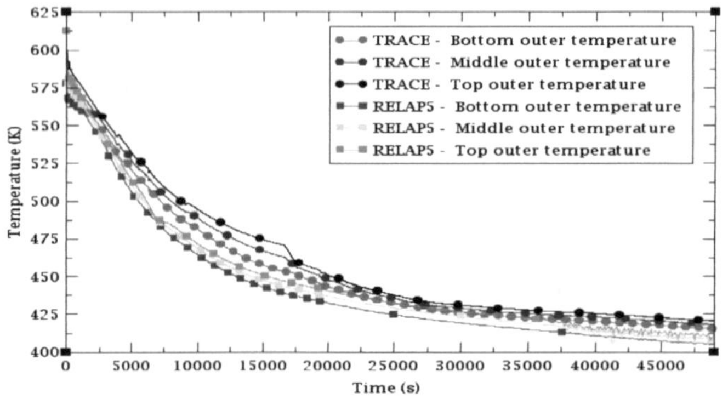

Figure 14. Heated rod surface temperature – hot rod.

Conclusions

The simulation of the SBO allowed to understand the phenomena occurring in the transient and to verify that the plant is suitable to cope with this kind of accident. No particular or critical situation occurred and the system slowly cools-down. Anyway, notwithstanding the difference in the codes and modeling, results are comfortably similar in trend and values, providing a further confirmation of the SPES3 facility design choices. In the future, the simulation with the two codes of other cases, included in the test matrix, will provide further information for a deeper investigation of the phenomena and the possible optimization of the system.

Acknowledgments

The authors appreciate the efforts and support of the colleagues of the SIET Laboratories, as well as the financial support of the Italian Minister of Economic Development through the agreement ENEA-MSE.

References

[1] R. Ferri, C. Congiu: Conceptual design of the SPES3-IRIS facility. SIET 01 334 RT 07 Rev.1. Piacenza (I), September 5th, 2008.

[2] G. D. Storrick: IRIS integral system test specification. Westinghouse Electric Company STD-AR-08-01 Rev.2, June 2010.

[3] RELAP5 code manual. NUREG/CR-5535/Rev.1 Idaho National Engineering Laboratory (USA), 2001.

[4] R. Ferri, C. Congiu: SPES3-IRIS facility nodalization for RELAP5 Mod.3.3 code and steady state qualification. SIET 01 423 RT 08 Rev.0. Piacenza (I), January 30th, 2009.

[5] R. Ferri, A. Achilli, G. Cattadori, F. Bianchi, P. Meloni: Design, experiments and RELAP5 code calculations for the PERSEO facility. Nuclear Engineering and Design 235 (2005) 1201-1214.

[6] TRACE V5.0, Theory and User's manuals, 2008. Division of System Analysis, Office of Nuclear Regulatory Research, U.S. Nuclear Regulatory.

[7] Symbolic Nuclear Analysis Package (SNAP) Users Manual, April 2007. Applied Programming Technology, Inc. 240.

[8] F. Castiglia, et al, 2010, "TRACE input model for SPES3 facility", CIRTEN-UNIPA RL-1208/2010, Palermo, Italy, July.

[9] P. Chiovaro, et al, 2011, "Analysis of the SPES3 direct vessel injection line break by using TRACE code", XXIX Congresso UIT sulla Trasmissione del Calore, Torino, Italy, June 20-22.

Journal of Physics: Conference Series **547** (2014) 012036 doi:10.1088/1742-6596/547/1/012036

Analysis of the influence of the heat transfer phenomena on the late phase of the ThAI Iod-12 test

B Gonfiotti and S Paci

University of Pisa, Dipartimento di Ingegneria Civile e Industriale, Largo Luciano Lazzarino 1, 56126, Pisa, Italy

Email: bruno.gonfiotti@for.unipi.it,

Abstract. Iodine is one of the major contributors to the source term during a severe accident in a Nuclear Power Plant for its volatility and high radiological consequences. Therefore, large efforts have been made to describe the Iodine behaviour during an accident, especially in the containment system. Due to the lack of experimental data, in the last years many attempts were carried out to fill the gaps on the knowledge of Iodine behaviour. In this framework, two tests (ThAI Iod-11 and Iod-12) were carried out inside a multi-compartment steel vessel. A quite complex transient characterizes these two tests; therefore they are also suitable for thermal-hydraulic benchmarks. The two tests were originally released for a benchmark exercise during the SARNET2 EU Project. At the end of this benchmark a report covering the main findings was issued, stating that the common codes employed in SA studies were able to simulate the tests but with large discrepancies. The present work is then related to the application of the new versions of ASTEC and MELCOR codes with the aim of carry out a new code-to-code comparison vs. ThAI Iod-12 experimental data, focusing on the influence of the heat exchanges with the outer environment, which seems to be one of the most challenging issues to cope with.

1. Introduction

During the last years many researches were carried out [1] to investigate aerosol behaviour during a severe accident (SA) in a Nuclear Power Plant (NPP). In particular these tests investigated the nuclear aerosol formation, transport and interaction with the various containment surfaces. Moreover, also their influence on thermal-hydraulics of primary circuit and containment system has been investigated [2]. A quite large part of these tests focused on the Iodine, being one of the major contributors to the external source term during a SA for its volatility and high radiological consequences [3].

As part of these efforts, the ThAI Iod-12 test [4] was performed. The test treats mainly the Iodine transport, deposition and resuspension from steel surfaces. The test is characterized by an Iodine release in a large steel-vessel under stratified conditions. Following, an helium injection breaks these stratified conditions leading to a mixing atmosphere and then to a mixed atmosphere. Therefore, thanks to the particular thermal-hydraulics conditions achieved during the transient, this test can also be used as thermal-hydraulics benchmark.

ThAI Iod-12 was originally released as a numerical benchmark [4] under the SARNET2 framework [5] in order to investigate the coupling between containment thermal-hydraulic and iodine codes, and to evaluate the capability of the SA codes to simulate the Iodine-steel interactions. At the end of the

benchmark (2012) a report covering the main findings was issued [6], stating that the employed codes, common used in SA studies, were able to simulate the test but large discrepancies were also present, especially on thermal-hydraulics and in Iodine modelling. For these causes model improvements and then further investigations were suggested.
Two of the main SA codes employed in the SARNET-2 framework were ASTEC and MELCOR [7]. Both codes have shown large thermal-hydraulics discrepancies during the benchmark, therefore the scope of this work is to build a new nodalisation capable to correctly simulate the entire thermal-hydraulics ThAI-12 transient and then utilize it with the latest ASTEC and MELCOR code versions. In particular the heat transfer phenomena towards the outer environment is of main concern, for its influence on the local atmospheric temperatures, and thus on the local relative humidity (r.h.). Both the local atmospheric temperatures and the local relative humidity influence the Iodine behaviour, therefore a proper thermal-hydraulics prediction is needed also for a correct evaluation of the Iodine behaviour.

The present article is structured into four parts. The first part describes the ThAI facility and the Iod-12 test. The second one describes the proposed common nodalisation. The third one describes the main thermal-hydraulics results, and finally in the fourth part of the paper the results of the study are reported.

2. The ThAI facility

The ThAI facility, acronym for Thermal-hydraulic Hydrogen, Aerosol and Iodine, is an experimental facility located at Eschborn (D). It is a coupled-effects test facility, simultaneously investigating stratification, convection and wall condensation as well heat conduction and storage in structures. Tests performed are mainly transients governed by the thermal inertia of the containment vessel. The facility is also characterized by a wide and complex instrumentation, allowing the determination of the thermal-hydraulic conditions inside each compartment. Thanks to its characteristics ThAI can be considered as an intermediate step between a separate-effect facility and a prototypical one.

3. Vessel description

The main ThAI component is a cylindrical steel vessel (Figure 1) with a total free volume of 60 m^3 (height: 9.2 m. Maximum diameter: 3.2 m). The vessel could resist to pressures and temperatures up to 14 bar and 180 °C and its free volume can be subdivided into six zones: the sump (1.7 m^3), the bottom area (6.6 m^3), the lower and the upper annulus (11.4 m^3 and 15.4 m^3 respectively), the dome area (17.7 m^3) and the inner cylinder area (6.05 m^3).

The lower and upper vessel surfaces are composed by 3 cm of AISI 316 Ti, 12 cm of mineral wool and 1 mm of Aluminum, except the vessel lid which is made in St-35 steel instead of AISI 316 Ti (mineral wool and aluminum thicknesses remain the same). The thermal insulation in the sump and dome areas is ~0.14 W/m·K
The external cylindrical vessel surface is composed by three heating/cooling jackets (h/c jackets) characterized by an oil moving helicoidally around the vessel circumference. The h/c jackets are composed (from the inner side to the outer side) by ~2.2 cm of AISI 316 Ti, a gap of ~1.7 cm for the carrier oil, 6 mm of AISI 316 Ti, 12 cm of mineral wool and 1 mm of Aluminium. The thermal insulation for the vertical walls is 0.1 W/m·K and the thermal conductivity of the carrier oil is 1.0 W/m·K. An electrical heater can also provide heat to the sump water.
As addendum, three intermediate decks of AISI 316 Ti were installed inside the vessel to enhance its spatial subdivision during the test. The upper deck closes the inner cylinder at top and reduces the flow area from the upper annulus to the dome area, leaving opened two sections of 0.123 m^2. The intermediate deck reduces the flow area between the upper and the lower annulus, leaving two sections of 0.198 m^2 and 0.325 m^2. The lower deck reduces the flow area from the lower vessel zone to the inner cylinder leaving a section of 0.176 m^2. Other important vessel components are the condensate collectors dividing the lower and the upper annulus, and the condensate gutters that drain condensate water out of the vessel.

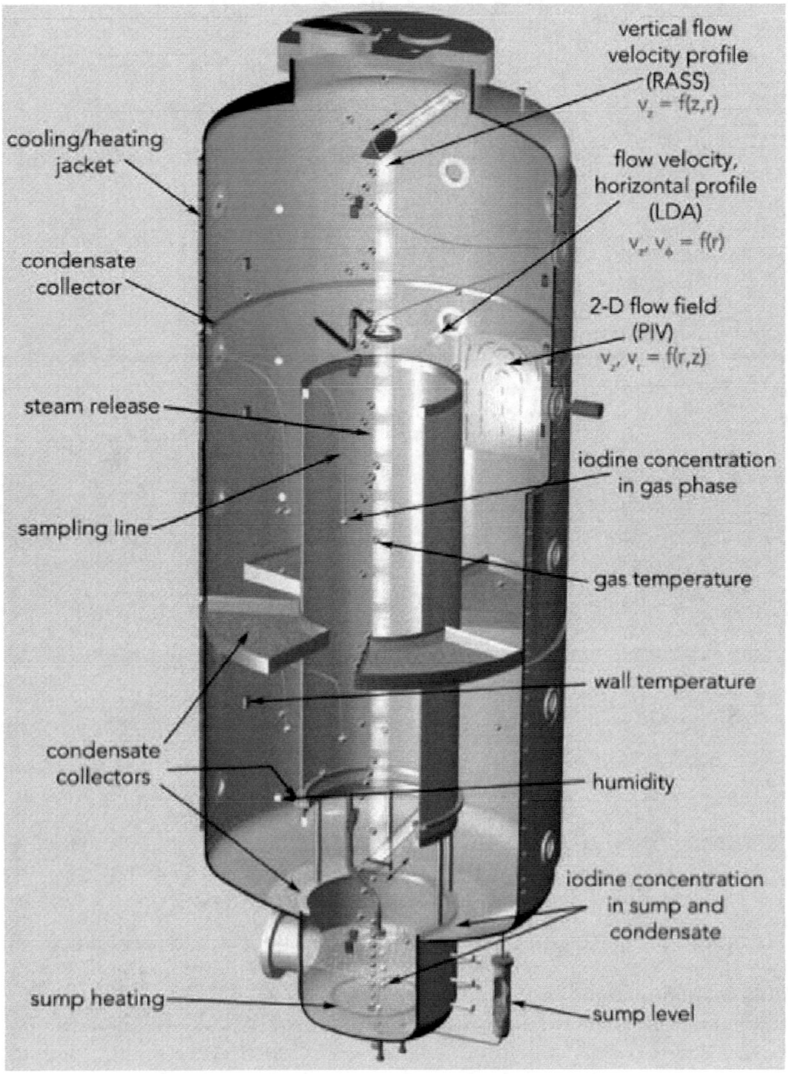

Figure 1. Position of the main vessel fittings and sensors.

4. The Iod-12 test

The Iod-12 test is characterized by injections of Iodine, Helium and steam as well as of energy by the external h/c jackets and the sump electrical heater. The test can be subdivided into eight phases, but only 6 were simulated, as requested in the SARNET2 benchmark specification [4]. In table 1 the main events occurring during the transient are reported.

Journal of Physics: Conference Series **547** (2014) 012036 doi:10.1088/1742-6596/547/1/012036

Table 1. Main events occurring during the Iod-12 test.

Time	Event
	Phase 2 – Setting of test conditions
-1.0 h	Start of the simulation
-0.32 h to -0.23 h	Injection of 0.335 Nm^3 of helium
	Phase 3 – Iodine injection
0.0 h to 0.1 h	First Iodine injection in the dome area (h =8.4 m) with air as carrier gas
0.17 h to 0.21 h	Second Iodine injection (same location and carrier gas)
1.75 h	Upper and middle h/c jackets switched off.
	Phase 4 – Establishing a mixed vessel atmosphere
2.62 h to 9.63 h	Injection of saturated steam (109 °C) in the upper zone of the inner cylinder area at a rate of 30 g/s, and then following increased to 34 g/s (4.0 h) and decreased to 4.3 g/s (4.82 h – 5.75 h). Furthermore the steam injection is re-increased to 6 g/s (6.95 h) and ended at 9.63 h.
2.68 h	Middle and lower h/c jackets switched of full cooling.
3.7 h	Activation of the sump electrical heater. Initially at 13.4 kW and then reduced to 3 kW (5.7 h).
4.92 h to 5.22 h	Injection of 563 g of Helium in the bottom area.
5.48 h	Upper h/c jacket switched on.
	Phase 5 – Mixed vessel atmosphere
9.67 h	Deactivation of the sump electrical heater.
9.68 h	H/c jackets thermal oil circuit switched off, but a limited cooling effect is presents due to the thermal oil natural circulation.
	Phase 6 – Rest period
-	-
	Phase 7 – Iodine resuspension
24.4 h to 30 h	Upper h/c jacket switched on full heating initially at 39 kW and then reduced to 3 kW at 25.68 h.

5. Common nodalisation

The ThAI facility has been modelled (figure 2) employing 66 control volumes, 60 for the vessel (3 volumes for the sump zone, 6 in the bottom area, 8 for the lower annulus, 16 volumes for the upper annulus, 14 in the dome area and 13 for the inner cylinder zone), 5 for the drain discharge volumes and 1 simulating the outer environment. The dome is vertically subdivided into 5 layers, while the upper and the lower annulus into 4 and 2 vertical layers respectively.
These control volumes are connected with 136 atmospheric junctions, necessary for the simulation of the mass and heat exchanges among the various vessel zones. Moreover, in ASTEC, 41 drain junctions were also employed to simulate the condensate water drain, connecting the upper and lower annulus and the outer volumes of dome, inner cylinder and bottom area to their respective discharge volumes. On contrary, MELCOR has not a proper drain junction model, then the film tracking model was employed to simulate the condensate water route to the external volumes.

The vessel walls have been modeled with 136 heat structures, 76 simulating the h/c jackets (24 the lower, 30 the middle and 22 the upper h/c jacket). The carrier oil has been simulated as a "solid slab" instead of a "flowing fluid", as requested in the benchmark specifications.

This quite detailed nodalisation has been developed to allow the investigation of the natural circulation phenomena among the main ThAI vessel zones.

Journal of Physics: Conference Series **547** (2014) 012036 doi:10.1088/1742-6596/547/1/012036

In the following two cases for each code will be reported, differing for the external boundary conditions imposed in the input data. The cases for each code are slightly different due to the different conditions that can be imposed in the input decks. The four cases are here listed:

- the "ASTEC-K" is characterized by an outer wall temperature and an outer heat transfer coefficient calculated by the code;
- the "ASTEC-T" is characterized by an outer heat transfer coefficient forced at 5 W/m^2K and by an outer wall temperature set to 20 °C;
- the "MELCOR-K" is characterized by an outer heat transfer coefficient forced to 5 W/m^2K. In this case the outer wall temperature is calculated by the code;
- the "MELCOR-T" is characterized by and imposed outer wall temperature set to 20 °C. In this case the outer heat transfer coefficient is calculated by the code.

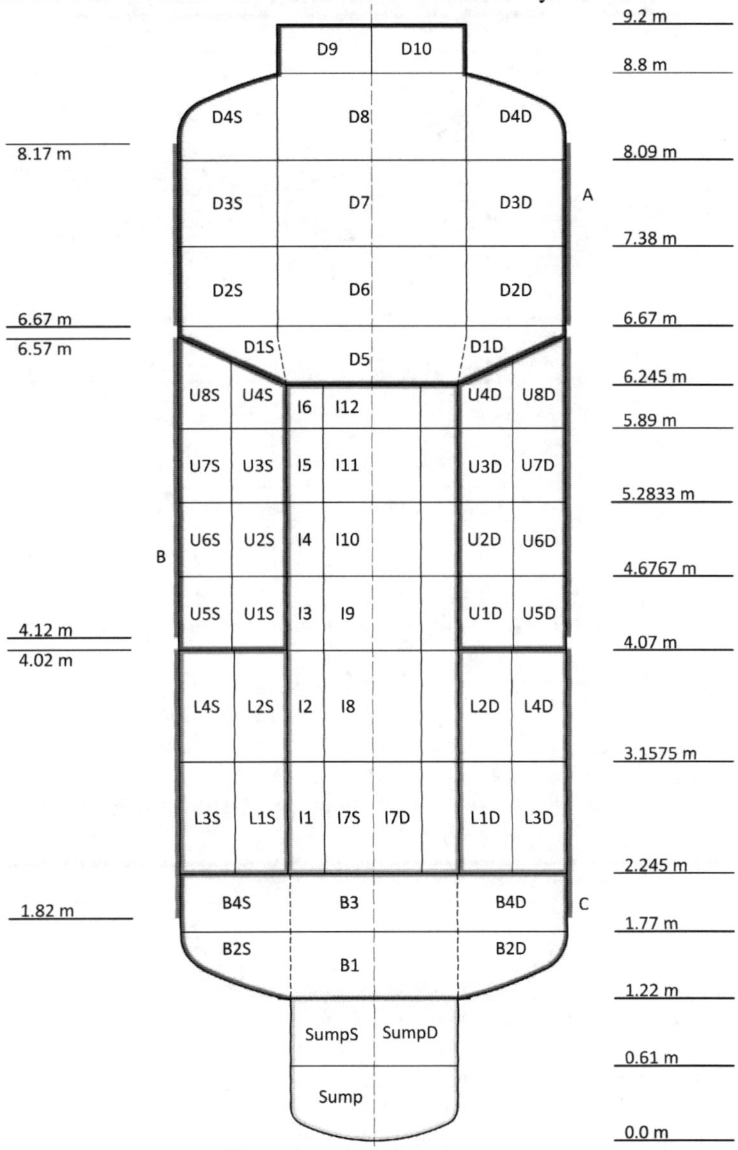

Figure 2. Nodalisation scheme.

6. Results

In the following the main thermal-hydraulics results of the ThAI Iod-12 test are reported. In particular, the analysis is mainly focused on the total pressure, the local atmosphere temperatures, and the local relative humidity, which are the most important parameters to achieve a good Iodine behaviour simulation [7].

6.1. Total pressure

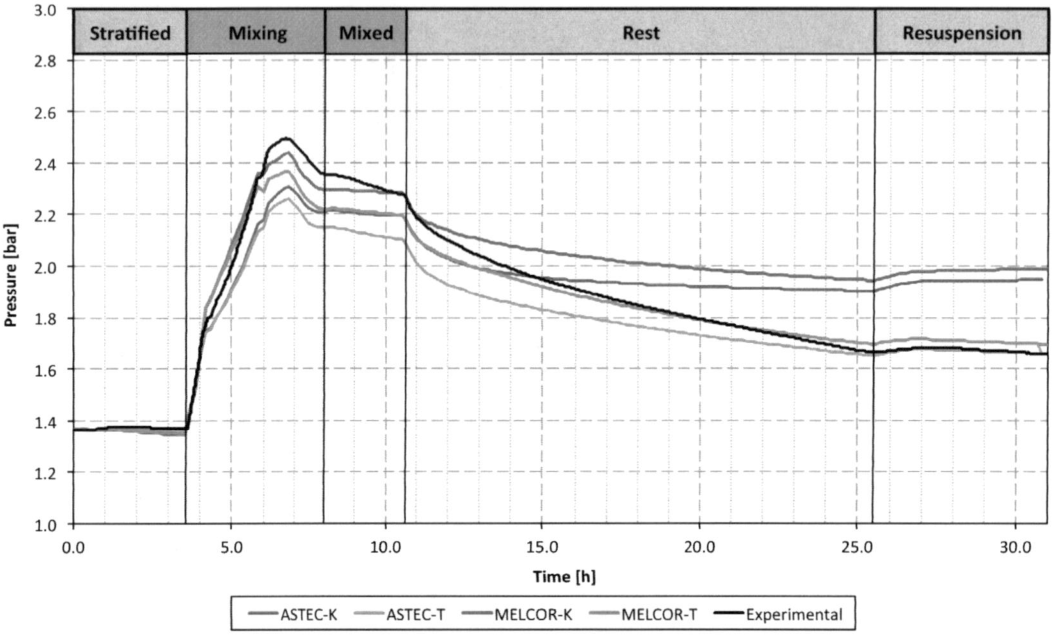

Figure 3. Total pressure in the dome area (7.7 m).

The total pressure was measured at 7.7 m inside the dome area (figure 3). As shown, the total pressure trend is similar in all the four cases till the end of the mixed phase, while in the following phases a quite large difference is shown.

A slightly pressure peak underestimation during the mixing phase characterizes the four cases. This difference is around 0.2 – 0-3 bar for the ASTEC cases, while for the MELCOR cases this underestimation does not exceeds 0.15 bar. An insufficient heat exchange with the outer environment during the late phases of the tests affects the ASTEC-K and MELCOR-K cases, leading to a total pressure overestimation of about 0.3 – 0.4 bar. On contrary, the ASTEC-T and the MELCOR-T cases follow the experimental trend without major discrepancies.

As conclusion it can be stated that both the ASTEC-K and the MELCOR-K cases are unable to predict the experimental trend, while ASTEC-T and MELCOR-T are in good agreement with the experimental data.

6.2. Dome temperature

The atmospheric temperature has been measured at various heights during the Iod-12 test. In figure 4 the atmospheric temperature inside the vessel dome at 8.4 m is reported.

The four cases show a similar trend till the begin of the rest phase, while in the following the influence of the imposed boundary conditions starts to be relevant. The MELCOR-K cases present the maximum temperature difference (8 °C) among the calculated data and the experimental data during the first 12 h of the test. In the rest and the resuspension phases both the ASTEC-K and the MELCOR-K cases show a quite large overestimation of about 35.0 – 40.0 °C. On contrary, both the ASTEC-T and the MELCOR-T cases follow without major discrepancies the experimental trend. However, during the final resuspension phase also these two cases present a slight overestimation of about 8.0 °C for the MELCOR-T case, and of 15.0 °C for the ASTEC-T case. This final discrepancy is probably due to the wrong temperature redistribution inside the vessel, being the total pressure correctly predicted during this phase (see above paragraph).

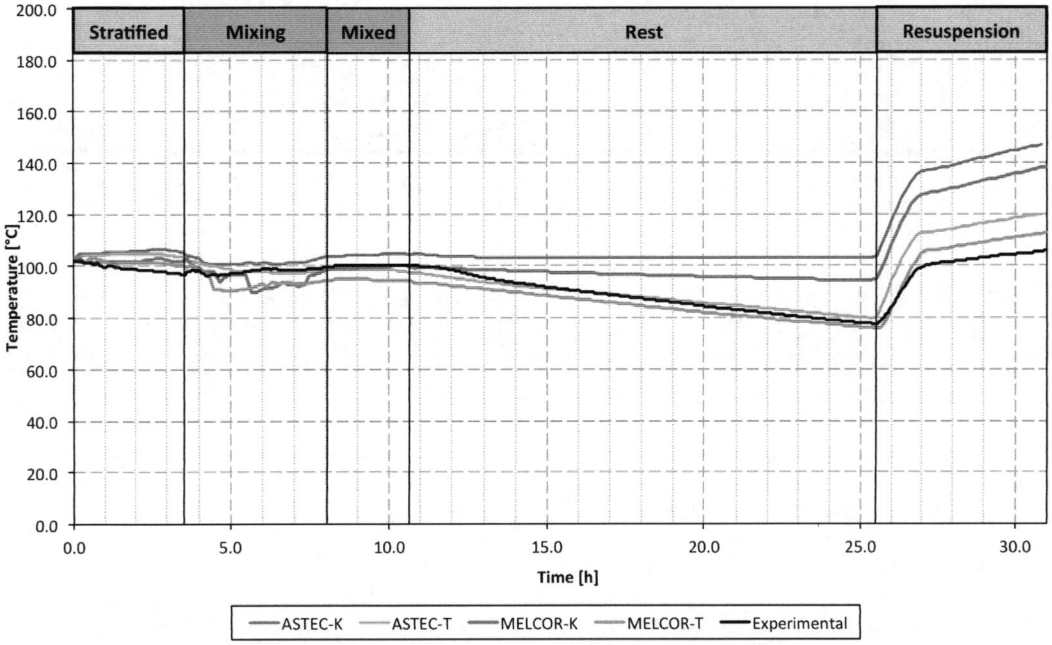

Figure 4. Atmospheric temperature inside the dome area (8.4 m).

6.3. Inner cylinder temperature

In Figure 5 the atmospheric temperature inside the inner cylinder at 5.6 m is shown. The atmospheric temperature inside the inner cylinder zone is probably one the best indicator of the correct mass and energy redistribution among the inner and outer vessel zones.

As shown in the graph all the 4 cases are able to predict extremely well the temperature trend till the end of the mixed phase, while on the rest and the resuspension phases some discrepancies are reported. In detail, the MELCOR-T and the MELCOR-K cases suffer for an insufficient heat exchange toward the outer zones of the vessel, leading to a temperature overestimation during the rest phase which influences also the final resuspension phase. Moreover, the trend of the MELCOR-T case is continuously decreasing instead of being “flat”.

On contrary, the ASTEC-K and the ASTEC-T cases are closer to the experimental data for the entire transient. A slight overestimation is shown in ASTEC-K case starting from 14.0 h, leading to a total error of 10 °C at the end of the resuspension phase, while the ASTEC-T case does not suffers major discrepancies during the entire test. As conclusion it could be stated that the ASTEC-T is the only case in which the redistribution of mass and energy among the vessel zones is well predicted.

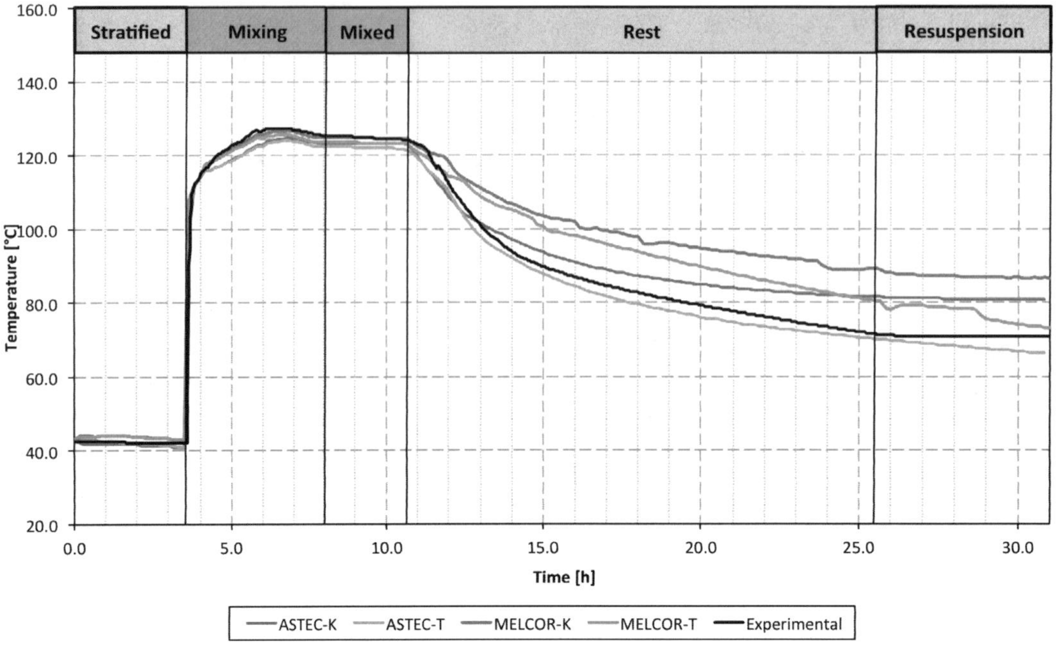

Figure 5. Atmospheric temperature inside the inner cylinder area at 5.6 m.

6.4. Relative Humidity

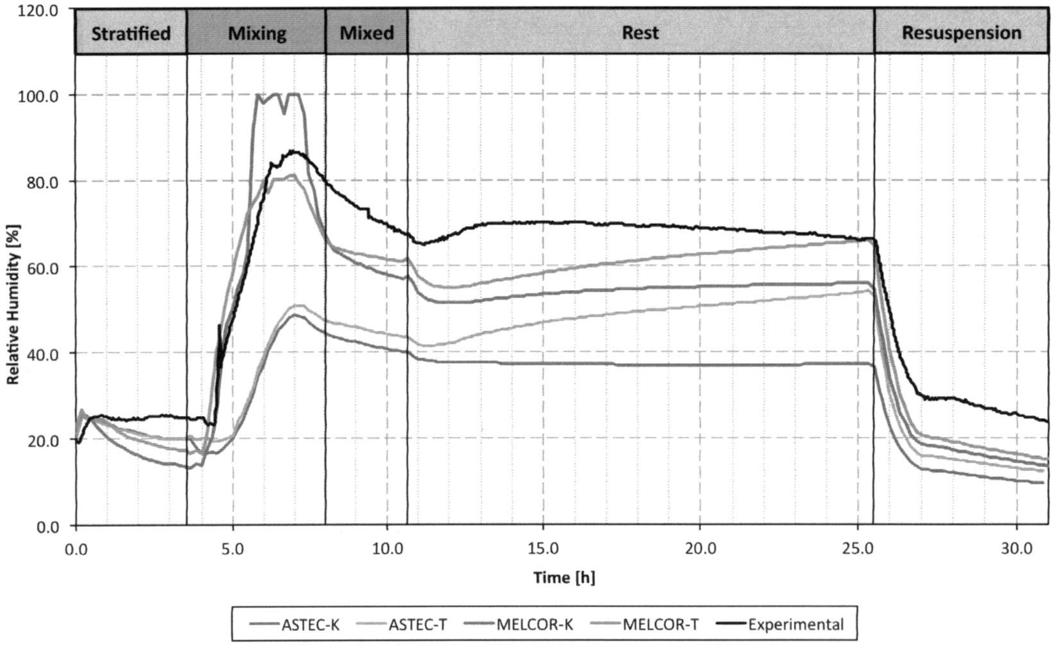

Figure 6. Relative humidity inside the dome area (8.4 m).

The relative humidity has been measured at three eights during the test. In Figure 6 the relative humidity at 8.4 m inside the dome area is reported.

The differences among the cases here reported are quite large. The ASTEC cases are both characterized by a complete underestimation of the r.h., especially in the mixing, the mixed, and the rest phases. This underestimation is around 40% for the ASTEC-K cases, while for the ASTEC-T the magnitude of this underestimation changes during the various phases of the test, from around 40% in the mixing phase, to about 10% at the end of the rest phase. Moreover, in the ASTEC-T case the r.h. trend during the rest phase is different compared to the experimental one. Therefore, it should be stated that both the ASTEC cases are unable to correctly predict the r.h. values and trend during the Iod-12 test.

On contrary, the MELCOR cases are closer to the experimental data values and trend, but large discrepancies can be highlighted during specific phases. The mixing phase is well predicted by the MELCOR-T case, while the MELCOR-K case is characterized by a global overestimation leading to an almost saturated condition from ~6.0 h to ~7.0 h. During the mixed phase both cases shown the same behaviour, but in the rest phase the trends became different, in particular the MELCOR-K trend becomes similar to that of ASTEC-K case, and the MELCOR-T case becomes similar to that of the ASTEC-T case. However, in the MELCOR cases the global underestimation respect to the experimental data is lower, reaching and error not larger than 20%. Finally, the resuspension phase is characterized by larger underestimation in all the four cases, therefore, it could be stated that except for the MELCOR-T case all the other three cases are unable to correctly predict the r.h. trend in the Iod-12 test.

6.5. Helium concentration

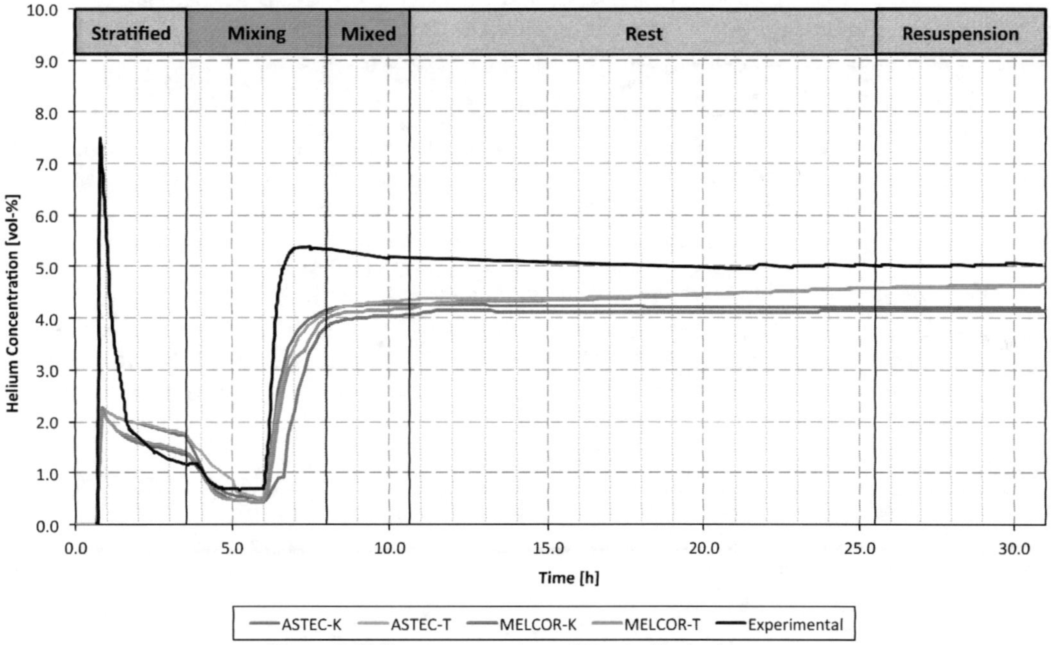

Figure 7. Helium concentration at 8.7 m inside the dome area.

The helium concentration has been measured at various heights during the Iod-12 test. In figure 7 the helium concentration at 8.7 m is shown. It should be highlighted as the experimental data provided are characterized by several uncertainties [4]. First, they refer to a steam-free atmosphere, therefore the concentrations shown after 5 h are slightly overestimated. Second, the initial peak is a local effect; hence its simulation is quite difficult. Therefore, the comparison between calculated and experimental data should be based on overall trends instead of a pure values comparison.

As shown, the four cases here reported show similar trends during the first 15.0 h, but in the latest hours of the test the ASTEC-T and the MELCOR-T cases report a continuous helium concentration increase, while the ASTEC-K and the MELCOR-K trends are similar to the experimental one.

7. Conclusions

As conclusion of this work it can be stated that the new versions of the ASTEC and MELCOR codes are able to predict the thermal-hydraulics transient of the ThAI Iod-12 test, but proper boundary conditions should be provided. In particular, it seems that the best choice is to let the code calculates the heat transfer coefficient instead of imposing an user-value. Moreover, it should be highlighted that heat transfer coefficient has a preeminent influence on the overall thermal-hydraulics results, as clearly shown in figures 3 and 4. Nevertheless, a good thermal-hydraulics simulation is needed for proper Iodine calculations, because the main phenomena involving Iodine are strongly influenced by the local thermal-hydraulics conditions. In particular the Iodine deposition and resuspension mechanisms are strongly influenced by the local r.h. values and by the local wall temperatures, therefore these two parameters should be as closer as possible to the experimental data.

References

[1] Sehgal B R et al. 2012 *Nuclear Safety in Light Water Reactors* (Great Britain: Elsevier)

[2] Allelein H, Auvinen A, Ball J, Güntay S, Herran L E, Hidaka A, Jones A V, Kissane M, Powers D and Weber G 2009 State-of-the-art report on nuclear aerosols (NEA/CSNI report)

[3] Clément B, Cantrel L, Ducros G, Funke F, Herranz L E, Rydl A, Weber G and Wren C 2007 State of the Art Report on Iodine Chemistry (NEA/CSNI report)

[4] Weber G 2011 Specification of the SARNET-2 WP8 THAI Benchmark – Part 2: Multi-Compartment Iodine Test Iod-12 (SARNET report)

[5] Van Dorsselaere J P, Auvinen A, Beraha D, Chatelard P, Journeau C, Kljenak I, Miassoedov A, Paci S, Tromm T W and Zeyen R 2012 The European Research on Severe Accidents in Generation 2 and 3 Nuclear Power Plants *Sci. Technol. Nucl. Ins.* **2012**

[6] Weber G, Herranz L E, Bendiab M, Fontanet J, Funke F, Gonfiotti B, Ivanov I, Krajewski S, Manfredini A, Paci S, Pelzer M and Sevón T 2012 SARNET2 WP8 Benchmark on the THAI Iodine Multi-Compartment Tests Iod-11 and Iod-12 (SARNET report)

[7] Weber G, Herranz L E, Bendiab M, Fontanet J, Funke F, Gonfiotti B, Ivanov I, Krajewski S, Manfredini A, Paci S, Pelzer M and Sevón T 2013 Thermal-hydraulic - Iodine chemistry coupling: Insights gained from the SARNET benchmark on the THAI experiments Iod-11 and Iod-12 *Nuclear Engineering and Design* **265** pp 95–107

Journal of Physics: Conference Series **547** (2014) 012037 doi:10.1088/1742-6596/547/1/012037

Numerical-experimental analyses by Hot-Wire method of an alumina cylinder for future studies on thermal conductivity of the fusion breeder materials

R. Lo Frano[1], M. Moscardini[2], D. Aquaro[1]
[1]DICI- University of Pisa, Pisa- Italy
[2]Karlsruhe Institute of Technology -Karlshrue- Germany

rosa.lofrano@ing.unipi.it

Abstract.
The determination of the thermal conductivity of breeder materials is one of the main goal in order to find the best candidate material for the fusion reactor technology.
Experimental tests have been and will be carried out with a dedicated experimental devices, built at the Department of Civil and Industrial Engineering of the University of Pisa. The methodological approach used in doing that is characterized by two main phases strictly interrelated each other: the first one focused on the experimental evaluation of thermal conductivity of a ceramic material, by means of hot wire method, to be subsequently used in the second phase, based on the test rig method, to determine the thermal conductivity of pebble bed material. To the purpose, two different experimental devices have been designed and built. This paper deals with the first phase of the methodology.
In this framework, the equipment set up and built to perform Hot wire tests, the ceramic material (a cylinder of alumina), the experimental procedure and the measured results obtained varying the temperature, are presented and discussed. The experimental campaign has been lead from 50°C up to 400°C.
The thermal conductivity of the ceramic material at different bulk temperatures has been obtained in stationary conditions (detected on the basis of the temperature values measured during the experiment). Numerical analyses have been also performed by means of FEM code Ansys©. The numerical results were in quite good agreement with the experimental one, confirming also the reliability of code in reproducing heat transfer phenomena.

1. Introduction

The knowledge of the effective thermal conductivity of pebble beds, to be used as breeder material in fusion reactor, is of meaningful important to extract thermal power, as it is widely known in the scientific literature [1]-[7]. Breeding blanket material, in form of sintered block and pebble bed (among the candidate ceramics they are lithium orthosilicate, lithium titanate, etc.), is considered one of the most promising fusion blankets and worldwide efforts have been devoted to its R&D.

Pebble beds are used in many applications, such as nuclear reactor fuel rods, thermal insulation, heat exchangers, chemical catalysts, automotive catalytic converters, and recently, fusion blankets. Ceramic materials have some problems, such as uncertainty of thermal conductance at breeder/structure interface, cracking due to high thermal stresses, creep, swelling etc., therefore an accurate prediction of the thermal conductance at its interface is required.

The pebble beds in fusion blanket offers several advantages, primarily due to the bed characteristics (e.g. size and shape of pebbles, pressure of cover gas, and bed packing) to be tailored to obtain the required thermal characteristics or to be adjusted to match the effective thermal conductivity [8][9]. Accordingly the effective thermal conductivity and the interface thermal conductance of the ceramic pebble beds are key thermal properties to be well measured and characterized: several parameters may influence them as the thermal conductivity of pebbles and filling gas, gas pressure, bed porosity, and pebbles' deformation. In literature it is also possible to find a lot of application of hot wire method for determining the thermal conductivity of solid materials.

Specifically, in respect to [10] and [11], where a thin wire, embedded in the pebble bed, a heated up and the wire surface temperature is measured as a function of time, this study uses a heating rod operated in steady-state condition aiming also at minimizing the accuracy of results that e.g. for deformed ceramic breeder and beryllium pebble beds is questionable [12].

In this framework an extensive experimental activity aiming at the evaluation of thermo-mechanical characterization of pebble beds, as a function of the average bed temperature in the range of 50÷500°C by adopting test ring method and hot wire technique, as described in [8] and [13], has been done and is still ongoing at the Department of Civil and Industrial Engineering (DICI) of the University of Pisa. The methodology used in doing that is based on:

1) the experimental evaluation of thermal conductivity of a ceramic material, by means of the hot wire method;
2) the adoption of test rig method to determine the thermal conductivity of pebble beds.

The effective thermal conductivity of the ceramic pebbles, namely λ_{PB}, is hence measured as function of a known conductivity of an alumina disc (λ_A) (used as reference value) and temperature differences across the alumina disc [13] and the bed, respectively ΔT_A and ΔT_{PB}.

This paper deals with point 1) of the above methodology.

In what follows, it will be presented and discussed:

- the experimental device used to determine the thermal conductivity of the alumina by Hot-Wire method;
- the experimental procedure;
- the results (experimental value of λ_A) obtained varying the furnace temperature from 20° to 500°C;
- the numerical results, obtained by means of FEM code Ansys©, and their comparison with the experimental ones.

2. Hot Wire Method (HWM): description of experimental device

The experimental equipment, consists of the following main parts: the furnace; the cylinder of alumina, which was composed of three cylinders in series as shown in Figure 2; the thermal resistance; the Fan-Inverter system and the Data Acquisition System (DAS).

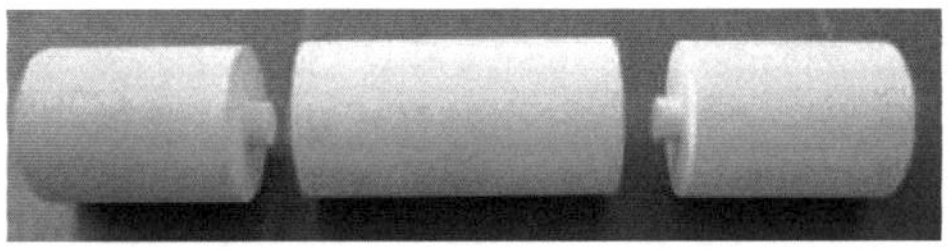

Figure 1- Alumina sample used in 'hot wire test' for determining the thermal conductivity

The alumina specimen, namely Alubit 90 (of same material of that used in "hot plate method with guardian rings" tests), is 35 mm in diameter and 280 mm length. The cylinder of alumina has three holes (Figure 2) of which one central through, to accommodate the electrical resistance, and the other two radials with the bottom end, to insert the thermocouples. These latter are positioned respectively at 7.75 and 21.2 mm from the centre of the specimen.

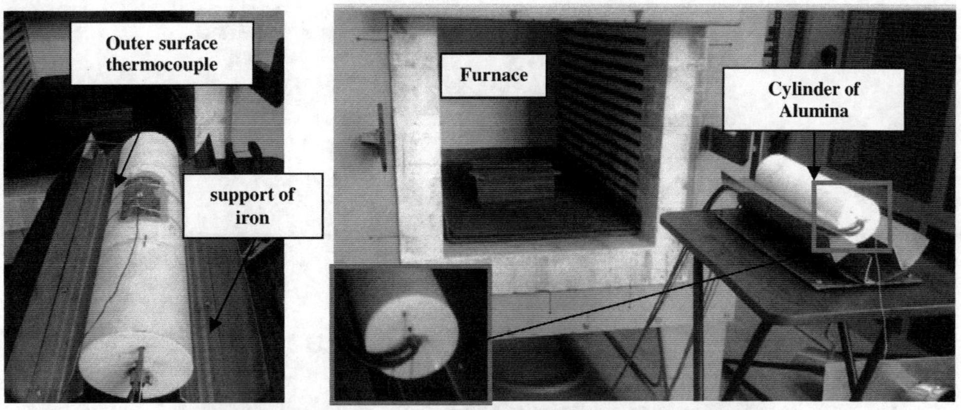

a)

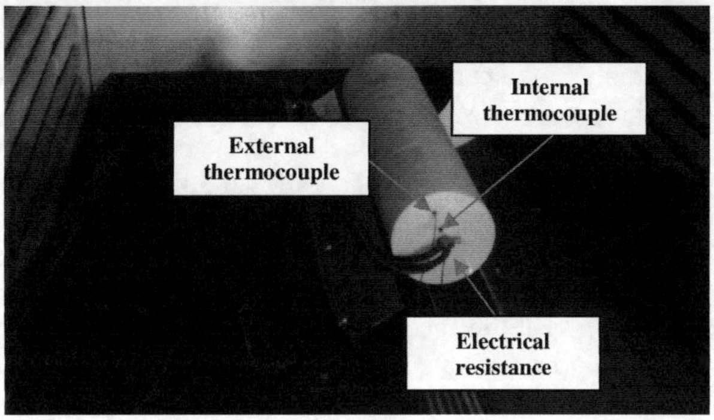

b)

Figure 2 a, b - Experimental equipment

The two thermocouples, type K, have been equipped with a final cap in aluminium in order to ensure the adherence to the walls of the holes.
In addition, on the outer surface of the specimen it was glued a thin copper plate, on which was welded a thermocouple that has allowed to control/measure the surface temperature of the specimen.

It is also worthy to note that the cylinder is resting on a support of iron, in this way, the alumina will not be in direct contact with the surface of the furnace in which it will be placed. Additionally in order to properly determine the thermal conductivity of the alumina specimen, the power generated by the electrical resistance has been removed by forced convection (Figure 3).

To the purpose a fan, duly supported to guarantees the parallelism of the tube with the ground, has been connected to an inverter to obtain a more precise regulation: pipe connected to the fan conveys the air directly on the outer surface of the alumina.

During the tests, a data acquisition system (National Instruments-SCXI 1000) directly connected to the PC via Lab View© interface) allowed to control and measure the temperatures of:

- alumina (on the external surface and inside the specimen);
- furnace;
- ambient air.

The thermocouples were of course calibrated, before the execution of the tests: the calibration was carried out every 10°C from 30 to 450°C in order to cover the temperature range foreseen in the wire tests (50 ÷ 400°C).

Figure 3 - View of the Fan-Inverter system

2.1. *Tests execution*

After having placed the alumina specimen in the furnace and connected the acquisition system and the ventilation system, the furnace was closed and heated up to the fixed temperature.

The power generated by the thermal resistance has been removed by forced convection, in this way the two thermocouples reach the steady state condition: a difference of 4°C is observed from each other.

The thermal conductivity of the alumina specimen was therefore determined at stationary conditions: the temperature gradient arisen across the longitudinal direction of the specimen was measured by the internal and external thermocouples for at least 5 minutes. The thermal conductivity (λ_A) is therefore calculated accordingly to the following equation:

$$\lambda_A = \frac{\phi_0}{2\pi\Delta T L}\ln\left(\frac{r_2}{r_1}\right) \tag{1}$$

Where Φ_0 is the power of the heat source; ΔT the temperature difference measured by the thermocouples between the inner and outer of the alumina; L is the length of the source which corresponds with the length of the cylinder of alumina; while r_1 and r_2 are the radial distances where internal and external thermocouples are positioned (see Figure 2).

2.2. *Experimental results*

The tests were carried out in the temperature range 50÷400°C, with increment of 50°C. The temperature behaviour measured, as an example, at 300°C, vs. time is represented in Figure 4: the red box indicates the stationary conditions, while, of course, in the first part (up to ~700 s) of the diagram the temperatures are different. Results in Figure 4 show that under an average temperature of 300°C, an average power of 217.1 W and a temperature difference between the internal and external thermocouple of 13°C, the thermal conductivity of alumina is equal to 9.41 W/m°C.

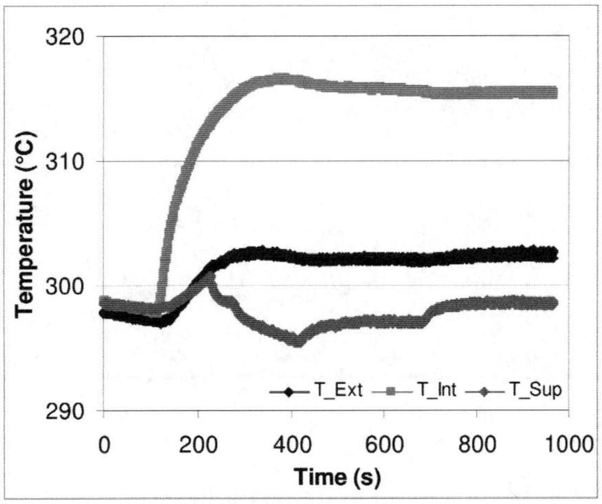

Figure 4 – Hot wire test results at 300°C: temperature measured across the alumina specimen

The wire tests were repeated several times; for each test campaign the calibration of thermocouples were also repeated. The thermal conductivity calculated on the basis of the hot wire tests are showed for different temperature in Figure 5. Analysing the λ_A behaviour, it is possible to observe that the conductivity decreases along with the increase of the test temperature; the calculated maximum deviation resulted about 1.3% (experimental data have been interpolated with a third-order polynomial curve). Finally taking into account the error related to acquisition system, the mean error of the measure resulted less than 3.5%.

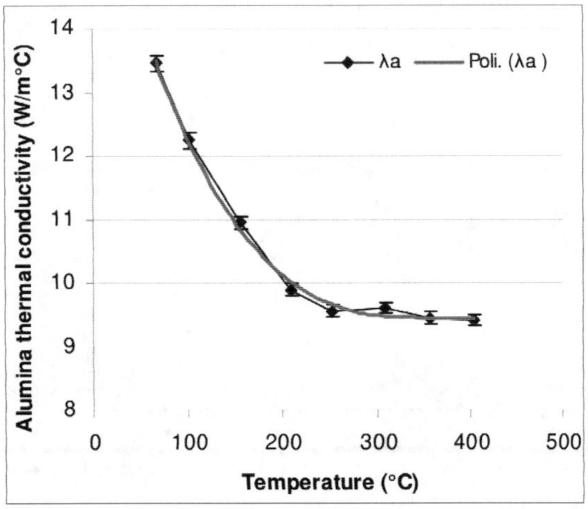

Figure 5 - Values of the thermal conductivity vs. temperature

Moreover the thermal conductivity values of the alumina, experimentally determined, were compared to the literature ones (Figure 6): a good agreement is observed at 400°C, while for temperatures above or below 400°C a relevant discrepancy appears. An explanation of that could be possibly relied on the different conditions based on which literature data has been determined (no accurate description of the procedure adopted); another relevant aspect to think about is however the lack of data available in literature.

Journal of Physics: Conference Series **547** (2014) 012037 doi:10.1088/1742-6596/547/1/012037

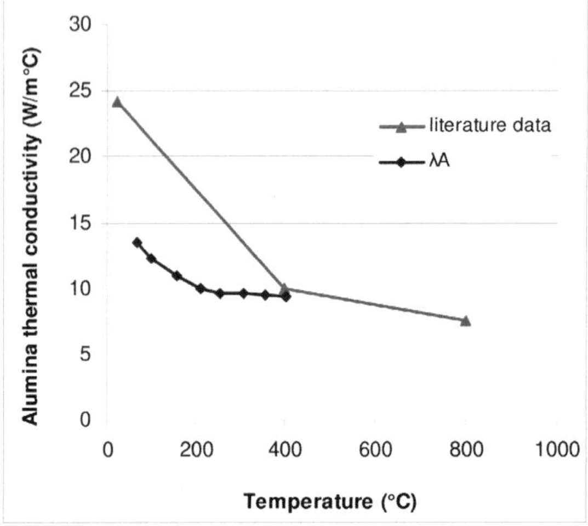

Figure 6 - Comparison between the literature and experimental values of the thermal conductivity of alumina

3. Numerical simulations

Numerical simulations (both steady state and transient) were also performed in order to validate the experimental hot wire tests and, in turn, the measurements of the thermal conductivity. In what follows, are briefly described the FEM simulations carried out along with the assumed initial and boundary conditions (Figure 7 a). The FEM model of the experimental device (1:1 scale) has been implemented by using four nodes axial-symmetric elements, as shown in Figure 7 b.

Input data entered to the code, characterizing the alumina, are:

- Thermal conductivity experimentally determined by tests (see previous Figure 5);
- Density = 3940 Kg/m^3;
- Specific heat capacity = 0.8 J/KgK.

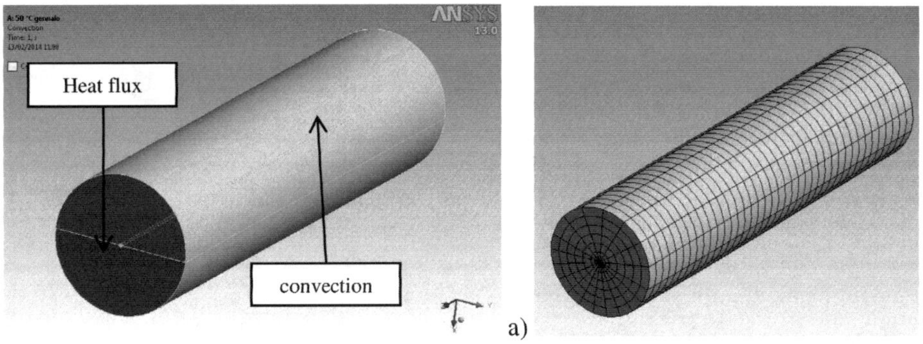

Figure 7 – 3D (a) and discretized (b) FEM model

3.1 Steady-state thermal analysis

The initial and boundary conditions reproduce the hot wire tests conditions, as follows:

- the initial temperature refers to the uniform temperature reached by the alumina specimen;
- the heat flow value, applied in the hole of lodging resistance, is the average value of power obtained from each test in the last five seconds divided by the surface of the hole;

• the convection on the outer surface of the specimen (Figure 7 a) is calculated by dividing the average power for the outer surface and temperature difference -ΔT- between that of the alumina wall and that of the cooling fluid (air conveyed by the fan on alumina).

Figure 8 show that the results obtained experimentally are very similar to the numerical ones; in the graph are represented the radial behaviour (a thermal path) of the thermal conductivity of the alumina.

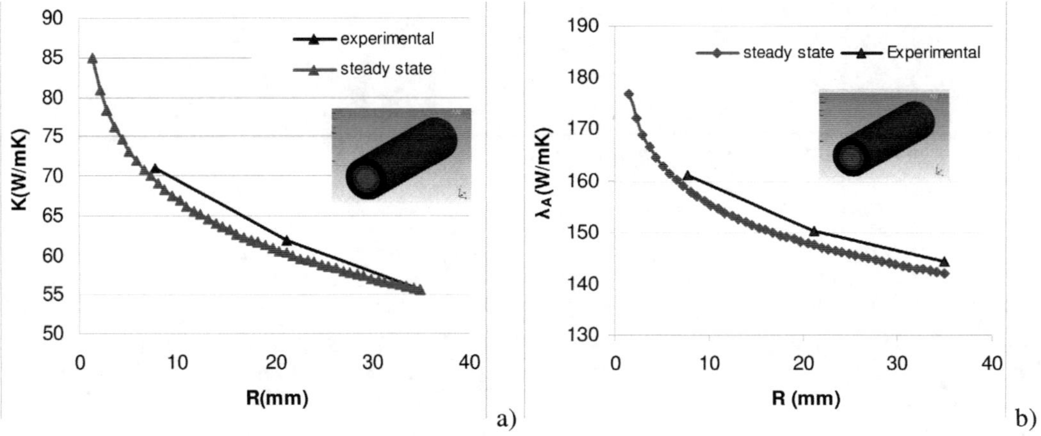

Figure 8 – Radial behaviour of the thermal conductivity of the alumina at 50°C (a) and 150°C (b)

3.2 Transient thermal analysis

The transient analyses have been carried out on the model of Figure 7 by implementing the same initial and boundary conditions, used for the steady state analysis, and the thermal steps summarized in Table 1, which represent the heating conditions (initial steady state, the ramp representing the ignition of the resistance, the final steady state) observed during the wire tests.

Table 1- Initial conditions of thermal transient simulation

Step	Time (s)	HTC (W/m$^{2\circ}$C)	Heat flux (W/m^2)	T_{blulk} (°C)
0	0	10	0	60
1	10	10	0	60
2	40	101,6	1,3551e+005	21,38
3	340	101,6	1,3551e+005	21,38

Figure 9 show the results obtained for an arbitrary thermal path along the radius of the alumina specimen: as for the previous analysis, the experimental and numerical results are in good agreement (difference of about 1°C until 150 ° C).

This good agreement allowed therefore to validate the used code.

In addition, in Figure 10 it is represented the comparison between the measured and calculated temperature behaviours for a hot wire test with oven temperature of 50°C and 150°C.

The steady state is reached in the first case after 150 s while in the latter case after 300 s: this different time scale is conditioned by the time required by the oven to reach the temperature test condition.

It is worthy to note that the stationary condition in the numerical simulation is reached immediately due to assumed initial conditions (the thermal interaction is simulated with an effective heat transfer coefficient, see Table 1); while as regards the experimental data, stationary condition was obtained by varying the air flow rate, for this reason, the graphs behaviour have different times.

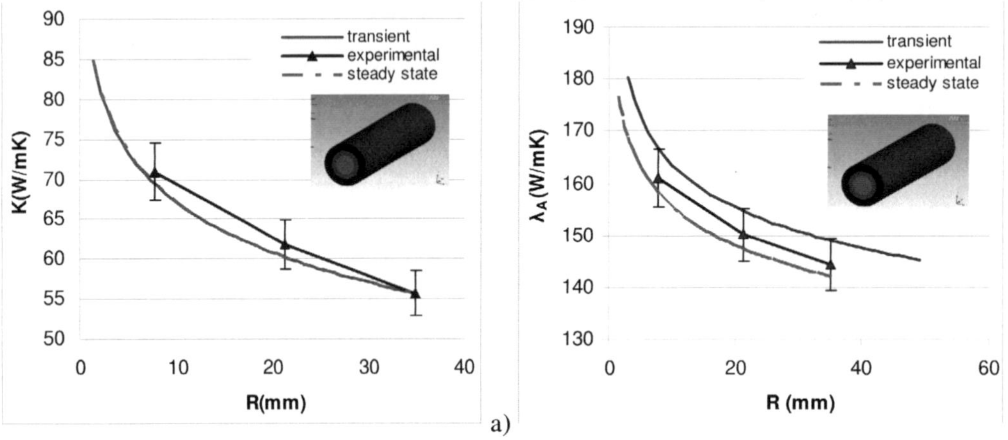

Figure 9 – Radial behaviour of alumina thermal conductivity: 50°C (a) and 150°C (b)

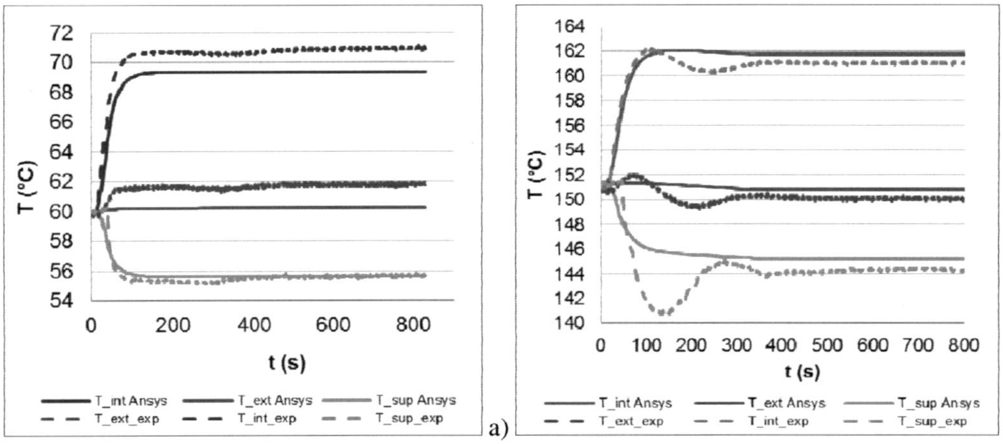

Figure 10 – Comparison between measured and calculated temperature: 50°C (a) and 150°C (b)

4. Conclusion

Breeding blanket material, in form of sintered block and pebble bed, is considered one of the most promising fusion blankets and worldwide efforts have been devoted to its R&D.

In this paper the thermal conductivity of the alumina, to be used as reference thermal conductivity in the test rig method, has been determined by performing hot wire tests at different temperatures. The experimental activity was carried out at the Lab. Scalbatraio of the Dep. DICI of the University of Pisa by using a dedicated device and an adequate DAS.

The thermal conductivity of an alumina specimen (of cylindrical shape) was determined at stationary conditions by means of the temperature gradient measured across the longitudinal direction of the specimen for at least 5 minutes. The tests were carried out in the temperature range 50÷400°C, with increment of 50°C. The results show that for an average temperature of 300°C, an average power of 217.1 W and a temperature difference between the internal and external thermocouple of 13°C, the thermal conductivity of alumina is equal to 9.41 W/m°C.

Numerical (post-test) simulations were also carried out to validate hot wire test results by performing both steady state and transient analyses.

The experimental results indicated that:

1) the thermal conductivity λ_A decreases along with the increase of the test temperature; the calculated maximum deviation resulted about 1.3% (experimental data have been interpolated with a third-order polynomial curve).
2) λ_A, experimentally determined, was compared to the literature value of the thermal conductivity: a very good agreement is observed (Figure 6) at 400°C, while for temperatures above or below 400°C it was not possible to make any comparison because of the lack of data available in the literature;
3) the comparison between the experimental and numerical results, in terms of radial behaviour of the thermal conductivity or radial temperatures, highlights a very good agreement with a discrepancy less than 5%.

The alumina thermal conductivity will be used as reference value for the application test rig method for the evaluation of thermo-mechanical characterization of ceramic pebble beds of interest for the design and analysis of fusion breeding blankets (activity currently ongoing).

Future experimental tests will be carried out increasing the temperature test condition up to 800°C in order to the required thermal characteristics to be used, subsequently, in the test rig method to measure the effective thermal conductivity of pebble beds.

References

[1] Enoeda M et al., 2001, Effective Thermal Conductivity Measurement of the Candidate Ceramic Breeder Pebble Beds by the Hot Wire Method", *Fusion Tech.* **39**, 612
[2] Abou-Sena A, Ying A, Abdou M, 2003, Experimental Investigation and Analysis of the Effective Thermal Properties of Beryllium Packed Beds, *Fusion Science and Techology* **44**, 79
[3] Reimann J, Hermsmeyer S, 2002, Thermal conductivity of compressed ceramic breeder pebble beds, *Fusion Engineering and Design*, **61-62** pp 345-351
[4] Enoeda M et al.. 1998, Effective thermal conductivity measurements of the binary pebble beds by hot wire method for the breeding blanket, *Fusion Technology* **34** November
[5] Ying A et al., 2012, Status of ceramic breeder pebble bed thermo-mechanics R&D and impact on breeder material mechanical strength, *Fusion Engineering and Design* **87** pp 1130-1137
[6] Aquaro D, Zaccari N, 2006, Experimental and numerical analysis on pebble beds used in an ITER Test Module Blanket *Fus. Eng. Des.* **81** 707-712
[7] Aquaro D, Zaccari N, 2007, Experimental and numerical analyses on Li4SO4 and Li2TiO3 Pebble bed used in a ITER test module blanket *J. Nucl. Mat.* **367-370** 1293-1297
[8] Lo Frano R et al., 2014, Ceramic breeder for fusion nuclear reactors: thermo-mechanical tests on alumina pebble beds, *Fusion Engineering and Design* **01**
[9] Abou-Sena A et al., 2007, Experimental measurements of the effective thermal conductivity of a lithium titanate (Li2TiO3) pebbles-packed bed, *J. M. Processing Technology* **181** pp 206-212
[10] Reimann J et al., 2006, Thermal conductivity of compressed beryllium pebble beds, *Fusion Engineering and Design* **81** pp 449-454
[11] Tanigawa H et al., 2009, Thermal Conductivity Measurement with Silica-Coated Hot Wire for Li4SiO4 Pebble Bed, *J. Nuclear Science and Technology*, **46**, 6 pp 553-556
[12] Mandal D et al., 2012, Experimental measurement of effective thermal conductivity of packed lithium-titanate pebble bed, *Fusion Engineering and Design*, **87** pp 67-76
[13] Aquaro D, Lo Frano R, 2014, Experimental evaluation of thermal conductivity of ceramic pebble beds *J. Phys.: Conf. Ser.* **501** 012001
[14] Hammerschmidt U,2003, A quasi-steady state technique to measure the thermal conductivity, *Intl. J. Thermophysics*, **24**, pp 234-248
[15] ANSYS© code, 2013, *User guide*

32nd UIT (Italian Union of Thermo-fluid-dynamics) Heat Transfer Conference IOP Publishing
Journal of Physics: Conference Series **547** (2014) 012038 doi:10.1088/1742-6596/547/1/012038

Onset of instability due to variable viscosity and dissipation in a plane porous channel

Antonio Barletta, Michele Celli

Department of Industrial Engineering, Alma Mater Studiorum Università di Bologna, Viale Risorgimento 2, 40136 Bologna, Italy.

E-mail: antonio.barletta@unibo.it, michele.celli3@unibo.it

Abstract. The effect of viscous dissipation is considered on modelling the fully-developed heat transfer in a parallel plane channel filled with a saturated porous medium. The basic Darcy's flow in a regime of forced convection is analysed insofar as the variability of fluid viscosity with temperature is taken into account. The thermal boundary conditions at the impermeable channel walls are described by assuming external convection with a constant heat transfer coefficient, *viz.* by imposing Robin conditions for the temperature as parametrised through the Biot number. The emergence of a singular behaviour in the basic velocity and temperature profiles is found when the Péclet number and the variable viscosity parameter are large enough as to imply a failure of the linear fluidity model. A linear stability analysis of the basic parallel flow is carried out to detect the reaction of the system to small-amplitude external perturbations. Different odd or even normal modes of the longitudinal type are studied. It is shown that no instability arises until the parametric condition for the emergence of the singularity is approached. An argument to predict the behaviour of normal modes of oblique type is eventually presented.

1. Introduction

Viscous dissipation is an important effect in heat transfer processes involving the flow of Newtonian or non-Newtonian fluids with a large viscosity and relatively small thermal conductivity. Such fluids display, in most cases, large viscosity variations with temperature. Examples of fluid flows with variable viscosity and dissipation are frequent also in fluid saturated porous media, where viscous dissipation is not only an effect arising close to boundary walls, as for Navier-Stokes flows, but it arises also in the bulk region of the seepage flow. The reason is that viscous dissipation in Navier-Stokes flow is set off by the square strain, while for Darcy's flows its intensity is directly proportional to the local square velocity [1, 2].

It is well established that viscous dissipation and variable viscosity may cause a thermal instability in the flow [2–10]. There are basically two physical causes for the instability, *viz.* the variability of viscosity with temperature and the effect of thermal buoyancy [2, 10]. While the former cause was recognised several years ago by Joseph [3], the role played by the thermal buoyancy has been identified only recently [9]. A wide work has been carried out in recent years on this topic, as surveyed by Barletta [2], dealing either with Navier-Stokes flows and with seepage flows in saturated porous media. On the interplay between the effect of variable viscosity and that of variable density (thermal buoyancy) in triggering the instability, an analysis has been recently carried out by Barletta and Nield [10]. In this paper, the instability of the fully developed parallel flow in a horizontal porous channel was proved to arise when the Péclet number exceeds a critical value, that depends both on the Gebhart number and on the slope parameter of the viscosity law. The configuration analysed by Barletta and Nield [10] is such

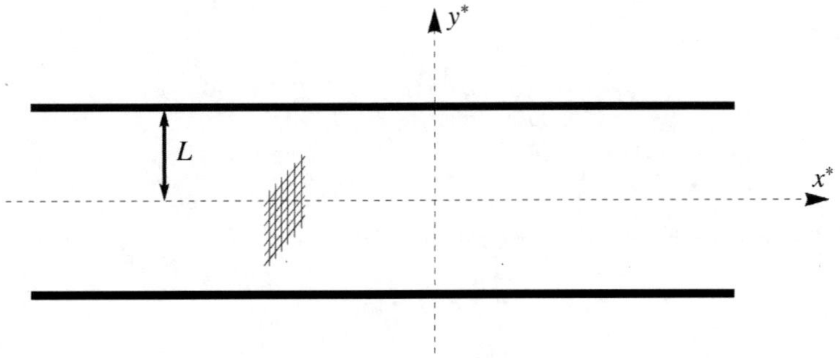

Figure 1. Two-dimensional sketch of the porous channel.

that the upper channel wall is isothermal and impermeable, while the lower wall is impermeable and adiabatic.

Our aim is to develop further the analysis of Barletta and Nield [10], in order to study a configuration where the boundary conditions are symmetric instead of asymmetric, and to focus our attention on the role played by the variable viscosity. To this end, we will assume flow conditions of forced convection, so that the thermal buoyancy will be neglected. The symmetric channel boundaries will be assumed impermeable and subject to Robin temperature conditions, so that a generic process of heat transfer to the external environment is envisaged, reducing to the purely isothermal case when the Biot number tends to infinity.

2. Governing equations

The analysis developed in this paper is relative to a plane porous channel with half-width L (see Fig. 1). The channel walls, $y^* = \pm L$, are considered impermeable and cooled up by an external fluid environment at temperature T_0, with a constant heat transfer coefficient h. The porous medium is homogeneous and isotropic; it is saturated by a fluid under local thermal equilibrium conditions. Darcy's law is satisfied and the fluid density is taken to be constant, so that the buoyancy force can be neglected.

Coordinates x^* and z^* define the streamwise and the spanwise directions, respectively. The velocity field $\boldsymbol{u}^*$ has components (u^*, v^*, w^*) along the (x^*, y^*, z^*) axes. Time is denoted by t^*, temperature field by T^*, and pressure field by p^*. Asterisks denote dimensional fields, coordinates and time. Dimensionless quantities are defined through the scaling

$$\frac{1}{L}\,(x^*, y^*, z^*) = (x, y, z), \quad \frac{\alpha}{\sigma L^2}\,t^* = t, \quad \frac{K}{\mu_0 \alpha}\,(p^* - p_0) = p,$$
$$\frac{L}{\alpha}\,\boldsymbol{u}^* = \frac{L}{\alpha}\,(u^*, v^*, w^*) = (u, v, w) = \boldsymbol{u}, \quad \rho_0 c K\,\frac{T^* - T_0}{\mu_0 \alpha} = T, \quad \frac{\mu_0}{\mu(T^*)} = \eta(T), \tag{1}$$

where α is the average thermal diffusivity of the saturated porous medium, p_0 is the reference pressure, ρ_0 is the reference fluid density evaluated at temperature T_0, μ_0 is the reference dynamic viscosity evaluated at temperature T_0, c is the specific heat of the fluid, K is the permeability, and σ is the heat capacity ratio of the saturated porous medium. The latter quantity is defined as the ratio between the average volumetric heat capacity of the saturated medium and the volumetric heat capacity of the fluid. In Eq. (1), the change of the fluid viscosity with temperature is accounted for by expressing the temperature-dependent dynamic viscosity $\mu(T^*)$ through a dimensionless function, $\eta(T)$, of the dimensionless temperature T. Function $\eta(T)$ is called fluidity [3]. By definition, fluidity is such that $\eta(0) = 1$, and $\eta(T) > 0$.

If we take into account the effects of viscous dissipation and variable viscosity, the local mass, momentum and energy balance equations can be written in a dimensionless form as

$$\nabla \cdot \boldsymbol{u} = 0, \tag{2a}$$

$$\boldsymbol{u} = -\,\eta(T)\,\nabla p, \tag{2b}$$

$$\frac{\partial T}{\partial t} + \boldsymbol{u} \cdot \nabla T = \nabla^2 T + \eta(T)^{-1} \boldsymbol{u} \cdot \boldsymbol{u}. \tag{2c}$$

Equations (2) lead to a pressure-temperature formulation,

$$\nabla \cdot [\eta(T)\,\nabla p] = 0, \tag{3a}$$

$$\frac{\partial T}{\partial t} - \eta(T)\,\nabla p \cdot \nabla T = \nabla^2 T + \eta(T)\,\nabla p \cdot \nabla p. \tag{3b}$$

The boundaries $y = \pm 1$ are impermeable with a convective condition to the external fluid environment, as sketched in Fig. 1, so that we may write

$$y = \pm 1: \qquad \frac{\partial p}{\partial y} = 0, \quad \frac{\partial T}{\partial y} \pm BT = 0, \tag{4}$$

where

$$B = \frac{hL}{\lambda} \tag{5}$$

is the Biot number, and λ is the average thermal conductivity of the saturated medium. We mention that $B \to \infty$ describes the limiting case of a perfectly isothermal wall, while $B \to 0$ yields a perfectly insulated wall.

3. Basic flow

A stationary solution of Eqs. (3) and (4) exists such that the pressure gradient is a parallel vector field directed along the $x-$direction, and such that the temperature depends only on y. This basic solution is given by

$$\nabla p_b = (P, 0, 0), \tag{6a}$$

$$\frac{\mathrm{d}^2 T_b}{\mathrm{d}y^2} = -P^2 \eta(T_b), \quad \left(\frac{\mathrm{d}T_b}{\mathrm{d}y} \pm BT_b\right)_{y=\pm 1} = 0, \tag{6b}$$

where P is a dimensionless constant expressing the Péclet number associated with the imposed pressure gradient along the $x-$direction, and the subscript "b" denotes the basic solution.

The basic solution, Eqs. (6), is fully determined if and only if the fluidity function $\eta(T)$ is prescribed. Assuming that viscous dissipation does not cause very large temperature changes over the channel cross-section, a linear fluidity model is a reasonable assumption [3],

$$\eta(T) = 1 + a^2 T, \tag{7}$$

where a is a positive constant parameter. The positive slope of the fluidity function, Eq. (7), implies that the viscosity of the fluid is a decreasing function of temperature, which is the behaviour experimentally observed for liquids. We mention that the linear fluidity model, Eq. (7), can be considered as a first-order approximation of the exponential, or Nahme, model [11].

On account of Eq. (7), Eqs. (6) yield

$$\nabla p_b = (P, 0, 0), \tag{8a}$$

$$T_b(y) = \frac{B\cos(aP) - aP\sin(aP) - B\cos(aPy)}{a^2[aP\sin(aP) - B\cos(aP)]}, \tag{8b}$$

$$\eta_b(y) = \frac{B\cos(aPy)}{B\cos(aP) - aP\sin(aP)}, \tag{8c}$$

Journal of Physics: Conference Series **547** (2014) 012038 doi:10.1088/1742-6596/547/1/012038

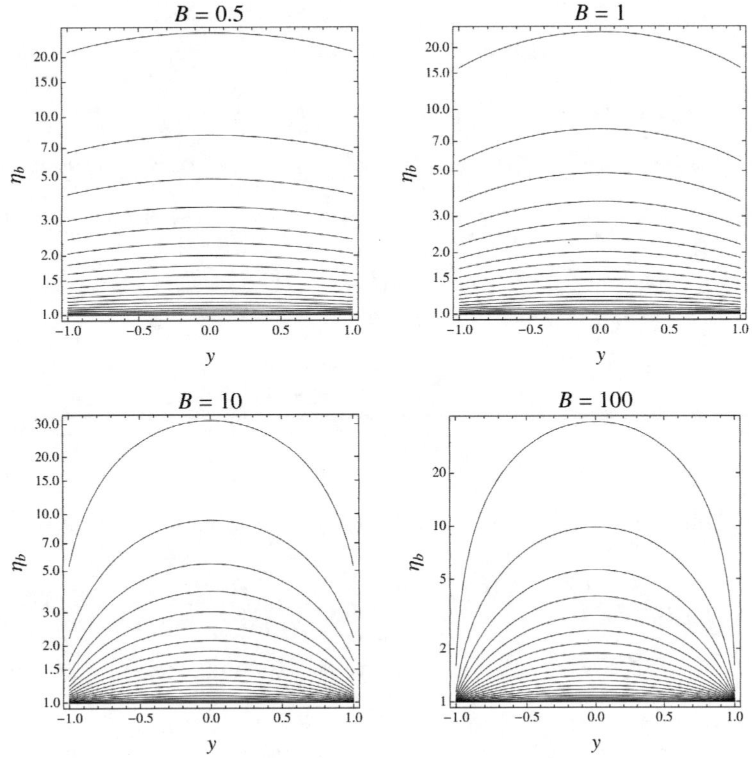

Figure 2. Basic flow. Plots of $\eta_b(y)$ for different products aP. Each frame corresponds to a given Biot number, B, and aP increases from bottom to top. For $B = 0.5$, aP ranges from 0.02 to 0.64 in steps of 0.031; for $B = 1$, aP ranges from 0.02 to 0.84 in steps of 0.041; for $B = 10$, aP ranges from 0.02 to 1.4 in steps of 0.069; for $B = 100$, aP ranges from 0.02 to 1.53 in steps of 0.0755.

where $\eta_b = \eta(T_b)$.

For a given Biot number, B, Eqs. (8) imply that the basic solution may be singular, and hence unphysical, when aP is a root of the equation

$$aP \sin(aP) - B \cos(aP) = 0. \tag{9}$$

In the special case $B \to \infty$ (isothermal duct wall), the smallest singular value of aP is $aP = \pi/2$. On gradually reducing the Biot number, the smallest singular value of aP decreases approaching 0 when $B \to 0$. In fact, when the Biot number vanishes the numerator and denominator at the right hand side of Eq. (8b) are simplified and T_b achieves a uniform distribution, with value $-1/a^2$. This uniform distribution is allowed only as a consequence of the variable viscosity model. When $a = 0$ (uniform viscosity), no stationary flow solution with a vanishing axial temperature gradient is possible. However, the uniform temperature distribution for the adiabatic case, $T_b = -1/a^2$, is a definitely unphysical limiting solution as it represents an identically vanishing fluidity $\eta_b = 0$ and, hence, an infinite viscosity.

Plots of the fluidity $\eta_b(y)$ for different values of aP, below the lowest singularity, and with given Biot numbers are reported in Fig. 2. We mention that, on account of Eqs. (2b) and (6a), these plots are also illustrative of the qualitative trend of the basic velocity profile, given by $u_b(y) = -P\,\eta_b(y)$.

4. Stability analysis

When the basic state is perturbed,

$$p = p_b + \varepsilon\,\hat{p}, \quad T = T_b + \varepsilon\,\hat{T}, \tag{10}$$

with a small amplitude parameter $|\varepsilon| \ll 1$, the system of governing equations (3) and (4) can be rewritten as

$$\eta_b(y)\nabla^2\hat{p} + \eta_b'(y)\,\frac{\partial\hat{p}}{\partial y} + a^2 P\,\frac{\partial\hat{T}}{\partial x} = 0, \tag{11a}$$

$$\frac{\partial\hat{T}}{\partial t} - \eta_b(y)P\,\frac{\partial\hat{T}}{\partial x} - \frac{1}{a^2}\,\eta_b(y)\eta_b'(y)\,\frac{\partial\hat{p}}{\partial y} = \nabla^2\hat{T} + a^2P^2\,\hat{T} + 2P\eta_b(y)\,\frac{\partial\hat{p}}{\partial x}, \tag{11b}$$

$$y = \pm 1: \qquad \frac{\partial\hat{p}}{\partial y} = 0, \quad \frac{\partial\hat{T}}{\partial y} \pm B\hat{T} = 0, \tag{11c}$$

4.1. Normal modes

Eqs. (11) will be solved by expressing $\hat{p}$ and $\hat{T}$ as linear combinations of normal modes, given by

$$\hat{p} = f(y)\,e^{st}\,e^{i(k_x x + k_z z)}, \qquad \hat{T} = g(y)\,e^{st}\,e^{i(k_x x + k_z z)}, \tag{12}$$

where (k_x, k_z) is a pair of real parameters expressing the components of the wave vector with wavenumber $k = (k_x^2 + k_z^2)^{1/2}$. Moreover, $s = s_r - i\,\omega$ is a complex parameter, with s_r and $-\,\omega$ denoting the real and imaginary parts of s, respectively. The condition $s_r > 0$ yields instability, $s_r = 0$ neutral stability, and $s_r < 0$ denotes linear stability. The angular frequency ω is given by the negative imaginary part of s.

On account of Eq. (12), Eqs. (11) yield

$$\eta_b(y)\left(f'' - k^2 f\right) + \eta_b'(y)\,f' + ik_x\,a^2P\,g = 0, \tag{13a}$$

$$g'' - \left[k^2 - a^2P^2 + s - ik_x\,\eta_b(y)P\right]g + 2ik_x\,P\eta_b(y)\,f + \frac{1}{a^2}\,\eta_b(y)\eta_b'(y)\,f' = 0, \tag{13b}$$

$$y = \pm 1: \qquad f' = 0, \quad g' \pm B\,g = 0. \tag{13c}$$

5. Longitudinal modes

A special class of normal modes is that of perturbations expressed through Eq. (12) with $k_x = 0$ and $k_z = k$. These modes are $x-$independent, or streamwise invariant. With $k_x = 0$, a dramatic simplification of Eqs. (13) occurs, namely

$$\eta_b(y)\left(f'' - k^2 f\right) + \eta_b'(y)\,f' = 0, \tag{14a}$$

$$g'' - \left(k^2 - a^2P^2 + s\right)g + \frac{1}{a^2}\,\eta_b(y)\eta_b'(y)\,f' = 0, \tag{14b}$$

$$y = \pm 1: \qquad f' = 0, \quad g' \pm B\,g = 0. \tag{14c}$$

Solution of Eqs. (14) can be sought analytically by recognising that Eq. (14a) does not contain h, so that it can be solved independently of Eq. (14b) with the boundary conditions $f'(\pm 1) = 0$. This yields $f(y) = \textit{constant}$, where the constant is arbitrary, when $k = 0$. On the other hand, when $k > 0$, we get $f(y) = 0$. No other solutions of Eqs. (14a) and (14c) exist, as it is proved in Appendix A. Thus, Eq. (14b) is satisfied either with even modes

$$g_n(y) = \cos(\gamma_n y), \qquad n = 0,\ 1,\ 2,\ \ldots\ , \tag{15}$$

such that γ_n are the roots of

$$B\cos(\gamma) - \gamma\sin(\gamma) = 0, \tag{16}$$

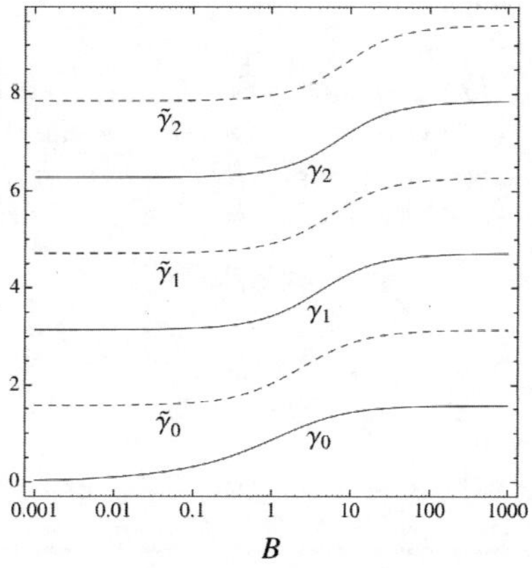

Figure 3. Longitudinal even and odd modes: plots of γ_n and $\tilde{\gamma}_n$ versus B with $n = 0,\ 1,\ 2$.

in increasing order, or with odd modes

$$g_n(y) = \sin(\tilde{\gamma}_n y), \qquad n = 0,\ 1,\ 2,\ \ldots\ , \tag{17}$$

such that $\tilde{\gamma}_n$ are the roots of

$$B \sin(\tilde{\gamma}) + \tilde{\gamma} \cos(\tilde{\gamma}) = 0, \tag{18}$$

in increasing order. A dispersion relation must hold expressed by

$$\begin{cases} a^2 P^2 = k^2 + \gamma_n^2 + s, & \text{for even modes,} \\ a^2 P^2 = k^2 + \tilde{\gamma}_n^2 + s, & \text{for odd modes.} \end{cases} \tag{19}$$

The imaginary part of s vanishes, *i.e.* $\omega = 0$, as it is proved in the following.

By taking into account that $f' = 0$, we multiply Eq. (14b) by $\bar{g}$, where $\bar{g}$ is the complex conjugate of g, and integrate over $y \in [-1, 1]$. Thus, by employing integration by parts for the first term, we obtain

$$\bar{g}(1)\, g'(1) - \bar{g}(-1)\, h'(-1) - \int_{-1}^{1} |g'|^2\, dy + (a^2 P^2 - s - k^2) \int_{-1}^{1} |g|^2\, dy = 0, \tag{20}$$

while use of Eq. (14c) allows further simplification to

$$B \left[|g(1)|^2 + |g(-1)|^2 \right]^2 + \int_{-1}^{1} |g'|^2\, dy - (a^2 P^2 - s - k^2) \int_{-1}^{1} |g|^2\, dy = 0. \tag{21}$$

In order to satisfy Eq. (21), both the real and the imaginary parts of its right hand side must be zero. As for the imaginary part, one obtains

$$\omega \int_{-1}^{1} |g|^2\, dy = 0, \tag{22}$$

which implies either $\omega = 0$ or g identically zero. Only, the former alternative is allowed, as the latter would lead to no perturbation of the basic flow. Thus, the dispersion relation (19) can be rewritten as

$$\begin{cases} a^2 P^2 = k^2 + \gamma_n^2 + s_r, & \text{for even modes,} \\ a^2 P^2 = k^2 + \tilde{\gamma}_n^2 + s_r, & \text{for odd modes.} \end{cases} \tag{23}$$

Journal of Physics: Conference Series **547** (2014) 012038 doi:10.1088/1742-6596/547/1/012038

As a by-product of this argument, one can infer that the roots, γ_n, of Eq. (16), as well as the roots, $\tilde{\gamma}_n$, of Eq. (18) are real.

Instability arises when $s_r > 0$. On account of Eq. (23), this means $a^2P^2 > \gamma_n^2$ for even modes, or $a^2P^2 > \tilde{\gamma}_n^2$ for odd modes. The most restrictive among these inequalities being $a^2P^2 > \gamma_0^2$, which corresponds to the even mode with $n = 0$ and $k = 0$. The trend of γ_n^2 and $\tilde{\gamma}_n^2$ versus B is plotted in Fig. 3 with $n = 0,\ 1,\ 2$. Thus, the neutral stability condition is defined by

$$a^2P^2 = \gamma_0^2. \tag{24}$$

Comparison between Eq. (9) and Eq. (16) allows one to reckon that the threshold for instability is at the lowest singularity of the basic solution, Eqs. (8). This means that the basic flow is linearly stable to longitudinal modes as long as the product $a\,P$ is smaller than its lowest value satisfying Eq. (9). Asymptotic values of γ_0 are attained in the limiting cases of $B \to 0$ and $B \to \infty$, where they coincide with 0 and $\pi/2$, respectively.

6. Three-dimensional modes

The analysis of instability to longitudinal modes revealed that the onset of instability is for a finite product aP. The use of a linear fluidity law is legitimate as long as the slope parameter a is small enough, namely $a \ll 1$. Thus, the regime of instability is to be considered as one where simultaneously $a \ll 1$ and $P \gg 1$, so that $aP \sim O(1)$. This finding implies a drastic simplification of Eqs. (13), where Eq. (13a) is well approximated by Eq. (14a). The proof given in Appendix A allows us to conclude that $f = 0$, as for the oblique modes $k_x \neq 0$ and, hence, $k > 0$. Thus, Eq. (13b) can be approximated as

$$g'' - \left[k^2 - a^2P^2 + s - ik_x\,\eta_b(y)P\right] g = 0, \tag{25a}$$

with the term $ik_x\,\eta_b(y)P\,g$ being the dominant contribution. As a consequence, Eq. (25) implies that $g = 0$ as well. The conclusion is that, in a physically sensible regime for instability where $a \ll 1$ and $P \sim O(a^{-1})$, no instability to oblique normal modes is possible.

7. Conclusions

The stability analysis of forced convection with viscous dissipation in a plane parallel porous channel has been carried out. Fully developed conditions have been assumed for the basic flow. Viscous dissipation acts as source of the temperature gradient within the flow region, while variable viscosity yields a thermal coupling between velocity and temperature in the momentum balance, thus expressing the physical cause of a possible flow instability. The stability analysis has been carried out under the assumptions that Darcy's law holds, that the law expressing the change of fluidity *vs.* temperature is linear, and that the boundary walls are impermeable while undergoing heat transfer to an external environment (Robin temperature conditions).

The linearised perturbation equations have been solved analytically for normal modes of the longitudinal type. Oblique normal modes have been shown to be ineffective, as for the onset of instability, in a physically sensible regime where the slope parameter of viscosity *vs.* temperature is very small, and the Péclet number is very large.

The instability of the basic flow to longitudinal normal modes arises for the same conditions leading to a singular behaviour of the basic flow solution. Hence, the main physical fact emerging from our stability analysis is that the basic forced convection flow with viscous dissipation remains stable in the whole parametric domain where the viscosity model adopted to describe it is physically conceivable. We mention that a similar conclusion was drawn by Joseph [3] on studying the dissipation instability of a basic Couette or Poiseuille flow in a clear fluid. As suggested by Joseph's analysis, what is expected when the linear fluidity model is replaced by more realistic models (Nahme or Arrhenius) is that the basic solution implies dual flows (see also Magyari and Barletta [12] for a second-degree polynomial fluidity model). Joseph [3] proved that the principal branch of the dual flows is stable, while the second branch is not. We expect that a similar behaviour is to be found also for the flow in a porous channel examined in this paper.

Journal of Physics: Conference Series **547** (2014) 012038 doi:10.1088/1742-6596/547/1/012038

We mention that our stability analysis refers just to the thermally developed regime of forced convection. The onset of instability in the thermal entrance region could imply more restrictive conditions for stability, depending on the inlet conditions assumed for the flow at the entrance cross-section.

References

[1] Nield D A and Bejan A 2013 *Convection in Porous Media (4th edition)* (Springer, New York)

[2] Barletta A 2014 *International Journal of Thermal Sciences* (in press, corrected proof) doi: 10.1016/j.ijthermalsci.2014.02.009

[3] Joseph D D 1965 *Physics of Fluids* **8** 2195–2200

[4] Sukanek P C, Goldstein C A and Laurence R L 1973 *Journal of Fluid Mechanics* **57** 651–670

[5] Ho T C, Denn M M and Anshus B E 1977 *Rheologica Acta* **16** 61–68

[6] Yueh C S and Weng C I 1996 *Physics of Fluids* **8** 1802–1813

[7] White J M and Muller S J 2002 *Journal of Fluid Mechanics* **462** 133–159

[8] Nouar C and Frigaard I 2009 *Physics of Fluids* **21** 064104

[9] Barletta A, Celli M and Rees D A S 2009 *International Journal of Heat and Mass Transfer* **52** 337–344

[10] Barletta A and Nield D A 2012 *Physics of Fluids* **24** 104102

[11] Nahme R 1940 *Archive of Applied Mechanics* **11** 191–209

[12] Magyari E and Barletta A 2007 *Heat and Mass Transfer* **44** 251–259

Appendix A.

A convenient way to express Eq. (14a) with the boundary condition given by Eq. (14c) is

$$[\eta_b(y)\, f']' - k^2\, \eta_b(y)\, f = 0, \qquad \text{(A.1a)}$$

$$y = \pm 1: \qquad f' = 0. \qquad \text{(A.1b)}$$

We multiply Eq. (A.1a) by $\bar{f}$, where $\bar{f}$ is the complex conjugate of f, and integrate over $y \in [-1, 1]$. Thus, we obtain

$$\int_{-1}^{1} \bar{f}\, [\eta_b(y)\, f']'\, dy - k^2 \int_{-1}^{1} \eta_b(y)\, |f|^2\, dy = 0. \qquad \text{(A.2)}$$

Integration by parts for the first term of Eq. (A.2), yields

$$\eta_b(1)\, \bar{f}(1)\, f'(1) - \eta_b(-1)\, \bar{f}(-1)\, f'(-1) - \int_{-1}^{1} \eta_b(y)\, |f'|^2\, dy - k^2 \int_{-1}^{1} \eta_b(y)\, |f|^2\, dy = 0. \qquad \text{(A.3)}$$

By employing the boundary condition, Eq. (A.1b), we further simplify to

$$\int_{-1}^{1} \eta_b(y)\, |f'|^2\, dy + k^2 \int_{-1}^{1} \eta_b(y)\, |f|^2\, dy = 0. \qquad \text{(A.4)}$$

By recalling that $\eta_b(r) > 0$, we conclude that Eq. (A.4) can be satisfied only with $f(r) = constant$, when $k = 0$, or with $f(r) = 0$, when $k > 0$.

Journal of Physics: Conference Series **547** (2014) 012039 doi:10.1088/1742-6596/547/1/012039

Onset of Absolute Instability Induced by Viscous Dissipation in the Poiseuille-Darcy-Bénard Convection of a Newtonian Fluid

P. V. Brandão[1], L. S. de B. Alves[1] and A. Barletta[2]

[1]Programa de Pós-Graduação em Engenharia Mecânica, Universidade Federal Fluminense, Rua Passo da P átria 156, Bloco E, São Domingos, Niterói, RJ 24210-240, Brazil.
[2]Dipartimento di Ingegneria Industriale, Alma Mater Studiorum Università di Bologna, Viale Risorgimento 2, I-40136 Bologna, Italy.

E-mail: `leonardo.alves@mec.uff.br`

Abstract. The present paper investigates the transition from convective to absolute instability induced by viscous dissipation. As far as the authors are aware, this is the first time such a study is reported in the literature. Its framework is provided by the Poiseuille-Darcy-Bénard convection of a Newtonian fluid. We found the same behaviour observed in the absence of viscous dissipation whenever the Gebhart number is smaller than $Ge < 0.95$, which is the stabilising effect of the cross flow. When $0.95 < Ge < 4.31$, weak cross flows still stabilise the onset of absolute instability but stronger cross flows destabilise it. For a stronger viscous dissipation, i.e. $Ge > 4.31$, the cross flow always destabilises this onset. The latter two conditions create a scenario where viscous dissipation is capable of inducing a transition to absolute instability in the absence of wall heating, i.e. with a zero Rayleigh number.

1. Introduction

Buoyancy induced natural convection is an important phenomenon that has been widely studied over the past century. In the present paper, we focus on a special subset of this problem that takes place in a fluid-saturated porous medium. It has important applications in geophysical and engineering research and technology, including aquifers in permeable rocks, underground pollutant dispersion, solar energy collectors, and so on. A detailed discussion of many possible applications and their respective literature review can be found elsewhere [1, 2]. The onset of convection through buoyancy was first postulated more than a hundred years ago [3], although it was later found that natural convection in this particular study was induced by surface tension gradients at the interface between liquid and air [4, 5]. Applications for this phenomenon abound in the areas of film cooling, crystal growth, electronic equipment cooling in microgravity, and so on. More information about this interface tension gradient induced natural convection can also be found elsewhere [6, 7].

This classical driver for natural convection known as buoyancy works as follows: Adverse density gradients force gravity to push higher density fluid downwards against lower density fluid, which moves upwards due to mass conservation. Density gradients can be imposed by one of several traditional mechanisms, such as internal heating as well as solid boundaries with differentially prescribed temperatures and/or concentrations, including any variations where their gradients are prescribed instead. Viscous dissipation has also been considered a possible mechanism for the generation of density gradients. Its influence on the heat transfer and fluid dynamics of the flow within a saturated a porous medium has been investigated for the first time over twenty years ago [8]. However, it took many years for this topic to be further considered by other researchers [9, 10]. It was only recently that viscous dissipation has been found responsible for mixed convection in porous media flows [11, 12]. In the presence of a strong enough through flow, viscous dissipation generates a nonuniform temperature distribution in an incompressible fluid, which, in turn, creates the necessary density gradients for buoyancy to initiate mixed convection. Several linear stability analyzes, such as the ones performed in these two former studies, have been performed since then to uncover similar onsets of mixed convection. They include porous media with lateral confinement [13], Darcy-Hadley flows [14] as well as Darcy-Bénard convection with through flow of viscoelastic fluids [15]. Similar studies have also been performed for viscous dissipation induced thermal instabilities in the Couette [16] and Poiseuille [17] flows of clear fluids.

One important drawback of the aforementioned studies is their use of a temporal stability analysis alone, which is only able to predict the onset of convective instabilities. Today it is well known that the onset of mixed convection might appear as an absolute instability instead of a convective one in the presence of through flow [18]. The onset of convective instability is determined with a temporal stability analysis, yielding the critical point beyond which a disturbance will grow in time in a reference frame that travels with the base flow. When observed from a laboratory (stationary) reference frame, it manifests as spatial growth. If the control parameter is increased beyond its first critical value, disturbances might become strong enough to overcome the base flow and travel upstream. This second critical point represents the onset of absolute instability, manifesting through the appearance of a pinching point in a spatial stability analysis, which is a saddle point formed between upstream and downstream travelling modes in the complex wave number map. The need to distinguish between spatial and temporal disturbance growth to identify the convective or absolute nature of the instability was first proposed in the context of plasma instabilities [19]. This is accomplished by analyzing the evolution and asymptotic behavior of wave packets initially subjected to an impulse excitation, which was originally done for boundary-layers [20]. A complete methodology for this analysis was soon developed for local and parallel shear flows subject to two-dimensional [21] as well as three-dimensional [22] instabilities. However, the latter is a quite nontrivial task when the dispersion relation does not have an algebraic form and there is no well established approach in the literature to deal with the numerical difficulties that arise when this is not the case [23].

The present paper is an extension of earlier work on the onset of convective instabilities in mixed convection within a fluid-saturated porous medium [11, 15]. On the one hand, the former study considered viscous dissipation as the only source of base flow temperature gradients. Furthermore, they derived an algebraic dispersion relation, but considered transverse and longitudinal rolls as separate two-dimensional disturbances. On the other hand, the latter study considered both viscous dissipation and differentially heated boundaries simultaneously as sources of base flow temperature gradients. In addition, it considered an Oldroyd-B fluid model, although a Newtonian fluid was analyzed as a limiting case. However, it derived a differential dispersion relation for three-dimensional disturbances, but solved this equation numerically. In the present paper, an algebraic dispersion relation is derived for the Newtonian fluid limiting case and used to determine its onset of absolute instability.

2. Mathematical Model

Two horizontal plane walls at $z^* = 0$ and h^* confine the fluid-saturated porous medium. They allow slip, are impermeable and maintained at uniform but different temperatures. In this configuration, x^* and y^* are the stream wise and span wise coordinates in the $\hat{\mathbf{i}}$ and $\hat{\mathbf{j}}$ directions, respectively. According to an incompressible flow model subject to the Oberbeck-Boussinesq approximation, it is possible to write conservation of mass as

$$\nabla \cdot \mathbf{u}^* = 0 \quad , \tag{1}$$

conservation of momentum through the Darcy's law as

$$\frac{\mu^*}{\kappa^*}\,\mathbf{u}^* = \mathbf{F}^* \quad , \tag{2}$$

and conservation of energy as

$$\sigma^* \frac{\partial T^*}{\partial t^*} + \mathbf{u}^* \cdot \nabla T^* = \varkappa^* \, \nabla^2 T^* + \frac{\mathbf{F}^* \cdot \mathbf{u}^*}{\rho_h^* \, c^*} \quad , \tag{3}$$

where the drag force driving the flow is defined as

$$-\mathbf{F}^* = \nabla \mathcal{P}^* + \rho_h^* \, \mathbf{g}^* \, \beta^* \, (T^* - T_h^*) \quad , \tag{4}$$

and the term $\mathbf{F}^* \cdot \mathbf{u}^*$ is the thermal power per unit volume generated by viscous dissipation. In the above equations, $\mathbf{u}^*$ is the velocity vector, μ^* is the dynamic viscosity, κ^* is the permeability, t^* is the time coordinate, T^* is the temperature, ρ_h^* is the fluid density at the reference temperature T_h^*, c^* is the specific heat of the fluid, σ^* is the volumetric heat capacity of the saturated porous medium divided by $\rho_h^* \, c^*$, $\varkappa^*$ is the effective thermal difusivity, $\mathcal{P}^*$ is the gauge pressure with respect to the hydrostatic pressure, $\mathbf{g}^* = -g^* \, \hat{\mathbf{k}}$ is the gravity acceleration vector, where g^* is the gravity acceleration and $\hat{\mathbf{k}}$ is the unit vector associated with the z^* coordinate and β^* is the fluid thermal expansion coefficient.

This flow is also subject to the following boundary conditions

$$z^* = 0 \quad \rightarrow \quad w^* = 0 \quad \text{and} \quad T^* = T_0^* \quad , \tag{5}$$

$$z^* = h^* \quad \rightarrow \quad w^* = 0 \quad \text{and} \quad T^* = T_h^* \quad , \tag{6}$$

where (u^*, v^*, w^*) are the three velocity components associated with the (x^*, y^*, z^*) coordinates in the $(\hat{\mathbf{i}}, \hat{\mathbf{j}}, \hat{\mathbf{k}})$ unit vector directions of $\mathbf{u}^*$, respectively, u_0^* is a uniform stream wise velocity, T_0^* is the prescribed lower wall temperature and T_h^* the prescribed upper wall temperature.

2.1. Dimensionless Model

Equations (1) to (6) can be written in dimensionless form using

$$\mathbf{u} = \frac{\mathbf{u}^*}{\varkappa^*/h^*} \quad , \quad \mathbf{F} = \frac{\mathbf{F}^*}{(\varkappa^* \, \mu^*)/(h^* \, \kappa^*)} \quad , \quad \mathbf{x} = \frac{\mathbf{x}^*}{h^*} \quad ,$$

$$t = \frac{t^*}{\sigma^* \, (h^*)^2/\varkappa^*} \quad , \quad \mathcal{P} = \frac{\mathcal{P}^*}{\mu^* \, \varkappa^*/\kappa^*} \quad \text{and} \quad T = \frac{T^* - T_h^*}{T_0^* - T_h^*} \quad , \tag{7}$$

as the scaling, leading to the following dimensionless parameters

$$Ra = \frac{\rho_h^* \, g^* \, \beta^* \, (T_0^* - T_h^*) \, \kappa^* \, h^*}{\mu^* \, \varkappa^*} \quad , \quad Ge = \frac{g^* \, \beta^* \, h^*}{c^*} \quad \text{and} \quad Pe = \frac{u_0^* \, h^*}{\varkappa^*} \quad , \tag{8}$$

which define the Rayleigh, Gebhart and Péclet numbers, respectively. They become

$$\nabla \cdot \mathbf{u} = 0 \quad , \tag{9}$$

for mass conservation,

$$\mathbf{u} = \mathbf{F} = Ra\, T\, \hat{\mathbf{k}} - \nabla \mathcal{P} \quad , \tag{10}$$

for Darcy's law and

$$\frac{\partial T}{\partial t} + \mathbf{u} \cdot \nabla T = \nabla^2 T + \frac{Ge}{Ra}\, \mathbf{F} \cdot \mathbf{u} \quad , \tag{11}$$

for energy conservation, subject to dimensionless boundary conditions

$$z = 0 \quad \rightarrow \quad w = 0 \quad \text{and} \quad T = 1 \quad , \tag{12}$$

$$z = 1 \quad \rightarrow \quad w = 0 \quad \text{and} \quad T = 0 \quad . \tag{13}$$

2.2. Base Flow

A known steady-state of Eqs. (9) to (13) is given by

$$\mathbf{u}_b = \mathbf{F}_b = Pe\, \hat{\mathbf{i}} \quad , \quad T_b(z) = 1 - z + \frac{Ge\, Pe^2}{2\, Ra}\, (1 - z)\, z \quad \text{and}$$

$$\mathcal{P}_b(x, z) = \mathcal{P}_0 - Pe\, x + \frac{Ra}{2}\, (2 - z)\, z + \frac{Ge\, Pe^2}{12}\, (3 - 2\, z)\, z^2 \quad , \tag{14}$$

where $\mathcal{P}_0$ is a reference pressure and the subscript b means base flow, since Eq. (14) is the stationary and parallel solution whose linear stability is being investigated in the present paper.

3. Algebraic Dispersion Relation

3.1. Linear Disturbances

In order to do so, small disturbances are superposed to base flow (14) using

$$\begin{aligned} \mathbf{u}(x, y, z, t) &= \mathbf{u}_b(z) + \epsilon\, \mathbf{u}_d(x, y, z, t) \quad , \\ T(x, y, z, t) &= T_b(z) + \epsilon\, T_d(x, y, z, t) \quad \text{and} \\ \mathcal{P}(x, y, z, t) &= \mathcal{P}_b(x, z) + \epsilon\, \mathcal{P}_d(x, y, z, t) \quad , \end{aligned} \tag{15}$$

where $\mathbf{u}_d = (u_d\, , v_d\, , w_d)$, ϵ is the disturbance amplitude parameter and the subscript d means disturbance. Substituting relations (15) into Eqs. (9) to (11) leads to new systems of equations at different orders in ϵ. The base flow equations that lead to solutions (14) are recovered at order zero, i.e. by collecting all terms of $O(\epsilon^0)$, whereas the linear disturbance equations

$$\nabla \cdot \mathbf{u}_d = 0 \quad , \tag{16}$$

$$\mathbf{u}_d = \mathbf{F}_d = Ra\, T_d\, \hat{\mathbf{k}} - \nabla \mathcal{P}_d \quad , \tag{17}$$

$$\frac{\partial T_d}{\partial t} + \mathbf{u}_b \cdot \nabla T_d + \mathbf{u}_d \cdot \nabla T_b = \nabla^2 T_d + \frac{Ge}{Ra}\, (\, \mathbf{F}_b \cdot \mathbf{u}_d + \mathbf{F}_d \cdot \mathbf{u}_b\,) \quad , \tag{18}$$

for conservation of mass, momentum and energy, respectively, are found at order one, i.e. by collecting all terms of $O(\epsilon^1)$. The same can be done to boundary conditions (12) and (13), which leads to

$$z = 0 \quad \rightarrow \quad w_d = T_d = 0 \quad \text{and} \tag{19}$$

$$z = 1 \quad \rightarrow \quad w_d = T_d = 0 \quad , \tag{20}$$

for the linear disturbances.

3.2. Normal Modes

Disturbances are now assumed to have a wave like behavior in the homogeneous directions, which allows them to be decomposed as Fourier modes in the form of

$$\begin{aligned} \mathbf{u}_d(x,y,z,t) &= \hat{\mathbf{u}}(z)\,\exp[\,i\,(\alpha\,x+\beta\,y-\omega\,t)\,] \quad , \\ T_d(x,y,z,t) &= \hat{T}(z)\,\exp[\,i\,(\alpha\,x+\beta\,y-\omega\,t)\,] \quad \text{and} \\ \mathcal{P}_d(x,y,z,t) &= \hat{\mathcal{P}}(z)\,\exp[\,i\,(\alpha\,x+\beta\,y-\omega\,t)\,] \quad , \end{aligned} \tag{21}$$

where the hat symbol $\hat{}$ indicates a normal mode amplitude varying in the non-homogeneous direction and $i=\sqrt{-1}$. Substituting relations (21) into Eqs. (16) to (18) leads to

$$i\,\alpha\,\hat{u}(z)+i\,\beta\,\hat{v}(z)+\hat{w}'(z)=0 \quad , \tag{22}$$

from linear disturbance mass conservation Eq. (16),

$$i\,\alpha\,\hat{\mathcal{P}}(z)+\hat{u}(z) = 0 \quad , \tag{23}$$

$$i\,\beta\,\hat{\mathcal{P}}(z)+\hat{v}(z) = 0 \quad \text{and} \tag{24}$$

$$\hat{\mathcal{P}}'(z)+\hat{w}(z) = Ra\,\hat{T}(z) \quad , \tag{25}$$

from linear disturbance momentum conservation Eq. (17) and

$$i\,(Pe\,\alpha-\omega)\,\hat{T}(z)+T_b'(z)\,\hat{w}(z)=\hat{T}''(z)-(\alpha^2+\beta^2)\,\hat{T}(z)+2\,\frac{Ge\,Pe}{Ra}\,\hat{u}(z) \quad , \tag{26}$$

from linear disturbance energy conservation Eq. (18). Equations (22) to (26) can be combined to generate a single fourth-order ordinary differential equation for the linear disturbance cross stream velocity component, given by

$$\begin{aligned} &\hat{w}''''(z)-i\,(\,Pe\,\alpha-2\,i\,(\,\alpha^2+\beta^2\,)-\omega\,)\,\hat{w}''(z)-2\,i\,\alpha\,Ge\,Pe\,\hat{w}'(z)+ \\ &(\,\alpha^2+\beta^2\,)\,(\,i\,Pe\,\alpha+(\alpha^2+\beta^2)-i\,\omega+Ra\,T_b'(z)\,)\,\hat{w}(z)=0 \quad , \end{aligned} \tag{27}$$

which is subject to boundary conditions

$$z=0 \quad \rightarrow \quad \hat{w}=\hat{w}''=0 \quad \text{and} \tag{28}$$

$$z=1 \quad \rightarrow \quad \hat{w}=\hat{w}''=0 \quad , \tag{29}$$

where the impermeable wall boundary conditions are complemented by the additional zero second derivative boundary conditions, derived from the isothermal wall boundary conditions through relation

$$\hat{T}(z)=\frac{(\,\alpha^2+\beta^2\,)\,\hat{w}(z)-\hat{w}''(z)}{Ra\,(\,\alpha^2+\beta^2\,)} \quad , \tag{30}$$

obtained by combining Eqs. (22) to (25). Equations (27) to (29) are collectively known as the dispersion relation. They must be solved to identify the eigenvalues, given by α, β and ω, that characterize the present problem as functions of the governing parameters Pe, Ge and Ra.

3.3. Generalized Integral Transform Technique

The search for an absolute instability onset has a large computational time cost when it relies solely on a graphical verification of the pinching condition. However, this cost can be greatly reduced when the zero group velocity conditions can be employed beforehand to find possible saddle points. Since these conditions are necessary but not sufficient, the pinching condition still must be verified graphically. Nevertheless, it is significantly less costly to do so if the possible saddle point location is already known. Arguably the major difficulty restricting the use of this alternative approach is the differential form of the dispersion relation. Hence, the next step in the present procedure is to reduce Eqs. (27) to (29) to an algebraic form. This is achieved here through the Generalized Integral Transform Technique [24, 25], which is an extension of Separation of Variables and the Classical Integral Transform Technique [26].

In order to generate this algebraic dispersion relation, an eigensystem must be defined for the integral transformation procedure. An adequate system for the present problem can be obtained from homogeneous equation

$$\psi_m''''(z) = \lambda_m^4\, \psi_m(z) \quad \text{with} \quad m = 1\,,\, 2\,,\, 3\,,\, \dots\,,\, N_Z \quad , \tag{31}$$

subject to homogeneous boundary conditions

$$z = 0 \quad \rightarrow \quad \psi_m = \psi_m'' = 0 \quad \text{and} \tag{32}$$

$$z = 1 \quad \rightarrow \quad \psi_m = \psi_m'' = 0 \quad , \tag{33}$$

leading to normalized eigenfunctions

$$\tilde{\psi}_m(z) = -\frac{\sinh[\,\lambda_m\,]\,\sin[\,\lambda_m\, z\,]}{\sqrt{N_m}} \quad , \tag{34}$$

where their respective eigenvalues are defined as

$$\lambda_m = m\,\pi \quad , \tag{35}$$

and norm is defined as

$$N_m = \frac{\cos[\,2\,\lambda_m\,] + \cosh[\,2\,\lambda_m\,]}{4} - \frac{1}{2} \quad . \tag{36}$$

Having established the eigensystem to be used for the integral transformation procedure, it is now possible to define the inverse/transform pair

$$\hat{w}(z) = \sum_{m=1}^{N_Z} \tilde{w}_m\, \tilde{\psi}_m(z) \quad \text{and} \tag{37}$$

$$\tilde{w}_m = \int_0^1 \tilde{\psi}_m(z)\, \hat{w}(z)\, dz \quad , \tag{38}$$

where the integral transformed cross stream velocity component defined in Eq. (37) forms the new algebraic variable set with N_Z terms. Since eigensystem (31) to (36) belongs to the Sturm-Liuville class, the summation in Eq. (38) is convergent. Hence, N_Z can be chosen high enough to yield a solution for the eigenvalues α, β and ω with the desired user prescribed accuracy.

Substituting base flow (14) into disturbance Eq. (27), multiplying the result by normalized eigenfunction (34) and integrating it over the dimensionless domain leads to

$$\begin{aligned}
&\int_0^1 \tilde{\psi}_m(z)\,\hat{w}''''(z)\,dz - i\left(Pe\,\alpha - 2\,i\left(\alpha^2+\beta^2\right) - \omega\right)\int_0^1 \tilde{\psi}_m(z)\,\hat{w}''(z)\,dz \,- \\
&2\,i\,\alpha\,Ge\,Pe\int_0^1 \tilde{\psi}_m(z)\,\hat{w}'(z)\,dz - \left(\alpha^2+\beta^2\right)Ge\,Pe^2\int_0^1 z\,\tilde{\psi}_m(z)\,\hat{w}(z)\,dz \,+ \\
&\frac{1}{2}\left(\alpha^2+\beta^2\right)\left(Ge\,Pe^2 + 2\,i\left(Pe\,\alpha - \omega - i\left(\alpha^2+\beta^2-Ra\right)\right)\right)\int_0^1 \tilde{\psi}_m(z)\,\hat{w}(z)\,dz = 0 \quad ,
\end{aligned} \tag{39}$$

where the first integral can be split by parts to yield

$$\int_0^1 \tilde{\psi}_m(z)\,\hat{w}''''(z)\,dz = \int_0^1 \tilde{\psi}_m''''(z)\,\hat{w}(z)\,dz = \lambda_m^4\int_0^1 \tilde{\psi}_m(z)\,\hat{w}(z)\,dz = \lambda_m^4\,\tilde{w}_m \tag{40}$$

upon substitution of Eqs. (31) and (38). Using this result as well as inverse and transform definitions given by Eqs. (37) and (38), respectively, Eq. (39) can be re-written as

$$\sum_{n=1}^{N_Z} A_{m,n}\,\tilde{w}_n = 0 \quad , \tag{41}$$

with integral transform coefficient matrix being defined as

$$\begin{aligned}
A_{m,n} = &\left(\lambda_m^4 + \frac{1}{2}\left(\alpha^2+\beta^2\right)\left(Ge\,Pe^2 + 2\,i\left(Pe\,\alpha - \omega - i\left(\alpha^2+\beta^2-Ra\right)\right)\right)\right)A_{m,n}^{(0)} \\
&-\left(\alpha^2+\beta^2\right)Ge\,Pe^2\,A_{m,n}^{(1)} - 2\,i\,\alpha\,Ge\,Pe\,A_{m,n}^{(2)} - i\left(Pe\,\alpha - 2\,i\left(\alpha^2+\beta^2\right)-\omega\right)A_{m,n}^{(3)}
\end{aligned} \tag{42}$$

which is itself dependent on eigenfunction integral coefficients

$$A_{m,n}^{(0)} = \int_0^1 \tilde{\psi}_m(z)\,\tilde{\psi}_n(z)\,dz = \delta_{m,n} \quad , \tag{43}$$

$$A_{m,n}^{(1)} = \int_0^1 z\,\tilde{\psi}_m(z)\,\tilde{\psi}_n(z)\,dz \quad , \tag{44}$$

$$A_{m,n}^{(2)} = \int_0^1 \tilde{\psi}_m(z)\,\tilde{\psi}_n'(z)\,dz \quad \text{and} \tag{45}$$

$$A_{m,n}^{(3)} = \int_0^1 \tilde{\psi}_m(z)\,\tilde{\psi}_n''(z)\,dz \quad , \tag{46}$$

where $\delta_{m,n}$ is the Kronecker delta and the remaining coefficients yield analytical expressions.

Equation (41) will only have a nontrivial solution when

$$\det[\,A_{m,n}\,] = 0 \quad , \tag{47}$$

for a fixed value of N_Z, which is the final form of the algebraic dispersion relation. A possible onset of absolute instability is then identified using the zero group velocity conditions

$$\frac{\partial \det[\,A_{m,n}\,]}{\partial\alpha} = 0 \;\text{ with }\; \frac{\partial\omega}{\partial\alpha} = 0 \;\text{ and }\; \frac{\partial \det[\,A_{m,n}\,]}{\partial\beta} = 0 \;\text{ with }\; \frac{\partial\omega}{\partial\beta} = 0 \quad , \tag{48}$$

as additional equations in the system of three equations for the three unknowns α_A, β_A and ω_A, which identify a possible saddle point location. Its behavior is then analyzed graphically to verify if it satisfies the pinching condition, confirming it as an onset of absolute instability.

4. Results and Discussion

4.1. Single Term Approximation

It is useful to first investigate an approximate solution of Eqs. (47) and (48) when the symmetric matrix (42) is generated with $N_Z = 1$ term only. In this case, Eq. (47) yields

$$\pi^4 + \pi^2(i\,Pe\,\alpha + 2\,(\alpha^2+\beta^2) - i\,\omega) + (\alpha^2+\beta^2)(i\,Pe\,\alpha + \alpha^2 + \beta^2 - i\,\omega - Ra) = 0 \quad , \tag{49}$$

while the zero group velocity conditions in Eq. (48) yield

$$i\,Pe\,(\,\pi^2 + 3\,\alpha^2 + \beta^2\,) + 2\,\alpha\,(\,2\,\pi^2 + 2\,(\,\alpha^2+\beta^2\,) - i\,\omega - Ra\,) = 0 \quad \text{and} \tag{50}$$

$$\beta\,(\,2\,\pi^2 + i\,Pe\,\alpha + 2\,(\,\alpha^2+\beta^2\,) - i\,\omega - Ra\,) = 0 \quad , \tag{51}$$

respectively. Two important conclusions can be drawn from these simplified relations. The first one is their Gebhart number independence of Eq. (49) and, hence, Eqs. (50) and (51) as well, i.e. the transition to absolute instability does not vary with Gebhart when $N_Z = 1$. Since the above relations are expected to be increasingly more accurate as the Gebhart number decreases, such a conclusion is not entirely unexpected as it is true as well for the onset of convective instability. The second conclusion is the existence of an onset of absolute instability for transverse modes, which arises from the fact that $\beta = 0$ is an obvious solution of Eq. (51). In fact, we were not able to find an onset of absolute instability for either oblique modes or longitudinal modes, independent of the value of N_Z utilize.

4.2. Convergence Analysis

A convergence study for the onset of absolute instability has been performed and shown in Fig. 1, which presents the critical Rayleigh number relative error as a function of the Péclet number with $Ge = 5$. This particular case was selected because it was the highest Gebhart number studied and convergence worsens when this parameter is increased. Furthermore, the error for each solution generated with N_Z terms was estimated using the solution generated with $N_Z + 1$ terms. It is clear in this figure that graphically converged curves can be obtained using $N_Z = 5$ terms in the summation series, since maximum relative errors are smaller than 1%. Hence, all results presented next were obtained with this number of terms. Furthermore, only transverse modes are shown in the present paper for the reason discussed in the previous subsection.

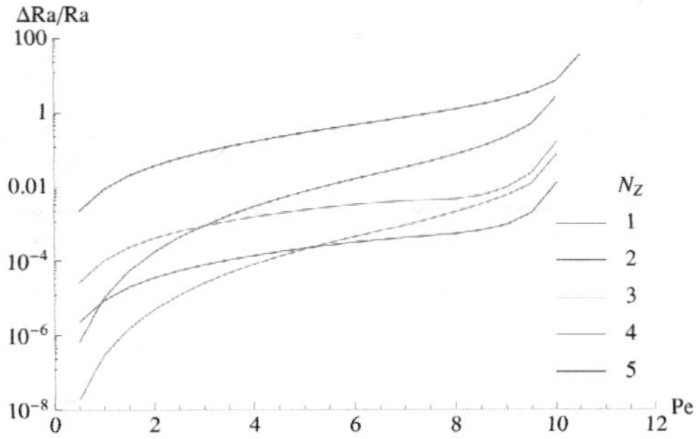

Figure 1. Relative error of the critical Rayleigh number for the onset of absolute instability with $Ge = 5$ as a function of Pe for different N_Z.

4.3. Numerical Solution

Figures 2 shows both onsets of convective and absolute instability, through the critical Rayleigh number and frequency, for transverse modes as functions of the Pe with $Ge = 0.01$, 0.1, 1, 2 and 5. When $Ge = 0.01$, both onsets agree with reported data in the literature for the case without viscous dissipation within the reported Péclet range [27]. In this limiting scenario, the onset of convective instability is not affected by the cross flow whereas the onset of absolute instability is always stabilized by it. When viscous dissipations is present, the cross flow always destabilises the onset of convective instability. Furthermore, this effect gets stronger as the Gebhart number increases. On the other hand, this scenario changes at the transition to absolute instability. Cross flow destabilisation is only observed for weak viscous dissipation. Beyond a certain threshold, estimated here to be $Ge_{c_1} \simeq 0.95$, weak cross flows stabilise the onset of absolute instability but stronger cross flows destabilise this onset. In addition, beyond a second threshold, estimated here to be $Ge_{c_1} \simeq 4.31$, the cross flow is always a destabilising factor for the onset of absolute instability. This behaviour leads to the peculiar condition where a transition to absolute instability can be observed even without a temperature difference between horizontal walls, i.e. even when $Ra = 0$. Finally, these transverse modes are always oscillatory in the presence of cross flow. Although their frequency does not vary significantly with the Gebhart number for weak cross flows, an increase in this parameter increases the critical frequency associated with both onsets for strong cross flows.

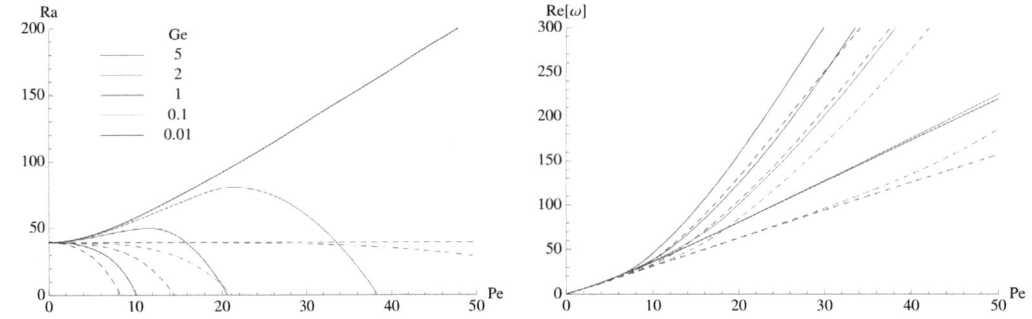

Figure 2. Critical Rayleigh number (left) and frequency (right) for the onsets of convective (dashed lines) and absolute (solid lines) instability as a function of Pe for different Ge.

All results presented so far were obtained using the zero group velocity conditions in Eqs (48), in addition to the dispersion relation (47). However, these conditions are necessary but not sufficient for the determination of an onset of absolute instability. In order to prove they indeed represent such an onset, a graphical analysis of the saddle point formation is required, showing a merger between originally upstream and downstream propagating modes. Figure 3 presents this analysis for $Pe = 30$ and $Ge = 1$, showing the simultaneous pinching of the saddle point $\alpha = 4.86559 - 5.86233\,i$ and $\beta = 0$ in both complex wave number maps. A few other points were similarly analysed and the same behaviour was observed. Hence, we expect this to be the case for all data points shown in Fig. 2.

5. Conclusions

The present paper considered the effects of viscous dissipation on the onset of absolute instability for a Newtonian fluid flowing inside a porous medium subjected to differentially heated horizontal walls. Three different regimes were identified. The first one is the small viscous dissipation regime, where the zero viscous dissipation behaviour is observed. On the other hand, the second regime is the average viscous dissipation one, where weak cross flows stabilise this onset but

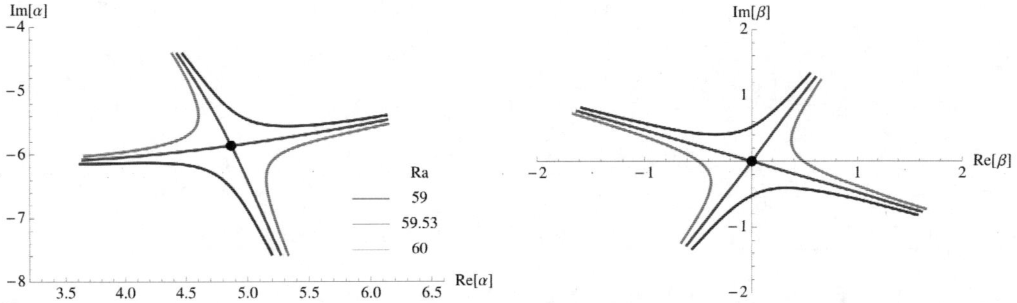

Figure 3. Simultaneous pinching point in stream wise (left) and transverse (right) complex wave number maps. Black dot indicates the saddle point located at $\alpha = 4.86559 - 5.86233\,i$, $\beta = 0$, $Ra = 59.5294$ and $\omega = 201.085$ for $Pe = 30$ and $Ge = 1$.

strong cross flows destabilise it. Finally, the cross flow always destabilises this onset in the third regime with strong viscous dissipation.

Additional studies will be pursued to verify the existence of additional transversal modes and well as the non existence of longitudinal or oblique modes. Furthermore, disturbance stream function behaviour will be investigated for positive and negative Rayleigh numbers. These studies will also include an extension to non-Newtonian fluids.

References

[1] Kaviany M 1995 *Principles of Heat Transfer in Porous Media* 2nd ed (New York: Springer – Verlag)
[2] Nield D A and Bejan A 2006 *Convection in Porous Media* 3rd ed (New York: Springer)
[3] Bénard H 1900 *Rev. Gén. Sci. Pures Appl.* **12** 1261–1309
[4] Block M 1956 *Nature* **178** 650
[5] Pearson J R A 1958 *Journal of Fluid Mechanics* **4** 489–500
[6] Simanovskii I B and Nepomnyashchy A A 1993 *Convective Instabilities in Systems with Interface* (Amsterdam: Gordon and Breach Science Publishers S.A.)
[7] Nepomnyashchy A A, Velarde M G and Colinet P 2001 *Interfacial Phenomena and Convection* (*Monographs and Surveys in Pure and Applied Mathematics* vol 124) (Boca Raton: Chapman & Hall/CRC)
[8] Nakayama A and Pop I 1989 *International Communications in Heat & Mass Transfer* **16** 173–180
[9] Murthy P V S N 1998 *Heat and Mass Transfer* **33** 295–300
[10] Nield D A 2000 *Transport in Porous Media* **41** 349–357
[11] Barletta A, Celli M and Rees D A S 2009 *International Journal of Heat and Mass Transfer* **52** 337–344
[12] Barletta A and Rees D A S 2009 *International Journal of Heat and Mass Transfer* **52** 2300–2310
[13] Barletta A and Storesletten L 2010 *International Journal of Thermal Sciences* **49** 621–630
[14] Barletta A and Rees D A S 2012 *Physics of Fluids* **24** 22
[15] de B Alves L S, Barletta A, Hirata S and Ouarzazi M N 2014 *International Journal of Heat and Mass Transfer* **70** 586–598
[16] Barletta A and Nield D A 2010 *Journal of Fluid Mechanics* **662** 475–492
[17] Barletta A, Celli M and Nield D A 2011 *Journal of Fluid Mechanics* **681** 499–514
[18] Huerre P and Monkewitz P A 1990 *Annual Review of Fluid Mechanics* **22** 473–537
[19] Briggs R J 1964 Electron-stream interaction with plasmas Research Monograph 29 MIT
[20] Gaster M 1968 *Physics of Fluids* **11** 723–727
[21] Huerre P and Monkewitz P A 1985 *Journal of Fluid Mechanics* **159** 151–168
[22] Brevdo L 1991 *Journal of Applied Mathematics and Physics* **42** 911–942
[23] Suslov S A 2006 *Journal of Computational Physics* **212** 188–217
[24] Cotta R M 1993 *Integral Transforms in Computational Heat and Fluid Flow* (Boca Raton, FL: CRC Press)
[25] Cotta R M (ed) 1998 *The Integral Transform Method in Thermal and Fluids Science and Engineering* (New York: Begell House, Inc.)
[26] Özisik M N 1993 *Heat Conduction* 2nd ed (New York: Wiley Interscience)
[27] Hirata S C and Ouarzazi M N 2010 *Physics Letters, Section A: General, Atomic and Solid State Physics* **374** 2661–2666

Journal of Physics: Conference Series **547** (2014) 012040 doi:10.1088/1742-6596/547/1/012040

Integral Transform Analysis of Poisson Problems that Occur in Discrete Solutions of the Incompressible Navier-Stokes Equations

Daniel J. N. M. Chalhub, Leandro A. Sphaier*, Leonardo S. de B. Alves

Laboratory of Theoretical and Applied Mechanics – LMTA
Department of Mechanical Engineering – TEM/PGMEC
Universidade Federal Fluminense, Niterói, Rio de Janeiro, Brazil

E-mail: lasphaier@id.uff.br

Abstract. The present work presents an alternate method for solving the poisson equation for calculating the pressure field that appears in many discrete numerical solvers of the incompressible Navier-Stokes equations. The methodology is based on a pressure-correction scheme with a mixed approach that employs Integral Transform Technique for the calculation of the pressure field from a given discrete velocity field. Two solution schemes are analyzed, these being the single transformation and the double transformation. The poisson equation is solved with the two different schemes using a prescribed source term to simulate the discrete data that could arise in the solution process of the momentum equation and an numerical results are presented. An error analysis of these results show that the single-transformation scheme is computationally superior to the double transformation, and that good convergence rates can be obtained with few terms in the series. Moreover, it was also verified that the series solution employed for the Poisson equation maintains the original spatial order of the discretization.

1. Introduction

A major difficulty in the numerical simulation of incompressible flows is the fact that velocity and pressure are coupled by the incompressibility constraint. To overcome this difficulty in time-dependent viscous incompressible flows, fractional step methods, which are also referred as projection methods, were developed. These methods can be classified into three classes [1], namely pressure-correction methods, velocity-correction methods, and consistent splitting methods. The most attractive feature of projection methods is that, at each time step, one only needs to solve a sequence of decoupled elliptic equations for the velocity and the pressure, making it very efficient for large-scale simulations. In spite of the advantages of projection methods, the employed decomposition is intrinsically second-order accurate, hence limiting the application of higher-order approximations.

Pressure-correction schemes are time-marching techniques composed of two sub-steps for each time step: the pressure is treated explicitly or ignored in the first sub-step and is corrected in the second one [2]. The most common methodology to obtain a pressure-correction equation involves combining the momentum and continuity equations by taking the divergence of the former and substituting the latter where necessary, effectively generating a Poisson-type equation for determining the pressure field from a given velocity distribution. Naturally, the correction scheme is performed in an iterative fashion until mass conservation is satisfied at each time-iteration. As a result, a major computational cost associated with pressure-correction methods resides in the required sub-iterations per time step. At each sub-iteration, a Poisson equation needs to be solved, which consumes large amounts of CPU time. To make matters worse, the convergence rate of common iterative algorithms for this purpose, such as the Jacobi and Gauss-Seidel, rapidly decreases as the mesh is refined [3]. However, this issue can generally be remedied by employing Multigrid methods, which are designed to exploit the inherent differences of the error behavior among meshes of different size [4].

In the realm of analytical methods, the Integral Transform Technique [5] has been used for the solution of a variety of problems. The method deals with expansions of the sought solution in terms of infinite orthogonal basis of eigenfunctions, keeping the solution process always within a continuous domain. The resulting system is generally composed of a set of uncoupled differential equations which can be solved analytically. A few works have implemented a mixed approach using the Integral Transform Technique and other discrete schemes. Most of these employ the Generalized version of the technique (GITT) [6], which can be used to non-transformable problems in general, including those with non-linear effects. Cotta and Gerk [7], employed the integral transform method in conjunction with second-order-accurate explicit finite-differences schemes, to handle a class of parabolic-hyperbolic problems. Guedes and Ozisik [8, 9] analyzed unsteady forced convection in laminar flow between parallel plates, solving the problem with a hybrid scheme that combines the GITT with second-order finite differences. More recently, Castelões and Cotta. [10], employed a partial integral transformation strategy in periodic convection in micro-channels and Naveira-Cotta et al. [11] used a similar approach in a conjugate conduction-convection problem. In theses studies, one of the spatial variables was handled automatically by a numerical PDE solver, while the other was handled by integral transformation. Very recently, different from these previous studies, a new approach was employed in [12], where integral transformation and upwind discretization techniques were applied within a same spatial variable.

This work is focused on an alternative numerical scheme for solving the unsteady incompressible Navier-Stokes equations with primitive variables in three dimensions. The methodology is based on pressure-correction schemes using a mixed approach based on the Integral Transform Technique. This work proposes the use of the Integral Transform Technique to find the pressure dependence on the discrete velocity field analytically, which potentially eliminates the need for a pressure-velocity decomposition. In this paper, the analysis of the Integral Transform solution of the Poisson equation for pressure field with a given discrete source term is analyzed.

2. Problem Formulation

The governing equations are the traditional form of the Navier-Stokes Equations for incompressible flow:

$$\frac{\partial \boldsymbol{v}}{\partial t} + \boldsymbol{v} \cdot \nabla \boldsymbol{v} = -\frac{1}{\rho} \nabla p + \nu \nabla^2 \boldsymbol{v} + \boldsymbol{f}, \quad \text{for} \quad \boldsymbol{x} \in \mathcal{V} \quad \text{and} \quad t \geq 0, \tag{1a}$$

$$\nabla \cdot \boldsymbol{v} = 0, \quad \text{for} \quad \boldsymbol{x} \in \mathcal{V}, \tag{1b}$$

in which equation (1a) is the momentum conservation equation and equation (1b) is the mass conservation equation, also called the incompressibility constraint.

The proposed methodology to solve the Navier-Stokes equations is based on the projection methods for incompressible flows [1]. The first step of the methodology is to apply the divergence operator on equation (1a) and use the continuity equation (1b) to obtain a Poisson equation for determining the pressure field:

$$\frac{1}{\rho}\nabla^2 p = \nabla \cdot \boldsymbol{f} - \nabla \boldsymbol{v} : (\nabla \boldsymbol{v})^{\mathrm{T}}, \quad \text{for} \quad \boldsymbol{x} \in \mathcal{V}, \tag{2}$$

where ρ is the fluid density, $\boldsymbol{f}$ is the body force vector (in acceleration units), $\boldsymbol{v}$ is the velocity vector, p is the pressure and $\mathcal{V}$ is a general domain volume. With this procedure, the continuity equations can be replaced by equation (2), which needs to be solved together with the momentum equation (1). With the given formulation, one needs to specific pressure boundary conditions at all boundaries. In this work, normal zero gradients for pressure at the boundaries will be used for illustrational purposes:

$$(\nabla p \cdot \boldsymbol{n})_{\partial\mathcal{V}} = 0, \tag{3}$$

where $\partial\mathcal{V}$ is the boundary of the general domain volume.

2.1. Initial Value Problem Discretization

The next step of the methodology is the time discretization. There are many ways to discretize the problem in time [13], such as methods like Forward Euler, Backward Euler, BDF methods, among others. If a Forward Euler Method is employed, the time-discretized version of equations (1a) and (2) become:

$$\boldsymbol{v}^{l+1} = \Delta t \left(-\frac{1}{\rho}\nabla p^l + \nu\,\nabla^2 \boldsymbol{v}^l - \boldsymbol{v}^l \cdot \nabla \boldsymbol{v}^l + \boldsymbol{f}^l \right) + \boldsymbol{v}^l, \tag{4a}$$

$$\frac{1}{\rho}\nabla^2 p^l = \nabla \cdot \boldsymbol{f}^l - \nabla \boldsymbol{v}^l : \left(\nabla \boldsymbol{v}^l\right)^{\mathrm{T}}, \tag{4b}$$

where, in order to find the velocity $\boldsymbol{v}^{l+1}$, it is necessary to know the pressure field p^l.

2.2. Poisson Equation Solution

For this work, the two dimensional case in cartesian coordinates will be considered. For these assumptions, the equation (4b) is simplified to the following form:

$$\frac{\partial^2 p(x,y,t)}{\partial x^2} + \frac{\partial^2 p(x,y,t)}{\partial y^2} = \rho[h(x,y,t) - g(x,y,t)], \tag{5a}$$

$$\left(\frac{\partial p(x,y,t)}{\partial y}\right)_{y=0} = 0, \qquad \left(\frac{\partial p(x,y,t)}{\partial y}\right)_{y=H} = 0, \tag{5b}$$

$$\left(\frac{\partial p(x,y,t)}{\partial x}\right)_{x=0} = 0, \qquad \left(\frac{\partial p(x,y,t)}{\partial x}\right)_{x=L} = 0, \tag{5c}$$

in which:

$$g(x,y,t) = \left(\frac{\partial u}{\partial x}\right)^2 + 2\frac{\partial u}{\partial y}\frac{\partial v}{\partial x} + \left(\frac{\partial v}{\partial y}\right)^2, \qquad h(x,y,t) = \frac{\partial f_x}{\partial x} + \frac{\partial f_y}{\partial y}. \tag{5d}$$

The Classical Integral Transform Technique (CITT) [5] is then used for the purpose of solving the Poisson equation (4b). Two approaches are used for this purpose: CITT transforming in one direction (single transformation) and CITT transforming in two directions (double transformation), as described next.

3. CITT with Single Transformation

In order to establish the transformation pair, the pressure field is written as function of an orthogonal eigenfunctions obtained from the following auxiliary eigenvalue problem known as the Helmholtz classical problem, where $X_n(x)$ are the eigenfunctions and λ_n are the eigenvalues.

$$\frac{\mathrm{d}^2 X_n(x)}{\mathrm{d}x^2} + \lambda_n^2 X_n(x) = 0, \qquad X'(0) = 0, \qquad X'(L) = 0, \tag{6}$$

which has the following solution:

$$X_n(x) = \cos(x\lambda_n), \qquad \text{with} \qquad \lambda_n = \frac{\pi n}{L}, \qquad \text{for} \qquad n = 0, 1, 2, 3, \ldots \tag{7}$$

Now, the transformation pair can be defined:

$$\text{Transformation} \quad \Rightarrow \quad \bar{p}_n(y,t) = \int_0^L p(x,y,t)\, X(x)_n \, \mathrm{d}x, \tag{8a}$$

$$\text{Inversion} \quad \Rightarrow \quad p(x,y,t) = p_0(y,t) + \sum_{n=1}^{\infty} \frac{X_n(x)\bar{p}_n(y,t)}{N_n}, \tag{8b}$$

where the norms N_n are defined by:

$$N_n = \int_0^L X_n^2 \, \mathrm{d}x \tag{9}$$

which gives $N_n = L/2$ for $n \neq 0$ and $N_n = L$ for $n = 0$.

The final solution is given by a decomposition of the pressure field in two parts: the average pressure in the x direction p_{avg}, and the modified pressure p_{mod}:

$$p(x,y,t) = p_{\text{avg}}(y,t) + p_{\text{mod}}(x,y,t), \tag{10}$$

in which p_{avg} comes from the solution of the eigenproblem when $\lambda = 0$ and p_{mod} comes from the solution when $\lambda \neq 0$, in other words:

$$p_{\text{avg}}(y,t) = \frac{p_0(y,t)}{N_0}, \qquad p_{\text{mod}}(x,y,t) = \sum_{n=1}^{\infty} \frac{X_n(x)\bar{p}_n(y,t)}{N_n}. \tag{11}$$

3.1. Solution for p_{mod} ($\lambda \neq 0$)

The integral transformation of the governing differential equation is derived by applying the operator $\int_0^L (\bullet)\, X_n \, \mathrm{d}x$ to equation (5), obtaining the following transformed Poisson equation:

$$\frac{\partial^2 \bar{p}_n(y,t)}{\partial y^2} - \lambda_n^2 \bar{p}_n(y,t) = \rho \bar{h}_n(y,t) - \rho \bar{g}_n(y,t), \tag{12a}$$

$$\left(\frac{\partial \bar{p}_n(y,t)}{\partial y}\right)_{y=0} = 0, \qquad \left(\frac{\partial \bar{p}_n(y,t)}{\partial y}\right)_{y=H} = 0, \tag{12b}$$

where $\bar{g}_n$ and $\bar{h}_n$ are the transformed versions of g and h:

$$\bar{g}_n(y,t) = \int_0^L g(x,y,t) X_n \, \mathrm{d}x, \qquad \bar{h}_n(y,t) = \int_0^L h(x,y,t) X_n \, \mathrm{d}x, \tag{13}$$

and equation (12) admits a closed-form analytical solution:

$$\bar{p}_n(y,t) = e^{-y\lambda_n}\left(-\frac{1}{2}\cosh(y\lambda_n)\operatorname{csch}(H\lambda_n)\,e^{(H+y)\lambda_n}\int_0^H \frac{\rho e^{-y\lambda_n}}{\lambda_n}\left(\bar{h}_n(y,t)-\bar{g}_n(y,t)\right)\mathrm{d}y + \frac{1}{4}\left(e^{2y\lambda_n}+1\right)\left(\coth(H\lambda_n)-1\right)\int_0^H \frac{\rho e^{y\lambda_n}}{\lambda_n}\left(\bar{g}_n(y,t)-\bar{h}_n(y,t)\right)\mathrm{d}y + \frac{1}{2}\int_0^y \frac{\rho e^{y'\lambda_n}}{\lambda_n}\left(\bar{g}_n(y',t)-\bar{h}_n(y',t)\right)\mathrm{d}y' + \frac{1}{2}e^{2y\lambda_n}\int_0^y \frac{\rho e^{-y'\lambda_n}}{\lambda_n}\left(\bar{h}_n(y',t)-\bar{g}_n(y',t)\right)\mathrm{d}y'\right) \tag{14}$$

To find the actual solution for modified pressure p_{mod}, the inversion formula is used, equation (11). By observing equations (13), one notices integrals of the discrete variables u, v, f_x and f_y. In order to compute these integrals, the following integral separation is proposed:

$$\int_0^L \Lambda(u,v,f_x,f_y)\,\mathrm{d}x = \sum_{q=1}^{i_{\max}} \int_{x_{q-1}}^{x_q} \Lambda(u,v,f_x,f_y)\,\mathrm{d}x \tag{15}$$

where Λ is a general function of u, v, f_x and f_y. Then, to compute the integrals analytically, a Taylor expansion is used to expand the variables u, v, f_x and f_y in each subdomain:

3.2. Solution for p_{avg} ($\lambda = 0$)

In order to obtain the transformed differential equation for $\lambda = 0$, a similar process is done, leading to the following the transformed equation:

$$\frac{\partial^2 \bar{p}_0(y,t)}{\partial y^2} = \rho\,\bar{h}_0(y,t) - \rho\,\bar{g}_0(y,t), \tag{16a}$$

$$\left(\frac{\partial \bar{p}_0(y,t)}{\partial y}\right)_{y=0} = 0, \qquad \left(\frac{\partial \bar{p}_0(y,t)}{\partial y}\right)_{y=H} = 0, \tag{16b}$$

where:

$$\bar{g}_0(y,t) = \int_0^L g(x,y,t)\,\mathrm{d}x \qquad \text{and} \qquad \bar{h}_0(y,t) = \int_0^L h(x,y,t)\,\mathrm{d}x. \tag{17}$$

The equation (16) admits a general analytical solution in the following form:

$$\bar{p}_0(y,t) = \int_0^y \int_0^{y''} \rho\left(\bar{h}_0(y',t) - \bar{g}_0(y',t)\right)\mathrm{d}y'\,\mathrm{d}y'' + c_1 y + c_2. \tag{18}$$

By applying the boundary conditions, one arrives to the following equations:

$$c_1 = 0, \qquad 0 = \int_0^H \rho\left(\bar{h}_0(y,t) - \bar{g}_0(y,t)\right)\mathrm{d}y + c_1, \tag{19}$$

such that c_1 must be zero and the integral also must be zero:

$$\int_0^H \rho\left(\bar{h}_0(y,t) - \bar{g}_0(y,t)\right)\mathrm{d}y = 0. \tag{20}$$

One can easily prove that the integral above indeed equals zero due to the momentum conservation law. With this information at hand, the solution of the transformed differential equation is achieved:

$$\bar{p}_0(y,t) = \int_0^y \int_0^{y''} \rho\left(-\bar{g}_0(y',t) + \bar{h}_0(y',t)\right) \mathrm{d}y'\, \mathrm{d}y'' \tag{21}$$

Then, the same integral separation (equation (15)) and Taylor series expansions are used to derive analytically the coefficients $\bar{h}_0$ and $\bar{g}_0$:

3.3. Discrete Derivatives

In order to solve the pressure problem, the discrete derivatives of u, v, f_x and f_y must be calculated. In this work, a second order central differencing scheme is used inside the domain and second-order the backward/forward (depending of the boundary) differencing scheme is used at the boundaries.

4. CITT with Double Transformation

In a similar manner as done for the single transformation scheme, one needs to establish the transformation pair. In order to obtain that for this approach, two eigenvalue problems are defined. The eigenvalue problem associated with the x direction is the same as given by equations (6), whereas the problem associated with the y direction is given by:

$$\frac{\mathrm{d}^2 Y_n(y)}{\mathrm{d}y^2} + \beta_m^2 Y_n(y) = 0, \qquad Y'(0) = 0, \qquad Y'(H) = 0, \tag{22}$$

which has the following solution:

$$Y_n(x) = \cos\left(y\beta_m\right), \qquad \text{with} \qquad \beta_m = \frac{\pi\, m}{H}, \qquad \text{for} \qquad m = 1, 2, 3, \ldots, \tag{23}$$

and the transformation pair can then be defined as:

$$\text{Transformation} \quad \Rightarrow \quad \bar{\bar{p}}_{n,m}(t) = \int_0^H \int_0^L p(x,y,t)\, X(x)_n Y(x)_m \,\mathrm{d}x\, \mathrm{d}y, \tag{24}$$

$$\text{Inversion} \quad \Rightarrow \quad p(x,y,t) = \sum_{n=0}^{\infty}\sum_{m=0}^{\infty} \frac{\bar{\bar{p}}_n(t)\, X_n(x)\, Y_m(y)}{\mathrm{Nx}_n\, \mathrm{Ny}_m}, \tag{25}$$

where the norms Nx_n and Ny_m are defined by:

$$\mathrm{Nx}_n = \int_0^L X_n^2 \,\mathrm{d}x \qquad \text{and} \qquad \mathrm{Ny}_m = \int_0^H Y_m^2 \,\mathrm{d}y \tag{26}$$

Applying the operator $\int_0^H \int_0^L (\bullet)\, X_n Y_m \,\mathrm{d}x\, \mathrm{d}y$ to the Poisson equation, the following transformed Poisson equation is obtained:

$$-(\lambda_n^2 + \beta_m^2)\, \bar{\bar{p}}_{n,m}(t) = \rho\left(\bar{\bar{h}}_{n,m}(t) - \bar{\bar{g}}_{n,m}(t)\right). \tag{27}$$

This equation is a simple uncoupled algebraic systems where $\bar{\bar{g}}_{n,m}$ and $\bar{\bar{h}}_{n,m}$ are written as:

$$\bar{\bar{g}}_{n,m}(t) = \int_0^H \int_0^L g(x,y,t)\, X_n\, Y_m \,\mathrm{d}x\, \mathrm{d}y, \qquad \bar{\bar{h}}_{n,m}(t) = \int_0^H \int_0^L h(x,y,t)\, X_n\, Y_m \,\mathrm{d}x\, \mathrm{d}y, \tag{28}$$

and has a direct solution given by:

$$\bar{\bar{p}}_{n,m}(t) = \frac{\rho\left(\bar{\bar{g}}_{n,m}(t) - \bar{\bar{h}}_{n,m}(t)\right)}{\lambda_n^2 + \beta_m^2}, \tag{29}$$

such that the final solution is obtained by using the inversion formula, eq. (25).

The greatest advantage of this approach is that it requires a lot less analytical effort and the final solution is simpler and more compact. But the final solution has a double summation that can increase computational cost. In order to minimize this cost, one can use a reordering scheme, switching from the double summation to a single one. This is done by associating pairs (n, m) with a single index k in by organizing pairs that promote lower values of $(\lambda_n^2 + \beta_m^2)$ in ascending order. With the reordering scheme one arrives at the following expression for the pressure distribution:

$$p(x,y,t) = \sum_{k=0}^{\infty} \rho \, \frac{\bar{\bar{g}}_{n(k),m(k)}(t) - \bar{\bar{h}}_{n(k),m(k)}(t)}{\lambda_{n(k)}^2 + \beta_{m(k)}^2} \, \frac{X_{n(k)}(x)\, Y_{m(k)}(y)}{\mathrm{Nx}_{n(k)}\, \mathrm{Ny}_{m(k)}}. \tag{30}$$

5. Results

The results presented in this paper are intended to evaluate the feasibility of using a CITT poisson solver with the discrete source terms that appear in algorithms that use pressure-based projection methods, comparing both single and double transformation schemes. For all cases herein presented $L = 1$, $H = 1$, $\rho = 1$ and $\mu = 1$ were used and the source term of the Poisson equation (5) was assumed to be of the following form:

$$\Big[\rho(h(x,y,t) - g(x,y,t))\Big]_{i,j} = -\sin\left(\frac{2\,\pi\,y_j}{H}\right) - \sin\left(\frac{2\,\pi\,x_i}{L}\right), \tag{31}$$

which was chosen in this form to ensure that relation (20) is satisfied.

A comparison of computational cost is done for the two transformation schemes presented in this work. In order to compare the CITT performance, a fixed mesh is used and different truncation orders for the summations are computed, so only the CITT error is captured. The CITT error is calculated using the following formula:

$$\epsilon_{i,j}^{\mathrm{CITT}}(n_{\max}) = \mathrm{abs}[p_{i,j}(n_{\max}) - p_{i,j}(n_{\max} + 5)] \tag{32}$$

Figure 1 presents a comparison of the error versus CPU time for single and double transformation strategies. As can be seen, the single transformation has better performance results, running the code in less CPU time and generating smaller errors. In other words, the final performance of the CITT single transformation overcomes the bigger effort in the analytical manipulation, which suggests that this is a better approach.

Once the single transformation scheme was shown to have better performance, from now on, the results for the solution of the Poisson equation for the pressure field is carried-out only using the single transformation scheme. With these results, the convergence will be analyzed and discussed. In order to analyze the convergence of the single transformation scheme, an absolute error in the x direction is defined as:

$$\epsilon_{i,j}^{x}(i_{\max}) = \mathrm{abs}[p_{i,j}(i_{\max}, 1024) - p_{i,j}(2i_{\max}, 1024)], \tag{33}$$

in which ϵ is calculated at the mesh point (i, j) and for a mesh size $(i_{\max}, 1024)$. Similarly, for the y direction:

$$\epsilon_{i,j}^{y}(j_{\max}) = \mathrm{abs}[p_{i,j}(1024, j_{\max}) - p_{i,j}(1024, 2j_{\max})], \tag{34}$$

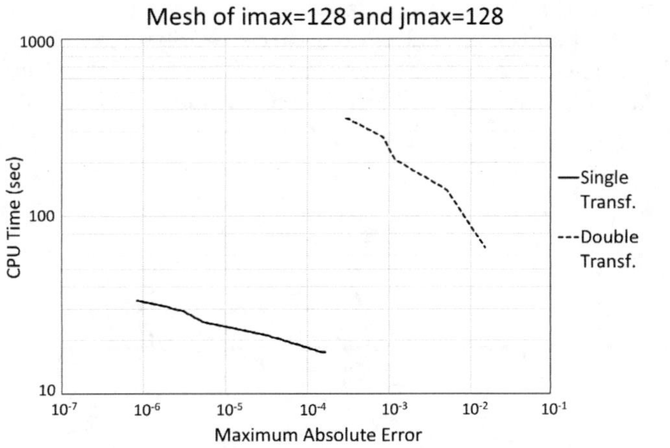

Figure 1. Comparison of the computational cost of CITT using single transformation and CITT using double transformation for a mesh $i_{\max} = 128$ and $j_{\max} = 128$.

where ϵ is calculated at the mesh point (i, j) and for a mesh size $(1024, j_{\max})$. For the calculation of the error in the x-direction, a refined mesh of 1024 divisions in y-direction was used in order to isolate the x error and minimize the effects of the y convergence. The same 1024 mesh divisions in y was used for the error in the x-direction. All meshes utilized in this work have divisions of the form 2^k, thus, one could take maximum advantage of parallel computing.

A great amount of data was generated with the proposed algorithm, and an analysis showed that the error distribution is almost insensitive to the variation of $n_{\max}$ showing that CITT has a very good convergence rate. Hence, for simplicity, the presented results show data for a fixed $n_{\max} = 10$. Figure 2 shows the distribution of the absolute error $\epsilon^x_{i,j}(64)$ while figure 3 shows the distribution $\epsilon^y_{i,j}(64)$. As can be seen from the former, the error oscillates in the x-direction.

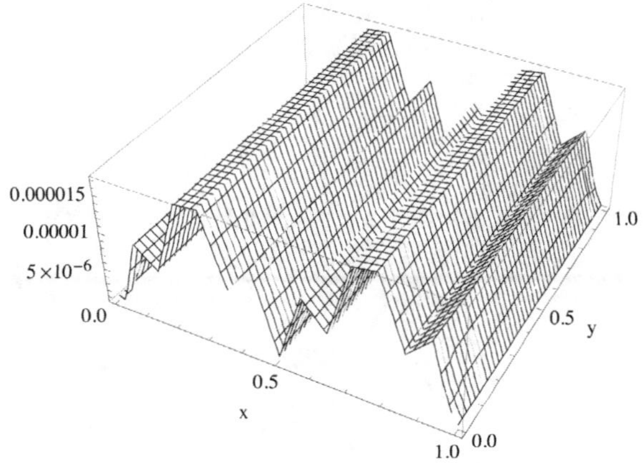

Figure 2. Absolute error variation inside the domain for $n_{\max} = 10$, $\Delta x = 2^{-6}$ and $\Delta y = 2^{-10}$.

This effect is explained by the nature of the Integral Transform Technique, since it is based on oscillatory eigenfunctions and the equation was transformed in the x-direction. Another point to be highlighted is that the eigenfunctions satisfy the boundary conditions. As a result, the error is zero at $x = 0$ and $x = 1$. The error behavior does not change with the variation Δx, the only notable change being the magnitude of the results, that naturally, decays as Δx is decreased. Figure 3 also displays the error behavior for the mesh in the y-direction, showing that it increases along with the y-coordinate. This phenomenon is due to the exponential feature of

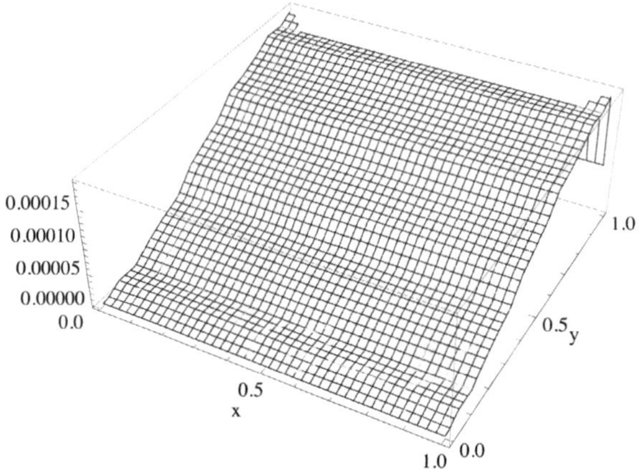

Figure 3. Absolute error variation inside the domain for $n_{\max} = 10$, $\Delta x = 2^{-10}$ and $\Delta y = 2^{-6}$.

the ODE solution obtained after the transformation of the problem. This exponential behavior of the solution magnifies the error while y increases. Similar observations to those pointed out in figure 2, in respect to Δx, can be done here about the variation of Δy.

In order to illustrate the convergence of the solution with the variation of Δx, figure 4 shows a plot of the maximum absolute error with the variation of the mesh size Δx. As one can observe, the convergence order is about 2, which was expected since all approximation made in the mathematical formulation were of this order. Figure 5 shows the convergence of the absolute error with Δy. Although the solution seems to have a higher order for the poorer refined meshes, the order stabilizes at 2 when more refined meshes are implemented.

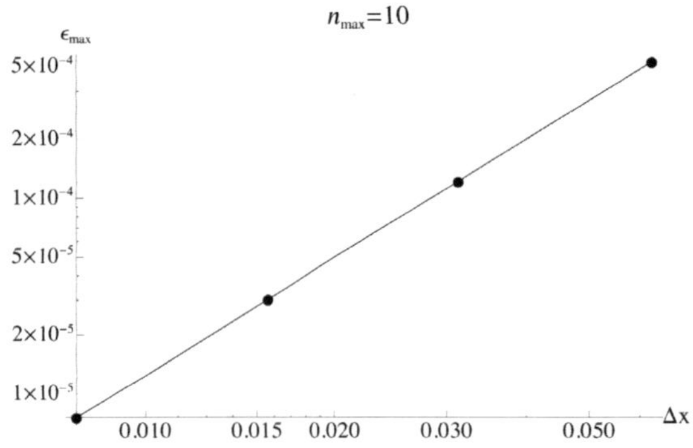

Figure 4. Maximum absolute error vs. Δx for $\Delta y = 2^{-10}$.

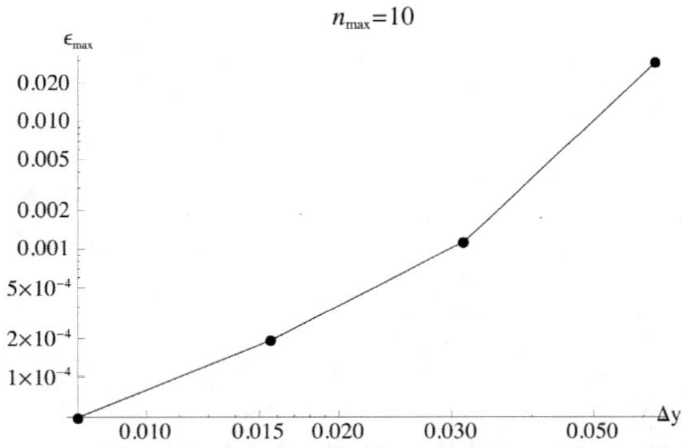

Figure 5. Maximum absolute error vs. Δy for $\Delta x = 2^{-10}$.

6. Summary and Conclusions

The main goal of the proposed work was the development of a new methodology using the Integral Transform Technique in a semi-analytical fashion for the Poisson equation, allowing an explicit expression for the pressure p, that then can be input in the momentum equations and thus be solved using a numerical technique for initial value problem. The solution of the Poisson equation using this semi-analytical approach was accomplished in this work using two different schemes: CITT single transformation and CITT double transformation. The comparison between both schemes showed that the double transformation has poorer performance in comparison with the single transformation scheme. The error convergence rate was shown and it was possible to see that the spatial approximation order matched the expected value of 2 in both x and y directions. It was also observed that the Integral Transform Technique had a very good performance, converging with very few terms in the series.

7. Acknowledgements

The authors would like to acknowledge the financial support provided by CAPES, CNPq, FAPERJ, and the Lemann Foundation.

References

[1] Guermond J L, Minev P and Shen J 2006 *Comput. Methods Appl. Mech. Engrg.* **195** 6045
[2] Ferziger J H and Perić M 1999 *Computational Methods for Fluid Dynamics* 2nd ed (New York, NY: Springer – Verlag)
[3] Versteeg H K and Malalasekera W 2007 *An Introduction to Computational Fluid Dynamics, The Finite Volume Method* 2nd ed (Pearson Prentice Hall)
[4] Wesseling P 1992 *An Introduction to Multigrid Methods* (New York: John Wiley & Sons)
[5] Mikhailov M D and Özişik M N 1984 *Unified Analysis and Solutions of Heat and Mass Diffusion* (New York: John Wiley & Sons)
[6] Cotta R M 1993 *Integral Transforms in Computational Heat and Fluid Flow* (Boca Raton, FL: CRC Press)
[7] Cotta R M and Gerk J E V 1994 *Numerical Heat Transfer - Part B* **25** 442–448
[8] Guedes R O C and Ozisik M N 1994 *International Journal Heat and Mass Transfer* **37** 2699–2706
[9] Guedes R O C and Ozisik M N 1994 *Int. J. Heat and Fluid Flow* **15** 116–121
[10] Castellões F V and Cotta R M 2006 *Progress in Computational Fluid Dynamics* **6** 321–326
[11] Naveira C P, Lachi M, Cotta R M and Padet J 2009 *International Journal of Heat and Mass Transfer* **52** 112–123
[12] Chalhub D J M N, Sphaier L A and de B Alves L S 2013 *Numerical Heat Transfer - Part B* **63** 167–187
[13] Moin P 2001 *Fundamentals of Engineering Numerical Analysis* (New York, NY: Cambridge University Press)

Journal of Physics: Conference Series **547** (2014) 012041 doi:10.1088/1742-6596/547/1/012041

Nusselt number correlations for simultaneously developing laminar duct flows of liquids with temperature dependent properties

Stefano Del Giudice, Stefano Savino and Carlo Nonino

Dipartimento di Ingegneria Elettrica, Gestionale e Meccanica, Università degli Studi di Udine,Via delle Scienze 206, 33100 UDINE, Italy

E-mail: stefano.delgiudice@uniud.it, stefano.savino@uniud.it, carlo.nonino@uniud.it

Abstract. New correlations, suitable for engineering applications, for the mean Nusselt number in the entrance region of circular tubes and square ducts with uniform heat flux boundary conditions specified at the walls are proposed. These correlations are obtained on the basis of the results of a previous parametric investigation on the effects of temperature dependent viscosity and thermal conductivity in simultaneously developing laminar flows of liquids in straight ducts of constant cross-sections. In these studies, a finite element procedure has been employed for the numerical solution of the parabolized momentum and energy equations. Viscosity and thermal conductivity are assumed to vary with temperature according to an exponential and to a linear relation, respectively, while the other fluid properties are held constant. Axial distributions of the mean Nusselt number, obtained by numerical integration from those of the local Nusselt number, are used as input data in the derivation of the proposed correlations. A superposition method is proved to be applicable in order to estimate the Nusselt number by considering separately the effects of temperature dependent viscosity and thermal conductivity. Therefore, for each of the considered cross-sectional geometries, two distinct correlations are proposed for flows of liquids with temperature dependent viscosity and with temperature dependent thermal conductivity, in addition to that obtained for constant property flows.

1. Introduction

In many laminar duct flows, entrance effects on fluid flow and heat transfer must be taken into account, since the total length of the duct is comparable with that of the entrance region. Moreover, the development of the velocity and temperature fields can be strongly affected by temperature dependence of fluid properties. If the fluid is a liquid, the relative variations of viscosity with temperature are the most relevant, while those of thermal conductivity are, in general, much smaller, and those of density and specific heat capacity are almost negligible [1, 2]. As far as velocity distribution and pressure drop are concerned, the main effects of temperature dependent fluid properties can be retained even if only viscosity is considered to vary with temperature, while the other properties are assumed constant [1, 3]. Instead, if heat transfer characteristics, as the Nusselt number, have to be evaluated, also the temperature dependence of thermal conductivity must be taken into account [2].

In previous articles, we carried out systematic analyses of the effects of temperature dependent viscosity and thermal conductivity on forced convection in simultaneously developing laminar flows of liquids in straight ducts with uniform wall heat flux boundary conditions [4, 5]. In particular, we obtained axial distributions of the local value of the peripherally averaged Nusselt number for ducts of different cross-sections. We assumed that viscosity decreases with increasing temperature according to an exponential relation, while thermal conductivity varies linearly with temperature. Suitable dimensionless parameters, namely, the viscosity and thermal conductivity Pearson numbers, were used to quantitatively express temperature dependence of the corresponding properties. A finite element procedure was employed for the step-by-step solution of the parabolised momentum and energy equations in a domain corresponding to the cross-section of the duct [6]. Finally, a superposition method was proved to be applicable in order to obtain accurate estimates of the local Nusselt number by considering separately the effects of temperature dependent viscosity and thermal conductivity [4, 5].

In this paper, first we use previous numerical results for the local Nusselt number for circular and square cross-sectional geometries [5] to obtain by numerical integration the corresponding axial distributions of the mean, i.e., longitudinally averaged, Nusselt number. Then, on the basis of such distributions, we obtain very accurate correlations, suitable for engineering applications. Actually, taking advantage of the above mentioned superposition method, we propose for each of the considered cross-sectional geometries a correlation valid for constant property flows and two distinct correlations for flows of liquids with temperature dependent viscosity and with temperature dependent thermal conductivity. Finally, we demonstrate that, by properly assembling these correlations, accurate estimates of the mean Nusselt number can be obtained in wide ranges of operative conditions. No correlations are presented here for pressure drop parameters since, as demonstrated in previous works [4, 5], the influence of thermal conductivity variation on pressure drop is almost negligible if compared with that of viscosity variations. Therefore, the correlations for pressure drop parameters already proposed by the authors for liquids with temperature dependent viscosity and constant thermal conductivity [7] can be used for all liquids with temperature dependent properties.

2. Statement of the problem

The laminar forced convection in the entrance region of straight ducts of constant cross-sections with uniform wall heat flux $q''_w > 0$ (fluid heating) is studied. Since the liquid heating is assumed to begin at the duct inlet, uniform values of the axial velocity component u and of the temperature t (i.e., $t = t_e$, $u = u_e = \overline{u}$ and $v = w = 0$) are specified as the appropriate inlet conditions, being $\overline{u}$ the average axial velocity and v and w the transverse velocity components. The heat flux (Neumann) boundary condition $q'' = k\,\partial t/\partial n = q''_w$ is applied at the solid walls, where k is the thermal conductivity and n is the outward normal to the boundary.

Since the fluids considered here are liquids, the dynamic viscosity μ is assumed to decrease with increasing temperature according to the widely used exponential formula [3]

$$\mu = \mu_e \, \exp[-\beta\,(t - t_e)] \tag{1}$$

where μ_e is the value of μ at t_e and $\beta = -(\mathrm{d}\mu/\mathrm{d}t)/\mu = \text{const}$ is positive. The thermal conductivity k can be assumed both to increase or decrease with increasing temperature, depending on the fluid considered, according to the linear relation

$$k = k_e\,[1 + \alpha\,(t - t_e)] \tag{2}$$

where k_e is the value of k at t_e and $\alpha = (\mathrm{d}k/\mathrm{d}t)/k_e = \text{const}$ can be both positive or negative. By means of simple manipulations, equations (1) and (2) can be cast in the following dimensionless forms

$$\frac{\mu}{\mu_e} = \exp(-\mathrm{Pn}_\mu\, T) \tag{3}$$

$$\frac{k}{k_e} = 1 + \mathrm{Pn}_k\, T \tag{4}$$

where $T = (t - t_e)\, k_e/(q''_w\, D_h)$ is the dimensionless temperature, D_h is the hydraulic diameter, $\mathrm{Pn}_\mu = \beta\, q''_w\, D_h/k_e$ is the viscosity Pearson number (representing the ratio of the characteristic process temperature difference $q''_w\, D_h/k_e$ to the characteristic temperature difference $1/\beta$ that can produce appreciable viscosity variations) and $\mathrm{Pn}_k = \alpha\, q''_w\, D_h/k_e$ is the thermal conductivity Pearson number (representing the ratio of the characteristic process temperature difference $q''_w\, D_h/k_e$ to the characteristic temperature difference $1/\alpha$ that can produce appreciable thermal conductivity variations).

It is worth noting that, since the density ρ and the specific heat c are constant, the local Reynolds number $\mathrm{Re} = \rho\, \overline{u}\, D_h/\mu$, the local Prandtl number $\mathrm{Pr} = \mu\, c/k$ and the local Péclet number $\mathrm{Pe} = \mathrm{Re}\,\mathrm{Pr} = \rho\, c\, \overline{u}\, D_h/k$ all vary with temperature because of the variations of μ, of the ratio μ/k and of k, respectively. Therefore we have $\mu/\mu_e = \mathrm{Re}_e/\mathrm{Re}$ and $k/k_e = \mathrm{Pe}_e/\mathrm{Pe}$, where Re_e and Pe_e are the Reynolds and Péclet numbers evaluated at t_e. Moreover, since the viscosity of liquids decreases with increasing temperature ($\beta > 0$), in the case of fluid heating we have $\mathrm{Pn}_\mu > 0$ and $\mathrm{Re}_e/\mathrm{Re} < 1$, while $\mathrm{Pn}_\mu = 0$ and $\mathrm{Re}_e/\mathrm{Re} = 1$ refer to constant viscosity fluids ($\beta = 0$). Instead, since the thermal conductivity can either increase ($\alpha > 0$) or decrease ($\alpha < 0$) with increasing temperature, we have $\mathrm{Pn}_k > 0$ and $\mathrm{Pe}_e/\mathrm{Pe} > 1$ in the first case and $\mathrm{Pn}_k < 0$ and $\mathrm{Pe}_e/\mathrm{Pe} < 1$ in the second one, while $\mathrm{Pn}_k = 0$ and $\mathrm{Pe}_e/\mathrm{Pe} = 1$ refer to constant thermal conductivity fluids ($\alpha = 0$).

As already pointed out, as the fluid temperature rises along the duct, the viscosity decreases while Re increases. Therefore, to ensure laminar flow conditions, the local Reynolds number Re_b evaluated at the bulk temperature is only allowed to reach the maximum value $(\mathrm{Re}_b)_{max} = 2,000$, corresponding to the maximum value x_{max} of the axial coordinate x, whereupon computations are stopped. Therefore, taking into account equation (3) and the appropriate heat balance for the duct, the following expression for the maximum value X^*_{max} of the dimensionless axial coordinate $X^* = x/(D_h\, \mathrm{Pe}_e)$ can be obtained [4, 5, 7]

$$X^*_{max} = \frac{1}{4\,\mathrm{Pn}_\mu} \ln\left[\frac{(\mathrm{Re}_b)_{max}}{\mathrm{Re}_e}\right] \tag{5}$$

Numerical results of interest for the present study concern the axial distributions of the peripherally averaged local Nusselt number $\mathrm{Nu} = \overline{h} D_h/k_e$, where $\overline{h}$ is the peripherally averaged local convective heat transfer coefficient defined as

$$\overline{h} = \frac{q''_w}{\overline{t}_w - t_b} = \frac{k_e}{D_h\,(\overline{T}_w - T_b)} \tag{6}$$

In equation (6), $\overline{t}_w$ and t_b are the peripherally averaged wall temperature and the mean bulk temperature, respectively, and $\overline{T}_w$ and T_b are their dimensionless forms. According to equation (6), the peripherally averaged local Nusselt number can be expressed as

$$\mathrm{Nu} = \frac{1}{\overline{T}_w - T_b} \tag{7}$$

Axial distributions of the longitudinally averaged Nusselt number, defined as [1]

$$\overline{\mathrm{Nu}} = \frac{1}{x}\int_0^x \mathrm{Nu}\,\mathrm{d}x = \frac{1}{X^*}\int_0^{X^*} \mathrm{Nu}\,\mathrm{d}X^* \tag{8}$$

can be obtained by means of an appropriate numerical integration rule.

3. Numerical procedure

If the effects of axial diffusion can be neglected ($\mathrm{Re}_e > 50$ and $\mathrm{Pe}_e > 50$) and there is no recirculation in the longitudinal direction, steady-state flow and heat transfer in straight ducts of constant cross-sections are governed by the continuity and the parabolized Navier-Stokes and energy equations [8, 9]. These equations are not reported here for lack of space, but they are reported elsewere together with the boundary conditions specified on the boundaries of the computational domain [5, 7].

A finite element procedure for the analysis of the forced convection of fluids with temperature dependent properties in the entrance region of straight ducts [6, 7, 10] is used to solve the model equations. The adopted procedure is based on a segregated approach which implies the sequential solution of the momentum and energy equations on a two-dimensional domain in the case of three-dimensional geometries and on a one-dimensional domain in axisymmetric problems, corresponding to the cross-section of the duct. A marching method is then used to move forward in the axial direction. The pressure-velocity coupling is dealt with using an improved projection algorithm already employed by one of the authors (C.N.) for the solution of the Navier-Stokes equations in their elliptic form [11]. The procedure has already been validated, with reference to both constant and temperature dependent property fluids, by comparing heat transfer and pressure drop results with existing literature data for simultaneously developing laminar flows in straight ducts [6, 10, 12, 13].

4. Numerical results

In this study, two cross-sectional geometries are considered, namely circular and square, chosen as representative of axisymmetric and three-dimensional duct geometries, respectively. The corresponding computational domains, defined taking into account existing symmetries, are one-dimensional and two-dimensional for the circular and the square cross-sections, respectively. Details of finite element domain and time discretisations are reported elsewere [4, 5].

The same value of the Reynolds number $\mathrm{Re}_e = 100$ is assumed in all the computations. Instead, the values $\mathrm{Pr}_e = 5, 20$ and 100 of the Prandtl number at t_e are selected to take into account the behaviours of different liquids. The corresponding values of the Péclet number at t_e are $\mathrm{Pe}_e = 500, 2000$ and 10 000. The values of the dynamic viscosity Pearson number $\mathrm{Pn}_\mu = 0, 1, 2$ and 4 are considered to account for reasonable viscosity temperature dependences. Thus, for the assumed Re_e, the maximum values of the dimensionless axial coordinate given by equation (5) are $X^*_{max} = 0.7489, 0.3745$ and 0.1872 for $\mathrm{Pn}_\mu = 1, 2$ and 4, respectively. For each nonzero value of Pn_μ eight values of Pn_k are selected (four positive and four negative), giving the corresponding values of the ratio $\mathrm{Pn}_\mu/\mathrm{Pn}_k = \beta/\alpha = \pm10, \pm20, \pm40$ and ±80. Thus, on the whole, the values $\mathrm{Pn}_k = 0, \pm0.0125, \pm0.025, \pm0.05, \pm0.1, \pm0.2$ and ±0.4 are considered.

The effects of temperature dependent properties (viscosity and thermal conductivity) on heat transfer can be illustrated by comparing the local Nusselt number $\mathrm{Nu}_{\mu k}$, obtained for given nonzero values of Pn_μ and Pn_k, with the corresponding local Nusselt number Nu_c, computed for simultaneously developing constant property flows ($\mathrm{Pn}_\mu = \mathrm{Pn}_k = 0$). Therefore, we can assume the value of the ratio $(\mathrm{Nu}_{\mu k} - \mathrm{Nu}_c)/\mathrm{Nu}_c = (\mathrm{Nu}_{\mu k}/\mathrm{Nu}_c) - 1$ as a measure of the effects of both temperature dependent viscosity and thermal conductivity. Axial distributions of the ratio $\mathrm{Nu}_{\mu k}/\mathrm{Nu}_c$ with reference to the dimensionless coordinate X^*, obtained by means of the numerical procedure described in Section 3 for the same cross-sectional geometries and values of dimensionless input parameters considered in this paper, are reported elsewere [4, 5]. The conclusion reached there is that both temperature dependent viscosity and thermal conductivity can have comparable effects on the Nusselt number, according to the values of Pn_μ and Pn_k. Moreover, it has been demonstrated that if the effects of temperature dependent viscosity and thermal conductivity are considered separately to compute Nu_μ (under the assumptions of temperature dependent viscosity and constant thermal conductivity, i.e., $\mathrm{Pn}_\mu > 0$ and $\mathrm{Pn}_k = 0$)

and Nu_k (under the assumptions of temperature dependent thermal conductivity and constant viscosity, i.e., $\mathrm{Pn}_k \neq 0$ and $\mathrm{Pn}_\mu = 0$), a superposition method is applicable in order to obtain approximate values of the Nusselt number $\mathrm{Nu}_{\mu k}$, according to the relation

$$\mathrm{Nu}_{\mu k} - \mathrm{Nu}_c \cong (\mathrm{Nu}_\mu - \mathrm{Nu}_c) + (\mathrm{Nu}_k - \mathrm{Nu}_c) \tag{9}$$

with an accuracy which can be considered satisfactory in most situations. With reference to the ratios $\mathrm{Nu}_{\mu k}/\mathrm{Nu}_c$, $\mathrm{Nu}_\mu/\mathrm{Nu}_c$ and $\mathrm{Nu}_k/\mathrm{Nu}_c$, equation (9) can be recast in the form [4, 5]

$$\frac{\mathrm{Nu}_{\mu k}}{\mathrm{Nu}_c} \cong \left(\frac{\mathrm{Nu}_{\mu k}}{\mathrm{Nu}_c}\right)' = \frac{\mathrm{Nu}_\mu}{\mathrm{Nu}_c} + \frac{\mathrm{Nu}_k}{\mathrm{Nu}_c} - 1 \tag{10}$$

where $(\mathrm{Nu}_{\mu k}/\mathrm{Nu}_c)'$ is the approximate value of $\mathrm{Nu}_{\mu k}/\mathrm{Nu}_c$ given by the superposition method. Therefore, the values of the differences $(\mathrm{Nu}_\mu/\mathrm{Nu}_c)-1$ and $(\mathrm{Nu}_k/\mathrm{Nu}_c)-1$ approximately measure the separate effects of temperature dependent viscosity and thermal conductivity, respectively.

In this work, previous numerical results concerning axial distributions of the ratios $\mathrm{Nu}_{\mu k}/\mathrm{Nu}_c$, $\mathrm{Nu}_\mu/\mathrm{Nu}_c$ and $\mathrm{Nu}_k/\mathrm{Nu}_c$ [5] have been used to obtain, by means of a suitable numerical integration rule according to equation (8), the corresponding distributions of the ratios $\overline{\mathrm{Nu}}_{\mu k}/\overline{\mathrm{Nu}}_c$, $\overline{\mathrm{Nu}}_\mu/\overline{\mathrm{Nu}}_c$ and $\overline{\mathrm{Nu}}_k/\overline{\mathrm{Nu}}_c$. Then, it has been verified that the following relation, obtained by combining equations (8) and (9), still holds true with an accuracy which can be considered satisfactory in most situations

$$\frac{\overline{\mathrm{Nu}}_{\mu k}}{\overline{\mathrm{Nu}}_c} \cong \left(\frac{\overline{\mathrm{Nu}}_{\mu k}}{\overline{\mathrm{Nu}}_c}\right)' = \frac{\overline{\mathrm{Nu}}_\mu}{\overline{\mathrm{Nu}}_c} + \frac{\overline{\mathrm{Nu}}_k}{\overline{\mathrm{Nu}}_c} - 1 \tag{11}$$

In equation (11), $(\overline{\mathrm{Nu}}_{\mu k}/\overline{\mathrm{Nu}}_c)'$ represents the approximate value of $\overline{\mathrm{Nu}}_{\mu k}/\overline{\mathrm{Nu}}_c$ given by the superposition method. Axial distributions of the ratio $(\overline{\mathrm{Nu}}_{\mu k}/\overline{\mathrm{Nu}}_c)'$ for the highest values of Pn_μ, i.e., $\mathrm{Pn}_\mu = 2$, and $\mathrm{Pn}_\mu = 4$, different values of Pn_k and $\mathrm{Pr}_e = 20$ are reported in figures 1 and 2 for circular tubes and square ducts, respectively, to allow the comparison with the corresponding distributions of $\overline{\mathrm{Nu}}_{\mu k}/\overline{\mathrm{Nu}}_c$. As can be seen, even for the highest values of Pn_μ and the highest/lowest values of Pn_k the dashed curves, representing axial distributions of the ratio $(\overline{\mathrm{Nu}}_{\mu k}/\overline{\mathrm{Nu}}_c)'$, are very close to the solid ones giving the numerical solutions for $\overline{\mathrm{Nu}}_{\mu k}/\overline{\mathrm{Nu}}_c$,

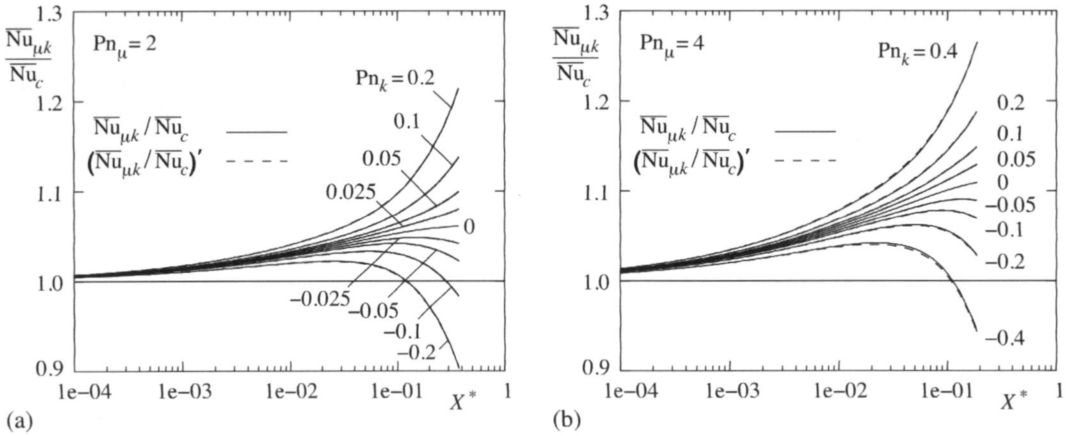

Figure 1. Comparison of axial distributions of the ratios $\overline{\mathrm{Nu}}_{\mu,k}/\overline{\mathrm{Nu}}_c$ and $\left(\overline{\mathrm{Nu}}_{\mu,k}/\overline{\mathrm{Nu}}_c\right)'$ for simultaneously developing laminar flows in circular tubes with $\mathrm{Pr}_e = 20$, different values of Pn_k and: (a) $\mathrm{Pn}_\mu = 2$ and (b) $\mathrm{Pn}_\mu = 4$.

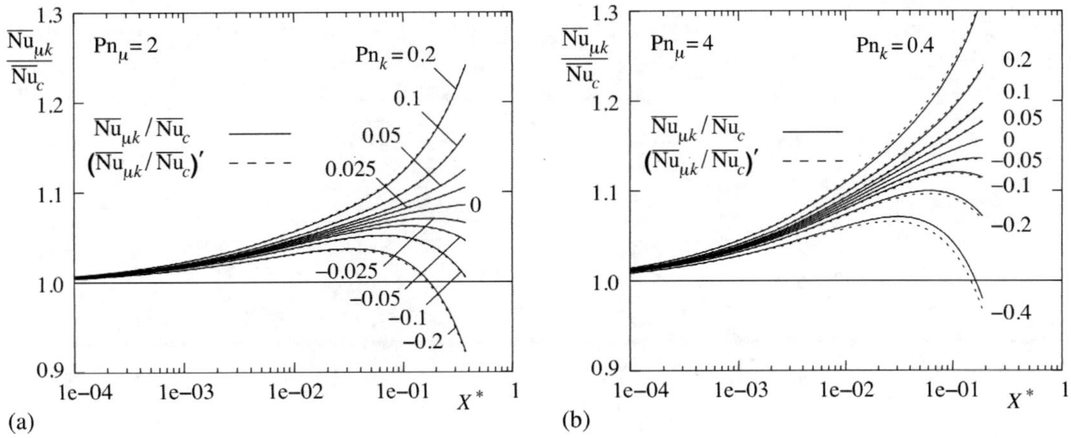

Figure 2. Comparison of axial distributions of the ratios $\overline{\mathrm{Nu}}_{\mu,k}/\overline{\mathrm{Nu}}_c$ and $\left(\overline{\mathrm{Nu}}_{\mu,k}/\overline{\mathrm{Nu}}_c\right)'$ for simultaneously developing laminar flows in square ducts with $\mathrm{Pr}_e = 20$, different values of Pn_k and: (a) $\mathrm{Pn}_\mu = 2$ and (b) $\mathrm{Pn}_\mu = 4$.

thus confirming the validity of equation (11). Moreover, the maximum positive and negative relative errors in the approximation of $\overline{\mathrm{Nu}}_{\mu k}/\overline{\mathrm{Nu}}_c$ by means of $(\overline{\mathrm{Nu}}_{\mu k}/\overline{\mathrm{Nu}}_c)'$ are $\varepsilon^+_{max} = 0.15\%$ and $\varepsilon^-_{max} = -0.28\%$ for circular ducts and $\varepsilon^+_{max} = 0.42\%$ and $\varepsilon^-_{max} = -1.39\%$ for square ducts in the range $10^{-4} \leq X^* \leq X^*_{max}$ with $\mathrm{Pr}_e = 5$, 20 and 100 and different values of Pn_μ and Pn_k. Therefore, we can conclude that the agreement between computed and approximate results is very good.

5. Correlations for the mean Nusselt number

Taking advantage of equation (11), a correlation for $\overline{\mathrm{Nu}}_{\mu k}$ has been obtained by assembling the different correlations for $\overline{\mathrm{Nu}}_c$, $\overline{\mathrm{Nu}}_\mu/\overline{\mathrm{Nu}}_c$ and $\overline{\mathrm{Nu}}_k/\overline{\mathrm{Nu}}_c$ obtained from computed results for flows of liquids with constant property ($\mathrm{Pn}_k = \mathrm{Pn}_\mu = 0$), temperature dependent viscosity ($\mathrm{Pn}_k = 0$ and $\mathrm{Pn}_\mu > 0$) and temperature dependent thermal conductivity ($\mathrm{Pn}_\mu = 0$ and $\mathrm{Pn}_k \neq 0$), respectively.

5.1. Correlations for the mean Nusselt number for flows of liquids with constant properties

The results obtained for circular and square cross-sectional geometries under the assumption of constant property flow ($\mathrm{Pn}_k = \mathrm{Pn}_\mu = 0$) have been used to obtain correlations for the mean Nusselt number $\overline{\mathrm{Nu}}_c$ as a function of X^* in the form

$$\overline{\mathrm{Nu}}_c = (\mathrm{Nu}_c)_{fd} + \frac{a\,(X^*)^{-m}}{1 + b\,(X^*)^{-n}} \tag{12}$$

where $(\mathrm{Nu}_c)_{fd}$, i.e., the asymptotic value of Nu_c for fully developed conditions, the coefficients a and b and the exponent m only depend on the cross-sectional geometry, while the exponent n also depends on the Prandtl number Pr_e according to the relation

$$n = n_1\,\mathrm{Pr}_e^s - n_2\,\mathrm{Pr}_e \tag{13}$$

whose coefficients n_1 and n_2 and the exponent s only depend on the cross-sectional geometry. The appropriate forms of equations (12) and (13) for circular and square cross-sections, valid

Table 1. Correlations for $\overline{\mathrm{Nu}}_c$ corresponding to equations (12) and (13) for circular and square cross-sections.

Cross-section	Correlations for $\overline{\mathrm{Nu}}_c$
circular	$\overline{\mathrm{Nu}}_c = 4.3636 + \dfrac{0.065\,(X^*)^{-1.3}}{1 + 0.10\,(X^*)^{-n}}$ $n = 0.761\,\mathrm{Pr}_e^{0.0224} - 0.000109\,\mathrm{Pr}_e$
square	$\overline{\mathrm{Nu}}_c = 3.0874 + \dfrac{0.065\,(X^*)^{-1.3}}{1 + 0.11\,(X^*)^{-n}}$ $n = 0.755\,\mathrm{Pr}_e^{0.0255} - 0.000136\,\mathrm{Pr}_e$

for $5 \le \mathrm{Pr}_e \le 100$ and $10^{-4} \le X^* \le X^*_{max}$, are reported in table 1. Since, for $\mathrm{Pr}_e = 5$, 20 and 100, the maximum positive and negative relative errors in the approximation of $\overline{\mathrm{Nu}}_c$ by means of equation (12) are $\varepsilon^+_{max} = 0.80\%$ and $\varepsilon^-_{max} = -1.29\%$ for circular ducts and $\varepsilon^+_{max} = 2.20\%$ and $\varepsilon^-_{max} = -1.31\%$ for square ducts, we can conclude that the agreement between computed and approximate results is very good.

5.2. Correlations for the mean Nusselt number for flows of liquids with temperature dependent viscosity

The results obtained for circular and square cross-sectional geometries for $\mathrm{Pn}_k = 0$ and $\mathrm{Pn}_\mu > 0$ have been used to obtain correlations for the ratio $\overline{\mathrm{Nu}}_\mu/\overline{\mathrm{Nu}}_c$ as a function of X^* in the form

$$\frac{\overline{\mathrm{Nu}}_\mu}{\overline{\mathrm{Nu}}_c} = 1 + \left[\left(\frac{\mathrm{Nu}_\mu}{\mathrm{Nu}_c}\right)_{fd} - 1\right]\{1 - \exp\left[-a\,(X^*)^m - b\,(X^*)^n\right]\} \tag{14}$$

In equation (14), $(\mathrm{Nu}_\mu/\mathrm{Nu}_c)_{fd}$ is the asymptotic value for fully developed conditions of the ratio $\mathrm{Nu}_\mu/\mathrm{Nu}_c$, which can be evaluated by means of the very accurate formula

$$\left(\frac{\mathrm{Nu}_\mu}{\mathrm{Nu}_c}\right)_{fd} = 1 + d_1\,\mathrm{Pn}_\mu + d_2\,\mathrm{Pn}_\mu^2 \tag{15}$$

whose coefficients d_1 and d_2 depend on the cross-sectional geometry. Besides, still with reference to equation (14), the coefficient b and the exponent n only depend on the cross-sectional geometry, while the coefficient a and the exponent m also depend on the viscosity Pearson number Pn_μ and on the Prandtl number Pr_e, respectively, and can be expressed as

$$a = a_1 + a_2\,\mathrm{Pn}_\mu \tag{16}$$

$$m = m_1\,\mathrm{Pr}_e^{-s} \tag{17}$$

where the coefficients a_1, a_2 and m_1 and the exponent s only depend on the cross-sectional geometry. For engineering application, the appropriate forms of equations (14) to (17) for circular and square cross-sections, valid for $5 \le \mathrm{Pr}_e \le 100$, $1 \le \mathrm{Pn}_\mu \le 4$ and $10^{-4} \le X^* \le X^*_{max}$, are reported in table 2. For $\mathrm{Pr}_e = 5$, 20 and 100, the maximum positive and negative relative errors in the approximation of $\overline{\mathrm{Nu}}_\mu/\overline{\mathrm{Nu}}_c$ by means of equation (14) are $\varepsilon^+_{max} = 0.14\%$ and $\varepsilon^-_{max} = -0.21\%$ for circular ducts and $\varepsilon^+_{max} = 0.31\%$ and $\varepsilon^-_{max} = -0.28\%$ for square ducts. Therefore, we can conclude that the agreement between computed and approximate results is very good.

Journal of Physics: Conference Series **547** (2014) 012041 doi:10.1088/1742-6596/547/1/012041

Table 2. Correlations for $\overline{\mathrm{Nu}}_\mu/\overline{\mathrm{Nu}}_c$ corresponding to equations (14) to (17) for circular and square cross-sections.

Cross-section	Correlations for $\overline{\mathrm{Nu}}_\mu/\overline{\mathrm{Nu}}_c$
circular	$\frac{\overline{\mathrm{Nu}}_\mu}{\overline{\mathrm{Nu}}_c} = 1 + \left[\left(\frac{\mathrm{Nu}_\mu}{\mathrm{Nu}_c}\right)_{fd} - 1\right]\left\{1 - \exp\left[-a\,(X^*)^m - 2.0\,(X^*)^{0.6}\right]\right\}$ $\left(\frac{\mathrm{Nu}_\mu}{\mathrm{Nu}_c}\right)_{fd} = 1 + 3.39\cdot 10^{-2}\,\mathrm{Pn}_\mu - 9.2\cdot 10^{-4}\,\mathrm{Pn}_\mu^2$ $a = 2.62 + 0.084\,\mathrm{Pn}_\mu$ $m = 0.48\,\mathrm{Pr}_e^{-0.076}$
square	$\frac{\overline{\mathrm{Nu}}_\mu}{\overline{\mathrm{Nu}}_c} = 1 + \left[\left(\frac{\mathrm{Nu}_\mu}{\mathrm{Nu}_c}\right)_{fd} - 1\right]\left\{1 - \exp\left[-a\,(X^*)^m + 3.0\,(X^*)^{0.8}\right]\right\}$ $\left(\frac{\mathrm{Nu}_\mu}{\mathrm{Nu}_c}\right)_{fd} = 1 + 4.74\cdot 10^{-2}\,\mathrm{Pn}_\mu - 1.0\cdot 10^{-3}\,\mathrm{Pn}_\mu^2$ $a = 6.79 + 0.044\,\mathrm{Pn}_\mu$ $m = 0.57\,\mathrm{Pr}_e^{-0.050}$

5.3. Correlations for the mean Nusselt number for flows of liquids with temperature dependent thermal conductivity

The results obtained for the considered cross-sectional geometries for $\mathrm{Pn}_\mu = 0$ and $\mathrm{Pn}_k \neq 0$ have been used to obtain correlations for the ratio $\overline{\mathrm{Nu}}_k/\overline{\mathrm{Nu}}_c$ as a function of X^* in the form

$$\frac{\overline{\mathrm{Nu}}_k}{\overline{\mathrm{Nu}}_c} = 1 + a\,\mathrm{Pn}_k + \left(b_1\,\mathrm{Pn}_k + b_2\,\mathrm{Pn}_k^2\right) X^* \tag{18}$$

where the coefficient a can be expressed as

$$a = a_1\,\mathrm{Pr}_e^d\left\{1 - \exp\left[-a_2\,(X^*)^n\right]\right\} \tag{19}$$

In equations (18) and (19) the coefficients a_1, a_2, b_1 and b_2 and the exponent d only depend on the cross-sectional geometry and on the sign of Pn_k, while the exponent n also depends on the Prandtl number Pr_e according to the relation

$$n = n_1\,\mathrm{Pr}_e^{-s} \tag{20}$$

where the coefficient n_1 and the exponent s only depend on the cross-sectional geometry. For engineering application, the appropriate forms of equations (18) to (20) for circular and square cross-sections, valid for $5 \le \mathrm{Pr}_e \le 100$, $0.0125 \le |\mathrm{Pn}_k| \le 0.4$ and $10^{-4} \le X^* \le X^*_{max}$, are reported in table 3. For $\mathrm{Pr}_e = 5$, 20 and 100, the maximum positive and negative relative errors in the approximation of $\overline{\mathrm{Nu}}_k/\overline{\mathrm{Nu}}_c$ by means of equation (18) are $\varepsilon^+_{max} = 0.53\%$ and $\varepsilon^-_{max} = -0.21\%$ for circular ducts and $\varepsilon^+_{max} = 0.30\%$ and $\varepsilon^-_{max} = -0.17\%$ for square ducts. Therefore, we can conclude that the agreement between computed and approximate results is very good.

Table 3. Correlations for $\overline{\mathrm{Nu}}_k/\overline{\mathrm{Nu}}_c$ corresponding to equations (18) and (19) for different cross-sections and signs of Pn_k.

Cross-section	Pn_k	Correlations for $\overline{\mathrm{Nu}}_k/\overline{\mathrm{Nu}}_c$
circular	>0	$\frac{\overline{\mathrm{Nu}}_k}{\overline{\mathrm{Nu}}_c} = 1 + a\,\mathrm{Pn}_k + \left(1.97\,\mathrm{Pn}_k - 0.238\,\mathrm{Pn}_k^2\right) X^*$; $a = 0.032\,\mathrm{Pr}_e^{0.031}\,\{1 - \exp[-28\,(X^*)^n]\}$; $n = 0.60\,\mathrm{Pr}_e^{-0.024}$
	<0	$\frac{\overline{\mathrm{Nu}}_k}{\overline{\mathrm{Nu}}_c} = 1 + a\,\mathrm{Pn}_k + \left(1.90\,\mathrm{Pn}_k - 0.399\,\mathrm{Pn}_k^2\right) X^*$; $a = 0.033\,\mathrm{Pr}_e^{0.035}\,\{1 - \exp[-28\,(X^*)^n]\}$; $n = 0.60\,\mathrm{Pr}_e^{-0.024}$
square	>0	$\frac{\overline{\mathrm{Nu}}_k}{\overline{\mathrm{Nu}}_c} = 1 + a\,\mathrm{Pn}_k + \left(1.98\,\mathrm{Pn}_k - 0.295\,\mathrm{Pn}_k^2\right) X^*$; $a = 0.052\,\mathrm{Pr}_e^{0.024}\,\{1 - \exp[-18\,(X^*)^n]\}$; $n = 0.60\,\mathrm{Pr}_e^{-0.037}$
	<0	$\frac{\overline{\mathrm{Nu}}_k}{\overline{\mathrm{Nu}}_c} = 1 + a\,\mathrm{Pn}_k + \left(1.92\,\mathrm{Pn}_k - 0.716\,\mathrm{Pn}_k^2\right) X^*$; $a = 0.053\,\mathrm{Pr}_e^{0.034}\,\{1 - \exp[-18\,(X^*)^n]\}$; $n = 0.60\,\mathrm{Pr}_e^{-0.037}$

5.4. Estimation of the mean Nusselt number for flows of liquids with temperature dependent properties by means of the proposed correlations

The correlations proposed in the previous sections can be used to predict the value of $\overline{\mathrm{Nu}}_{\mu k}$. In fact, according to equation (11), we can write

$$\overline{\mathrm{Nu}}_{\mu k} = \overline{\mathrm{Nu}}_c\,\frac{\overline{\mathrm{Nu}}_{\mu k}}{\overline{\mathrm{Nu}}_c} \cong \overline{\mathrm{Nu}}_c\left(\frac{\overline{\mathrm{Nu}}_{\mu k}}{\overline{\mathrm{Nu}}_c}\right)' = \overline{\mathrm{Nu}}_c\left(\frac{\overline{\mathrm{Nu}}_\mu}{\overline{\mathrm{Nu}}_c} + \frac{\overline{\mathrm{Nu}}_k}{\overline{\mathrm{Nu}}_c} - 1\right) \tag{21}$$

where $\overline{\mathrm{Nu}}_c$, $\overline{\mathrm{Nu}}_\mu/\overline{\mathrm{Nu}}_c$ and $\overline{\mathrm{Nu}}_k/\overline{\mathrm{Nu}}_c$ are given by equations (12), (14) and (18), respectively. The maximum absolute values $|\varepsilon|_{max}$ of the relative error in the approximation of $\overline{\mathrm{Nu}}_{\mu k}$ by means of equation (21) for $\mathrm{Pr}_e = 5$, 20 and 100 and different values of Pn_μ and $\mathrm{Pn}_\mu/\mathrm{Pn}_k$ are reported in table 4 for circular and square cross-sectional geometries. As can be seen, the agreement between computed and approximate results is very good.

6. Conclusions

New correlations, suitable for engineering applications, for the mean Nusselt number in the entrance region of circular tubes and square ducts with uniform heat flux boundary conditions specified at the walls have been proposed. These correlations have been obtained on the basis of the results of a previous parametric investigation on the effects of temperature dependent viscosity and thermal conductivity in simultaneously developing laminar flows of liquids in straight ducts of constant cross-sections. In these studies, a finite element procedure has

Journal of Physics: Conference Series **547** (2014) 012041 doi:10.1088/1742-6596/547/1/012041

Table 4. Maximum absolute values $|\varepsilon|_{max}$ (%) of the relative error in the approximation of $\overline{\mathrm{Nu}}_{\mu k}$ by means of equation (21) for simultaneously developing flows in circular and square ducts with different viscosity and thermal conductivity Pearson numbers.

Cross-section	Pn_μ	$\mathrm{Pn}_\mu/\mathrm{Pn}_k$							
		-10	-20	-40	-80	80	40	20	10
circular	1	1.26	1.26	1.26	1.26	1.25	1.26	1.25	1.25
	2	1.25	1.23	1.23	1.22	1.22	1.22	1.21	1.21
	4	1.29	1.23	1.20	1.19	1.17	1.16	1.15	1.11
square	1	1.67	1.67	1.67	1.67	1.67	1.66	1.66	1.66
	2	1.59	1.60	1.61	1.61	1.61	1.61	1.61	1.62
	4	1.45	1.40	1.43	1.44	1.46	1.47	1.57	1.76

been employed for the numerical solution of the parabolized momentum and energy equations. Viscosity and thermal conductivity have been assumed to vary with temperature according to an exponential and to a linear relation, respectively, while the other fluid properties are held constant. The temperature dependences of viscosity and thermal conductivity have been quantitatively expressed by the corresponding Pearson numbers. Axial distributions of the mean Nusselt number, obtained by numerical integration from those of the local Nusselt number, have been used as input data in the derivation of the proposed correlations. A superposition method has been proved to be applicable in order to estimate the Nusselt number by considering separately the effects of temperature dependent viscosity and thermal conductivity. Therefore, for each of the considered cross-sectional geometries, two distinct correlations have been proposed for flows of liquids with temperature dependent viscosity and with temperature dependent thermal conductivity, in addition to that obtained for constant property flows.

Acknowledgments

This work was funded by MIUR (PRIN/COFIN 2009 project).

References

[1] Shah R K and London A L 1978 (New York: Academic Press)
[2] Herwig H 1985 *Int. J. Heat Mass Transfer* **28** 423
[3] Kakaç S 1987 *Handbook of Single-Phase Convective Heat Transfer* ed S Kakaç, R K Shah and W Aung (New York: Wiley) chapter 18
[4] Del Giudice S, Savino S and Nonino C 2010 *Proc. 28th UIT Heat Transfer Congress (Brescia, Italy)* p 231
[5] Del Giudice S, Savino S and Nonino C 2012 Entrance and temperature dependent property effects in laminar duct flows of liquids (to appear)
[6] Nonino C, Del Giudice S and Comini G 1988 *Numer. Heat Transfer* **13** 451
[7] Del Giudice S, Savino S and Nonino C 2011 *J. Heat Transfer, Trans. ASME* **133** 101702
[8] Patankar S V and Spalding D B 1972 *Int. J. Heat Mass Transfer* **15** 1787
[9] Hirsh C 1988 vol 1 (New York: Wiley) p 70
[10] Nonino C, Del Giudice S and Savino S 2006 *Int. J. Heat Mass Transfer* **49** 4469
[11] Nonino C 2003 *Numer. Heat Transfer, Part B* **44** 61
[12] Del Giudice S, Nonino C and Savino S 2007 *Int. J. Heat and Fluid Flow* **28** 15
[13] Nonino C, Del Giudice S and Savino S 2007 *J. Heat Transfer, Trans. ASME* **129** 1187

Journal of Physics: Conference Series **547** (2014) 012042 doi:10.1088/1742-6596/547/1/012042

Energy and water vapor transport across a simplified cloud-clear air interface

L. Gallana, S. Di Savino, F. De Santi, M. Iovieno, D. Tordella

Dipartimento di Ingegneria Meccanica e Aerospaziale, Politecnico di Torino, Corso Duca degli Abruzzi 24, 10129 Torino, Italy

E-mail: `luca.gallana@polito.it`, `daniela.tordella@polito.it`

Abstract. We consider a simplified physics of the could interface where condensation, evaporation and radiation are neglected and momentum, thermal energy and water vapor transport is represented in terms of the Boussinesq model coupled to a passive scalar transport equation for the vapor. The interface is modeled as a layer separating two isotropic turbulent regions with different kinetic energy and vapor concentration. In particular, we focus on the small scale part of the inertial range of the atmospheric boundary layer as well as on the dissipative range of scales which are important to the micro-physics of warm clouds. We have numerically investigated stably stratified interfaces by locally perturbing at an initial instant the standard temperature lapse rate at the cloud interface and then observing the temporal evolution of the system. When the buoyancy term becomes of the same order of the inertial one, we observe a spatial redistribution of the kinetic energy which produce a concomitant pit of kinetic energy within the mixing layer. In this situation, the mixing layer contains two interfacial regions with opposite kinetic energy gradient, which in turn produces two intermittent sublayers in the velocity fluctuations field. This changes the structure of the field with respect to the corresponding non-stratified shearless mixing: the communication between the two turbulent region is weak, and the growth of the mixing layer stops. These results are discussed with respect to Large Eddy Simulations data for the Planetary Boundary Layers.

1. Introduction

Warm clouds as stratocumuli swathe a significant part of earth's surface and play a major role in the global dynamics of atmosphere by strongly reflecting incoming solar radiation – thus contributing to the Earth's albedo – so that an accurate representation of their dynamics is important in large-scale analyses of atmoshperic flows [1]. They are controlled by the tight interplay between radiative driving, turbulence, surface fluxes, latent heat release, and entrainment. Among them, the mixing and entrainment processes at the cloud top have been identified as fundamental to determine the internal structure of warm clouds, so that a clear and complete understanding of their physics is required [2]. As pointed out by Malinowsky *et al.* [3], data from most field campaigns and large-eddy simulations are too poorly resolved to allow to infer the details of the interfacial layer, even if they indicate that, in order to allow for entrainment, a high level of turbulence must be present. For this reason, in this work we study the local transport through a clear air/cloud interface through DNS (Direct Numerical Simulation). As our focus is on the dynamics of the smallest scales of the flow which influence the microphysics of warm clouds, we have simulated an idealized configuration to understand, under controlled conditions, some of the basic phenomena which occur at the cloud interface

Journal of Physics: Conference Series **547** (2014) 012042 doi:10.1088/1742-6596/547/1/012042

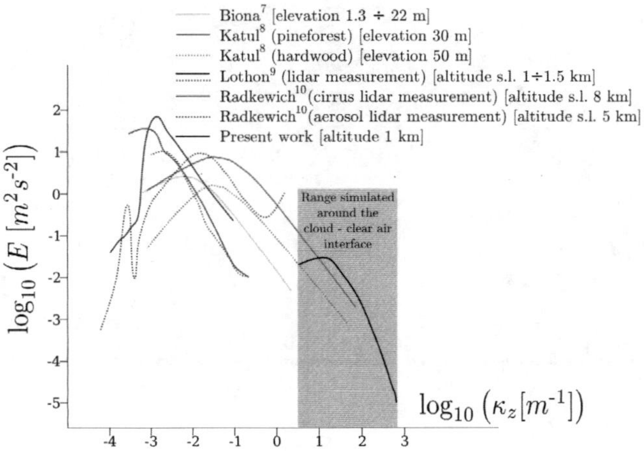

Figure 1. Kinetic energy spectra. Present study: black solid line, representing the small inertial and dissipative range. Colored lines: spectra from in-situ atmospheric observations [7, 8, 9, 10] where the kinetic energy injection and low wave-number inertial range are measured.

over length scales of the order of few meters. In these conditions, we solve scales from few meters down to few millimeters, that is, we resolve only the small scale part of the inertial range and the dissipative range of the power spectrum in a small portion ($6\ m \times 6\ m \times 12\ m$) of the atmosphere across the clou - clear air interface. This allows us to investigate the dynamics of entrainment which occurs in a thin layer at the cloud top, which a smaller scale with respect ot the scale explicitly resolved in large-eddy simulations of clouds [4]. In this preliminary work, we focus on two concomitant aspects of the cloud top mixing layer: the effect of the presence of a stratification and of a turbulent kinetic energy gradient. We do not consider the wind shear neither the phenomena linked to the processes of evaporation and condensation and radiative cooling which are important in conditions of buoyancy reversal [5, 6]. Therefore, our simulations were performed by applying the Boussinesq approximation to the Navier-Stokes momentum and energy equations together with a passive scalar transport equation which models the water vapor transport.

The details of the physical problem we have considered and of the governing equations are given in section 2. Section 3 contains a selection of our main results about intermittency, energy redistribution and entrainment. Conclusion remarks are in Section 4.

2. The physical problem

We consider the interface between clear air and cloud in a $6\ m \times 6\ m \times 12\ m$ parallelepipedic domain. Compared to in situ measurement of the atmospheric energy spectra, as shown in figure 1, we are able to simulate the small scales, that is, the highest wavenumbers in the inertial range and the dissipative range. As shown in figure 2, the system is composed by two homogeneous and isotropic turbulent regions that interact through a mixing layer, whose initial thickness is nearly equal to the integral scale of the present simulations ℓ, here assumed equal to $3 \cdot 10^{-1}$ m. It should be noted that the integral scale in our simulations turns out to be nearly equal to the Taylor microscale of the turbulence in the Planetary Boundary Layer (PBL) where the integral scale ranges from 10 to 10^3 m, as can be seen from figure 1 and [7, 8, 9, 10]..

The two isotropic regions that we simulate have a different kinetic energy and we assumed that the kinetic energy is higher in the cloud than in the external region. The root mean square of the velocity inside the cloud is $u_{rms} = 0.2$ m/s, and the energy ratio between the cloud energy E_1 and the external region energy E_2 is equal to 6.7. This energy ratio is of the same order of the ones measured in warm clouds (see, e.g., [3]) and, furthermore, it allows us to compare our results with experiments on shearless mixing (see [11, 12]) in absence of any stratification. In our simulations the Taylor microscale Reynolds number Re_λ of the higher energy region is

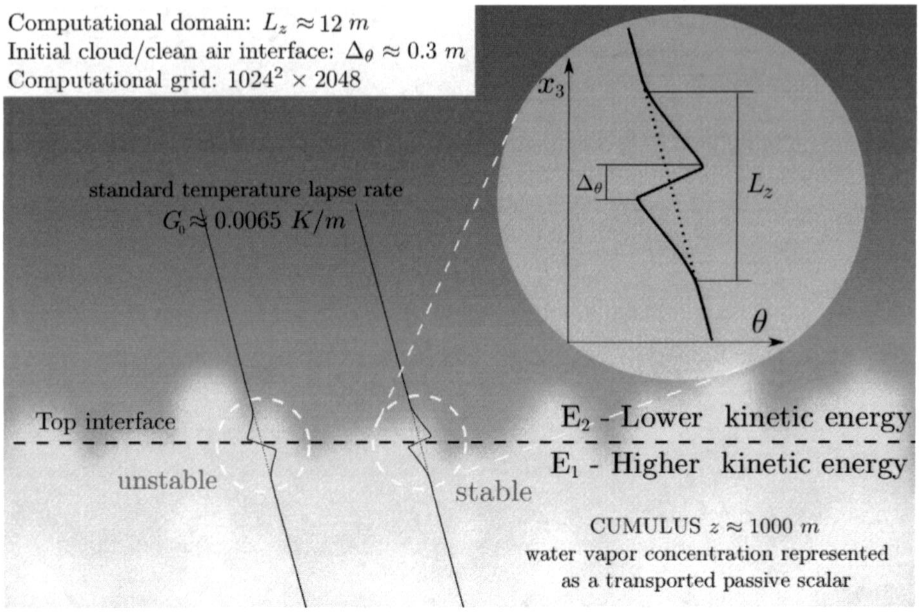

Figure 2. Scheme of the initial conditions. E_1 is the mean initial turbulent kinetic energy of the bottom region (inside the cloud), E_2 of the upper region (outside the cloud). For the top cloud mixing here presented we consider $E_1/E_2 = 6.7$. The water vapor is initially present only inside the cloud (bottom region). The zoom in the yellow circle is an example of initial temperature perturbation θ' of the standard boundary layer lapse rate.

equal to 250. Buoyancy is taken into account through a local perturbation θ' in the profile of the temperature distribution θ inside the troposphere, and the water vapor concentration χ is considered as a passive scalar. The Prandtl (Pr= 0.74) and the Schmidt number ($Sc = 0.61$) considered refers to an altitude of 1000 m s.l.

The initial conditions for the temperature perturbation is described in figure 2 and in table 1. The ratio between inertial and buoyancy forces is expressed by the Froude number Fr, based on the maximum gradient within the initial interface, which in our simulations ranges from 31.2 (negligible stratification) to 0.62 (strong stratification).

We use the continuity, momentum and energy balance equations within the Boussinesq approximation, which holds for small temperature variations [13]:

$$\nabla \cdot \mathbf{u}' = 0 \tag{1}$$

$$\frac{\partial \mathbf{u}'}{\partial t} + (\mathbf{u}' \cdot \nabla)\, \mathbf{u}' = -\nabla \frac{\tilde{p}}{\rho} + \nu \nabla^2 \mathbf{u}' + \alpha \mathbf{g} \theta' \tag{2}$$

$$\frac{\partial \theta'}{\partial t} + \mathbf{u}' \cdot \nabla \theta' + u_3 G = \kappa \nabla^2 \theta' \tag{3}$$

$$\frac{\partial \chi}{\partial t} + \mathbf{u}' \cdot \nabla \chi = d_\chi \nabla^2 \chi, \tag{4}$$

where the temperature θ is composed as the sum of a fluctuation $\theta'(\mathbf{x}, t)$, a static component $\tilde{\theta}(x_3) = Gx_3$, and a reference constant temperature θ_0, $\tilde{p} = p + \alpha g x_3 (\theta_0 + Gx_3/2)$ is the total hydrodynamic pressure (where p is the fluid-dynamic pressure, α the thermal expansion coefficient, g the gravity acceleration), u' is the velocity fluctuation and χ is the vapor

Table 1. Initial stratification level parameters. G is the maximum gradient of θ, expressed in terms of the standard troposphere lapse rate $G_0 = 0.0065\ Km^{-1}$; $N_{ci} = \sqrt{\alpha\theta_0 g \frac{d\theta}{dx_3}}$ is the characteristic Brunt-Väisälä frequency of initial condition. The Froude number Fr$= \frac{u'_{rms}}{N_{ic}\ell}$ and the Reynolds Buoyancy number $\text{Re}_b = \frac{\varepsilon N^2}{\nu}$ give a measure of the order of magnitude of the buoyancy forces compared with the inertial terms (here ε is the initial energy dissipation rate and ν is the kinematic viscosity).

G	ΔT $[K]$	N_{ic} $[s^{-1}]$	Fr	Re_b
$2G_0$	4.0e-3	2.13e-2	31.2	7
$30G_0$	6.0e-2	5.24e-2	12.7	112
$100G_0$	2.0e-1	1.50e-1	4.4	273
$500G_0$	1.0e0	3.35e-1	1.8	833
$5000G_0$	1.0e1	1.06e0	0.62	2635

concentration of the air - water vapor mixture present in the cloud, here considered as a passive scalar. The constant κ and d_χ are respectively the thermal and water vapor diffusivity.

The simulations are performed using our home produced computational code that implements a pseudospectral Fourier-Galerkin spatial discretization and an explicit low storage fourth order Runge-Kutta time integration scheme. Evaluation of non-linear (advective) terms is performed through the 3/2 de-aliased method [14]. The initial conditions for the velocity field are obtained by a linear matching of two different isotropic homogeneous turbulent fields (that are randomly generated, respecting physical conditions imposing spectra, solenoidality, integral scale and kinetic energy)[15]. The grid has 1024×1024×2048 points, and allows to capture all the turbulent scales from the greatest (integral scale ℓ) to the smaller (Kolmogorov scale η). The computational code uses a distributed memory paradigm through the MPI libraries: the simulation were performed at the TGCC Curie supercomputer within the PRACE project n° RA2011050773 for a total of 3 million cpu-hours [16].

3. Results

According to the ratio between buoyancy force and kinematic forces (that are advection and diffusion), the evolution of the system can be split in two main stages. As long as the ratio remains small, there are no significant differences with respect to a non-stratified case. On the contrary, as the stratification perturbation level become higher, buoyancy effects are no more negligible: differences are present from both a quantitative and qualitative point of view. These considerations can be observed through the statistical analysis of the simulated fields. The statistics are computed by averaging the variables in the planes normal to the mixing direction (with a sample of 1024 × 1024 data-points), focusing on the variation along the vertical (non-homogeneous) direction. The effects of the different stratification levels are clearly visible on the second order moment of velocity, temperature, and vapor as shown in figure 3 (a - b - c). When the stratification level is mild ($Fr > 4$) there are no relevant differences with respect the neutrally buoyant flow, while significant differences appears for intense stratification ($Fr < 2$). In particular, in correspondence of the local temperature perturbation, the formation of a layer with a pit of kinetic energy can be observed. The presence of such a layer deeply changes the physics of the system, because in this situation two interfaces are produced. The first interface, (which is present also in the absence of stratification), now separates the high turbulent energy region from the pit, while the second one (not present without stratification) separates the low

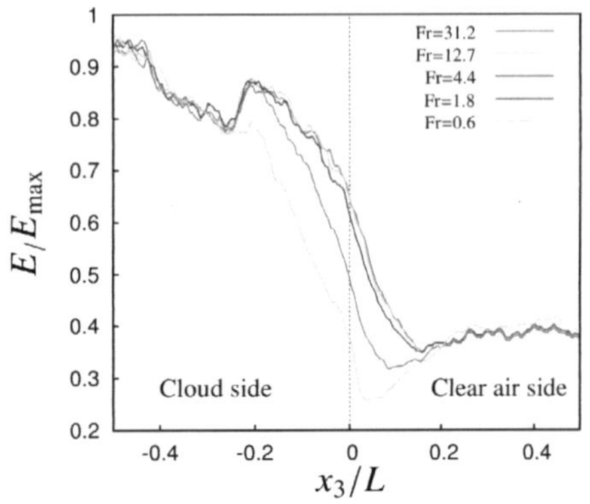

(a)

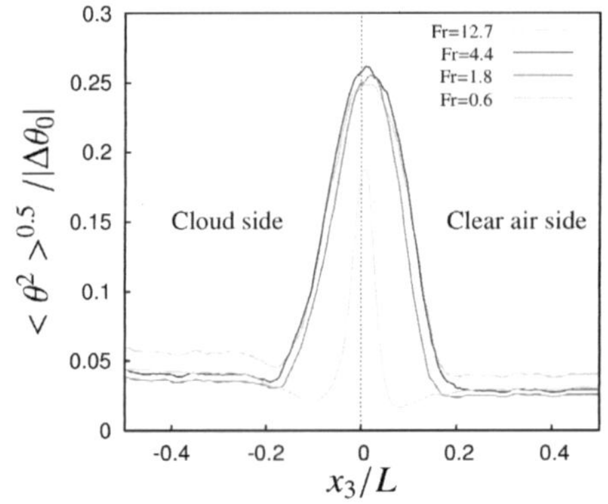

(b)

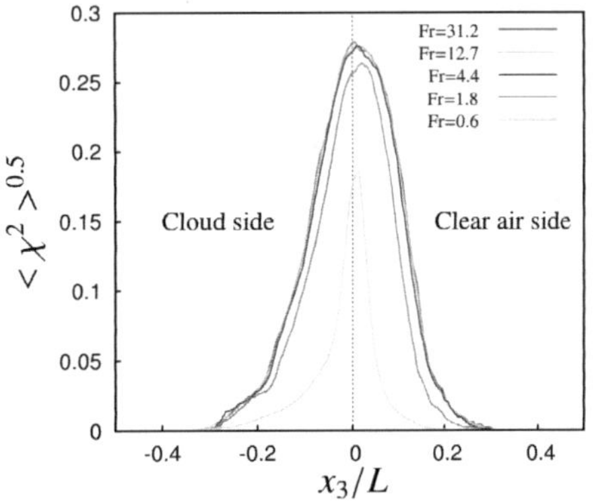

(c)

Figure 3. Spatial distribution across the mixing (i.e. along the coordinate x_3) of the kinetic energy (a), temperature variance (b) and water vapor concentration variance (c) at $t/\tau \approx 6$ for different levels of stratification.

turbulent energy region from the center of the mixing layer. Therefore, a strong stratification induces a physical separation between the two external regions, greatly decreasing the interaction between them. Both interfaces present an intermittent behavior, as shown in figure 4 (a - b) by skewness and kurtosis distribution (respectively third and fourth moments normalized with the local variance). In fact, two peaks of skewness and kurtosis can be observed in the highly stratified case: one is placed inside the cloud, and the other is placed close to the position of the intermittency peak in case of absence of stratification, see data at $Fr = 0.6$ in figure 4 (a). Observing the magnitude of the kurtosis maximum in figure 4 (b), it can be noted that the peak inside the cloud reaches values as high as 4, that are about the 10% larger than when the stratification is milder.

The behavior of the temperature and the water vapor fields appear qualitatively analogue as far as the shape of the statistical distributions are concerned, but are quantitatively different. See, for instance, the temperature and the vapor concentration peaks in the skewness and kurtosis distribution, shown in figure 4 (c - d - e - f).

It can be also observed that higher levels of stratification produce a relevant reduction of intermittency in the flow, with a drop of about 70% in the peaks of skewness and kurtosis. The interaction between the two regions aside the interface is greatly reduced, so the fluctuation at the sides of the mixing layer are damped, preventing the formation of the intermittent layer typical of the passive scalar transport [17, 18]. This large reduction of the skewness and kurtosis values is coupled with a slight increase in the higher order moments of the vertical velocity

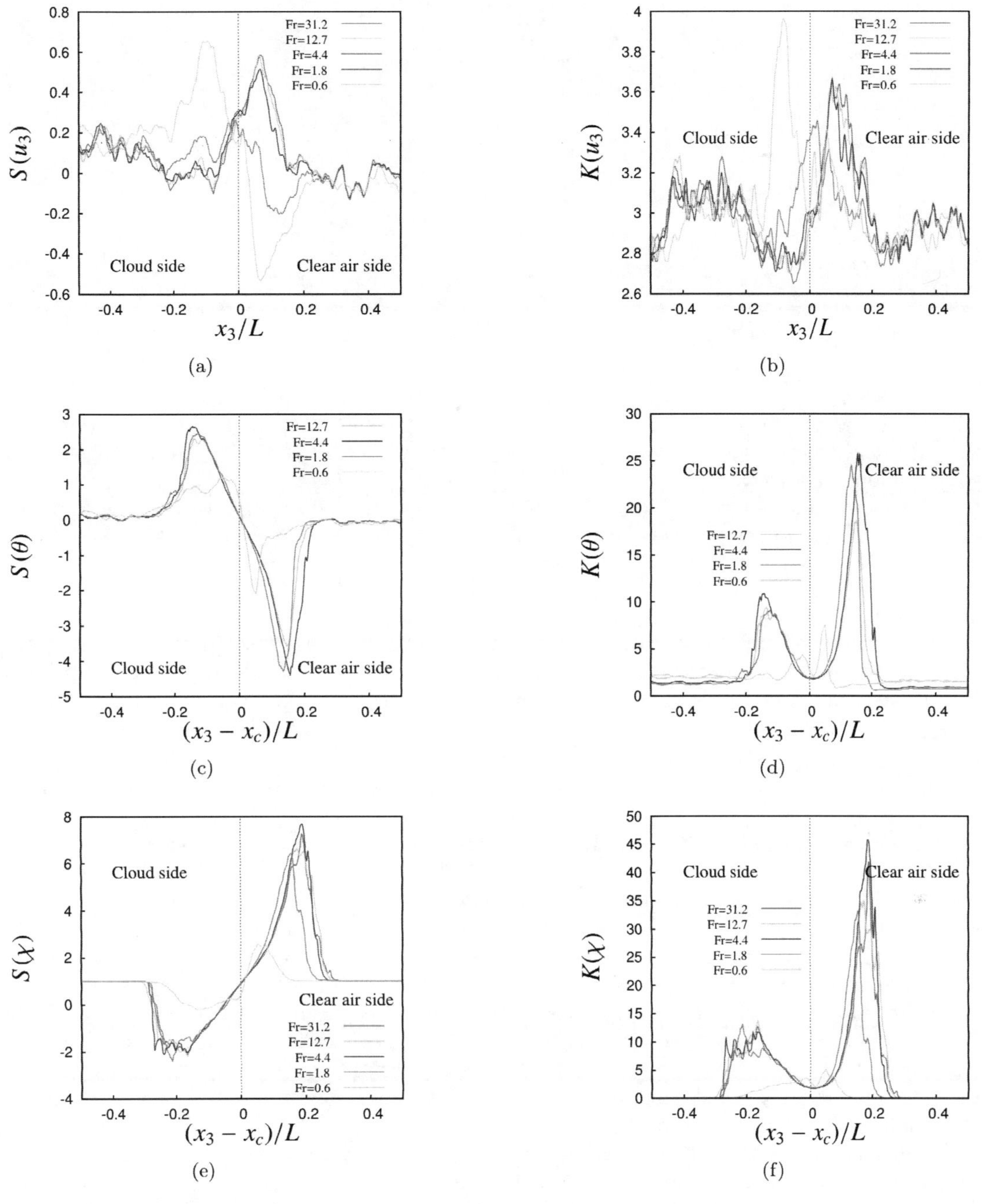

Figure 4. Spatial distribution across the mixing (i.e. along the direction x_3) of the skewness and the kurtosis of vertical component of the velocity (a - b), temperature (c - d) and water vapour concentration (e - f) at $t/\tau = 6$ for different levels of stratification.

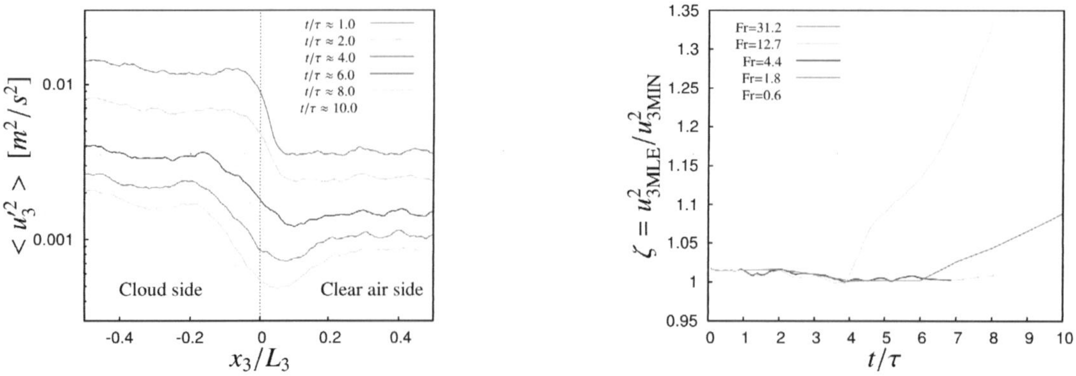

(a) (b)

Figure 5. (a) Temporal evolution of the spatial distribution along the vertical direction x_3 of the turbulent kinetic energy when Fr=1.8. A pit of energy appears after about 6 eddy turnover times in correspondence of the original interface.
(b) Temporal evolution of the ratio ζ between the mean vertical velocity variance in the lower energy (clear air side) region $u_{3,MLE}$ and the minimum value of the vertical velocity $u_{3,MIN}$. When this ratio departs from 1, a pit of kinetic velocity appears as shown in (a).

– for sufficiently strong stratification – which is in fair agreement with the trend observed in [19, p. 158]. Moreover, observing both figures 3 and 4, in case of stratification with $Fr < 2$, one can see that the thickness of the mixing layer is reduced. See th next section for more details.

3.1. The onset of a kinetic energy pit

As shown in figure 5 (a), in case of high stratification level, in the center of the domain – where the initial temperature step is placed – the onset of a layer with a kinetic energy lower than both the external regions can be seen. This layer can be considered as a pit of kinetic energy. Varying the stratification intensity, the genesis and the evolution of such pit can be measured by considering the ratio ζ between the variance of vertical velocity in the low energy region u^2_{3MLE} (mean value) and in the center of the pit u^2_{3MIN} (where the variance reaches its minimum); the temporal evolution of ζ is shown in figure 5 (b). Qualitatively similar results have been observed in the large-eddy simulations of stratocumulos-topped planetary boundary layer carried out by several physics of the atmosphere research group, as described in Moeng *et al.* 1996 [20, fig.7(c) and 8(a) at pag. 11]. In particular, in the case of sufficiently strong stratification ($Fr \approx 0.4$ and $Re_\lambda \approx 5500$), the trend of our vertical velocity variance which can be seen in figure 8(a), is analogue to those observed in the LES carried out by the NCAR group (Deardoff TKE model [21]) and the WVU (ARAP TKE model [22]).

A visualization of such phenomenology is represented in figure 6, where the vertical velocity fluctuations in a vertical slice of the domain are represented using an elevation plot (where such elevation is proportional to the square of u_3). In presence of a mild stratification, $Fr = 12$, even after 8 time scale, there is a smooth mixing layer between the high (left) and the low (rigth) energy regions. The differences in the case of strong stratification, with $Fr = 0.62$, after 4 time-scales (b) and 8 time scales (c), are clearly visible: a separation layer is present in the center of the domain, that becomes even more evident as the time pass by.

As said, the presence of the pit generates a physical separation between the two external turbulent regions and damps the mixing, thus reducing the exchange of information. As a

Journal of Physics: Conference Series **547** (2014) 012042 doi:10.1088/1742-6596/547/1/012042

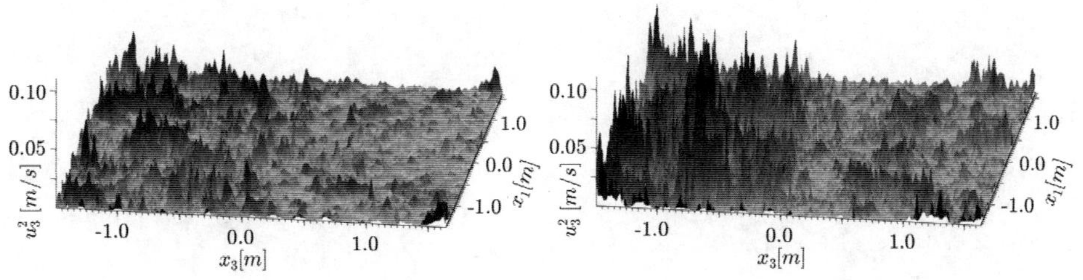

(a) (b)

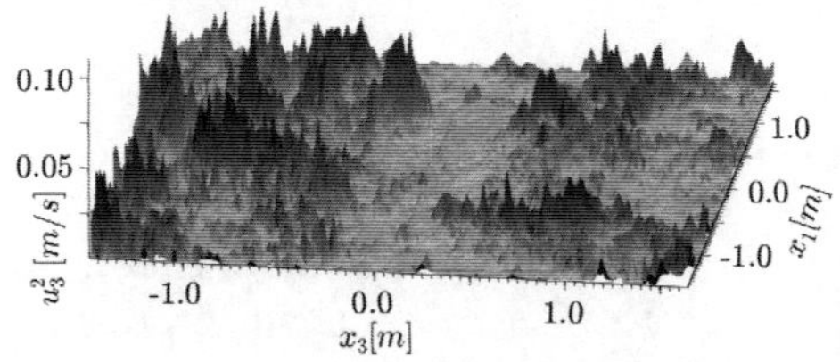

(c)

Figure 6. Visualization of the vertical component of the velocity in a vertical plane. The elevation is proportional to the square of vertical velocity fluctuations and the colors respect the velocity directions (blue for downward , red for upward) in various case: $Fr = 12$ after 8 eddy turnover times (a) and for Fr=0.62 after 4 and 8 eddy turnover times (panels b, c respectively). In the last two panels it can be observed the formation of the pit of kinetic energy.

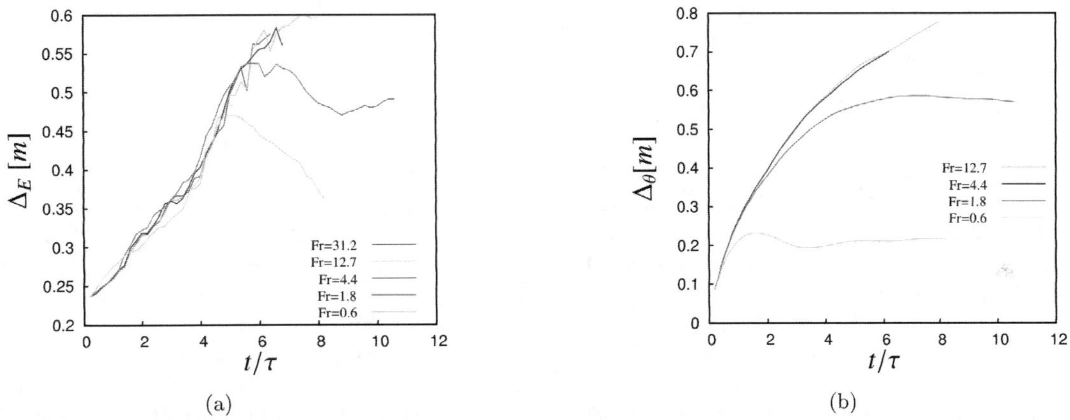

(a) (b)

Figure 7. Thickness of the mixing layer of velocity (a) and temperature (b). For both fields, the mixing layer thickness is the same as defined in [18].

consequence, there is a saturation of the thickening of the mixing layer Δ_E, see figure 7 (a).

Looking to the temperature mixing layer thickness Δ_θ, shown in figure 7 (b), it can be seen that, for strong stratification, the thickening stops approximately after the same amount of times scale required by Δ_E. In that case, contrary to what observed for the kinetic energy, the thickening does not stop suddenly, but rather with a transient that lasts a couple of time scales.

3.2. Entrainment

The entrainment of clear air inside the cloud is an important aspect of the top cloud interface as it concurs in the evaporation/condensation of droplets inside a cloud [1].

Journal of Physics: Conference Series **547** (2014) 012042 doi:10.1088/1742-6596/547/1/012042

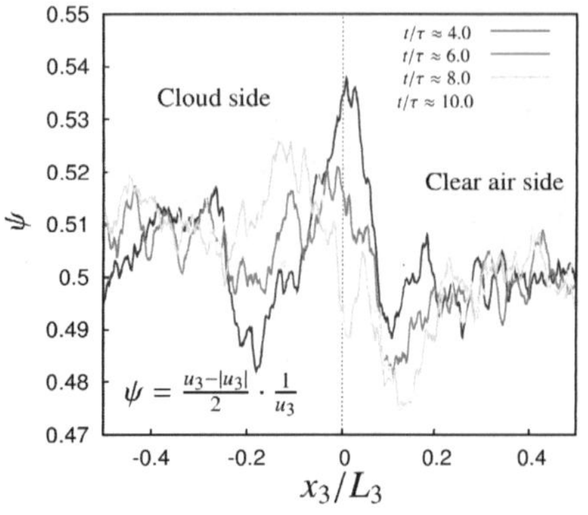

(a)

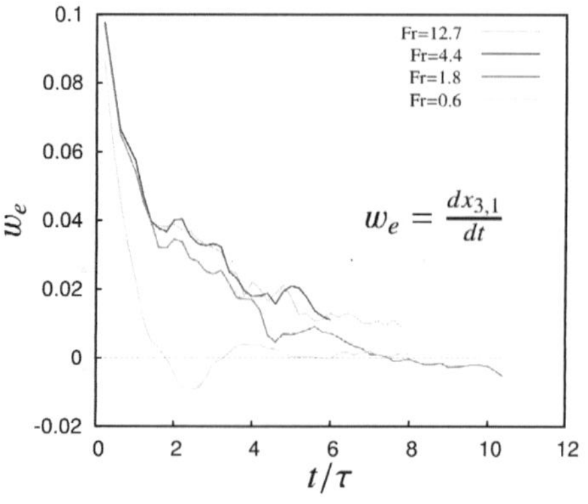

(b)

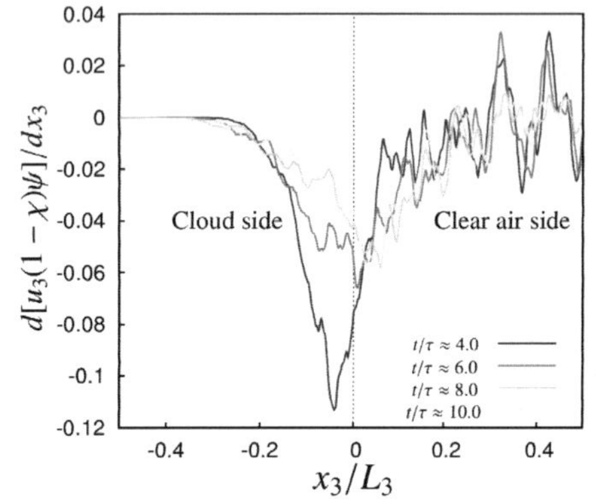

(c)

Figure 8. Evolution of the entrainment across the top cloud. Fraction ψ of downwards velocity (a) and vertical variation of the mean flux of clear air into the cloud (c) when with Fr= 1.8 Evolution of entrainment velocity w_e normalized with the high kinetic energy E_1 root mean square (b), see figure 2.

In any plane parallel to the interface, in absence of a mean velocity, only downward velocity fluctuations can transport clear air into the cloud. Their presence can be represented by a marker function ψ that is equal to 1 where u_3 is negative, and 0 otherwise. Its average in each horizontal plane, shown in figure 8 (a), shows a small deviation from the mean value of 0.5 which would be observed in an homogeneous and isotropic flow. This implies that upward and downward fluctuations are almost equally distributed; the maximum departure from an homogeneous distribution is about 4%, and the spatial distribution of ψ looks like the one which has been observed in the third order moment of the velocity, see figure 4 (a).

The entrainment of clear air is responsible of the growth of the cloud. In fact, the velocity $w_e = dx_{3,i}/dt$ of the cloud top interface ($\langle x_{3,i} \rangle$ is the mean vertical position of the cloud top, defined as the position where the mean vapor concentration is 25%) has often been used as a parameter to measure the entrainment rate [5, 4]. In figure 8 (b), for different levels of stratification, the temporal decay of w_e is represented. In presence of weak stratification, that is Fr larger than 4, its value gradually decreases with an almost exponential law, due to the decay of the kinetic energy. On the contrary, when the stratification is stronger, the damping of w_e is much faster, and the entrainment vanishes after few times scale, when the presence of the pit of kinetic energy substantially reduces the flux of clear air inside the cloud.

Figure 8 (c) shows the vertical derivative of the downward flux of clear air when $Fr = 1.8$ As the flow evolves, the downward flux reduces and its derivative, which represents the net variation of $1 - \chi$ at a given instant, rapidly tends to zero inside the cloud. This implies that the entrainment of clear air is confined to a thin interfacial layer.

4. Conclusion

In this work we have carried out numerical simulations on the transport of energy and scalars in a turbulent shearless mixing layer associated to temporal perturbation of the temperature lapse rate across the clear air - cloud interface. The perturbation locally introduces a stable stratification. This idealized configuration models some of the phenomena which are present in the kinetic dynamics of the cloud and clear air interaction, namely those linked to turbulent mixing and entrainment.

We have shown that this flow configuration develops an horizontal layered structure characterized by a sublayer – with a kinetic energy lower than both the external regions – which acts as a barrier for the transport between the cloud and the external ambient. In our transient simulations, this flow structure appears when the buoyancy terms becomes of the same order of magnitude of the inertial ones, therefore the time needed for this transition becomes shorter when the stratification is more intense. Once buoyancy dominates and the new flow regime is reached, we observed two highly intermittent regions with opposite kinetic energy gradients. As a direct consequence, the entrainment is damped.

The results here presented correspond to a preliminary analysis of the DNS run campaign carried out under the PRACE Project n° RA2011050773 [16]. We observed a fair agreement with few outcomes from LES simulation of the cloud topped Planetary Boundary Layer. This support the use of LES and subgrid modeling applied to these kind of very complex atmospherical systems but at the same time can offer a detailed description of smaller scales.

Acknowledgement

We acknowledge PRACE for awarding us access to resource Curie based in France at TGCC.

Bibliography

[1] Wood R 2012 *Mon. Weather Rev.* **140** 2373–2423 ISSN 0027-0644
[2] Gerber H, Frick G, Malinowski S P, Jonsson H and Khelif D 2013 *J. Geohys. Res. D* **118** 12094–12109
[3] Malinowski S P, Gerber H, Jen-La Plante I, Kopec M K, Kumala W, Nurowska K, Chuang P Y, Khelif D and Haman K E 2013 *Atmos. Chem. Phys.* **13** 12171–12186 ISSN 1680-7316
[4] Moeng C 2000 *J. Atmos. Sci.* **57** 3627–3643 ISSN 0022-4928
[5] Mellado J P 2010 *J. Fluid Mech.* **660** 5–36 ISSN 0022-1120
[6] Mellado J P, Stevens B and Schmidt H 2014 *J. Atmos. Sci.* **71** 1040–1057 ISSN 0022-4928
[7] Biona C, Druilhet A, Benech B and Lyra R 2001 *Agric. Forest Meteor.* **109** 135–141 ISSN 0168-1923
[8] Katul G, Geron C, Hsieh C, Vidakovic B and Guenther A 1998 *J. Appl. Meteor.* **37** 1533–1546 ISSN 0894-8763
[9] Lothon M, Lenschow D H and Mayor S D 2009 *Bound. Lay. Meteor.* **132** 205–226 ISSN 0006-8314
[10] Radkevich A, Lovejoy S, Strawbridge K B and Schertzer D 2008 *Quart. J. Roy. Meteor. Soc.* **134** 317–335
[11] Veeravalli S and Warhaft Z 1989 *J. Fluid Mech.* **207** 191–229
[12] Tordella D and Iovieno M 2011 *Phys. Rev. Lett.* **107** 194501
[13] Drazin P and Reid D 1981 *Hydrodynamic stability* (Cambridge (UK): Cambridge University Press)
[14] Iovieno M, Cavazzoni C and Tordella D 2001 *Comp. Phys. Comm.* **141** 365–374
[15] Tordella D and Iovieno M 2006 *J. Fluid Mech.* **549** 429–441
[16] Tordella D, Iovieno M, De Santi F, Di Savino S and Gallana L 2011 Fluid turbulence: self and passive scalar diffusion. application to stably stratified flows URL `http://www.prace-ri.eu/prace-4th-regular-call/`
[17] Ma B and Warhaft Z 1986 *Phys. Fluids* **29**(10) 3114–3120
[18] Iovieno M, Di Savino S, Gallana L and Tordella D 2014 *J. Turb.* **15** 311–334 ISSN 1468-5248
[19] Qiu X, Huang Y x, Lu Z m and Liu Y l 2009 *App. Mat. and Mech.* **30** 153–162 ISSN 0253-4827
[20] Moeng C, Cotton W, Bretherton C, Chlond A, Khairoutdinov M, Krueger S, Lewellen W, MacVean M, Pasquier J, Rand H, Siebesma A, Stevens B and Sykes R 1996 *Bull. Am. Meteorol. Soc.* **77** 261–278
[21] Moeng C 1986 *J. Atmos. Sci.* **43** 2886–2900
[22] Sykes R, Lewellen W and Henn D 1990 *Mon. Wea. Rev.* **118** 363–373

Natural convection in asymmetric triangular enclosures heated from below

O.M. Kamiyo[1], D. Angeli[2], G.S. Barozzi[3,*] and M.W. Collins[4]†

[1]Department of Mechanical Engineering, University of Lagos, Lagos, Nigeria.
[2]DISMI – Dipartimento di Scienze e Metodi dell'Ingegneria, Università di Modena e Reggio Emilia, via Amendola 2, I-42122 Reggio Emilia, Italy
[3]DIEF - Dipartimento di Ingegneria "Enzo Ferrari", Università di Modena e Reggio Emilia, via Vignolese 905, I-41125 Modena, Italy
[4]School of Engineering and Design, Brunel University, Uxbridge, Middlesex, UB8 3PH, UK

*E-mail: giovanni.barozzi@unimore.it

†Prof. Michael Wesley Collins passed on 2014 August 24th. Among a number of scientific topics he cultivated in his brilliant academic career, convection heat transfer was the first and the most beloved.

Abstract. Triangular enclosures are typical configurations of attic spaces found in residential as well as industrial pitched-roof buildings. Natural convection in triangular rooftops has received considerable attention over the years, mainly on right-angled and isosceles enclosures. In this paper, a finite volume CFD package is employed to study the laminar air flow and temperature distribution in asymmetric rooftop-shaped triangular enclosures when heated isothermally from the base wall, for aspect ratios (AR) $0.2 \leq AR \leq 1.0$, and Rayleigh number (Ra) values $8 \times 10^5 \leq Ra \leq 5 \times 10^7$. The effects of Rayleigh number and pitch angle on the flow structure and temperature distributions within the enclosure are analysed. Results indicate that, at low pitch angle, the heat transfer between the cold inclined and the hot base walls is very high, resulting in a multi-cellular flow structure. As the pitch angle increases, however, the number of cells reduces, and the total heat transfer rate progressively reduces, even if the Rayleigh number, being based on the enclosure height, rapidly increases. Physical reasons for the above effect are inspected.

1. Introduction

The study of natural convection in rooftop enclosures is gaining importance for various environmental applications. In roof design, the attic space is often given paramount consideration because its thermal characteristics have great influence on the conditions of the space directly below it. In tropical climates, both humid and arid, conventional types of roof construction suffer from excessive mid-day overheating due to the high solar radiation incident on the surface area. Solar radiation absorption can actually cause the roofing sheets to become very hot in the midday sun. A low-pitched roof, common in the tropics, is particularly prone to trap heat in the attic, and a significant amount of the cooling load in residential and industrial buildings is the result of heat transfer across the ceiling from the attic [1]. Also, in some rural areas, agricultural produce are sometimes kept in the rooftops of residences either for drying or for storage. It is, therefore, desirable to have a thorough knowledge of the flow patterns and heat transfer characteristics of the attic space under realistic conditions.

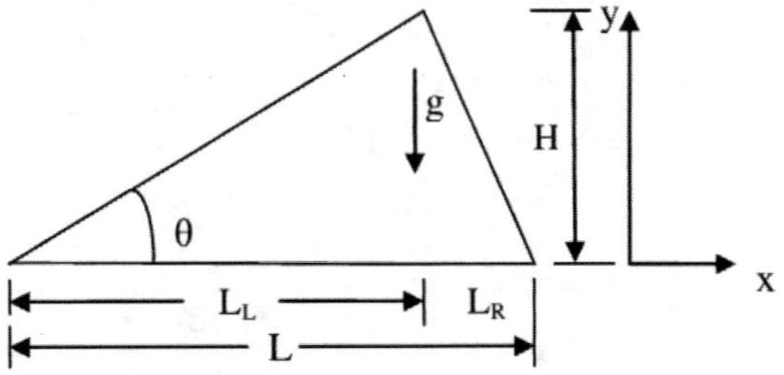

Figure 1. Physical Model

Natural convection heat transfer and fluid flow in enclosed spaces has been studied extensively in recent years especially in response to energy-related applications. The earlier numerical study was carried out by Gershuni et al. in 1974 [2]. In their work, they investigated specific features of fluid motion in an enclosure with bottom heating. Studies carried out by Poulikakos and Bejan [3] for a right-triangular enclosure indicate single cell fluid circulations for Rayleigh numbers (Ra) up to 10^4, and for values of the aspect ratio (AR) between 0.02 and 1.0. Ridouane and Campo [4] observed that, for $10^2 \leq Ra \leq 10^5$, the flow bifurcation is time-dependent. Sahar et al. [5] studied natural convection in tilted isosceles triangular enclosures with discrete bottom heating for $10^3 \leq Gr \leq 10^6$, $0.5 \leq AR \leq 1.0$, $Pr = 0.7$, and inclination angles from 0° to 60°. The investigation showed that the mean Nusselt number decreases as the heated strip enlarges, and increases along with the inclination angle. Kent [6] used the finite volume method to study triangular enclosures for both summer and winter conditions within the range $10^3 \leq Ra \leq 10^5$ for $15° \leq \theta \leq 75°$. It was observed that, for winter conditions and at small pitch angles, increasing Ra resulted into a multicellular flow structure. Further review on this subject can be found in Kamiyo et al. [7].

It is worthy of note that most of the studies on triangular enclosures have been restricted to isosceles or right-angled shapes. Reports on scalene shape are uncommon, despite that many buildings in different climes are of that shape. In this paper, ANSYS FLUENT©, a finite volume CFD package, is employed to study the flow field and heat transfer in rooftop-shaped asymmetric triangular cavities when heated from the base wall (winter condition). As in all the investigations mentioned previously, this study is restricted to steady laminar flow conditions ($Ra \leq 5.25 \times 10^7$).

2. Computational details

A long air-filled ($Pr = 0.71$) triangular enclosure with a triangular cross-section is considered as shown in Fig. 1. The enclosure extension in the direction perpendicular to the cross-section is assumed more than double its width so that the flow features can be assumed to be two-dimensional [8].

The horizontal base, depicting a ceiling being heated by the warm space below it, is assigned a uniform temperature T_H. The pitched roof is considered to be all at the same cold temperature, T_C. Four pitch angles, 14°, 25°, 35°, and 45°, representing an aspect ratio range $0.2 \leq AR \leq 1.0$ were simulated. The enclosure aspect ratio, AR, is defined here as the ratio of the enclosure height, H, to the left fraction of the base, L_L (as from Fig. 1). The base length and the base length ratio, L_L/L_R, were kept unchanged in all the analysis. The Rayleigh and Prandtl numbers are defined

$$Ra = \frac{g\beta(T_H - T_C)H^3}{\nu\alpha} \quad ; \quad Pr = \frac{\nu}{\alpha}$$

Here, g is the gravitational acceleration module, and α, β, and ν, designate the thermal diffusivity, the thermal expansion coefficient, and the kinematic viscosity of air, respectively.

Leaving the base-length unchanged, results in a progressive increase in the cavity height for increasing the pitch angle. This, in combination with a fixed temperature difference between the hot and cold walls, constrains the Rayleigh number to increase steadily with the pitch angle. In the four cases considered, the Rayleigh number ranges between 8.06×10^5 and 5.29×10^7. The four geometries considered and the corresponding Ra-values are reported in Table 1.

PITCH ANGLE (θ)	14°	25°	35°	45°
BASE LENGTH RATIO (L_L/L_R)	3:1	3:1	3:1	3:1
ASPECT RATIO (AR =H/L_L)	0.2	0.5	0.7	1.0
RAYLEIGH NUMBER (Ra)	8.06×10^5	5.27×10^6	1.78×10^7	5.29×10^7

Table 1. Non-dimensional characterization of the enclosures, and Rayleigh number values

The governing equations for buoyancy-driven laminar flow under steady-state conditions are conservation of mass, momentum and energy. Subject to the Boussinesq approximation, they are written

$$\frac{\partial u}{\partial x}+\frac{\partial v}{\partial y}=0$$

$$\frac{u\partial u}{\partial x}+\frac{v\partial u}{\partial y}=-\frac{1}{\rho}\frac{\partial p}{\partial x}+\nu\left(\frac{\partial^2 u}{\partial x^2}+\frac{\partial^2 u}{\partial y^2}\right)$$

$$\frac{u\partial v}{\partial x}+\frac{v\partial v}{\partial y}=-\frac{1}{\rho}\frac{\partial p}{\partial y}+\nu\left(\frac{\partial^2 v}{\partial x^2}+\frac{\partial^2 v}{\partial y^2}\right)+g\beta(T-T_C)$$

$$\frac{u\partial T}{\partial x}+\frac{v\partial T}{\partial y}=\alpha\left(\frac{\partial^2 T}{\partial x^2}+\frac{\partial^2 T}{\partial y^2}\right)$$

Here, x and y designate the horizontal and vertical coordinates, and u, v, T, p are the field variables: horizontal and vertical velocity components, temperature, and pressure.

The validity of the Boussinesq approximation in the present context is corroborated by Gray and Giorgini [9], since they reported that the results would deviate of at most 3% from the corresponding non-Boussinesq solution. Also, in the process of validating their code, Ridouane et al. [10] re-examined the benchmark numerical study of natural convection of air in a square enclosure carried out by de Vahl Davis [11]. Results obtained without enforcing the Boussinesq approximation were just 2% different from the benchmark solution. Such a level of accuracy is generally considered to be acceptable for natural convection flow simulations.

The computational domain coincides with the physical domain. The triangular shape of the domain makes it difficult to use a structured grid; use was therefore made of unstructured triangular meshes.

The steady-state flow field and heat transfer predictions were carried out using the finite volume based ANSYS FLUENT© code (version 14). The SIMPLE algorithm was employed for solving the pressure and velocity coupling, while the PRESTO discretization scheme was enforced for pressure. The QUICK scheme was adopted for space discretization of the momentum and energy equations. The use of the above schemes, combined with body-fitted meshes having the corners inflated with nodes tightly clustered, is expected to guarantee good accuracy in the prediction of the local and mean Nusselt numbers. About 10^4 grid elements were used for the final simulations, after a series of grid-independence experiments. For example, results were obtained for 6344, 9196 and 12,438 elements, for the 14°–pitch triangle. For the sake of exemplification, the grid for case θ = 35° is shown in Fig. 2. Convergence criteria were fixed at 5×10^{-6} for the continuity residual, and at 10^{-7} for the residuals of the momentum and energy equations.

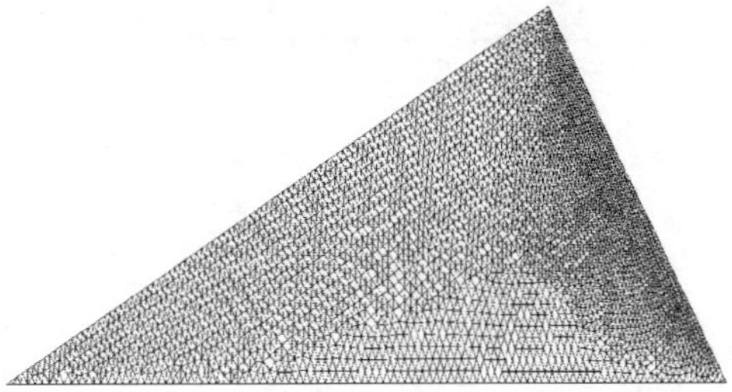

Figure 2. Computational grid for the 35°–pitch triangle.

3. Results and discussion

Results are given in terms of streamline and isotherm representations in Figs. 3 and 4, respectively. The stream-function is scaled using $U=\sqrt{g\beta(T_H-T_C)H}$ as the reference velocity, and H as the scale length. Isotherms are for the non-dimensional temperature $\Theta=(T-T_C)/(T_H-T_C)$ and, therefore, range from zero to one. The x-coordinate is scaled with the base length, L, with the origin in correspondence of the triangle vertex $(-0.75\le x/L\le +0.25)$.

The results in Fig. 2 show that, over the relatively high Ra-range covered by this numerical experiment, the flow is always characterized by the presence of multiple buoyancy-driven vortices. Fig. 3, correspondingly, demonstrates that the thermal field is convection-dominated even at the lowest Ra-value, with no evidence of thermal diffusion. In all the cases the flow-field is led by a main central vortex rotating clockwise with a relatively high velocity. This drives smaller counter-rotating cells, whose number and strength changes drastically as the pitch angle increases. Up to ten vortices, whose size progressively reduces towards the lower corners, can be observed in the shallow 14°-pitch cavity (Fig. 2.a). These reduce to two large cells and three smaller vortices for the 25°-pitch enclosure. The vortex structure remains similar for the 35°-pitch enclosure, where, however, a single vortex resists at the left lower corner, while two rolls appear at the upper corner. In the 45°-pitch enclosure (Fig. 2.d), the two large counter-rotating cells observed in (Fig. 2.c) merge to form a larger vortex occupying the lower part of the space, while smaller cells are formed towards the top and bottom-left corners. It is observed that the multi-cellular flow structure is now vertical, as opposite to the one observed for the lowest pitch-value. In each enclosure, the velocity of rotation decreases as the cell size reduces.

This result is similar to that of Holtzman et al. [12] who presented flow visualization results from experiments performed in a smoke-filled isosceles triangular enclosure heated from the base wall, to show that, as Ra increases for a given geometry, the flow pattern becomes multi-cellular and the number of counter-rotating cells increases.

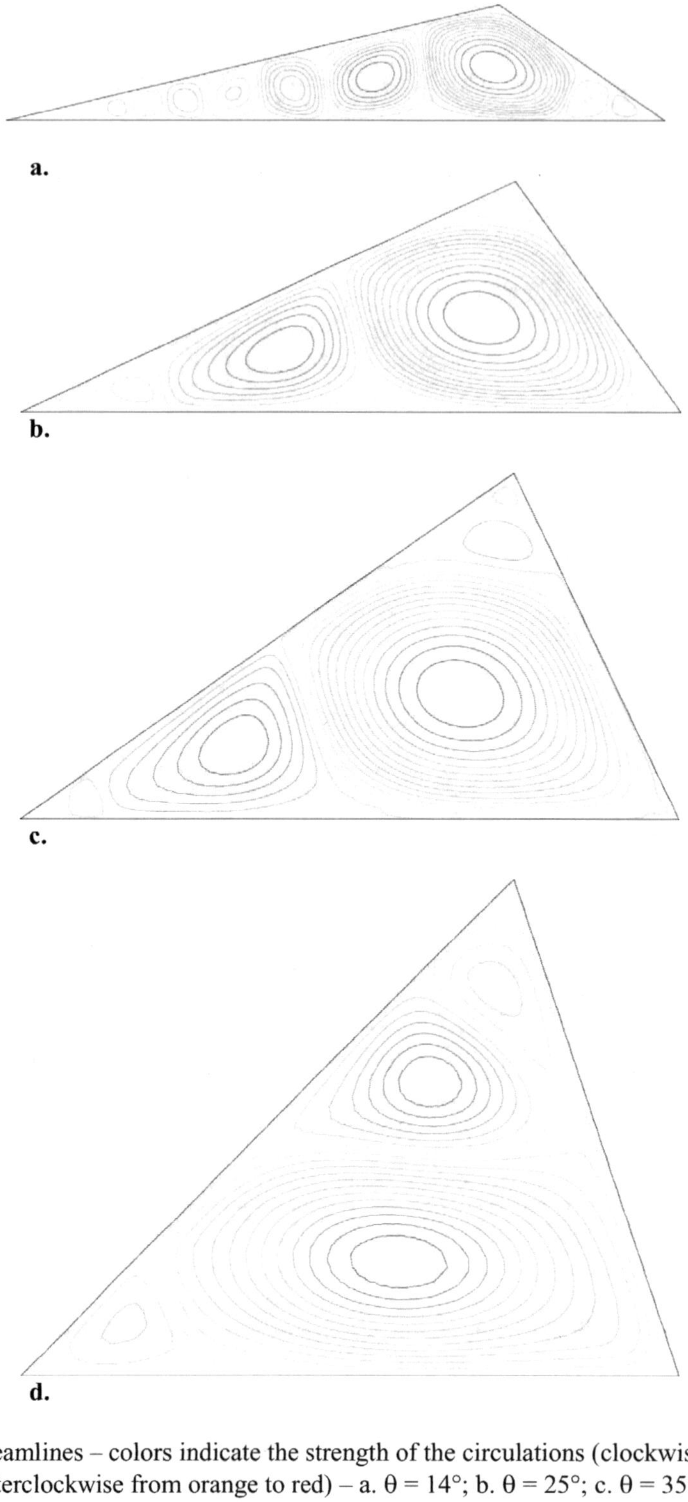

Figure 3. Streamlines – colors indicate the strength of the circulations (clockwise from blue to yellow – Counterclockwise from orange to red) – a. $\theta = 14°$; b. $\theta = 25°$; c. $\theta = 35°$; d. $\theta = 45°$.

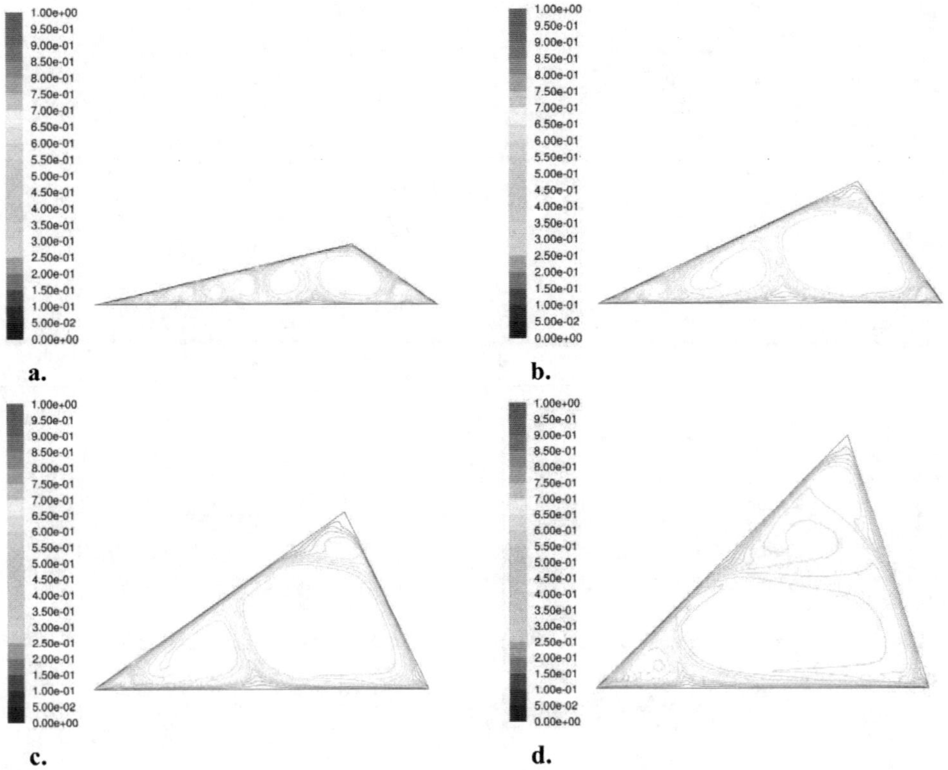

Figure 3. Isotherms $\Theta = (T - T_C)/(T_H - T_C)$ – a. θ = 14°; b. θ = 25°; c. θ = 35°; d. θ = 45°.

The strength and number of the vortices definitely controls the transport processes within the enclosures, as confirmed by the thermal fields in Fig. 3. Isotherms indicate the presence of hot thermal plumes rising from the base side and of cold jets leaving the cold upper sides. These correspond to regions included between two counter-rotating rolls, where the convective transport attains its maximum efficiency. As a consequence, the number of thermals progressively reduces for increasing the pitch angle. The multi-cellular flow within the 14°-pitch enclosure, Fig. 3.a, results in a thorough mixing of the fluid, to which corresponds a relatively high value of the mean temperature of air. In addition, high temperature gradients are observed at the periphery of the cells and along the walls. The temperature distribution within the enclosure becomes progressively more uniform as the aspect ratio and the Rayleigh number increase. This effect is directly related to the formation of big vortices whose central part remains practically isothermal and at temperatures of order 0.3-0.4.

The local values of the heat transfer coefficient, h_x, are reported in Fig. 4, in terms of the local Nusselt number

$$Nu_{x,L} = \frac{h_x L}{\lambda}$$

where λ is the thermal conductivity of air. In the Nusselt number definition, the base length, L, was preferred to the cavity height, H, to facilitate the direct comparison of results relevant to different geometries, but having a common base-length. In Fig. 4, data for the hot and cold walls are both referred to the x-coordinate.

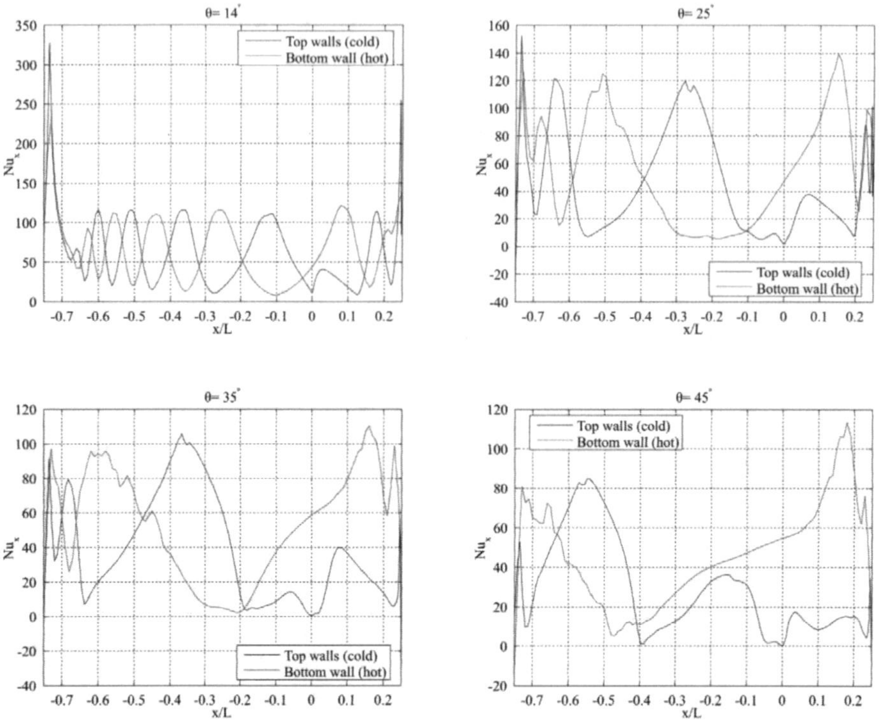

Figure 4. Local Nusselt number distributions along the bottom (red), and top (blue) enclosure walls.

The Nusselt number distributions along the bottom and top surfaces confirm that the heat transfer rate is directly linked to the attachment and detachment processes of the thermal plumes at the solid walls. In particular, as expected, maxima and minima in $Nu_{x,L}$ closely correspond to thermal-jet impact points, and to detachment saddle points, respectively. This is particularly well evident for the 14°-pitch in Fig. 4, showing a quite well ordered oscillatory sequence. This is no longer the case for the highest pitch-angle values, 35° and 45°, where the trends become irregular and a limited number of relatively high peak-value is observed. A minimum in $Nu_{x,L}$ is always found at the upper corner, since air stagnates in its correspondence. The heat transfer rate drops to zero, or quasi-zero, at the detachment points, where air at the wall temperature is driven on the separation area by two counter-rotating vortices, thus creating a region of almost-zero temperature gradient over it. In most of the cases high Nu-values are found in the vicinity of the bottom corners, mainly due to the closeness of the hot and cold walls. These extreme values are however of little practical significance in terms of the overall heat transfer efficiency of the system. This is well documented by the values of the mean Nusselt number over the bottom wall in Fig.5. Results are presented for two alternative Nusselt number definitions, assuming the base length, L, and the cavity height, H, as the reference length, respectively

$$Nu_{av,L} = \frac{h_{av}L}{\lambda} \quad ; \quad Nu_{av,H} = \frac{h_{av}H}{\lambda}$$

Here, h_{av} is the mean value of h_x over L. While it is obvious that the two definitions are equivalent, they have different informative contents in terms of interpretation of the heat transfer results.

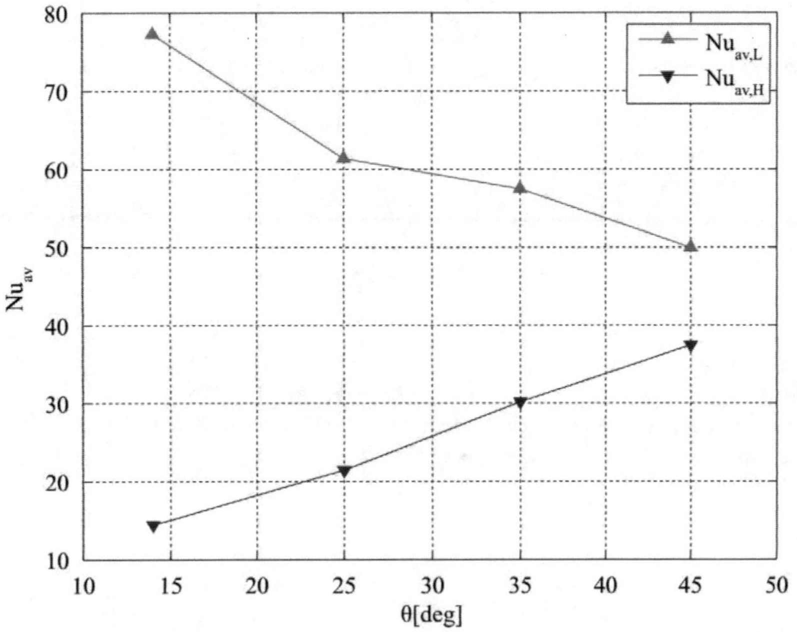

Figure 5. Averaged Nusselt-number values for the bottom enclosure wall.

The increasing trend of $Nu_{av,H}$ with θ actually corresponds to expectations, since Ra changes of two orders of magnitude passing from the 14° to the 45°-pitch angle cases. The $Nu_{av,H} - \theta$ plot is quasi-linear, and this results in an almost linear dependence of $Nu_{av,H}$ on $Ra^{1/3}$. However, the increasing trend of $Nu_{av,H}$ with the Rayleigh number might lead to the wrong conclusion that the system's heat transfer efficiency increases for increasing θ. That this is actually not the case is clearly shown by the decreasing trend of $Nu_{av,L}$. This parameter directly reflects the total heat transfer rate, and, for cases of equal base-length, indicates that the thermal power driven to the cold walls undergoes a 23% reduction passing from the 14°- to the 25°-pitch case, reducing further for increasing θ.

The practical significance of the above results is that the reduction of the number of the buoyancy-driven rolls and their progressive vertical alignment for increasing the pitch-angle provide a corresponding reduction of the heat transfer rate, in spite of the increase in the buoyancy forces indicated by the Rayleigh number.

4. Conclusion

Two-dimensional, steady, laminar natural convection of air contained in a long, horizontal asymmetric triangular enclosure was investigated for bottom heating conditions, while varying the pitch-angle and keeping unchanged the enclosure base-length and the upper vertex coordinate. The results show that, over the range covered by the numerical experiment, heat is transferred from the hot base wall to the cold inclined walls by a multi-cellular flow structure. As the pitch angle increases, the number of cells is seen to reduce, and the cells tend to pass from a horizontal alignment to a vertical one. Since the Rayleigh number increases with the third power of the enclosure height, it would be expected that the higher the pitch angle is the higher is the heat transfer rate. Results however demonstrate that the effect of the pitch angle is just the opposite, since, in fact, the above-mentioned reorganization of the buoyancy-driven circulations invariably produces a reduction in the heat transfer coefficients for increasing θ. The work therefore provides qualitative directions for the choice of a roof shape, in view of the control of heat losses through triangular attic spaces.

References

[1] ASHRAE Handbook: Fundamentals 2005, SI Edition, ASHRAE Inc., USA
[2] Gershuni GZ, Zhukhovitskiy EM and Shvartzblat, DL 1974 *Hydrodyn.* **7** 89
[3] Poulikakos D and Bejan A 1983 *J. Heat Transfer* **105** 652
[4] Ridouane EH and Campo A 2006 *Proc. 9th AIAA/ASME Joint Thermophysics Heat Transfer Conference*, California.
[5] Sahar G, Tofiqul I, Saha S and Quamrul I 2007 *Thammasat Int. J. Sci. and Tech.* **12** 24
[6] Kent EF 2009 *Proc. IMechE C* **223** 1157
[7] Kamiyo OM, Angeli D, Barozzi GS, Collins MW, Olunloyo VOS and Talabi SO 2010 *Appl. Mech. Rev.* **63**
[8] Penot F and N'Dame A 1992 *Proc 3rd UK National Conf. and 1st European Conf. on Thermal Science*, Birmingham, UK
[9] Gray DD and Giorgini A 1976 *Int. J. Heat Mass Transfer* **19** 545
[10] Ridouane EH, Campo A and McGarry M 2005 *Int. J. Thermal Sci.* **44** 944
[11] De Vahl Davis G 1983 *Int. J. Num. Meth. Fluids* **3** 249
[12] Holtzman GA, Hill, RW and Ball, KS 2000 *J. Heat Transfer* **122** 485

Journal of Physics: Conference Series **547** (2014) 012043 doi:10.1088/1742-6596/547/1/012043

In memory of

Prof. Michael Wesley Collins

This short commemoration is dedicated to Prof. Michael W. Collins, as an outstanding researcher in thermal sciences, a passionate promoter of education in sciences, and a very good friend of Italy and UIT (Italian Union of Thermo-Fluid Dynamics).

Michael left us on 2014 August 23rd, after a short, but relentless disease. He had been born in Dorchester, Dorset, UK, in 1939. He got the BA in Engineering Science at the University of Oxford, and run most of his academic career at City University, in London, where he served as Lecturer, Senior Lecturer, and Professor, up to his retirement in 1998. At City University he first got the Ph.D. in Mechanical Engineering, and the title of Doctor of Science in 1990.

Besides, Michael Collins was Visiting Professor at Fudan University, Shanghai & Shanghai University of Science & Technology, at London South Bank University, and at Brunel University, West London, from 2005.

He was appointed Honorary Fellow of UIT in 1991 and Copernicus Visiting Scientist at the University of Ferrara, Italy, in 2006. He was awarded the Knights Cross, Order of Merit for services to Polish Science, in 1991, and the Busk Prize of the Royal Aeronautical Society in 1992.

The above public recognitions of the scientific value of Michael Collins are the results of his untiring research activity, resulting in an incredible number of scientific contributions, around 400, including papers in archival journals, conference keynotes and papers, contributions to volumes.

Heat transfer by convection was his first, and never abandoned, field of interest, even if his open minded view of science led him soon to extend his studies to other bordering topics, such as computational thermofluids, applied mathematics (Michel was Chartered Mathematician as well as Chartered Engineer), medical flows, non-invasive measurement methods and data processing, joint engineering & biological studies.

Michael Collins collaboration with the Italian heat transfer community dates back to 1983, when he was appointed Contract Professor at University of Bologna, and did never interrupt up to his death. We remember him as a member of the Scientific Committee and as an invited speaker in several UIT Conferences, as well as the supervisor of a number of Italian Ph.D. students. Under this latter aspect, the role of Michael Collins in indicating novel and frontier research topics to our scientific community deserves special mention.

Giovanni Sebastiano Barozzi and Walter Grassi

Experimental study of cooling enhancement using a Fe_3O_4 magnetic nanofluid, in an applied magnetic field

F D Stoian and S Holotescu

Faculty of Mechanical Engineering and Research Center for Engineering of Systems with Complex Fluids, Politehnica University Timisoara, 1 Mihai Viteazu Bv., 300222 Timisoara, Romania

E-mail: floriana.stoian@upt.ro

Abstract. This paper presents the results of an experimental study that envisaged the evaluation of the cooling capabilities of a transformer oil based magnetic nanofluid with the solid volume fraction of magnetite nanoparticles equal to 0.0162, in an AC applied magnetic field (f = 50 Hz). The heating and cooling regimes of a coil immersed in the magnetic nanofluid were compared to that corresponding to the base fluid (transformer oil). The results of our study indicate that the temperature rise rate of the magnetic nanofluid is lower than that corresponding to the transformer oil and a lower stationary temperature is obtained in the coil core, where the magnetic flux density is the largest.

1. Introduction

Researches regarding the use of nanofluids in thermal applications carried out worldwide are based primarily on the findings that their effective thermal conductivity is enhanced relatively to that of the carrier liquid [1,2]. This offer in turn the possibility to obtain a larger heat transfer coefficient in those devices and equipments where the heat transfer fluid is replaced with an application-compatible nanofluid. The magnetic nanofluids (widely known as ferrofluids or magnetic liquids) have an additional property – they are responsive to the application of a magnetic field, which allows for the possibility of controlling the flow and the convective heat transfer [3-5]. Magnetic nanofluids heat transfer related research is ranging from heat removal in magneto-fluidic devices for maintaining the desired magnetic properties, to their use as heat transfer fluids [3-7]. A field of research that gained promising results up to date is that of cooling and insulation of power transformers and other types of electromagnetic devices using magnetic nanofluids [7-12]. In this case, the magnetic field generated by the windings of a power transformer is used also to generate the flow of the magnetic nanofluid (the magneto-convection), thus enhancing the heat removal from the windings [11,12].

The aim of this work was to evaluate the cooling potential of a magnetic nanofluid for use as cooling medium in a power transformer. To accomplish this, the cooling of a coil powered by a 50 Hz AC power supply using a specially prepared Fe_3O_4 transformer oil based magnetic nanofluid [13], was investigated experimentally.

2. Materials

A magnetic nanofluid (MNF) sample and its carrier fluid – the transformer oil (TO-40A, producer MOL) were used in this experimental study. The MNF consisted of a stable colloidal dispersion of magnetite (Fe_3O_4) nanoparticles in TO-40A, with oleic acid (Vegetable oleic acid, 65-88 %, Merk) as surfactant, and was prepared using an established procedure developed by Bica et al [5,14].

The volume fraction of magnetite nanoparticles was determined based on the measured values of the transformer oil density and magnetic nanofluid density (indicated in table 1) and magnetite density, taken equal to 5180 kg m^{-3}. The resulting volume fraction is Φ = 1.62 %. The details of the preparation procedure and characterization of the magnetic, electric, rheological and thermal properties of the MNF are given in [13]. A summary of the characteristic properties for the MNF and TO-40A, at 25 °C, is given in Table 1.

Table 1. Physical properties.

	ρ kg m^{-3}	η mPa s	k W m K^{-1}	c_p J kg^{-1} K^{-1}	β K^{-1}	ε F m^{-1}	M_s A m^{-1}
Magnetic nanofluid (MNF)	937	18.5	0.133	1743.39	6.62 10^{-4}	2.25 ε_0	2785.3
Transformer oil (TO-40A)	867	17.5	0.127	1915.12	7.15 10^{-4}	2.2 ε_0	-

The properties in table 1 are as follows: ρ is the density, η –dynamic viscosity, k – thermal conductivity, c_p - heat capacity, β – thermal expansion coefficient, ε – dielectric constant, M_s - saturation magnetization.

3. Experimental Setup

The cooling potential of the fluids was tested in an experimental setup presented in figure 1. A coil (4) of R = 680 Ω was placed in the center of the measurement cell – a cubic vessel (1) with Plexiglas walls of size 60 x 60 x 60 mm^3. The coil has the windings disposed on a rectangular plastic housing (3) and the internal air core dimensions are 15 mm x 15 mm x 18 mm. The coil housing is set on four legs on the bottom of the cell such that the fluid can circulate from the coil core to the outside bulk. The test fluid (2) was consecutively, TO-40A and MNF. The same volume of fluid, V_f = 130 cm^3 was used. The coil was powered by a 50 Hz AC power supply (7), the working voltage being set at 40 V. Six temperature measurement points were established from the external side of the cell wall to the center of the coil core, as follows: the temperature of the external side (T_1) and of the internal side (T_2) of the measurement cell wall, the temperature on the external wall (T_4) and internal wall (T_5) of the coil, the temperature in the fluid (T_3) at the middle point between T_2 and T_5, and the temperature in the fluid in the central axis of the coil core (T_6). All measurement points are in the same horizontal plan. The six T – type thermocouples (Omega GmbH, 0.5 mm, Teflon insulated, standard error of ±1°C or ±0.7%) were introduced through plastic tubes inserted in a foam lid, placed on top of the measurement cell (not drawn in the schematic, for simplification reasons). Each thermocouple was connected through a NI SCC TC-01 module to a data acquisition system (6) consisting of a NI DAQ PCMCIA 6062E card and the NI SC 2345 connector block. The acquisition system was connected to a laptop computer (8) and the temperature measurements were recorded using the software package LabView 8.6. The sampling rate was 1 sample/s.

The coil magnetic flux density, B, was measured with a FW Bell 5080 Gaussmeter, before the heat transfer experiments. Using the axial and transversal Hall probes, the magnetic flux density was measured in several point of interest in the measurement cell for the applied voltage, as follows: the middle of the coil core axis (A), the top center of the core coil (B), the point situated 15 mm above the top center of the coil (C) and the point situated 15 mm above the top of the core corner (D), as indicated in figure 2. The obtained values are listed in Table 2.

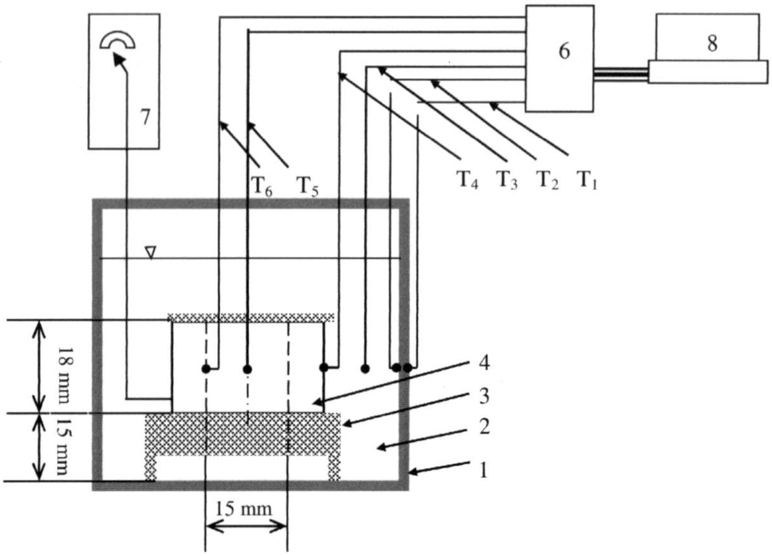

Figure 1. Experimental setup

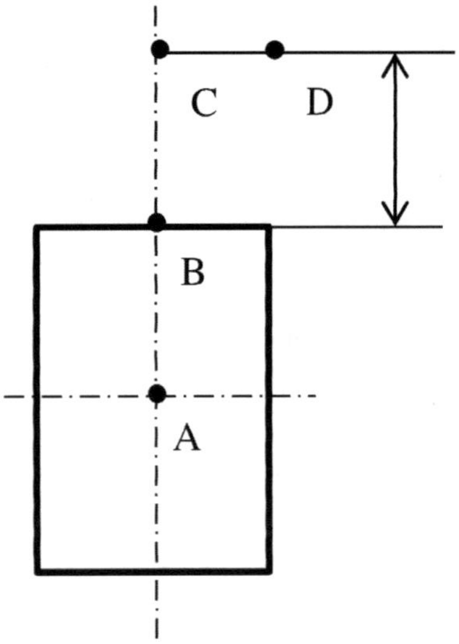

Figure 2. Indication of the magnetic flux density measurement points in the coil core and above it

Table 2. Measured magnetic flux density values.

Point	B [mT]
A	8.1
B	4.9
C	1.4
D	1.0

Using Vizimag software for visualising the magnetic field with the characteristics of the coil and measured magnetic field data for calibration, it was obtained a model for the magnetic field lines and the magnetic flux density contours of the coil, which are presented in figure 3. It can be observed that the effect of the magnetic field generated by the coil is limited to the coil core and its vicinity.

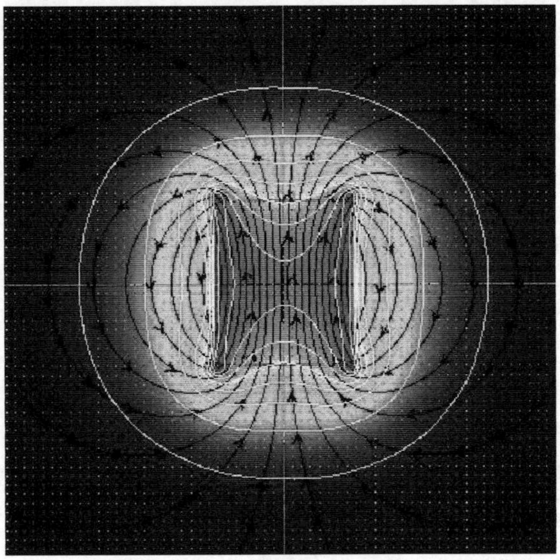

Figure 3. Model of the coil magnetic field lines (black lines with arrows) and the flux density contours (white lines)

4. Results and Discussion

The measurement session for each fluid was divided into three periods: (I) – a non-stationary heating regime characterised by a temperature rise rate, (II) – a stationary heating regime, and (III) cooling regime characterised by a temperature decrease rate.

In each experiment, the temperatures measured by the six thermocouples were continuously recorded by the data acquisition system, starting from the room temperature (28±1.0 °C) and until the end of the cooling regime. During the measurement session, the start of the stationary heating regime was established as the beginning of the period when the temperatures varied less than the standard error (that is less than ±1.0 ° C). The stationary heating regime was attained after about 130 minutes from the start of the experiment. After a stationary period of at least 10 minutes, the power supply was turned off and the experimental system entered the cooling regime until it returned at the initial temperature.

The results of the temperature measurements were compared for the two tested fluids and are presented in figures 4 to 7. Figures 4 and 5 show that, when comparing the stationary values of temperatures $T_1 - T_6$ for the two test fluids, the main differences are obtained for T_5 - the temperature measured in the coil core. The average values of the measured temperatures in the stationary regime are presented in table 3.

Table 3. Average temperatures at the measurement points, in the stationary heating regime.

Working fluid	$T_{1,avg}$ [°C]	$T_{2,avg}$ [°C]	$T_{3,avg}$ [°C]	$T_{4,avg}$ [°C]	$T_{5,avg}$ [°C]	$T_{6,avg}$ [°C]
TO-40A	35.7	36.7	39.4	39.8	40.6	39.4
MNF	35.3	37.6	39.1	40.4	37.2	40.7

Journal of Physics: Conference Series **547** (2014) 012044 doi:10.1088/1742-6596/547/1/012044

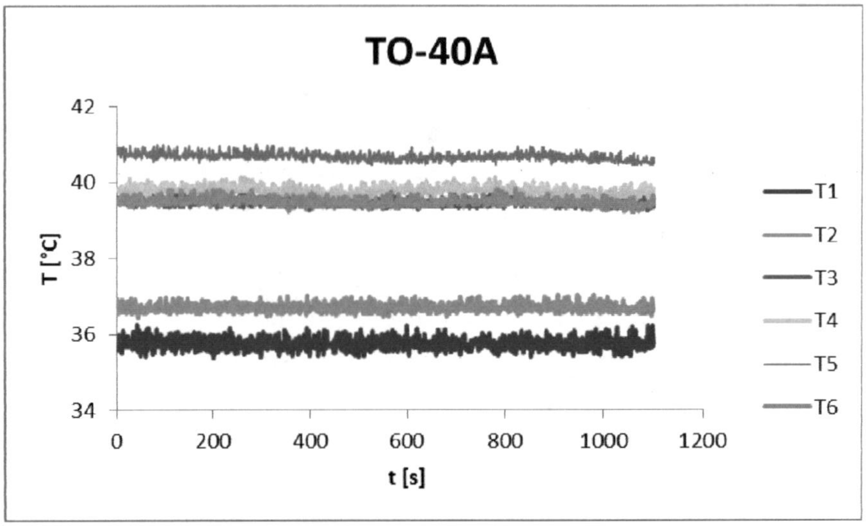

Figure 4. Stationary heating regime for the transformer oil

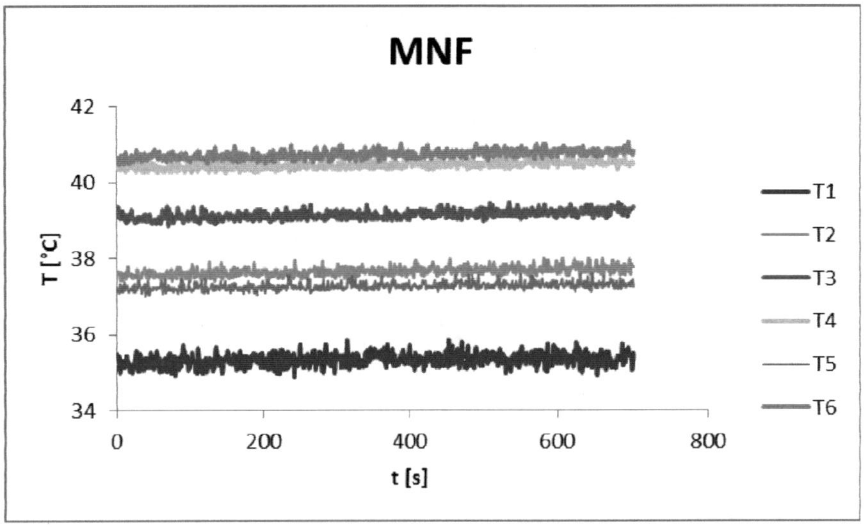

Figure 5. Stationary heating regime for the magnetic nanofluid

The coil external and internal walls temperatures (T_4 and T_6) are slightly higher in the case of magnetic nanofluid test. A possible explanation might be the adherence of some nanoparticles at the coil walls. Due to the low value of the magnetite volume fraction and the low magnetic field outside the coil, there is no significant difference between the corresponding average temperatures in the fluid outside the coil ($T_{3,avg}$). The only clear difference for the temperature measured in the fluid was inside the coil core ($T_{5,avg}$). Here, a difference of 3.4 °C in the corresponding values was obtained. Thus, the effect of the magneto-convection could be observed only in the coil core region, characterised by larger values of the magnetic field density (as indicated by measurements and simulation).

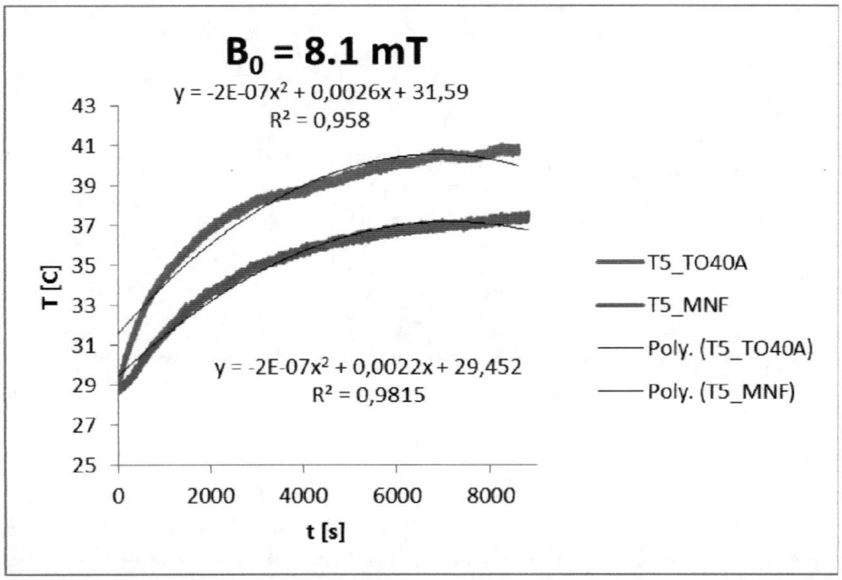

Figure 6. Comparison of heating curves in the central section of the coil core

To study further the heat transfer inside the coil core, the heating curves for TO-40A and MNF at the temperature measurement point T_5 (axis of the coil core) and their polynomial trend lines are compared in figure 6. The magnetic flux density at the measurement point is the one corresponding to the core centre (point A in figure 2). The results of the cooling process using both test fluids at the same point (T_5) are compared in figure 7. The temperature rise rate was calculated from each trend line equation displayed in figure 6 and the temperature decrease rate was calculated from the trend line equations displayed in figure 7. The obtained results are presented as a function of temperature in figure 8. The temperature rise rate of MNF for the explored temperature range is lower than that of the TO-40A. Also, the stationary regime of MNF it is approached at a lower temperature compared to TO-40A case, as observed from figure 5. This result is due to the magneto-convection, a combined effect of the addition of nanoparticles dispersed in the transformer oil, the temperature gradient inside the nanofluid and the applied external field.

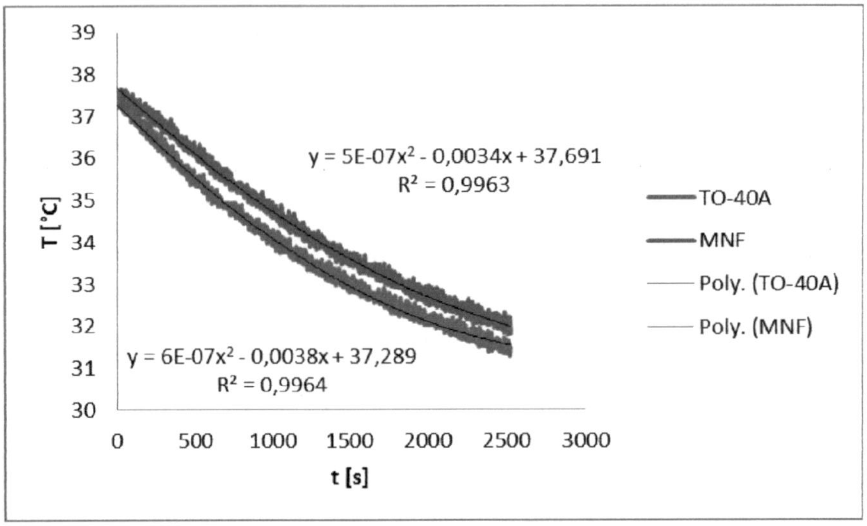

Figure 7. Comparison of the cooling curves in the central section of the coil core (with the coil being powered off)

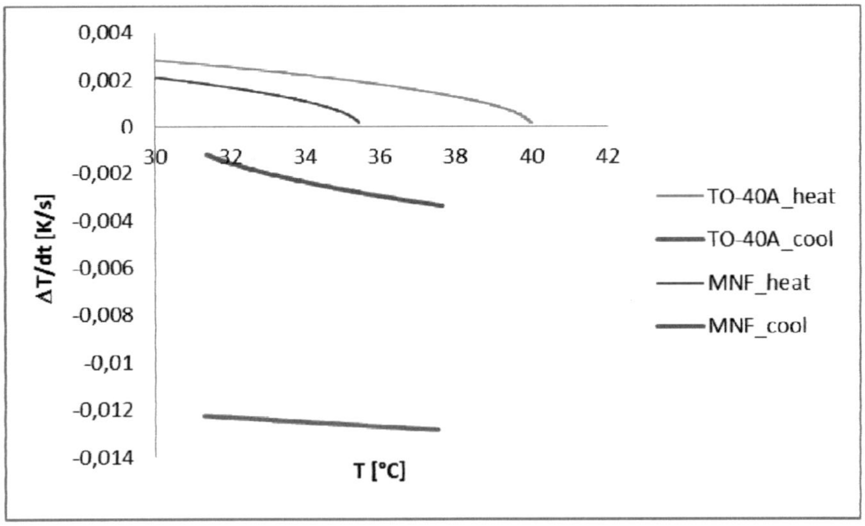

Figure 8. Temperature rise rate (subscript -"heat") and temperature decrease rate (subscript – "cool") in transformer oil and magnetic nanofluid

The MNF temperature decrease rate for the explored temperature range is larger than that corresponding to the TO-40A. That is, due to higher viscosity of magnetic nanofluid compared to that of transformer oil and the absence of the magnetic field, MNF has a lower heat transfer rate.

5. Conclusions

The cooling of an air cored coil powered by an AC source (f = 50 Hz) with magnetic nanofluid and, for comparison, with transformer oil (its base fluid) have been investigated. The obtained results indicated that the magneto-convection inside the coil core led to a decrease of the temperature rise rate and of the stationary temperature at the related measurement point (T_5). If the magnetic field is low or it is off, the heating and cooling temperature data are similar. The tested magnetic nanofluid has potential to be used as cooling medium in a power transformer, if magneto-convection is generated by the transformer windings magnetic field in the bulk of the fluid. The experimental results will be used as reference for a numerical study dedicated to the exploration of other geometrical configurations for the experimental cell in order to increase the magneto-convection outside the coil.

Acknowledgements

Funding for this work was provided by the Executive Agency for Higher Education, Research, Development and Innovation Funding (UEFISCDI), Ministry of National Education (Romania), FP7 MNT-ERA.NET Project MAFINCO No. 7-018/2009-2011 and Project PN II 157/2012-2016.

References

[1] Saidur R, Leong K Y, Mohammad H A 2011 *Ren. Sustain. Energy Rev.* **15** 1646–1668
[2] Philip J, Shima P D 2012 *Adv. Colloid Interface Sci.* **183-184** 30-45
[3] Berkovski B, Bashtovoi V (Eds.) 1996 *Magnetic Fluids and Applications Handbook* Begell House Inc. Publishers ISBN 1567000622
[4] Borbath T, Bica D, Potencz I, Vékás L, Borbath I and Boros T 2010 *Proc. of 25th IAHR Symposium on Hydraulic Machinery and Systems,* IOP Conf. Series: Earth and Environmental Science 12 012105
[5] Vékás L, Bica D, Avdeev M 2007 *China Particuol.* **5**(1-2) 43-49
[6] Nkurikiyimfura I, Wang Y, Pan Z 2013 *Renew. Sust. Energ. Rev.* **21** 548-561
[7] Segal V, Hjotsberg A, Rabinovich A, Nattrass D, Raj K 1998 *IEEE Int. Symp. on Electrical Insulation (Arlington, Virginia, USA)* Vol. 2 p 619
[8] Chiesa M, Das S K 2009 *Colloids and Surfaces A: Physicochem. Eng. Aspects* **335** 88-97
[9] Yamaguchi H, Zhang X-R, Niu X-D, Yoshikawa K 2010 *J. Magn. Magn. Mater.* **322** 698-704
[10] Kopčanský P, Tomčo L, Marton K, Koneracká M, Potočová I, Timko M 2004 *J Magn. Magn. Mater.* **272** 2377-2378
[11] Morega A M, Morega M, Pîslaru-Danescu L, Stoica V, Nouraş F, Stoian F D 2010 *12th International Conference on Optimization of Electrical and Electronic Equipment – OPTIM (Brasov, Romania),* IEEE Xplore p 140
[12] Pislaru – Danescu L, Morega A M, Morega M, Stoica V, Marinica O, Nouras F, Paduraru N, Borbath I, Borbath T 2013 *IEEE Transactions on Industy Applications* **49**(3) 1289-1998
[13] Stoian F D, Holotescu S, Taculescu A, Marinica O, Resiga D, Timko M, Kopcansky P, Rajnak M 2013 *Proc. 8th Int. Symp. on Advanced Topics in Electrical Engineering - ATEE (Bucharest, Romania)* IEEE Xplore p 1-4.
[14] Bica D 1995 *Romanian Reports in Physics* **47** 65-272

Journal of Physics: Conference Series **547** (2014) 012045 doi:10.1088/1742-6596/547/1/012045

A simplified analytical model of radiative heat transfer in open cell foams

A Andreozzi[1], N Bianco[1], S Cunsolo[1], V Naso[1] and M Oliviero[2]
[1] Dipartimento di Ingegneria Industriale, Università degli studi Federico II, 80 P.le Tecchio, 80125, Napoli, Italy
[2] Istituto per i materiali compositi e biomedici, Consiglio Nazionale delle Ricerche, 1 P.le Fermi, 80055, Portici (Napoli), Italy

E-mail: nicola.bianco@unina.it

Abstract. A simplified one-dimensional analytical model of radiative heat transfer in foams is presented, based on the idea of dividing the porous material into layers at the pore level and then modeling each layer of the porous material as an equivalent semi-transparent, absorbing and reflecting plane. Compared to existing models, the model proposed in this paper has the advantage of explicitly accounting for the geometry of the foam and the radiative energy fluxes, at the same time ensuring self-consistency and offering the computational lightness of analytical models, without sacrificing the mathematical simplicity of the formulation. Using a regular cubic lattice representation and assuming diffuse radiation, straightforward analytical correlations are derived to evaluate the characteristics both of single layers of foam and of finite thickness samples, accounting for various boundary conditions. The predictions of the model are in good agreement with experimental data taken from the literature.

1. Introduction

Open cell metal and ceramic foams are being widely adopted in many heat transfer areas, such as compact heat exchangers, fire barriers and volumetric absorbers in concentrating solar systems receivers. In many heat transfer applications thermal radiation is the predominant heat transfer mode in open cell solid foams [1].

Radiative behavior of open cell foams has been investigated using analytical, numerical and experimental methods [2].

Most analytical models in literature are based on independent scattering in randomly dispersed media in geometric optics regime. The foam is modeled as a random dispersion of particles, whose contributions are summed up to obtain the effective radiative properties. This approach, originally proposed by Glicksmann et al. [3], who modeled the foam as a set of dodecahedral cells, was also followed by Kuhn et al. [4-5], and Doermann and Sacadura [6]. Coquard and Baillis [7], Coquard et al. [8], and Loretz et al. [9] extended these results by considering models with different cell and strut shapes.

Some alternative approaches have been attempted by a number of authors. Kamiuto [10] calculated radiative properties of the foam on the basis of Mie scattering theory. De Micco and Aldao [11] proposed a model based on treating the foam as a series of high-optical density and low-optical density layers. The model of Zhao et al. [12] is one of the very few that try to explicitly account for the cell organization of the foam, though using a simplified cubic lattice. Contento et al. [13] modified the model proposed in [12] in order to improve its accuracy.

In this paper, following an approach similar to that used by Zhao et al. [12], an alternative analytical model is proposed. Compared to existing models, the model proposed in this paper has the advantage of explicitly accounting for the geometry of the foam and the radiative energy fluxes, at the same time ensuring self-consistency and offering the computational lightness of analytical models, without sacrificing the mathematical simplicity of the formulation. Some applications are shown and, finally, a comparison of predictions with experimental data taken from the literature is made.

2. Model and analysis

As a premise, it is worth summing up the assumptions made in the formulation of the model:

- surface reflectivity independent of wavelength and incidence angle;
- geometric optics approximation;
- diffuse reflection;
- the foam is modeled as a uniform cubic lattice;
- the model is one-dimensional.

2.1 Idealized cell structure

The microstructure of ideal open celled foams is constituted by cells with 12-14 square, pentagonal and hexagonal faces. In the present study reference is made to the simplified cubic representation proposed by Bhattacharya et al. [14]. The structure is depicted in figure 1.

Unit cells are characterized by the cell diameter, d_p, and the ligament diameter, d_0, that can be correlated with the characteristic parameters of foams, such as porosity, φ, and pores per inch, *PPI*.

In order to ensure an unchanged porosity in the simplified cubic representation, the following correlation has to hold

$$d_0 \approx 2\left(\frac{1-\varphi}{3\pi}\right)^{1/2} d_p \tag{1}$$

Additionally, the cell diameter is chosen so that the cubic cells have the same average diameter as the real cells. This can be achieved by using directly measured values of the cell diameter or, alternatively, when only technical specifications are available, the following approximate correlation can be used

$$d_p \approx 25.4\,(PPI)^{-1}\;[mm] \tag{2}$$

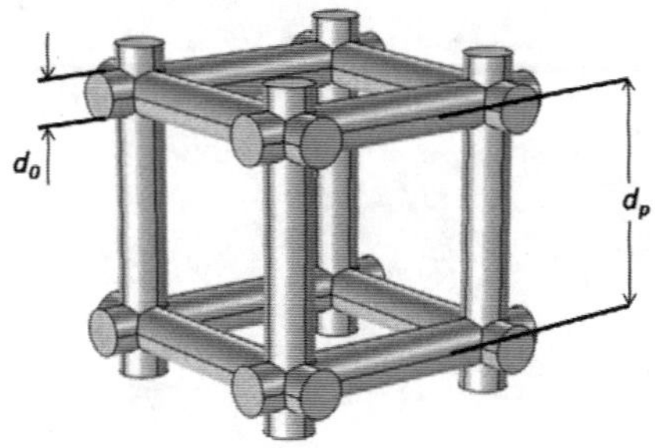

Figure 1. Structure of an open celled foam.

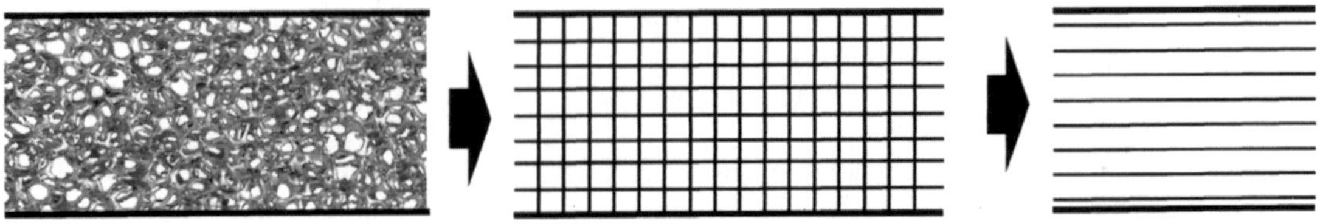

Figure 2. Partition of the cubic lattice into parallel planes.

2.2 Equivalent planes

After simplifying the original foam structure the assumed cubic lattice can be partitioned into parallel planes, each of them corresponding to a layer of cubes, by means of the process sketched in figure 2.

Reference being made to an indefinite single layer of cells, the top and bottom planes, along with the enclosed solid matter, constitute an optical cavity, as depicted in figure 3.

The optical cavity can now be used to calculate the equivalent characteristics of the plane. The view factor, F_{12}, between the opposite planes 1 and 2 can be evaluated once the view factors between the opposite faces A and B, F_{AB}, and between the contiguous orthogonal faces A and C, F_{AC}, are known. To this aim, one can first notice that the view factor between the two planes is equal to the view factor between a single face and the opposite plane, which can be calculated considering both direct radiation and radiation through side faces.

In all practical foams geometry F_{AB} and F_{AC} assume a nearly common value, given by the following correlation [15]

$$F_{AB} \approx F_{AC} \approx \frac{2}{\pi X^2}\left[\ln\left(\frac{(1+X^2)^2}{1+2X^2}\right)^{1/2} + 2X(1+X^2)^{1/2}\tan^{-1}\left[X(1+X^2)^{-1/2}\right] - 2X\tan^{-1}X\right] \quad (3)$$

where $X = 1 - d_0/d_p$.

It is also worth noticing that equation (3) can be approximated as

$$F_{AB} \approx F_{AC} \approx 0.26\,X - 0.063 \quad (4)$$

for all typical X values of open cell solid foams.

Equation (4) allows to evaluate F_{12}, F_{S1}, F_{1S} and F_{SS}, by making use of reciprocity and cavity properties

$$F_{12} = F_{AB} + 8\,F_{AB}^2\ \ (1 - 3F_{AB})^{-1} \quad (5)$$

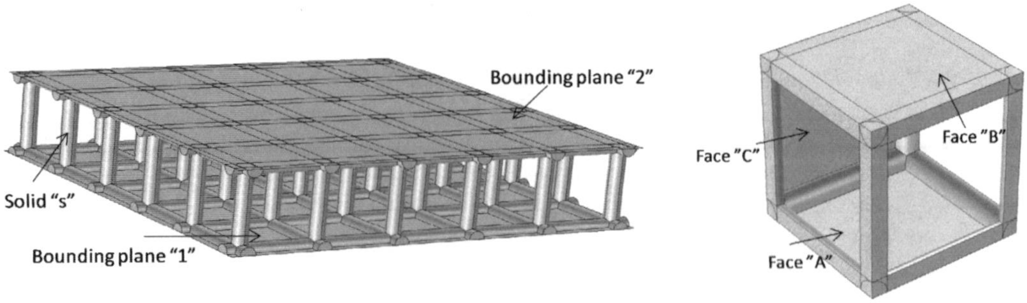

Figure 3. Sketch of the optical cavity.

$$F_{1S} = 1 - F_{12} \tag{6}$$

$$F_{S1} = F_{1S}(A_1/A_S) \tag{7}$$

$$F_{SS} = 1 - 2\,F_{S1} \tag{8}$$

The characteristics of a plane optically equivalent to the cell layer considered can then be determined from the amount of emitted, reflected and transmitted radiation. The solid surface is considered as an opaque grey body, characterized by emissivity ε and reflectivity $\rho = 1 - \varepsilon$. Reflection is assumed to be diffuse and multiple reflection effects are taken into account.

The plane is characterized in terms of equivalent emissivity, ε_{eq}, reflectivity, ρ_{eq}, and transmissivity, τ_{eq}. The following correlations are obtained

$$\varepsilon_{eq} = \varepsilon\, F_{1S}(1 - \rho\, F_{SS})^{-1} \tag{9}$$

$$\rho_{eq} = \rho\, F_{1S}\, F_{S1}(1 - \rho\, F_{SS})^{-1} \tag{10}$$

$$\tau_{eq} = F_{12} + \rho_{eq} \tag{11}$$

2.3 Plane grouping

Using the aforementioned correlations, any given thickness of foam can be treated as a series of planes with given optical properties. In some applications, it can be useful to introduce a further simplification and determine a single plane optically equivalent to a group of n contiguous planes. This is accomplished by using recursive relations for the emissivity, ε_p^n, reflectivity, ρ_p^n, and transmissivity, τ_p^n, which are hereby presented

$$\varepsilon_p^n = \varepsilon_{eq} + \varepsilon_{eq}\left[\tau_{eq}\,\rho_p^{n-1}\left(1 - \rho_{eq}\,\rho_p^{n-1}\right)^{-1}\right] + \varepsilon_p^{n-1}\left[\tau_{eq}\left(1 - \rho_{eq}\,\rho_p^{n-1}\right)^{-1}\right] \tag{12}$$

$$\rho_p^n = \rho_{eq} + \tau_{eq}^2\,\rho_p^{n-1}\left(1 - \rho_{eq}\,\rho_p^{n-1}\right)^{-1} \tag{13}$$

$$\tau_p^n = \tau_p^{n-1}\,\tau_{eq}\left(1 - \rho_{eq}\,\rho_p^{n-1}\right)^{-1} \tag{14}$$

It should be remarked that the values of ε_p^0, ρ_p^0 and τ_p^0 should be chosen such as to fit the effective boundary condition. For example, if the boundary is a flat metal plate, its emissivity and reflectivity can be set equal to ε_p^0 and ρ_p^0, whereas if the boundary is an open surface τ_p^0 should be set equal to 1.

The three successions are convergent and the values ε_p^∞, ρ_p^∞ and τ_p^∞ depend only on ε_{eq}, ρ_{eq} and τ_{eq}, i.e. the optical characteristics of a sufficiently thick foam sample depend only on the characteristics of the foam. The above characteristics can be obtained either by iteration or using the following correlations

$$\varepsilon_p^\infty = \varepsilon_{eq}\left\{\left[1 + \rho_p^\infty(\tau_{eq} - \rho_{eq})\right]\left(1 - \rho_{eq}\,\rho_p^\infty - \tau_{eq}\right)^{-1}\right\} \tag{15}$$

$$\rho_p^\infty = \left\{1 + \rho_{eq}^2 - \tau_{eq}^2 - \left[\left(1 + \rho_{eq}^2 - \tau_{eq}^2\right)^2 - 4\,\rho_{eq}^2\right]^{1/2}\right\}\left(2\,\rho_{eq}\right)^{-1} \tag{16}$$

$$\tau_p^\infty = 0 \tag{17}$$

Journal of Physics: Conference Series **547** (2014) 012045 doi:10.1088/1742-6596/547/1/012045

2.4 Two flux model

A simple two-flux model [15] can be now implemented to manage radiation in the layered structure (figure 4).

Since, in the following, correlations will be given in the form of reverse recurrences, a reverse form of equations (12 - 14) is presented

$$\varepsilon_p^n = \varepsilon_{eq} + \varepsilon_{eq}\left[\tau_{eq}\,\rho_p^{n+1}\left(1-\rho_{eq}\,\rho_p^{n+1}\right)^{-1}\right] + \varepsilon_p^{n-1}\left[\tau_{eq}\left(1-\rho_{eq}\,\rho_p^{n+1}\right)^{-1}\right] \tag{18}$$

$$\rho_p^n = \rho_{eq} + \left[\tau_{eq}^2\,\rho_p^{n+1}\left(1-\rho_{eq}\,\rho_p^{n+1}\right)^{-1}\right] \tag{19}$$

$$\tau_p^n = \tau_p^{n+1}\left[\tau_{eq}\left(1-\rho_{eq}\,\rho_p^{n+1}\right)^{-1}\right] \tag{20}$$

As it is shown in figure 4, the calculation of radiosities J_y and J_{-y} is performed considering three contributions:

a. the heat flux irradiated from the plane in the chosen direction;
b. the heat flux irradiated from the plane in the opposite direction and reflected by the boundary plane;
c. the heat flux irradiated from the rest of the domain in the chosen direction that passes through the boundary plane.

The radiosity J_y can be written as

$$J_y = \varepsilon_{eq}\,\sigma\,T^4 + \varepsilon_{eq}\left[\tau_{eq}\,\rho_{p,y}^0\left(1-\rho_{eq}\,\rho_{p,y}^0\right)^{-1}\right]\sigma\,T^4 + J_y'\left[\tau_{eq}\left(1-\rho_{eq}\,\rho_{p,y}^0\right)^{-1}\right] \tag{21}$$

It is worth noticing that equation (21) can be easily generalized in the form of a reverse recursion. Then the equations for the calculation of the radiosities in both directions can be written as

$$J_y^n = \varepsilon_{eq}\,\sigma\,T_n^4 + \varepsilon_{eq}\left[\tau_{eq}\,\rho_{p,y}^n\left(1-\rho_{eq}\,\rho_{p,y}^n\right)^{-1}\right]\sigma\,T_n^4 + J_y^{n+1}\left[\tau_{eq}\left(1-\rho_{eq}\,\rho_{p,y}^n\right)^{-1}\right] \tag{22}$$

$$J_{-y}^m = \varepsilon_{eq}\,\sigma\,T_m^4 + \varepsilon_{eq}\left[\tau_{eq}\,\rho_{p-y}^m\left(1-\rho_{eq}\,\rho_{p,-y}^m\right)^{-1}\right]\sigma\,T_m^4 + J_{-y}^{m+1}\left[\tau_{eq}\left(1-\rho_{eq}\,\rho_{p,-y}^m\right)^{-1}\right] \tag{23}$$

where J_y^N and J_{-y}^M are the first elements of the recursion and are chosen to represent the irradiation coming from boundary planes. For opaque plates with emissivity ε_p^N and ε_p^M, J_y^N and J_{-y}^M are equal to

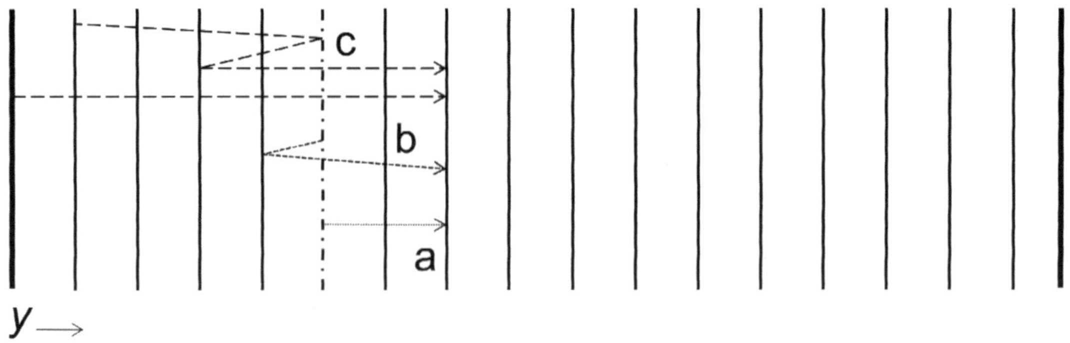

Figure 4. Schematic illustration of contributions to radiosities.

$$J_y^N = \varepsilon_p^N \, \sigma \, T_N^4 \tag{24}$$

$$J_{-y}^M = \varepsilon_p^M \, \sigma \, T_M^4 \tag{25}$$

For an open boundary with the irradiations I_0^N and I_0^M one can simply write

$$J_y^N = I_0^N \tag{26}$$

$$J_{-y}^M = I_0^M \tag{27}$$

One can remark that the recursions can be truncated, giving approximated solutions. It has been found that a 10 term approximation gives accurate results, with an error less than 5%, in most practical cases.

Finally the net radiative heat flux, q_r, can be determined, taking into account reflection effects, using the following correlation

$$q_r = \left[J_y^0\left(1 - \rho_{p,-y}^0\right) - J_{-y}^0\left(1 - \rho_{p,y}^0\right)\right]\left(1 - \rho_{p,y}^0 \, \rho_{p,-y}^0\right)^{-1} \tag{28}$$

3. Applications

In the following some applications of the model to problems commonly found in literature will be presented, along with some useful simplifications.

3.1 Evaluation of the radiative conductivity

The equations presented in the previous section can be used to predict the radiative conductivity, k_r, on the basis of a linear, one-dimensional modeling of the phenomenon. The radiative conductivity is currently used when account is to be taken of the radiation effects [1].

A uniform and linear thermal gradient $\mathrm{d}T/\mathrm{d}y$ is imposed on the foam, which takes the form of a constant $\Delta T = \mathrm{d}T/\mathrm{d}y \, (d_p)$ between contiguous planes, and then k_r is written as

$$k_r = q_r(\mathrm{d}T/\mathrm{d}y)^{-1} \tag{29}$$

If the foam material can be considered black and the boundary effects can be neglected ($N > 20$, $M > 20$), equations (22), (23) and (28) greatly simplify, and an approximate analytical solution can be derived

$$k_r = 4 \, d_p \frac{1 + \tau_{eq}}{1 - \tau_{eq}} \, \sigma \, T^3 \tag{30}$$

A numerical study has been carried out, which showed that emissivity has minor effects on predicted radiative conductivity, down to very low values ($\varepsilon < 0.05$). Therefore, equation (30) can be used for the determination of k_r in thick foam samples.

3.2 Simplified treatment of external sources of radiation

In some cases (for example, in volumetric solar receivers) internal energy transfer may be negligible compared to the heat transfer with an external source. For the sake of simplicity, a domain indefinitely extending in one direction can be considered.

In this case the net incoming radiative heat flux across the n planes can be calculated as

$$q_r^n = I_0\left[\tau_p^n(1-\rho_p^\infty)\right]\left(1-\rho_p^n\,\rho_p^\infty\right)^{-1} \tag{31}$$

with $\tau_p{}^0 = 1$ and $\rho_p{}^0 = 0$.

The radiative heat flux absorbed by the solid contained in the nth plane, q_{in}^n, is

$$q_{in}^n = q_r^n - q_r^{n+1} \tag{32}$$

Then, the amount of heat flux radiating from plane n and irradiating outside of the foam, q_{out}^n, can be calculated as

$$q_{out}^n = \tau_p^n\left[1+\tau_{eq}\,\rho_p^\infty\left(1-\rho_{eq}\,\rho_p^\infty\right)^{-1}\right]\left(1-\rho_p^n\,\rho_p^\infty\right)^{-1}\varepsilon_{eq}\,\sigma\,T^4 \tag{33}$$

Finally, the difference between the incoming and the outgoing heat flux gives the net absorbed heat flux

$$q_{net}^n = q_{in}^n - q_{out}^n \tag{34}$$

This term can be easily rewritten as a volumetric internal energy generation term in the energy equation

$$S_r^n = \left(q_{in}^n - q_{out}^n\right)\left(d_p\right)^{-1} \tag{35}$$

3.3 Characterization as optically thick media

To obtain the extinction, β, absorption, κ, and scattering, σ_s, coefficients from the model proposed in the present paper, the above correlations developed for the evaluation of the radiative conductivity, k_r, are coupled with the well known Rosseland approximation to obtain the radiative conductivity [1].

Under the Rosseland approximation we can write

$$k_r = \left[16\,(3\,\beta_R)^{-1}\right]\sigma\,T^3 \tag{36}$$

where β_R is the Rosseland extinction coefficient. Coupling equation (36) with equation (30) we obtain

$$\beta_R = 4\left(1-\tau_{eq}\right)\left[3d_p\left(1+\tau_{eq}\right)\right]^{-1} \tag{37}$$

Under the assumption of grey and diffuse surfaces, one can finally write

$$\beta = \beta_R \tag{38}$$

$$\kappa = (1-\rho)\beta \tag{39}$$

$$\sigma_s = \rho\,\beta \tag{40}$$

4. Validation of the model

The model was validated by comparing its predictions with experimental data of radiative conductivity presented by Zhao et al. [16]. Experimental data presented in [16] was used to set boundary conditions of the problem. The small thickness (30 mm) of the samples used suggests that effects of the boundaries may be not negligible. The nickel coated copper plates enclosing the samples were modeled as gray flat plates with an emissivity ε_b = 0.05. Data on the microscopic features of the materials was taken from another work from Zhao et al. [17], where SEM imaging was used to accurately characterize the significant process parameters.

The average radiative conductivity, $k_{r,\ av}$, was calculated by simulating coupled conductive-radiative heat transfer in ANSYS Fluent.

The values predicted by the herein proposed model and the experimental data of the radiative conductivity as a function of the temperature, for different values of the Pore Per Inch and the porosity, are reported in figure 5. The figure shows that the model well predicts the temperature dependence of the radiative conductivity exhibited by experimental data. The agreement is rather good, with maximum deviations about 25%.

It may be interesting to focus the attention on a peculiar phenomenon that emerged analyzing data from Zhao et al. [16]. We noticed that for small thicknesses of the foam samples, the effect of boundary emissivity on the value of radiative conductivity averaged along the sample thickness is quite large. The average radiative conductive as a function of boundary emissivity, at 750 K, for a 30 mm thick sample, *30 PPI*, for 90% and 95% porosities, is reported in fig.6. The aforementioned effect is more marked for the higher porosity foam. At very low emissivities of the boundary the 90% porosity foam shows higher average radiative conductivity, as depicted in fig. 6. This is likely caused by high boundary reflectivity "bottlenecking" radiation heat transfer at the boundary, causing energy to be transferred conductively: less porous foams are then advantaged due to their significantly higher effective solid conductivity.

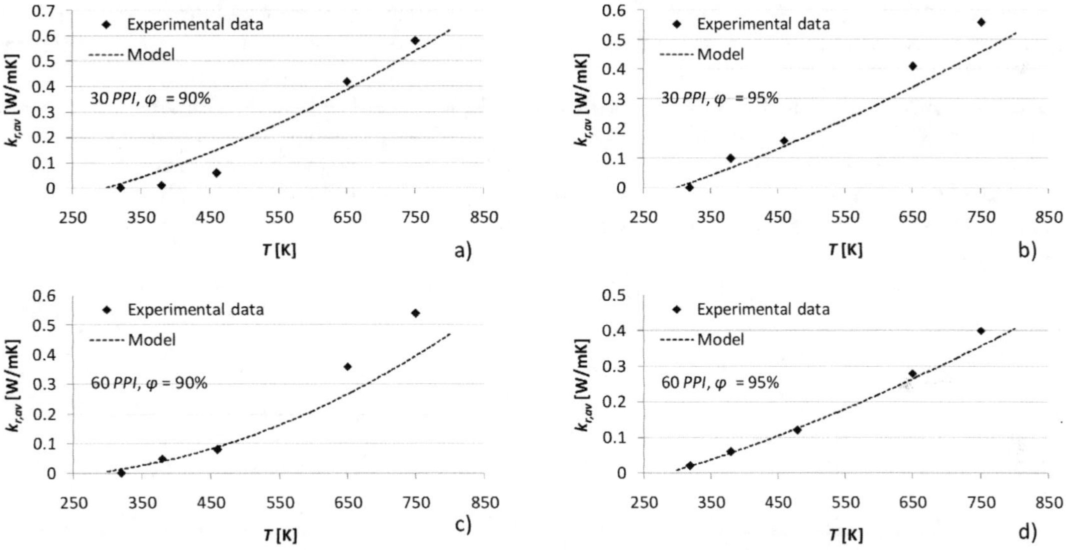

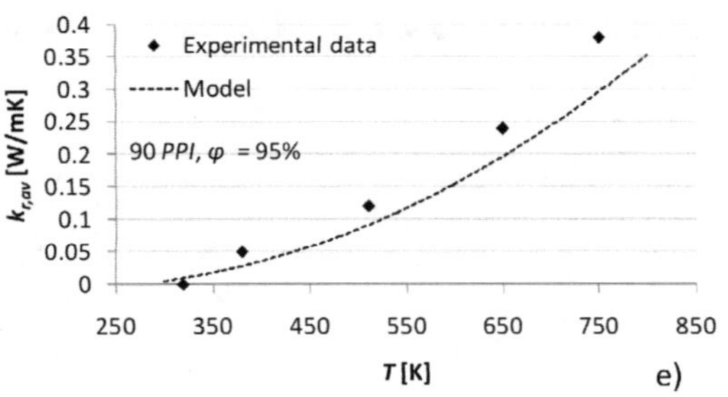

Figure 5. Predicted and experimental data of the radiative conductivity *vs* the temperature:
a) 30 *PPI* and $\varphi = 90\%$;
b) 30 *PPI* and $\varphi = 95\%$;
c) 60 *PPI* and $\varphi = 90\%$;
d) 60 *PPI* and $\varphi = 95\%$;
e) 90 *PPI* and $\varphi = 95\%$.

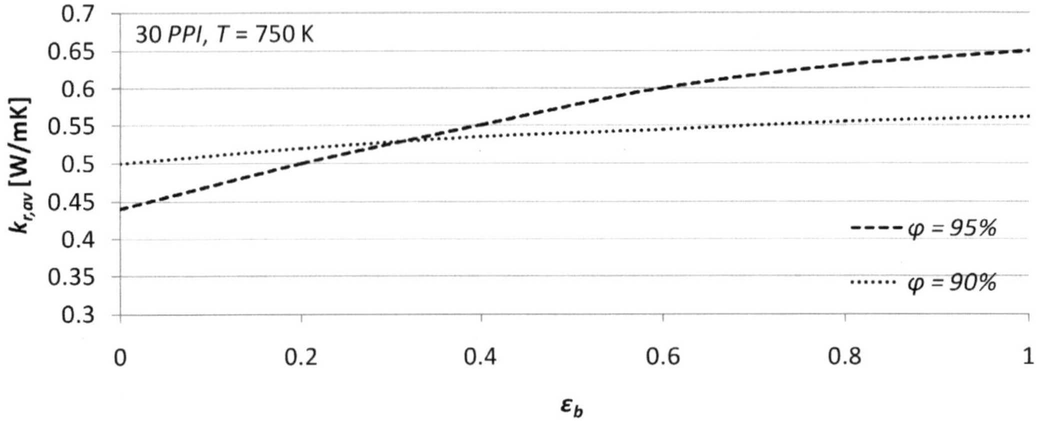

Figure 6. Average radiative conductivity *vs* boundary plates emissivity, for a 30 mm thick sample at $T = 750$ K, with 30 *PPI*, $\varphi = 90\%$ and $\varphi = 95\%$.

5. Conclusions

A simplified one-dimensional analytical model of radiative heat transfer in foams is presented. The approach adopted by the model is alternative to the methodology, most frequently used in literature, based on independent scattering from randomly distributed particles and, although in a very simplified fashion, explicitly accounts for the cell-like structure of the foam. The model first reduces the foam to a set of optically equivalent planes and then uses a two-flux like method to model radiation.

Correlations have been proposed to simplify the application of the model, and some examples have been provided, with reference to problems frequently encountered in literature. Finally, the model has been validated comparing its prediction with experimental data from Zhao et al. [16], showing good agreement.

While still based on relatively coarse approximations of the geometric structure of the foam, the proposed model can effectively be used as an alternative analytical reference, with the added advantage of an explicit treatment of radiation energy fluxes and geometry.

6. Nomenclature

A_i	Area of the *i* surface
d_p	Cell diameter
d_0	Ligament diameter
F_{ij}	View factor
I	Irradiation
J	Radiosity
k_r	Radiative conductivity
PPI	Pores Per Inch
q	Heat flux
S_r^n	Volumetric internal energy generation
T	Temperature
y	Cartesian coordinate
Greek symbols	
β	Extinction coefficient
β_R	Rosseland mean extinction coefficient
ΔT	Temperature difference
ε	Emissivity
κ	Absorption coefficient
ρ	Reflectivity
σ	Stephan-Boltzmann constant
σ_s	Scattering coefficient
τ	Transmissivity
φ	Porosity
Subscripts	
av	Average
b	Boundary plates
eq	Equivalent
in	Incoming
m	Refers to *m*th plane
n	Refers to *n*th plane
net	Net
out	Outgoing
p	Grouped planes
r	Radiative

Journal of Physics: Conference Series **547** (2014) 012045 doi:10.1088/1742-6596/547/1/012045

y	Refers to y axis
0	Refers to 0th plane
Superscripts	
m	Refers to mth plane
M	Refers to the Mth plane
n	Refers to nth plane
N	Refers to the Nth plane
0	Refers to 0th plane
∞	Grouping succession limit

7. References

[1] Coquard R, Rochais D and Baillis D, 2012 *Fire technol.* **48.3** 699-732
[2] Zhao C Y 2012, *Int. J. Heat Mass Transfer* **55.13** 3618-32
[3] Glicksmann L R, Schuetz M A and Sinofsky M J 1987 *J. Heat Transfer* **109** 809-12
[4] Kuhn J, Ebert H P, Arduini Schuster M C, Buttner D and Fricke J 1992 *Int. J. Heat Mass Transfer* **35** 1795-1801
[5] Kuhn J, Placido E and Arduini Schuster M C 2005 *Infr. Phys. Technol.* **46** 219-31
[6] Doermann D and Sacadura J F 1996 *J. Heat Transfer* **118** 88-93
[7] Coquard R and Baillis D 2006 *J.Heat Transfer* **128** 538-49
[8] Coquard R, Baillis D and Quenard D 2009 *J. Heat Transfer* **131** 012702.1- 012702.10.
[9] Loretz M, Coquard R, Baillis D and Maire E 2008 *J. Quant. Spectr. Radiat. Transfer* **109** 16-27
[10] Kamiuto K 1997 *JSME Int. J. Series B* **4** 577-82
[11] De Micco C and Aldao C M 2005 *J. Polymer Science B: Polymer Physics* **43.2** 190-2
[12] Zhao C Y, Tassou S A and Lu T J 2008. *Int. J. Heat Mass Transfer* **51** 929-40
[13] Contento G, Oliviero M, Bianco N and Naso V 2014 *J. Thermal Sciences* **76** 147-54
[14] Bhattacharya A, Calmidi V V and Mahajan R L 2002 *Int. J. Heat Mass Transfer* **45.5** 1017-31
[15] Howell J R, Siegel R and Menguc M C 2010 *Thermal Radiation Heat Transfer*, 5th ed., , Taylor and Francis/CRC, New York
[16] Zhao C Y, Lu T J, Hodson H P and Jackson J D 2004 *Mat. Science Eng. A* **367.1** 123-31
[17] Zhao C Y, Kim T, Lu T J and Hodson H P 2001 Micromechanics Centre & Whittle Lab, Department of Engineering, University of Cambridge

Journal of Physics: Conference Series **547** (2014) 012046 doi:10.1088/1742-6596/547/1/012046

A comparison of nanofluid thermal conductivity measurements by flash and hot disk techniques

Bernardo Buonomo[1], Laura Colla[2], Laura Fedele[2], Oronzio Manca[1], Lorenzo Marinelli[1]

[1]Dipartimento di Ingegneria Industriale e dell'Informazione, Seconda Università degli Studi di Napoli, Via Roma, 29, 81031 Aversa (CE), Italy
[2]Consiglio Nazionale delle Ricerche, Istituto per le Tecnologie della Costruzione, Corso Stati Uniti, 4, 35127 Padova, Italy

E-mail: lorenzo.marinelli@unina2.it

Abstract. The conversion into nanofluids is considered a suitable solution to increase the heat transfer efficiency of such fluids. Several theories with an emphasis on different thermal nanofluid mechanisms have appeared to predict enhanced conductivity measurements. There are many ways to measure the thermal conductivity of fluids. Some researchers argued that the anomalous *k* enhancement data are caused by inaccuracies of thermal measurement methods. In this paper, measurements on thermal conductivities of nanofluid mixtures (alumina/water) by means of two different methods are accomplished, *i.e.* the flash and the hot disk technique. In the first method, a NETZSCH model LFA 447 NanoFlash is employed, while in the second one a Hot Disk model TPS 2500 S is used. A comparison between the results obtained from the different measurement techniques is done. Two-step method is used to prepare nanofluids with a nanoparticles volumetric concentration from 0.1% to 4%. Each mixture, at assigned volumetric concentration, is treated with a sonicator for different times and thermal conductivity is measured in the range of temperature from 20°C to 50°C. Moreover, for assigned volumetric concentration and sonication, the stability analysis is performed and thermal conductivity measurements are carried out to determine the effect of sonication time. Results show the thermal conductivity dependence on sonication time, and an asymptotic value is evaluated for each volumetric concentration.

1. Introduction

Heat transfer fluids, such as water, mineral oil, and ethylene glycol, always have an important role in many industrial processes. The poor heat transfer properties of these common fluids hamper their effectiveness as heat exchangers. The conversion into nanofluids is considered a suitable solution to increase the heat transfer efficiency of such fluids. Ever since the report of the abnormal thermal conductivity *k* enhancement of nanofluids by Choi [1], many researchers have tried to explain the mechanisms leading to extraordinarily high thermal conductivity. Accordingly, several theories with an emphasis on different thermal nanofluid mechanisms have appeared to predict enhanced conductivity measurements. The predictive models were developed using the experimentally nanofluid measured thermal conductivity data, provided by many research groups.

There are many ways to measure the thermal conductivity of fluids, such as the cylindrical cell method, temperature oscillation method, steady-state parallel-plate method, 3ω method, thermal constants analyser method, thermal comparator method, and transient hot-wire or hot disk method. However, the experimental findings have been controversial and theories do not fully explain the mechanisms of elevated thermal conductivity. Some researchers argued that the anomalous k enhancement data are caused by inaccuracies of thermal measurement methods. Nanofluids are composed of a common fluid, such as water, oil or glycols and solid particles of nanometric dimensions made of various materials. Nanofluids attract the scientific and industrial interest because of their particular properties, above all the thermal conductivity that is the most studied property.

Since 1995, when Choi [1] advanced the concept of nanofluid, showing substantial increase of heat transported in suspensions of copper or aluminum nanoparticles in water and other liquids, many studies have been done on nanofluids [2,3]. The large application of nanofluids has determined the interest in the numerical simulations and experimental investigations, in both laminar and turbulent regimes, as reviewed in Terekhov *et al.* [4], Sarkar [5], Chandrasekar *et al.* [6] and Hussein *et al.* [7]. These new suspensions are very complicated fluids, difficult to product, handle and study. Many parameters affect nanofluid behaviour amongst which nanoparticles size, shape and concentration, the presence of chemical agents often necessary to ensure stability, and, certainly, the method of preparation. One method frequently used to obtain a stable dispersion of nanoparticles in the fluid consists in breaking particle-particle interactions using an ultrasonic probe. The effectiveness of this method is confirmed by several works [8,9,10,11,12], which often indicate an optimal sonication time. Due to the difficulties in replicating equivalent nanofluids and controlling all the preparation parameters and the test conditions, the wide literature results are often controversial. For example, Buongiorno *et al.* [13] and Utomo *et al.* [14] do not observed anomalously high thermal conductivity enhancement.

In this study, water-based nanofluid containing Al_2O_3 nanoparticles was analysed at a volume fraction ranging between 0.1% and 4%. All the concentrations were sonicated, for a time range from 30 min to 180 min, in order to found an optimal sonication time. Diameter of nanoparticles in suspension was measured using the Dynamic Light Scattering (DLS) technique. Sonication time is effective in the reduction of nanoparticles size up to 120 min. Further investigations has allowed to assess the stability of properties such as the viscosity and the thermal conductivity over time, in the presence of different concentrations of nanoparticles [15-19]. Thermal conductivity of nanofluids at all the concentrations were measured using laser flash and hot disk methods, in a temperature range between 25°C and 65°C, and a comparison between the two sets of data was done.

2. Experimental

2.1. Materials and preparation

The water-based nanofluid containing Al_2O_3 nanoparticles was purchased by Alfa Aesar at a mass concentration 50%. A surfactant, not specified by Alfa Aesar, was present at a concentration less than 1%. The nanofluid was diluted using bidistilled water, in order to obtain the other volume fractions (0.1%, 0.5%, 2.0%, 3.0% and 4.0%) by means of a precision balance KERN 440-45 N, with a maximum uncertainty of 0.2 g). The first step was to prepare nanofluids at 2, 3 and 4 vol.%. Each composition was sonicated for 30, 60, 120, 150 and 180 min, using a Hielscher UP 400S sonicator (frequency 24 kHz, maximum power 400 W).

Then, the nanofluid, resulting more stable, was used to obtain the other two compositions, 0.1 and 0.5 vol.%, adding bidistilled water. The maximum uncertainty on the volumetric concentration is 0.5% for the nanofluid mixture at 0.1%.

2.2. Nanofluid stability characterization

The Dynamic Light Scattering (DLS) technique was used to analyse the average size of nanoparticles in suspension within the base-fluid, by means of a Zetasizer Nano ZS (Malvern). The instrument measures the mean diameter of nanoparticles reading the back-scattered light at 173° and all the tests were performed at 25°C. This instrument can detect particle size from 0.6 nm to 6 μm using the DLS process, with a declared accuracy better than ± 2%. A sample of fluid is put into a proper cell, which is illuminated by a laser and the particles scatter the light which is measured using a detector. The particles move randomly under Brownian motion and their speed is used to determine the particles dimension.

2.3. Instruments

In order to make a comparison, two different instruments, based on different techniques, are used for the thermal conductivity measurements on the same samples.

The first instrument is the NETZSCH model LFA 447, based on the "nano-flash" technique. A pulse of electromagnetic energy heats a sample side and the temperature increases as a function of time. It is evaluated on the opposite face of the same sample by means of an infrared detector, as shown schematically in figure 1a. The energy, released from the Xenon flash lamp (wavelength: from 150 nm to 2000 nm), can be adjusted using a software in terms of voltage and pulse length. The pulse width is adjustable between 0.06 ms and 0.3 ms. The higher the thermal diffusivity of the sample, the faster the energy increase on the other side. The flash method by means of LFA 447 was employed for fluid and pastes for the first time in [20]. The optimal volume of liquid in the sample holder was obtained by a calibration of the system. In fact, it can be filled in a wrong way as illustrated in figure 1b and 1c where the overfilled sample holder and the optimal volume of liquid in the sample holder, are schematically shown, respectively. The drawing of the sample holder employed in the measurements in the NanoFlash NETZSCH LFA 447 device is reported in figure 1d.

The calibration to estimate the fluid volume is obtained evaluating the thermal diffusivity of bidistilled water as a function of the fluid volume at 25°C, as shown in figure 2. It was found that the optimal volume is equal to 0.055 ml. This value is in agreement also with the nominal volume of the sample holder [20].

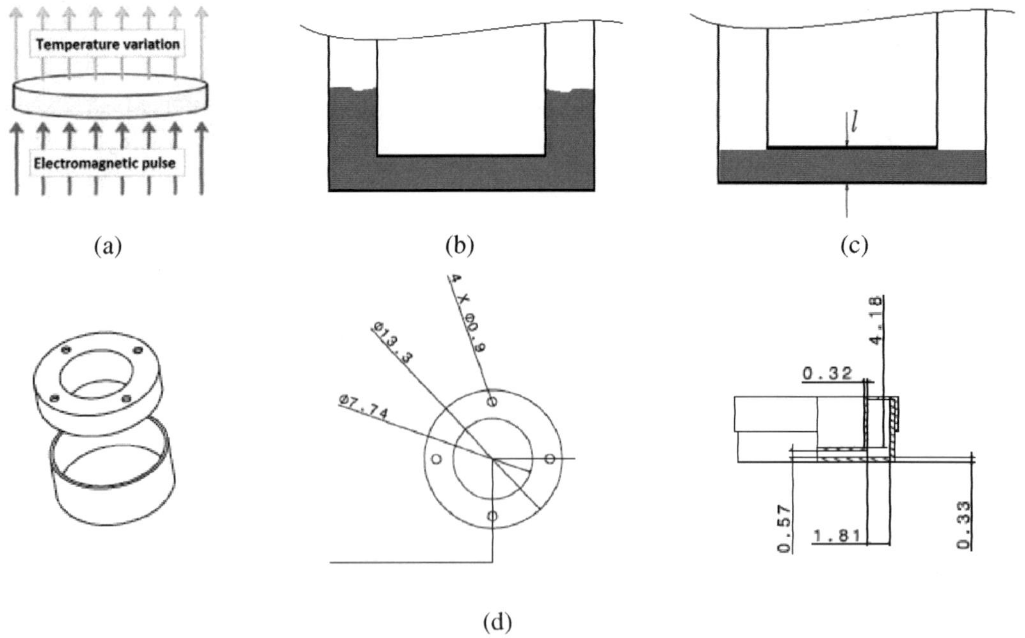

Figure 1. NanoFlash NETZSCH LFA 447 system: (a) working principle, (b) overfilled sample holder, (c) optimal volume of liquid in the sample holder (d) drawing of the sample holder, size in mm.

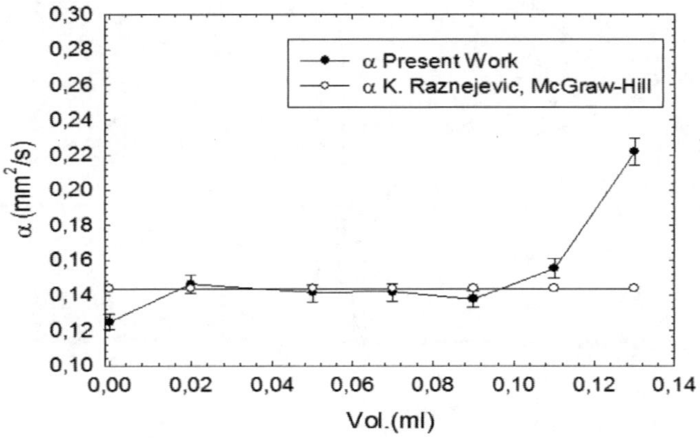

Figure 2 Thermal diffusivity as a function of the volume of fluid in the sample holder for bidistilled water at 25°C and comparison with the literature data.

Temperature profile as function of time, which is detected by the infrared sensor, depends on thermal diffusivity of the material, solid or liquid, as indicate in [21-23]. In figure 3a, an ideal response curve is reported. The evaluation of thermal diffusivity is carried out estimating the half maximum temperature rise and the corresponding time, t_{50}, as schematically shown in figure 3b. In the simplest one dimensional adiabatic model the thermal diffusivity is calculated as

$$a = 0.1388 \frac{d_s^2}{t_{50}} \quad (1)$$

where a is the thermal diffusivity, ds is the sample holder diameter and t_{50} is the time at which the half maximum temperature rise corresponds. The following equation

$$\lambda(T) = a(T)c(T)\rho(T) \quad (2)$$

allows to obtain the thermal conductivity of the liquid sample.

In figures 3c and 3d the effective response curves with bidistilled water at 25°C (b) and 65°C (c) are shown. The uncertainty of the "Nanoflash LFA 447" system for the thermal diffusivity measurements is 3% as indicated by the datasheet. The tests were repeated on three different samples and on each sample temperature ranged between 25°C and 65°C. For each sample and at assigned temperature the run was repeated ten times, *i.e.* the Nanoflash system carried out ten shoots for the assigned temperature. For each temperature the mean value was evaluated and the corresponding variance resulted less than ±3%.

The second instrument is a Hot Disk model TPS 2500 S, that uses a transient plane source method. The main part of the instrument is the sensor, shown in figure 4a, made of a double spiral of thin Nickel, immersed in the nanofluid, contained in a proper box. The sensor works as a continuous plane heat source and, at the same time, as temperature sensor. Actually, through the sensor, a power input is supplied to the sample and the increase of temperature is measured by the electrical resistance of the sensor itself. According to the Fourier's law of heat conduction, the increase in temperature can be calculated as

$$\Delta T(\tau) = \frac{Q}{\pi^{1.5} r \lambda} D(\tau) \quad (3)$$

where

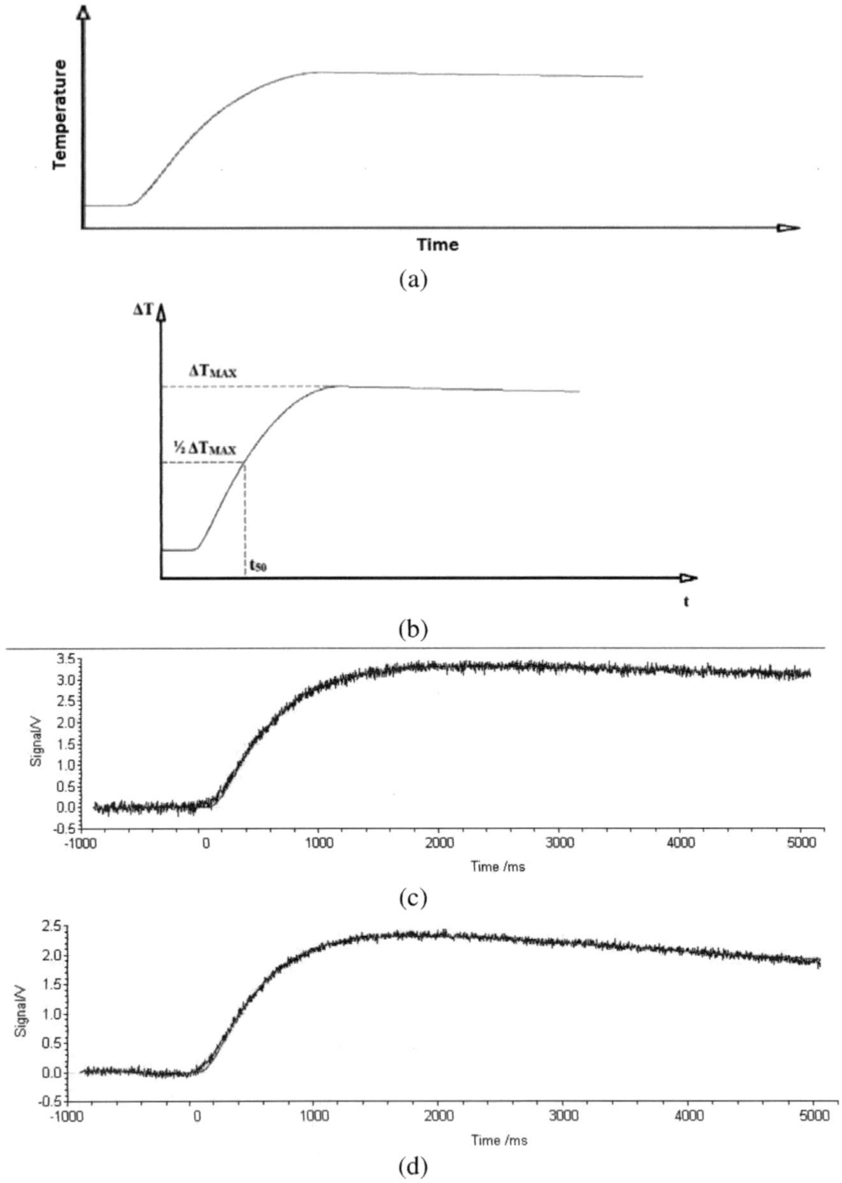

Figure 3. Signal carried out from the LFA 447: (a) theoretical response curve; (b) response curve and time to evaluate the thermal diffusivity, (c) and (d) effective response curve at (c) 25°C and (d) 65°C.

$$\tau = \sqrt{\frac{t\alpha}{r^2}} \tag{4}$$

and r is the probe radius, Q is the heat power, supplied as electric power, and D is a dimensionless time function of τ. Without natural convection, fitting the line given by equation (3) to the experimental data, its slope gives $1/\lambda$.

Figure 4b shows the Hot Disk model TPS 2500 S set up. This instrument can measure a wide range of materials, after suitable sample preparation and choosing the proper sensor diameter. The thermal

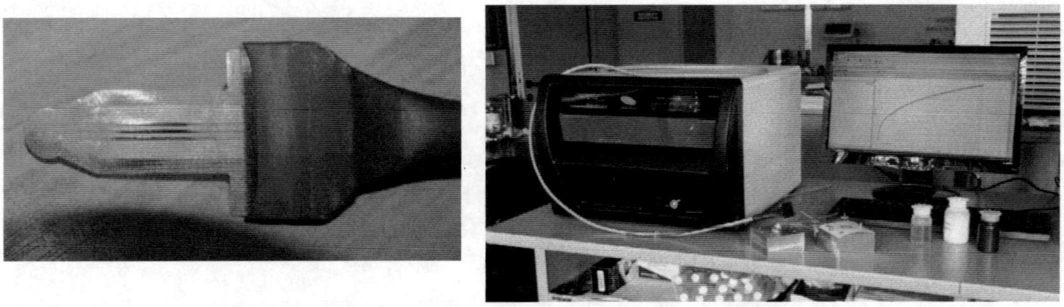

(a) (b)

Figure 4. TPS 2500 S device: (a) sensor, (b) Hot Disk setup.

conductivity which can be detected ranges from 0.005 W/m·K to 500 W/m·K, over a wide temperature range. The declared instrument uncertainty is 5%. However, before measuring nanofluids, thermal conductivity of a well known fluid, as bidistilled water, was measured at ambient pressure in the temperature range between 10°C and 70°C to test the sensor of the instrument and to evaluate the instrument accuracy. All the measured data were compared literature data [24], obtaining an absolute average deviation less than 1%, well within the 5% accuracy declared by the constructor.

3. Results

3.1.Stability

Stability of nanofluid was studied considering two important parameters: the nanoparticles volume concentration and the sonication time. The figure 6 shows the particle size distribution, according to the intensity detected by the Zetasizer, for the water-Al_2O_3 nanofluids at the volume concentration of 2% and 4%. In the figure 7, the nanoparticles size is represented as a function of the sonication time for nanofluid at several concentrations. Al_2O_3 diameter decreases after 30 min sonication, for all the compositions, and slightly continues to decrease until 120 min of sonication. However, further sonication seems to have no effects on nanoparticle size. At a sonication time of 120, 150 and 180 min, nanoparticles diameter remains constant, indicating that there is not an optimal sonication time, but a value beyond which the further sonication is unnecessary. This value is the same for all the concentrations. Nanofluids at a sonication time 180 min were chosen as samples to measure thermal conductivity.

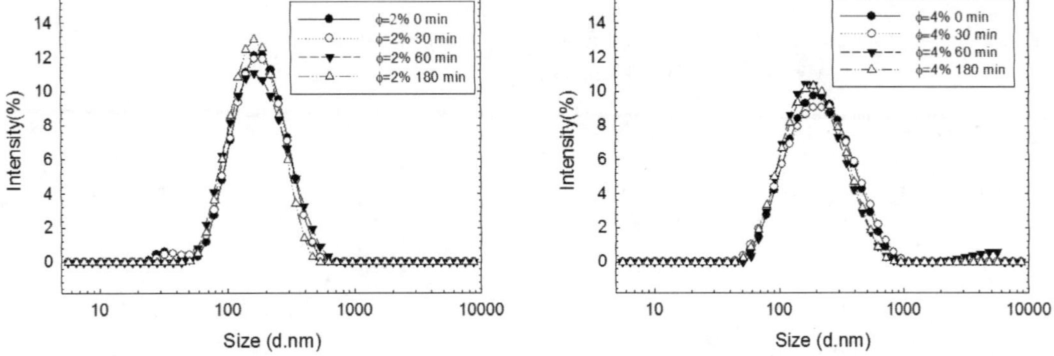

(a) (b)

Figure 6. Particle size distribution for the water Al_2O_3 nanofluids at different sonication times for volume concentration of (a) 2% and (b) 4%.

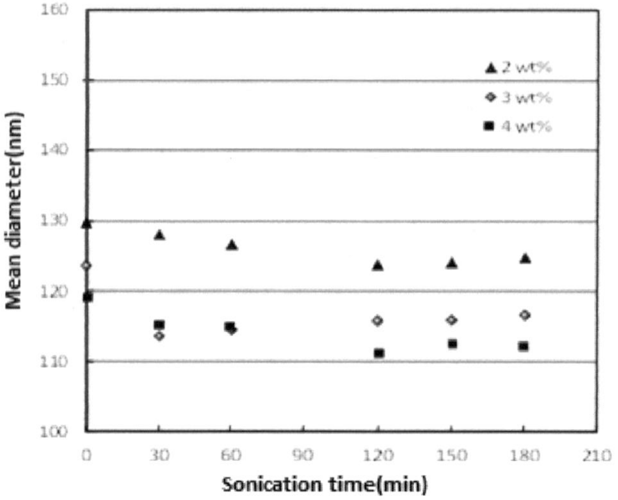

Figure 7. Mean diameter of nanoparticles at different sonication times at considered volume concentrations.

3.2.Thermal conductivity comparison

Thermal conductivity measurements were performed in a temperature range between 25°C and 65°C, with step of 10°C, at ambient pressure. Tests on pure water were initially made in order to verify the accuracy of the instrument, using a referential fluid and in order to compare base-fluid and nanofluid properties. Results indicate good agreement with literature data [24], all the deviations being within 2%, in the entire temperature range and for both the used instruments.

In table 1, thermal conductivity data are reported, for both the laser flash and the hot disk methods. Nanofluid was studied in a volume concentration between 0.1 and 4 vol.%. Deviations between the two methods are below or similar to the experimental uncertainty, the maximum deviation being 5.4% at the higher volume fraction. It confirms the goodness of the measurements and indicates that these different techniques are able to properly characterize the nanofluid. Thermal conductivity increases with nanoparticles concentration and with temperature.

In figure 8, the enhancement in thermal conductivity with respect to that of the base-fluid is reported as a function of the temperature, from 25°C to 65°C. Thermal conductivity is very similar to that of the base-fluid for nanofluid at temperature up to 45°C and low concentrations, while at higher temperatures, nanofluid thermal conductivity is always higher than water thermal conductivity. As expected, the maximum enhancement was found for the nanofluid at 4 vol.%, resulting 13.3% and 10.2% with nano flash and hot disk methods, respectively.

4. Conclusions

Stability of water-based nanofluid containing Al_2O_3 nanoparticles was studied considering the nanoparticles volume concentration at 1.0%, 2.0%, 3.0% and 4.0% and different sonication time, between 0 and 180 min. Increasing the sonication time, Al_2O_3 diameter decreases until 120 min of sonication, for all the compositions. However, further sonication seems to have no effects on nanoparticle size.

Thermal conductivity measurements were performed on nanofluid at 0.1, 0.5, 2, 3 and 4 vol.%, in a temperature range between 25°C and 65°C, using laser flash and hot disk methods and a comparison between the two sets of data was done. Good agreement between the two methods was found. Thermal conductivity increases with nanoparticles concentration and with temperature. The maximum enhancement, *i.e.* 13.3% and 10.2% for laser flash and hot disk method, respectively, was found for the nanofluid at 4 vol.%.

Table 1. Thermal conductivity data.

$$\Delta\% = \frac{\lambda_{hotdisk} - \lambda_{nanoflash}}{\lambda_{nanoflash}} \cdot 100$$

T [°C]	***Thermal Conductivity*** [W/mK]		*Δ%*
	Nano flash	*Hot disk*	
		0.1%	
25.0	0.612	0.610	-0.4
35.0	0.631	0.635	0.7
45.0	0.649	0.657	1.1
55.0	0.677	0.680	0.3
65.0	0.712	0.695	-2.4
		0.5%	
25.0	0.614	0.614	-0.1
35.0	0.635	0.636	0.1
45.0	0.653	0.658	0.7
55.0	0.687	0.682	-0.8
65.0	0.721	0.692	-4.0
		2%	
25.0	0.619	0.619	0.0
35.0	0.641	0.643	0.3
45.0	0.662	0.667	0.5
55.0	0.699	0.683	-2.4
65.0	0.725	0.699	-3.7
		3%	
25.0	0.626	0.636	1.6
35.0	0.649	0.656	1.0
45.0	0.678	0.684	0.9
55.0	0.711	0.702	-1.4
65.0	0.732	0.693	-5.4
		4%	
25.0	0.636	0.642	0.9
35.0	0.653	0.666	2.1
45.0	0.691	0.696	0.7
55.0	0.721	0.707	-2.0
65.0	0.747	0.726	-2.7

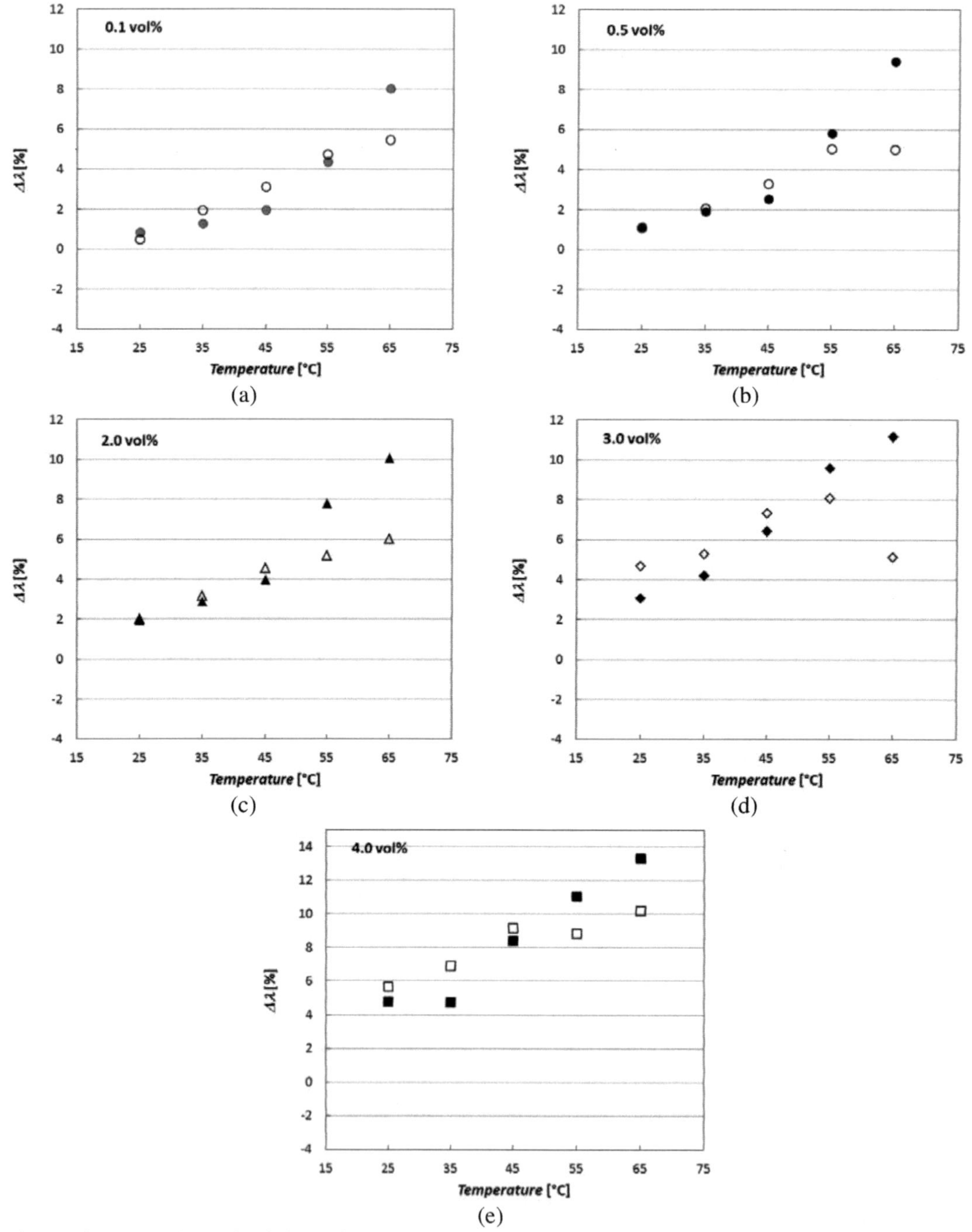

Figure 8. Thermal conductivity enhancement at 0.1(a), 0.5(b), 2(c), 3(d) and 4 vol%(e). Full symbol for nano flash method, empty symbols for hot disk method.

Acknowledgements

The authors want to acknowledge Mrs. A. Barizza and Mr. M. Scattolini for their help in this work.

References

[1] Choi S U S 1995 *ASME FED* **231** 99.
[2] Goharshadi E K, Ahmadzadeh H, Samiee S and Hadadian M 2013 *Phys. Chem. Res.* **1** 1.
[3] Yu W, France D M, Routbort J L and Choi S U S 2008 *Heat Transfer Eng.* **29** 432.
[4] Terekhov V I, Kalinina, S V and Lemanov, V V 2010 *Thermoph. Aeromec.* **17** 157.
[5] Sarkar J, 2011 *Renew. Sust. energy Rev.* **15** 3271.
[6] Chandrasekar M, Suresh S and Senthilkumar T 2012 *Renew. Sust. Energy* **16** 3917.
[7] Hussein A M, Sharma K V, Bakar R A and Kadirgama K 2014 *Sust. Energy* **29**, 734.
[8] Behnaz T, Abbas A, Majid S A and Mehdi A N 2012 *Pow. Tech.* **217** 171
[9] Ruan B and Jacobi A M 2012 *Nanos. Res. Letters* **7**:127.
[10] Ramis M K, Jawaz P and Shebeer A R 2012 *Intl Journal of Eng. Science and Technology.***07** 3514.
[11] Sonawane S S, Khedkar R S. and Wasewar K L. 2013 *Journal of Exper. Nanoscience* DOI: 10.1080/17458080.2013.832421.
[12] Hong TK and Yang HS 2005 *Journal of the Korean Physical Society* **47** 321.
[13] Buongiorno J *et al*, 2009 *Journal of Applied Physics* **106**, 094312.
[14] Utomo A T, Poth H, Robbins P T and Pacek A W 2012 *Int. Journal of Heat and Mass Transfer*, **55**, 7772.
[15] Ji H L et *International Journal of Heat and Mass Transfer.*
[16] Juan C Y, Feng C L, Wen W Z, Yu R H and Bao C J 2012 *Int. Journal of Heat and Mass Transfer* **55** 3160.
[17] Lee J H, Hwang K S, Jang S B, Lee B H, Kim J H, Choi S U S and Choi C J 2008 *Int. J. Heat Mass Transfer* **51** 2651.
[18] Le P Z, Bu-X W, Xiao-F P, Xiao Z D and Yong P Y 2010 *Hindawi Publishing Corporation Advances in Mechanical Eng.*, ID 172085.
[19] Pastoriza M. J, Casanova C, Páramo R, Barbés B, Legido J L and Piñeiro M M 2009, *Journal of Applied Physics* **106**, 064301.
[20] Blumm J, Lindemann A and Min S 2007, *Thermochimica Acta* **455** 26.
[21] Blumm J and Opfermann J 2002, *High Temperature High Pressure* **34** 515.
[22] Manual 2006, NanoFlash NETZSCH LFA 447.
[23] Sato Y and Taira T 2006, Optics Express **14** 10528.
[24] Lemmon E W, Huber M L, McLinden M O 2010 *Ref. Fluid Thermodynamic and Transport Prop. (REFPROP).*

AUTHOR INDEX

AUTHOR INDEX